Preface

At first glance, this book may inspire a certain uneasiness, even alarm, on the part of the reader—and rightly so, since we are confronted with so many ravening jaws and poison fangs, needle-sharp teeth, claws and talons for shearing through hair and hide, pincers for crushing and rending the flesh of lesser creatures. All of this nevertheless provides us with a true composite portrait of life itself, which, to put it rather unoriginally, can only be sustained in death.

This ongoing drama is not the theater of cruelty that we humans have often made it out to be, even though there surely must be some animals that are so misguided or degraded as to kill for pleasure. When a fox gets into the henhouse and slaughters all the inmates, we are still not sure that he does so purely because he has been infected by the panic terror of his victims, and there are some domestic carnivores that seem to take pleasure in the massacres that they commit, though perhaps this is part of what we have taught them.

For the biologist, there are neither good nor evil beings in the living world, nothing but honest co-participants, each of which plays out its assigned role in the natural order of things. There are no such invidious distinctions to be drawn between the peaceful herbivore and the fearsome predator that seeks its life, the delicate insect that feeds only on nectar and the hunting wasp that stalks a victim to be eaten alive by its offspring. If there were no carnivores, then the vegetarians would have vanished from the earth long ago, though the paradox here is less real than apparent. We have only to recall what happened some years ago in an American wilderness preserve, when well-meaning conservationists became determined to save the deer by killing off the skulking packs of wolves and coyotes that preyed on them. Not too long after this campaign of extermination was successfully carried out, the deer had overbrowsed their winter range and were dying of starvation and disease—aggravated in many cases by the congenital weaknesses that had proliferated in the absence of selective constraints.

The meat eater is the vegetarian's inevitable complement, at every level of organization of the animal kingdom. The stalker's strategy is matched against the quarry's skill in taking flight or seeking cover, the speed of the pursuer against the speed of the pursued. The whole range of these subtle maneuverings constitutes a kind of dialogue between one species and another, an enactment of the tremendous forces that permeate the living world, recreating life even as they are destroying it. This is a contest in which the odds tend to favor the survival of all contestants, for the good of all, and a struggle in which nieiher side will ever completely triumph: There would be no more screech owls if they caught every mouse or vole they went after, since the stock of game that was available to them would quickly be exhausted.

This highly original work gives us a detailed account of a great many such relationships between predator and prey species, each of which is characterized by a unique set of behavior patterns on either side and has a highly specific effect on the environment as a whole. And beyond cataloguing so many of the wordless dramas that are played out in this way, by day as well as by night and all over the living world, this book also serves as an eloquent piece of special pleading—it pleads the predators' case for them and argues compellingly for their restoration to their rightful place in that world. We must let them live because they are above and beyond all of our morality, and because each one of them gives evidence of the perfection that has been achieved by living organisms as the result of three and a half billion years of the slow and patient process of evolution.

I am delighted that more than seventy zoologists have pooled their efforts here in order to allow us all to share in their knowledge, and that their message is being disseminated by a publisher that is well known for the quality of its productions. It may be thanks to their efforts, at least in part, that the lion, the tiger, the bear, and the wolf, as well as the hawk, the shark, the venomous snake, the tiger beetle, and the digger wasp will finally be allowed to carry on with their game of life and death—or, as we might as well say, with their game of life—without undue interference from us.

Jean Dorst,
French Academy

Introduction

From *Acanthrometra* to the zorilla and from agama to *Xantusia,* passing by way of the raccoon and the lobster (which we tend of think of as being more preyed upon than preying), all of these living creatures feed upon one another. It is enough to make the latter-day disciples of Rousseau, as they leaf through the pages of this book, give up their last remaining illusions about a universe in which everything is equipped with claws, pincers, stingers, tentacles, and mandibles.

Scientists take the position that the difference between the tiger and the pygmy shrew or between the eagle and the wood warbler is only one of scale. As far as the general public is concerned, however, a decorative creature like the sea anemone or the harmless cochineal insect, which the French call "the cows of the Good Lord," is not a predator, even though the latter devotes its entire brief life to the wholesale slaughter of aphids. But it is an article of faith among our own species that a predator can be recognized by its sharp teeth or its cruel beak, and a predator is by definition a detroyer, a kind of animal bandit that lives by its depredations—since the distinction between *predation* and *depredation* often becomes somewhat blurred.

This idea is so deeply engrained that it requires a genuine effort to even begin to question its validity. I remember that I began to be troubled by doubts during the Second World War, when game animals continued to proliferate in Europe even though humans had been obliged to give up hunting (and thus the normal campaign of destruction waged against raptors and small carnivores) almost entirely. Doubt later turned to conviction—of the truth of the opposite viewpoint, that is—when I was traveling through the more sparsely settled regions of several continents, where there was an astonishing abundance of life in spite of the presence of all sorts of predators (human beings largely excepted, of course).

With the subject so much obscured by prejudice and unconscious fear, a phenomenon as complex as predation can only be encompassed by a rigorous scientific study that will enable us to understand its real impact on the animal species involved. The entomologists and ornithologists have been the pioneers in this respect, though it is true that the subjects of their investigations are somewhat more readily accessible than is the case with the mammalogists, herpetologists, and ichthyologists. The habits of a preying mantis, a digger wasp, or a bird that brings food back to its nestlings on a regular schedule are easier to observe and record than those of a wary nocturnal hunter like the beech marten, or a viper or a pike in its natural habitat.

The Russians, the Americans, and the Canadians were the first to take a serious interest in the larger carnivores, since the fur trade was still a matter of considerable economic interest to them and they wanted to be able to continue it on a more rational basis. The European countries began to follow their lead more recently, generally at the behest either of hunting enthusiasts who hoped to be able to measure the impact of predators on game animal populations (and thus to provide themselves with a warrant for their destruction) or, alternatively, of ecologists who were eager to restore these unjustly maligned creatures to their rightful place at the head of the terrestrial food chain. In France, it has only been in the past few years that a new generation of scientists has begun to come to grips with this problem on behalf of such institutions as the Institut National de la Recherche Agronomique, Office National de la Chasse ("National Bureau of the Hunt"), and the Centre National de la Recherche Scientifique.

The adoption of so-called capture-recapture methods of animal recognition marking and especially biotelemetry (long-term tracking of individual animals that have been outfitted with radio-transmitting collars) has provided us with a great deal more information then ever could have been obtained from the most painstaking visual observations and inferential (i.e. anatomical) evidence of predator behavior. From the very earliest data it was clear that population densities of predatory species are far lower than those of prey species, though, as we shall see shortly, there are some predator species whose numbers are subject to considerable fluctuations in response (though not necessarily in proportion) to rapid population growth (or decline) on the part of the prey species.

Thus, there may be an average of 2–5 weasels or 1–11 stoats per square kilometer, but the comparable figure for the water voles on which they feed is 10,000–40,000. It is also worth noting that since a water vole normally weighs 80 g and a stoat 130 or 250 g (typical weights for the female and the male respectively), then the stoat can get along quite comfortably by catching just one or two voles per day. The other Western European carnivores are present in even smaller densities, if at all: 1–1.3 foxes, 0.3–1 badger, 0.1–0.2 European wild cats, 0.3–1 marten, 0.4–0.5 polecats per square kilometer.

Apart from illustrating the remarkable complexity and variability of the relationships between predators and prey animals, these observations have also enabled us to make a preliminary distinction between two basic types of predator, namely the specialists, which prey on a particular kind, sometimes a particular species, of animal, and the generalists or opportunists, which prey on many kinds of animals and feed additionally (or even primarily) on plant material. This basic distinction appears to exist throughout the animal kingdom; the cochineal insect, for example, is a specialist that preys exclusively on aphids, whereas the praying mantis is a generalist that will seize on and devour virtually any insect that crosses it path.

The weasel and the stoat are both specialists, great destroyers of small rodents and of voles in particular. Their small size (20–50 cm and 50–120 g in the case of the weasel, 30–40 cm and 100–300 g in the case of the stoat), flattened skull and pointed muzzle, and sinuous, snakelike bodies make them particularly adept at pursuing rodents through their tunnels and galleries in the ground or under the snow. Studies conducted in different countries and different habitats (cultivated plains, mountainous or wooded areas) have shown that between 58 and 86 percent of the weasel's normal diet is made up of small rodents, with the remaining fraction consisting of shrews and moles, hares and rabbits, birds, and insects. In the case of the stoat, the percentage of rodents (and especially voles) may be as high as 90 or even 100 percent in regions where they are particularly abundant. In the British Isles, on the other hand, where voles are much less prevalent than on the Continent, the stoat has learned to rely primarily on rabbits, which are not native to the British Isles. (Perhaps surprisingly, the rabbit is an "exotic" import from the Mediterranean Basin and was first introduced into Northern Europe as a game or domestic animal during the Middle Ages.)

The polecat is the rabbit catcher par excellence in Western Europe, at least in those regions where their ranges overlap, and its prowess at sniffing out rabbits in their burrows was so much admired by man that it was taken into his service; the polecat is the ancestor of the domesticated ferret. In the wild, the polecat is often found in low-lying, humid areas and thus also consumes prodigious quantities of frogs, toads, and other amphibians. In my own country, the European wild cat (which looks very much like a large, thick-coated housecat) is the leading vole specialist; it swallows its prey whole, and its stomach has been found to contain as many as 20 or 25 water voles.

Generalist predators of Western Europe are somewhat more numerous, ranging in size from the beech and pine marten to the brown bear and including the fox, the badger, and the genet (a relative of the mongoose brought from Africa by the Moors). Their diet contains a very large proportion of vegetable material (dry or pulpy fruit, berries, bulbs, and shoots): up to 60 percent in the case of the beech and the pine martens and as much as 80 percent in the case of the almost-extinct brown bears of the Pyrenees. Small rodents are also a staple of the generalist predator's diet, comprising between 20 and 75 percent (depending on the season and the habitat) of all the animal prey consumed by the martens, which also feed on mole and shrews, hares and rabbits, birds, frogs and toads, reptiles, and invertebrates of various sorts. Earthworms typically make up about 40 percent of the stomach contents of the badger, along with a roughly equivalent amount of vegetable forage in addition to rodents, insects, frogs, and toads.

The specialist and the generalist predtors can often be distinguished by their reproductive patterns as well, since it is only the specialists that are so highly dependent on a particular prey species that their own population growth are subject to the same cyclical "boom-and-crash" fluctuations as the prey speices, whereas the population levels of generalist predators are not subject to these populations and remain relatively stable. The classic examples of these "boom-and-crash" partnerships include the Canadian lynx and the snowshoe rabbit (varying hare), the Arctic fox or the great white owl and the lemming, and, in more temperate regions, the stoat and the water vole. These cycles, which are determined by the relative abundance of the prey animal's food sources and thus primarily by climatic factors, are not strictly seasonal, and the upward curve of the cycle often continues for several years. In the Jura Mountains, the density of the stoat population was subject to a sixteenfold increase over a two-year period in response to an even more rapid population boom on the part of the voles.

The reproductive capacity of specialized predators is greater than that of the generalists, allowing them to rapidly adjust their numbers in order to take advantage of such a windfall—or alternatively, to compensate for their losses in the wake of the inevitable crash of the prey animal population. Starting over again with a relatively small number of mating pairs, the population can rapidly reconstitute itself (allowing for a slight time lag) in time to catch the next upward trend when food resources become more abundant again. A population of specialist predators that is decimated by some other agency than famine thus stands a good chance of recovering, whereas a population of generalist predators with a relatively low reproductive rate may not be able to withstand a comparable level of persecution, with the result that the species may be exterminated from the region altogether (as the pine marten and the polecat have been from the British Isles).

Recently developed techniques of animal recognition marking, and especially biotelemetry, have also given us a great deal of information on the movements of individual animals in a particular locality. As far as stoats and weasels are concerned, for example, a given area is likely to support three different populations—permanent residents, which are observed (or trapped and then released) in the same locality at various times throughout the year; transients, which are only observed; in a given locality on a single occasion, and seasonal residents that occupy different territories at certain parts of the year.

The individual hunting territories of all of these predators are quite extensive, ranging from several tens or hundreds of hectares for the mustelids and the fox to several thousand or several tens of thousands of hectares for larger predators such as the lynx, the wolf, and the bear. All of these predators cover considerable distances in their search for prey and only return to hunt in the same locality at widely spaced intervals. They energetically defend their territory against intruders of their own species, even killing and eating them on occasion. All the mustelids attack one another on sight, as does the wild cat, which also attacks fox cubs, while the lynx has acquired a taste for the flesh of either wild or domestic cats, foxes, and small dogs. Nonfatal encounters of this kind may also be responsible for the spread of a number of virulent diseases to which wild and domestic carnivores are equally susceptible, notably canine distemper, and the raptors may also become involved in this interspecific melee, with the eagle, eagle owl, goshawk, screech owl, and barn owl swooping down, according to their sizes and abilities, to prey on foxes, cats, martens, stoats, and weasels.

In other words, and in contrast to our more traditional notions, we have learned that the predator's place in its environment is quite precarious, not even taking account of the extent to which it is systematically persecuted by man and his four-footed auxiliaries. Everything that we have learned so far (of which the above discussion only represents the sketchiest summary) has made it clear that their populations are very sparsely distributed and the extent of their "depredations" very limited, even against species such as the voles that are enormously abundant during the peak periods of their long-term population cycles. In such cases, the toll of these creatures taken by predators may have the effect of accelerating and possibly prolonging a downturn in the rodent population cycle that has already been triggered by other causes, but is ultimately not very significant.

Intensive studies of winged as well as terrestrial predators and more general observations of entire ecosystems have led us to conclude that predation is purely a marginal phenomenon, one of a great many constraints that affect certain species and are clearly less significant in any case than the environmental and climatic factors that determine the available food supply. On the other hand, predators undoubtedly play a selective role in eliminating individual prey animals that are not as well adapted for survival, and they help to control the spread of certain diseases, notably among gamebirds and game animals that have been bred in captivity in order to be released during the hunting season and that are often congenitally weak or diseased.

Ordinary logic would suggest that, with any particular exertions on their behalf, it would at least make sense to relent a little in the ancient campaign of persecution that we have waged against these predators. On the other hand, ordinary logic and even the soundest scientific conclusions are rarely sufficient to prevail against "common-sense" notions that may originally have been formulated in the days when the carnivore really was a serious competitor, even sometimes a predator, of *Homo sapiens* and have rarely been called into question since then.

Pierre Pfeffer,
Research Director,
French National Center for
Scientific Research (CNRS)

Acknowledgments

The editors would like to acknowledge the information and other invaluable assistance provided by the following organizations, all of which are located or have their headquarters in Paris unless otherwise indicated: Centre National de Recherche Scientifique, Conservatoire National du Littoral (Rochefort), Ecole Normale Supérieure, Institut National de la Recherche Agronomique, Museum National d'Historie Naturelle, Université Paris VII, Association des Amis des Renards et autres Puants, Fédération Française des Sociétés de Protection de la Nature, Fonds d'Intervention pour les Rapaces (La Garenne), Ligue pour la Protection des Oiseaux (Rochefort), Office National de Chasse, Office pour l'Information Ecoentomologique (Guyancourt), Rassamblement des Opposants à la Chasse, Société Française pur l'Étude et la Protection des Mammifères (Bohallard), Société Française de la Protection de la Nature, Société Herpétologique de France, Fonds Mondial pour l'Environnement (World Wildlife Fund).

A

AARDVARK *Orcyteropus afer* 1

1
Sub.: Vertebrata
Cl.: Mammalia
O.: Tubulidenta
F.: Orycteropodidae

The aardvark (an Afrikaans word meaning "earth pig") is so called because of its elongated, piglike snout, though in other respects it more closely resembles the giant anteater of South America; this is an example of the evolutionary convergence of two different lineages, since the two species are not at all closely related. The aardvark is the only extant representative of a primitive order of mammals, the Tubulidentata, which are also thought to have been the ancestors of the ruminants; fossil specimens have been discovered in Europe, southern Asia, and Madagascar, but the species is now found exclusively in sub-Saharan Africa (excluding Madagascar), where it is presumed to have originated.

▶ The aardvark digs its burrow in the vicinity of a number of large termite mounds. It hunts by night, singly or in pairs, and rarely ventures very far from its burrow as long as the supply of termites remains plentiful. It always follows the same trail to the termite mounds, though it never feeds at the same mound on two consecutive days. It begins by scooping out a hole about 40 cm in diameter at the base of a termite mound, then, when a gang of termites turns up to repair the damage, the aarvark picks them off with its long, adhesive tongue or literally inhales them by pressing its snout against one end of a termite tunnel that has been partially exposed. The aardvark seems to find most of its food, however, by ambushing columns of termites in the course of their nocturnal migrations.

◀ The aardvark's main natural enemies are the lion, the leopard, and the hyena; it is also hunted by man in certain parts of Africa. When threatened, the aardvark takes refuge in its burrow, partially blocking up the entrance with earth; if it is caught out in the open, it will attempt to dig itself in, or it may attempt flight, hoping to elude its pursuer with its first few loping bounds, since it is not very quick on its feet. If neither of these maneuvers is successful, the aardvark attempts to defend itself by rolling over on its back and brandishing its powerful front claws.

P.A.

AARDWOLF *Proteles cristatus* 2

2
Sub.: Vertebrata
Cl.: Mammalia
O.: Carnivora
F.: Protelidae

The aardwolf is the only representative of the family Protelidae. Curiously, though its closest relatives are the mongooses and civets (Viverridae), the aardwolf looks a great deal like the striped hyena. It is about 80 cm in length, with a bushy tail that measures an additional 30 cm; its coat is light-colored with black stripes, and the hair grows very thick along its back, in the manner of the hyena's mane. The aardwolf is found on the open savannahs of eastern and southern Africa; it sometimes digs its own burrow, more frequently makes its home in the abandoned burrow of an aarvark, and it marks the boundaries of its territory with the secretions of its anal glands.

▶ The aardwolf feeds primarily on termites, as well as on other insects and their larvae and on birds' eggs. It searches for termites that are just setting out on their nightly foraging expeditions, lapping them up with its sticky tongue in the manner or the aardvark or the anteater; it is incapable of breaching the walls of a termite mound. The aardwolf feeds only between the hours of 5 p.m. and midnight, and spends the daylight hours either sunning itself by the entrance to its burrow or curled up asleep inside. It sometimes feeds opportunistically on small mammals as well as on carrion.

◀ The aardwolf is very wary and rarely ventures more than a short distance from its burrow. The flesh of the aardwolf is sometimes consumed by man, and packs of dogs that are trained to hunt jackals often fail to make a distinction between a jackal and an aardwolf. The systematic destruction of termite mounds on farmland or rangeland deprives this species of its livelihood and invariably results in its disappearance from the region.

The aardwolf's defensive tactics are mainly passive. Like other canids, it raises its hackles (which are especially luxuriant) to make itself appear larger than it is, and also secretes a foul-smelling musky substance if an attack appears imminent. Some authorities believe that the aardwolf's strong resemblance to the striped hyena has been enhanced by natural selection and is thus an instance of protective mimicry, which allows this harmless insectivore to trade on the reputation of one of the most formidable carnivores of the African savannah. P.A.

3
Sub.: Vertebrata
Cl.: Aves
O.: Coraciadiformes
F.: Bucerotidae

ABYSSINIAN GROUND HORNBILL
Bucorvus abyssinicus 3

The bill of *Bucorvus abyssinicus* lacks the spectacular proportions of the more celebrated hornbills of the genus *Buceros* and is adorned with fleshy excrescences called *caruncles,* which continue to grow larger throughout the bird's lifetime. The Abyssinian ground hornbill is about 1.1 m long and weighs between 3.5 and 4 kg. It becomes sexually mature at the age of 3 and has been known to survive to the age of 12 in captivity. This species is found on the steppes and savannahs of Subsaharan Africa, particularly in the vicinity of baobab trees, where it builds its nest.

▶ Ground hornbills are so called because, though perfectly capable of flying, they usually prefer to get around by hopping along the ground. They mate for life, and since the female does not undergo the lengthy confinement to which other female hornbills are subject (see RED-BILLED HORNBILL, below), both parents are free to join in the hunt for food for their nestlings. Hornbills feed on locusts, crickets, and other large insects, snails, frogs, lizards, snakes, and the young of other ground-dwelling birds; like all large birds, they also eat fruit, both wild and cultivated (including peanuts in the latter category). They swallow their food whole but quickly spit up such indigestible items as fruit pits and the sclerotized shells of insects. They are attracted to grass and forest fires, since they make a habit of preying opportunistically on insects and other small creatures fleeing the approaching flames.

◀ This hornbill is large enough to be safe from most predators, though the nestlings are not brought up in the high-security environment enjoyed by hornbill chicks of other species. The hornbill is said to be respected by Central African agriculturalists because it preys on snakes, locusts, and crickets—and is also revered by some cultures as a model of conjugal devotion.

The hornbill family is comprised of 14 genera, 55 species in all, most of which feed primarily on fruit. Some of the larger arboreal species include the rhinoceros hornbill *(Buceros rhincoeros)* of the Malay Peninsula and Borneo and the black-casqued hornbill *(Ceratogymna atrata)* of the equatorial jungles of Africa; the rufous hornbill *(Buceros hydrocorax)* of the Philippines is also known as the calao, a name that is sometimes (though perhaps less properly) applied to all hornbill species. J.-F. T.

4
Sub.: Protozoa
Sup.cl.: Actinopoda
Cl.: Acantharia
O.: Arthracantha

Acanthometra pellucida 4

This protozoan has a spherical body from .02 to .08 mm in diameter supported by an array of 20 minute spines, or spicules, composed of strontium sulphate; the arrangement of the spicules is invariable: 4 around the "poles" of the creature's body, 8 around the tropics, and 4 around the equator. The internal cytoplasm (endoplasm) has a yellow-green tint caused by the presence of pigmented bodies called zooxanthellae (in this case, algae of the family Prymnesiphysceae) and is encased in a capsular membrane (so-called microfibrillary packing) pierced with pores through which extrudes an incomplete covering of external cytoplasm (ectoplasm).

A delicate external membrane, or pellicule, called the periplasmic cortex, a highly organized microfibrillary network, is stretched between the points of the spicules like the fabric of a tent. The periplasmic cortex is attached to the spicules by means of extensible fibrils called myomenes, the contraction and relaxation of which permit the position of the cortex to be raised or lowered with respect to the endoplasm. Finally, numerous axopodia, delicate wands of cytoplasm, extend outward from the body of the cell, generally in the vicinity of the spicules.

Acanthometra pellucida is very abundant between April and November in subtropical, tropical, and equatorial waters from a depth of 0 to 200 m; it is rarely found in winter. Off the bar of Villefranche-sur-Mer, for example, in the south of France, 60 to 80 percent of microorganisms present in a sample skimmed off the surface on a hot summer day (when the water temperature reaches 26–28°C) will typically consist of members of this species.

▶ Unlike those of its relatives the Heliozoa, the axopodia of *Acanthometra* are neither very adhesive nor very sensitive and play only a limited role in the capture of larger prey; they are primarily useful in conveying the smaller organisms (nanoplankton) and clumps of organic debris on which *Acanthometra* feeds toward its interior. The surface of the periplasmic cortex is covered with a tracery of little channels characterized by convulsive cytoplasmic activity. Prey adheres to the surface of the cortex and is then rafted back into the subcortical space (into the ectoplasm, in other words) and thence into the endoplasm, where the process of digestion begins.

◀ Tiny crustaceans called copepods will occasionally make a meal of an acantharian, leaving nothing but the bare spicules behind, though this behavior has not been observed very often and may be only be characteristic of organisms that have been scooped up together in a plankton net. Tunicates, or salps (barrel-shaped marine creatures related to sea squirts), are often collected with a fairly large number of acantharians in their buccal chambers, and may thus be presumed to feed on them on a regular basis.

Most acantharians live in the same climatic zones and have much the same feeding habits as *Acanthometra pellucida*. The species *Acanthochiasma rubescens*, however, is well provided with adhesive axopodia, which enables it to go trolling for copepods, and a few of the larger species *(Acanthoplegma, Astrolonche)* are able to engulf large, very fast-moving ciliata *(Tintinides* and *Strombidium)* of 20–30 microns in length; the periplasmic cortex ruptures and propels the prey into the subcortical space, where it is immediately ensnared by fingerlike emanations from within as soon as it makes contact with the ectoplasm. The captive protozoans are digested on the spot, by a process that may take several hours.

J.F.

Acineta tuberosa 5

5
Sub.: Protozoa
Cl.: Ciliophora
O.: Suctorida
F.: Acinetidae

Acineta tuberosa is a suctorian (a member of an order of sessile protozoans) that generally lives as an ectocommensal (a creature that lives on the surface of another organism without feeding on the body of its host) on the gills and appendages of certain marine crustaceans. *A. tuberosa'* s monocellular body measures 90–100 μm in length and is attached to the host by means of a long stalk (peduncle); the retractile tentacles that are the chief distinguishing feature of the order Suctoria are found in clusters near the apex of the cell. *A. tuberosa* reproduces by budding; the larvae emerge from the internal budding pouch as free-swimming ciliates, then quickly select a suitable crustacean host, attach themselves by secreting a permanent stalk, shed their cilia, and sprout tentacles with which to trap and ingest their prey.

▶ This species feeds primarily on free-swimming marine ciliates (notably *Strombidium). A. tuberosa*'s clublike tentacles are tipped with tiny prehensive organs called haptocysts, which spring out and penetrate the cell wall of any protozoan that brushes against the ends of the tentacles as it moves through the water. If the prey animal is considerably larger than the predator, it may be able to escape, but more frequently the prey only succeeds in impaling itself on the adjoining haptocysts as it struggles to tear itself away. Eventually the cilia stop beating and the prey is immobilized.

The cell wall of the prey is ruptured by the initial contact with the haptocysts, and *A. tuberosa* ingests the cytoplasm of the captive ciliate through a regular network of slender tubules that run down to the base of the tentacles and are connected to one or more digestive vacuoles inside the cell. The membrane that covers the club-like tip of each tentacle immediately forms a network of capillarial connections with these tubules, and the tentacle partially retracts as the cytoplasm of the prey animal is gradually siphoned off into the interior of the predator. (The name of the order Suctoria is derived from the Latin *suctus*, "sucked dry," "depleted").

◀ It seems likely that *A. tuberosa* is preyed on in a similar fashion by other suctorians, perhaps other acetinid species, but this has yet to be confirmed by direct observation.

Other suctorians are found in both fresh and salt water; some remain free-swimming throughout their lives, but most attach themselves to the exterior (or even the interior) of a commensal host. Some species have a hard external skeleton called a lorica ("coat of mail"); the arrangement of the tentacles is often irregular, and a number of species (including the common *Dendrocometes paradoxus)* have branching tentacles. The reproductive process is also subject to various modifications: the larvae may develop inside an internal pouch, as is the case with *Acineta tuberosa*, or a swelling may appear on the surface of the cell from which one or more buds quickly develop. J.G.

6
Sub.: Arthropoda
Cl.: Insecta
O.: Hemiptera
F.: Saldidae
S.f.: Aepophilinae

Aepophilus bonnairei 6

This small insect measures only about 3 mm in length; its head is very dark, its eyes bright red, its antennae are fairly long, and its beak (rostrum) consists of 4 jointed segments. The thorax is reddish, and the walking legs (which are not adapted for swimming) are yellowish; the adbomen is brownish and covered with a fuzz of golden hairs, and the vestigial wing covers are located at the rear of the thorax. The sexes are similar in appearance, though the male is slightly more elongated.

Aepophilus bonnairei is found exclusively in the intertidal zone (primarily toward the lower edge, at approximately the same level as the periwinkle and beds of brown seaweed [fuscus]) and is active at low water. When the tide is high, *Aepophilus* retreats into fissures and tiny cavities in the rocks and stones along the shore, and is thus able not only to continue breathing but even to remain completely dry while the waters rise around it. This species, the sole representative of the subfamily Aepophilinae, is found all along the shores of the Eastern Atlantic and the English Channel, from Britain and Ireland down to Morocco.

▶ Not a great deal is known of *Aepophilus*'s feeding habits, though it appears to prey primarily on marine worms that are left stranded in the seaweed beds at the lower edge of the intertidal zone at low tide. It uses its beak to inject the prey animal with its gastric juices, continuing to do so until its tissues are completely liquefied, then ingesting the dissolved nutrients through its beak (external, or preoral, digestion).

◀ The pseudoscorpion *Obisium maritimum* is one of several predacious insects found in the same habitat that is thought to feed on *Aepophilus bonnairei.*

The Salididae, collectively known as shore bugs, comprise roughly 230 species worldwide, of which only 8 inhabit the intertidal zone; the remainder are found along the banks of fresh or brackish bodeis of water, and are substantially different in their habits and morphology. J.-L.D.

AESCULAPIAN SNAKE *Elaphe longissima* 7

This handsome snake has a narrow head and slim, greenish-bronze body dotted with small white scales. It is not common in any one region but found sporadically throughout Central and Southern Europe and in Asia Minor up to a height of 1500 m. It is primarily arboreal, though it is also capable of burrowing and is well adapted to a variety of habits. It avoids very humid areas and will only take to the water on an extremely hot day or when otherwise driven by necessity. In early fall, it takes refuge from the cold in a burrow, underneath a stump, or in some other sheltered spot and will not emerge again until March or April.

The conventional assumption that this species served as the model for the coiled serpent on the medical caduceus, sacred to Aesculapius (Asklepios), the classical god of medicine, has been questioned in recent years and other candidates have been proposed, including the Montpellier snake (*Malpolon monspessulamus),* the leopard snake *(Elaphe situla)*, and the four-lined snake *(Elaphe quatuorlineata)*. However, the fact that isolated populations of *Elaphe longissima* are still found in the vicinity of mineral springs frequented by the Romans, apparently having been deliberately introduced there, does suggest that this species has been identified with the cult of Aesculapius at least since late antiquity.

▶ The Aesculapian snake is sometimes seen making its way across a sunny patch of ground in the morning or early afternoon, or perhaps moving more purposefully through the dried grass in search of prey. It is a very proficient climber and attacks its prey from ambush, remaining motionless until the prey animal is within range (giving rise to the notion that it somehow "fascinates" its victims before it strikes) and then abruptly extending the forepart of its body, jaws agape, to seize its prey around the middle.

It kills its prey by suffocation, wraping its coils around it and constricting its body, very much in the manner of a python or a boa; it is sometimes claimed that constrictors also crush their victims and break their bones as an aid to digestion, but this is not the case. The process of ingestion requires no more than a few minutes and takes place by means of the usual alternate movements of the snake's jaws. Sixty percent of the diet of the Aesculapian snake consists of small insectivores, especially shrews of the genus *Sorex*, *Neomus*, or *Crocidura*, an addditional 30–35% of sitting birds and nestlings, which it catches on the ground as well as in trees, and the remainder of lizards and reptiles' eggs. The young of this species feed on lizards and insects, especially grasshoppers.

◀ The Aesculapian snake is preyed on by certain diurnal raptors (the short-toed eagle, the common buzzard), turkeys, chickens, and other free-ranging domestic fowl, and wild boar as well as a fellow colubrid, the Montpellier snake. Like the watersnakes of the genus*Natrix*, it can emit a foul-smelling liquid from its cloacal gland in order to discourage predators; many specimens are moderately aggressive and will not hesitate to inflict a painful bite, though the Aesculapian snake will often allow itself to be handled quite freely once it has grown accustomed to captivity and the company of human beings.

Elaphe longissima might conceivably be mistaken for its congener the ladder snake *(Elaphe scalaris)*, which is found all along the Mediterranean coastline, since *E. scalaris*'s ladderback markings are not always that strongly contrasted. The four-lined snake *(E. quatuorlineata)*, found in Italy, Greece, and Yugoslavia, also bears some resemblance to the Aesculapian snake.

J.C.

7
Sub.: Vertebrata
Cl.: Reptilia
O.: Squamata
S.o.: Ophidia
F.: Colubridae

AFRICAN CIVET *Viverra civetta* 8

8
Sub.: Vertebrata
V.: Mammalia
O.: Carnivora
F.: Viverridae

This catlike carnivore is typically about 70 cm long, plus an additional 35 cm for the tail. Its coat is grayish or yellowish with black spots, and it has hackles, or erectile guard hairs, all up and down its back. This particular civet species, sometimes called civet cat, is found virtually everywhere in Africa except for desert regions and in the northeast. It makes its home in a crevice in the rock, a hollow tree, or a burrow abandoned by some other creature, and marks its territory with the secretions of its anal glands (of which more later).

▶ The civet feeds on hares and rabbits, rodents, birds, amphibians, reptiles, and insects as well as certain fruits. It is a solitary nocturnal hunter that stalks a prey animal that is already in motion, gives chase, and fells it with a series of bites, approaching successively closer to the head, and by raking with the claws on its forefeet.

◀ The civet is competes successfully with a variety of larger and more formidable nocturnal carnivores, which is able to evade thanks to the special keenness of its hearing and sense of smell. Man is the civet's only serious enemy, since the musky secretions of its anal glands (an individual civet produces about 20 g per week) is widely as used as ingredient in soaps and perfumes.

Other notable members of the civet family include the lesser Oriental civet *(Viverricula indica)* of Southern Asia, which is also a soure of commercial civet musk, the palm civet (genus *Paradoxorus)* of Southern India, one of a number of civet species found on the subcontinent (also called "toddy cat" because of its fondness for palmtree sap), and the fossa *(Cryptoprocta ferox)* of Madagascar, a civetlike creature that (if not already extinct) is regarded by many authorities as the rarest of living carnivores. P.A.

9
Sub.: Vertebrata
Cl.: Reptilia
O.: Squamata
S.o.: Ophidia
F.: Colubridae

AFRICAN EGG-EATING SNAKE
Dasypeltis scabra 9

This colubrid species typically measures between 60 and 70 cm, and only exceptionally exceeds a meter in length. This species ranges across North Africa as far east as Egypt, and a relict population was recently discovered in the extreme southwestern corner of Morocco. Its congeners, mainly found on the savannahs and arid steppes of sub-Saharan Africa, also include at least one one forest-dwelling species.

▶ *Dasypeltis* feeds exclusively on birds' eggs; it is a marauding hunter that searches for nests along the ground as well as in low-lying bushes, sampling the air with its flicking tongue and transferring the organic molecules that adhere to it to the olfactory cells in the lining of Jacobson's organ. As soon as a nest has been located, *Dasypeltis* taps the eggs with its snout at either the round or the pointed end, then suddenly opens its jaws very wide, and begins to swallow it—or rather, the egg remains stationary while the snake moves forward with alternate movements of its jaws and engulfs it.

With the egg securely lodged in its esophagus, just behind its head, *Dasypeltis* sets to work on the shell. The vertebrae have a number of well-developed bony projections that protrude into the gullet in a kind of sawtooth formation, so that when *Dasypeltis* begins to undulate its spinal column, the egg is pressed against these protruding spikes, the shell cracks, and its liquid contents are propelled down the gullet and into the stomach. *Dasypeltis* immediately closes its glottis to prevent the egg from feeding back into the gullet, then spits out the shell fragments onto the ground. Not surprisingly, *Daspyletis's* peak periods of predatory activity correspond to the nesting seasons of the various birds whose eggs its feeds on in a particular locality. Though there are a great many snakes, as well as monitor lizards, that feed opportunistically on birds' eggs, the various *Dasypeltis* species, like the Gila monster of the American West, are the only creatures to do so exclusively in their respective habitats.

The predators of *Dasypeltis* include small carnivores, monitor lizards, birds of prey, and other snakes.

Certain species are similar in coloration to common night adders of the genus *Causus*; a related species, *Elachistodon westermanni*, is commonly referred to as the Indian egg-eating snake. D.H.

10
Sub.: Vertebrata
Cl.: Aves
O.: Charadriiformes
F.: Jacanidae

AFRICAN LILY-TROTTER
Actophilornis africana 10

The African lily–trotter, or African jacana, is 30 cm long and weighs about 150 g; it is found in flooded meadows and on the surface of lakes and pools and along riverbanks that are overgrown with floating vegetation. Strongly territorial during the nesting season, the lily–trotter is gregarious at other times of year and sometimes congregates in very large flocks on certain marshes. The lily–trotters (Jacanidae) are so called because their toes are enormously elongated—the toenails of the species in question may be as long as 7 cm—enabling them to walk on waterlilies and indeed to spend their entire lives on a frail, shifting platform of floating vegetation without breaking through the surface. The African lily-trotter is primarily diurnal, though it is occasionally active at night as well.

▶ The lily-trotter feeds on aquatic insects and larvae that it finds swimming on the water or over the submerged tops of water lilies; it also seeks its prey by turning over the edges of water lilies and investigating the foliage and the drifts of vegetable debris along the riverbank. It feeds on mollusks, small fish, and the seeds of aquatic plants.

◀ Several species of rails, coots, and sultanas are found in the the same habitat as the African lily-trotter, but none of these are equipped to live permanently in this particular habitat. At the first sign of danger, lily-trotter chicks jump into the water and conceal themselves underneath a lilypad while the male stands guard; their parents sometimes tuck them under their wings and carry them around with them. There are no predators that prey specifically on the African lily-trotter, and the inaccessibility of its habitat also protects it from human hunters.

The family Jacanidae comprises 7 species altogether, found throughout the tropical regions of the world: The name *jacana* is of South American origin, and perhaps the most remarkable member of the family is the long-tailed jacana *(Hydrophasianus chiurgus),* an Indo-Malaysian species noted for its plumage and its elaborate nuptial display. J.-F. T.

AFRICAN SUN-GREBE *Podica senegalensis* 11

This shy African waterbird, also called the African finfoot, is about 56 cm long; it lives along lakeshores, on the banks of watercourses lined with overhanging vegetation, and in mangrove swamps. Its body rides low in the water when it swims, and it seems equally at home in a placid or swift-moving current. It conceals its nest among the drifts of leaves and sticks that are deposited along the high-water mark on the bank—and generally conceals itself in the foliage along the bank and is rarely seen by man. The sun-grebe shares some of the anatomical characteristics of the coot and the rail but seems more like a cross between a grebe and a cormorant as far as its external appearance is concerned.

▶ The sun grebe feeds on small fish and crustaceans as well as insects, which it catches on the bank as well as in the water. It appears to have no specific predators and is the only species to exploit its particular niche; its habitat is rarely frequented and thus virtually unaffected by man.

This curious family consists of 3 separate species of pantropical distribution, each of which has been assigned to a different genus. In addition to the African sun-grebe, the South American sun-grebe *(Heliornis fulica)* and the masked sun-grebe *(Heliopsis personata)* of Southeast Asia are found in similar biotopes and are virtually identical in their habits. J.-F. T.

11
Sub.: Vertebrata
Cl.: Aves
O.: Gruiformes
F.: Heliornithidae

Ageniaspis fuscicollis praysincola 12

This tiny encyrtid wasp is quite common in olive and citrus groves all around the Mediterranean basin, since these are the exclusive habitats of its 2 parasitic hosts, *Prays oleae* and *Prays citri* (small moths of the family Yponomeutidae). The body of the adult wasp is solid black and rather chunky (relative to its overall body length of only 1 mm). Throughout most of its range, the male of this species of encyrtid is extremely rare, though it has been described by entomologists, and reproduction is primarily parthenogenetic (i.e. all the genetic material of the offspring is provided by a single, female parent).

▶ The encyrtid wasp is endophagous, which means that it devours its prey from within, a parasitic feat that it accomplishes in the following manner: The adult wasp normally deposits a single egg, secured by means of a slender stalk called a peduncle, inside the egg case of the host. This species is polyembryonic, which means in this case that the egg develops into a so-called embryonic chain, with each link in the chain subsequently differentiated into an individual larva. The larvae complete their development in isolation inside the body cavity of the moth, which finally succumbs either before or after the larva's nymphal molt but invariably after the construction of the silky cocoon that the wasp inhabits during its nymphal stage is already well underway.

Thus, the larvae come to maturity (the process is called nymphosis) while lodged in individual compartments inside the body of the host, or beneath the surface of the moth cocoon or the outer shell of the chrysalis (which exhibits a characteristically swollen appearance in consequence). The adult wasps finally emerge by breaking a hole through the body case of their host. In this species, an average of 10 larvae will develop from a single embryo, a comparatively modest rate of polyembryonic differentiation, though it sometimes happens that 2 eggs will be deposited, hence 2 embryonic chains will develop simultaneously, inside the body of a single moth.

◀ During the larval and nymphal stages, the encyrtid is virtually immune from direct predation, which appears to be the advantage of this peculiar mode of existence. It is possible that the larvae may sometimes suffer from competition with the larvae of other endophagous species

12
Sub.: Arthropoda
Cl.: Insecta
S.cl.: Hymenopteroidea
O.: Hymenoptera
S.o.: Chalcidoidea
F.: Encyrtidae

inside the body of a single moth, though no specific indications of this have so far been forthcoming.

Another subspecies, *A. fuscicollis S. Str.*, is parasitic on a related moth, *Yponomeuta malinella*, and may produce as many as 100 offspring from a single embryonic chain. The genus *Ageniaspis* includes several other species, all parasitic on moths, but the family Encyrtidae is numerous and inflicts itself on a variety of aphids, scale and cochineal insects, as well as moths.

Y.A.

13
Sub.: Vertebrata
Cl.: Amphibia
O.: Anura
F.: Bufonidae

AGUA TOAD *Bufo marinus* 13

The agua toad, or marine toad, is found in open terrain and in the vicinity of human settlements in tropical America, from southern Texas down to central Brazil; it will not attempt to penetrate a dense tropical forest, but makes its way from one clearing or settlement to the next by following roadways and the banks of rivers and streams. This species has been deliberately introduced into many other regions (the French Antilles, Australia, Hawaii, New Guinea, Taiwan, Fiji) to prey on insects that are destructive to sugar cane, though the mainland variety is the largest of these farflung geographical races, and in fact the largest toad in the world.

In French Guiana, where the agua toad is known as *crapaud boeuf*, or "ox toad," the females may attain a length of 25 cm, though the males never exceed 13 or 14 cm; one individual, which was collected by the present writer and lived out its days in the vivarium of the Jardin des Plantes in Cayenne, is undoubtedly the world recordholder at 25.5 cm. The female deposits a clutch of as many as 35,000 eggs in a pool, puddle, or other body of stagnant water; the eggs, like the agua toad itself, contains an alkaloid that is highly toxic even to man.

▶ The agua toad feeds primarily on ants, beetles, and other terrestrial arthropods, notably leaf-cutter ants of the genus *Atta*, which are frequently found in manioc patches in French Guiana. It also eats snails in certain regions and can handle quite large grasshoppers and centipedes *(Scolopendra)* of 12 cm, half as long as itself, though unlike other tropical frogs (Ranidae) and toads (Bufonidae) of comparable size, it feeds almost exclusively on invertebrates. The stomach contents of 350 agua toads collected on Guadeloupe and Martinique included the remains of only a single batrachian.

The agua toad often hunts for insects attracted by the lights of human dwellings and has been known to make devastating attacks on beehives that have not been raised off the ground; the bees make practically no attempt to sting the toad, which in any case appears to be impervious to their attacks. The agua toad catches its prey with its protractile tongue, and the predatory mechanism is triggered by the approach of a moving object of appropriate size (see COMMON TOAD for additional details, since the agua's predatory technique is similar to that of *Bufo bufo).*

Alexander (1964) mentions agua toads feeding on dog food, and the present writer was acquainted with a tame agua that lived in a yard with a cat and had become accustomed to eating from the cat's dish. In each case, the toads may have initially been attracted by the presence of flies or other insects that had settled on the food. It is interesting to note that the toad observed by the present writer engaged in a series of seemingly purposeless yawning and lateral rubbing motions (comparable to the hand motions normally made by toads while pushing a freshly caught prey animal into their mouths) before it began to eat the cat food—apparently as a kind of substitution response, like the meaningless, "nervous" pecking motions sometimes made by a rooster in the midst of an encounter with a rival, since the normal predatory mechanisms had been thwarted by the absence of animate insect prey.

◀ Juvenile agua toads may fall prey to various snakes, raptors, wading birds, and small mammals, though the large size and the well-developed parotid venom glands of the adults appear to be a formidable deterrent to predators. When approached by a larger animal, the agua toad raises itself up on all four feet, inflates its body by swallowing air, displays its venom glands, and emits a loud puffing or panting sound. As with *Bufo bufo*, the agua toad sometimes engages in intraspecific head-butting and tongue-flicking duels combined with aggressive displays of the type just mentioned.

The predatory behavior of *Bufo paracnemis* of southern Brazil and the other large toads of tropical America is probably comparable to that of *Bufo marinus*. J.L.

ALGERIAN SKINK *Eumeces algerensis* 14

14
Sub.: Vertebrata
Cl.: Reptilia
O.: Squamata
S.o.: Sauria
F.: Scincidae

This large, heavy-bodied lizard sometimes attains a length of 45 cm (about half of which is accounted for by its tail). It is yellowish-gray in color with transverse red stripes, and its longevity (in captivity) may exceed 20 years. It is found in the Maghreb, from Morocco to Tunisia and from the seacoast to the arid pre-Saharan plateau as well as in the Atlas Mountains, though rarely above an altitude of 1500 m. It prefers flat terrain more or less covered with brushwood and isolated boulders, and it frequently takes shelter underneath stumps or in hedges of prickly pear *(Opuntia)*. In more arid terrain, it digs its own burrow in the banks of a oued (dry creekbed) or between the roots of a tamarisk or other shrub.

▶ The Algerian skink is active during the day , though in summer when the sun is at its height, it retreats to its burrow, which is as cool and moist as the terrain permits. It is a marauding hunter, feeding on insects and other arthropods, which it locates by sight, and especially on snails, which it locates by their smell, then wedges the shell against a small rock or into an irregularity in the ground and cracks it with its powerful jaws. In captivity, it flourishes on a diet of newborn rodents.

◀ This species exhibits pronounced sexual dimorphism; the males are much larger than the females and are aggressively intolerant of one another, though considering the sparseness of available food resources, this appears to be an ecologically enlightened attitude on their part. The Algerian skink does not compete with smaller, faster lizards like the acanthodactyls (lacertids) or with burrowers like the sand skinks.

In ascending order of frequency, the principal predators of this species include raptors, other lizards, small mammals, and reptiles, particularly the Montpellier viper. Too ungainly in its movements to flee from predators, it is still capable of inflicting a powerful bite.

Other *Eumeces* species are found in North America, western India, and Southeast Asia. D.H.

ALLIGATOR SNAPPING TURTLE *Macroclemys temmincki* 15

15
Sub.: Vertebrata
Cl.: Reptilia
O.: Chelonia
S.o.: Cryptodira
F.: Chelydridae

This North American aquatic turtle measures more than 120 cm in length (the carapace alone measuring 50–60 cm) and typically weighs about 100 kg (maximum 180 kg). Its dark-colored irregular shell is marked with a cryptic pattern of bumps and ridges; its long tail is also studded with scaly tubercules, and its enormous head is covered with numerous ragged excrescences—a remarkable morphology that, still more remarkably, recalls that of its earliest known ancestors, which emerged during the Triassic.

The alligator snapping turtle is found in the deep waters of lakes and rivers, generally along a muddy bottom that is covered with vegetation, and it rarely leaves the water except during the breeding season. The female deposits its 20–50 spherical eggs in a nest scooped out to a depth of 35–50 cm, and the hatchlings initially measure 45–50 mm and will have attained only about a quarter of their eventual length (30 cm) by the age of 10; typical longevity is in excess of 60 years. This species is found throughout the Mississippi Basin, from Iowa down to the Gulf, ranging westward as far as Texas and eastward to Florida.

▶ During the day, the alligator snapping turtle hunts from ambush, periodically opening its mouth to display its slightly forked bright red tongue, which looks very much like a worm and stands out in especially bold relief against the dull gray backdrop of the turtle's palate. The ability of this appendage to attract curious fish has been verified in the laboratory, and when the potential victim ventures over the threshold to claim this prize, the alligator snapping turtle snaps its great hooked beak shut with truly astonishing speed, killing or maiming the prey animal and swallowing the fragments almost instantaneously.

A larger prey animal may be seized in the turtle's beak and torn to pieces with its front claws. At night, which is when this species is generally active, it hunts by more conventional means, tracking its prey by smell and feeding on fish, amphibians, turtles, snakes, mollusks, crustaceans, and essentially any aquatic creature that crosses its path.

◀ The juveniles probably fall prey to alligators on occasion, and the egg clutches are frequently dug up and devoured by raccoons. The flesh of the alligator snapping turtle is consumed locally, though this species is more frequently hunted by man in revenge for its depredations against freshwater fish stocks. The alligator snapping turtle will vigorously defend its territory against trespassers of its own kind, and when attacked, it raises itself up on all fours, darts out its head,and snaps at anyone or anything within range. The larger specimens are fully capable of removing a human finger, and its long claws can be formidable weapons as well. Alternatively, the alligator snapping turtle may attempt to make itself less appetizing to a potential predator by releasing an unpleasant musky odor from its anal scent glands or by expelling urine from its cloacal pouch.

The family Chelydridae consists of only three monospecific genera. The SNAPPING TURTLE *(Macroclemys temmincki)* of North America is smaller, less specialized in its feeding habits, and more extensive in its range than the alligator snapping turtle; the third member of the family is the big-headed turtle *(Platysternon megacephalum)*, which is found in mountain torrents in India, China, and Southeast Asia. Its carapace measures only 20 cm, and its head, as its name suggests, is disproportionately large. This species feeds on mollusks and is noted for its climbing abilities, though in other respects its behavior is more typical of other aquatic turtles than that of either of its American cousins.

R. Bo.

16
Sub.: Vertebrata
Cl.: Amphibia
O.: Urodela
F.: Salamandridae

ALPINE NEWT *Triturus alpestris* 16

The males of this species measure between 80 and 100 mm, the females between 100 and 120 mm. The range of this species is much more extensive than its name suggests, since it is found throughout much of Central and Eastern Europe (from northern and eastern France to the Carpathians and from southern Denmark down to northern Italy and central Greece) as well as in the Alps up to an altitude of 2500 m. It tends to remain in the water for a larger part of each year than other European newts and is generally found in, or in the vicinity of, a pool or small pond, slow-moving stream, or a mountain lake. Some individuals remain in the water all year round; those that remain on land for the winter bury themselves in the ground and go into hibernation.

▶ During its aquatic phase, the Alpine newt preys primarily on the larvae and pupae of chironomids and beetles and on the larvae of crane flies, cladocerans (water fleas), and ostracods; during its terrestrial phase, it preys on earthworms and small arthropods. In the water, the newt initially locates its prey with its binocular vision, then approaches fairly close to the prey animal and pauses for some time while it confirms its initial identification by olfactory means. Since it ingests its prey by aspiration, it also spends some time is making certain that the prey animal is "in the pipeline" (i.e. correctly aligned with its mouth and gullet), carefully adjusting its position either with lateral movements of its head or by pushing against the substrate with its mouth.

As with the Pipidae (cf. BOETTGER'S CLAW-FOOTED TOAD), the Alpine newt ingests its prey by abruptly depressing the floor of its mouth as it opens its jaws, thus creating a relatively strong current that, in effect, sweeps the prey animal down its gullet. The stream of water is expelled and sucked in again several times to ease the prey animal's passage through the esophagus, and in the case of a larger prey animal, the process of swallowing is assisted by movements of the tongue and by retracting the undersides of the eyeballs into the pharyngeal cavity.

The Alpine newt's predatory reflexes may be triggered by the scent of a dead prey animal; this species is also equipped with a series of lateral line detectors, similar to those of fish, which are sensitive to vibrations produced in the water by moving prey animals. In the case of the Alpine newt, these vibrations can be detected within an area that extends only a few millimeters in front of the creature's snout and about a centimeter on either side of its body. In blind newts, the predatory reflex may be evoked by vibrations alone or by direct physical contact with the prey. The prey may be initially located by means of visual, olfactory, or vibratory stimuli, but the newt will not actually approach the prey animal unless the olfactory or vibratory stimulus is confirmed by a visual sighting and it will not attempt to ingest the prey unless its identity can be confirmed by olfaction.

During the Alpine newt's terrestrial phase, its predatory reflexes are essentially triggered by visual cues provided by moving prey animals, and the olfactory receptors and lateral line detectors play a less important role. This species can be induced to attack a moving lure that reflects only 2–5 percent more or less light than the background.

Until recently it was thought that tailed amphibians (urodeles, including newts and salamanders among others) seized their prey directly in their jaws and the role of the tongue was primarily as described above, to help propel the prey animal toward the back of the mouth and down the gullet after it had already been caught. However, Severtsov's filmed records (1971) of the terrestrial feeding behavior of Asian land salamanders (*Ranodon siberius* and *Hynobius keyserlingui*) and European newts (*Triturus cristatus* and *T. vulgaris*) have made it clear that these urodeles catch prey animals on land by flicking out their tongues somewhat in the manner of the higher anurans: the upper surface of the tongue flips down and adheres to the prey animal's body (typically for an interval of 0.06–0.1 second) before the tongue is rapidly retracted (0.04–0.1 second), drawing the prey animal after it, and it is only at this point that the jaws snap shut, immobilizing the prey.

◀ In lakes or streams, trout, minnows, and chub are the chief predators of the Alpine newt. The numerous terrestrial predators that habitually feed on anurans will normally also feed on newts when circumstances permit.

Salamanders (*Salamandra salamandra* and *S. atra*) invariably feed on land and, in contrast to the anurans, locate their prey with a simple movement of their heads. Since they lap up insects and other small invertebrates with their tongues, they need not take the time to align their upper bodies with respect to the quarry, and the last-moment olfactory confirmation of the identity of the prey animal is rare. Salamanders will attack only moving prey animals, and the predatory reflex can be evoked with an inanimate lure that is moving at a rate of 2 cm/sec. J.L.

AMERICAN ALLIGATOR *Alligator mississippiensis* 17

17
Sub.: Vertebrata
Cl.: Reptilia
O.: Crocodilia
F.: Alligatoridae

This powerful but rather unassertive predator may attain a length of 6 m and sometimes lives to the age of 60. It lives in the coastal wetlands of the southeastern United States—swamps, bayous, inland creeks and mudflats—sometimes venturing out into the brackish water at the mouth of a river. The alligator's jaws are wider but perhaps not quite as strong as those of the Nile crocodile.

▶ Baby alligators live on insects and crustaceans at first, gradually making the transition to a diet of frogs and small fish, and finally a more substantial adult diet of larger fish, supplemented by birds and small mammals on occasion. Statistics compiled at commercial alligator farms suggest that during their first year of life alligators eat from 8 to 10% of their body weight every day, and the so-called feed-to-flesh ratio (ratio of the total amount of feed the animal consumes to total weight gain over a given period) is a remarkably low 2.5. By maturity, or the age of 5, these figures will have dwindled to a somewhat less impressive 1% of bodyweight consumed daily, 10 pounds of feed for every pound of increased bodyweight.

In the wild, alligators live in small groups and go out to hunt at dusk, emitting deep-bass grunts and rumbles (50–60 Hz) and slapping the surface of the water with their tails to frighten the fish. This technique is generally effective, since the fish dart out in all directions, many of them swimming straight into the gaping jaws of an alligator. When the alligators have temporarily depleted their fishing grounds, they move on to another pond or pool by night, which will be known for the duration of their tenancy as an "alligator hole."

The alligator is otherwise reluctant to make any expenditure of energy without being virtually guaranteed of a reasonable return. Thus, it only takes the trouble of catching such smaller prey—frogs, snakes, and insects—as venture within its reach, and rarely goes after large wading birds or quadrupeds. The alligator is easily intimidated by loud noises and aggressive behavior; humans are advised to shout loudly and smack the surface of the water with sticks if it is necessary to approach them for any reason—the method used by stockherders who have to drive a herd of cattle or a string of horses across an alligator-infested stream.

◀ Its smaller South American cousin, the caiman, is sometimes taken by anacondas, but the adult American alligator's only natural enemy is man. Now a protected species, it was once intensively hunted for its hide and was exterminated throughout a large portion of its former range during the early years of this century; in the early morning, when it lies sluggishly on a mudbank, digesting its evening meal, the alligator is particularly vulnerable to attack. The current vogue for alligator hide and even alligator flesh (prompted by the recent resurgence of interest in traditional Cajun dishes) is sustained largely by commercial alligator farms.

The eggs and young of the American alligator are sometimes eaten by raptors, bobcats, cougars, and other large predators. The female alligator stands guard over its nest and will not hesitate to attack any creature that approaches it. Normally, however, the alligator prefers to assume a defensive posture when approached—opening its mouth very wide and hissing menacingly. If attacked, it generally retreats, and only when cornered will it attempt an aggressive counterattack.

The Chinese alligator *(A. sinensis)* lives in brackish water at the mouth of the Yangtze River and is a much smaller (1.5 m) animal, its body length being less than that of its tail. It feeds on fish, amphibians, small mammals, and aquatic birds, and never attacks larger prey. The caiman is also smaller than the American alligator (too small to be considered dangerous to man) though otherwise similar in its appearance and habits. It has been introduced into Asia, where it is also raised for its hide on commercial farms. M.D.

AMEIVA *Ameiva ameiva* 18

18
Sub.: Vertebrata
Cl.: Reptilia
O.: Squamata
S.o.: Sauria
F.: Teiidae

This South American lizard, also known as the jungle runner, is somewhat arbitarily included with the larger members of its family—notably the meter-long teju, *Tupinambis nigropunctatus*—in the subgroup Macroteiidae. Even though *Ameiva* measures only 45 mm from snout to vent (i.e. excluding the tail) when it emerges from the egg and no more than 164 mm at maturity (135 mm in the case of the female), the lizard's tail may be twice as long as its body, though it is rarely intact in adult specimens, particularly the males. Brown at birth, the young *Ameiva* develops greenish patches on top of its head and on its foreparts, the rest of the trunk and tail remaining brown mottled with beige; this color scheme is reversed in adults of both sexes (i.e. brown head and foreparts, the rest green).

A. ameiva is common from Panama to Brazil (excluding the western slopes of the Andes). It is especially partial to warm, open spaces with sparse or low-lying vegetation, both natural and manmade—dry savannah, sand dunes, riverbanks, dirt roads and village streets, dumping grounds, logging cuts, etc. It also lives along the banks of watercourses in the Amazonian forests, particularly in the temporary clearings created by fallen trees.

▶ *Ameiva* is active only during the hottest part of the day, and then only on clear days when direct sunlight reaches the ground It spends the rest of its time in a burrow that it digs in open country, taking the precaution of covering up the entrance behind it, and even after leaving its burrow, it still requires a brief warmup period before it can begin its hunting sweep. This consists of a noisy reconnaissance of the layer of dead leaves and organic debris on the floor of the clearing, which it turns over with its muzzle and strong front claws. It will venture into the underbrush at the edge of the forest and other shady regions, pausing occasionally in a patch of sunlight when its body temperature begins to drop.

Ameiva seeks its prey with frequent transverse movements of its head, continually flicking its tongue, which is equipped with olfactory sensors. It also hunts by sight, and can spot a moving insect at a distance of several meters. It feeds on grasshoppers, spiders, caterpillars and cocoons, ants, earthworms, snails, and other small creatures that inhabit the uppermost strata of the topsoil. During the rainy season, *Ameiva* can go for days on end without leaving its burrow, though it will emerge shortly after the sunlight strikes the floor of its clearing, continuing its hunting sweeps as long as the sunlight lasts, and finally returning to its burrow with a full belly.

◀ *Ameiva* can run remarkably fast when disturbed (the younger ones with only their hindlegs touching the ground) and undoubtedly derives some benefit from the fact that few other creatures are active at all during the hottest part of the equatorial day. Exceptions include a number of small diurnal raptors that find a perch at the edge of the forest and seek their prey on the floor of the adjoining clearings, ground-dwelling snakes (notably the boa constrictor and other Colubridae), and *Ameiva*'s larger cousin, the teju, which also hunts on open ground, feeding on bird's eggs and small vertebrates. While in its burrow, *Ameiva* is almost totally immoblized by the nocturnal temperature drop, but we do not know whether any predator has learned to take advantage of this fact.

Close relatives in the same or adjoining regions of South America (but with a preference for slightly different terrain types) include *Kentropyx calcaratus* and *Cnemidophorus lemniscatus*. (Species that share the same geographic range, by the way, are said to be sympatric; those that share the same local habitat are said to be syntopic). *K. calcaratus* also hunts in small treeless patches in the forest, but its delicate forelimbs are much better adapted to climbing than rummaging through the topsoil. It catches most of its prey well above ground level and will sometimes jump entirely free of the ground to snag a flying insect.

C. lemniscatus is found along the Atlantic coast of Brazil and along the banks of the Amazon and a few of its larger tributaries. It digs its burrow in sandy soil and hunts among the tall grasses or the roots of shrubs, feeding on locusts, grasshoppers, and flying insects. Like *Ameiva*, it is quick to take possession of cleared areas opened up by man; its capacity for settling new territories is perhaps enhanced by the fact that this species is also capable of parthenogenetic reproduction. J.-P. G.

AMOEBA *Amoeba proteus* 19

19
Sub.: Protozoa
Sup.cl.: Rhizopoda
Cl.: Lobosea
Fl.: Amoebina

The amoeba is a protozoan, or one-celled organism, one of the first of its kind to be identified (1755). *Proteus*, like all amoeba species, moves from place to place and ingests its prey by extruding pseudopodia, "false feet," or long lobate extensions of cellular material; thus the amoeba's shape, if it can be said to have one, is irregular and constantly changing. *Proteus* has several of these locomotor appendages; the anterior pseudopod may be described as scalloped or slightly fringed while the rest are smooth.

Fully extended, *proteus* may be as long as 300 to 600 μm and can be distinguished from other amoeba species by the dipyramidal crystalline structure of its nucleus (i.e. in the form of two pyramids set base to base). It reproduces by fission, the life cycle of each "individual" being about 20–30 hours under optimal conditions. The amoeba can also encyst itself, taking refuge inside an envelope of mucilaginous cellular secretions when conditions are extremely unfavorable. *Proteus* is a cosmopolitan species, found in freshwater ponds, lakes, marshes, and bogs, frequently on the bottom or adhering to the surface of plants.

▶ The amoeba is an omnivore, feeding on bacteria and other microorganisms, unicellular or filamentous algae, grains of starch and other vegetable debris, other protozoans, especially cilates such as *Paramecium* and *Tetrahymena*, and even minute multicellular organisms (metazoa), notably certain free-swimming nematode worms. The well-known method by which the amoeba engulfs its prey without benefit of mouth, teeth, limbs, or other specialized structures is known as phagocytosis. When *proteus* encounters a paramecium, for example, the prey is immobilized (perhaps as a result of some chemical secretion), and additional pseudopodia begin to form on either side of it, finally merging and enclosing the paramecium in a vacuole, which is lined with a membrane made up of material from the amoeba's cell wall. The vacuole acts as a sort of temporary stomach into which enzymes are released so that the process of digestion can begin. This predation cycle is probably triggered by some sort of molecular interchange between the amoeba and its prey, since it only seems to occur when the two are actually in contact.

An amoeba has been shown to be capable of ingesting an average of 28 tetrahymenas (each measuring about 40 x 15 μm) in a single day; when food was withheld during the previous day, the amoeba attempted to compensate for this by ingesting 46 tetrahymenas. Such vigorous phagocytic activity and the formation of so many digestive vacuoles may require the amoeba to regenerate as much as 50% of its bodily surface every hour. Other laboratory studies have shown that *proteus* is omnivorous to a truly indiscriminate degree, since it can be induced to ingest latex pellets and other inert material.

Nevertheless, the quality of its diet seems to influence the duration of the cellular cycle. With *Tetrahymena* as its exclusive prey, the amoeba divides every 24 hours; with *Paramecium,* the figure is more like 50 hours, and somewhere in the neighborhood of 100 hours with flagellate protozoa of the genus *Chiromonas.* An encysted amoeba is able to go without food of any kind for a very long period.

◀ Studies of population succession in a medium congenial to the development of unicellular life forms (hay steeped in water, for instance) have shown that the initial proliferation of bacteria is succeeded by certain flagellate protozoan species, then by ciliata, then by amoebas. In nature, a succession population of amoebas would undoubtedly be reduced in short order by a predatory microfauna consisting of rotifers, crustaceans, and worms.

There are a great many varieties of free-swimming amoebas (as opposed to parasitic amoebas, Endamoebidae), most of them very difficult to distinguish from one another. Some are covered with tiny scales, others (Testacea) are encased in a mineral shell or organic capsule. Most of them, like *proteus,* are not harmful to man, and may even do him some good in their own small way by discouraging the proliferation of bacteria.

Several species of the genera *Naegleria* and *Acanthamoeba* that develop as free-swimming species in warm water (particularly in poorly maintained swimming pools) occasionally find their way into the human body, where they may be responsible for certain very serious forms of encephalitis. Of the exclusively parasitic varieties, the most notorious is *Entamoeoba histolytica,* an organisim only about 20 μm in diameter (hence, about one-fifteenth the size of *Amoeba proteus)* that lives in the human intestinal tract and may cause such disorders as amoebic dysentery and intestinal amibiasis, from which pulmonary and hepatic complications sometimes result.

J.-P. M

ANGLERFISH *Lophius piscatorius* 20

20
Sub.: Vertebrata
Cl.: Osteichthyes
S.cl.: Teleostei
O.: Lophiiformes
F.: Lophiidae

A large, ungainly fish, weighing between 5 and 10 kg though only 40 to 60 cm long, the anglerfish has an enormous, flat, wedge-shaped head and extremely sharp teeth; its body is camouflaged with an assortment of raglike excrescences resembling shreds of seaweed. The anglerfish is found in tropical and temperate waters throughout the world; in the Atlantic, it ranges from the Gulf of Guinea as far north as the Barents Sea. The anglerfish spends most of its time half-buried in the sand or mud on the bottom (down to a depth of 100 m). It can crawl very slowly along the bottom by using its sturdy pectoral fins as "walkers" but very rarely ventures out into open water. The anglerfish's extremely static mode of existence uses up little oxygen, so that its gill openings are very small and the movements of its gill slits scarcely perceptible.

▶ The anglerfish attracts its prey by vibrating the small wormlike "lure" that is attached to the first of the six spiny vanes of its dorsal fin, situated well forward on its body. The frequency and amplitude of the vibrations generally increase when the potential prey approaches, then decrease if the prey veers off or appears to be alarmed by this display. Once the prey is within striking distance, the anglerfish opens its enormous mouth and swallows it whole—

so quickly that to the unaided human eye it may seem that the prey has inexplicably disappeared from view.

The anglerfish's teeth point backward as a precaution against escape, since the anglerfish's prey is often as large as or larger than itself and still capable of putting up a vigorous resistance to being swallowed. Ingestion is a protracted process; the stomach contents of an anglerfish sometimes equals a third of its own weight. The anglerfish feeds primarily on bottom-dwellers like itself, though it willingly devours any creature that takes an interest in its lure—crabs, conger eels, sea robins, rays, dogfish; in coastal waters, the anglerfish sometimes preys on diving birds as well.

◀ We know of no predators that prey specifically on the anglerfish, which seems to be well protected by its camouflage outfit and its sedentary habits. Despite its fairly repulsive exterior, the anglerfish makes very good eating and is caught commercially with trawls (dragnets) or baited longlines in quantities of about 9000 metric tons per year. In European fishmarkets (and more recently, in North American ones as well), the anglerfish is customarily relieved of its head and skin and sold under the name of its more presentable relative, the *lotte*, or monkfish.

There are a dozen anglerfish species altogether, not counting the numerous members of the related families *Antennaridae* and *Ococephalidae*, all of which depend on camouflage and a vibrating fishing lure to entrap their prey (cf. ANTENNARIUS CHIRONECTES, below).

J.-Y. S.

Anonconotus alpinus 21

21
Sub.: Arthropoda
Cl.: Insecta
S.cl.: Orthoptera
O.: Ensifera
S.o.: Tettigonidoidea
F.: Tettigoniidae

This Alpine grasshopper is most notable for its degenerate wings, which are only 2 mm long; the male's body barely exceeds 22 mm in length, the female's 37 mm (including ovipositor). It is dark- or olive-green in color (like many species found in mountainous regions) its legs are short, and its antennae only slightly longer than its body. The male's elytra are also greatly reduced in size; it is still capable of stridulating, though the sound is very faint. This species is common in the coastal mountain ranges of Southern Europe, up to an altitude of 300 m; in July and August, the adults are most conspicuous while hopping across the stone walls that separate one meadow from the next. The larvae are similar in appearance to the adults.

▶ *Anonconotus* is an omnivore, browsing on vegetable material (avidly drinking dew and other liquids exuded by plants) and, like its tettigoniid cousin the WART-BITER *(Decticus verrucivorous)*, feeding on recently dead or disabled insects that it catches along the ground. *Anonconotus* detects its insect prey by sight and, if the prey is still mobile, *Anonconotus* is capable of at least a short predatory pounce before seizing it with its forelegs and devouring it with its powerful grinding mandibles.

◀ The predators of this species are unknown.

R.C.

Antennarius chironectes 22

22
Sub.: Vertebrata
Cl.: Osteichthyes
S.cl.: Teleostei
O.: Lophiiformes
F.: Antennariidae

Antennarius, a genus of deep-sea anglerfishes, is found throughout the Indian Ocean, including the eastern coast of South Africa. It attains a maximum length of 20 to 25 cm, and it characteristically conceals itself amid the the waving fields of wrack grass on the sea floor, in the recesses of a coral reef, or in the clefts and crevices of a submarine rock, particularly one that is overgrown with algae or with sponges. It is a solitary creature, always profoundly distressed by the proximity of another *Antennarius;* since it is ill-equipped for combat, it will attempt to scare away the intruder with an assortment of menacing transformations—puffing up its body, changing color, quivering violently—before darting forward to jostle its adversary or throw it off balance. The loser in these contests signals defeat by turning very pale and hobbling away. Long floating strings, or "rafts," of unfertilized eggs are sometimes found in aquariums, but nothing is known of *Antennarius'* (or any of the other deep-sea anglers') reproductive behavior in the wild.

▶ *Antennarius* may sometimes feed on crustaceans, annelid worms, and other bottom-dwellers, but its typical prey consists of fish that live in open water or bottom-feeding fish that occasionally come up to feed in open water. Unlike other fish, which have to move their fins in order to maintain their position in the water, *Antennarius* is able to remain motionless for long periods by grasping an outcrop of rock or coral or a strand of algae or seaweed with its weblike pectoral fins, uncannily like little clenched fists in appearance, which also enable it to "walk" in short, halting steps across the substrate.

In keeping with its overall strategy of camouflage and concealment, *Antennarius* is highly variable in coloration, though it may take the chromatophores in its skin several hours to adjust to a dramatic change in surroundings. In such cases, it often prefers to creep slowly along the bottom until it can find a more appropriate backdrop. As noted, *Antennarius* can also vary its appearance by inflating itself or by turning pale, and its body is covered with ragged excrescences that resemble shreds of seaweed or other seaborne debris. To complete its camouflage outfit, its gills have been reduced to two tiny apertures, and if necessary, it is able to breathe through only one of its gills at a time—whichever is facing away from its potential prey.

As with other deep-sea anglers, the anterior vanes of *Antennarius'* dorsal fin (which is located very far forward on the fish's body) has been modified into a complicated and highly realistic fish lure at the end of a semirigid stalk. The precise form of the lure varies greatly from one species to the next; in the case of *Antennarius chironectes,* it may resemble a clump of tiny worms or perhaps the delicate branchiate (gill) filaments of certain annelid worms (genus *Spirographa).* Normally, the stalk remains motionless, but as soon as a fish of the appropriate size appears within *Antennarius*'s visual field of 30 to 40 cm, the stalk begins to vibrate with a jerky and irregular rhythm, a very convincing simulation of a writhing clump of worms that has drifted down from the surface toward the ocean floor.

If the fish comes closer to investigate, *Antennarius* "plays" its quarry by alternately agitating the lure at the end of the stalk and allowing it to remain motionless for a short period. At each interval, the stalk bends a little bit farther forward, bringing the lure a little closer to *Antennarius*'s enormous mouth—which, like the rest of *Antennarius,* is to all intents and purposes invisible, having merged with the surface features of an adjoining rock or a chunk of coral with a luxuriant coat of seaweed.

When the fish is within effective range (about 10 cm), *Antennarius* opens its mouth and sucks in a large quantity of water—and the fish along with it—then clamps its jaws shut. The rear section of the fish frequently protrudes, since the quarry is usually presented head first due to the hydrostatic resistance of the tailfins. In such cases, the prey may be at least as large as *Antennarius* and generally begins to struggle violently, which is why it is essential for *Antennarius* to remain firmly attached to its perch. When the anterior portion of the prey has been devoured (and only then), *Antennarius* is able to relax its deathgrip in order to swallow the rest; the entire process may take as long as two hours. An anglerfish that is 20 cm long may catch 5 or 6 fish that are each 5 cm long in the course of a 24–hour period, but after it has caught a fish as long as itself, it will eat nothing else for the next 48 hours.

◀ Camouflage and concealment is *Antennarius'* best defense against predators as well, since both fight and flight are equally out of the question. In general, it can only propel itself in short hops of 20 to 30 cm in open water, but when severely threatened, it can project itself either forward or backward for a distance of about 50 cm by violently expelling water from its gills chambers or its mouth (the narrowness of the gill openings serve to accentuate the propulsive effect of this maneuver). More frequently, when a fish approaches that is too big for it to swallow, *Antennarius* simply stops moving and further retards its rate of respiration, which is already barely detectable. In spite of these precautions, it is still sometimes scooped up by indiscriminate bottom-feeders like sharks and rays, as well as other anglerfish and larger or more agile individuals of its own species.

In general, however, even when deprived of the opportunity to camouflage itself (in an aquarium, for example), *Antennarius* does not seem to be very tempting to other predators, though there may be hostile intraspecific encounters of the type described earlier; these seem to be more in the nature of territorial disputes than attempted homophagia. *Antennarius* is collected for the benefit of tropical-fish enthusiasts, but only in relatively modest numbers; the principal cause of mortality in the wild is probably the modification of its habitat by dredging and other catastrophes.

Other deep-sea anglers vary considerably in details of coloration, the arrangement of the lure, etc.; the spots of *A. moluccensis,* for example, appear to suggest an algae-encrusted rock, the stripes of *A. phymatoides,* fringes of seaweed. *Histrio histrio,* the tiny sargassum fish, which inhabits the Sargasso Sea, is virtually impossible to distinguish from a clump of sargassum weed, and its pectoral fins are very well adapted to a life of crawling and clambering through this highly resistant medium, though older individuals are sometimes found on the ocean bottom. Note that while all deep-sea anglers are sometimes collectively referred to as *anglerfish,* the term is frequently reserved for the more familiar *Lophius piscatorius* of the North Atlantic (q.v.). M.D.

ANT LION *Euroleon nostras* 23

The adult ant lion can be distinguished from a large dragonfly (libellulid), which it otherwise resembles, by its clublike antennae and its pointed wingtips. Its wingspread is 55–65 mm. The adult ant lion's body is slender and highly elongated; it has four transparent wings reinforced by a tracery of dark veins and interspersed with dots of brownish pigment. When at rest, the ant lion folds its wings obliquely over its body like the sloping sides of a tent. It remains motionless during the daylight hours, clinging to a plant stalk or a leaf; if disturbed, it will take to the air briefly and then alight again a short distance away. The males are attracted by artifical light. The lifespan of the adult varies from several days to several weeks, though the larval stage lasts for one or two years.

▶ The adult ant lion hunts at night, making its way from one plant stem to the next and feeding on small homopterans (true bugs). The larva hatches out in summer and spends its first winter in hibernation; it does not feed at all during this period, but in the spring, it digs a funnel-shaped pit in dry, sandy soil, often in a pine–forest clearing, that is protected from the rains by an overhanging rock or tree-trunk or by the eaves of a building. Head upward, it backs itself into the ground, moving at first in a wide spiral that gradually tapers to a point and tossing out the loose dirt with a flick of its flat, wedge-shaped head. The completed pit is smoothly conical, and the inward tilt of its walls corresponds to what miners call the angle of repose for that particular variety of sand or soil (i.e. the greatest angle at which the walls can be pitched without sliding down or caving in spontaneously). The diameter of the pits constructed by older larvae is about 8 cm, the depth about 5 cm.

When the larva has backed itself into the ground in this manner, the top of its head (including a cluster of 7 little eyes on each side) and its two curved pincers are still visible at the bottom of the funnel. Because of the steep angle at which the walls are pitched and the dry and friable consistency of the soil, ants and other small insects that wander into the funnel immediately begin to slide toward the bottom. If their rate of descent is too slow to suit the ant lion larva, it will try to dislodge them by hurling little pellets of chewed-up dirt. When the insect completes its irreversible slide toward the bottom of the funnel, is is seized in the ant lion's large curved mandibles.

23
Sub.: Arthropoda
Cl.: Insecta
S.cl.: Nevropteroidea
O.: Planipennia (Neuroptera)
F.: Myrmeleonidae

There is a canal-like groove that runs the length of the ant lion's mandibles, which—in conjunction with a corresponding groove in the maxillae—forms a hollow tube when the two sets of mouthparts are clamped together. The larva's mouth is still fused together, so it cannot ingest its food in the more conventional way. Instead, it uses this tube to inject its prey with a mixture of paralyzing toxins and digestive fluids and, a short time later, to ingest the liquefied tissues of the prey, then tosses the indissoluble remnants out of the pit. The gut of the ant lion larva is also not connected to its rectum, so that solid wastes continue to accumulate inside its body case and are only disposed of after the larva breaks out of its nymphal cocoon and emerges as an adult, though liquid wastes can still be eliminated through the malpighian ducts, organs analogous to the kidneys, even during the larval stage.

◀ The parasitic larva of the chalcid wasp *Hybothorax graffi* Ratz. normally feeds on the paralyzed larva or the nymph of the ant lion.

There are some 1200 ant lion species worldwide, grouped into some 300 different genera; the wingspread of the largest of them may exceed 17 cm. A great many, perhaps the majority of them, do not trap insects in the manner described—as many as 13 out of 18 of the common Western European ones, for example, including *Acanthoclisis baetica* and the one of the larger ant lions, *Palapares libelluloides*. G.B.

24
Sub.: Arthropoda
Cl.: Insecta
S.cl.: Hymenopteroidea
O.: Hymenoptera
S.o.: Chalcidoidea
F.: Aphelinidae

Aphelinus asychis 24

This species of chalcid wasp is found in hot, humid regions throughout the Mediterranean Basin as well as in India; it has recently been introduced into the Americas in the hope that it will prove to be an effective non-chemical means of controlling aphids of the families Aphidinae, Myzinae, and Macrosiphinae, which include many species that feed on alfalfa and cereal crops.

The adult is a typical chalcid wasp (sometimes referred to as a "chalcid fly" because of its very small size, about 1 mm, though it is a member of the order Hymenoptera rather than Diptera): it has a compact body, almost veinless wings, short, crooked antennae, and (for the most part) a black body and a yellow abdomen. At a congenial temperature of 20–25°C and with a plentiful supply of aphids, *Aphelinus*'s lifespan is about a month.

▶ *Aphelinus* is both a predator and a parasite; the female inserts its ovipositor into the aphid's body case and feeds on its body fluids (known as hemolymph), a process that takes 6 to 10 minutes before the paralyzed aphid is drained completely dry. On some occasions, the female contents itself with a brief exploratory incision, with no permanent consequences for the aphid. However, when the local aphid population has reached a certain density, the female *Aphelinus* introduces a single egg into the aphid's "bloodstream." The embryo develops in about 3 days; at the end of another 6 to 10 days—and with nothing left of the aphid host but the shell, or cuticle—the larva secretes a substance from its labial gland that impregnates the cuticle and turns it black (except for the area around the head, which becomes transparent).

This substance also acts as an adhesive, which secures the aphid "mummy" to a leaf or stalk; at the end of 18 days (counting from the deposition of the egg and at a temperature of 20°C), the adult *Aphelinus* cuts a ragged-edged hole in the lower part of the cuticle with its mandible and emerges from confinement. Mating may take place immediately on emergence, but the female does not lay its eggs until some time later. When conditions are favorable, the female *Aphelinus* may produce as many as 200 offspring.

◀ *Aphelinus*' principal natural enemies are so-called hyperparasites, tiny wasps (in this case), notably of the genus *Asaphes*, whose larvae feed on other parasitic larvae. The adult deposits its egg inside the aphid mummy, and the hyperparasite larva invariably devours the developing *Aphelinus* nymph (never the other way around).

A great many species of the family Aphelinidae are parasitic on hompterans (scale insects and whiteflies as well as aphids) and a certain number of these have been enlisted by man as agents of biological pest control (cf. APHYTIS, below). The blackened aphid mummies created by *Aphelinus* furnish a graphic, perhaps even a poetic image of parasitism at work; these can be readily distinguished from the cocoons of the family Aphidiides (see APHIDUS MATRICARIAE, below), since the *Aphelinus* mummy preserves the original shape of the aphid's body case whereas the *Aphidus* mummy is greatly swollen and distorted in order to accommodate the nymph in its cocoon. J.-M. R.

25
Sub.: Arthropoda
Cl.: Insecta
S.cl.: Diptera
O.: Nematocera
F.: Cecidomyiidae

Aphidoletes aphidimyza 25

This is a species of cecidomyiid, or gall midge, an elegant little fly whose entire body is only about 2 mm long. The adults' nuptial flight takes place at dawn or at dusk, and shortly afterwards the female lays about 100 eggs, either individually or in small groups, in proximity to an aphid colony. The eggs hatch out 3 or 4 days later; the activities of the larvae are described below. Prior to undergoing nymphosis, the larva drops to the ground and burrows a few cm into the soil, where it weaves a silky cocoon from which the adult cecidomyiid will emerge in 1–3 weeks. In this way, several generations of midges will live and die in the space of a single year.

▶ This species is classified as a predacious rather than parasitic gall midge, which refers exclusively to the feeding habits of the larva. The lifespan of the adult is very short, and it consumes nothing at all except perhaps a few droplets of moisture or honeydew (the sugary fluid secreted by aphids). The larva begins to feed immediately after hatching; it is known to have a taste for at least 60 different species of aphids. It literally finds the chink in their armor by inserting its hooked harpoonlike mouth parts into the intersegmentary membrane that protects the aphids' joints. The saliva of the gall midge larva contains a toxin that immobilizes the aphid as well as enzymes that will liquefy and "predigest" its tissues in the usual way. The aphid's flexible body shell deflates fairly quickly, like a collapsed balloon, but the shell remains in place, the stylets still firmly affixed to the veins in the leaf. *Aphidoleptes aphidimyza* has no known parasites or predators. R.C.

Aphidus matricariae 26

Aphidus is not an aphid *(Aphis)* but a cosmopolitan genus of parasitic wasps belonging to the ichneumon fly family. The larva is endoparasitic (a strategy that is also known as *parasitoidism,* which is to say that the living body of the host, in this case an aphid, provides both shelter and nourishment for the developing larva—and that the parasitic activities of the larva are invariably fatal to the host). The adult is an elegant little (2 mm) brown wasp with long flexible antennae. It feeds on nectar and honeydew (the sticky syrup secreted by aphids and other insects), and at a controlled laboratory temperature of 15–20° C lives only for a week or two at the most; it has been observed that *Aphidus* populations are usually about 60% female. Since *Aphidus* is destructive to many different varities of aphid, it has recently been the object of some scientific interest, especially with regard to its potential as an agent of biological pest control, though unlike *Aphelinus asychis* (see above) it has not yet been actively recruited into the service of commercial agriculture.

▶ After mating, the female *Aphidus* seeks out a colony of aphids, typically of the genus *Aphis* or *Myzus*. It deposits a tiny egg (0.1 mm) in the aphid's abdominal cavity by means of its short ovipositor. At an optimal temperature of 20°C, the larva emerges from the egg after 3 and a half days. The cells of the serosa, the membranous container in which the larva is enclosed, grow very quickly and are referred to as "giant cells"; the aphid's fatty tissues and developing embryos begin to break down in the presence of the *Aphidus* larva, the products of which are immediately resorbed by the giant cells of the serosa.

Each of the 3 initial stages of development takes a single day, and during this period, the larva takes in nutrients through this membrane by osmosis. By the fourth stage, the larva has developed mandibles and embarks on a more aggressive policy of slowly devouring its host's internal organs until at the end of 36 hours, there is generally nothing left of the aphid but the dry body case, though an aphid that is parasitized at an advanced stage of its development may be able to hold out long enough to permit the emergence of several *Aphidus* larvae.

The larva makes a ventral incision in the body case and begins to spin its cocoon, which is roughly spherical in shape and tightly covered by the aphid's body case; the cocoon inside which *Aphidus* undergoes its final transformation (nymphosis) is golden in color, like raw silk, and attached to a leaf or a stalk by a cable of silken filaments that pass through the incision in the aphid "mummy." The cocoon should be complete by the eighth or ninth day, and the imaginal molt occurs about 5 days later, whereupon the adult *Aphidus* cuts an escape hatch in the coccon with its mandibles and crawls out to spend the rest of its brief, vegetarian existence in the open air.

Clearly, it may be advantageous for a parasitic species like *Aphidus* to eliminate its host on an individual basis without

26
Sub.: Arthropoda
Cl.: Insecta
S.cl.: Hymenopteroidea
O.: Hymenoptera
Sup.f.: Ichneumonoida
F.: Aphidiidae

exterminating it as a species. A comparative study of *Aphidus matricariae* and one of its principal victims, *Myzus persicae*, revealed that under controlled conditions (in a greenhouse kept at an optimal temperature of 20°) both species were impressively prolific, with the female aphid producing an average of 60 offspring, the wasp as many as 300. The larval cycle of the aphid lasts for only 9 days as compared to 14 in the case of the wasp, but it seems clear, at least according to this study conducted in 1978, that the parasitoidal strategy of the wasp can act as an effective curb on a burgeoning aphid population that theoretically doubles every 3 days. Thus, on May 16, only 1.6% of the aphids examined were found to have been parasitized; by the end of the week, the figure had risen to 3%, with the wasp population vector rising sharply to achieve a quite respectable figure of 33% by the end of the following week (May 30).

◀ We do not know whether any predators feed habitually on the short-lived adult *Aphidus;* the cocoon seems to offer effective protection against predators, though not against hyperparasites such as *Dendrocerus* (family Ceraphronidae), *Asaphes, Pachyneuron,* and *Courone* (family Ptepromalidae), which deposit their eggs in the cocoon. The developing larvae are external parasites (ectoparasites) on *Aphidus*. The larvae of a number of cynipid wasps that deposit their eggs inside the aphid's body cavity *(Alloxysta, Phaenoglyphis)* are endoparasites that attack the *Aphidus* larva and may eventually dispose of one another as well. In the spring, the aphids emerge first, followed closely by *Aphidus* and other primary parasites, followed a short time later by the hyperparasites, which require a slightly higher temperature for their development.

All the members of the family Aphidiidae are solitary wasps that are parasitic on aphids (of all varieties except for the families Phylloxeridae and Adelgidae) during the larval stage. The adults are inconspicuous, and the cocoons provide the only visible evidence of their presence in the midst of an aphid colony. These vary in color from straw yellow to a deep maroon or even black in the case of the genus *Ephedrus* ; ichneumon flies of the genus *Praon* spin their cocoons directly beneath rather than inside the dessicated "mummy" of the aphid, creating a decorative little ensemble that looks something like an aphid's shell dislayed on a silken pedestal.

J.-M. R.

Aphytis mytilaspidis 27

27
Sub.: Arthropoda
Cl.: Insecta
S.cl.: Hymenopteroidea
O.: Hymenoptera
S.o.: Chalcidoidea
F.: Aphelinidae

Another member of the family Aphelinidae, *Aphytis* preys on diaspid scale insects, notably the oystershell scale *(Lepidosaphes ulmi),* which is generally found in fruit orchards. The adult *Aphytis* has a yellow body and is only about 1 mm long, the female slightly smaller than the male. *Aphytis* is very active during the daylight hours, constantly scurrying around in search of a parasitic host. When alarmed, it hops rather than flies away, and at night it conceals itself on the underside of a leaf.

▶ The adult *Aphytis* feeds primarily on honeydew and plant nectar, a diet that is sometimes varied (most frequently during the period just before the female deposits its eggs) with a vampire feast of hemolymph. In this latter case, the female penetrates the armored shell and the dorsal tegument of the scale insect with its ovipositor in order to extract the body fluids; alternatively, *Aphytis* may penetrate the shell but not the dorsal tegument in order to deposit its egg on the scale's dorsal surface or (exclusively in the case of the oystershell scale) among the eggs of the insect host.

The *Aphytis* larva is oval and semitransparent with segementation appearing by the time the second or third larval stage is complete. The larva progressively consumes its host's body tissues in the usual way, so that only the shell and a few other inedible body parts remain by the time the pupa has developed. When the egg is deposited among the eggs of *Lepidosaphes ulmi,* it generally succeeds in devouring most of a single clutch (the oystershell scale breeds only once a year) by the time it is about to enter the pupal stage. The dorsal surface of *Aphytis* 's all-black pupa may sometimes be seen on the underside of the shell of the scale insect; a round perforation in the latter means that an adult *Aphytis* has already gnawed its way through the shell.

◀ *Aphytis* does not appear to have any particular predators, but in spite of the beneficial role it plays as a tireless exterminator of scale insects, it may have been inadvertently eradicated in many orchards due to its extreme susceptibility to various chemical fungicides. The use of more selective chemical sprays has allowed the natural ratio of predator (or parasite) and prey populations to be restored.

There are a great many *Aphytis* species, most of which (though not all have been classified thus far) are destructive to scale insects and would undoubtedly be of use in biological pest control. *Aphytis mytilaspidis,* which is found all over Europe, has already been introduced into California from France in an effort to provide a biological control on the fig scale *(Lepdosaphes ficus)* that has turned out to be remarkably successful. *Aphytis* has become widely dispersed and has almost totally eradicated this particular scale insect species in its new habitat. C.B.

ARAPAIMA *Arapaima gigas* 28

The arapaima (also called pirarucu) lives in the Guiana Highlands and the Amazon Basin; it sometimes exceeds 4 m and 150 kg, and is one of the largest of all freshwater fish (perhaps second only to the European sheatfish, *Silurisglanis,* which frequently weighs twice as much). It is found in shallow rivers that frequently flood their banks and are rich in aquatic plants; it prefers sunlit to shady areas or water that is too acidic, where it would have difficulty finding prey, but it can live in oxygen-poor water since it uses its swimming bladder as an auxiliary organ of respiration.

The female lays its eggs on a patch of river bottom from which all plant life has been removed; the male watches over the young fry. The cephalic glands of the arapaima secrete a subtance whose exact composition is unknown but which is believed to serve as some sort of territorial marker, either to ward off intruders or to furnish a rallying point for its young brood.

▶ The arapaima feeds almost entirely on other fish, primarily cichlids, large characids, and catfish; it is certainly large enough to catch birds and small mammals but apparently does not do so, unlike other outsize predatory fish. The young arapaima eats worms, insects, and the young fry of other species, as well as one another. The greenish- gray coloration of the adult often allows it to pass unnoticed as long as it remains motionless among the waterweeds. It patrols its territory incessantly and relies primarily on speed (and on a very quick takeoff) to catch its prey, which it detects by sight or vibration and seizes in its protractile mouth.

◀ Adult arapaima are preyed on by caimans, anacondas, and man; their gray-green coloration may be of some assistance in protecting them against all 3 of these predators. Young arapiama are eaten by catfish and the larger cichlids, matamata turtles, and snakes. It was estimated in 1978 that a medium-sized Indian village might easily consume 10,000 arapaima over a 3 month period, so there is probably some danger of overfishing in populated areas; it is said, for example, that specimens measuring over 2 meters are now very difficult to come by. The dried flesh of the arapaima is used as currency by some Indian groups; the bony tongue (whence the family name, Osteoglossidae) is sometimes used as a file or grater, and the skin, scales, bones, bladder, and other organs are all exploited as well. Arapaima are caught with spears or with nets, often with the assistance of a natural or artificial dam or fish weir.

A smaller relative, the arowana *(Osteoglossum bicirrhosum),* about 60 cm in length, appears to accompany the arapaima on its hunting forays, benefiting from the panic created in shoals of fish by seizing an occasional straggler. The arowana has a flexible, laterally flattened body, two senstitive barbels on its chin, like a catfish, and a very wide mouth, angled up toward the surface, which it where it normally finds its prey; it feeds on insects and other fish, sometimes quite large ones, and is adept at concealing itself in the water weeds to avoid being devoured by its hunting companion, the arapaima.

Two other large osteoglossid fish, *Scleropages formosus* of Malaysia and *S. leichardi* of Australia (sometimes called the barramunda), are similar to the arapaima in their habits; both are currently endangered, and the Malaysian species is thought to be very close to extinction. M.D.

28
Sub.: Vertebrata
Cl.: Osteichthyes
S.cl.: Teleostei
O.: Osteoglossiformes
F.: Osteoglossidae

29
Sub.: Vertebrata
Cl.: Osteichthyes
S.cl.: Teleostei
O.: Perciformes
F.: Toxotidae

ARCHERFISH *Toxotes jaculator* 29

The archerfish generally measures between 15 and 18 cm; maximum length is 24 cm. It is found in turbid or brackish coastal waters, in coastal marshes, mangrove swamps, and occasionally in the lower reaches of coastal rivers and in freshwater estuaries. Its range extends from India and southeastern China through Malaysia and the Philippines to northern Australia. The juveniles live in schools, generally in fresh water, whereas the adults tend to be solitary and are generally found in brackish water or (exclusively during the breeding season) on coral reefs. The longevity of this species is unknown.

▶ The archerfish preys almost exclusively on low-flying insects and on surface-dwelling aquatic insects and occasionally on bottom-dwelling mollusks and crustaceans. For a fish to catch flying insects obviously requires a certain technique: The archerfish's mouth tapers into a longitudinal groove through which, when the aperture is partially blocked with the tongue, a powerful stream of water can accurately be projected for a distance of up to 1.5 m. The pressure is supplied by the contractions of the fish's gill flaps (opercula), and the force of the stream is more than sufficient to dislodge an insect from an overhanging leaf or branch, or to disable a flying insect by drenching its wings with water and sending it tumbling down to the surface.

Though its marksmanship is generally very good, it can also unleash a volley of up to 5 or 6 shots in fairly rapid succession if it fails to hit the target with its first shot. It stations itself directly beneath the target with the tip of its muzzle protruding just above the surface and its eyes still underwater (in other words, it has learned to compensate for the effects of refraction). Young archerfish begin to spit out unaimed jets of water when they are just a few centimeters long, though a fairly lengthy process of conditioning is required before they are able to shoot with sufficient force and accuracy to bring down a flying insect. The adult archerfish also uses its hydraulic weapon to dislodge mollusks and crustaceans that have buried themselves in the mud; it sometimes catches aquatic insects on the surface by pouncing on them from below, and in all of these cases, the archerfish's sense of sight is its primary means of prey location.

◀ The archerfish is undoubtedly preyed on opportunistically (rather than specifically) by a variety of other creatures (turtles, snakes, wading birds). Its defensive tactics are purely passive—schooling behavior in the case of the juveniles, the vertical stripes on its body (which encourage predators to misestimate the fish's actual size and position), and its predilection for turbid or cloudy water—and it is not known to use its archery as a deterrent against predators. The archerfish is a very popular aquarium fish, though individuals bred in captivity rarely exceed 10 cm in length.

There are 5 species of archerfish; the predatory behavior of *Toxotes chatereus* is identical to that of *T. jaculator*, as described above. J. Y.-S.

30
Sub.: Vertebrata
Cl.: Mammalia
O.: Carnivora
F.: Canidae

ARCTIC FOX *Alopex lagopus* 30

This polar species, somewhat smaller than the red fox, may attain a length of 45–65 cm and a weight of up to 9 kg, though 2.5–5 kg is closer to the average; its tail measures as much as 30 or 40 cm. Its summer coat is light brown, with distinctly lighter patches at the throat and along the belly; its winter coat varies from bluish-gray to solid white. The arctic fox is the only small mammal that lives exclusively above the Arctic Circle and does not move southward with the approach of winter; it is found in Canada, Greenland, northern Scandinavia, and the Soviet Union. It digs a burrow in clay or sandy soil and sometimes makes its den in a crevice between two boulders or a crevasse in the ice or snow.

▶ The arctic fox patrols its hunting grounds, which are always at some remove from its burrow or den, by day as well as by night, feeding on eggs, nestlings, and small mammals (lemmings) during the warmer months and frequently on carrion, especially the remains of beached seals and other marine mammals. In winter, though it is obviously incapable of bringing down a large herbivore like a reindeer or musk ox, it still follows the herds in the hope of coming upon a fresh carcass or the remains of a herd animal that has been brought down by wolves. In hunting for rodents under the snow, the arctic fox is guided by its sense of smell and hearing; it moves noiselessly over the surface, pounces on its quarry, and seizes it in its strong jaws. Frequently if the prey is small enough, it will be swallowed whole; if not, the leftovers may be buried under the snow to keep for another occasion.

◀ The harshness of the arctic winter and the scarcity of food are the chief causes of mortality, especially among the cubs. The fur of the arctic fox, particularly its winter coat, is greatly prized, and the species itself has no natural fear of man and is not a very elusive quarry; as a result, it is currently being exterminated from some parts of its range. A few individuals may be taken by wolves, but this is largely accidental, since the artic fox is by far the swifter and more agile of the two canids.

P.A.

ARCTIC FULMAR *Fulmarus glacialis* 31

31
Sub.: Vertebrata
Cl.: Aves
O.: Procellariiformes
F.: Procellariidae

The arctic fulmar measures between 42 and 46 cm, with a wingspread of 106–113 cm, and weighs between 500 g and 1000 g. This species reaches sexual maturity between the ages of 6 and 12; adult life expectancy is about 18 years, and a record longevity of 34 years has been reported in two different cases. The fulmar spends much of its life on the open ocean, coming ashore to nest in large, noisy cliffside colonies long the coasts of Greenland, Iceland, Scotland, and the Faeroes. Over the past thirty years, nesting fulmars have also been reported on a few rocky islets as far south as Normandy and Brittany.

The fulmar is an excellent swimmer and flies very low over the water, skimming over the troughs and crests of the waves with outstretched wings, occasionally regaining momentum with a few sharp wingbeats. Like many seabirds, however, it is ungainly in its movements on land and has difficulty taking off from a level surface; this is not a serious problem, however, since the fulmar generally frequents vertical rock faces and cliffsides when it comes ashore.

▶ The fulmar preys on a variety of pelagic organisms (jellyfish, crustacean larvae, cephalopods, fish hatchlings), which it skims off the surface with its beak; it also feeds on the refuse discarded by fishing boats.

◀ The osprey, the great skua, and several species of pelagic gull are the chief predators of the adult fulmar; in addition, it is estimated that 45 percent of the fulmar's eggs and 15 percent of the nestlings are devoured by gulls and crows. (The fulmar will attempt to discourage any intruder on the nesting site by spitting out a foul-smelling stomachic secretion.) Since it finds its food on the open sea, the fulmar is not troubled by any particular competitors.

In former times, the inhabitants of isolated fishing communities on the North Atlantic (notably the island of St. Kilda in the Hebrides) organized yearly expeditions to relieve the fulmars of their eggs and chicks; the flesh of the chicks was prized as a distinctively high-flavored delicacy, though the bodies of the adults were more suitable for rendering into lamp oil. This practice persisted until the 1940s along the coasts of Iceland and Greenland, when it was discovered that, in addition to eggs, squabs, and blubber, the fulmar nesting colonies were also an abundant source of infectious psittacosis.

The order Procelliformes comprises the albatrosses and shearwaters as well as a great many species of petrel and the closely related fulmars; most petrels and fulmars are relatively small birds, a notable exception being the southern giant petrel *(Macronectes giganteus)* of the Antarctic, which is about the size of an albatross; it feeds on carrion and sometimes preys on penguin colonies as well.

ARCTIC LOON *Gavia arctica* 32

32
Sub.: Vertebrata
Cl.: Aves
O.: Gaviiformes
F.: Gaviidae

The arctic loon, or black-throated diver, measures between 58 and 70 cm, with a wingspread of 113–127 cm, and weighs between 2 and 2.75 kg. It is sexually mature by the age of 2 or 3; record longevity in nature is 27 years. The arctic loon inhabits the arctic and subarctic tundras of Eurasia and North America (and less frequently, the taigas, or grasslands, further south). It nests along the shores of lakes and ponds where the water is clear and deep, and outside the breeding season, it is found in coastal waters or on the open ocean. In Europe, this species sometimes nests as far south as northern Scotland, and winters along the coasts of the Atlantic, the Baltic or (in the case of Siberian populations) along the shores of the Black Sea; in North America, it alternates between the coastal waters and the inland lakes of Alaska and northern Canada.

Loons spend up 99 percent of their lives on the water and, like many aquatic birds, are ungainly in their movements on land—so much so that the arctic loon is incapable of venturing more than about a meter from the shore. The grebes and the loons (more frequently called divers in Europe) are very similar in appearance, particularly in silhouette, as well as in their mode of life, and were often grouped together in former classifications.

▶ The arctic loon is a voracious feeder and virtually no marine or aquatic organism of suitable size (ranging from tiny invertebrates to creatures as large a duckling or a codfish) will escape its notice; European populations feed primarily on trout, carp, perch, and bleak in fresh water and on herrings, sprats, and smelts in salt water, as well as on mollusks, frogs, and crayfish and other small crustaceans. The loons are acknowledged to be second only to the penguins at swimming and diving, and though they are capable of abrupt vertical descents of up to 30 m, the loon's dive usually describes a flatter trajectory, extending downward to a depth of only 3–6 m and forward for as much as 500 m; it usually remains submerged for 3–6 minutes.

The loon reconnoiters its hunting ground by poking its head below the surface, and if it catches sight of a fish or other prey animal, it slips noiselessly into the water and sets off in pursuit. If the prey animal is too large to be swallowed whole, the loon literally shakes it to pieces by brandishing it in its beak and then dipping it back into the water from time to time. The nestlings are fed at first on small invertebrates and begin to fish for themselves when they are about 60 or 70 days old.

◀ The loon is quick and agile enough to avoid most predators simply by ducking below the surface and remaining submerged for several minutes, though a sitting bird will use its powerful notched beak to defend its nest against such predators as the skunk, the fox, and even (in exceptional cases) the polar bear. While in its winter quarters, the loon is very wary of humans.

A nesting pair will defend a territory of 50–150 hectares of open water against interlopers of their own species, using vocal and gestural signals to coordinate their movements. Outside the breeding season, and directly before migrations in particular, artic loons will assemble in substantial flocks of 20–40 individuals, thus exploiting the food resources of the habitat with maximal efficiency (and shutting out competitors of other species). The chief competitor of the arctic loon, at least in certain regions, is the red-throated loon or diver *(Gavia stellata)*, though this species is generally found on smaller bodies of water and has found an alternate solution to the problem of competition by ranging much farther from its nesting area in search of its food.

The genus *Gavia* consists of only four species, which are remarkably homogeneous in behavior and appearance: the common loon, or great northern diver *(Gavia immer)*, the yellow-billed loon, or white-billed diver *(G. adamsii)*, both found in the North American and Eurasian Arctic and sub-Arctic regions, as well as the arctic loon and the red- throated loon mentioned above. J.-F. T.

Argyrodes gibbosus 33

33
Sub.: Arthropoda
Cl.: Arachnida
O.: Araneidea
F.: Theridiidae

Agyrodes gibbosus is the Southern European representative of a genus of semitropical spiders. It has a triangular yellow or silvery abdomen and is only a couple of millimeters long; it spins its web, which is also very small, in the immediate outskirts of the web of a larger araneidan, frequently *Cyrotophora citricola*, which, thanks to its inconspicuous size and coloration, it is able to do with some impunity. The female lays its eggs in the middle of summer and wraps them in a silken cocoon that is suspended by a single strand from the edge of the web. The vasiform (vaselike) shape of the cocoon is characteristic of this species. The young spiders hatch during the autumn and immediately set off on their own.

▶ *Argyrodes* feeds on small insects that it catches in its own web as well as those that are rejected as too small by its larger neighbor (an instance of commensalism). *Argyrodes* is able to clamber around freely on the web of the host spider without setting off vibrations of sufficient magnitude to announce itself to the latter as a potential prey animal.

◀ Specimens of several*Agyrodes* species have been found in the nests of hunting wasps, apparently destined as food for the wasp larvae.

J.-F. C.

ARROW WORM *Sagitta setosa* 34

34
Sub.: Chaetognatha
Cl.: Sagittoridea
O.: Aphragmophora
S.o.: Ctenodontina
F.: Sagittidae

The arrow worm is a typical representative of the phylum Chaetognatha, which takes its name from the clusters of hook-shaped filaments that are found on either side of the creature's head (chaetos = 'silk,' gnathos = 'jaw'). The chaetognath's body is transparent, elongated, and divided into three separate segments (head, trunk, and tail) by transverse septa. It has either one or two pairs of lateral fins and a single pair of caudal fins to propel itself through the water (though no other appendages), and the digestive tube extends all the way to the posterior end of the trunk.

Sagitta measures about 1 cm in length and weighs 2 mg. This species is an important constituent of the marine plankton in the Eastern Atlantic (including the North Sea, Irish Sea, and the Channel, as well as the French coast down to the mouth of the Loire), the Western Mediterranean, the Adriatic, Black Sea, and elsewhere. It is not found on the open sea, though it is very abundant in coastal waters, including inlets and estuaries, and can tolerate a very low level of salinity. During the day, arrow worms generally congregate at a depth of 5–40 m, with the adults forming the upper layer, the juvenile and immature individuals on the bottom, though this entire population moves up to the surface to feed at night. Maximum longevity varies seasonally from several weeks to up to 3 months.

▶ *Sagitta setosa* feeds on small and medium-sized planktonic organisms, primarily on copepods *(Acartia, Centropages, Clausocalanus . . .)*, which make up the greater part of the marine plankton layer in these waters, and to a lesser extent on cladocerans *(Evadne, Podon)*, tunicates *(Oikopleura)*, and other chaetognaths. The adults are sometimes capable of ingesting large prey animals, including the larvae of decapod crustaceans and fish; the juveniles feed primarily on the nauplia (larvae) of copepods and Tintinnidae *(Stenosemella)*.

Sagitta does not actively seek its prey; it maintains a vertical position in the water, head upward, and bobs up and down slowly, moving through a distance of several tens of centimeters. The entire surface of its body is liberally covered with long rigid filaments that serve as tactile receptors and inform it of the location and direction of any small prey animal in its immediate vicinity. As the result of a slight (often <1 cm) but very rapid course adjustment, *Sagitta* ensnares the prey animal in one of the fanlike arrays of hook-shaped prehensive filaments located on either side of its head; these filaments are made of sclerotized protein and provided with numerous tiny teeth, which act to restrain the prey until it can be absorbed into *Sagitta's* mouth.

Sagitta setosa feeds very little during its daily period of submersion, and almost all of the biochemical processes involved with the reproductive cycle of this hermaphroditic organism are closely synchronized with its intensive nightly feeding sesssions. The final phase of sperm production and the exchange of gametes (mating) both occur in the early evening, whereas the peak periods in the growth of the ovocytes and the vitelline sac that nourishes the embryos tend to continue throughout the night. In the laboratory, *Sagitta selosa* has shown itself to be capable of going without food entirely for up to a week, though after its initial reserves are exhausted, this quickly results in the atrophy of the reproductive organs and an alarming decrease in body length.

◀ Arrow worms are preyed on by a variety of larger planktonic organisms, including ctenarians *(Pleurobranchia)*, scyphozoans *(Liriope, Chrysaora)*, siphonophores *(Apolemia)*, and polychaetes *(Tomopteris)*. Cannibalism is fairly prevalent among laboratory specimens, considerably less so in nature, unless prey animals are in unusually short supply.

The almost perfect transparency of its body is *Sagitta's* best defense against predators that locate their prey visually, though this does not always protect it from certain shrimps *(Pasiphea, Sergetes)* and euphausids, as well as herrings, shad, and anchovies. *Sagitta's* flight response is not very effective, since it merely involves a succession of short hops of rapidly diminishing length and velocity. The adaptive value of *Sagitta*'s daily migrations into deeper water is all the more apparent when we take note of the fact that partially digested food in the digestive tube may be visible as a sort of opalescent blur, and the mature eggs are opaque and readily visible, so that Sagitta loses its best defense against surface predators that hunt by sight during this culminating phase of its digestive and reproductive cycles.

Fewer than a hundred chaetognath species have been classified thus far, and they are all quite similar in their morphology. All of them are carnivorous, and though a few spend all or some portion of their lives on the bottom, the vast majority are found in the plankton layer, thus on or fairly near the surface, either in coastal waters *(like Sagitta)* or on the open ocean. Most of these species feed on copepods for the most part, though some are more specialized, such as *Khronitta,* which feeds on appendicularians (a variety of tunicate), and *S. hexoptera,* which seems best adapted to preying on other chaetognaths. Chaetognaths are found from the tropics to the polar seas, and thus, simply by virtue of their ubiquity and abundance, play an important part in stabilizing the population levels of planktonic crustaceans throughout the oceans of the world S.D.

35
Sub.: Vertebrata
Cl.: Reptilia
O.: Squamata
S.o.: Ophidia
F.: Viperidae

ASP VIPER *Vipera aspis* 35

The asp viper, or European asp, measures up to 70 cm in length and is readily recognizable by its turned-up snout; though there may be considerable variation in color (beige, gray, brown, orange, brick-red), it generally has a dark zigzag stripe running down the back, and some populations found in mountainous regions are entirely black. The asp viper ranges throughout southern France, southwestern Belgium, western Germany, Switzerland, and Italy, and has been reported at an altitude of 2500 m in the Pyrenees. It is generally found in brushwood or on stony ground, along wooded slopes, in uncultivated fields, and under hedges.

Between about October 15 and March 15 (depending somewhat on the region, the altitude, and the sex of the individual) the asp viper remains buried at a depth of several centimeters in a burrow or underground tunnel or tucked away inside a rotten stump or a stone wall. It will not reemerge until its body temperature reaches an optimal level of 26–32°, and it continues to regulate its temperature throughout its active season by alternately sunning itself and then retiring into the shade of a patch of vegetation. Toward the end of August or the beginning of September, the females give birth to 6–12 live young, which typically measure about 20 cm.

▶ The venom of the newborn vipers is already potent enough to dispatch a baby mouse, and they begin to hunt for themselves either shortly after birth or at the end of their first winter (which only about 50 percent will survive). The asp viper's dentition is of the solenoglyphic type, which means that its venom is injected through a channel in the center of its two hollow fangs rather than through a groove in the outer surface. The fangs are ankylosed to the jawbone, and when the asp viper opens its mouth extremely wide (almost at an angle of 180°), the fangs are tilted forward and the venom is injected under pressure into the body of the prey animal.

The asp viper's venom contains several different enzymes that break down protein, so that its effects are primarily lytic (tissue-destroying), though there also a mild neurotoxic effect as well. Apart from immobilizing and eventually killing the prey animal, the venom also serves as a kind of concentrated "super saliva," so that the process of digestion effectively begins even before the viper has swallowed its victim. Digestion also proceeds relatively quickly, so that the asp viper can digest a prey animal weighing about a third as much as itself in 4 or 5 days, excluding parts such as hair and keratinized scales.

The asp viper conceals itself in ambush, then lunges out and sinks its fangs into the prey animal. It pulls them out almost immediately, and the prey animal staggers off to die a short distance away. The viper retrieves its quarry by olfaction and begins to swallow it, generally starting with the head, though a very small prey animal may be swallowed immediately after it has been bitten. Juvenile and adult rodents (voles, vole rats, mice) and shrews comprise about 95 percent of the diet of the asp viper, with lizards and snails furnishing an additional 1 or 2 percent.

The diet of the young vipers includes a fairly high proportion of small lizards as well as a smaller proportion of newborn mice and other small mammals. By the time it is 1 or 2 years old, the asp viper will have begun to concentrate on adult mammals, which, being more mobile and more conspicuous in their movements, are undoubtedly easier for it to track down. In captivity, asp vipers can be maintained on their natural diet, though they will accept prey animals that have already been killed (including those that have been frozen and thawed). The adults, and especially the juveniles, can also subsist on the internal organs of fish, cattle, and horses. ◀ The chief predators of the asp viper include the short-toed eagle, buzzards and harriers, corvids, pheasants, and domestic fowl; wild boar and free-ranging hogs also feed occasionally on this species, and the European hedgehog is noted for its ability to defeat the viper in single combat by ramming it with its spines. The flesh of the asp viper is never eaten by humans, but just as whiskey kegs in the American West were reported to have been flavored with rattlesnake heads, the body of an asp viper is still believed to impart similar virtues to the family moonshine jug in certain rural areas of Europe.

When threatened, the asp viper will first attempt to flee, and if thwarted in this attempt, it curves back the forepart of its body into the shape of an S and begins to hiss noisily. It continues to follow the movements of the aggressor with its eyes, and strike very quickly and without warning. If the asp viper is approached from behind and pinned to the ground with a forked stick or some similar object, the viper will immediately turn its head and strike at the stick.

A number of distinct *Vipera* species have populated Europe since the middle of the Tertiary. The adder *(V. berus)* is more northerly in its range than the asp viper, and apart from the HORNED VIPER, the LEVANTINE VIPER, and ORSINI'S VIPER, the other more common European and Near Eastern species include Lataste's viper *(V. latasti)*, found in the Iberian peninsula and North Africa, the sand viper *(V. ammodytes)* of the northeastern Mediterranean region, and the mountain viper *(V. xanthina)* of eastern Anatolia.

J.C.

ASSASSIN BUG *Rhodnius prolixus* 36

36
Sub.: Arthropoda
Cl.: Insecta
O.: Hemiptera
F.: Reduviidae

This species of assassin bug (family Reduviidae) is an intermediate host of the protozoan *Trypanosoma cruzi,* the organism that causes a debilitating and occasionally fatal illness in humans known as Chagas's disease, named for C. Chagas of the Oswaldo Cruz Institute in Rio de Janeiro, who first discovered this relationship in the early years of this century. It has been suggested that Charles Darwin was a lifelong sufferer from Chagas's disease, which he is thought to have contracted in Brazil while serving as naturalist aboard *H.M.S. Beagle* (1831–36).

Rhodnius prolixus ranges from Mexico to Brazil, and is the most common vector of Chagas's disease in Colombia, Venezuela, and Central Africa; the disease is primarily spread by a related species, the conenose *Triatoma megista,* in Brazil. Both the adult and the larva can be distinguished by its long cylindrical head and elongated body; the beak consists of three jointed segments and is kept tucked back beneath the head when not in use.

▶ *Rhodnius prolixus* remains completely immobile by day and actively seeks its prey by night, either by flying or by crawling along the ground. In the newly settled backlands of Brazil, Venezuela, Colombia, and the Guianas, where *R. prolixus* is particularly common, it is often found in settler's shacks, thatched huts, and similar structures of this kind. It is suspected that this species originally fed on other insects, as it continues to do occasionally, before its territory was invaded by warm-blooded vertebrates, though it has been recovered from the burrows of anteaters as well; it is also known to feed on the blood of opossums and armadillos.

The sensory hairs on the antennae of *R. prolixus* can detect not only only the scent of humans and domestic animals but also the heat given off by them as well. *R. prolixus*'s beak is tipped with 4 sharp pipettes (stylets), with which it pierces the skin and ingests the blood of its victims. Its bite is almost painless, and the host rarely awakens while *R. prolixus* gorges itself on its blood for as long as half an hour at a time, consuming up to 10–12 times its own weight. It takes several days for *R. prolixus* to absorb its blood meal into its system.

◀ It is suspected that the eggs of *R. prolixus* are sometimes parasitized by wasp larva of the family Scelionidae, and there are several species of ants, notably the Argentine ant *(Iridomyrex humilis),* that feed on both the larvae and the adults. Significant effort has also been made to exterminate this species from regions inhabited by humans or their livestock. R.C.

37
Sub.: Echinodermata
Cl.: Asteroidae
O.: Phanerozonia
F.: Astropectinidae

Astropecten irregularis 37

Together with its several subspecies, this starfish species is commonly found in the North Sea and in the Eastern Atlantic, from Morocco to the Lofoten Islands, as well as in the Mediterranean. It lives primarily in coastal waters out to a depth of 30 m, but is sometimes found beyond the continental shelf, even at a depth of 1000 m, and generally in areas where the bottom is covered with fine sand—in which *Astropecten* likes to bury itself—or with organic detritus or with beds of wrack grass or posidonia. Reproduction takes place in the spring, and *Astropecten*'s arms begin to develop in the free-swimming planktonic (bipinnaria) stage. *Astropecten* is a primitive starfish with several anatomical peculiarities: its tube feet lack suction cups, for example, and its digestive tube lacks an anus.

▶ All the members of the genus *Astropecten* are carnivores that swallow their prey whole and expel the indigestible portions afterwards; they feed on gastropod and pelycopod mollusks (notably *Spisula* and *Natica),* which are very numerous in the offshore sands, as well as on sipunculid and polychaete worms, crustaceans (amphipods, isopods, cumaceans), and other echinoderms (especially irregular sea urchins and brittle stars). *Astropecten* is a relatively dainty feeder, and very little sediment is found among its stomach contents. It detects its prey by means of chemoreceptors at the tips of its arms, and when confronted with a choice among several mollusks of different metabolic intensities, it will select the mollusk with the rapidest metabolism (i.e. that will probably be the first to open its shell spontaneously). In the fine sand layer on the bottom, it feeds primarily on benthic larvae and generally seeks for larger prey in the deeper layers of the substrate.

It does not fully extrude its stomach, like other starfish, to digest its prey extraorally, but it will partially extrude its stomach to scoop up smaller marine organisms like a shovel; in the case of larger prey, the ambulacral feet help to propel the prey animal toward the starfish's mouth, which is highly extensible. The expulsion of mollusk shells and other indigestible debris also requires a modest extrusion of the stomach lobes. In spite of being a voracious feeder, *Astropecten* is capable of going as long as 14–18 months without food—or rather without live food, since starfish are continually absorbing minute organic particles from the water.

◀ We have little information on this subject, but it seems likely that *Astropecten* larvae are sometimes devoured by flatfish, the starfish's principal competitor in the fine sand layer, as soon as they descend to the bottom, possibly by other starfish as well. Attacks on *Astropecten* by large crabs and lobsters have also been reported.

Most of the numerous *Astropecten* species are found in tropical waters; *A. irregularis* and *A. auranciacus* are the most common off the coasts of Europe.

A.G.

38
Sub.: Mollusca
Cl.: Gasteropoda
O.: Mesogastropoda
Sup.f.: Atlantoidea (Heteropoda)

Atlanta lesueuri 38

The superfamily Heteropoda consists of a group of sea snails that have adapted to a free-swimming existence on the surface of the ocean. The muscular foot has been transformed into a powerful fin, and the eyes are very well developed. *Atlanta* has a flat, transparent spiral shell; it lives in the plankton layer (0 to 100 m below the surface), typically measures 1 to 3 mm, and is found all over the world, though nowhere in any great abundance. The female lays a string of eggs; the planktonic larva lives on plankton and propels itself through the water by means a ciliated "sail."

▶ *Atlanta* feeds on other pelagic mollusks and their larvae, including members of its own species. It locates its prey by sight and propels it toward its mouth by moving its fin back and forth; then it seizes the prey with the suction pad attached to the fin and devours it with the help of a kind of sawblade of chitonous material called the radula, which is located at the entrance to its mouth and is equipped with several rows of teeth.

Members of 2 other heteropod genera attain quite respectable sizes: 26 cm in the case of *Pterotrachea*, sometimes called the "sea elephant," 46 cm in the case of *Carinaria*. *Carinaria* has a bulky, gelatinous body; the shell of *Pterotrachea* has entirely disappeared. Both are found in all the world's oceans and consume large quantities of other planktonic organisms. Fish are attracted by the heteropod's conspicuous black eyes and dark visceral mass (which comprises the digestive and ovarian tubes) and feed on them avidly. J.-P. T.

ATLANTIC COD *Gadus morrhua* 39

This sizable marine fish generally attains a length of 110 cm and weighs 15 kg, with a possible maximum 150 cm and 40 kg. It is abundant in many regions of the North Atlantic, invariably above 45° N lat. and generally in cold (2–10° C) coastal waters at a depth of 500–600 m. Several geographic races can be distinguished by their habitats and modes of life, a few of which—notably those living in certain lochs and fjords—are sedentary, though most carry out substantial migrations between their feeding and spawning grounds. The eggs are laid in open water, float upward, and are caught by the currents on the surface; the hatchlings do not return to deeper water until they are about 5–6 months old. The adults generally stay close to the bottom, though some choose to remain in open water, especially certain very old individuals that cover vast distances in pursuit of herring shoals and other schooling fish.

▶ The juveniles prey on planktonic crustaceans at first, then move on to larger prey as they themselves grow larger—polychaetes, mollusks, and echinoderms as well as herrings, capelins, sand launces (ammodytes), and other small fish. Bottom-dwelling species such as the grenadier (q.v.) are often recovered from the stomachs of the larger adults, and the migratory races of the Atlantic cod assemble in large schools to follow the shoals of herring and capelin, of which they consume substantial quantities. The sensory organs along the cod's lateral line detect vibrations in the water created by other fish. A school of cod hunts cooperatively, first encircling a herring shoal and then moving in simultaneously for the attack. The Atlantic cod tends to eat considerably less during the spring spawning season that at other times of the year.

◀ The natural enemies of the Atlantic cod include sperm whales, killer whales, sharks, and seals *(Phoca)*. Schooling behavior makes it more difficult for a predator to target a particular individual, which might well serve as a deterrent to all the predators just mentioned except the sperm whale. The Atlantic cod's normal cruising speed is about 3–5 kph, but it is capable of attaining speeds up to 9 kph and probably does so while actively pursuing prey or when being pursued by an enemy.

The Atlantic cod also supports a very important fishery, with an annual catch estimated at 1.5 million metric tons. The fish are caught with dragnets, purse seines, and other types of nets, primarily in the Barents Sea and off the coasts of Iceland, Greenland, and Newfoundland; nonmigratory coastal populations are fished with stationary fish weirs and fish traps as well as with hook and line. In the North Sea, conservation measures prohibit the taking of fish less than 30 cm long. Cod is sold fresh, dried and salted (bacalao), and in frozen filets; the scraps are processed into fish meal, and of course a celebrated home remedy and nutritional supplement is extracted from the liver of the Atlantic cod.

The Greenland cod *(Gadus ogac)* is about

39
Sub.: Vertebrata
Cl.: Osteichthyes
S.cl.: Teleostei
O.: Gadiformes
F.: Gadidae

50 cm long; it has grown rarer in recent years and seems to be giving way to the Atlantic cod in the coastal waters of Greenland. The polar cod *(Boreogadus saida)* is only about 40 cm long; it lives just below the surface, in the region of floating pack ice near the edges of the polar icecap, and feeds primarily on plankton. Its predators include the Atlantic cod, seals, toothed whales (Odontoceti), and even seabirds on occasion. J.-F. M.

40
Sub.: Vertebrata
Cl.: Osteichthyes
S.cl.: Teleostei
O.: Pleuronectiformes
F.: Pleuronectidae

ATLANTIC HALIBUT *Hippoglossus hippoglossus* 40

The Atlantic halibut is the largest species of flatfish and one of the largest deep-sea bony fishes. The record specimen measured 4.7 m and weighed 330 kg, but on the average, the males are between 1.5 and 1.8 m long and weigh about 50 kg; the females are larger (2.2–3 m) and quite a bit heavier (100–150 kg). The males are sexually mature at the age of 10–14, by which time they will have attained a length of about 1 m and a weight of 15–20 kg; the females mature at 10–11, by which time they will be slightly longer (1.2 m) and about twice as heavy (30 kg), this latter proportion remaining roughly constant for all age groups. Longevity exceeds 30 years (even 100 years, according to some authorities). The several halibut species are found in cold water, normally above 45° N, throughout the Northern Hemisphere. Young halibut are found at a depth of 35–70 m, though the adults may descend to a depth of 1000 m. The halibut is a bottom-dweller, though, unlike the other flatfishes (Pleuronectiformes), it normally swims in a vertical position and thus raises itself above the level of the bottom.

▶ A voracious carnivore, the halibut hunts by sight and by detecting vibrations in the water; it feeds on many different species of fish (Gadidae, Clupidae, Pleuronectiformes) as well as cephalopods (octopus as well as squid) and, during the summer, on the larger crustaceans. It spends the winter in the depths of the ocean, where it feeds on abyssal shrimps and other creatures, whose position is detected either indirectly (by means of vibrations in the water) or directly, with the help of the dim light from the bioluminescent organs of certain abyssal fish and other creatures.

◀ The adult halibut's size and vertical swimming position—its bulbous eyes are mounted on top of its head, and it can easily descend into the depths if attack is threatened from above—serve as sufficient deterrent to most deep- sea predators, though the young are undoubtedly preyed on by codfish and their numerous relatives (Gadidae) as well as by other flatfish. After the introduction of trawler fishing, the North Atlantic halibut grounds were seriously depleted during the first half of this century; the annual catch had declined by as much as two thirds during the period 1914–27 and is currently about 2000 metric tons, shared primarily between Canada and Norway.

Norway is the only country that still maintains a large-scale commercial halibut fishery during certain seasons in which especially stringent conservation measures continue to be in force; no fish smaller than 50 cm, for example, may be legally taken at any time. Canada and the United States have had some success in reviving the Pacific fishing grounds by delimiting a number of restricted zones which are only open to commercial fishing during very brief periods and in which the size of the catch is strictly limited. Various attempts to raise halibut on experimental fish farms are currently underway, but none has proved successful so far. Halibut are sold fresh, frozen, and dried, the latter form known commercially as räkling.

The Pacific halibut, only slightly smaller than *H. hippoglossus,* is *H. stenoleptis.* Two other species are also found in the North Atlantic, the Greenland halibut *(Reinhardtius hippoglossus),* which never exceeds 1.2 m in length and weighs about 45 kg, and the long rough dab *(H. platessoides),* which is only about 50 cm long. J.-M. R.

41
Sub.: Vertebrata
Cl.: Osteichthyes
S.cl.: Teleostei
O.: Perciformes
F.: Scombridae

ATLANTIC MACKEREL *Scomber scombrus* 41

This species is found on both sides of the Atlantic as well as in the Mediterranean, the Black Sea, the North Sea, the western Baltic, and (in years when the water is particularly warm) in the White Sea as far north as Murmansk. Populations that actually live in the Atlantic grow larger on the average (35–40 cm by the age of 5–9 years; specimens measuring up to 50 cm and weighing 500 g have been reported) than those found in the Mediterranean (30 cm). The Atlantic mackerel is sexually mature at the age of 3 or 4 (by which time it will have already attained a length of 28–30 cm); average life expectancy is on the order of 10 years. This species lives and feeds on the surface from the spring through the fall, then descends into deeper water, sometimes as deep as 250 m.

▶ Mackerel hunt in shoals, making a broad meandering sweep through the water while each individual fish keeps its mouth open wide, straining out zooplankton and other marine organisms with its gill rakers. The mackerel shoals begin to feed on zooplankton as soon as they return to the surface in the spring; after spawning, which occurs in May or June, depending on the region and the temperature, small shoals of mackerel begin to hunt more actively for larger prey, including sand launces, sprats, and other smaller members of the herring family. The pelagic juveniles feed initially on zooplankton. In September, the mackerel make their way down to the depths again and generally do not feed at all throughout the winter. The mackerel is an excellent streamlined swimmer, capable of attaining speeds up up to 11 kph at least for short bursts, though it swims at a speed of about 3 kph while it is searching for food and only at about 1 kph during its seasonal migrations.

◀ The mackerel's potential predators include the tuna, dolphin, tope shark *(Galeorhinus galeus)*, gilthead *(Sparus auratus)*, and other large pelagic fish, as well as the dogfish, which is found in deeper waters. Flight is its best defense, and the density of the mackerel shoal at least assures the survival of the group if not of any particular individual. The mackerel has no swim bladder, which makes it possible for it to dive or to ascend toward the surface very quickly. Its coloration—greenish back and shiny white belly—provide it with a measure of protection against avian predators (herring gulls and mews) as well as pelagic predators that might be lurking in deeper water. Man, of course, is a major predator of the mackerel, which is caught in dragnets, purse seines, and drift nets; the European trade in fresh, smoked, and canned mackerel alone accounts for about 150,000 metric tons.

The Spanish mackerel *(Scomber maculatus)* is also found in European waters; its habits are essentially identical to those of the Atlantic mackerel, though it is less abundant and rarely ventures to the north of the English Channel. Two other species, *S. concolor* and *S. commersoni*, are also found off the Pacific coast of North America.

J.-M. R.

ATLANTIC PUFFIN *Fratercula arctica* 42

42
Sub.: Vertebrata
Cl.: Aves
O.: Alciformes
F.: Alcidae

The Atlantic puffin is 28–31 cm long, with a wingspread of 57–60 cm, and weighs between 300 and 450 g. Average longevity is probably about 10 years, and the puffin does not become sexually mature until the age of 4 or 5. Like the razor-billed auk, the puffin is a gregarious bird, roaming the waters of the North Atlantic in dense flocks and coming ashore only to mate and rear its young; it comports itself more skilfully and gracefully on land than the other members of the auk family (Alcidae). It establishes its nesting colonies on seacliffs and small offshore islands, where a puffin pair digs a burrow, to a depth of nearly a meter, to receive their single egg. The puffins of the North Atlantic nest from Spitzbergen down to Newfoundland and Brittany; the yearly migration of the puffins of the North Sea to the coast of Portugal and the Mediterranean remains somewhat mysterious. This species is thought to number at least 10 million mating pairs altogether.

▶ The puffin feeds primarily on small fish, sprats and sand launces, and less frequently on crustaceans, marine worms, and small mollusks. It pokes it head under the water to locate its prey, then dives down after it; it may bring as many as 20 fish back to its chick in a single trip, all neatly centered in its brightly colored beak. The puffin chick eats as many as 30 or 40 fish and, unlike the young razor-billed auk, which learns to fish from its parents, the puffin chick does not venture out to sea for the first time until its parents have already abandoned the colony.

◀ The Atlantic puffin may sometimes have to dispute the ownership of its burrow with a wild rabbit, but it gets along quite amicably, and often shares the same fishing grounds, with other members of the auk family. In some areas, the puffin may be preyed on by the gyrfalcon or the peregrine.

The history of the puffin colonies on Sept-Iles, off the coast of Brittany, may serve to illustrate the changing fortunes of the species during the twentieth century. They presented an easy target to market gunners from the mainland prior to 1920, when the colonies were first protected by law. By 1950, the puffin population had risen to 7000 pairs, and then precipitously declined as a result of catastrophic oilspills and the pervasive hydrocarbon pollution of the Breton coastline during the 1970s; fewer than 400 pairs remain—fewer than the threshold population required for these colonies to maintain themselves. Puffins are still caught for food by fishermen in the North Atlantic; the adults are taken in nets and the chicks dug out of their burrows by the Icelanders and the Faeroese.

Two true puffins and several related species are found along the shores of the North Pacific, ranging between Kamchatka and Alaska: the horned puffin *(Fratercula corniculata)*, the tufted puffin *(Lunda cirrhata)*, the rhinoceros auklet *(Cerorhinca monocerata)*, the crested auklet *(Aethia cristatella)*, whiskered auklet *(A. pygmaea)*, least auklet, *(A. pusilla)*, and parakeet auklet *(Cyclorrynchus psittacula)*. The first 3 of these species feed on fish, the last 4, which are considerably smaller, on marine invertebrates. J.-F. T.

43
Sub.: Vertebrata
Cl.: Osteichthyes
S. cl.: Teleostei
O.: Salmoniformes
F.: Salmonidae

ATLANTIC SALMON *Salmo salar* 43

This species typically measures between 50 and 100 cm and weighs between 2 and 20 kg; maximum size is 150 cm and 50 kg in the case of the males and 120 cm and 20 kg in the case of females. It is found in the coastal waters of the North Atlantic, from New England to Labrador and from the Bay of Biscay to the Baltic, as well as in coastal rivers and streams of Europe (from Spain to Iceland, including Ireland) and North America.

The young salmon, or parrs, are born in the cold, oxygen-rich waters of coastal rivers and streams, where they remain for 1–5 years (depending on the region); now called smolts, they begin to migrate downstream and out into the open ocean. The markings of the young salmon change as they grow older; the parrs can be distinguished by the lighter strips along their sides, the smolts by their silvery backs. The young adult salmon, or grilses, live as pelagic predators for 3 or 4 years before embarking on the long migrations that take them back to the streams and rivers in which they were born. Most Atlantic salmon die shortly after spawning, so that the typical lifespan is about 6–8 years, though aquarium specimens have been known to survive to the age of 13. There are relict freshwater populations in several inland lakes (which presumably once had an outlet to the sea) in both Europe and North America.

▶ In fresh water, the parrs feed initially on fish hatchlings and small invertebrates—including gammarid crustaceans and insect larvae, including caddis fly, stone fly, dragonfly (libellulid), and mosquito larvae—and later on worms, chub, minnows, and other small fish. When the smolts have returned to the sea but are still in coastal waters, they feed on shrimps, copepods, sticklebacks, sand launces, and other small fish. On the high seas, the grilses prey on shoals of herring, sprats, and sardines, as well as on large planktonic crustaceans.

The adult Atlantic salmon is a solitary hunter and a rapid swimmer that hunts by sight during the day and actively seeks out shoals of smaller fish (generally at a depth of several hundred meters); at night, it stations itself in the midst of a shoal of fry or plankton, so it can browse freely and more or less at random throughout the night. The Atlantic salmon's flesh is tinted pink by the carotenoid pigments of small planktonic crustaceans. Adult salmon will sometimes follow a school of fish for thousands of kilometers and are occasionally encountered off the coasts of Greenland, though they live off their accumulated reserves (e.g the lower jaw is resorbed) and feed very little in the course of their return migration to their freshwater spawning grounds.

◀ In fresh water, eels feed on salmon roe, and diving birds, perch, pike, and trout feed on the hatchlings. Pelagic predators of the adult salmon include seals, sharks and other large fish, dolphins, and cormorants, and the adults are also preyed on by otters, bears, and eagles during their return migration. Salmon are caught in trawls, weirs, and nets, and during the middle decades of this century, populations were severely reduced (even exterminated in some cases) by overfishing, pollution of coastal streams, the installation of locks and hydroelectric facilities that obstructed the salmon's migrations, etc. In more recent years, this trend is begining to be reversed in some areas as the result of intensive antipollution campaigns, installation of fish ladders, restocking of streams with hatchery-bred fry, and heavy restrictions on commercial fishing in coastal waters and estuaries.

The genus *Onchorhynchus* comprises 6 species of Pacific salmon, all quite similar in their habits to the Atlantic salmon.

J.-Y. S.

Aubria subgillata 44

44
Sub.: Vertebrata
Cl.: Amphibia
O.: Anura
F.: Ranidae

This large semiaquatic frog, 50–100 mm long, is found in the forests of equatorial Africa from the Gulf of Guinea to the Uele River in central Zaire; it spends the daylight hours buried in moist earth at a depth of about 50 cm and only comes out at night to search for food. The species name *subgillata* refers to the spots that are found on the lower part of the body.

▶ Knopffler (1976) was able to observe the predatory behavior of this species, groups of which will assemble around the edges of a small pool or puddle; these are filled with water all year round. The frogs gobble up various prey animals as they emerge from the water, but what is most notable about this is much of their diet consists of several species of killifish (genus *Epiplatys,* family Cyprinodontidae), small, minnowlike fish that are noted for their ability to portage from small pool to another in a series of short hops over the ground. Of 169 specimens of *Aubria subgillata* examined by Knoepffler, as many as 103, or 62 percent, had most recently fed on *E. macrostyma* and *E. sheldjukoi* (plus as an additional 2 percent on killifish belonging to the genus *Aphysemion).* The remainder had fed on insects and other arthropods (15. 5 percent), on other frogs of their their own species (18 percent), on *Hymenochirus boettgeri,* a species of aquatic frog (3 percent), and on mollusks (1 percent).

J. L.

AVOCET *Recurvirostra avosetta* 45

45
Sub.: Vertebrata
Cl.: Aves
O.: Charadriiformes
F.: Recurvirostridae

The avocet is a long-legged wading bird with a long curved bill, white plumage with a black cap, black patches on its back and along the edges of its wings; its legs are gray. It measures 42 cm from beak to tail, has a wingspread of 71 to 75 cm, and weighs between 320 and 400 grams. It is found in Eurasia and Africa, and is protected everywhere in Europe, so that its numbers, greatly reduced in this century, have begun to increase in recent years. The avocet has recently returned to Great Britain after an absence of almost a century, and there are about a thousand breeding pairs left in France, 800 of them in the marshy southern region known as the Camargue.

The avocet is found invariably in brackish coastal wetlands, salt marshes, coastal lagoons, or tidal mudflats dotted with pools of clear water. The avocet is gregarious and diurnal; during courtship, individuals of both sexes participate in a curious round dance that sometimes results in an orgy of mass copulation. About half the European avocets make the winter migration to Africa, the rest preferring to remain where they are.

▶ The avocet feeds on small crustaceans and other invertebrates that it scoops up by "beating" the water in front of it with vigorous, saberlike strokes of its bill, which it flicks rapidly back and forth, keeping it slightly open in order to entrap its prey. As it as gets out of its depth on the mudflats, it continues its hunting sweeps while swimming. In subtropical regions, the avocet is frequently found in association with the pink flamingo *(Phoenicopterus ruber),* which feeds on similar prey in similar surroundings, though it employs a very different food-gathering technique.

◀ The avocet chicks are supervised by the entire colony and agressively defended against such predators as crows, gulls, and hunting cats. The adult avocet adopts a less aggressive maneuver to discourage predators, pretending to be sick or even dead; this seems to be effective with species that prefer to feed on live prey, though not with rats, which are less fastidious or perhaps less easily deceived. Avocet eggs are also sometimes eaten by rats when the hummocks on which the avocets build their nests are accessible from shore.

There are 3 other avocet species, which are found along the seacoast of North America *(R. americana)* and Australia *(R. novaehollandiae)* and on the brackish lakes of the Andean altiplano *(R. andina)*. The avocet belongs to the stilt family of shorebirds, which includes another denizen of the high plateau, the red-beaked curlew *(Idiorhyncha struthersii)* of the Himalayas, which has bluish-gray plumage and the characteristic curve-tipped beak and lives at a height of about 3000 m above sea level.

J.-F. T.

B

46
Sub.: Arthropoda
Cl.: Insecta
S.cl.: Pterygota
O.: Hemiptera
S.o.: Hydrocorisa
F.: Notonectidae

BACKSWIMMER *Notonecta glauca* 46

This common aquatic insect is about 14–16 mm long. Its body is elongated and very compact, dorsally convex and ventrally flattened (though note that it is the backswimmer's ventral surface that normally faces upwards), and it is black and yellow in color with patches of brown. Its eyes are so large that they touch the edge of thorax, and its beak, composed of four jointed segements, is fairly long; the backswimmer's sting can be very painful if it is picked up with bare hands.

The backswimmer's first two pairs of legs are used for prehension of prey as well as for holding the stalks of plants and the various other props that the insect clings to while at rest. The swimming legs are very long, slightly flattened, and fringed with numerous fine hairs, and the backswimmer sculls itself over the surface with great facility, though rarely for more than a few centimeters at a time; while at rest, the swimming legs are tucked toward the forepart of its body.

The backswimmer is active both by night and by day, though it generally remains stationary with its head just below the surface and its abdomen tilted upward into the air. It breathes air directly, and the entrances to the trachea (stigmata) are kept clear by a cordon of water-repellent hairs. The backswimmer is found in permanent or impermanent bodies of stagnant water, including ponds, ditches, puddles, and swimming pools (where it can become a serious nuisance during the breeding season). This particular species ranges throughout Europe and North Africa. The larvae are similar to the adults in appearance, though their bodies are less elongated; the wings are initially distinct and gradually take on their adult form over the course of several molts.

▶ Both the larva and the adult feed voraciously on a number of aquatic creatures, including some that are substantially larger than themselves (fry, tadpoles), and a swarm of backswimmer larvae can wreak serious havoc in a fish hatchery. Their normal prey consists of dragonfly, mayfly, and mosquito larvae, both the larvae and the adults of various beetles and true bugs (hemipterans), as well as copepods, cladocerans, ostracods, and other crustaceans, and mollusks. The backswimmer is not very energetic in its movements until it detects the vibrations made by a prey animal skimming across the surface of the water; then with a powerful stroke of its swimming legs, it launches itself forward to intercept the prey, which it seizes with its first two pairs of legs. It inserts its beak into the body of the prey and assimilates it by external digestion.

◀ The predators of *Notonecta glauca* include the diving beetle *(Dytiscus)*, dragonfly larvae (mainly Aeshnidae), water scorpions and water stick insects *(Ranatra)*, as well as certain aquatic birds. *Notonecta unifasciata* and other large tropical species are commonly consumed by man in Mexico and elsewhere.

The family Notonectidae comprises severalhundredspeciesworldwide.

J.-L. D.

BADGER *Meles meles* 47

47
Sub.: Vertebrata
Cl.: Mammalia
O.: Carnivora
F.: Mustelidae

The common European badger (the North American variety is *Taxidea taxus)* is 85 to 90 cm long, including 15 to 20 cm for the tail, and stands about 30 cm high at the shoulder; it weighs from 15 to 18 kg, sometimes as much as 25 kg. With its short legs and chunky body, pointed muzzle and slightly turned-up nose, the badger seems rather bearlike in appearance; its neck is fairly long and made to appear longer than it actually is by the two black stripes on either side of its head that run from the muzzle to the base of the neck via the eyes and ears.

The badger's underbelly is darker than its back and flanks; its upper body is pepper-and-salt, since each of the hairs in its coat is dark in the middle and light gray, even white at either end. Its bushy tail is lighter in color, and its paws and belly are at least partially concealed by its long fur. The badger reaches maturity after its first year and normally lives to the age of 15. It is found throughout Europe (except for the northern tip of Scandinavia and Corsica, the Balearics, and Sardinia) up to a height of 2000 m.

It is active at dusk and during the night and lives in woods and forests (mixed or deciduous), particularly in the vicinity of clearings or cultivated land. It digs its den, which it generally ends up sharing with a fox or other creatures, in a hillside, often not far from water. Since a badger's den is an elaborate underground complex that may have as many as 50 separate entrances (though 5 or 6 is closer to the average), it prefers to dig in dry calcareous soil. It spends a great deal of its time replacing its bedding, which consists of dried leaves and grass, and in extending and elaborating its tunnel system; the dirt that it removes from these tunnels is piled up in a conical mound at the entrance to its den.

Badgers usually live in small family groups; the young, from 2 to 5 in number, are born in January or February (on the Mediterranean coast) or in March (in Scandinavia). Adult badgers are fairly inactive during the winter, though they do not go into deep hibernation; their winter sleep is light and frequently interrupted, and their tracks can often be seen in the snow, particularly where they go down to drink.

▶ The badger is omnivorous, and roots and rhizomes, bulbs, grass, ripe fruit, and certain mushrooms comprise a large part—perhaps the larger part—of its diet. Hainard observed a badger cracking nuts with its teeth, after tidily cleaning off the shells; badgers also have a special fondness for young corn and have sometimes been persecuted by farmers as a result. Badgers also feed on a variety of small animals: dung beetles—which they search for by turning over dried cowpats—earthworms, snails and mollusks, frogs, toads, lizards, wasps' and bumblebees'nests, the eggs of ground-nesting birds, moles, small rodents, and young rabbits, which they dig out of their burrows. Since the European wild rabbit population has been severely reduced by outbreaks of an epidemic viral disease called myxomatosis, the vole appears to have replaced the rabbit as the badger's principal animal food.

The seasonal variety of the badger diet, based on analysis of the stomach contents of a single individual in each case, may be indicated by the following observations: in June, a kilo of cherries, 50 crickets, a fieldmouse, a lizard; in August, a viper, a mouse, and a quite a few grapes (Hainard); in September, a very large quantity of june bug grubs and nothing else (Heim de Balsac). In locating its animal prey and vegetable foodstuffs, the badger seems to depend almost exclusively on its sense of smell, the only one of its senses that is particularly acute.

◀ In Northern Europe, the lynx occasionally preys on badgers, and badger cubs are sometimes taken by the eagle owl. The badger is pursued by man for a variety of reasons: for its fur and for its bristles (from which various kinds of brushes are made), also, perhaps primarily, because it competes with sportsmen by digging up rabbit's nests and destroying bird's nests on the ground as well as wreaking a modest amount of destruction in cornfields and vineyards. Finally, in regions where rabies is endemic among wild animals, lethal gas is pumped into foxes' dens to prevent the spread of the disease; this may be an even more effective means of disposing of badgers, however, since those who administer this procedure are often incapable of distinguishing between a badger's den and a fox's (and in any case, as noted earlier, the two are sometimes one and the same). To make matters worse (for the badger), the fox often takes to its heels at the approach of human beings, whereas the badger tends to respond to danger by retreating into the innermost recesses of its den.

The European badger has recently encountered two exotic competitors in the form of the familiar raccoon *(Procyon lotor)* of North America and the raccoon dog *(Nyctereutes procyonoides)* of Japan and Eastern Asia. Raccoons, descendants of specimens that have escaped from captivity, are now quite widely distributed in Western Europe, and the raccoon dog, originally introduced into the region around Moscow for the benefit of the commercial fur trade, has slowly made its way across Central Europe over the last 50 years and has just begun to be seen in France. The raccoon dog—a relative of the domestic dog rather than the raccoon—is actually somewhat similar in appearance to the badger, though its tail is longer and bushier and the dark markings on its head are very different. F.T.

48
Sub.: Vertebrata
Cl.: Reptilia
O.: Squamata
S.o.: Ophidia
F.: Elapidae

BANDED KRAIT *Bungarus fasciatus* 48

The banded krait, or pama, is a large venomous snake with yellow and black rings, a round head, and a triangular body when viewed in cross-section. It is generally between 120 and 175 cm long, and occasionally attains a length of 210 cm; this species is native to Bengal, South China, Vietnam, and the Indonesian archipelago, where it is usually found in wet or humid areas (marshes, rice paddies, the shores of ponds, the banks of streams, roadside drainage ditches) in mountainous terrain up to a height of about 2300 m.

▶ The krait is a member of the cobra family, and its dentition is of the same basic type as the Egyptian cobra (q.v.). Its fangs are extremely short, but its venom is highly toxic, causing death by paralysis of the central nervous system and thus of the muscles that assist in respiration. It feeds almost exclusively on snakes—wolf snakes *(Lycodon)*, rat snakes *(Ptyas)*, cobras, and other kraits—very occasionally on other creatures, such as lizards, amphibians, fish, and small mammals (shrews and rodents).

This species is strictly nocturnal and shuns the light; it emerges from its burrow at dusk and is particularly active after a rainfall. It locates its prey by chemoreception, following the chemical trails left along the ground by other snakes, and confirms its location by touch (i.e. with the tip of its forked tongue). It usually seizes a prey animal around the middle of the body and maintains a tenacious grip while inserting its venomous fangs, working its way forward to swallow its prey head first, a process that may require the best part of an hour in the case of a good-sized snake and which involves alternate motions of the left and right hinges of the banded krait's jaws.

◀ If dislodged from its lair in daylight by a potential predator or curious investigator, the banded krait is totally sunstruck *(heliocatalepsy* is the technical term for this condition) and incapable of defending itself with its venomous fangs (Hediger). Instead, it stiffens the forepart of its body and tucks its head underneath; if the aggressor persists, this rigor extends to the posterior part of its body so that, with its tail tucked underneath as well, it assumes roughly the shape of a figure eight. The banded krait is probably preyed on by other snakes, presumably while in this heliocataleptic stupor for the most part; it competes with a number of other snake-eating (ophidaphagous) reptiles, none of which occupies precisely the same ecological niche (with respect to size, circadian rhythms, habitat, etc.).

There are several other krait species *(Bungarus multicintus, flaciceps . . .)* that are similar in their habits. The venom of the Javanese krait *(B. javanicus)*, at least in theory, is among the deadliest of any snake's. The common krait *(B. cerulaeus)* of India and Pakistan is often found in the vicinity of human settlements; it is not particularly aggressive toward human beings, but its bite is almost invariably fatal. D.H.

BANDY-BANDY *Vermicella annulata* 49

49
Sub.: Vertebrata
Cl.: Reptilia
O.: Squamata
S.o.: Ophidia
F.: Elapidae

An Australian representative of the cobra family, the bandy-bandy has a cylindrical body with alternate black and white rings; the boundary between its head and neck is ill defined. It is generally about 40 to 60 cm long and may attain a length of 90 cm in exceptional cases. Its average lifespan is unknown. It is found all over Australia, particularly in regions of loose, sandy soil; it spends most of its time underground or hidden under rocks or beneath a layer of vegetable debris, occasionally making an appearance on the surface on a warm night.

▶ The bandy-bandy is nocturnal and feeds almost exclusively on blind burrowing snakes of the genus *Typhlops*, perhaps occasionally on burowing skinks as well. It pursues *Typhlops* through its tunnels, and locates its prey by smell. It has venomous fangs and the typical proteroglyphic dentition (see EGYPTIAN COBRA) of the family Elapidae.

◀ It seems likely that the bandy-bandy is preyed on by nocturnal raptors and by other snakes; the young may also be eaten by large insects and other arthropods. When it is disturbed, the bandy-bandy adopts a remarkable defensive posture, raising the middle of its body about 10 cm off the ground in one or more semicircular loops, which probably has the effect of making it seem much larger than it really is, perhaps also making it more difficult to grasp by its head or its tail. The venom of the bandy-bandy is too weak to inflict any serious damage on a large animal (its mouth is said to be too small for it to bite a human being), but this behavior and its conspicuous coloration may serve as a reminder to smaller predators that the bandy-bandy is not to be trifled with. It is the only Australian snake that occupies its particular ecological niche (a burrowing snake that feeds on other snakes) and thus does not compete directly with any other species. D.H.

BARN OWL *Tyto alba* 50

50
Sub.: Vertebrata
Cl.: Aves
O.: Strigiformes
F.: Tytonidae

The barn owl is a common species in both the Old and New World between latitude 40° north and 40° south. It prefers flat open terrain (cultivated or fallow fields, wastegrounds, and woodlots) to thickly forested or mountainous terrain. This species measures 30–39 cm with a wingspread of 91–95 cm; it weighs between 290 and 355 g and reaches maturity by the end of its first year; average longevity probably does not exceed 2 years, though a captive barn owl is known to have lived to the age of 18.

▶ Though sometimes active in the daylight hours (while feeding a batch of nestlings or when food is scarce), the barn owl usually conceals itself until sunrise in a hollow tree, a hayloft, a crevice in the rocks, a belfry, abandoned building, or some other dark, secluded spot. In Europe, it feeds primarily on shrews, fieldmice, voles, occasional rats and young rabbits, and other small mammals, as well as bats, frogs, and insects. In some localities, birds furnish a substantial portion of the barn owl's diet, and there has been one report of a barn owl catching fish—not just skimming along the surface with its talons but actually diving down into the water. Different types of prey may be represented in different proportions in the diet of an individual owl in different seasons and localities.

The barn owl appears to patrol its hunting grounds in a regular pattern, flying several meters above the ground, the sound of its flight muffled by the specialized soft feathers along the edges of its wings. It may also station itself on a fencepost, a branch, or the roof of a building, and when a suitable prey animal approaches, it plummets downward, its talons outstretched, legs extended, and impales the prey with its long curved nails, then quickly seizes it in its beak. Normally, the barn owl makes use its sense of sight as well as hearing, but laboratory observations have confirmed that its sense of hearing alone is sufficiently keen to lead it to its prey in total darkness. A small prey animal may be swallowed whole, a larger one torn into pieces and taken back to the young ones in the nest or carried off to be eaten later in a more secluded spot.

◀ The barn owl is not systematically preyed on by any other species, though an occasional owlet or adult may be captured by a marten, cat, or other small carnivore. The barn owl depends on the same food sources as a number of other nocturnal raptors, but their hunting territories and behaviors as well as favored nesting sites are sufficiently different that these other species cannot truly be said to be competitive with the barn owl.

Until comparatively recent times, the barn owl suffered greatly from human ignorance and negligence—caught in pole traps intended for "destructive" raptors that were not disarmed at night, or even nailed to the barn door as a sort of charm against rodents or as a superstitious talisman against misfortunes of every kind. Nowadays automobiles and hunting rifles in the hands of vandals still take their toll of *Tyto alba,* not to speak of the destruction of nesting sites, the indiscriminate use of insecticides, and other such practices. However, in spite of past and present injustices, barn owls will still occupy a nesting box set out for them by humans.

The genus *Tyto* (subfamily Tytonidae) consists of 9 different species, all basically similar to *Tytus alba,* found in different parts of the world. The related subfamily Phodilinae has only 2 species, one of which is found in India and Southeast Asia, the other in Central Africa. Both appear to seek their prey along watercourses in dense tropical rainforests, feeding on small mammals, birds, amphibians, insects, and perhaps on fish as well; the African species, *Phodilus prigoginei,* is only known from a single specimen captured in the vicinity of Lake Tanganyika in 1951. The snowy owl, or harfang *(Nyctea nycta* syn. *scandiaca)* is also found in both hemispheres though much more restricted in its range than the barn owl—it normally lives on the Arctic tundra of Canada and Siberia, ranging down into more southerly regions (Scotland, Scandinavia, and the United States) in colder weather. J.-P.R.

51
Sub.: Vertebrata
Cl.: Aves
O.: Passeriformes
F.: Hirudinidae

BARN SWALLOW *Hirundo rustica* 51

The barn swallow is 20 cm long, with a wingspread of 33–34 cm, and weighs only about 20 g. With the exception of the arctic regions, this species is found throughout the Northern Hemisphere (the North American susbspecies is *Hirundo rustica erythrogaster);* it remains in Western Europe from April through October, and spends the rest of the year in subsaharan Africa, sometimes as far south as the Cape of Good Hope. It occupies a variety of perches during the day, including rooftops, dead trees, and of course telephone wires and power lines, and sometimes returns in substantial numbers to spend the night as well. The barn swallow is so called because it builds its nest under the eaves of barns and other farm buildings, but it does not form nesting colonies in the strict sense of the term.

▶ The barn swallow is a swift and effortless flier, and rarely sets foot on the ground except to gather mud to build its nest. It spends most of its time on the wing, snapping up small flying insects (flies, mosquitoes, dragonflies, flying ants, etc.) as well as the small spiders that use their gossamer webs to ride the air currents like hot-air balloonists; it swallows its prey immediately (the noun and the verb *swallow* seem to be etymologically unrelated, by the way) and sometimes catches insects on the ground as well. The nestlings are fed on little pellets of insects stuck together with saliva.

◀ The hobby (a small falcon) as well as the screech owl and the tawny owl occasionally prey on the barn swallow, as do the domestic cat and the beech marten; when one of these predators shows itself, a flock of swallows will begin circling around it, emitting cries of alarm. Nestlings are often victimized by blood-sucking mites such as *Protocalliphora caerulea, Dermonyssus gallinae,* and other parasites. Barn swallows are still hunted in Italy and North Africa during their fall migrations, and elsewhere more parsimonious techniques of barn construction (doing away with broad overhanging eaves) and the use of pesticides are supposed to have caused a decline in certain populations, but in general, man and his activities seem to have had little effect on this species.

The family Hirundidae consists of 80 species altogether, which are found in every region of the globe. European swallows of the genus *Delichon* are called martins. The house martin *(Delichon urbica)* is a gregarious species found in populated areas. Other common martins are assigned to different genera: The sand martin *(Riparia riparia)* builds its nest in a burrow in a sandbank, and nests of the the crag martin *(Ptyonoporgne rupestris)* adheres to sheer rock faces and cliffsides. Both species nest in colonies. J.-P. R.

BASKING SHARK *Cetorhinus maximus* 52

52
Sub.: Vertebrata
Cl.: Chondrichthyes
O.: Lamniformes
F.: Cetorhinidae

This enormous, harmless creature is found throughout the world's oceans, though more frequently in temperate waters, particularly off the coasts of Ireland and Southern California. Next to the whale shark, the basking shark is the largest fish in existence, typically measuring between 9 and 11 m in length and weighing 3–4 metric tons; maximum size is probably 15 m and 5 metric tons. Longevity is in excess of 10 years.

The basking shark swims slowly (4 kph) through the plankton layer, often with its dorsal fin and even the upper fluke of its caudal fin projecting above the surface. Some individuals are solitary; others congregate in schools numbering as many as 50 or 200. At the end of a very long gestation period (possibly on the order of 2 years), the female gives birth to a single live offspring that already measures between 1.5 and 2 m; basking sharks reach sexual maturity when they have grown to about half their full size (5–6 m).

▶ The basking shark feeds exclusively on plankton, and since the nutritive value of plankton is fairly slight and the basking shark expends up to 660 calories per hour merely by propelling its massive body through the water, it has been calculated that that an individual measuring 6.5 m has to filter at least 1500 cu m of water through its gill slits every hour in order to obtain those 660 calories. The gill slits of the basking shark extend all the way from its head to the median ventral line, and its branchial arches are elaborated into a series of comblike gill rakers, each of which is equipped with as many as 1200 long, thin filaments (known as branchitenia) measuring between 5 and 10 cm. The whale shark's internal filtration system is much more efficient than that of the baleen whales, by the way.

The slow cruising speed of the basking shark not only minimizes the expenditure of energy, it also keeps eddies from forming around its mouth that would disrupt the flow of planktonic organisms through the gill rakers. These organisms consist primarily of microscopic crustaceans, particularly copepods those of the genus *Calanus,* as well as jellyfish, roe, and the larvae of pelagic crustaceans. Since plankton disappear from temperate waters in the winter, the basking shark is obliged to live off its accumlated reserves, which includes the gill filaments, or branchitenia, which are resorbed during the winter and regenerated in the spring.

◀ It is conceivable that young, sick, or wounded basking sharks could fall prey to killer whales, large pelagic sharks, or even schools of tuna. The liver of the basking shark accounts for about 10 percent of its body weight, or 300 kg on the average. One specimen contained 1500 l of oil, which is used in the manufacture of leathergoods. In the coastal waters of Ireland and Norway, basking sharks are still hunted with harpoons from wooden whaleboats. Since a pair of basking sharks produce a single offspring, at most, every 2 years, it seems unlikely that the species will be able to support an intensive fishery for much longer, and it is possibly already on its way to extinction. J.-Y. S.

BATH SPONGE *Spongia officinalis* 53

53
Sub.: Porifera
Cl.: Demospongea
O.: Dictyoceratida
F.: Spongiidae

There are 5000 different species found in all the seas and oceans of the world, but this is *the* sponge as far as most of us are concerned. The bath sponge is abundant in the Mediterranean, especially off the coasts of Tunisia, Greece, Crete, and Syria at a depth of 5–100 m; it is also found in the Caribbean, especially off the coasts of Florida and the Bahamas. The sponges that we use to wash ourselves, our cars, and our domestic surroundings, if not synthetic, are actually the skeletons of *Spongina officinalis* or some other comon species, composed of anastomosed (elaborately interconnected) fibers of an organic substance known as spongine. The living sponge is a spherical mass about 50–80 cm in diameter, its shiny gray or black surface interrupted at intervals by the small conical protuberances at the tips of the skeletal fibers.

The sponge is unique in the animal kingdom insofar as it has evolved neither organs nor tissues in the usual sense and consists entirely of an open-circuit water-filtration system. Seawater enters through small orifices (50 μ in diameter) and is pumped through a network of inhalant canals into the choanocytary chambers, hollow spherical cavities lined with specialized cells called choanocytes, each consisting of a collar and a flagellum, or whiplike vibrating stalk. The beating of the flagella keeps the water circulating through the system; it is pumped back to the surface through a network of exhalant canals, each of which terminates in a relatively large opening (1–5 mm in diameter) called an osculum. The workings of this system can readily be verified, either in the aquarium or in situ, by releasing a quantity of india ink or fluorescent dye in the vicinity of the inhalant orifices; a thin stream of colored water will be expelled through the oscula about 30 seconds later.

▶ The bath sponge feeds on bacteria, protozoa, and other microorganisms, as well as other organic particles dissolved or suspended in the water that it filters through its body. These are absorbed by the collars at the base of the choanocytes, then first transferred to digestive vacuoles, or phagosomes, and finally to cells called archeocytes, in which the process of digestion takes place.

◀ The body of *Spongia officinalis* is a genuine microcosm, the habitat of many lesser creatures. These consist of worms and small crustaceans for the most part; along with *Trypton spongicola* (q.v.), the cirriped *Acasta spongites* is a barnaclelike crustacean that spends its entire life in a shallow burrow that it excavates for itself in the sponge's body, with only the cirri, the jointed filaments with which it gathers its food from the incoming and outcoing current, protruding above the surface. A secondary advantage of this arrangement is that it offers absolute security from predators (excluding man, of course), since *Spongia officinalis* is inedible.

The water-retentive properties of the sponge have probably been exploited by man since prehistoric times; a bath sponge figures among the insignia of an official of pharoah's household, and it was certainly in evidence in the Roman baths as well as in the bathhouses of medieval towns and castles, though less so during the relatively bathless centuries of the Renaissance and early modern period (15th—18th centuries). Because of its prominence in the Gospel accounts of the Crucifixion, it also served various medical and semiliturgical functions (such as scrubbing the high altar) during the Middle Ages. Over the last 30 centuries, sponges have been harvested at successively greater depths by waders, boatmen, and divers, some equipped with sponge hooks or tridents, most recently with weighted diving suits and scuba gear.

When the exhaustion of the accessible sponge beds seemed imminent at the beginning of this century, it was discovered that artifical sponge beds could be created by grafting; earlier experiments in raising sponges from larvae had proved unsuccessful. A healthy mature sponge can be cut into cubes about 7 or 8 cm on each side, propped up on a support, and harvested in 4 or 5 years, by which time it will have grown to commercial size. Vast artificial sponge beds were established in the Gulf of Mexico and in the waters off Japan, though these, along with many naturally occuring beds, were destroyed by a worldwide epidemic in 1938–39. Since then, the demands of the industrial and domestic markets have largely been satisfied by artificial sponges made of cellulose and other substances. The annual harvest of the Greek sponge fishery still exceeded 100,000 kg in 1960, though subsequent harvests have been rapidly diminishing as a result of widespread pollution in the Aegean.

Hippospongia communis, found in the same waters as *Spongia officinalis,* is harvested commercially and is used for rubbing down horses and for more demanding industrial chores. A few species of sponge are found in fresh waters, and some, unlike *S. officinalis,* are attacked by predators, especially those that shelter pigmented algae, *Zooxanthellae,* in their external tissues; thus, *Verongia aerophoba* and *Petrosia ficiformis* are specifically preyed on by a small nudibranch mollusk, *Tylodina perversa,* and a gastropod, *Peltodoris atromaculata,* respectively.

These are isolated cases, however, and in general the sponge has secured an ecological niche for itself by means of the wholesale consumption of microorganisms that have been passed up by other creatures (and in some instances, have already passed through them). The sponge also contributes to the purity of the water in which it lives by ingesting a certain quantity of bacteria, though, like other marine organisms, it cannot survive in an environment where bacteria are too numerous, especially if other pollutants are present as well. N.B.-E.

BIBRON'S AGAMA *Agama bibroni* 54

54
Sub.: Vertebrata
Cl.: Reptilia
O.: Squamata
S.o.: Sauria
F.: Agamidae

This small lizard grows to a length of about 30 cm. Its lifespan in the wild is unknown, but some individuals have survived for over 5 years in captivity. It is quite common in the desert regions of North Africa, from the shores of the Mediterranean to the Sahel, including Tunisia, Morocco, and the interior of Mauritania. The species is readily distinguished by the rows of spiny scales on either side of its head, right behind the eardrums. The females' backs are adorned with transverse bands of lemon yellow and vermillion, their throats with blue and white stripes; their heads are bright blue. The males' heads are grayish-white, their backs normally gray-green (sometimes bluish during the mating season) with cream-colored spots. Young agamas are invariably gray.

In the Sahara, the agama prefers the sort of rocky, boulder-strewn outcroppings that are known as jebels. It climbs and leaps from rock to rock with great agility, and digs its burrow underneath a boulder, in a sloping bank, or in the edge of a ravine that is exposed to the sunlight; frequently it will also find shelter in a crevice or in the space between two boulders. The agama is diurnal in its habits. In summer, it leaves its burrow at daybreak; in spring and autumn, it prefers to wait until the sun has warmed up the ground sufficiently. Its periods of activity are greatly diminished in winter, but it does not undergo true hibernation.

▶ Bibron's agama lives a solitary life, spending much of its time stretched out on top of a boulder surveying its domain. It feeds primarily on insects—small beetles (family Tenebrionidae), locusts (Pamphagides), ants, and their larvae—and spiders. It also captures small nocturnal lepidopetera, which it probably finds under rocks during daytime periods of inactivity (Gauthier, 1967), as well as their larvae. The passage of a swarm of migratory locusts through the agama's territory occasionally provides it with a seasonal banquet of memorable proportions. Bons (1969) has discovered that agamas normally consume a substantial quantity of vegetable matter (notably *Anvillea radiata),* no doubt a primary source of moisture.

The agama actively stalks its prey, which it detects by sight. After a short period of observation, it darts out its tongue, which is covered with sticky mucus, and entraps the prey, then seizes it in its mouth and swallows it in a single gulp. The dentition of *Agama bibroni* is of a type called acrodontic (meaning that the teeth are specialized in certain ways); among others, it has two very sharp incisors that can pierce the thick chitin shells of beetles, locusts, and other insects. The quantity of insect prey that is ingested by a particular agama varies with the season and current activity level of the individual animal.

◀ Bibron's agama is a favorite food of a number of small mammals (including the desert fox, fennec, zorilla, and wild cat) and other reptiles (the monitor lizard, horned viper, and others). On the other hand, its reactions are very quick, its flight responses sometimes remarkably so; it is often able to escape all but the fleetest of its predators. When it is running flat out, with its tail sticking straight up in the air, it may appear that its front feet are not touching the ground at all.

The agama is rarely hunted by the nomads of the Sahara. Gast (1968) reports that the inhabitants of the Ahagar region of Algeria consider its flesh to be poisonous. On the other hand, according to Champault D. (1963), two agama species *(A. bibronis* and *mutabilis)* are relentlessly hunted down by the inhabitants of some other regions of the Sahara as soon as their tracks are sighted. Once captured, the animal's throat is cut and its body is left lying on a rock, belly up. This last detail is important, since the reason that is given for this curious custom is that the *bubress* (local name for the agama) is said to have sworn that Allah would never be permitted to see his belly. Since Allah, of course, sees everything in the end, these tribesmen have taken it upon themselves, like the Furies in a Greek tragedy, to mete out this symbolic punishment for the agama's blasphemy and presumption.

As just mentioned, a close cousin, *Agama*

mutabilis, is also found in the Sahara. *Mutabilis* prefers a slightly different terrain, the rocky flatlands known as regs and hamada. It is slightly smaller than *bibronis* (25 cm) and lacks the scaly spines behind the eardrums; both sexes are more or less uniformly gray in color with well-defined dark green or brownish-black spots on their backs. The males' throat pouches turn a very bright blue if the animal is excited or subjected to a rapid temperature increase. *Mutabilis* does not dig a burrow for itself, and usually finds shelter underneath a rock or a clump of brush during the hottest part of the day. This species can survive for very long periods without food, and is often obliged to do so in summer when insects are scarce.

Tourneville's agama *(A. flavimaculata tournevillei)* inhabits the so-called erg terrain type, which is characterized by massive sand dunes. This agama is thin-bodied, also about 25 cm long; as with *A. mutabilis,* the male is equipped with a throat pouch that turns a bright Prussian blue in similar circumstances. The insect diet of Tourneville's agama is somewhat more extensive, both in quantity and in kind, than that of its cousins. It climbs with great agility through the brush, where it catches flying insects (butterflies, bees and wasps, flies) and devours their larvae or cocoons; it also eats ants and termites, small beetles, and various sand-dwelling larvae.

Agama agama is especially common in the forests and savannahs of central Africa, even in the vicinity of human settlements, since it is very bold and these areas frequently afford a ready supply of insects. During the mating season, the male provides a remarkable display of color—bright red head, blue body, black and orange tail. The female is brown with green spots on her head. A study undertaken by Vernon A. Harris (1966) in the vicinity of Ibadan, Nigeria, has shown that this species lives almost exclusively on ants. R.V.

BIBRON'S BURROWING VIPER
Atractaspis bibronii 55

55
Sub.: Vertebrata
Cl.: Reptilia
O.: Squamata
S.o.: Ophidia
F.: Colubridae

Atractaspis, formerly assigned to the family Viperidae (true vipers), is now considered to be a colubrid genus, which comprises the African mole vipers, or burrowing vipers. This species measures between 40 and 60 cm. Its body is smooth and cylindrical and uniform in color, ranging from brown to black; its tail is short, its eyes are small, and its pupils are round; there is no clear differentiation between its neck and conical head, which has broad scutes on its underside and a single beak-like scale (frontal shield) at the tip of the snout. *Atractaspis bibronii* is a nocturnal burrowing snake, found in Central and East Africa, that is sometimes encountered on the surface after a rain. Its burrow consists of a short (30–40 cm) tunnel that leads to a larger chamber under a rock.

▶ *Atractaspis bibronii* feeds on blind snakes (both *Typhlops* and *Leptotyphlops)* and amphisbaenians as well as on skinks *(Feylinia)* and other lizards, small rodents, and shrews. It tracks its prey by olfaction and by sensing vibrations in the walls of the burrows and tunnels in which it seeks its prey. It seizes the quarry in its jaws and injects its venom into its body while it is already in the process of swallowing it, sometimes pressing the prey animal against the wall of the burrow or even (since its neck is remarkably flexible) against its own flanks for additional leverage. The dentition of the burrowing viper is unlike that of other venomous snakes: it has two very long fangs rooted in the lower jaw, though it is incapable of opening its mouth wide enough for them for to be fully erected; its venom glands are highly developed and in some *Atractaspis* species even extend into the forepart of the body.

◀ In spite of its peculiar dentition, which is unlike that of any other venomous snake, the burrowing viper can still be dangerous to humans or other large mammals, since it can still inject its venom through one of its long protruding fangs without opening its mouth, which it does by jerking up its head and bringing it back down with a rapid stabbing motion. If it is surprised on the surface, its usual reaction is to make for the entrance to the nearest tunnel or to play dead. It is also quite bad-tempered and quick to strike when approached or handled, and because of the extreme flexibility of its neck vertebrae, it can still twist around and strike with one of its fangs even if it is seized just behind the head.

The burrowing vipers are discreet in their habits, and little else is known about them;

their chief predators are probably other burrowing snakes.

Several *Atractaspis* species are found in tropical Africa; *A. engaddensis* is found in Israel. *Calemelaps,* a genus of venomous African colubrids, are also burrowers and probably close competitors of *Atractaspis.*

D.H.

BICHIR *Polypterus bichir* 56

This primitive freshwater fish measures up to 60–80 cm in length. It has a flattened head, an elongated body, an almost continuous fringe of stubby, squared-off fins (pinnules) along its backbone, and an armored plating of thick, close-set scales along its flanks. The bichir is found in the Upper Nile, Lake Rudolf, Lake Chad, and the associated rivers sytems as well as in several coastal rivers of West Africa; when not searching for prey, it normally spends the day at the bottom of a deep pool along the bank, concealed in a clump of aquatic vegetation.

▶ The bichir hunts at dusk as well as in daylight; it is not a proficient swimmer, and feeds almost exclusively on insect larvae, small fish and small frogs as well as foraging for worms along the bottom. Its hunting technique is distinctive, more reminiscent of a cat stalking its prey than a pike or a perch. It uses its pectoral fins like crutches to propel itself slowly along the bottom, comes to an abrupt halt when confronted by obstacles, then moves forward again, just as cautiously as before, pausing frequently to check its bearings by raising its head and sniffing the water.

The bichir seems to detect its prey primarily with the aid of its well-developed nostrils, which are mounted on short stalklike projections. It also has a row of sensitive barbels on either side of its snout, and its eyes appear to play a relatively minor role in prey location. Once it has arrived within striking distance of its prey, it rapidly propels itself forward by means of wriggling, eel-like undulations of its body, while its dorsal pinnules are pressed flat against its spines and its pectoral fins pressed back against its flanks. It catches the prey animal in its jaws, and swallows it whole.

◀ The bichir has a number of passive defenses at its disposal, including flight, protective, coloration, and concealment, and appears to have no systematic predators. The dark stripes on its sides mimic the stems of aquatic plants, and when threatened, it will attempt to take refuge in a dense clump of vegetation along the bank (naturally using the streamlined high-speed swimming technique described above rather than stumping along on its pectoral fins). Its body armor is thick enough to resist the sharp teeth of larger fish, and its dorsal pinnules are ribbed with sharp spines, which, like those of the perch or stickleback, have an equally discouraging effect on predators that are accustomed to swallow their food whole. The bichir is locally consumed by man, though it is not exploited commercially; its flesh is very bony but excellent-tasting, particularly when the fish is baked over the coals in its own armor.

The genus *Polypterus* comprises approximately ten species, all found in tropical Africa. In some taxonomies, the bichirs and the reedfish *(Calamoichthys calabaricus)* are grouped together in the order Cladistia; the feeding behavior of the reedfish is identical to that of the bichir, but is body is more elongated and it tends to be nocturnal in its habits.

J.-Y. S.

56
Sub.: Vertebrata
Cl.: Osteichthyes
S.cl.: Chondrostei
O.: Polypteriformes
F.: Polypteridae

BLACK-BANDED SEA KRAIT *Laticauda laticauda* 57

57
Sub.: Vertebrata
Cl.: Reptilia
O.: Squamata
S.o.: Ophidia
F.: Hydrophiidae

Less specialized than the other sea snakes (Hydrophidae), the sea kraits *(Laticauda)* have retained the broad ventral plates (scutes) that enable them to crawl up on shore, and their nostrils are laterally rather than dorsally mounted as is the case with the other Hydrophidae. The males measure between 80 and 100 cm, the females between 100 and 120 cm, though the females may exceptionally attain a length of up to 160 cm; the laterally flattened tail, which serves as a rudder and represents about 9–13 percent of total body length (proportionately longer in the males), is the sea krait's only visible adaptation to marine life. The body takes the shape of a slightly flattened cylinder and is marked with approximately thirty black bands separated by lighter bands of equal width.

This species is found throughout much

of the Indo-Pacific region, from southern Japan to northern New Zealand and westward to the Bay of Bengal. Sea kraits frequently come ashore, most often on small tropical islands, not only to mate and lay their eggs but also to slough their skins and even digest their meals. In tropical waters, colonies of several dozen black-banded or yellow-lipped sea kraits *(L. colubrina)* may be found in caves, recesses in the rock, or simply curled up together under a bush.

▶ Sea kraits feed on fish that are thinner (though not necessarily smaller) than themselves, especially on eels, whose elongated shape and lack of sharp protruding fins makes them that much easier to swallow. They hunt along rocky substrates and coral reefs, and detect their prey by smell, making use of Jacobson's organ exclusively, which is able to function underwater whereas olfactory epithelium is almost entirely sealed off in all sea snakes. The eel is thus easily tracked to its lair in a crevice among the rocks or coral; the sea krait seizes its quarry in its jaws and disables it by injecting a potent neurotoxin through its poison fangs, which are similar in location and design to those of the cobra.

The sea krait maintains its grip on the prey until the venom takes effect; the snake initially sinks its fangs into whatever part of the eel's body is accessible to it, then works its jaws along the eel's body until it reaches the head; if the eel is particularly large, the snake may release its hold momentarily in order to get a fresh grip on the eel's snout with its jaws. Then, as snakes normally do when attempting to ingest a large, thick-bodied prey animal, the sea krait slowly advances forward to engulf the prey with alternate lateral motions of its jaws, rather than swallowing it the way a mammal would. The sea krait identifies its victim's head solely by "taste" (with the aid of its darting tongue and Jacobson's organ) rather than by sight.

Sea snakes have never been observed to hunt at night, though it seems quite likely that they do so. The process of digestion may take several days or even weeks, depending on the size of the prey animal and the external temperature; the optimal temperature for this process is 28°C, which is rarely attained in the water except at the height of summer, so sea kraits generally come ashore to digest their food in a sunny spot on the beach. Whether or not their stomachs are full, however, they are entirely indifferent to the terrestrial lizards and small mammals that scurry past them, and frequently clamber over them, as they lie coiled up on the shore.

◀ Several species of shark and other large fish appear to be the only marine predators that are not deterred by the sea krait's deadly venom. Sea kraits are frequently attacked by fish-eating eagles either while on shore or while swimming on the surface, and their remains are often found in the vicinity of the eyries. It seems probable that gulls and other shorebirds feed on the very young sea kraits, which are hatched on shore.

It is curious that whereas the sea krait is quite wary and quick either to take flight or to prepare to strike when approached in the water, it seems to lose these instincts of self-preservation when it comes up on shore. A snake that appears to be alarmed by the slightest ripple in the water will make only the sketchiest and most perfunctory attempts at self-defense when it emerges from the water and will allow itself to be handled with impunity as soon as it has crawled one or two meters from the shoreline.

As noted earlier, sea kraits are often found in substantial numbers along tropical beaches, and are regarded in some areas as a harmless curiosity and in others a resource to be exploited ruthlessly while the supply lasts. In the Philippines, sea kraits have been slaughtered for their skins, and the flesh is consumed locally or dried and processed to be used as animal feed; there is also some demand for the caul fat that surrounds the gall bladder, which plays a role in the traditional East Asian pharmacopeia. At a conservative estimate (based on figures for processed snakeskins exported from Japan), it appears that several hundred thousand sea kraits of various species are killed every year for commercial purposes. Sea krait populations on several Philippine islands are known to be severely threatened, though as yet there are no legal restrictions or controls on this traffic.

Other sea snakes, which are ovoviviparous and never come ashore (and may not even be capable of doing so), are basically similar to the sea kraits in their predatory strategy, including several species that are sometimes found in the lower reaches of large coastal rivers. *Pelamys platurus,* the only pelagic species, preys on small carrion-eating fish—which it attracts by floating lifelessly on the surface. The dentition of *Emydocephalus* is poorly developed, and these species feed exclusively on fish eggs (Voris, 1966). Many species of sea snakes are more aggressive than the sea kraits and have been held responsible for a number of human fatalities (particularly common among fishermen hauling in their nets by hand) in the Indo-Pacific region.

H.S.C.

BLACK-BILLED MAGPIE *Pica pica* 58

This handsome corvid measures 41–42 cm, with a wingspread of 48–53 cm, and weighs about 210 g on the average. In Europe, the magpie is commonly encountered on flatlands dotted with occasional groves of trees (including cultivated fields and woodlots) and even in small villages and on the fringes of urban areas. It is less commonly seen in mountainous terrain (below 1500 m), and it avoids dense forests and marshland. The black-billed magpie ranges throughout much of Europe (though it is fairly rare in parts of Southern Europe), Asia, North Africa, and western North America.

The magpie is a sedentary, gregarious bird. It is heavy and ungainly in flight and normally prefers to get around on foot; during the winter, magpies sometimes congregate in rookeries containing as many as 200 individuals. This species reaches sexual maturity in its first or second year; perhaps contrary to appearances, the social life of the magpie is quite harmonious, but nesting pairs prefer to keep to themselves during the breeding season.

▶ About 60 percent of the magpie's diet consists of animal prey or carrion, the remainder of seeds and grain, fruit, and a variety of other vegetable material. In Western Europe, the black-billed magpie feeds primarily on insects and their larvae (cockchafers, rosechafers, rove beetles, dung beetles, butterflies, moths, and caterpillars, crickets, and grasshoppers) as well as on spiders, snails, slugs, earthworms, lizards, small rodents and shrews. In spring, the magpie is an inveterate nest robber, preying especially on the eggs and nestlings of finches, blackbirds, and ringdoves, which build their nests in gardens, parks, and other sparsely wooded areas that are typically frequented by the magpie.

The black-billed magpie catches most of its prey on the ground, seizing the larger creatures with its claws and dispatching them with vigorous blows of its beak. Like other corvids, the magpie is accustomed to stuff the leftovers into its cheek pouches and carry them off to a private cache in a hollow tree or elsewhere. Though the reasons for this behavior are still obscure, the magpie's notorious fondness for "bright glittering objects" has done little to improve its reputation among humans as a skulker, a loiterer, and an all-round undesirable character.

◀ Even though it flies rather poorly, the magpie is wary and quick to go to cover; it thus avoids such common aerial predators as the sparrowhawk, goshawk, peregrine, and eagle owl except on very rare occasions. The magpie protects its nest against predators by constructing a barricade of thorns across the top. The magpie's nest is large and sturdily built (and, at least according to legend, richly furnished with stolen objets d'art) and is sometimes preempted by crows while still under construction in the early spring; abandoned nests from previous years are frequently occupied by long-eared owls and the smaller falcons. The enduring human prejudice against the magpie is still expressed in a variety of ways—eggs poisoned with strychnine are still being set out as bait and (contrary to the ethics of the hunting field) sitting birds are still being picked off with rifles—but none of these activities seems to have prevented the black-billed magpie from maintaining, and even increasing, its numbers.

J.-F. T.

58
Sub.: Vertebrata
Cl.: Aves
O.: Passeriformes
F.: Corvidae

BLACK BULLHEAD *Ictalurus melas* 59

59
Sub.: Vertebrata
Cl.: Osteichthyes
S.cl.: Teleostei
O.: Siluriformes
F.: Ictaluridae

A native of the Eastern United States, this catfish was introduced into several European rivers about a century ago and is now found all over Europe. North

American black bullheads typically measure 45 cm and weigh up to 2 kg; European black bullheads measure about 33 cm and weigh about 450 g. This species is nocturnal, and, like other catfish, it is a bottom-dweller that does surprisingly well in oxygen-poor and even polluted water.

The black bullhead is normally found in the backwaters of slow-moving rivers and streams as well as in ponds and lakes, particularly where the bottom is muddy, and can even survive for some time while buried in the mud at the bottom of a dried-up pool or ditch. In the breeding season, the adults scoop out a nursery for the hatchlings on the bottom, and they are noted for the vigilance with which they supervise their developing offspring.

▶ The hatchlings make their way upward to the plankton layer, where they congregate in considerable numbers, feeding on copepods, *Daphnia*, and chironomid larvae. The older juveniles are solitary and forage along the bottom for larvae (Diptera, mayfly), oligochaetes, and mollusks; when they have grown to a length of about 15 cm, they are ready to start feeding on fish and crayfish, though later on, when they have attained a length of about 25 cm, they tend to revert to zooplankton to some extent. The black bullhead is especially fond of crayfish and on the roe of other fish, though it is very flexible in adjusting its feeding habits in accordance with its current circumstances.

The 8 senstive barbels on the bullhead's snout are studded with olfactory receptors, and it probes with these as well as with its mouth while foraging along along a muddy or sandy substrate. The sensory organs of the lateral line probably play some role in catching fish and other free-swimming prey, as the bullhead can sense variations in electroconductivity created by the passage of other creatures through the water; however, it lacks the electric organs used by the electric catfish and the electric eels to stun their prey. The bullhead's sense of hearing is very good; a set of small bones collectively known as the weberian apparatus and corresponding to the ossicles in the human ear are connected by ligaments to the swim bladder, which is thus pressed into service as a resonating chamber that picks up and amplifies the vibrations transmitted along the ossicles.

◀ The black bullhead is so diversified in its feeding habits that it cannot really be considered a competitor of the pike or the perch or any of the other large fish that occupy a similar position in the aquatic foodchain. Though it feeds voraciously on roe, it feeds on other fish in substantially smaller numbers than these other species do. Its spines and aggressive behavior keep it safe from aquatic predators, though badgers, otters, and bears (lacking the esthetic or hygienic scruples that deter many human fishermen) feed avidly on its flesh. Like many bottom-feeding fish, it is considered to be inedible in many regions, though more often on the basis of local custom than empirical evidence. The black bullhead does not present much of a challenge to sports fishermen, and is particularly easy to catch in a spring freshet or a stream swollen with floodwater. F.M.

BLACK-HEADED GULL *Larus ridibundus* 60

60
Sub.: Vertebrata
Cl.: Aves
O.: Charadriiformes
F.: Laridae

This handsome bird, also known as the laughing gull, measures 33–39 cm, with a wingspread of 84–100 cm, and weighs between 210 and 300 g. It becomes sexually mature at the age of 2 and may live as long as 25 years. During the spring, the black-headed gull is commonly found in flooded marshes or on lakes that are substantially overgrown with aquatic vegetation or, less commonly, on sandbanks in rivers, coastal dunes, and offshore islands. Northern European populations spend the winter in southwestern Europe or in North Africa whereas the native populations may be found all along the coastline during the winter as well as in the vicinity of large cities and inland lakes and rivers. There they establish noisy, crowded dormitories on sandbars and the shores of lakes and descend in enormous flocks on their winter feeding grounds—plowed fields, landfills, sewer outlets, etc. Extremely gregarious (though snappish and irascible), the black-capped gull performs all of its essential activities—feeding, nesting, and migrating—en masse. The French population is currently estimated at about 20–30,000 pairs and is growing rapidly.

▶ The black-headed gull is an omnivore, though as much as 90 percent of its diet consists of animal prey. During the spring, it feeds on earthworms, cockchafers, flies, and mosquitoes, caterpillars, dragonflies, frogs and tadpoles, mollusks, fish, and even fieldmice; it sometimes robs the nests of other birds as well, devouring both the eggs and the nestlings. The vegetable portion of its diet consists of seeds and green leaves.

During the winter, the black-headed gull feeds on human organic wastes of various sorts as well as on sick and injured fish, which it either retrieves from the surface or as the result of a brief underwater reconnaissance. Black-capped gulls will sometimes gather in enormous numbers to feed on swarming gnats or on flying ants during their nuptial flight. Ordinarily they search for prey along riverbanks or in wet or flooded fields, where they will snatch up any small creature that can be caught without too much trouble. They have evolved a number of labor-saving strategems, such as pressing their toes down into a mudbank to force sandworms up to the surface and searching for worms and grubs turned up along a freshly plowed furrow. Like other gulls, the black-capped gull feeds its chicks by disgorging part of its most recent meal, whatever that might happen to be.

◀ The black-headed gull's gregarious instincts and irascible temperament work together to protect it from such predators as the fox, domestic dog, reed buzzard, and crow, at least in the majority of instances. The chicks often fall victim to pike, crows, and small mammals, and the adults are sometimes taken in flight by peregrine falcons.

Colonies of black-headed gulls tend to replace smaller and more specialized members of the family Laridae, especially the black tern *(Chlidonias nigrus)* and other tern species. In localities where food is put out for gulls or other waterbirds during the winter, the black-capped gull may also enter into competition with mute swans, coots, and ducks. Like the herring gull, this species has quickly learned to profit from man and his activities, notably modern methods of agriculture and waste disposal. Some contamination of egg clutches with chlorinated hydrocarbons (pesticide residues) has been reported, but this does not appear to have significantly interfered with the expansion of this species all along the coasts of Western Europe.

The family Laridae comprises 6 genera, 43 species in all, and is morphologically quite homogeneous. Gulls belonging to the genus *Larus* (the smaller Atlantic species may be known as mews or kittiwakes) weigh between 120 and 2000 g. Audouin's gull *(L. audouini)* is now severely threatened, with only about 2000 pairs still occupying its nesting sites in the Balearics and other small islands in the Western Mediterranean. The familiar herring gull *(L. dominicanus;* q.v.), common on both sides of the Atlantic, has increased its numbers by a factor of 10 in some areas and may pose a threat to the exploitation of coastal resources by humans (the harvesting of shellfish, for example).

J.-F.T.

BLACK IGUANA *Ctenosaura similis* 61

61
Sub.: Vertebrata
Cl.: Reptilia
O.: Squamata
S.o.: Sauria
F.: Iguanidae

The body of this large tropical iguana is slender, laterally compressed, and sometimes exceeds a meter in length. Its legs and digits are well developed, its tail is sometimes twice as long as the rest of its body, and its color is usually a drab shade of brown or green; the coloration of the juvenile is sometimes more intense than that of the adults.

The black iguana is found throughout Central America and in virtually every available habitat, including arid savannahs and grasslands, rough, uneven ground littered with fallen rocks and treetrunks or broken with rocky outcroppings, and valleys covered with thick brush. Primary rain forest and mountainous uplands are the only terrain types in the region that are not frequented by the black iguana. In wet terrain it confines itself to dry, open ground and is more arboreal in its habits than in dry terrain. It has enthusiastically adapted to manmade biotopes—cultivated fields, gardens, and the cleared land at the outskirts of a village—and is often found in conjunction with herd animals in pastures. Stone walls, hedges, fences and corrals, and refuse heaps may also furnish the black iguana with shelter or merely a place to sun itself. The black iguana digs its own burrow, 1–2 m in length, which it may end up sharing with other creatures (snakes, rodents, other lizards, and arthropods). Diurnal and heliophilic, it is frequently active in the morning, coming out to warm itself in the sun for increasingly longer periods, though it returns to its burrow at midday; it may come out once again in the evening, but spends the night (and the entirety of any day when the sun is not shining) in the shelter of its burrow.

▶ The black iguana feeds primarily on leaves, flowers, fruits, and even seeds and grains, as well as on insects (grasshoppers, crickets, beetles, caterpillars, butterflies, bees, wasps, flies, bugs); it catches terrestrial insects by scratching or digging in the ground; the flying insects are probably swallowed accidentally while it is browsing on flowers or foliage.

The black iguana also feeds, though to a still lesser extent, on frogs, young lizards (including young iguanas), small birds, rodents, and lizard's eggs. In the vicinity of human settlements, it often feeds on human excrement as well. The juvenile iguanas are markedly arboreal in their habits, and a larger percentage of their diet consists of insects and other animal prey. The iguana detects its prey primarily by sight, occasionally by smell (as in the case of insect eggs or pupae); it seizes the prey with its jaws, chews it hastily, and swallows. Larger prey may be pressed against the ground to make it easier to swallow.

◀ The iguana will defend its burrow or its favorite basking ground against encroachments by other iguanas, but the individual iguana's territory has no strictly defined boundaries. Females never engage in territorial disputes, and the squabbles of the juveniles rarely result in serious injury. The territory of individual males usually overlaps with or includes the territories of several females and juveniles, and the males seem to feel an active distaste for one another's company; however, this is generally manifested in a sequence of ritualized aggressive postures that fall short of actual combat. The iguana is an indolent, home-loving creature and rarely wanders more than a few meters from its burrow, perhaps a few tens of meters in the case of the full-grown males, though males in search of a mate and females searching for food or a suitable place to lay their eggs will often venture farther afield.

The young iguanas are frequently devoured by birds and reptiles of various species, but the original predators of the adult iguana, primarily birds of prey and carnivorous mammals, have been exterminated in most of the regions it inhabits. Local cultivators, on the other hand, will often hunt iguanas as food or to prevent them from destroying their crops; the iguana is commonly hunted with dogs, dug out of its burrow, or shot with a rifle or a slingshot.

There are several other *Ctenosaura* species, one of which is found along the southern borders of the United States. The iguanas of the Caribbean, more primitive in their morphology, are assigned to the genus *Cyclura* and are now restricted to Haiti and the Leeward Antilles, where they are extensively hunted for food. As with the black iguana, the diet of the juvenile contains a higher proportion of animal prey, though all the Caribbean iguanas are strictly terrestrial in their habits. The male rhinoceros iguana *(Cyclura cornuta)* is notable for the fatty pouches on either side of its body and a series of hornlike protuberances along its snout. A related species, *Conolophus subcristatus*, is the "land iguana" of the Galapagos Islands (terrestrial rather than arboreal) and appears to be exclusively vegetarian as well. D.H.

BLACK MAMBA *Dendroaspis polylepis* 62

62
Sub.: Vertebrata
Cl.: Reptilia
O.: Squamata
S.o.: Ophibia
F.: Elapidae

The black mamba—which is not really black, but brown, olive-green, or grayish—is the largest venomous snake in Africa, and the second largest in the world, after the king cobra. Its body is muscular and covered with smooth scales; its head is long and narrow, and its eyes are large, with rounded pupils. Average body length is between 2 and 3 m, though some individuals may attain a length of 4.3 m and live to the age of 20. A denizen of the open savannah, the black mamba is found in dry terrain, sparsely wooded or covered with brushwood, in East and Central Africa; it is extremely rare in West Africa and is never found in dense tropical forest.

▶ Like all the Elapidae (cobras and their kin), the dentition of the black mamba is proteroglyphic, which primarily means that it has two long fangs in the front of its upper jaw, each of which is supplied by a venom gland. The venom canal found in other elapids is only partially enclosed and thus takes the form of a groove, though it fulfills the same function and in a similar manner. The mamba's fangs are between 3 and 7 mm long, proportionally larger than those of other elapids, and may be erected by a pivoting motion of the upper jaws—in this respect resembling the solenoglyphic dentition of the Viperidae (see DIAMONDBACK RATTLESNAKE)—whereas the fangs of the cobra are permanently erect.

The mamba's venom is more toxic than any of the cobras' with the exception of the Cape cobra, *Naja nivea*. It contains neurotoxins and causes death primarily by paralyzing the muscles responsible for respiration; in the case of a human, this can occur within several hours if not within several minutes. The black mamba feeds on homeothermic (warm-blooded) vertebrates for the most part—birds and especially mammals, including hyraxes, rats, squirrels, and other rodents, shrews, and bats—as well as on lizards (skinks, agamids), particularly when young. It also eats bird and lizard eggs and sometimes eats other snakes of quite considerable size. A mamba measuring 2.9 m was found to have just swallowed a black-and-white cobra measuring 2.25 m. The black mamba is active during the daytime and occasionally at dusk, most often between the hours of 10 a.m. and 4 p.m.

The black mamba makes its home in an abandoned burrow, a hollow tree, or crevice in the ground, even an empty bird's nest, which it leaves only to hunt or find a mate and otherwise as infrequently as possible. It climbs trees easily but is not really arboreal; the adult prefers to find its food on the ground. It travels quite rapidly (see below) with the first third or quarter of its body raised high off the ground, its tongue darting quite far out of its mouth, the tip curving alternately upward and downward. The olfactory data that it collects in this way may be of help in determining its trajectory, though the mamba locates specific prey essentially by sight.

Once the quarry has been sighted, the mamba approaches rapidly, sometimes halting briefly from time to time; it often attacks its quarry from one side and, unlike the cobra, does not pause to allow its venom to flow into the wound but strikes rapidly several times. Death is almost instanteous in the case of a small animal, and the mamba flicks the animal's body with its tongue in order to locate the head, which it invariably swallows first. It normally digests its prey rapidly; 8–10 hours suffices for a creature the size of a rat.

◀ A large adult mamba has little to fear from predators, but the young can easily be gobbled up by diurnal raptors and the secretary bird as well as by other snakes; a mamba measuring 72 cm was found in the stomach of a sand snake *(Psammophis sibilans)*. The mongoose and other viverrids (suricate, genet), celebrated for their resistance to the cobra's venom, and the wart hog, which is at least partially protected by a layer of subcutaneous fat, may also feed on young mambas. Crocodiles and large fish may occasionally catch a mamba that is swimming across a watercourse—thus, a mamba measuring 2 m was found in the stomach of a 3 kg brindle bass (a kind of freshwater perch). Large numbers of black mambas are also killed by grassfires, or as a result of them, since marabou storks and other predatory birds are often standing by to snap up the fleeing survivors.

The aggressiveness of the black mamba, though often exaggerated, is certainly worthy of comment. Normally a mamba will head for its burrow when disturbed; a mamba that is surprised while crawling along a branch will drop to the ground and flee, its head held up in its normal vigilant attitude. However, it may strike at anyone that interferes with its retreat, deliberately or not; if it is prevented from escaping, it flattens out its neck, a little like a cobra, and goes over to the attack, frequently veering off to one side, as it does with prey animal, in order to strike from the side. The black mamba also exhibits territorial behavior during the mating season and may launch an unprovoked attack on any intruder in its territory.

The maximum speeds attained by the black mamba under these conditions are naturally very difficult to measure accurately; estimates generally range between 12 and 17 kph, and a figure of 32 kph attested by one authority may accordingly be regarded with some skepticism. Luckily, the black mamba usually makes use of its agility to seek the safety of its burrow when alarmed, and fatal attacks on humans are fairly rare. However, because of its size and speed, the toxicity of its venom, and the unpredictability of its tactics and its temper, the black mamba must be considered the most dangerous of African reptiles.

Three species of green mamba are also found in tropical Africa, *D. virdis, augusticeps,* and *jamesoni*. The largest of the 3, *D. jamesoni,* may exceed 3 m in length; the other two species measure between 1.8 and 2 m. Though all 3 species are equally comfortable in the water or on the ground, they tend to be more arboreal in their habits than the black mamba. They feed primarily on birds and their eggs, on chameleons, agamids, geckos, and other lizards, and tree frogs as well as on small mammals and other snakes. Green mambas are unaggressive towards humans, though they sometimes kill cattle by biting them on the head and neck. D.H.

63
Sub.: Vertebrata
Cl.: Aves
O.: Podicipediformes
F.: Podicipedidae

BLACK-NECKED GREBE *Podiceps nigracollis* 63

This species measures about 28–31 cm, with a wingspread of 53–58 cm, and weighs betwen 300 and 400 g. It is found in Europe, western North America, and parts of Asia and Africa. Rather timid and more restricted in its habitat that the great crested grebe (q.v.), the black-necked grebe is found only on freshwater ponds and open marshes with abundant foliage along their banks and clean water in which insect larvae and other aquatic creatures are likewise abundant. The black-necked grebe is a gregarious species, and nests in colonies intermingled with those of gulls of various species; in the marshes of the Sologne, along the Loire in central France, the black-necked grebe is associated with the laughing gull, for example, which provide the grebe with some measure of protection against predators. Longevity is unknown, but the black-necked grebe becomes sexually mature in its second year of life.

▶ In the spring, the black-necked grebe feeds primarily on insects and their larvae (mosquitoes, libellulids, diving beetles), mollusks and small crustaceans, and amphibians. In winter, it feeds primarily on small perch, chub, gobies, and other small fish. The grebe searches for insects swimming on the surface or plucks them out of the air. It is a remarkably agile diver and underwater swimmer, though it forages at a depth of only 1.5–2 m; maximal duration of the dive is 50 seconds. It can reverse directions very quickly while swimming underwater in order to keep pace with the evasive maneuvers of the schools of small fish that it feeds on; it swallows its catch before surfacing, thus minimizing the possibility of it meal being shared with a seagull.

Other potential competitors include the little grebe, or dabchick *(P. rudicollis),* whose range overlaps with that of the black-necked grebe in Western Europe, though it is not nearly as particular in its choice of nesting sites, and the horned or Slavonian grebe *(P. auritus),* before which the black-necked grebe is currently retreating in Northern Europe and Central Asia.

The genus *Podiceps* (syn. *Colymbus)* comprises 6 different species, of which the little grebe is, not surprisingly, the smallest; it feeds primarily on mollusks and is markedly gregarious in its behavior. The largest extant grebe is the western grebe *(Aechomophorus occidentalis)* of North America, best known for its mating dance, an exuberant web-footed gallop across the surface of a pond or lake. Rolland's grebe *(Rollandia rollandi),* a small species native to Patagonia, is sometimes found in association with the black-necked swan *(Cygnus melanocoryphus),* particularly in ponds that are choked with water weeds, which would normally prevent the grebe from diving. The swan's neck clears out a swath of open water as it feeds on the bottom and the grebe follows along in its wake, diving down to forage in the mud.

J.-F.T.

64
Sub.: Vertebrata
Cl.: Mammalia
O.: Cetacea
F.: Balaenidae

BLACK RIGHT WHALE *Eubalaena glacialis* 64

This cetecean (member of an order of marine mammals that includes both whales and porpoises) attains a maximum length of 17 m, an average length of 14 m, and weighs between 40 and 50 metric tons; the females are slightly larger than the males.

The black right whale is generally found in temperate waters, and the species has been further classified into 3 regional subspecies: *E. g. glacialis* of the North Atlantic, *E. g. japonica* of the North Pacific, and *E. g. australis* of the Southern Hemisphere.

▶ The black right whale feeds exclusively on copepods, or microplankton—tiny crustaceans no more than 3 or 4 mm long. The baleen, the thin whalebone strips that strain these tiny creatures out of the water, is often longer than 2 m and is arranged in 2 rows along the upper jaw that do not quite meet in front, leaving a triangular aperture through which water flows when the whale plows openmouthed through the dense subsurface "pastures" of copepods in the North Atlantic. When a sufficient quantity of copepods has collected on the baleen strainers, the whale closes its mouth, expels the remaining water by raising its tongue, and swallows the copepods; this method of food-gathering differs from that of the rorquals (another group of baleen whales that includes the finback and the blue whale), which take in water in great gulps rather than in a continuous stream.

The black right whale was virtually exterminated in European waters in the 16th century, but prior to that time, the whales migrated to their spring basking grounds to feed on the rich pastures of copepods in the waters off Iceland, the Faeroes, and the Hebrides; in the autumn, they returned to their breeding grounds in the deep, warmer waters of the Bay of Biscay and the mid-Atlantic in the vicinity of the Azores.

◀ The black right whale is sometimes attacked by packs of killer whales, though no scientific observations of this phenomenon have ever been made. The species was being hunted by the Basques in the Bay of Biscay as early as the 11th century, and by the 16th century, European whalers were obliged to shift their attentions to the bowhead, or Greenland, whale of the Arctic seas. The black right whale was brought under the protection of an international whaling convention in 1936, by which time the species had been hunted to the brink of extinction throughout its range. Today there is still a small population in the southern portion of the Northwest Atlantic and perhaps a few thousand individuals of this species in the waters of the Southern Hemipshere.

The Greenland whale *(Balaena mysticetus)* has the same feeding habits as the black right whale; in addition to copepods, it feeds on small pteropod mollusks (free-swimming sea snails) of 4 to 5 mm. The baleen strainers may be as long as 4 m in this species but are narrower than those of the right whale. D.R.

BLACK SKIMMER *Rhynchops nigra* 65

65
Sub.: Vertebrata
Cl.: Aves
O.: Charadriiformes
F.: Rhynchopidae

This shorebird has a dark-brown back with a white underside and white around its eyes and the edges of its wings. Its feet are webbed and bright red; its tail is forked, and its bill is black with an orange streak at the base. It is typically about 45 cm long with a wingspread of 85 cm. The black skimmer is found in sandy coastal regions of North America, from Canada to Argentina on the Atlantic coast, from California to Chile on the Pacific.

▶ The black skimmer feeds on small fish, crustaceans, insects, worms, and other small creatures that are found on or near the surface of the water. It is so called because it skims over the water with its lower mandible, which is longer than the upper one, cutting through the water like a plowshare. As soon as it encounters a prey animal, it snaps its bill shut and swallows its prize; the bladelike lower mandible fits into the upper one like a knife in a knife grinder, which allows the skimmer to make off with even the slipperiest of prey. To enhance the efficiency of this formidable if rather ungainly apparatus, the skimmer leaves a luminous trail behind it when it plows through the water at night; this attracts small marine organisms, which in turn attract fish, which are snapped up by the skimmer. It practises the same technique while wading through deep water.

◀ The black skimmer has no specific predators or competitors; it is not hunted by man, though local populations may be reduced due to the destruction of nesting sites.

Three skimmer species comprise the entire family Rhynchopidae. *Rhynchops flavirostris* lives on the seacoasts and the shores of inland watercourses of Egypt and Subsaharan Africa; *Rhynchops albicollis* is found in India, Burma, and Indochina. All three are very similar in their habits.

J.-P. R.

66
Sub.: Vertebrata
Cl.: Aves
O.: Picoides
F.: Indicatoridae

BLACK-THROATED HONEY GUIDE *Indicator indicator* 66

This small woodpeckerlike (piciform) bird is about 20 cm long, with a wingspread of 98–119 cm; its life expectancy is probably fairly long, as with most tropical birds, to compensate for its low reproductive rate. It is found in open, sunlit wooded terrain and on savannahs, up to an altitude of 2500 m, from Senegal to Eritrea down to the Cape. The honey guide lays its eggs in the nests of various other birds, primarily those that build their nests in cavities in trees (barbets, magpies, kingfishers, bee-eaters, and others); the nestling is equipped with a diamond-shaped excrescence on its upper and lower mandible that serves the function of a shovel blade and enables it to rid the nest of unhatched eggs and its foster siblings of the host species. The adult's plumage is drab; it is solitary (though occasionally seen in small, scattered flocks) and inconspicuous in its habits (with exceptions noted below).

▶ The honey guide is an insectivore, feeding primarily on bees and wasps, and is so called because its forms an impromptu symbiotic partnership with mammals of many different species (notably a badgerlike African carnivore called the ratel, as well as the leopard, the rhinoceros, and man, among others). Its sense of smell is well developed, and it is probably by this means that it finds its way to a hive of wild bees. It attracts the attention of its mammalian accomplice by chattering continuously and flying off in the direction of the hive, pausing occasionally to wait for its larger and slower partner to catch up. When the latter has broken open the hive and robbed it of its honeycomb, the honey guide swoops down to feed on the honey and larvae and especially on the beesewax of the honeycomb.

The honey guide is the beneficiary of another symbiotic partnership—its intestines contain bacteria that can break down beeswax into digestible fatty acids. The honey guide is even reported to feed on beeswax candles on occasion, and Africans usually leave a piece of the honeycomb behind as a sort of finder's fee for the bird that led them to the hive. The honey guide also catches bees and wasps in flight; it has a very thick skin that protects it from stings and it emits a strong musky odor that acts as an insect repellent.

The honey guides are classfied into 4 diferent genera, 17 species in all; most of them are parasitic. One exception is *Indicator xanthonotus,* found along the southern slopes of the Himalayas, which builds its own nest. Apart from one other Asian species, the honey guides are native to Africa.

J.-F. T.

67
Sub.: Vertebrata
Cl.: Aves
O.: Charadriiformes
F.: Recurvirostridae

BLACK-WINGED STILT *Himantopus himantopus* 67

This elegant-looking European shorebird measures about 37 cm from beak to tail, with a wingspread of 69–74 cm, and weighs between 140 and 180 g; it is found in shallow water along the banks of freshwater lagoons and open marshes where the bottom is muddy and the vegetation not too thick. The clamorous nesting colonies are frequently displaced by the fluctuating water level of the marshes. The black-wined stilt spends the winter months in tropical Africa. There are just a few hundred nesting pairs in France, both on the Atlantic and Mediterranean coastlines and in the inland marshes; the adult birds generally live long enough to hatch out several broods of chicks, which helps to compensate for the very high mortality of the latter.

▶ The black-winged stilt feeds on aquatic insects and their larvae—diving beetles, flies and mosquitoes, libellulids (large dragonflies), and hydrophilid (diving) beetles—as well as worms, freshwater crustaceans, and small frogs. It seizes its prey on the surface of the water or forages along the muddy bottom with the long beak while its head is completely immersed; perched on its enormously long legs like a tiny stork, it able to seek its prey in much deeper water than other shorebirds that inhabit the same biotopes and thus has no direct competitors.

◀ The black-winged stilt makes its nest on the ground or on a tuft of marshgrass, and the entire colony will take up the hue and cry and flap off in pursuit of the crow, buzzard, or fox that attempts to raid and make off with the chicks. The adults birds have no natural enemies, are protected throughout most of their range, and are rarely molested by man, but like other shorebirds, their survival may be threatened by the draining of marshlands and similar land reclamation projects—especially so in the case of the black-winged stilt, since the entire European population is concentrated in a very small number of nesting areas.

The closely related black-necked stilt *(H. mexicanus)* is a seasonal resident of the Pacific coasts of North and South America, from Oregon to Peru; the Australian white-headed stilt is assigned to a different genus*(Cladorryhynchusleucocephalus)*.

J.-F.T.

BLIND SNAKES *Typhlops, Leptotyphlops* 68

Though these two families of primitive snakes (each represented by a single genus) appear to be descended from different evolutionary lineages, they are very similar in their habit and appearance. All are eyeless burrowing snakes, sometimes known as worm snakes, that have cylindrical bodies covered with smooth scales of uniform size (i.e. they lack the projecting ventral scutes used for locomotion by surface-dwelling snakes). The largest species, *Typhlops punctatus*, found in Africa, may attain a length of 80 cm, but 15–20 cm is a more typical length for its congeners, and the smallest Leptotyphlopidae are no longer than 7 or 8 cm and no bigger around than a pencil lead. Most blind snakes are found in tropical forests and a few in arid regions; both genera are widely distributed throughout the tropical and subtropical regions of the globe and, for obvious reasons, are invariably found in friable rather than sandy soil.

▶ Though they are clearly adapted to a fossorial (burrowing) rather than a terrestrial mode of life, the blind snakes can still move quite rapidly along the ground and are often encountered on the surface in damp weather, though it is not clearly whether they are actively seeking for prey or simply taking a shortcut over a patch of ground that is not well suited for burrowing. Blind snakes feed primarily on termites and the larvae of other insects and occasionally on other invertebrates; they are often found with their stomachs literally bulging with termites, all consumed in the course of a single feeding. Blind snakes appear to detect their prey by smell, but the relative contributions of the different olfactory organs (Jacobson's organ and the olfactory epithelium) have not been established; both are highly developed in both genera.

Previous investigators had been puzzled by the absence of of identifiable remains in the stomachs of some of the smaller worm snakes until Smith (1957) was able to observe the feeding behavior of a Mexican species *(Leptotyphlops phenops)* in a terrarium. *L. phenops* does not swallow a termite whole, but rather seizes the abdomen in its jaws and bites down vigorously, thus projecting the termite's internal organs into its mouth without requiring it to swallow the the chitinous exoskeleton.

◀ There are a large number of semifossorial snakes, many of them venomous, that prey primarily on other burrowing snakes, including *Typhlops* and *Leptotyphlops.* (It is perhaps worth noting that terrestrial, aquatic, and arboreal species tend to prey less frequently on other snakes, perhaps simply because they are presented with a wider choice of potential prey animals). Predators of this first type include the false coral snakes *(Anilius)* of South America, the pipe snakes *(Cylindrophis, Anomachilus)*, sunbeam snake *(Xenopeltis unicolor)*, and coral snakes *(Maticora, Calliophis)* of India and Southeast Asia, the coral snakes *(Micrurus, Micruroides, Leptomicrurus)* of North America, the mole vipers *(Actraspis)* of southern Africa, and the BANDY-BANDY *(Vermicella annulata)* of Australia.

Of these species, the mole vipers prey primarily and the bandy-bandy almost exclusively on *Typhlops.* Blind snakes are obviously vulnerable to attack by a great other many nocturnal predators during their occasional forays up to the surface—possbily including frogs, toads, and even large predacious insects in the case of some of the

68
Sub.: Vertebrata
Cl.: Reptilia
O.: Squamata
S.o.: Ophidia
F.: Typhlopidae et Leptotyphlopidae

smallerLeptotyphlops species—but the semifossorial species just mentioned are the only creatures that prey systematically on *Typhlops* and *Leptotyphlops*. H.S.G.

69
Sub.: Vertebrata
Cl.: Reptilia
O.: Squamata
S.o.: Sauria
F.: Anguidae

BLINDWORM *Anguis fragilis* 69

The blindworm, or slowworm, is a European legless lizard, the only member of the family Anguidae to be found in Western Europe. The adults are generally brown or gray and rarely exceed 50 cm in length; the juveniles are golden or silvery in color with darker undersides and a dark median stripe. The blindworm (which is not actually blind, incidentally) is found throughout Europe, with the exception of Ireland and southern Spain, and is generally encountered in moist or humid areas, such as meadows, hedgerows, lawns and gardens, and forest tracks, though it is not particularly fond of running water. It has been reported up to an altitude of 2000 m.

▶ The blindworm is inconspicuous in its habits, tends to be most active at dusk, and remains in hibernation from October until March or April. It feeds on earthworms *(Lumbricus, Allobophora)*, slugs *(Arion, Limax)*, snails *(Helix, Cepea*, etc.), woodlice *(Oniscus, Porcellio, Armadillidum)*, caterpillars, cockchafer and other insect larvae, and occasionally on adult beetles and small spiders. It moves rapidly through the leaf-mold layer or forest litter in search of prey, which it identifies by means of several exploratory flicks of its tongue. It seizes the prey animal and shakes it sharply and rapidly from side to side; its teeth are sharp and backward-pointing, which facilitates the ingestion of prey animals.

There is some seasonal variation in the size and frequency of its meals; it seems likely that, like many reptiles living in temperate regions, it feeds most actively in the spring and autumn to compensate for or to prepare for its lengthy hibernation. It is also worth noting that the blindworm pauses frequently to drink and, apart from the fact that it consumes a substantial numbers of earthworms, does yeoman service on behalf of the farmer and the gardener.

◀ Most large European birds, including diurnal raptors, domestic fowl, and corvids (rooks, magpies), will prey on the blindworm, as will a number of mammals, including wild boar and free-ranging hogs, badgers, cats, and probably hedgehogs and a number of other small terrestrial carnivores as well. The Montepellier snake and the dark-green snake also feed avidly on the blindworm, which is completely unaggressive and has no more effective means of defense than rapid flight (frequently expedited by its ability to detach its tail by autotomy).

While being handled by a human being, the blindworm wriggles and contorts its body vigorously; its fine smooth scales make it extremely difficult to hold on to, though its body is perfectly dry. It may also attempt to free itself by striking out with the pointed tip of its tail, which is not especially painful to humans but may be a more effective deterrent in the case of smaller predators.

The blindworm is roughly the same size and color as a young smooth snake *(Coronella)*, though as a very general rule, snakes and lizards can be distinguished by the fact that the lizard has movable eyelids. Similarly, there is a small Mediterranean skink found in southwestern France, the three-toed seps *(Chalcides chalcides)*, whose legs are extremely tiny and thus might readily be mistaken for a legless lizard at first glance—were it not for the fact that the three-toed seps moves with such remarkable speed and agility over the carpet of grass or moss on which it makes its home that it is unlikely to be seen at all except by a trained naturalist. J.C.

70
Sub.: Arthropoda
Cl.: Crustacea
O.: Decapoda
F.: Portunidae

BLUE CRAB *Callinectes sapidus* 70

This is one of the largest members of the family Portunidae (swimming crabs); the shell of the male, slightly larger than that of the female, may be as wide as 20.9 cm. The blue crab becomes sexually mature after it has moulted its shell 18 or 20 times, which usually takes between a year and 18 months altogether. Since copulation can occur only while the female's shell is soft, the male uses one pair of its legs to carry the female around with it for several days, pressed against the underside of its shell (a procedure known as "cradle-carrying"), until the female is ready to moult. The female stores the male's sperm in its body for several months and lays its eggs during the following spring or summer; the number of eggs laid by a single female is quite considerable, varying between 900,000 and 2,100,000. The eggs hatch in 15 days at an optimal temperature of 26° C.

The larvae emerge in the open sea but complete their development in estuarine waters of low salinity; the juvenile and adult crabs can live in water of extremely low salinity and are occasionally found in large rivers as much as 300 k from the sea. The adult blue crab is not very particular about its suroundings—it can tolerate temperatures ranging from 3 to 35° C and salinity from 0 to 44 g/l—and is normally found on sandy or muddy bottoms out to a depth of 30 m all along the Eastern Atlantic coastline, from New England down to Uruguay. The blue crab is generally found only as an accidental species along the coasts of Europe, though it has been succesfully introduced into certain parts of the Eastern Mediterranean. Maximum longevity is probably 3 or 4 years.

▶ The planktonic larvae feed on coastal plankton; the adult blue crab, unlike most crabs, is an active and aggressive hunter, quick enough to outswim a fish on occasion, though as a rule it prefers to seek out whichever food source is the most plentiful and requires the least effort to obtain. Between 30 and 40% of the blue crab's diet consists of bivalves—primarily mussels, clams (including Mactridae and Tellinidae, *Donax),* and oysters. It feeds on gastropods, 15–20% of decapods, amphipods, and microcrustaceans, the decapods including mud crabs *(Panopeus)* as well as *Palaemonetes, Macrobrachium,* and other shrimps that live in fresh or brackish water, the amphipods including *Talorchestia, Gammarus, Corophium,* and the smaller crustaceans including ostracods, acorn barnacles *(Balanus),* and Mysidae. An additional 15–20% of the blue crab's diet consists of small fish and less than 5% of marine worms and bryozoans, insects (beetles, diptera, waterbugs, hymenopterans, dragonflies) as well as hydrozoans and other jellyfish.

To catch free-swimming creatures such as *Nereis pelagica,* jellyfish, or small fish, it conceals itself on the sandy bottom and springs out from ambush; it sifts through the sand in search of amphipods and clams, and sometimes hunts for snails and insects on the beach. Many, though by no means all, of the animals it feeds on are sessile or slow-moving; in the case of the marsh periwinkle *Littornia irrorata,* it swims over the submerged mudbanks to which these snails attach themselves, selects those with the thinnest shells, cracks them open with its claws (which takes about 30 seconds), and devours them on the spot.

◀ The species name of the blue crab, *sapidus,* means "good-tasting," and its flesh is greatly appreciated by a variety of different predators, including man. The eggs are eaten by tiny marine worms, and the larvae may be devoured by any one of the myriad creatures that feed on zooplankton; it is estimated that at the moment of its emergence from the egg, the blue crab larva's chances of surviving long enough to reproduce are quite literally a million to one. Juvenile crabs are preyed on by coastal fish and seabirds, and the adults—once their shells have attained a width of 10–15 cm, at least according to North American practice—are eaten by man. The worldwide annual catch is estimated at 58,000 metric tons, most of it taken in United States and Mexican coastal waters; the U.S. fishery is concentrated in Chesapeake Bay, North Carolina, and Florida.

The genus *Callinectes* comprises 14 other species, the majority of which are found in tropical waters of the New World and West Africa; they are very similar to the blue crab in their morphology and behavior but are not as extensively exploited by man. The family Portunidae includes a number of other champion swimmers, notably the Old World velvet crab *Portunus pelagicus,* which is often found sculling along very far out at sea. *Ovalipes guadalupensis,* found off the coast of Florida, ia another portunid that conceals itself in the sand and leaps out at its prey from ambush, very much in the manner of the blue crab. It is especially active at night and feeds on fish, crustaceans, annelids, and brittle stars, and occasionally on mollusks and seaweed. P.N.

BLUEFIN TUNA *Thunnus thynnus* 71

71
Sub.: Vertebrata
Cl.: Osteichthyes
S.cl.: Teleostei
O.: Perciformes
F.: Scombridae

The bluefin tuna is the largest member of the mackerel family (Scombridae). It reaches sexual maturity at the age of 5, when it weighs about 50 kg. A full-grown adult measures 1.5–2 m, though strictly speaking the bluefin keeps growing throughout its lifespan, up to a maximum size of about 2.6 m and 500 kg; typical longevity is 15–30 years. The bluefin is a migratory pelagic fish that travels in small schools of 25 or more that may include members of other *Thunnus* species. *Thunnus thynnus* is found on both sides of the Atlantic and is a prodigious long-distance swimmer, sometimes covering up to 200 km per day in the course of its migrations; fish caught and banded off the coast of Morocco have later been recovered in the Gulf of Mexico.

▶ The adult bluefin is not a particularly specialized predator, though it feeds primarily on herring, sardines, menhaden, and other schooling fish as well as on flying fish, eels, and squid ; it also preys occasionally on ling (a relative of the codfish), and the young fry of its own species. Young bluefins feed on the fry of sardines, anchovies, and herrings, and on pelagic crustaceans; since it grows very rapidly and thus needs to maintain a high body temperature, a young bluefin may have to consume up to a quarter of its own weight every day.

The bluefin is the fastest swimmer of any pelagic fish, sometimes attaining speeds of up to 20 kph. An individual fish will sometimes attack a compact shoal of sprats, cutting a wide swath through their ranks and returning to make repeated passes until the quarry has been dispersed (or devoured) completely. Bluefins usually assemble in small hunting packs to pursue adult mackerel or herrings, plunging energetically through the midst of the school, beating the water with their tails to stun or disable as many of the quarry as possible, then immediately returning to snap them up. The shape of the bluefin's flukes enables it to execute rapid high-speed turns in the water, which it is often compelled to do when pursuing a fast-moving quarry such as the mackerel. Finally, as with many pelagic predators, the bluefin's back is darker colored (in this case, blue-black) and its belly white, which enables it to approach a school of fish from either above or below and still remain fairly inconspicuous.

◀ The adult bluefin can outswim a shark, and its bipolar coloration affords an additional measure of protection against predators that track their prey visually, such as the killer whale. Immature bluefins are frequently caught by dolphins and mako sharks, killer whales, and adults of the their own species, with the result that they frequently congregate in schools, though for defensive rather than aggressive purposes.

Human beings, of course, are by the far the most important predators of this species, and bluefins are caught for sport as well as for food, with hook and line in deeper water and in stationary nets along the shore. Their flesh spoils relatively quickly, which is why canning facilities are often located right next to the fishing wharves and why fresh tuna, though increasingly in demand in recent years (to the extent that bluefin stocks have been dangerously reduced by overfishing), still remains something of a luxury.

The albacore *(Thunnus alalunga)* is a staple food fish in North America as is the yellowfin *(T. albacares)* in Japan. The bigeye *(T. obesus)*, the little tuna *(Euthynnus affinis)*, and the bonitos *(Gymnosarda, Sarda,* and others), though substantially smaller than the bluefin, are all very fast swimmers and essentially similar in their feeding habits. J.-Y. S.

BLUE-LINE TRIGGERFISH
Pseudobalistes fuscus 72

72
Sub.: Vertebrata
Sup.cl.: Osteichthyes
Cl.: Actinogiterygii
S.cl.: Teleostei
O.: Tetraodontiformes
F.: Balistidae

This fish is found in coastal waters of the Indian Ocean, including lagoons on the outer edges of coral reefs, to a depth of 50 m. It attains a maximum length of 65 cm, though the average is closer to 30–50 cm; its lifespan is in excess of 10 years.

▶ The blue-line triggerfish is relatively slow-moving, patrolling the sandy bottom or the surface of a rocky outcrop by undulating its well-developed dorsal and anal fins. It feeds primarily on crustaceans, mollusks, and sea urchins, as well as on dead or disabled fish and coral animals; its strong jaws are able to crack the hardest shell or carapace. The triggerfish locates its prey by directing a steady stream of water along the bottom, uncovering creatures that have concealed themselves in the sand or mud, or, in the case of venomous sea urchins of the genus *Centre-*

chinus, washing the sand out from underneath them and exposing the unprotected oral surface, which normally rests directly on the bottom. The small size of its mouth compels the triggerfish to devour its prey piecemeal, a nibble at a time, rather than swallowing it whole; unlike most predators, it does not take the precaution of devouring its prey head first.

◀ The name "triggerfish" or *balistes* (from the Latin word for crossbow) has nothing to do with this fish's hydraulic undermining technique. Instead, it refers to the fact that the triggerfish has a row of well-developed defensive dorsal spines, the first of which can be erected and then locked into position by inserting the second, wedge- shaped spine into a groove in the base of the first—a method comparable to the trigger-lock mechanism of a crossbow. The triggerfish is capable of inflicting a nasty gash with its dorsal spine, and it will not hesitate to attack a diver or even a shark. The triggerfish also uses its spine to wedge itself into crevices in coral reefs and to discourage potential predators by clacking it against its body and grinding its teeth at the same time.

Accordingly, the blue-line triggerfish has few natural enemies, with the exception of sharks and barracudas on those rare occasions when it ventures into open water. Triggerfish, especially the larger ones, are much sought after by tropical-fish fanciers; the methods customarily used to capture them are extremely destructive of both the triggerfish and its habitat, since about the only practical way of extracting a live triggerfish from a crevice in a coral reef is by breaking up the coral around it or by disabling the fish with a toxic preparation that frequently proves fatal as well, within a few weeks if not immediately.

Another triggerfish species, *Balistiapus undulatus*, catches sea urchins by picking them up by one of their spines and dropping them repeatedly until all their spines are broken, then quickly seizing the urchin by its relatively soft oral surface. If the urchin somehow succeeds in escaping, the triggerfish goes into a terrible snit, clicking its teeth together and swimming around in all directions. Experiments have shown that the triggerfish is capable of removing a glass bell jar set over a prey animal by lifting it up with its jaws. Some species, notably *Balistes erythrodon*, subsist on plant as well as animal food.

J.-M. R.

BLUE SHARK *Prionace glauca* 73

73
Sub.: Vertebrata
Cl.: Chondrichthyes
O.: Carcharhiniformes
F.: Carcharhinidae

This voracious predator is found all around the world between 45° N. lat. and 45° S. lat., provided the water temperature is between 10 and 17° C. It typically measures 3–4 m, attains a maximum length of 6 m, and weighs between 100 and 150 kg. The blue shark is solitary (though it occasionally congregates in small packs) and swims very rapidly, sometimes attaining speeds of 50–70 kph, and often directly below the surface with its dorsal fin clearly visible. Its broad flukes provide stability when swimming at high speeds, and since the shark's respiratory apparatus requires a continuous intake of water, it can never come to rest on the bottom.

Its pectoral fins are also very long, and, like many pelagic predators, its underside is white and its back and flanks are darker colored (slate blue in this case), so that the outlines of its body tend to blend in with the background whether it is seen from above or below. Its teeth are roughly pyramidal in shape, with one serrated edge, and, like those of other sharks, are continually being worn down and replaced.

▶ The blue shark follows the migrations of large schools of pelagic fish, but will also attempt to swallow virtually anything that creates vibrations in the water—by struggling on the surface or merely by offering resistance to the ocean swell. Thus, examinations of the contents of its stomach (which is rarely full, due to the extreme rapidity with which it digests its food) have revealed the usual inedible or macabre assortment of objects, including, in the case of this species, glass bottles and a drowned dog, still wearing its leash. It has even been determined that an eviscerated blue shark that is immediately returned to the water will feed avidly on its own internal organs if these are flung into the water after it. The bodies of humans have also been recovered from the stomach of the blue shark, though there is some question as to whether this species will attack a living human, rather than merely feeding on the bodies of those who have died by drowning or from other causes.

Normally, however, this species feeds almost continuously on herring and mackerel, and occasionally on cod and squid as well; its sense of smell is very keen, and it follows the molecular trail left by a school of fish like a bloodhound. Its cryptic coloration enables it to approach very close to its quarry, and, as noted earlier, it is also very quick to detect vibrations in the water. Thus, as soon as the school gets wind of its approach and begins to take flight, it is the change in the frequency of the vibrations emitted by the quarry that triggers the shark's attack response. The blue shark also feeds on the refuse discarded from ships at sea, and sometimes plays a useful role as a scavenger around fishing wharves and in commercial harbors.

◀ This species is of no great economic interest to man, though it is sometimes fished in Japanese waters, since shark's fin soup, one of the costliest of Oriental delicacies, is made from its pectoral fins. The skin of the blue shark is also tanned into a medium-grade leather and has also occasionally been used to make a kind of rough sandpaper.

The feeding habits of all pelagic sharks are very similar to those of the blue shark. The mako shark *(Isurus oxyrhynchus)* swims rapidly enough to catch swordfish and marlin and, like them, is considered among the most challenging of gamefish. The tiger shark *(Saleoscerdo articus),* which is 4–5 m long and is found in coastal waters throughout the tropics, feeds on an immense variety of marine organisms (and organic material), including crustaceans, sea turtles, seabirds, saltwater crocodiles, and has been known to swallow plastic garbage bags in their entirety; because it is found in shallow water, it is reputed to be dangerous to humans, particularly to swimmers and fishermen hauling in their nets. J.-Y. S.

74
Sub.: Vertebrata
Cl.: Aves
O.: Passeriformes
F.: Paridae

BLUE TITMOUSE *Parus caerulus* 74

This small Eurasian titmouse measures 11–12 cm, with a wingspread of 19–21 cm, and weighs only 9–16 g. It is found in deciduous forests and woodlots, parks, gardens, and orchards (though it avoids coniferous forests, particularly pine and Norway spruce) in Europe (south of Finland and northern Scandinavia), as well as in North Africa, the Canary Islands, Asia Minor, the Caucasus, and Iran. Northern European populations are largely sedentary, though some may migrate to southern Europe when food becomes scarce in the winter. Average life expectancy is scarely more than 2 or 3 years, and juvenile mortality is quite high. The blue titmouse lives in pairs or in flocks, depending on the season, and is a lively, busy bird, venturing out to investigate the tips of the slenderest twigs, though often with its claws securely hooked around an overhanging branch. It flies swiftly and is quite comfortable on the ground as well; it builds its nest in a natural or artificial cavity of some sort.

▶ In spring, the blue titmouse feeds primarily on insects and insect larvae, which it catches in its beak (caterpillars, cocoons, chrysalids, beetle larvae); during the winter months, when the blue titmouse often searches for food in small flocks which may include birds of other species (tree creepers, nuthatches), its insect diet is supplemented with a high proportion of seeds, berries, and other vegetable material. It hunts for food primarily in concealed or inaccessible areas where larvae and cocoons are likely to be found—at the tips of twigs and branches, under the bark of trees, and on the undersides of leaves and buds. The daily intake of this species is roughly equivalent to its own weight.

◀ There may be some rivalry for nesting sites with other cavernicolous species, notably the great titmouse *(Parus major);* the blue titmouse has no specific predators, though both of these species are among the 16 songbirds most frequently taken by the European sparrow hawk. The blue titmouse is unafraid of man and is readily attracted to bird feeders and birdhouses, which is fortunate, since its natural nesting sites in hedgerows and hollow trees have been steadily disappearing in recent years.

The 60 species of titmice are found in all parts of the globe except for South America, Madagascar, New Guinea, and Polynesia. Apart from the great titmouse, or great tit, mentioned earlier, other congeneric species include the Old World marsh titmouse *(Parus palustris)* and the crested titmouse *(P. cristatus)* as well as the New World tufted titmouse *(P. bicolor).* Long-tailed titmice of the genus *Aegisthos,* represented in Europe by *A. caudatus,* are noted for their elaborately camouflaged nests fashioned out of mosses and lichens.

The penduline titmouse *(Remiz pendulinus)* weaves a basketlike closed nest out of plant fibers and animal hair that hangs from the branch of a tree; this species once commonly nested in the marshes of the Camargue, in the South of France, but has grown increasingly rare since about 1965. The so-called bearded titmouse *(Panurus biarmicus)* builds its nest in dense reedbeds in various parts of Europe, though its congeners are found exclusively in East Asia; these species belong to the thrush family (Muscicapidae) and are thus not true titmice, which, along with the chickadees of North America, comprise the family Paridae. J.-P. R.

BLUE WHALE *Balaenoptera musculus* 75

75
Sub.: Vertebrata
Cl.: Mammalia
O.: Cetacea
F.: Balaenopteridae

Typically measuring 25 m in length (maximum length is closer to 32 m) and weighing up to 150 metric tons (in the case of Antarctic populations), the blue whale is the largest mammal that has existed. The female is a little larger than the male, and longevity for both sexes is on the order of 40 years. This species is cosmopolitan in its distribution, though it prefers deeper, colder water and is rarely found near the coast; at one time it a was more numerous in the Southern Hemisphere, though as a result of several decades of unrestricted whaling in those waters, this no longer appears to be the case.

▶ Like other baleen whales (Mysticeti, or "whalebone whales"), the blue whale feeds on dense shoals of krill and other small marine organisms by filtering seawater through the curtain of flexible bony plates, known as baleen or whalebone, that hangs from its upper jaw. The baleen curtain of the blue whale, like that of the other rorquals (Balaenoptera), is relatively short, well adapted to the task of screening out the small (2–7 cm) euphausid crustaceans that are the primary component of krill in Antarctic waters. The blue whale also feeds on small fish in certain regions.

It is not known for certain how the blue whale detects and tracks its prey; it is possible that it uses some sort of ultrasonic echolocation system, as porpoises do (though the evidence for this is lacking), or it may simply rely on its sense of hearing, since a large bank of krill actually produces a considerable volume of sound and may be audible for some distance underwater. The blue whale swims with its mouth agape, taking in as much water as can be contained in the oral cavity, the capacity of which can be greatly increased if necessary by straightening out the deep external folds in the gullet.

It pumps the water through the baleen curtain in its upper jaw by pushing upward with its tongue and its lower palate, so that the krill and other appropriate-sized organisms it contains are left up clinging to the bristles on the inner surface of the slats of baleen. A blue whale in Antarctic waters consumes something like 4 metric tons per day, or about 480 metric tons by summer's end. During its winter migrations to its breeding grounds in more temperate latitudes, the blue whale eats very little and essentially lives off its reserves.

◀ It is possible that very young calves might be taken by killer whales, but man is (or was) the only significant predator of the adult blue whale. This species could not be hunted effectively with hand-held harpoons from an open boat, particularly since the body of the blue whale, unlike that of the sperm whale, sinks very rapidly after the animal is killed. It was only during the second half of the 19th century, with fast steam-powered vessels equipped with harpoon guns and pumps to inflate the giant carcasses and prevent them from sinking, that the blue whale fishery was launched in earnest.

Whaling vessels of this type made their first appearance in Antarctic waters at around the turn of the century, and the modern factory-ship system was introduced in about 1905. In 1925, factory ships were first outfitted with a ramp leading down to the waterline at the stern, called a slipway, while permitted the entire carcass of a blue whale to be hoisted aboard the vessel. (Prior to that time, the blubber layer was stripped from the carcass while it was still floating in the water.) This technical innovation soon resulted in a dramatic increase in the annual catch—in 1930, about 30,000 blue whales were taken in Antarctic waters alone.

Whaling operations tended to concentrate on the blue whale, simply because the large size of the quarry increased the economic efficiency of these costly expeditions to the Antarctic, though the size of the annual catch soon began to diminish steadily until 1964, when the taking of blue whales was finally prohibited by international convention. It was estimated at that time that there were no more than 1000 or at most 1500 blue whales remaining in Antarctic waters (as compared with an original population of perhaps 200,000 worldwide) and that if whaling operations against this species were ever to be resumed, it would take at least 50 years before the population could be restored to a level that would tolerate commercial exploitation.

Of the remaining rorqual species, only the sei whale (Balaenoptera borealis) feeds in a somewhat different manner, taking in water gradually through the corners of its mouth (cf. black right whale) rather than in enormous gulps like the blue whale. The size and structure of the baleen (particularly the length and diameter of the bristly filaments that coat its inner surface) varies somewhat from one rorqual species to the next, in proportion to the size of the prey animals habitually taken by that species. The common rorqual (B. physalus) also feeds on small crustaceans in polar waters as well as on herrings, capelins, and other small gregarious fish in the North Atlantic, and the beaked rorqual (B. acutorostratus) of the North Atlantic also feeds primarily on herrings and capelins. The sei whale (see above) feeds on very small planktonic crustaceans, whereas Bryde's whale (B. edeni), an equatorial species, feeds on small fish almost exclusively.

D.R

76
Sub.: Vertebrata
Cl.: Amphibia
O.: Anura
F.: Pipidae

BOETTGER'S CLAWED TOAD *Hymenochirus boettgeri* 76

The males of this species measure 32–36 mm, the females 35–40 mm. The Pipidae are a small family of primitive tongueless frogs, one of several that are included in the suborder Aglossa. Boettger's clawed toad (or frog) is found in forest pools in Cameroon; it is nocturnal and conceals itself in the layer of decaying vegetation on the bottom during the day. It is entirely aquatic, and if its pool should happen to dry up, it buries itself in the ground.

▶ Clearly the predatory technique of this tongueless aquatic toad is bound to be quite different from that of the higher anurans. In fact, *Hymenochirus* catches its prey by aspiration, in much the same way as the ALPINE NEWT, creating a sudden inrush of water by retracting its hyolarynx and depressing the ceratohyal arch that forms the floor of its mouth. It feeds on tadpoles and other aquatic invertebrates, and if it catches a tadpole or other comparatively large prey animal, it allows the water it has swallowed along with it to drain out again before it closes its mouth (while gripping the prey firmly in its jaws); with smaller prey animals, it closes its mouth and lets the water run out between its lips.

◀ The predators of Boettger's clawed toad include a variety of larger anurans, water snakes, fish, and mammals.

This method of feeding by aspiration, with the current of water being produced by the pumping action of the hyolarynx, is shared with the other pipids, notably *Xenopus* and *Pipa*. They track their prey by olfaction, and thus will feed on motionless prey animals; the predatory response of the higher anurans is generally triggered by the movements of the prey (cf. COMMON TOAD). The African clawed toad *(Xenopus laevis)* may begin to seize the prey with its forefeet before it has caught in its jaws, whereas the common toad only raises its forefeet to its mouth after the prey has already been captured. The Surinam toad *(Pipa pipa)*, noted for its large size and its peculiar reproductive habits, is often caught with hook and line or in fishtraps, and can be maintained on a diet of earthworms in captivity.

J.L.

77
Sub.: Arthropoda
Cl.: Insecta
O.: Coleoptera
F.: Carabidae

BOMBARDIER BEETLE *Brachinus crepitans* 77

This particular species is 7 to 10 mm long, lives in fields and meadows in Western Europe, on Corsica, and in Transcaucasia, and is generally found under rocks or tufts of grass. In nature, the bombardier's average lifespan is probably 2 two years (2 consecutive summers), though captive specimens have survived for as long as 5 years.

▶ Little is known of the bombardier beetle's behavior in nature, though like the other carabids, or ground beetles, it probably feeds on a great variety of other insect species, including several that are destructive to crops. The bombardier is dormant during the colder months and only active during the summer; the larva lives in leafmold or the uppermost layer of topsoil and probably feeds on ticks and mites (acarians) and earthworms for the most part.

◀ Scientific interest in the bombardier beetle has been largely focused on its remarkable talent for defensive chemical warfare. At the approach of a potential predator, the bombardier discharges an acrid chemical spray from its anus; this discharge is accompanied by a distinctly audible explosion and produces an intense burning sensation in the mucous membranes of the predator—thus discouraging any further interest in the bombardier on the part of birds, amphibians, and mammals (excluding entomologists).

The bombardier secretes a liquid from its anal gland that volatilizes when it comes in contact with the air; this substance is stored in a special pouch near the anus and then, when circumstances warrant, transferred (in quantities on the order of 0.02 to 0.4 microliters) by means of muscular contractions to a sclerotized "ignition chamber"—where it is mixed with air, producing an explosive chemical reaction followed by a projectile discharge of chemical vapor heated to a temperature of about 100° C.

Latin scholars may be surprised to note that the smaller bombardier species *B. explodens* produces a much less violent explosion than *B. crepitans*, involving only 0.01 to 0.05 microliters of chemical repellent. It was thought for some time that *Brachinus* was the only carabid genus to have developed this talent, but it appears that every member of this very extensive beetle family has evolved some sort of advanced chemical defense. *Procrustes purpurascens*, for example, can project its stinging spray over a distance of 30 cm; ground beetles of the genus *Calathus* secrete a chemical repellent that contains a large proportion of formic acid. G.B.

BOTTLE-NOSED DOLPHIN *Tursiops truncatus* 78

78
Sub.: Vertebrata
Cl.: Mammalia
O.: Cetacea
F.: Delphinidae

The males of this familiar "porpoise" species measure up to 4 m in length, the females up to 3.6 m. (To oversimplify somewhat, the term *dolphin* is preferred for small ceteceans with pointed beaks, such as this one, *porpoise* being reserved for the blunt-nosed species). The handsome *Tursiops* is noted for its streamlined contours and as a fast-moving, playful swimmer, sometimes attaining speeds of up to 30 kph. It travels in large schools and its behavior is governed by complex patterns of social interaction. The gestation period for this species is 1 year, and the females are very solicitious of the newborn calves, which measure about 1 m in length and weigh up to 12 kg. *Tursiops truncatus* and two closely related species (possibly subspecies) are found throughout the Atlantic and the Pacific, most frequently in shallow coastal waters and in tropical or temperate regions.

▶ The diet of the bottle-nosed dolphin is typical of the smaller and medium-sized representatives of the family Dephinidae (dolphins and porpoises) and consists primarily of small surface fish such as herrings, sardines, and anchovies. *Tursiops* can consume enormous quantities of these schooling fish, and the otoliths (calcified concretions found in the inner ear) of more than 3000 recently caught fish have been recovered from the stomach of a single individual. *Tursiops* also feeds on squid and cuttlefish and to a lesser extent on mollusks and crustaceans.

Tursiops is quick and agile enough to catch isolated prey animals with little difficulty; schooling fish are encircled by an entire school of dolphins, or herded into the shallows or into the lee of some obstruction, and devoured en masse. *Tursiops's* sonar echolocation system (which enables it to track the size, velocity, and direction of moving prey by high-frequency impulses through the water and monitoring the returning echoes) is particularly accurate and is thus its primary means of prey detection. *Tursiops* has between 80 and 100 sharp conical teeth, all basically identical in form, which are suitable for gripping and restraining prey but not for chewing; consequently, prey animals have to be small enough to be swallowed whole.

◀ Sharks and killer whales are the chief marine predators of this species, though *Tursiops* may be able to drive off an attacking a shark by ramming it with its beak. Dolphins are frequently, if unintentionally, caught in drift nets—currently a source of high-level political tension between the United States and Japan—and the smaller ceteceans are also fished commercially by the Japanese in their home waters. In addition, *Tursiops* is the dolphin species that is most often taken alive for purposes of scientific research and/or in order to display its adaptability and remarkable performing skills in aquariums and oceanariums on shore. V. de F.

79
Sub.: Arthropoda
Cl.: Crustacea
O.: Decapoda
S.o.: Brachyura
F.: Xanthidae

BOXING CRAB *Lybia tesselata* 79

Lybia tesselata is a very attractive little tropical crab; the width of its shell barely exceeds 4 cm and the shell itself is crisscrossed with streaks of yellow, pink, and brown, forming a network of polygonal patterns (hence, *tesselata,* "patterned like a mosaic"). The boxing crab is named for its curious habit of carrying a sea anemone *(Sagartia, Bunodeopsis, Phellia)* in each of its claws at all times; the anemones, which rather resemble boxing gloves in function as well as in appearance, may be almost as large as the crab itself and are made to serve as both offensive and defensive weapons. The boxing crab is found in the tropical Indo-Pacific, between the east coast of Africa and Hawaii; it lives in crevices in coral reefs, particularly living coral, down to a depth of about 30 m.

▶ The boxing crab spends much of its time concealed in its crevice, emerging occasionally to search for food. It moves its chelipeds (front claws) slowly back and forth in order to assist the anemones in their primary task, which is that of finding food for the crab. Occasionally, a fish is only grazed by the anemone's tentacles and is able to escape, but most prey animals are killed instantly by the anemone's venom and drop lifeless to the bottom or are seized by the anemone's tentacles. In either case, however, the boxing crab impales the prey with the pointed tips of its first or second pair of walking legs and snatches it away from the anemone; the venom that has been absorbed by the prey animal does not seem to interfere with the crab's digestive processes in any way.

The boxing crab is also capable of catching prey animals that come into contact with the sensory hairs on its walking legs, seizing them in its mandibles without any assistance from the anemone. Occasionally, the crab will take both of the anemones in one of its claws, picking up little bits of seaweed and small, soft-bodied prey animals and transferring them to its mandibles with its unencumbered claw. The boxing crab's claws, however, have been considerably modified in order to serve as a proper perch for the anemones; they are equipped with fine toothlike projections, which allow the crab to hold the anemones quite securely, but are incapable of crunching the shells of mollusks or other calcareous material.

◀ When threatened, the boxing crab brandishes its anemones like boxing gloves and, if this intimidatory tactic is unsuccessful, flings them at the aggressor and retreats into its crevice in the coral. It also sometimes plays dead, like the hermit crab. These defensive maneuvers have been observed in captivity, but we do not know what sort of predators they might be employed against in nature. This species has been commercially exploited, on a modest scale, for the benefit of saltwater aquarium enthusiasts.

There are several related species, considerably rarer than the boxing crab, that make use of sea anemones (in exceptional cases, of nudibranchs and small holothurians) for the same purpose, but very little else is known of their habits.

P.N. and C.V.

80
Sub.: Echinoderma
Cl.: Ophiuroidea
O.: Ophiuridae
F.: Ophiotrichidae

BRITTLE STAR *Ophiothrix fragilis* 80

This species is very common in the eastern Atlantic, from Iceland to South Africa, as well as in the Mediterranean; it is found from just off the shoreline out to a depth of 1000 m, frequently in dense aggregations along the bottom (as many as 300 individuals per sq m) that many contain several layers of brittle stars stacked up on top of one another. *Ophiothrix fragilis* is generally found on a muddy substrate containing a high proportion of organic sediment, in a clump of wrack grass, or on the surface of one of the many sessile organisms, living or dead, that litter the ocean floor, including sponges, tunicates (specifically *Microcosmus),* hydrozoans, gorgonians. (An organism that lives on the surface of a plant or animal in this fashion is said to be epibiotic).

The sexes are differentiated, and the eggs are fertilized in August and September; the free-swimming larva (called an ophipluteus) has 8 arms, two of which are very long and act like outriggers, enabling the larva to stay afloat in the water until it is time for it to descend to the bottom and undergo its metamorphosis. When they arrive on the bottom, the juveniles attempt to attach themselves to another organism, most frequently an adult brittle star.

▶ Most frequently, the brittle star extends 2 or 3 of its long flexible arms into the water to collect plankton and other organic nutrients with its adhesive tube feet (podia); the brittle star's habit of congregating in dense clumps on the bottom greatly increases the efficiency of this process, and the nutrients that are collected in this manner are gradually transferred to the brittle star's mouth, which is located on the ventral surface of the central disk. The digestive tube is a simple pouch confined to the space between the arms on the central disk.

When swells or currents interfere with the orderly collection of nutrients suspended in the water, the brittle star retracts its arms and begins to crawl along the bottom, browsing on sponges, bryozoans, hydrozoans, and feeding on small vagile crustaceans; the arms of the brittle star are strong enough to detach bits of flesh from a dead animal, and brittle stars can frequently be observed in the process of feeding on the arms of other members of their own species, including the juveniles. Like other starfish, however, brittle stars are able to detach and regenerate their arms by autotomy.

◀ Like many other echinoderms, *Ophiothrix fragilis* frequently becomes the prey of large carnivorous starfish of the families Asteriidae and Luidiidae; it is also preyed on by a variety of bottom-feeding fish and large crustaceans, including the lobster. Some authorities maintain that lobsters are more numerous in areas where brittle stars are especially prevalent, though they also pose a serious inconvenience to commercial fishermen; a dense clump of brittle stars can add several tons to the weight of a dragnet in a matter of minutes, often making it very difficult to haul it back up to the surface.

The genus *Ophitohrix* comprises about a hundred species, more than any other genus of brittle star (class Ophiuroidea), and is very common throughout the oceans of the world (excluding the polar regions), though the physical similarity between any two of these species may not be particularly pronounced. In tropical seas, *Ophiothrix* is frequently found on coral reefs, and in the Mediterranean, the most abundant species is *O. quinquemaculata,* which is very closely related to (some would even say indistinguishable from) *O. fragilis.*

A.G.

BROWN BEAR *Ursus arctos* 81

81
Sub.: Vertebrata
Cl.: Mammalia
O.: Carnivora
F.: Ursidae

The brown bear ranges throughout the Northern Hemisphere, though it is becoming increasingly rare in Western Europe. The remnant population in the French Pyrenees, for example, has been reduced to 15 or 20, though brown bears are considerably more numerous in Eastern Europe, Scandinavia, and especially in European Russia. North American subspecies include the grizzly and the Kodiak bear, or Alaskan brown bear, which is the largest of living bears.

The male brown bear measures up to 2.5 m in length; the females are slightly smaller. Individuals vary considerably in bulk, between 60 and 350 kg in the case of the European subspecies. The brown bear spends the winter in its den, which is generally a cave or a cleft between two large boulders, sleeping on a bed of moss and dry leaves. Unlike smaller creatures such as the marmot or forest dormouse, the brown bear does not go into deep hibernation, and its body temperature remains at 29–34° C. The cubs are born in the female's den at the height of winter;

bearcubs are hairless, tiny, and extremely underdeveloped at birth, which gave rise to the ancient belief that they emerge as formless lumps of living matter that have to be "licked into shape" by their mother. Maximum longevity for this species is on the order of 35 years.

▶ The brown bear is omnivorous, depending primarily on its impressive strength and its excellent sense of smell to provide it with a wide assortment of foodstuffs. Its eyesight and hearing are both mediocre (unlike most carnivores, its external ears are not habitually held erect), and with its massive body and plantigrade gait (which is to say it puts its weight down on its heels when it walks like a human being, rather than on its toes like a lion or tiger), it is ill equipped to pursue a fast-moving prey animal. The brown bear's enormous canines can easily shear through flesh, though its flattened molars are more suitable for crushing and grinding grass, roots, bulbs, fruit, and other vegetable food, which furnish the greater part of its diet.

As well as for its size and strength, the brown bear is justly celebrated for its resourcefulness in finding food. To begin with, it is a notorious beehive raider; it starts out by overturning the hives and smashing them open, then picks up the frames one by one and delicately scoops out the beeswax and honey from every cell in the hive with the tips of its claws. It is also fond of sugary fruits and can often seen browsing in blueberry and huckleberry patches during the fall and winter. Occasionally a brown bear will use its claws to loosen a long strip of turf and then roll it up like a carpet in order to feed on the grubs and insects that are buried underneath. Brown bears also spend a great deal of time digging up the ground, partly in the hope of flushing a rodent or other burrowing creature, partly for the sheer enjoyment of it. Hillsides frequented by brown bears may be covered with numerous traces of this pastime.

The Kodiak subspecies lives on salmon during the spring and on roots and grasses during the rest of the year. The bears wade out into the ocean when the salmon begin to run and then follow them upriver, scooping up the salmon with their paws and flipping them onto the bank, then dextrously separating the top filet from the backbone, turning the salmon over, and repeating this operation on the other side. Kodiak bears will often station themselves at the edge of a waterfall, a stretch of rapids, a fish ladder, or any other spot where the fish will be at a particular disadvantage.

Brown bears occasionally (though very rarely) develop a taste for fresh meat and begin to prey on livestock, though most bears will turn carnivore only when their normal vegetable food is unavailable. The brown bear has been observed to feed on stags, wild boar, and European elk (moose), particularly those that are sick or otherwise enfeebled. Once it has brought down its quarry, it begins by devouring the viscera (including the udder of a female ungulate), then buries the rest of the carcass and comes back to retrieve it later. The brown bear also feeds occasionally on carrion.

◀ Bearcubs are occasionally attacked by wolves, but man remains the principal predator of this species. The female brown bear is noted for the ferocity with which it defends its cubs. It may advance on the intruder, hissing and snorting threateningly, or it may even attack without warning if the cubs are very close by. The female will not normally abandon its cubs, even to give chase to a fleeing adversary, though if there is time to send them up a tree while the adversary is still approaching, the female bear may either turn back to confront him alone or pretend to run away in order to lure him away from the cubs.

The North American black bear *(Ursus americanus)* is also very catholic in its tastes, feeding on frogs and crayfish as well as on roots and grasses, and very occasionally on livestock. This species is noted for its habit of plundering campgrounds, trailers, and isolated cabins, and is also very fond of sugar; its numbers have been substantially reduced in recent years by poaching—largely because the gall bladder of the black bear is especially prized as a remedy in traditional Chinese medicine.

The Asiatic black bear *(Selenarctos thibetanus)* feeds primarily on fruit, though it may occasionally bring down a creature as large as a calf or a small yak. The sloth bear *(Melursus ursinus)* of India and Sri

Lanka feeds on termites and other insects. Its teeth are rudimentary, and its feeding habits are fairly similar to those of the aardvark or the pangolin. It attacks a termite mound by digging a small hole in its base, then purses its protruding lips into a sort of funnel, inserts them into the hole and vacuums up the termites that have arrived to inspect the damage. The Malayan sun bear *(Helarctos malayanus)* is primarily a vegetarian, though it also feeds on termites, which it laps up with its tongue like an anteater. The spectacled bear *(Tremarctos ornatus)* lives in the Andes and is a strict vegetarian. J.-F. B.

BROWN RASCASSE *Scorpaena porcus* 82

This Mediterranean benthic fish, known to ichthyologists as the small-scaled scorpion fish, measures between 10 and 20 cm (rarely exceeding 25 cm) and has achieved some celebrity even in English-speaking countries as one of the key ingredients in bouillabaisse, the robust Provençal fish chowder that contains as many as 6 different species of fish and is heavily flavored with garlic and herbs. The brown rascasse is found along rocky coastlines (at depths ranging from 0 to 800 m) from the Mediterranean up to the Black Sea and the Bay of Biscay, and occasionally off the south coast of England. It is nocturnal and quite sedentary, rarely venturing far from the cleft or crevice in the rock in which it makes its home.

▶ Like many scorpion fish, the rascasse hunts from ambush, concealing itself under a rock or an overhanging ledge and darting out at small fish (Sparidae, Gobiidae) and crustaceans that pass within striking distance. The rascasse is quite voracious and swallows its prey at a single gulp, so that it is obliged to form a accurate visual estimate of the size of its victim before it strikes.

◀ Like other scorpion fish (including the LION-FISH), the rascasse is well protected against predators: It has a large head with jagged spines projecting from its gills, and both these spines and the vanes of its dorsal fin are associated with venom glands. The toxins produced by the brown rascasse are not dangerous to man, though they are sometimes said to have a pronounced sedative effect when consumed in the form of bouillabaisse. The skin of the rascasse is covered with irregular excrescences, reminiscent of shreds of seaweed or other aquatic plants, which not only have considerable camouflage value but also play a role in attracting prey animals. The brown rascasse is sometimes caught in trawls, more often with hook and line, and in spite of the high prestige enjoyed by bouillabaisse in the culinary world, this species does not support an organized commercial fishery.

The Mediterranean scorpion fish *(Scropaena scrofa)* may attain a length of up to 50 m and is found in the Eastern Atlantic, from Britain (rarely) to the south of Morocco, as well as in the Mediterranean. Its habits are similar to those of the brown rascasse, though it is also found on sandy substrates at depths ranging from 0 to 500 m. It is fished commercially with trawls in the Mediterranean. J.-M. R.

82
Sub.: Vertebrata
Sup.cl.: Osteichthyes
Cl.: Actinopterygii
S.cl.: Teleostei
O.: Scorpaeniformes
F.: Scorpaenidae

BURBOT *Lota lota* 83

This freshwater fish is usually about 40 cm long and weighs about 500 g at maturity, though individuals as long as 1 m and weighing between 25 and 30 kg have been reported from Siberia. The species is found throughout much of Eurasia (excluding Italy and Spain, for example) and North America (except south of of 45° N lat.) up to an altitude of 1200 m, particularly in lakes and in streams where the water is clear and the current fairly sluggish. The burbot is nocturnal and takes refuge under a rock or in a hole in the bank during the day. It is also the only member of the family Gadidae to live in fresh water, though there is one geographic race that inhabits the brackish shallows of the eastern Baltic.

▶ The hatchlings immediately begin to feed on plankton, and the young fry are extremely partial to mayfly larvae—which they root out from underneath stones and gravel—as well as mollusks and crustaceans. Older burbot are very active predators, preying (in the Old World) on perch, roach, and gudgeon as well as numerous aquatic arthropods, particularly crayfish; they also devour substantial quantities of the roe of other fish. The burbot is a good

83
Sub.: Vertebrata
Cl.: Osteichthyes
S.cl.: Teleostei
O.: Gadiformes
F.: Gadidae

swimmer, attaining speeds of about 10 kph, and it appears to rely primarily on olfaction rather than vision in locating its prey. With a density of 810,000 rods (for night vision) and 400 cones for (daytime vision) per sq mm, the burbot's retinas could be said to be about 40 percent as well endowed with neural receptors as are its olfactory centers.

◀ The burbot, especially in its juvenile stages, is preyed on by a number of other creatures, including pike, black bass, watersnakes (colubrids), herons, and ospreys. The burbot's fondness for fresh roe may extend to that of its own species; roe expelled by the females near the surface is often snapped up by the males as it descends toward the bottom. In the Old World especially, the burbot is caught with weirs and seines as well as with hook and line; its firm flesh is greatly appreciated by connoisseurs, and it is still more of a gourmet delicacy than an important commercial food fish. F.M.

C

84
Sub.: Arthropoda
Cl.: Crustacea
O.: Decapoda
S.o.: Brachyura
F.: Calappidae

CALAPPA *Calappa granulata* 84

Calappa is the name given to several species of box crab that live in European waters. This particular one is found in the Mediterranean and in the Atlantic between the Azores and the coast of Portugal; its pinkish, roughly spherical carapace is dotted with little red knobs, or tubercules (whence *granulata),* and can measure up to 12 cm across.

The carapace is rounded in front and wider in the back with two ledgelike extensions, under which its rather spindly legs can be safely concealed. When threatened, the box crab's well developed pincers fit together into a sort of shield, which it folds back to protect its carapace somewhat in the manner of a boxer protecting his face with his gloves. This species is found at a depth of between 30 and 150 m; it often buries itself in the sand with only its eyestalks and adjacent regions of its carapace protruding, while its maxillipeds (the set of mouth parts directly behind the mandibles and maxillae) are pressed tight against the front of its carapace to form a sort of snorkel tube. This species probably lives several years.

▶ The calappa detects its prey—primarily snails, tooth sheels (scaphopods), and other mollusks—by tapping the sand with its forelegs, then enticing the prey animal up to the surface by means of rapid, pattering movements of its legs. The calappa's pincers are asymmetrical, with one of them sharp and elongated, the other (often the right) provided with a conical toothlike projection on its movable "thumb" that strikes against two blunt projections on the fixed portion of the pincer. First, it seizes the prey between its pincers and carapace. Next, it rolls the mollusk from side to side for a moment, as if it were playing with it, before breaking a hole in the edge of the shell with the sharper of its two pincers; it inserts the movable portion of its other pincer until the "tooth" on the inside is lined up against the two blunt projections on the outside. Finally, it cracks the shell by exerting pressure at this point and extracts the mollusk with its maxillipeds or with the pointed tips of one of its legs. When the calappa is buried in the sand at the bottom of an aquarium, it will spring from concealment with startling rapidity when presented with a little piece of meat, shrimp, or fish; it probably catches hermit crabs and other nonsedentary prey animals in much the same way.

◀ The shield formed by the calappa's interlocking pincers seems to be an effective defense against predators, though like all crabs it is vulnerable after it molts its shell. We have no specific information on its predators—apart from the fact that calappas are frequently offered for sale in waterfront markets in the South of France. P.N.

85
Sub.: Vertebrata
Cl.: Reptilia
O.: Squamata
S.o.: Ophidia
F.: Colubridae

CALIFORNIA KINGSNAKE
Lampropeltis getulus californiae 85

This colubrid typically measures between 80 and 130 cm though it may attain a length of up to 2 m; its thick cylindrical body is generally light yellow with black or chestnut-colored bands. This subspecies is found in several Western states (including southern Oregon, California, southern Nevada, southwestern Utah, northwestern Arizona) and in northern Baja California as well as in a variety of different habitats, including forests, marshlands and the banks of streams and rivers, cultivated fields and meadows, praries and semidesert plains, and in mountainous regions up to an altitude of 2300 m. The kingsnake is diurnal throughout most of the year and especially active around dawn and dusk, though during the height of the summer it tends to leave its burrow primarily at night.

▶ The California kingsnake is eclectic in its feeding habits, though it appears to prey primarily on small mammals and lizards and occasionally on birds, other snakes, the eggs of birds and reptiles, and (in the case of the juveniles) on frogs and insects. All the kingsnakes are marauding hunters, seeking their prey along the ground as well as in underground galleries and among the branches of shrubs and bushes. The kingsnake locates moving prey animals by sight as well as by smell, stationary prey animals and the eggs of birds and reptiles by smell alone. Eggs and nestlings are devoured on the spot, and small mammals are seized in the kingsnake's jaws, then encircled with one or more of its coils, and killed by constriction.

The kingsnake sometimes crushes a prey animal against a rock, a branch, or some other hard surface, and dispatches rodents and other burrowing animals by pressing them against the walls of their burrows and smothering them. A kingsnake that is customarily fed live mice in captivity may seize one of its victims in its jaws while simultaneously encircling and crushing three or four more in its coils, though it nature it is rare for a constrictor to attempt to kill a prey animal by constriction without first catching it in its jaws. In either case, the muscular pressure of the kingsnake's constricting coils is sufficient to cause apoplexy in a small rodent within a few seconds; lizards are generally killed by suffocation, though the smaller species may simply be swallowed alive.

When the kingsnake gets wind of another snake, venomous or otherwise, it increases its speed until it has almost overtaken the quarry, then tries to clamp down its jaws somewhere (not always directly) behind the head. Next, the kingsnake wraps its body around the other snake's and gradually works its jaws toward the head; the kingsnake enjoys a high degree of immunity even to the venom of the rattlesnake and need not concern itself especially with the quarry's attempts to defend itself. Once the kingsnake has gotten a firm grip on the other snake's head, it attempts to crush it in its jaws while continuing to exert presure on its body by pressing it against the ground or squeezing it in its coils.

The kingsnake starts to swallow its prey, which is generally still living though considerably battered, by slowly engulfing it with alternate motions of its upper jaws. The kingsnake generally preys on snakes that are considerably smaller than itself, though in periods of intense predatory activity it is likely to bring down a quarry that it is too large for it to swallow. The kingsnake is not averse to feeding on its own kind, though it does avoid a genus of watersnakes *(Nutrix)* that defend themselves by secreting a foul-smelling substance from their cloacal glands.

◀ The various *Lampropeltis* species range from southern Canada down to the Equator; they are found in a variety of different habitats, and thus enter into competition with a great many more specialized local species. They have no specific predators, but occasionally fall victim to raptors and corvids, small carnivores and other reptiles, including, as we have just seen, other kingsnakes.

One of these species, the milksnake *(L. triangulum),* is sometimes known as the false coral snake because its tricolor rings appear to mimic, in hue if not in sequence, the distinctive red, yellow, and black pigmentation of the venomous North American coral snakes *(Micrurus, Micruroides, Leptomicrurus).* It was originally suggested that this tricolor pattern had originally developed in the coral snakes as a warning signal, comparable to the rattle of a rattlesnake, directed at potential predators or heavy-footed ruminants and was consequently adopted by the nonvenomous milksnake and several other kingsnake species (including *L. zonata).* The difficulty with this explanation, however, is that while the venom of the true coral snakes is highly toxic, the mouths of these snakes are very small and their dentition poorly developed. The sort of predators that might pose a threat to the milksnake would probably never have been bitten by a coral snake—and if they had, they almost certainly would not have survived, so the effects of this conditioning would have been lost on them.

The herpetologist Robert Mertens proposed an ingenious alternative explanation, namely that the ring patterns of *both* the harmless colubrid *Lampropeltis* and the highly venomous elapid *Micrurus* (et al.) had originally evolved in imitation of a second group of colubrids, *Erythrolamprus,* venomous snakes with opisthoglyphic dentition whose bite is painful but rarely fatal. Thus, the formative experience of having been bitten by *Erythrolamprus* might reasonably be expected to serve as an object lesson to the potential predators of both the milk snake and the coral snakes. This very special case of protective mimicry—in which the coloration of a group of harmless species as well as group of highly toxic but rather ineffectual species converges on that of a group of genunely harmful species—has come to be known as Mertensian mimicry. D.H.

86
Sub.: Vertebrata
Cl.: Mammalia
O.: Carnivora
F.: Canidae

CAPE HUNTING DOG *Lycaon pictus* 86

The Cape, or African, hunting dog may look something like a spotted wolf, as its scientific names suggest (Greek *lykaios,* "wolfish," Latin *pictus,* "spotted"), and with its large, erect, rounded ears, poowerful jaws, and massive head, it looks quite a bit like a hyena as well. It attains a maximal length of 1.4 m and a maximal weight of 25–35 kg ; its tail is slender and tapering and about 40 cm long. As with most canids, the average life expectancy for this species is between 10 and 14 years.

The Cape hunting dog was originally found in all but the desert regions of Africa and most frequently in the vicinity of the game herds on which it preys, though it also follows predators in the hope of picking up their leavings. Intensively hunted in recent years, it is increasingly disappearing from those regions in which it was once most prevalent.

▶ Like the dingo and the wolf, the Cape hunting dog runs down its prey in packs, thus supplying the needs of an entire social group; this strategy enables it to prey on such large herbivores as Thomson's gazelle and Grant's gazelle. A very large pack may be able to bring down a zebra or a gnu, though smaller packs are restricted to yearling gnus and (only in exceptional cases) to older, enfeebled adults.

The Cape hunting dog lives in a burrow, generally one that has been abandoned by some other creature; this is the only safe refuge for the cubs when the pack is off hunting, and a female member of the pack is always left behind to watch over them. The size of a pack's hunting territory is about 50 sq km on the average, varying somewhat with the density of the game animal population. The pack first sets off on the hunt between 6 and 7:30 in the morning, and then again between 6 and 7 in the evening, or as soon as the heat of the day has begun to abate. The entire pack begins to howl until the pack leader gives the signal for the chase to begin; the pack stealthily approaches a large game herd, which generally ignores them until they are quite close by, at which point the entire herd begins to flee.

As the chase begins in earnest, the pack leader sets a steady, loping pace (40 kph) that the Cape hunting dog can sustain for a very long time. When one of the herd animals begins to tire and veers off from the main body, the dog pack deploys itself in a semicircle to prevent it from rejoining the herd. The pack closes in on its quarry in a mass rush, and since it lacks the sharp claws and the muscular strength of the great hunting cats, the Cape hunting dog tries to weaken the quarry by tearing off bits of flesh with teeth, eventually bringing it down when its viscera are exposed.

The carcass of a single large ungulate will barely suffice to feed the entire pack, each member of which requires about 3 kg of meat per day. The carcass is divided up on the site of the kill. The dogs bolt down large chunks of meat, some of which will be regurgitated when the pack returns to the den in order to feed the cubs and the female "nanny" that has stayed behind to look after them; the pups will even approach the adults and try to compel them to vomit up the scraps of meat they have torn from the carcass. The regurgitated meat will be sampled by every member of the pack in a sort of food-sharing ritual that is of great importance in maintaining and reaffirming the social bonds that hold the pack together. An individual Cape hunting dog is lost if it ever becomes separated from its fellows, since the survival of the individual is absolutely dependent on the cohesion of the pack.

◀ The Cape hunting dog obviously shares its home on the savannah and the open grasslands of eastern and southern Africa with a great many other predators, including the lion, cheetah, leopard, and hyena. Of these, the hyena is perhaps the more formidable competitor, and a hyena pack will often attempt to relieve a hunting dog pack of its kill; only the more numerous hunting dog packs will be able to resist these depredations.

The Cape hunting dog may compete with other carnivores, but it is not preyed on by them. As noted earlier, this species has been pitilessly exterminated by man from many areas in which it once was common—as with the hyena, the hunting dog's appearance and behavior fail to inspire the same respect and admiration as the great hunting cats. Probably only the extreme fecundity of the females (the average litter consists of as many as 10–16 pups) has enabled the species to maintain its numbers in recent years. P.A.

CENTIPEDE 87

87
Sub.: Arthropoda
Cl.: Chilopoda
O.: Geophilomorpha
F.: Geophilidae

The adult *Necrophloephagus longicornis* measures between 40 and 46 mm; its body contains about 70 mg of living tissue in addition to about 18 mg of chitinous scales, etc. This species of centipede actually has between 41 and 57 pairs of legs, all of which are present in its earliest juvenile stages, since the juveniles are morphologically similar to the adults (so-called epimorphic development). In most of the Geophilidae, postembryonic development is completed in 1 or 2 years.

N. longicornis is a familiar species in Western Europe and throughout most of the Mediterranean Basin and has also been introduced into the eastern United States, Newfoundland, and the Atlantic island of St. Helena. It occurs in a variety of habitats from dense deciduous forests to sunny open meadows and subalpine pastures (up to an altitude of 1800–2000 m) as well as in the mouths and intermediate photic zones of caves; like many Geophilidae ("earth lovers"), it is part of the fauna of the topsoil, most frequently encountered under rocks and rotting wood and in the deepest layers of the forest litter.

▶ *N. longicornis* is a nonspecialized predator that feeds on a variety of other invertebrates, including enchtraeids (members of a family of oligochaete worms), insect or mollusk larvae, juvenile centipedes (geophilomorphs or lithiobiomorphs); the same holds true for most of the family Geophilidae, the larger members of which may even be able to dispatch an earthworm. The centipede actively searches out its prey, which it seizes in its venomous pincers (forcipules). This species also functions as a scavenger, feeding on the bodies of recently deceased invertebrates; the extent to which its diet consists of dead matter rather than living prey is currently unknown.

Information on the means by which it captures and pursues its prey is equally sketchy, though these seem to be comparable to the hunting techniques of other centipedes (see *Lithobius forficatus*, for example). This species, like all the Geophilidae, is eyeless, but the filiform shape of its body enables its to enter and explore the smallest cracks and crevices in the soil at close range.

During its early larval stages, *N. longicornis* is subject to predation by larger centipedes and other carnivorous arthropods; it also frequently falls prey to spiders and large ground beetles during its adult molts. For the most part, since it is rarely encountered on the surface, *longicornis* is inaccessible to vertebrates, though it may be devoured by all sorts of burrowing, tunneling, or foraging predators—amphibians, reptiles, birds, or mammals.

N. longicornis frequently shares its habitat with other geophilomorph species, with which it enters into competition for the same food resources; these include *Geophilus proximus, carpophagus,* and *electricus* as well as *Scolioplanes acuminatus* and *Chaerechelyne vesuviana.* All of these relatively large carnivorous arthropods, occupying a position at or near the top of the food chain (as far as the fauna of the topsoil and forest litter are concerned), play an important role in regulating the populations of other species farther down the chain. J.-J.G.

88
Sub.: Arthropoda
Cl.: Insecta
O.: Coleoptera
F.: Cleridae

CHECKERED BEETLE *Trichodes apiarius* 88

As we have seen, there are a great number of wasp and beetle species that ensure an ample supply of food for their larvae by laying their eggs in a burrow that is already occupied by the larva of another insect. This small European beetle has evolved a subtler and more elaborate parasitic strategy in order to take advantage of the complex social structure of the honeybee *(Apis mellifera),* leaf-cutter bee *(Megachile),* and other hymenopterans. The bright, motley-colored adult beetle feeds on pollen and lays its eggs inside the same flowers that are visited by worker bees, so that the eggs of the checkered beetle are transported back to the hive in the nectar sacs of the workers.

▶ The larva of *Trichodes apiarius* will devour almost everything that it encounters inside the hive, with the exception of the adult bees and the honeycomb itself; this includes the eggs, larvae, and nymphs of *Apis mellifera,* as well as the hive's reserve supplies of pollen and nectar. A beehive that has been parasitized by beetle larvae emits a characteristic high-pitched sound, quick unlike the normal buzzing of the honeybee, which has earned this species the common name *clairon des ruches,* or "beehive bugler," in French-speaking countries.

The nests of 6 different species of leafcutter bees are known to be parasitized by *Trichodes apiarius,* though in this case the bee larva is encased in a cocoon and the mouth of each cell in the nest is plugged with little bits of leaves. Here, it will generally take the beetle larva 5 or 6 days before it is able to gnaw its way through this protective structure and attack its prey.

Experiments have shown that the larva of the checkered beetle is able to go for as long as 27 days without food (probably longer in nature), though there is undoubtedly a high mortality among the eggs and first-stage larvae. Here the choice of parasitic host is not made by direct intervention on the part of the adult beetle but by a sort of statistical preselection procedure (certain species of hymenopterans are attracted to the blossoms of the particular flowers frequented by the adult beetle), which is presumably a less reliable way of determining that the beetle larva will be "adopted" by the most suitable host species. The infestation of the hive takes place during July and August, and the beetle larva requires 4 or 5 months to complete its development. It has been observed that larvae that feed on pollen develop less rapidly than those that feed on the larval forms of other insects.

◀ Another species of checkered beetle, *Trichodes alvearius,* shares the habitat of the "beehive bugler," though the adverse effects of direct competition between these two species are mitigated by the fact that *T. alvearius* is more eclectic in its choice of parasitic hosts—15 hymenopteran species altogether, of which 11 belong to the genus *Megachile.* As far as human beings are concerned, this competition may not seem like a serious problem in the case of *Megachile rotundata,* which builds its nest on the eaves and gutters of buildings and is widely regarded as a nuisance. However, other leaf-cutter species do play an important role in the pollination of alfalfa and other leguminous crops, and attempts have accordingly been made by the French National Institute for Agricultural Research (INRA) to reduce the extent of these depredations—either by providing nesting sites for leaf-cutter bees in areas less likely to be frequented by *Trichodes* (large fields rather than smaller ones, e.g.) or by inducing *Trichodes* to lay its eggs on baited strips of velcro that can easily be removed and destroyed. G.B.

89
Sub.: Vertebrata
Cl.: Mammalia
O.: Carnivora
F.: Felidae

CHEETAH *Acionyx jubatus* 89

The cheetah is capable of attaining speeds of 115 kph, at least for very brief periods, and is known to be the world's fastest mammal. This distinction is suggested by a number of features of its external anatomy, including its long breastbone, very long, delicate legs, elevated hindquarters, its remarkable hinged backbone, and the powerful musculature in the lumbar region. Maximum length for this species is 2.9 m (including the tail), and its weight normally varies between 42 and 65 kg. As with many other members of the cat family, longevity rarely exceeds 16 years. Newborn cheetahs have a thick mane, like a hyena's, running down the neck and back, which will have disappeared by the age of 3 months. The adult cheetah has a thin black mask around its eyes that continues downward on either side of its muzzle to the jawline and gives the animal a rather sinister expression.

As with many specialized carnivores, the range and habitat, and indeed the very existence, of the cheetah is closely linked to that of its favorite prey, the gazelle. Thus, it is found on grassy plains and savannah interspersed with stands of small acacias on which gazelles and other herbivores can browse; the cheetah requires tall grass in which to conceal itself while it closes with its prey, since it cannot maintain pursuit speed for a distance of more than 400 m or so. It takes 15 minutes to recover its wind. The cheetah is widely but sparsely distributed in Africa and is occasionally found on the Arabian Peninsula and in Southern Asia as far east as India, where it is now extinct in the wild. It is now most often encountered in the large game parks of Kenya.

▶ The cheetah has excellent vision and surveys its vast hunting park from a hillock or other vantage point; it is also shy and wary, and will not attempt to defend its territory against other predators, though it patrols continually throughout the daylight hours insearch of prey. The territories of neighboring bands of cheetahs may overlap to some small extent, but this is never a source of friction, since hunting bands prefer to avoid confrontations with other predators, including members of their own species. Each band spends two or three days in the same locality and always comes back to the same place to sleep at night; a complete circuit of the entire territory can thus be accomplished in a week or so. The cheetah's habit of scent-marking with its urine is not intended, as was formerly supposed, to stake a claim to a particular territory but merely to leave some record of its passage through a given locality; it may also be of some use in attracting a mate.

The cheetah detects and tracks its prey entirely by sight. Its spotted coat enables it to conceal itself perfectly in the tall grass and to approach a grazing herd of gazelles without being seen (this maneuver is referred to as the "masked approach"). On the other hand, it does not take any precautions to conceal its scent and makes no attempt to remain downwind of its quarry. The cheetah is very patient, however, and will repeat this strategem over and over until it succeeds.

If a gazelle turns its head in the cheetah's direction, the cheetah instantly freezes in its tracks and breaks into a trot only when it has selected its victim—generally the weakest member of the herd, which often keeps its distance from the rest. If the cheetah's stalking maneuver has been successful, there is little likelihood that its quarry will be able to escape it during this last phase of the operation. The cheetah picks up its pace as the herd takes flight, overtakes the gazelle and knocks it off its feet with a blow of its paw, which is equipped with curved, nonretractable claws, immobilizes it, and kills it very quickly by sinking its canines into its neck.

The cheetah generally hunts during the early hours of the morning and rarely at night. The female usually hunts with the male and drags the carcass back to the spot where the cubs are waiting for it, and then lies down to rest as they begin to eat. The cheetah is very choosy about its food; it does not eat any of the internal organs except for the liver and usually only eats part of the animal's flesh. It buries the leftovers as soon as it has finished eating, though these will be most probably disinterred by jackals and vultures.

The cheetah's nutritional requirements are relatively modest; an adult will eat about 1.3 kg of meat every day for every 10 kg of body weight, so that a 50 kg cheetah can get by quite comfortably on 6.5 kg of meat a day. This is one reason why the cheetah tends to prey on smaller animals, though it will not hesitate to attack larger quarry if the latter is particularly slow-moving or otherwise easy to catch; in either case, there will be plenty of leftovers for other carnivores. Male cheetahs sometimes hunt in bands of 3 or 4 and share the carcass among them. As noted, the cheetah preys primarily on antelopes, including Thomson's gazelles, Grant's gazelles, impalas, and young gnus, as well as a variety of other creatures including zebras, warthogs, hares, and guineafowl.

◀ The lion is the only creature that actually preys on the cheetah, though the leopard may attack in order to relieve it of its kill; the cheetah is understandably reluctant to dispute the possession of a kill with a pack of hyenas. Young cheetahs, captured in the wild, have been trained as coursing animals since antiquity, though it generally takes several years before the cheetah will able to run down an oryx or other game animal at the behest of its human master. Even sheikhs and rajahs may have difficulty in locating suitable candidates for this training, since the cheetah is extremely rare in Asia, and this picturesque sport is rarely practiced today.

P.A.

90
Sub.: Vertebrata
Cl.: Aves
O.: Anseriformes
F.: Anatidae

CHILEAN TORRENT DUCK *Mergenetta armata* 90

The Chilean torrent duck is a small diving duck between 38 and 42 cm long with a long stiff tail that lives along mountain torrents and fast-flowing streams in the Andes at an altitude of 1000 to 4000 m. The torrent duck is highly territorial; each pair occupies a 2 or 3 km stretch of riverbank and makes its nest in a crevice in the rocks.

▶ The torrent duck braves the coldest and most powerful mountain currents to collect insect larvae and freshwater snails that crawl along or affix themselves to rocks or conceal themselves in crevices at the bottom of whitewater pools; its bill is narrow and flexible, and a typical prey animal is the larva of *Rheoplila*, measuring between 0.5 and 3 cm long. Very little else is known of its habits.

◀ The torrent duck has no particular predators or competitors (as noted below, there are only two other species worldwide that occupy the same ecological niche). It is very shy and easily disturbed by man, and its future may be threatened by the construction of dams and reservoirs, the pollution of streams, etc., though such occurrences have thus far been comparatively rare in the high Andes.

There are several species of torrent duck, all found in South America from Chile to Colombia; the genus *Merganetta* is regarded by taxonomists as a smaller relative of the merganser *(Mergus).* Two other species, the blue duck *(Hymenolaimus malacorhyncus)* of New Zealand and the harlequin duck *(Histrionicus histrionicus),* which ranges between Iceland and eastern Siberia, pursue a similar hunting strategy throughout at least part of their lives.

J.-F.T.

91
Sub.: Vertebrata
Cl.: Mammalia
O.: Primates
F.: Pongidae

CHIMPANZEE *Pan troglodytes* 91

The chimpanzee is the most familiar and the most manlike of the anthropoid apes, with its smooth-skinned face, its large, flat ears, and its extensive repertoire of "all too human" seeming gestures and facial expressions. It varies in height between 1 and 1.30 m and weighs between 45 and 80 kg. Its pelt is black and rather sparse and covers its entire body; the hair on its chin turns gray in later life. The chimpanzee is a protected species and lives in a number of isolated enclaves ranging across Central Africa from Senegal in the west to the great lakes of Uganda and Tanzania in the east, with the greatest concentrations (quite probably) to be found around the headwaters of the Niger and in the Congo basin.

The information in this article is based on studies of a group of chimpanzees that had been reintroduced into a national forest in Gabon in 1963; two very extensive field studies of this remarkable animal have also been carried out in Tanzania (notably by Jane Goodall and her associates at Gombe Stream National Park on Lake Tanganyika). The Gabon field study involved chimpanzees living in thick equatorial jungle where the trees may attain a height of 50 m (and observations of the chimps' daily peregrinations in search of food are made correspondingly difficult) as well as on the nearly treeless savannah; in either case, the chimps foraged actively during the day and then retired to their nests in the treetops or simply stopped to rest for the night.

▶ The Gabon chimpanzees get most of their nutrition from plant material (the typical diet consists of 28% foliage, 68% fruit by bulk) but supplement their protein intake by catching and consuming ants, termites, scorpions, and other invertebrates. The sting of this variety of scorpion, which is generally found under the bark of trees, is no more harmful than that of a wasp; the scorpion will be dispatched with a few quick blows of the chimp's bare fist and then devoured, along a few shreds of bark, and chewed for quite a long time before being swallowed. The occasional capture of a young warthog or a young colobus monkey provides the chimpanzee band with its only significant source of fresh meat. Scraps of the flesh of a young warthog—a larger prize of this kind is generally shared among the chimpanzee band—are eaten with a mouthful of leaves as a sort of vegetable garnish. Birds' eggs and nestlings are also a great favorite with chimpanzees and potential nesting sites are systematically explored by certain individuals.

However, all this animal and insect prey constitutes only a very small part (0.2%) of the chimp's diet; ants (or termites on the savannah) may comprise as much as 4 or 5% of the chimp's total intake. When foraging in the trees, the chimpanzee band searches for small ant colonies on the undersides of leaves, which are eaten up, leaves and all, in the customary manner. Jane Goodall was able to observe a more sophisticated ant-catching technique devised by the chimps of Gombe: a long straight branch is stripped of its leaves and introduced into the entrance of an anthill. Inside the anthill, the intruding object is immediately attacked by large warrior ants, which cling ferociously to the stick and are devoured by the chimpanzee as soon as the stick is withdrawn from the anthill.

◀ The forest-dwelling chimpanzees of Gabon have no natural enemies to speak of. On the savannah, chimpanzees are occasionally preyed on by leopards and other large felids. When chimpanzees become aware of the presence of a leopard, most of the band takes flight, but sometimes one of the dominant males will hold his ground and attempt to deter the aggressor by hopping up and down and shrieking while brandishing a stick or leafy branch. Though international traffic in chimpanzees for scientific and other purposes has now been halted, man still remains the greatest threat to the chimpanzee's survival. P.A.

CHINESE PANGOLIN *Manis pentadactyla* 92

92
Sub.: Vertebrata
Cl.: Mammalia
O.: Pholidota
F.: Manidae

At first glance, the Chinese pangolin may appear to be a reptile rather than a mammal, since its entire body (except for its underbelly) is covered with thick, overlapping scales. Its body measures 50 or 60 cm, its tail an additional 40 cm, and is relatively common in wooded grasslands and well as more sparsely vegetated arid regions of South China, Nepal, and Indochina. It lives alone or in pairs and is primarily active at night, spending its days curled up in a ball inside its burrow or in some other sheltered spot.

▶ The Chinese pangolin feeds exclusively on ants and termites, and its range is coextensive with those of the two termite species that it particularly prefers, *Captotermes formosaurus* and *Cyclotermes formosaurus*. It feeding habits are similar to those of the aardvark. It begins by making a breach in the termite mound with its claws, then leans back on its haunches and supports its weight on its tail—rather like a sportsman perched on a shooting stick—while it explores the termites' tunnels with its tongue, which it periodically retracts to ingest its booty.

Apart from its extensible tongue, the digestive tract of the Chinese pangolin displays a number of other remarkable adaptations: Its salivary glands have grown to enormous size in order to provide a continuous flow of adhesive saliva, and since there is no longer room for teeth inside its long tubular snout, the lining of the pangolin's stomach is studded with toothlike keratinized projections, so that it begins to chew its food (by means of vigorous contractions of the well developed musculature of its stomach) only after it has swallowed it.

◀ The Chinese pangolin defends itself by rolling itself up and tucking its head under its tail, thus presenting such potential predators as the tiger and the leopard with an impenetrable armored ball. Some of the smaller pangolin species (see below) have been known to defend themselves by spraying an aggressor with their urine, which is extremely pungent and corrosive.

Six other pangolin species, all assigned to the genus *Manis,* are found in the Old World. Two arboreal species, the white-bellied tree pangolin *(M. tricuspis)* and the long-tailed tree pangolin *(M. tetradactyla),* are found in the rainforests of West and Central Africa. The terrestrial Cape pangolin *(M. temnincki)* is actually more prevalent in Central and East Africa, and the giant pangolin *(M. gigantea)* is found on the open savannahs as well as in tropical forests of West and Central Africa. The Indian pangolin *(M. crassicaudata)* is also found on Sri Lanka, and the Malaysian pangolin *(M. javanica)* in Burma, Indonesia, and Indochina as well as in Malaysia.

P.A.

93
Sub.: Annelida
Cl.: Polychaeta
S.cl.: Polychaeta errantia
F.: Nereidae

CLAM WORM *Nereis virens* 93

This large polychaete worm measures between 20 and 30 cm. Its back is iridescent with purplish highlights, and like all nereids, it is equipped with two large pincerlike jaws at the end of its trunk. The clam worm is normally encountered on tidal flats and mudbanks at low tide; it lives in relatively cold water and is found along the coast of the North Sea as well as of the North Atlantic, from the Eastern seaboard to the Arctic and, though less frequently, in California. The clam worm digs long galleries lined with mucus in the upper 10 cm of the soft sand below the tidal mark; it may live for 2 or 3 years, but reproduces only once during its lifetime and dies shortly afterward.

▶ As with certain other polychaetes, there is still some controversy among specialists concerning the feeding habits of this species. The earlier authorities maintained that the clam worm was an indiscriminate carnivore, feeding on any small marine organism that ventured within reach of its long protectile trunk, whereas some now believe it to be an omnivore, feeding on vegetable material as well. Various experiments have recently been conducted in the laboratory to test the growth rate and chemoreceptive capabilities of the clam worm, and it can certainly be induced to feed on small cockles and mussels that are left outside the entrance to its burrow; it has no difficulty in crushing their shells with its pincers. It will also accept crushed shellfish and the flesh of other polychaete species; all these edible tributes are immediately seized in the animal's pincers and dragged down into the burrow.

The larvae live off their internal reserves for some time, then, in the aquarium at any rate, begin to feed on bits of seaweed, though this vegetable diet must be supplemented with shellfish after 3 weeks. The esophagus of the clam worm secretes large quantities of proteolytic enzymes, which also suggests an adaptation to a carnivorous diet. Instances of cannibalism have been observed in the aquarium; the clam worm does not feed when the temperature drops below 3°C. This species is often used as live bait by fishermen, and, as these experiments suggest, could easily be raised in captivity.

◀ The clam worm rarely leaves its burrow except during the mating season, when the males take to the open water and are vulnerable to predation by various fish, though the females remain in their burrows. The carnivorous annelid *Aphrodite,* the sea mouse, sometimes preys on the clam worm, though man undoubtedly remains its most significant predator.

Nereis diversicolor is a very common, wide-ranging species that feeds on nematode worms and other small marine organisms; it also constructs a sort of strainer out of mucus at the entrance to its burrow on which organic particles are allowed to collect and then are periodically devoured by the worm. *N. diversicolor* is commonly used as live bait by fishermen in the South of France. The eggs settle on the bottom, and the larvae are semipelagic at first, though they shortly return to the bottom to dig their burrows and begin their predatory careers as soon as the jaws and pharynx are fully developed.

C.M.

94
Sub.: Arthropoda
Cl.: Crustacea
O.: Decapoda
S.o.: Caridae
F.: Palaemonidae

CLEANER SHRIMP *Periclimenes sp.* 94

The genus *Periclimenes* consists of about 100 species, mostly found on coral reefs in tropical waters, about 10 of which are known to remove parasites (as well as other small marine organisms) from the gills, skin, and even from the mouths of fish. Four species of cleaner shrimp are found in the Mediterranean, 2 of them ranging outward into the Eastern Atlantic. Cleaner shrimps are rather inconspicuous in size (10–40 mm) but quite brilliantly colored; almost all species possess a heart-shaped spot on the abdomen.

▶ The cleaner shrimp clearly runs a considerable risk by crawling around on the gills and inside the mouth of a carnivorous "client" perhaps 20 times its size. The client fish signals its peaceable intentions by swimming up to a "cleaning station" frequented by several shrimp, often in association with an invertebrate protector equipped with venomous spines or tentacles such as a sea urchin, anemone, cnidarian, or crinoid. When the fish appears at the cleaning station, the cleaner shrimp shows itself and wiggles its antennae; the fish responds by immobilizing itself on the seabed and presenting or exposing the part of its body (gills, fins, flanks) that has become infested with parasites.

The shrimp approaches cautiously, systematically inspecting the fish's body and removing the parasites with the little hooked pincers on its first and second pairs of walking legs and devouring them. The fish sometimes changes color in the course of the cleaning process, which may take several minutes; the shrimp's "client" species typically measure from 4 to 80 cm and belong to several different families: Chaetodontidae (butterfly fish), Serranidae (sea bass and their relatives), Mullidae (red mullets), Bothidae (a type of flatfish).

◀ Cleaner shrimp are preyed on by fish of the family Serranidae, Scianidae (croakers), and others, though never the same species that present themselves at the shrimp's cleaning stations. The shrimp's first defense against predation is to take refuge among the spines or stinging tentacles of its invertebrate host. In addition, the cleaner shrimp's stripes, pigment patches, and other gaudy markings, like those of certain insects and stomatopods (a type of crustacean), may repel potential predators by mimicking the pigmentation of other organisms that are venomous, foul-tasting, or otherwise unattractive to predators. This type of mimicry is very common.

Brachycarpus biunguiculatus, another member of the family Palaemonidae, is associated with sea urchins of the genus *Diadema,* is active only at night, and mainly removes parasites from damselfish of the genera *Chromis* and *Eupomacentrus.* This species is somewhat more cautious in its methods than *Periclemenes,* however; it will clean the scales but not the gills or mouths of its clients, and it will not clamber up on the side of the fish's body if it is impossible for it to recover all the parasites while standing securely on the bottom.

Lysmata grahami, a shrimp of the family Hippolytidae, removes parasites from at least 20 different species of reef-dwelling fish (Apogonidae, Chaetodontidae, Muraenidae, etc.) as well as spiny lobsters, though it feeds on other nonparasitic organisms as well. The spiny shrimp *Stenopus hispidus,* a species of wide circumtropical distribution, feeds on parasitic isopods and copepods that it removes from the bodies of fairly large fish. There are a number of other creatures that collect their food in this manner, including one species of crab and a fish of the goby family.

P.N.

COELACANTH *Latimeria chalumnae* 95

95
Sub.: Vertebrata
Cl.: Osteichthyes
S.cl.: Crossopterygei
O.: Coelacanthiformes
F.: Coelacanthidae

Latimeria chalumnae is a member of an order of bony fishes that was once thought to have been extinct for several hundred million years. Since the fortuitous discovery of a living coelacanth in 1938, this opinion has had to be revised, but since then very little has been learned of its behavior of this species in nature. It is found only a few hundred meters off the shores of the Comoro Island , though also at a depth of between 100 and 400 m, since the Comoros are a chain of volcanic seamounts rising abruptly out of a deep abyssal plain. The coelacanth is a large fish, about 1.8 m long, and weighs about 8 kg; longevity is at least 11 years. This is an ovoviparous species, which means that the female incubates its eggs (each about 10 cm in diameter in this case) in its genital tract and gives birth to live young about 30 cm long.

▶ The coelacanth lives in salt or brackish water (since at these depths there is probably some admixture of fresh water) amidst the submerged volcanic outcroppings and grottoes that comprise the upper slopes of the Comoro seamount. Its behavior has only been observed in nature for a total of several hours, but it appears to be a calm, rather sluggish fish, roughly resembling the codfish in its general demeanor—in other words probably sedentary and not a very strong swimmer.

The "gombessa" is caught by local fishermen on hooks baited with a mackerel-like fish called the roudi *(Prometichthys prometheus)*, which is found at a depths etween 100 and 750 m. It seems likely that the coelacanth also preys on this fish spontaneously in nature, though its hunting strategy has not been directly observed. It also seems probable it hunts from ambush, like the codfish, concealing itself behind a rock or inside a crevice and darting out to give chase to any smaller fish that ventures too close to its lair.

◀ Luckily, the flesh of the coelacanth is very oily and not particularly in demand in the fishmarkets of the Comoros. It is caught accidentally only when it strikes the bait intended for more desirable food fish fish called the ruvet, a larger relative of the roudi, which is found to a depth of 700 m and may attain a length of 2 m. The coelacanth is also known to be very combative; the type specimen discovered in 1938 also distinguished itself by biting the hand of the captain of the research vessel that had hauled it up to the surface, and it is no simple matter to land one of these creatures in an outrigger canoe. Apart from *Latimeria*, there is one other extant coelacanth genus, *Malania*, which is found in the waters off nearby Madagascar.

F.M.

Coleodactylus amazonicus 96

96
Sub.: Vertebrata
Cl.: Reptilia
O.: Squamata
S.o.: Sauria
F.: Gekkonidae

This tiny South American gecko weighs less than a gram and measures barely 40 mm when its tail is intact, which makes it one of the smallest of living vertebrates. It is a member of a group of New World geckos (recognized by some taxonomists as the subfamily Spherodactylinae) that lack adhesive pads on the undersides of their digits and whose claws, ill adapted for climbing, are encased in a sheath of skin.

Coleodactylus lives and seeks its prey within the deep layer of leafmold and organic litter in the forests of the Amazonian basin and the Guianas, preferably in the less humid regions and sometimes at the edge of a glade or clearing. It is active during the day, and though hardly conspicuous, it appears to be one of the most abundant of the many tiny creatures of the forest floor. In a world of many perils, it moves very deliberately, somewhat in the manner of a chameleon, carefully planting its feet down on the flat side of an adjacent leaf or gripping the edge with its toes before it will risk its next step.

▶ About half of *Coledactylus's* diet is known to consist of collembolans (minute centipedelike denizens of the forest litter) followed, in order of frequency, by mites, crustaceans, and ants, though virtually nothing is known of the means by which it captures this prey.

◀ Like other lizards of the tropical forest floor, *Coleodactylus* relies heavily on protective coloration as its primary defense against predators. Its upper body is pinkish-beige with a sprinkling of dark chromatophores that enable it to adjust its basic coloration from a sort of pale khaki to pinkish gray. It has pink patches on its knees and elbows that are bordered with black; its underbody is dirty white, and the ridge of scales along its backbone are rounded and closely overlapping. Its eyes are large and almost frontally mounted; its snout is short and concave in profile.

The predators that this elaborate camouflage outfit is intended to deceive probably include birds, diurnal reptiles, and perhaps tarantulas and other large spiders. When alarmed, *Coleodactylus* abandons cautions and attempts to escape by means of a series of rapid wriggling movements in which the lateral undulations of the trunk play an important part.

The range of at least 4 other members of the subfamily Spherodactylinae overlaps with that of *Coleodactylus amazonicus*. First, there are 2 species of the genus *Gonatodes*, whose snout-to-vent length (excluding the tail, i.e.) exceeds 35 mm and whose relatively long legs and functional claws enable them to crawl along the bark of trees and thus seek their prey above ground level, though primarily at the base of stumps and among accumulations of forest debris. The larger of these 2 species,

G. annularis, attains a length of about 51 mm and seems to feed primarily on insects, particularly hymenopterans; the smaller, *G. humeralis,* is about 38 mm long and feeds primarily on colembolans.

There are also 2 smaller species, similar in size to *Coleodactylus amazonicus* as well as with respect to the articulation of their vertebrae, the relative shortness of their limbs, and the vestigial character of their claws: *Pseudogonatodes guianensis* and *Lepidoblepharis heyerorum*. The second of these 2 species is only known to us through a few specimens, but it seems probable that both also prefer to feed on colembolans.

J.-P.G.

COLOMBIAN HORNED FROG *Cetatophrys calcarata* 97

97
Sub.: Vertebrata
Cl.: Amphibia
O.: Anura
F.: Leptodactylidae

Each of the "horns" of the Colombian horned frog, a species found in northern Venezuela as well as northern Colombia, is formed by a triangular prolongation of the eyelid (as is also the case with the horned viper) and thus merely a rigid flap of skin rather than a true horn. The skull of the horned frog, on the other hand, is highly ossified, and its jaws are very wide and powerful; all the South American horned frogs are noted for their aggressiveness, and unlike other anurans, they rarely back down or hop away from a fight.

▶ The Colombian horned frog is reported to prey on small vertebrates, including other frogs and small mammals. Murphy (1975) provides an interesting account of a predatory strategem attempted by a captive specimen (100 mm long) in a terrarium; four smaller frogs of a different species *(Pleurodema brachyops,* 30–35 mm) had been placed in an adjoining terrarium and were clearly visible through the glass. The horned frog sat facing the *Pleurodemas* and began to rapidly flex and wiggle its toes (the third toe is extremely long) while horizontally extending its left hind foot perpendicular to its body.

After several seconds, it slowly raised its leg into the air and rotated it to display the sole of its foot, which it then stroked slowly along its flank, alternately flexing and wiggling its toes all the while. Next, it brought its leg forward until the foot was level with its external ear and rotated the foot back and forth to display both the sole and the instep, finally extending its leg still further, so that the foot was now directly over its brow ridge, and maintaining it in that position. The frogs in the other terrarium seemed transfixed by this performance, as the horned frog's wiggling, waving toes first crept along the ground, then wafted slowly through the air, and finally came to rest just a short distance in front of them. At that point, the horned frog made an abrupt bound in their direction, and its predatory ruse would no doubt have been successful if the glass wall of the terrarium had not been there to restrain it.

◀ When threatened or attacked from the front, the Colombian horned frog inflates its body by swallowing air—and continues to do so until the danger is past; it may also emit a warning cry. When touched from behind, it will turn its head and attempt to bite. The young frogs are probably preyed on by small mammals, though the adult has no predators that we know of.

A related species found in Brazil is even larger (100 mm long) and more aggressive. If a stick or metal rod is extended toward it, it will jump up and clamp its jaws down on it, hanging suspended for several moments before dropping back to the ground. It will also jump up at animals much larger than itself, with unmistakably agggressive intentions. The horned frogs in Argentina are said bite the lips of grazing horses in this manner; they are also reported, perhaps more reliably, to feed on other frogs and lizards, small birds and small mammals.

J.L.

COMMENSAL ANEMONE *Adamsia palliata* 98

98
Sub.: Cnidaria
Cl.: Anthozoa
S.cl.: Hexactiniaria
O.: Actiniaria
F.: Actiniidae

This species of anemone is found along the Mediterranean and Atlantic coasts of Europe from Norway to the south of France; it lives below the tidal zone, generally at a depth of 30 to 100 meters, on sandy, muddy, or gravel bottom. Its body is cream-colored and equipped with numerous tentacles (as many as 470). Its most noteworthy characteristic is its very frequent association with the hermit crab *Pagurus prideauxi*. In the earlier stages of its life cycle, it may attach itself to a pebble, but thereafter it is invariably found on a discarded mollusk's shell that is inhabited by *Pagarus prideauxi* or, in very rare instances, by a hermit crab of another species.

The morphology of *Adamsia palliata* is originally determined by the shape of the shell selected by the hermit crab, and subsequently by the growth of the crab itself. The anemone secretes a basal cuticle that attaches it solidly to the shell, then progressively increases the size of the crab's living quarters, so that neither party to the arrangement is obliged to move as soon as the crab has outgrown the original shell, as is the case in other hermit crab—anemone partnerships in which the anemone does not secrete a cuticle.

▶ These two species are involved in a particular variety of symbiosis that is known as commensalism, from the Latin *commensalis,* which refers to those who share the same table. This is almost literally the case with *Adamsia palliata* and *Pagurus prideuxi;* the anemone's broad coronet of tentacles is situated just below and behind the mouth parts of the crab. The tentacles of the anemone are thus well positioned to divert a reasonable proportion of the hermit crab's prey to its own use, as well as to gather up the sedimentary remains of the hermit's meal that collect on the sea floor.

◀ The anemone's contribution to the partnership is to protect the crab against its predators: fish, cephalopods, and other crabs. The tentacles themselves, because of their downward-pointing position and specialized food-gathering function, are of no great use in this respect, but the body of the anemone contains a number of whiplike whitish filaments called acontia, which are liberally equipped with stinging, venomous nematocysts. The acontia float freely inside the enteric cavity of the anemone with one end anchored to the base of the mesenteric membranes; when disturbed or threatened by a predator, the anemone projects them outward through its oral orifice and the perforations in its body column (cinclides).

There are numerous examples of commensal relationships between anemones and animals belonging to other phyla. The most frequently cited example is that of Calliactis parasitica and a different species of hermit crab, Pagarus bernhardus (known to the French as "Bernard the Hermit"). This association is less exclusive than the one involving Adamsia palliata and Pagarus prideauxi: the anemone's morphological structure is not modified in accordance with the form of the shell chosen by the crab, and the crab may belong to any one of several different genera *(Dardanus, Pagurus, Paguristes).* Other symbiotic partnerships between hermit crabs and anemones have been reported all over the world, including the deepest ocean trenches.

The association between the clownfish *(Amphiprion)* and anemones of the genera *Stichodactyla* and *Stoichactis* constitutes another well-known instance of commensalism, in which the anemone is the primary predator and the clownfish picks up the scraps; also, in this instance, both parties assist each other by providing protection from their respective enemies. In the same way, the tropical anemone *Bartholomea* gives shelter to several species of crustaceans, one of which, the pistol shrimp *Alpheus armatus,* defends it against one of its predators, the polychete worm *Hermodice carunculata.* M.v.P

COMMON BUZZARD *Buteo buteo* 99

99
Sub.: Vertebrata
Cl.: Aves
O.: Falconiformes
F.: Accipitridae

This sort of buzzard is not a vulture, like the North American turkey buzzard, but a large European hawk, about 51 to 57 cm long with a wingspread between 115 and 140 cm and a weight of 500–1060 g (for the male) or 700–1350 g (for the female). One banded specimen survived for at least 25 years, but considering the very high mortality rate among buzzard chicks, the average life expectancy at birth is probably just a few years. The common buzzard is in fact one of the most common of European raptors and is at home in a variety of habitats between sea level and the treeline, including small groves of trees and sparsely wooded areas as well as cultivated fields and pastureland interspersed with hedges and woodlots.

▶ The familiar description of a hawk "making lazy circles in the sky" applies perfectly to the common buzzard, which often patrols its hunting grounds from the air and is, in fact, a large, heavy, indolent bird (called *la cossarde,* "the lazy woman," in French). It may also station itself on a tree limb or other convenient vantage point and simply wait for a vole or other rodent to turn up at the entrance to one of its tunnels, whereupon the buzzard swoops down on its prey in the classic divebomber posture with its talons outstretched. It sometimes hunts on the ground as well, looking for rodents, insects, and earthworms in meadows or plowed fields. It takes to the air if necessary to spy out rodents or other prey in the tall grass or underbrush (assuming brisk and favorable winds) and systematically explores the sides of roads in search of fresh carrion. It pillages crows' nests and sometimes relieves smaller raptors of their kills.

The common buzzard is an opportunistic hunter, feeding on sick or wounded animals as well as on carrion. Voles, rats, and rabbits (in some areas) comprise the greater part of its diet, though it also feeds on moles, shrews, and the smaller mustellids, supplemented in some areas by lizards, vipers and other snakes, frogs and toads. In the spring, it feeds on nestlings (crows, magpies, blackbirds, pheasants, grouse, partridge); in the summer on insects (carabids, geotrupes, and other large ground beetles, crickets, and locusts); and in winter on earthworms. It is worth noting in connection with the common buzzard's reputation as a "destructive" species that voles typically comprise between 40 and 80 percent of its diet, the young of pheasants, partridges, and other gamebirds between 1 and 3 percent. The nestlings are fed initially on pieces of meat that have been shredded up by their mother, but by the time they are 2 weeks old, they are quite prepared to swallow an entire vole. The adults require about 90–100 g of nourishment a day, and a family of buzzards typically consumes 15 to 30 rodents a day during the nesting season.

The catholicity of its tastes and the abundance of its prey permit the buzzard to coexist with other "mousing hawks" (diurnal or nocturnal raptors that feed primarily on rodents) without feeling the pinch of competition. A pair of buzzards defends its territory by performing aerobatic displays to warn off other buzzards; the size of the territory varies according to the nature of the terrain (and the density of the vole population), from a few hundred hectares in lowland meadows to as many as several thousand in mountainous or forest regions. It is essentially a diurnal hunter, though it may continue its less strenuous hunting activities (sitting on its perch and looking out for voles and other rodents) until nightfall.

◀ The nestlings are sometimes caught by the marten or European wild cat, the adults by the eagle-owl or, very rarely, by the peregrine falcon or goshawk. Other than that, the common buzzard's only natural enemies are violent storms and man. It is now widely protected, though too late to prevent its extermination in many parts of its former range as a "chickenhawk" or a destroyer of small game. In spite of legal protection and the buzzard's much more important role as a destroyer of agricultural pests, this campaign of poisoning and snaring still continues in a great many rural areas.

The genus *Buteo* consists of 29 species, from the rough-legged hawk *Buteo lagopus,* of widespread circumpolar distribution, to others very limited in their range, notably the Galápagos hawk, *Buteo galapagoënsis,* which is confined to a few tiny islets of the Galápagos archipelago. There are a number of other European raptors similar in size and roughly similar in apparance to the common buzzard: the honey buzzard *(Pernis aprivorus),* the goshawk *(Accipter gentilis),* the black kite *(Milvus migrans),* and the booted eagle *(Hieraetus pennatus).*

J.-F. T.

COMMON CHAMELEON *Chamaeleo chamaeleon* 100

100
Sub.: Vertebrata
Cl.: Reptilia
O.: Squamata
S.o.: Sauria
F.: Chamaeleonidae

The North African variety of this Old World lizard is less than 30 cm long, with its tail contributing about one fourth of its total length. It is found in Egypt, Morocco, the southernmost parts of Spain, and the Canary Islands. The chameleon lives in brush and shrubs in dry terrain; in semidesert regions, it is confined to palm groves and intermittent watercourses *(oueds)* where there are tamarisk bushes and oleanders growing.

The anatomy of the chameleon reveals a number of useful adaptations to the life of an acrobatic arboreal hunter. Its body is laterally compressed, its legs quite sturdy, its feet equipped with opposable clamplike claws, each composed of 2 or 3 toes joined together by a web of skin. Its long prehensile tail provides it with an additional point of balance; its head is massive, primarily due to the addition of a bony casque—somewhat reminiscent of the ceratopsian dinosaurs (*Triceratops*, e.g.)—that is roughly pyramidal in shape. Its eyes are extremely mobile and enclosed within a single fused eyelid of ball-turret design pierced with a small opening opposite the pupil; its eyes can rotate in their sockets independently of one another, which gives the chameleon the combined advantages of binocular and monocular vision (see below).

The chameleon is strictly diurnal; it feeds every day during its period of activity, after it has been warmed up by the sun in the early morning. It spends the hottest parts of the day in the shade and sleeps on a branch at night, firmly secured to its perch by its opposable claws. It remains dormant during part of the winter, curled up under a rock, in a hole between the roots of a tree, or in a hollow treetrunk.

▶ The chameleon's tongue is actually a long hollow tube that it uses to catch insects in a rather complex but strikingly effective manner. The inner wall of the tube fits snugly around a sticklike bony process (the *entoglossum),* coated with a lubricant fluid, that projects outward from the floor of the chamelon's mouth. When not in use, the tube is compressed into tightly folded accordion pleats by the contraction of the powerful longitudinal muscle that runs along its entire length. There is a second set of circular bands of muscle that run around the outside of the tube and act as the triggering mechanism for this remarkable implement of predation. When these circular bands contract, the tube slips off the bone by the relaxation of the longitudinal muscle and is then catapulted forward to strike the chameleon's insect prey—sometimes at a distance equal to the total length of the chameleon's body. The bulbous tip of the chameleon's tongue, apart from being coated with sticky mucous secretions, flattens out into a spoonlike receptacle that makes it more difficult for the insect to escape during the brief interval before the chamelon's tongue is whisked back inside its mouth by the contraction of the longitudinal muscle.

The chameleon will eat virtually any insect of reasonable size (excluding a few that have a very hard shell or a repulsive odor). It detects its prey exclusively by sight and stalks it very slowly and deliberately before attempting the maneuver just described. It can spot an insect, even one that remains motionless, from quite far away; each of its eyes can sweep through 180° in the horizontal plane and 90° in the vertical. It turns its head to get a binocular fix on the prey that will permit it to evaluate the distance. Then it begins to stalk, sometimes providing a display of slow-motion pantomime acrobatics as it moves from twig to twig. With the target in sight, it opens its mouth, protrudes its tongue slightly (while still folded up) to get it clear of its jaws, then projects it toward its prey. In the case of larger insects, it crushes them in its jaws before swallowing them.

◀ The chameleon moves very clumsily on the ground, which is probably where it is most vulnerable to attack by snakes and predatory birds; when compelled to move from one bush to another, it usually tries to do so with all possible speed. The chameleon's gift for protective mimicry has undoubtedly been exaggerated; it can change color, to some extent, in response to the color of its surrounding by expanding or contracting the pigmented granules in its skin, but such changes occur in response to stimuli from its autonomic nervous system that may also be determined by the temperature, intensity of sunlight, and the chameleon's prevailing emotional state (particularly fear or anger). Since few predators have good color vision, the subtlety of these fluctuations is probably wasted on them for the most part; the chameleon's slow-motion swaying walk as it makes its way along a branch, plus the jagged spinal crest and scaly projections that obscure the true outlines of its body, probably have greater camouflage value against a backdrop of wind-fluttered desert foliage.

When the chameleon is cornered or threatened, it opens its mouth very wide, inflates its body to the maximum, and pants loudly while rocking from one side to the other. It launches into this same threatening display whenever it encounters another member of its own species, irrespective of sex. It is the only arboreal lizard in North Africa, so it faces no competition from other species, and its striking appearance and behavior seem to have convinced the inhabitants of North and Central Africa that it is extremely dangerous to man (which is not the case). Its dried and powdered flesh is widely used as an ingredient in folk medicine.

The European variety of the common chameleon is found all around the Mediterranean basin, including the Greek isles. There are a great many other chameleon species living in Southern Asia and the Near East and especially in Africa and on Madagascar; the largest exceed 60 cm in length, large enough to catch other lizards and small birds. The heads of many chameleons are adorned with bony or scaly projections that can assume a variety of interesting forms. Jackson's chameleon (*Chamaeleo jacksoni*), for example, has three little horns, which greatly heighten the generic resemblance to a miniature triceratops. Smaller chameleons of the genera *Brookseia* and *Rhampoleon*, also found in Africa and on Madagascar, never exceed 10 cm in length and are less strictly arboreal in their habits. D.II.

COMMON COENAGRION *Coenagrion puella* 101

Coenagrion puella is one of many common damselfly species in Western Europe, all noted for the subtlety and variety of their coloration, that can be seen hovering over ponds and marshes or darting through the dense vegetation on the bank. The head is narrow and transversely elongated, the abdomen is filiform (pointed at both ends), the wings perfectly transparent. The male's body is adorned with black rings and intricate designs against an azure background; the female's body is more greenish in color, and the upper part of the abdomen is almost entirely black. The hues increase in intensity as the damselfly matures; females of the same species may exhibit very different patterns of coloration. Damselflies live along the banks of sluggish streams or stagnant bodies of water—anything from a ditch to the sheltered margins of a large pond. The larvae live among the water weeds; their bodies are elongated, terminating in 3 sets of feathery gills.

▶ Like all dragonfly larvae, the agrion larva is equipped with a "mask" or "mandibular arm," an appendage that is tipped with a set of pincers and is well adapted to the task of seizing its prey and conveying it to its mouth. Its initial diet consists of rotifers and tiny crustaceans (cladocerans, or water fleas, and copepods) and the larvae of long-legged flies (chironomids) and mayflies (ephemerids). Normally, the adult damselfly stations itself on a leaf or grassy stem and flies off in pursuit of passing insects, which it seizes with its front feet and brings back to its perch to devour and digest (though the first of these activities may also be accomplished while the damselfly is still in flight).

The damselfly's behavior is quite aggressive, particularly in warm weather, and it often will not hesitate to drive a much larger dragonfly (anisopteran) from its perch. The damselfly's insect prey may also include rather large butterflies, though it only eats the upper body, the wings and legs (after these have been detached from the thorax). The damselfly flies slowly, which enables it to maneuver very skilfully through the tangled vegetation at the edge of a pond in order to catch its prey on the wing or to escape from predators.

◀ The young larvae are eaten by carp, tench, pike, and especially bream, as well as efts (the aquatic phase of the newt), water scorpions, backswimmers, water bugs, and other insect predators, including older members of their own or related species. The water scorpion, which can easily dispose of much larger dragonfly larvae of the genera *Aeshna* and *Anax*, is an especially formidable foe; both larval and adult diving beetles of the family Dytiscidae prey on odonata larvae that live in stagnant water.

A crucial, and frequently terminal, stage in the damselfly's development is the imaginal molt, in the course of which which the imago (adult) sheds its larval body case but has not yet acquired the use of its wings and is highly vulnerable to predation. As the mature larva climbs up a grassy stalk on the bank preparatory to its final

101
Sub.: Arthropoda
Cl.: Insecta
O.: Odonata
S.o.: Zygoptera
F.: Coenagriidae

molt or as the imago struggles to emerge from the husk of its former self *(exuvium* is the term preferred by entomologists), they may also attract the attention of older adult damselflies, frogs and other amphibians, robber flies (Asilidae), and especially birds.

Among European birds, kingfishers, little egrets, bee-eaters, ducks, and herons are all important predators of the odonata and are capable of bringing down even the most powerful anisoptera; swallows, cuckoos, and hobbies (a species of small falcon, *Falco subbuteo)* also feed on the smaller zygoptera, including *Coenagrion puella*. Her proximity to the surface of the water as she lays her eggs makes the female damselfly highly vulnerable to predation by frogs and other amphibia, trout, pike, and other large fish. Spiders and robber flies feed on adult damselflies, and of the larger dragonflies, the Libellulidae (though not the Aeshnidae) also seem to have a predilection for their smaller cousins.

Seven different genera, comprising 33 species, of the family Coenagridae are represented in Western and Central Europe and North Africa, all roughly similar in appearance and hunting behavior. In France, for example, there are 12 different species of the genus *Coenagrion,* which differ not only in size, coloration, and other morphological details but also in their ecological preferences, which is characteristic of the broad geographical distribution of the zygoptera all over the world.

A.de R.

102
Sub.: Vertebrata
Cl.: Aves
O.: Charadriiformes
F.: Scolopacidae

COMMON CURLEW *Numenius arquata* 102

This sturdy shorebird, also called the Eurasian curlew, measures about 55 cm with a wingspan of 94–110 cm and weighs between 640 and 1100 g, which makes it the largest of European shorebirds. In its northern nesting grounds in Europe (including Great Britain) and Central Asia, the common curlew is found only in open areas with low vegetation both along the coast and further inland, including marshes, bogs, and low-lying meadows as well as dry scrublands; during its migratory flight and its winter sojourn in southern Africa, it seeks out estuaries and exposed mudbanks at low tide. This species is not very prolific, which implies that the typical curlew must live for a number of years past sexual maturity in order to keep population levels constant.

The common curlew is a monogamous, highly gregarious bird that nests in sparsely populated colonies but migrates in enormous flocks, though substantial populations also remain behind in Europe through the winter. Migratory flocks are sometimes observed in close association with those of other shorebirds, including godwits *(Limosa),* redshanks *(Tringa),* and sandpipers *(Calidris),* each of which exploits the food resources of the intertidal zone in a somewhat different manner; the common curlew will not associate with flocks of a kindred and thus more directly competitive species, the whimbrel, or lesser curlew *(Numenius phaeopus).*

▶ With its long legs, long neck, and its long, curved bill, the curlew is well adapted to seek nourishment in the topsoil layer, amid the vegetation of marshlands and cultivated fields, in shallow water, whether fresh or salty, and in coastal mudbanks, in which it probes with its bill. It sometimes swims on the surface after wading out to a sufficient depth, and its prey varies according to the nature of the terrain—worms, mollusks, crustaceans, insects and their larvae, fish, and frogs. Young curlews have relatively short beaks and initially learn to hunt in the company of their parents, feeding on much the same sort of prey.

◀ Curlews will actively defend their nests against intruders; the eggs and nestlings, in spite of their protective coloration, are occasionally devoured by crows, gulls, hawks, foxes, and other marauders. The adults are sometimes taken by the peregrine falcon.

The common curlew, though a timid and elusive quarry, is frequently hunted by man, and restrictions on hunting during the spring breeding season and the establishment of wildlife preserves in coastal wetlands would seem essential if this species, along with many others, is to coexist successfully with man. Fortunately, with the rapid disappearance of bogs, marshes, and other suitable habitats, the curlew has learned to some extent to find its food on cultivated land.

The genus *Numenius* comprises just 8 species and thus makes up a relatively small fraction of the extensive family Scolopacidae (sandpipers, woodcocks, snipe, godwits, etc.). Two species of curlew, consdierably more trusting and less evasive than the common curlew, have already been hunted to the very edge of extinction—the Eskimo curlew *(N. borealis)* of North America and the slender-billed curlew *(N. tenuirostris),* which nests in western Siberia and winters along the shores of the Eastern Mediterranean. All curlews are prodigious travelers, spending the breeding season in the Northern Hemisphere and wintering in the Southern; the Tahitian curlew *(N. tahitiensis),* for example, nests in Alaska and winters in Polynesia—after a transoceanic voyage of some 8000 kilometers. J.-F. T.

COMMON CUTTLEFISH *Sepia officinalis* 103

When sexually mature, this small cephalopod measures between 10 and 25 cm, and its chief distinguishing feature is its calcified internal shell, or cuttlebone, which is partitioned into small chambers; the squid and the cuttlefish can most readily be distinguished by the presence or absence of the cuttlebone. This species is found in the coastal waters of the Atlantic, the English Channel and the North Sea, as well as the Mediterranean. The eggs are deposited near the shore sometime during the spring and summer, and most cuttlefish migrate into deeper water in the fall; the seasonal migratory and breeding cycles of this species are more pronounced in the Channel and the North Sea than in the Mediterranean.

The hatchlings measure just over 1 cm (with the mantle measuring 6–9 mm) and, under optimal conditions, will reach sexual maturity in about 10 months. The young cuttlefish are able to bury themselves in the sand or other friable substrate at a very early age, and the skin of the mantle and ventral arms can also serve the function of sucker disks, enabling them to anchor themselves to a rock or other hard surface. The adult cuttlefish spend their days buried in the sand and come out to hunt at nightfall.

▶ The hatchlings continue to feed on the remnants of the internal yolk sac for several days, and at the same time they begin to catch small mysidaceans and very small shrimp (e.g. Palaemonidae, Crangonidae) by projecting their retractile tentacles. After a few more weeks, they will have learned to catch small crabs or small fish by rapidly throwing out and reeling in their tentacles, or, in the case of the crabs, simply by pouncing on them and seizing them with their arms.

The cuttlefish slices off bits of its prey with its horny beak and thus devours it piecemeal; in later life, it continues to feed primarily on decapod crustaceans and bony fish, as well as on isopods and other crustaceans, cephalopods, and mollusks, though in relatively reduced quantities. The cuttlefish tracks a prey animal prey by focusing its two well-developed eyes on it, and as with the other ten-armed cephalopods (Sepioidea), the sucker disks on its arms and tentacles are reinforced by rigid bony rings.

Several generations of *Sepia officinalis* have been successfully raised in the laboratory, since this species will accept a variety of food (which need not even be living prey if the handler simply jiggles it up and down in the water in order to stimulate the cuttlefish's predatory response) and in general has proved to be better suited than the other cephalopods to aquarium life.

◀ Camouflage is the cuttlefish's principal defense against predators. The cuttlefish normally remains buried in the sand when not actively seeking prey, and even on a firm substrate, the pigment cells (chromatophores) and the pimply excrescences (papillae) on the cuttlefish's skin can be dilated or contracted in order to produce a pattern of stripes, spots, and other configurations as well as a "pebble-grain" texture that can closely match the color and texture of the patch of bottom on which the cuttlefish is currently reposing.

If attacked by a predator while swimming in the open water, the cuttlefish may be able to escape by expelling a cloud of sepia ink, which serves to confuse the predator by presenting it with one or more phantom targets. As with other cephalopods, the cuttlefish's ink also contains enzymes that anesthetize the olfactory receptors of a fish or other pursuer when a

103
Sub.: Mollusca
Cl.: Cephalopoda
O.: Decapoda - Sepioidea
F.: Sepiidae

substantial quantity of ink is expelled. The ink sac is already present in the embryo as a differentiated pouch in the intestine, which develops into an organ in which significant quantities of melanin (black pigment) are produced and stored; the ink itself is dark brown in color.

Marine predators that are known to prey on the cuttlefish include sharks, rays, conger eels, hake, and serranids (sea bass, groupers), seals and other marine mammals, and diving birds; they will attack while the cuttlefish is searching for prey in open water rather than lurking on the bottom. The flesh of the cuttlefish is a gourmet delicacy, particulalry appreciated in Mediterranean countries, and this particular species is taken with trawls, drift nets, and squid jigs (clusters of baited hooks) in substantial quantities.

Sepia orbignyana and *S. elegans,* both smaller than *S. officinalis,* are also found in European coastal waters. The family Sepiidae (including *Sepia, Sepiella,* and *Hemisepius)* comprises almost a hundred species, which are found off the coasts of Africa, Asia, and Australia, though this family is not represented in the New World. S.v.B.

104
Sub.: Vertebrata
Cl.: Aves
O.: Anseriformes
F.: Anatidae

COMMON EIDER *Somateria mollissima* 104

This gregarious saltwater diving duck measures between 49 and 71 cm from beak to tail and weighs between 1.5 and 2 kg; its wingspread is between 80 and 110 cm. It reaches maturity at the age of 3; the longevity record for this species (in the wild) is 15 years and 7 months. The eider is found in the subarctic regions of the Northern Hemisphere, more specifically along rocky coastlines that are sheltered from the full force of the waves by reefs or offshore islands. The eider builds its nest in the low-lying vegetation near the shore; natives of more southerly regions are primarily aware of the eider in connection with the soft, fluffy down with which it lines its nest, used as a lightweight insulating material in quilts, down jackets, and other cold-weather gear. Approximately 2 million eider ducks spend the winter in Northern Europe, from Greenland to the Barents Sea; the North American subspecies is found along the eastern and northern coasts of Canada, from Labrador to Baffin Island.

▶ The eider is generally found in the vicinity of offshore mussel beds, and mussels and other mollusks (bivalves and univalves) may comprise as much as 80 percent of its diet. Eiders feed in flocks, both by day and night, generally diving down to take mussels at a depth of 15–20 m; the dive lasts between 25 and 60 seconds, and the mussels are swallowed whole before the eider has returned to the surface; the shells are crushed by the contractions of the muscular walls of the bird's stomach. The eider also feeds on crabs and other crustaceans, brittle stars, and fry; it catches crabs in shallow water, kills them by grasping them in its beak and shaking its head violently, and swallows them whole. The adult eider duck requires about 300 g of nourishment a day; the chicks are fed on small mollusks and crustaceans.

◀ The eggs and chicks are highly vulnerable to marauding skuas, glaucous gulls, and Arctic foxes, among other predators, and the eider frequently seeks protection by nesting in the vicinity of a colony of Arctic terns or even the eyrie of a predator like the great white owl. An injured or sick eider may occasionally be dispatched by an osprey, but in general the adult eider has no significant predators.

The common eider is a protected species and its numbers are increasing throughout northwestern Europe, though the plundering of its nests in springtime (in search of eggs rather than eiderdown) still continues in some areas. The collection of eiderdown from artificial nesting boxes is, quite literally, an important cottage industry along the coasts of Scotland and Iceland, so that local eider populations are further safeguarded by considerations of economic self-interest; humans are nevertheless responsible for the two principal threats to the eider's survival, namely pesticide residues in the marine food chain and hydrocarbon pollution of the surface waters.

The tribe Somateriini consists of two genera *(Somateria, Polysticta),* comprising 4 species altogether, all of circumpolar distribution. In the Southern Hemisphere, the steamer ducks (genus *Tachyres)* of Patagonia and the Falklands, sometimes grouped with the shelducks in the tribe Tadornini, occupy the same ecological niche as the eiders; 2 of the 3 steamer duck species have lost the ability to fly.

J.-F.T.

COMMON EUROPEAN SQUID *Loligo vulgaris* 105

105
Sub.: Mollusca
Cl.: Cephalopoda
O.: Decapoda-Teuthoidea
F.: Loginidae

This long-finned cephalopod, sometimes called the calmary, is assigned to the suborder Myopsida (meaning that its two eyes are covered with a translucent flap of skin, a sort of primitive cornea) and is of the type popularly known as *calmar* or *calamar* in the Mediterranean because its uncalcified internal skeleton is shaped like a slender tube or quill (Latin *calamus,* "reed pen"). The common European squid is found in the coastal waters of the Atlantic, the North Sea, the English Channel, and the Mediterranean from the shoreline to the edge of the continental shelf.

The newly hatched squid is only 5–6 mm long (dorsal length of the mantle = 2.7–3.0 mm), though it immediately takes up an active, free-swimming existence in the plankton layer, and begins to grow very rapidly, so that it attains sexual maturity (and a total body length of about 10 cm in the case of the males and 13 cm in the case of the females) during its first year of life. The seasonal cycle of migration from deep to shallow water and reproduction varies considerably throughout this species's considerable range, so that the female may lay its eggs at various depths at virtually any time of year—though this occurs most frequently in the spring and in shallow water. *Loligo vulgaris* is an active swimmer, and unlike other squids of the calamar type, it remains constantly in motion and is never found resting on the bottom as long as it remains in good physical condition.

▶ The young squid feeds on small crustaceans (copepods, opossum shrimp) and planktonic larvae; it is not capable of ingesting the shells of its larger victims (large with respect to own body length of a few millimeters), though it can break them open and avail itself of their contents. Squids hunt by sight, and as soon as the squid has fixed both its eyes on a potential prey animal, capture is almost instantaneous; the long retractile tentacles are used to entangle and haul in the prey, while the corona of shorter tentacles holds it in place while the squid tears it in pieces with its beak.

The adults and the larger juveniles feed on fish and on other cephalopods, including members of their own species. When the juvenile squid is well on its way toward maturity, it starts to live in shoals composed of individuals of roughly the same size, and an encounter between two shoals of different-sized squid is believed to result in a cannibalistic melee as the larger squid begin to prey on the smaller ones.

◀ The common European squid is extensively preyed on by a many different species of fish as well as certain seabirds and ceteceans (notably dolphins). Jigging for squid with a long string of hooks on a weighted line is practiced by both amateur and commercial fishermen on both sides of the Atlantic, though it is the Japanese that have perfected this technique on a truly industrial scale, equipping factory ships with squid jigs mounted permanently on winches, conveyor-belt fashion.

Loligo forbesi, which is particularly common off the coasts of the British Isles, is quite similar to *L. vulgaris* and is sometimes found in the same waters and in the same biotopes. *Loligo pealei* and *Doryteuthis plei* are the two most common species in North American waters.

S.v.B.

COMMON GECKO *Tarentola mauritanica* 106

106
Sub.: Vertebrata
Cl.: Reptilia
O.: Squamata
S.o.: Sauria
F.: Gekkonidae

This small Mediterranean gecko measures up to 15 cm in length and ranges from the Iberian peninsula to the Ionian Islands and throughout much of North Africa (including the Sahara, where it is found in palm groves and oases). *T. mauritanica*'s head is flattened, its body is rather stout, and both its tail and body are covered with whitish tubercules. Coloration varies considerably, but common geckos are often pinkish gray with brownish spots on the body and transverse brownish stripes on the tail. The eyes are large, with vertical pupils, and the tips of the toes are equipped with adhesive pads (each divided into a series of transverse bands, or lamellae) that enable this species to run up walls and across windowpanes and perform the various acrobatic feats for which it is particularly noted.

The common gecko is frequently found in human dwellings, particularly in stone walls and masonry (including the ramparts of walled towns), as well as in quarries, on cliffsides and rock faces, woodpiles, under the bark of cork oaks, in subterranean galleries and tunnels, and in a variety of other seleltered habitats. Longevity is on the order of 8 years.

▶ The comon gecko is fairly quick in its movements and prefers to feed on noctua moths, flies and mosquitoes, or crickets and locusts, though it will also snap up almost any small insect that it encounters. In human dwellings, it hunts for spiders in the corners of rooms, wood-boring insects among the rafters, and flying insects circling around bright lights. It generally stations itself in ambush, and when it catches sight of an approaching insect (at a distance of up to 1 m), it immediately rushes forward and seizes it in its jaws, then carries it off to some nearby place of safety to devour it.

◀ The common gecko defends its hunting territory, which is often quite strictly demarcated, against interlopers of its own species. Its ability to run up walls and smooth rock faces keeps its safe from potential predators as a rule, and it will defend itself aggressively, often attempting to bite, if it is picked up or handled. In some parts of its range, this species is regarded as an omen of good luck, in others as bad luck, but is is almost universally tolerated as a destroyer of household insects and is frequently kept as a pet.

A subspecies, *T. m. deserti*, is most prevalent on the high plateaus of southern Algeria; it is pale yellow, or somewhat lighter in color than *T. m. mauritanica*. *T. neglecta* is found among the dunes of the Sahara as well as along wadis (intermittent watercourses) and in oases, where it tends to be arboreal. Like *Stenodactylus petrii* (q.v.), *Tarentola* is commonly believed to be poisonous by desert dwellers, a notion that may have been inspired by its warty skin and rather sickly appearance, nocturnal habits, and by its fondness for the interiors of blackened stumps, the trunks of ancient shrubs, and other mildly sinister habitats.

Q.V.

COMMON HOGNOSE SNAKE
Heterodon platyhinos 107

107
Sub.: Vertebrata
Cl.: Reptilia
O.: Squamata
S.o.: Ophidia
F.: Colubridae

This thick-bodied North American snake normally measures between 60 and 90 cm and may occasionally attain a length of 1.20 m. It is found in the Eastern United States, between Florida and southern New Hamsphire and to the east of a line extending from Minnesota and South Dakota through Kansas and Oklahoma to Texas. It prefers rather dry, sandy terrain in mountainous regions.

▶ The adult hognose feeds almost exclusively on anurans, particularly on burrowing toads, which are found in the same sort of sandy terrain. It primarily makes use of its sense of smell to detect its prey, which it swallows whole. The juveniles feed on tadpoles, earthworms, and probably on insects as well.

◀ The hognose is noted for its two-tier defensive strategy, which is based entirely on bluff and dissimulation. If disturbed or attacked and unable to flee, this harmless snake begins by striking a very aggressive posture, flattening out its head or body, spreading its cervical ribs somewhat in the manner of a cobra, then inflating itself with air, hissing noisily, lunging forward and posturing as if to strike. If the adversary is undeterred by this, the hognose suddenly changes its tack and begins to play dead. It wriggles and twitches violently, as if seized with convulsions, then rolls over on its back with its mouth lolling open and its tongue protruding. It may remain in this position for several minutes, but if it is rolled back over into its normal position, it instantly and somewhat unconvincingly rolls over on its back again.

As a predator (as well as a performer) the hognose is without competitors; there are other North American snakes that feed almost exclusively on frogs and toads, but none are found in the same biotopes.

There are a number of congeneric species in the Central United States and Mexico that practice the same defensive strategy as the common hognose snake.

D.H.

COMMON KINGFISHER *Alcedo atthis* 108

108
Sub.: Vertebrata
Cl.: Aves
O.: Coraciiformes
F.: Alcedinidae

This colorful Old World bird measures about 15 cm, with a wingspread of 25–28 cm, and weighs between 30 and 44 g. The kingfisher is found by the clear waters of streams, lakes, and ponds, provided there are steep overhanging banks of clay, shingle, or sandy soil, since the kingfisher builds its nest at the end of a long shaft excavated in the bank. The range of this species is quite extensive, including Western Europe (except for the northern tips of Scotland and the Scandinavian peninsula), North Africa, parts of Asia, Indo-Malaysia, Japan, and certain Pacific islands. The com mon kingfisher is essentially sedentary, with local populations fluctuating in accordance with the severity of the winter. It raises two or three broods every year, and during the summer, nesting pairs may occasionally be observed in regions not normally frequented by kingfishers; conversely, if a kingfisher's pond or stream freezes over in the winter, it may fly off in search of open water, sometimes even along the seashore.

▶ Outside the nesting season, the kingfisher leads a solitary existence, darting back and forth over its fishing grounds, skimming just above the water and perching frequently on one of its favorite observations posts, usually a root, rock, fencepost, or the like. It feeds primarily on smaller fish (hatchlings of various species, minnows, bleak, gudgeon) as well as on tadpoles and small frogs, crustaceans, and aquatic insects (diving beetles, backswimmers, libellulids); it has difficulty swallowing prey longer than 10 cm.

When the kingfisher catches sight of a fish or other prey animal below the surface, it drops straight down into the water to intercept it and seizes it in its beak. It sometimes clubs a fish against the bank to subdue it and then begins to swallow it headfirst—sometimes tossing it up in the air and catching it until the fish comes down in the proper orientation.

◀ The common kingfisher has no specific predators and no competitors besides members of its own species; it will not tolerate any intrusion by other kingfishers on its fishing grounds, particularly during the nesting season. The brightly colored plumage of the commmon kingfisher was a staple of the millinery trade during the 19th century and its numbers were seriously reduced by plume hunters.

Today, the kingfisher may still be in some danger from fishermen and, to an even greater extent, from commercial fish farmers. The species has been systematicaly eradicated from certain areas; the degradation of their habitat as the result of other human activities and their by-products (diversion and canalization of streams and rivers, rectification of riverbanks, water pollution) has had an adverse effect on kingfisher populations in many others.

The family Alcedinidae originated in the tropical forests of the Indo-Malaysian region and now comprises 86 species in Africa, Eurasia, Oceania, Australia, and in the New World; a great many of them feed on fish, but the majority are terrestrial, preying on insects and other terrestrial creatures. Thus, the forest kingfisher *(Halcyon macleayi)*, found in wooded areas in Australia, feeds exclusively on insects. The stork-billed kingfisher *(Pelargopsis capensis)* of India feeds on lizards, crabs, and nestlings, and the hunting kingfishers *(Dacelo)* of Australia and New Guinea search out insects, reptiles, and birds; the largest of these, the kookaburra *(D. gigas)*, is the size of a crow. J.-P. R.

COMMON MUDSKIPPER *Periophthalmus vulgaris* 109

109
Sub.: Vertebrata
Cl.: Osteichthyes
S.cl.: Teleostei
O.: Perciformes
F.: Gobiidae

The common mudskipper is a small fish that never exceeds 15 cm in length and in found in costal mangrove swamps in Malaysia and Indonesia; the aerial roots of the mangrove are submerged at high tide, but as the tide begins to go out, the water filters back slowly through this labyrinthine root system, leaving substantial quantities of sediment behind. These almost inaccessible tidal mudbanks are the domain of the mudskipper, which spends much of its life out of water, above the tidal mark, and lives in troops of 20–50. The mudskipper scoops out a burrow for itself with its mouth, then transports each mouthful of mud to the edges of its territory, where it promptly spits it out, thus creating a dike several centimeters high that completely surrounds its burrow.

During the breeding season, the mudskipper excavates a sort of nursery for its offspring in the mud, the lower level of which always remains submerged. Its numerous and mainly very curious adaptations to life on shore are not apparent at first, and young mudskippers look very much like young gobies (both belong to the family Gobiidae), though they are about to undergo a remarkable metamorphosis. The adult's enormous bulging eyes are mounted on top of its head, like gun turrets, which provide it with a panoramic field of vision, and the oddity of the mudskipper's appearance (acknowledged by the genus name *Periophthalmus,* which means "crosseyed" in Latin) is intensified by its habit of alternately retracting one of its eyeballs into its head.

It was formerly thought that it simply did this because it was unable to focus both of its eyes at once; the real problem is that the mudskipper's gill slits begin to dry out and stick together (thus depriving the fish of oxygen) after a certain amount of time on shore. It keeps its gills aerated by trapping a little reservoir of seawater in its inner ear; when it retracts one eyes and presses down on this reservoir with the underside of its eyeball, this stirs up (and thus reoxygenates) the remaining water, lubricates the gill flaps, and restores its gills to normal functioning. It sometimes achieves the same effect by rolling over on one side and then the other while pressing its gill chambers against the mudbank.

The mudskipper's fins have evolved into jointed walking (or crawling) legs, consisting of an elongated, movable proximal segment and a digital segment that has been enlarged into a paddle-shapped flipper, though it is capable of far more vigorous locomotion: It twists its tail to one side, then flicks it back with an abrupt triggerlike movement, which sends it skittering or a considerable distance across the mudbank. The forepart of its body is studded with tiny bumperlike projections (papillae) that help to absorb the impact of repeated landings, and the skin of its belly is covered with a layer of thick protective callus that comes into play when it crawls along with its flippers at a more deliberate pace.

▶ The mudskipper feeds on virtually every species of insect, sandworm, and small crustacean that is found on the mangrove flats, and it has been suggested that it was the remarkable richness of this fauna that first inspired its ancestors to linger for a few more hours on shore. As long it remains above the tidal mark, the mudskipper enjoys a considerable advantage in speed and mobility over its potential prey, and it is interesting to note that when it catches sight of a prey animal that is heading toward the water, it will try to intercept it and herd it back up onto the mudbank.

The mudskipper can even clamber up onto the lower branches of a mangrove tree (which admittedly grow very close to the ground) in search of caterpillars and beetle larvae, and it can skip along the mud with sufficient speeds and accuracy that it can even pounce on a sandfly before it has a chance to escape. The mudskipper catches its prey in its jaws; its teeth are fairly sharp, but (as a final oddity) the movement of its jaws is hampered by the size of the gill reservoirs in its inner ear, which gives it the appearance of a child with a mouthful of bubblegum when it chews its food.

◀ Frogs and toads, colubrids, mangrove crabs, and various shorebirds all prey on the mudskipper when the tide is out. When the tide is in, it is vulnerable to the assaults of grunters *(Therapon),* groupers, and other large fish. Characteristically, it will attempt to escape from these aquatic predators by crawling back up on shore. Its habitat is inaccessible to all but the most determined naturalists, and human beings do not currently pose a threat to this species.

A related species, *P. chrysophilos,* is often found in association with the common mudskipper. The largest of the mudskippers, *P. schlosseri,* sometimes attains a length of 25 cm; the nursery in which it incubates its eggs may be up to a meter in diameter and the muddy rampart that surrounds its burrow up to 2m. The glodog *(Scartelaos viridis)* belongs to a closely related family; it lacks the mudskipper's shock-absorbing papillae, though it also has a layer of callus that protects its belly as it clambers about on shore. *Boleophthalmus pectruirostris* only emerges from the water at low tide and excavates a sort of wallow for itself in the mud to keep its skin from drying out. These last two species seek their prey in much the same way as the mudskippers (though with considerably less speed and agility) and are pursued by the same predators. M.D.

COMMON OCTOPUS *Octopus vulgaris* 110

110
Sub.: Mollusca
Cl.: Cephalopoda
O.: Octopoda
F.: Octopodidae

This is the most familiar representative of the family Octopodidae, which, unlike the other Octopoda (octopuses and paper nautiluses), are sedentary bottom-dwellers. Apart from all having eight tentacles, the Octopoda are further distinguished by the almost total absence of an internal skeleton or an external shell. *Octopus vulgaris* is found in shallow coastal waters on both sides of the Atlantic (from the southern United States to Central America as well as the English Channel and the Mediterranean) and Indo-Pacific. It has rarely been encountered beyond a depth of 150–200 m along the continental shelf, and in the spring especially, it approaches quite close to the shore and is often found at a depth of only a few meters.

The females lay their eggs in cavities or crevices in the rock, generally quite close to shore, guarding them jealously (so that they appear to be incubating them) until they hatch out. The hatchlings measure barely 3 mm, the dorsal length of the mantle is only 2 mm, and the tentacles are barely visible. After a brief free-swimming interlude of several weeks, the young octopuses descend to the bottom, where they will remain throughout most of their lives, even though they can swim quite well in the open water. The common octopus reaches sexual maturity at the age of 12–15 months, on the average; the smallest individuals measure only about 7–9 cm while the largest may attain a diameter of up to 3 m. The breeding season, which is correlated with a pattern of complex but not very extensive migrations, is quite lengthy, beginning in the spring and lasting until the fall.

▶ The common octopus generally hunts by sight, fixing its single large eye on its prey. The octopus hatchlings prefer to feed on planktonic crustaceans, notably the larvae of crabs, shrimp, and other decapods (and probably feed on suspended organic particles as well). Aquarium studies have demonstrated that they will actually attack any light-colored object (since the larvae on which they feed are virtually transparent) that appears to be moving under its own power, for example, bits of the muscle tissue or gonads of a crab that were dropped from the surface. Even newly hatched octopuses contested vigorously for scraps of food several times larger than themselves and continued to pursue them closely as they drifted down to the bottom of the tank.

By the time the young octopus's tentacles have grown to be as long as its mantle (for a total body length of about 1 cm), it begins to seek its prey along the bottom. Here obviously, the lack of internal skeletal support works greatly to the octopus's advantage, enabling it to squeeze into the narrowest cracks and crevices in search of shrimps, crabs, and other relatively large crustaceans. In general, shrimps of the genus *Palaemon* and crabs of the genus *Carcinus* are its prey of choice, though some local populations feed primarily on bivalves and gastropods. Like other carnivorous mollusks (cf. NECKLACE SNAIL), *Octopus vulgaris* is able to bore through the shells of bivalves with the aid of a specialized organ called the radula, which is studded with sharp toothlike projections and serves the function of a drill bit or a rasp; this process is expedited by a corrosive decalcifying agent in the octopus's saliva, which is also secreted by the radula.

The stomach contents of *Octopus vulgaris* generally include a rather modest proportion of fish and other cephalopods. The octopus is sometimes able to catch bottom-feeding prey animals by seizing them with its tentacles, or by dropping down on them from above and trapping them inside the circle of its tentacles, with the weblike intertentacular membranes pressed together to form an impregnable barrier. The suckers on the common octopus' tentacles, like those of all the Octopoda, are also very flexible.

The prey animal is first restrained by one or more of the octopus's tentacles, then totally immobilized by the poisonous secretions of the posterior salivary glands before being conveyed to the octopus' mouth (a complicated arrangement of folds and inlets officially referred to as the buccal mass). Depending on its size, the prey is either swallowed whole or cut into pieces by the octopus' horny beak; the radula plays the further role—analogous to that of the tongue in a bird or mammal—of transporting the prey, either in whole or in part, into the interior of the buccal mass, where it is propelled toward the stomach by the peristaltic action of the esophagus. The forepart of the small intestine is differentiated into a sort of pouch (called the crop) that is found directly in front of the octopus' stomach.

◀ The common octopus' archcompetitor is the moray eel, which, along with the European lobster, is involved in a three-cornered competition for the very limited space provided by the rocky clefts and crevices that all three species prefer to inhabit. Conger eels, dolphins, seals, and certain sharks also prey on the common octopus, which, like the cuttlefish (q.v.), often succeeds in evading these predators by releasing a murky cloud of sepia ink. As noted elsewhere, the octopus is rather evenly matched in its combats with the moray eel, particularly since the moray's olfactory receptors are immediately desensitized by the octopus's ink.

The genus *Octopus* comprises almost a hundred species, ranging in size from the tiny (*O. joubini*, found off the coast of the southeastern United States) to the enormous (*O. dofleini* of the Pacific). Species found off the coasts of Western European water include *O. macropus*, *O. salutii*, and *O. defilippi*. Octopuses of the genus *Eledone*, which also belong to the family Octopodidae, have only a single row of suckers on their tentacles, but are biologically quite similar to *O. vulgaris* in other respects. For a discussion of *O. vulgaris'* more distant cousins, the Argonautidae, see PAPER NAUTILUS. S.v.B.

COMMON OTTER 111

111
Sub.: Vertebrata
C.: Mammifera
O.: Carnivora
F.: Mustelidae

Almost all of the mustelids (weasels, martens, and their kin) display some adaptations to a semiaquatic mode of life, and the common otter, with its long, sinuous body, webbed feet, and thick, water-repellent fur, is thoroughly at home in fresh water. Its pelt is brown with a white bib at its neck; its body length varies between 65 and 85 cm, and its tail may be as long as 52 cm. Its weight varies between 7 and 15 kg. Though the average longevity of this species in the wild in unknown, some individuals have lived to the age of 19 years in captivity.

This species is found throughout Eurasia from India to the Arctic Circle, always in the vicinity of fresh or brackish water, usually along the shores of calm lakes or marshes that are well stocked with fish, or in large estuaries where there are accessible offshore islands. The otter digs its den ("couch" is the technical term) in a steep bank, consisting of a central chamber with a subsidiary tunnel on either side, one of which serves as a sort of airshaft, the other opening of which is underwater, as an entranceway.

The common otter rarely ventures farther than 200 m from its couch unless food is scarce, in which case it sometimes undertakes lengthy migrations either by water or overland. It goes out to hunt by night; in parts of Western Europe from which the species has not been exterminated by man (such as the Scottish Highlands or the marshes of Poitou), it may still require much patience and vigilance to catch a glimpse of a wild otter. During the day, it prefers to rest in its couch or sun itself on the bank.

▶ The common otter feeds largely but not exclusively on fish; it also catches crayfish, frogs, muskrats, rabbits, various waterbirds, and is especially fond of eating the eggs of birds that nest by the water. To catch fish, the otter stations itself on the bank, slips into the water when it catches sight of its quarry, and swims off in pursuit. The otter's eyes are fitted with special membranes that enable it to see underwater, and its whiskers (vibrissae) serve to extend its tactile senses through an additional 25 cm on either side of its muzzle.

The common otter can remain submerged for up to 7 or 8 minutes, and it sometimes rounds up whole schools of fish, driving them toward the shore, then digging them out with its muzzles and its paws when they take refuge in holes in the bank. The otter catches them one by one and tosses them up on the bank, making sure that no more remain in hiding before it climbs up on the bank to enjoy its meal. The otter will frequently toss a fish up in the air several times or juggle it back and forth between its paws before swallowing it whole. If the fish is a small one, the otter may pause to devour it while floating on its back and grasping it firmly between its paws, not taking the trouble to come ashore before it resumes the chase.

◀ Apart from man, the common otter has no significant predators, and there is ample written evidence testifying to the prevalence of this species during the 19th century. However, the common otter was hunted for sport, as well as for its pelt, until it had virtually disappeared from both France and Great Britain (where a special breed of dog, the otterhound, was developed for this purpose). The species has also been persecuted by fishermen as an undesirable competitor, whereas fishermen in China and India continue to use domesticated otters to drive schools of fish into their nets. The wild otters that have survived in Western Europe, mainly in remote or inaccessible areas, are no longer even capable of maintaining their current numbers because of water pollution, land development, and other catastrophic alterations to their environment; a census of local populations and habitats has been undertaken in France.

The giant otter of South America, morphologically identical to the common otter, may attain a length of between 1 and 1.5 m (including its tail, about 70 cm) and weigh as much as 24 kg. It lives exclusively in rivers that flow into the Atlantic and whose current is not too swift. The giant otter hunts in packs of as many as 20 individuals and is primarily active between dawn and dusk, but in other respects its habits and dietary preferences are similar to those of the common otter.

P.A.

COMMON PARTRIDGE *Perdix perdix* 112

112
Sub.: Vertebrata
Cl.: Aves
O.: Galliformes
F.: Phasianidae

The common partridge measures 28–30 cm, has a wingspread of 47–52, and weighs between 290 and 500 g; the average weight is about 400 g. A captive specimen survived to the age of 7, but, as with other gamebirds, life expectancy is the wild is extremely short. During the early Middle Ages, much of Western Europe was still thickly forested, and the common partridge (not especially common at that time) would mainly have been confined to scrub forests, heath, moors, alpine pastures, and other marginal terrain. The clearing of the forests for agriculture opened up a vast new domain for the partridge, and the topography of the old-style European countryside—the medieval patchwork of cultivated and fallow fields, hedges, meadows, and orchards—represented an ideal habitat.

▶ The common partridge is a sedentary, terrestrial bird that rarely strays far from its native patch of ground. It lives in small flocks consisting of no more than one or two family groups, and feeds on seeds and grasses *(Poa, Trifolium, Polygonum . . .)* for the most part, though the nestlings are fed almost exclusively on caterpillars and other larvae, cochineal insects, flies and mosquitoes, and small ground beetles during their first few weeks of life. The adult partridge pecks in the ground and hunts for insects on the leaves of low-lying plants, but arthropods, worms, and snails only comprise about 5–10 percent of the adult diet.

◀ The common partridge seems to coexist quite comfortably with other members of the pheasant family (red-legged partridges, quail, pheasants), though flocks of common partridge tend to avoid one another. A high proportion of partridge eggs are destroyed by crows, magpies, and beech martens, and the nestlings are also preyed on by stoats, weasels, and buzzards. The adults are occasionally taken by goshawks, peregrines, and golden eagles, but coccidosis and other epidemic diseases of fowl as well as blizzards, thunderstorms, and other natural calamities take a much greater toll, and of course hunting is by far the most significant cause of mortality in this species. Partridges are hunted with dogs and, in more traditional fashion, flushed from cover and driven toward the guns by a line of beaters, a technique known as the battue. According to the French National Hunting Office, 5.5 million partridges were shot during the 1974–5 season.

The common partridge is very prolific and until recently has been able to maintain its numbers in spite of epidemics, natural disasters, and the carnage of the hunting field. During the postwar era, however, the introduction of chemical herbicides and pesticides and the disappearance of fallow land, hedgerows, and other suitable habitats finally seems to have shifted the balance, making it far more difficult for the partridge to find shelter and to feed its young. Attempts to breed partridges in captivity and then release them into the wild have not been notably successful; apart from being ill equipped to survive on their own, the released birds have have introduced a number of new diseases into the habitat. In Czechoslovakia, the partridge population is known to have dwindled from approximately 5–6 million in 1935 to only 400,000 in 1963, and in France, the common partridge is found in its former numbers only in a few localities where prudent game management policies have been in effect for some time.

The pheasant family (Phasianidae) comprises 174 species, found in tropical as well as temperate regions. The red-legged partridge *(Alectoris rufa)*, mentioned earlier, is more southerly in its range than the common partridge, and the greater redleg *(A. graeca)* is found primarily in upland meadows and the sunlit slopes of the Alps.

J.-F. T.

113
Sub.: Vertebrata
Cl.: Mammalia
O.: Cetacea
F.: Phocenidae

COMMON PORPOISE *Phocena phocena* 113

The common porpoise is almost, though not quite, the smallest of the ceteceans, measuring only 1.8 m and weighing 55 kg. It is thick-bodied and solidly built, though still perfectly streamlined. It is found in coastal water and estuaries, particularly in the Northern Hemisphere, on both sides of the Pacific and the Atlantic (the coasts of Europe and Canada), the North Sea, the Adriatic, and, to a lesser extent, the Mediterranean. Atlantic populations sometimes migrate as far north as Greenland in summer, returning to the waters around the Tropic of Cancer for the winter. The common porpoise travels in small schools and often ventures fairly far upriver, though it is almost never found on the high seas. Its gestation period is quite lengthy, and the young are already 80 cm long when they are born.

▶ The common porpoise feeds primarily on herring, sardines, anchovies, mackerel, and other fish that congregate in shoals near the surface, as well as cod, which are found in deeper waters, cuttlefish, and crustaceans. In general, porpoises hunt in groups, diving frequently and remaining submerged for only about 5 minutes, in order to encircle and devour a shoal of fish. Like the other odontocetes (toothed whales), the common porpoise relies entirely on echolocation (sonar) and complex vocalization patterns not only to locate fish but also to coordinate the activities of the group during a roundup of this kind. The porpoise's vision is not very keen, and it has entirely lost its sense of smell.

The dentition of toothed whales differs greatly from that of other mammals; the porpoise has a great many conical pointed teeth (haplodonty), which are not much good at slicing off bits of fish or flesh, but are mainly useful in preventing the prey from escaping. Such skill is a high priority in the case of a slippery customer like a herring or a sardine; the porpoise swallows its food whole, without preliminary chewing.

◀ Sharks and killer whales prey avidly on the common propoise, and its flesh has occasionally been eaten by man; young porpoise especially was a favorite banquet dish in ancient Rome and medieval England.

The common porpoise is the most widespread member of the family (or subfamily) Phocaenidae, all of which are small (1.2–1.8 m) coastal ceteceans that feed on fish and cephalopods.

V.deB.

114
Sub.: Vertebrata
Cl.: Aves
O.: Charadriiformes
F.: Scolopacidae

COMMON REDSHANK *Tringa totanus* 114

The redshank, also called the red-legged snipe, is a member of the same family as the snipe, woodcock, and sandpiper; it has long orange-red legs, pointed wings, and a short tail, with an overall length of about 27 cm, a wingspread between 47 and 50 cm; it weighs between 102 and 143 g, and is found all along the coasts of Europe. It prefers the banks of shallow freshwater ponds and saltwater lagoons, salt meadows that are regularly mowed or grazed by livestock, and other marshy or tidal regions with low-lying vegetation, particularly large clumps of sedge and other aquatic vegetation. Even sedentary populations tend to prefer coastal regions, and migratory redshanks are found primarily along the shores of shallow bays or tidal estuaries; Scandinavian and other redshank populations winter in Southern Europe (including France) and in North Africa. This species is essentially solitary, though it sometimes migrates in small groups that may include shorebirds of other species; it is monogamous and nests in small colonies.

▶ In freshwater marshes, the redshank feeds on aquatic insects and their larvae, worms, mollusks, and other invertebrates. It feeds on tidal flats by probing with its beak for mollusks and small crustaceans and hunts for marine worms *(Nereis diversicolor)* in the turbid water.

◀ The redshank conceals its nest in a patch of thick waterweeds and valiantly defends its eggs and chicks against any potential predator. The peregrine falcon sometimes preys on the adult redshank. The redshank often nests in the vicinity of other shorebirds, including lapwings, black-winged stilts, and black-tailed godwits, though a nesting pair will vigorously defend its territory against intruders of its own species. During its migrations, it avoids the company of large flocks of sandpipers, godwits, oystercatchers, and other redshank species—all shorebirds that are found in the same intertidal habitat, though they are not actually competitors for the same food sources.

The redshank is protected in several nations of Northern Europe but the *"pied-rouge"* is still considered a legitimate gamebird in France, which is one of the reasons why the sedentary population in that country has probably been reduced to fewer than 1000 nesting pairs; the survival of local populations is also threatened by current changes in land use: the draining of salt marshes, discontinuance of the traditional practice of harvesting salt hay and grazing livestock on salt meadows; in some cases abandoned meadows have been invaded by reedbeds, which is not a congenial habitat for the redshank.

The family Scolopacidae numbers some 85 species, of which no more than 10 are regarded as true redshanks; 7 species are found along the shores of Western Europe. The ruff *(Philomachus pugnax)* is regarded as a kind of sandpiper in English-speaking countries and as a redshank *(chevalier)* in French-speaking ones; in fact, it is transitional between the two, and owes its curious scientific name ("pugnacious combat-lover") to the elaborately choreographed prenuptial battles engaged in by the males. J.-F. T.

COMMON ROLLER *Coracias garrulus* 115

115
Sub.: Vertebrata
Cl.: Aves
O.: Coraciiformes
F.: Coraciidae

This species typically measures 29 cm in length, with a wingspread of 62–67 cm, and weighs 140 g. Originally an Asian tropical bird, *Coracias garrulus* currently ranges into southern Spain, France (Languedoc and the Rhône delta), and Italy, and well up into Central Europe (as far north as the island of Gotland, off the coast of Sweden), as well as the Balkans and Central Asia. It seeks out open, low-lying terrain interspersed with woodlands (since it nests in hollow trees), and lives in isolated pairs in Western Europe, though it congregates in small flocks in the Balkans. The common name of this species in both English and French *("rollier d'Europe")* refers to its acrobatic flight patterns, including numerous sideslips, loop-the-loops, and nosedives; the short-legged *Coracias* is ungainly in its movements on the ground, however.

▶ *Coracias* feeds primarily on scarabs, cockchafers, grasshoppers, and ants, occasionally on lizards, amphibians, and even songbirds and small rodents. It hunting technique is similar to that of the shrikes—namely, it stations itself on a upright post or a branch, swoops down on a ground-dwelling insect, carries it back up to its perch, and swallows it immediately; it sometimes catches flying insects in midair as well.

◀ *Coracias* has no particular predators, though jackdaws and starlings may compete with it for nesting sites; the number of available nesting sites may also be reduced, of course, by the felling of dead trees, the clearing of woodlands, and other human activities.

The 25 species of rollers are distributed throughout the tropical and temperate regions of the Old World, and apart from the true rollers *(Coracias)*, a number of which are prevalent in Africa and Asia, there are several genera of ground rollers on Madagascar (notably *Atelornis*). The broad-billed rollers *(Eurystomus)* are found in China and Australia as well as on Madagascar.

J.-P. R.

116
Sub.: Vertebrata
Cl.: Mammalia
O.: Pinnipedia
F.: Phocidae

COMMON SEAL *Phoca vitulina* 116

The male measures between 1.3 and 1.95 m and weighs between 100 and 120 kg; the female measures between 1.2 and 1.7 m and weighs between 47 and 150 kg. There is considerable physical variation among local populations and geographic races, which are often referred to by different vernacular names and are collectively dispersed over a very wide area of the Northern Hemisphere, thus creating numerous problems for the the taxonomist. Coloration varies considerably within a general pattern characterized by darker spots of pigment, sometimes blending together in a continuous network against a lighter background.

European populations are found along the coasts of Ireland, Iceland, the Hebrides, Orkneys, and Shetlands, Norway, and the shores of Baltic, totalling some 50,000. An additional 43,000 seals are found along the coasts of Greenland and the Atlantic seaboard of North America; the largest group, comprising 315,000 seals, is found along the Pacific coast of Canada and the United States, which yields a (very approximate) worldwide total of 400,000 for the species.

The females reach sexual maturity between the ages of 2 and 5 years, the males between 3 and 6. The period in which the pups are born only lasts for 6 weeks altogether in any particular locality, though it may begin as early as January in some areas and end as late as September in others. The pups weigh about 10 kg at birth and are weaned at the age of 4–6 weeks. Maximum longevity is about 40 years.

▶ The common seal is an opportunistic predator that seeks its prey in shallow coastal waters, feeding on squid, cuttlefish, crabs, and shrimp as well as on codfish, flatfish, salmon (a particularly sore point with human fishermen), and a variety of other deep-sea, coastal, and migratory fish. A captive female weighing 60 kg will eat about 2 or 3 kg of fish per day. Smaller fish are swallowed whole, larger fish have their heads and tails removed, and very large fish, including salmon, are carried back to the surface, where the seal makes a more leisurely meal of them while floating on its back and clasping them securely in its front flippers. Young seals begin to feed on medium-sized crustaceans (crabs and prawns) before they are ready to catch fish.

◀ The newborn pups are occasionally taken by sea eagles and Arctic foxes, the adults by killer whales and polar bears. The commercial exploitation of this species is strictly controlled; about 16–22,000, most of them yearlings, are taken by human hunters every year. The disappearance or degradation of habitat and the pollution of coastal waters (particularly with mercury and chlorinated hydrocarbons) currently pose much more serious problems for this species.

M.P.

117
Sub.: Vertebrata
Cl.: Reptilia
O.: Squamata
S.o.: Sauria
F.: Scincidae

COMMON SKINK *Scincus scincus* 117

This North African skink, sometimes called the "sand fish" *(poisson de sable)*, typically measures 15–20 cm and never exceeds 30 cm in length. As its nickname implies, it is well adapted to moving rapidly through the desert sand; it body is stout and cylindrical but otherwise quite streamlined, its conical tail is rather short, and its flattened muzzle is prolonged by sharp plowlike ridges that extend beyond the lower jaw. Both its eyes and external ears are very small, but its legs are fully developed, and its toes are provided with a scalloped fringe of scales to assist it in burrowing through the sand. Several subspecies of *Scincus scincus* range all across North Africa, from Egypt to the Spanish Sahara, though it is only found in regions of shifting sand (particularly the high dunes known as ergs). Longevity in captivity is on the order of 4 or 5 years.

▶ A study conducted in the Great Western Erg in southern Algeria revealed that, in that particular region at least, common skinks are either exclusively herbivorous

(12 percent of cases), exclusively carnivorous (30 percent), or omnivorous (58 percent). The vegetable portion of their diet is derived primarily from the seeds of grasses, sedges, and spiny shrubs *(Genista)* that collect in drifts, along with other windblown debris, at the foot of the dunes; in the spring (April—early June), common skinks occasionally feed on the fallen blossoms of desert shrubs.

Their animal prey consists very largely of beetles, primarily scarabs and tenebrionids as well as some rather large carabids (such as *Scarites*). Common skinks also feed occasionally on scorpions, bristletails, spiders (including Solpugida), homopterans, cockroaches, and, in the case of the larger adults, on smaller lizards *(Acanthodactylus)*. Though cockroaches are very numerous in the dunes, they are rarely preyed on by the common skink, and insects such as *Anthia* that are equipped with noxious chemical weaponry are avoided altogether (Gauthier, 1967). The common skink locates its prey by sight, foraging in tunnels and burrows in the dunes, sometimes pursuing its prey along the surface, sometimes attacking from ambush.

The stomachs of the specimens examined in this study virtually all included a certain amount, often a substantial amount, of sand (and especially so in the case of those that fed primarily on insects). It is not clear whether this sand was swallowed involuntarily, along with the prey animal, or whether, like the stones in a bird's crop, it may be of some use in crushing up the tough body cases of insects or the seed capsules of desert shrubs. The common skink preys most actively in the spring, after it emerges from hibernation, which is also when food is most abundant; its intake is considerably reduced in August, then picks up again from the beginning of September until the beginning of October, when it goes into hibernation until March.

◀ The numerous predators of the common skink include the zorilla, the fennec, the sand fox, the jackal, the brown-necked raven, the desert monitor, the sand viper, the Egyptian cobra, and several species of desert racer. The skink's best defense against these predators is flight, though like most reptiles, it has little endurance and is quickly exhausted. It can bury itself in soft sand with astonishing rapidity and agility, though in less favorable terrain, it may attempt to escape from predators by curling up in a ball and playing dead.

The common skink is very alert and, since it disappears into the sand at the slightest provocation, rather difficult to catch. This species, which is relatively abundant in the high dunes, is consumed in substantial quantities by desert tribesmen, and Bedouin children quickly learn to anticipate its evasive tactics, so that in a single day an experienced skink hunter may bring back as many as 20–50 common skinks to the family campfire, where they are skinned, roasted over the coals, and eaten as a sort of hash, with dates. The ashes of the common skink, alleged to have an aphrodisiac effect, sometimes figure as an ingredient in various folk remedies, whence the common name of this species in French and German *(scinque des boutiques, Apothekerskink,* both meaning "drugstore skink") as well as its original scientific name *(Scincus officinalis,* "medicinal skink.")

There are a dozen congeners altogether, including *Scincus mitranus,* which ranges as far east as Pakistan, *S. conirostris* of Iran, *S. hemprichi* of Ethiopia, and *S. philbyi* (and 5 other species) of the Arabian peninsula; they all feed on small invertebrates and seek their prey primarily on the surface of the sand. R.V.

COMMON SOLE *Solea solea* 118

118
Sub.: Vertebrata
Cl.: Osteichthyes
S.cl.: Teleostei
O.: Pleuronectiformes
F.: Soleidae

This flatfish measures between 30 and 40 cm and weighs between 300 and 350 g. Typical longevity is about 8 years, but soles continue to grow throughout their lifetimes, so that a fish that survives to the age of 10 will probably measure 45 cm in length; exceptionally long-lived individuals may even attain a length of 65 cm and a weight of up to 3 kg by the age of 20, though this is rather rare. Like other flatfish (Pleuronectiformes), the common sole is a benthic (bottom-dwelling) species, generally found along sandy substrates at depths ranging from 0 to 150 m. It prefers a water temperature of 12–16°C and is generally found in shallow coastal waters in the spring and at greater depths during the winter. This species ranges throughout the Mediterranean and the Black Sea as well as the Eastern Atlantic, from Norway to Senegal.

▶ The common sole is a nocturnal predator for the most part, relying on its sense of smell to locate prey animals concealed in the sand along the bottom; it feeds on thin-shelled lamellibranchs (shellfish), crustaceans, and polychaetes and other marine worms. The sole also uses the stiff fringe formed by its caudal and dorsal fins, which extends virtually all the way around its body, as a tactile organ to alert it to the presence of prey animals encountered while burrowing through the sand. The adult sole ingests its prey by pursing its mouth into a sort of suction tube. The pelagic larva feeds on plankton, which it locates by sight, until its metamorphosis, whereupon it descends to the bottom and assumes the form and feeding habits of the adult.

◀ The chief predators of the common sole include the ling and the cusk (relatives of the codfish), the turbot (a flatfish), the conger eel, and the flapper skate and other *Raja* species. The sole's basic defense against these predators is concealment, either by displacing sand with the fringe of fins along its body and buring itself completely or, on more resistant surfaces, by mimicking the color of the substrate, which it does by dilating or contracting the chromatophores in its skin. This reaction, which is triggered by fear and other forms of violent stress, occurs when a visual mapping of the sole's immediate surroundings is transformed into a series of neural impulses relayed by the hypothalamus to the pituitary gland, which in turn secretes the hormones that stimulate the dilation or contraction of the individual pigment cells.

The common sole, better known in restaurants and fishmarkets as the Dover sole (Dover having been the point of origin of the freshest fish that were available in London during the pre-refrigeration era), is greatly esteemed for its firm, delicate-tasting white flesh. It is caught with trawls and seine nets, and the annual catch unloaded in European ports is equal to about 25,000 metric tons.

Other members of the sole family are found on both sides of the Atlantic as well as in the Pacific; other Eastern Atlantic and Mediterranean soles that are too small to be fished commercially include the sand sole *(Solea lascaris)*, the eyed sole and the thickback sole *(Microchurus)*, and the lenguadilla *(Dicologlossa cuneata)*. J.-M. R.

COMMON TERN *Sterna hirundo* 119

119
Sub.: Vertebrata
Cl.: Aves
O.: Charadriiformes
F.: Sternidae

Also known as the sea swallow *(hirundo* = 'swallow'), this small gregarious seabird measures 34–36 cm in length, with a wingspread of 73–77 cm, and weighs between 100 and 175 g. The common tern is found on banks of sand or gravel in riverbeds and along the shores of lakes and ponds as well as along the coast; it ranges inland from New England and the Maritimes to the Great Lakes and Hudson's Bay as well as all across the temperate regions of Eurasia, occasionally nesting at altitudes of up to 4000 m on the Tibetan plateau.

The common tern builds its nest in a shallow dugout on the ground ("scrape" is the ornithologist's term), which is occasionally concealed by vegetation. European populations leave the nesting colonies in July and spend the winter along the shores of the Mediterranean or on lakeshores and riverbanks in the interior of Africa, sometimes as far south as the Cape. North American populations winter along both the Atlantic and Pacific coasts, from California and the Middle Atlantic seaboard as far south as Peru and Argentina. A banded common tern is reported to have survived to the age of 27 years.

▶ Small squadrons of terns hunt for shoals of fish in shallow coastal waters. Once a shoal has been spotted from the air, the terns roll out and dive with their wings tucked back against their bodies and generally return to the surface moments after striking the water with a small silvery fish glinting in their beaks. The prey (in European waters, generally a bleak, gudgeon, or sand launce) may be swallowed on the spot or brought back to the colony; symbolic food offerings play an important part in the courtship ritual, and unlike gulls for example, the common tern feeds its nestlings on whole fish, presented to them one at a time, rather than on regurgitated pap.

◀ The common tern's catch is sometimes pirated by gulls and skuas as it returns from its fishing grounds, and the eggs and nestlings are highly vulnerable to predation by rats, cats, hedgehogs, scops owls *(Otus),* and gulls, though the eggs are even more likely to be destroyed by floods, coastal storms, spring tides and freshets, and other natural calamities. As noted elsewhere, the HERRING GULL has often supplanted the common tern on many of its inland and coastal nesting grounds. Terns were widely hunted and their nests pillaged by humans during the 19th century when this species was far more common in Europe (only about 1500 nesting pairs remain in France), but nowadays their numbers are more likely to diminish as they are deprived of suitable nesting sites by civil engineers and real estate developers engaged in the canalization and rectification of riverbanks, and construction of seashore residential and recreational areas.

The tern family (Sternidae) comprises 9 genera, though three quarters of all tern species are assigned to the genus *Sternus.* All terns are shorebirds and most are found in tropical or subtropical regions, a notable exception being the Arctic tern *(S. paradisaea),* which nests in the Arctic tundra of North America and Eurasia and winters on the edge of the Antarctic pack ice; its two annual migratory flights of 18,000 km are thus the longest made by any species of bird.

The genus *Chlidonias* includes several smaller, stockier terns, including the whiskered tern *(C. hybridus),* which nests in Spain and in the marshes of Central France as well as in India, Australia, and other subtropical regions of the world. The black tern *(C. niger)* is found in the American West and Midwest as well as in much of Europe and Central Asia. Both species feed primarily on insects and other small creatures that are caught in midair, on the leaves of aquatic plants, or on the surface of the water. J.-F. T.

COMMON TOAD *Bufo bufo* 120

The common garden toad of the north temperate zone is usually between 7 and 13 cm long, though in Southern Europe the larger females may grow to a length of 18 cm. The toad is generally inactive until dusk (crepuscular) and is found in gardens and flowerbeds and frequently in the vicinity of human dwellings, particularly farmyards. It buries itself in the ground for its long winter period of dormancy, and each spring the adults return to breed by their native pond or pool. The female lays two long ribbonlike strands of eggs, which are attached to the leaves of an aquatic plant. The tadpoles are black, and in June and July they metamorphose into little "toadlets" about 1 cm long. Typical lifespan is about 10 years; record longevity for this species was achieved by an English toad that survived for 36 years in captivity.

▶ The diet of the common toad consists of ants (62.9%), beetles (14.3%), earwigs (4.7%), true bugs (4%), spiders (2.7%), daddy longlegs (1.7%), myriapods (1.8%), and sowbugs (1.4%), with 81% of the beetles consisting of ground-dwelling or herbaceous species such as carabids, staphylinids (rove beetles), and weevils. Diurnal flying insects and insects that live on the leaves and stems of plants are unlikely to be encountered by *Bufo bufo,* which is nocturnal and strictly terrestrial.

The toad has been variously acclaimed by farmers and gardeners for catching slugs and beetles (more deservedly, it would seem, in the case of the beetles than the slugs) and condemned by beekeepers for eating bees. Juvenile toads are almost immediately broken of the habit of eating bees by direct conditioning—the unpleasant experience of being stung repeatedly on the inside of its mouth. This is a lesson, however painfully acquired, that is occasionally forgotten by older toads, and the large adult females appear to be less susceptible to the bees' venom and quick enough with their tongues to neutralize most of their attackers in flight. In cases where a beehive has been placed directly on the ground rather than an elevated platform, large female toads have been known to station themselves on the little ledge at the entrance to the hive and devour a large proportion of its inhabitants.

The toad has an extremely wide field of vision, and predatory behavior is triggered by a visible movement of an insect or other appropriately sized animal. The toad turns first its head and then its entire body in the direction of the movement (what be-

120
Sub.: Vertebrata
Cl.: Amphibia
O.: Anura
F.: Bufonidae

havioral scientists call the *taxis of orientation),* though there may be a brief delay between the initial perception and the response. Usually the prey is not yet within reach of the toad's protracile tongue, so the toad begins to execute a series of cautious stalking maneuvers (collectively called the *taxis of locomotion),* continually adjusting its position and orientation in accordance with the movements of the prey.

When the toad is within striking distance, it raises itself up on outstretched toes, towering over its victim, and remains in this position for a moment or two until the prey makes a movement of some kind, at which point the toad shoots out its tongue and entraps the prey. The toad's tongue is not strong enough to "reel in" a large insect, which it seizes in its jaws, sometimes with the assistance of its hands, and swallows. Like every other animal, of course, the toad is not a complete automaton that inevitably exhibits a particular behavioral taxis in response to a given stimulus. Hunger is the initial stimulus that prompts the toad to leave its burrow on a hunting expedition, and as the toad's appetite is progressively satisfied, it becomes indifferent to the larger insects that might have aroused the complete panoply of predatory instincts a few hours earlier.

◀ Curiously, the toad may exhibit either predatory or defensive behavior toward a creature of a certain size (this threshold appears to be quite precisely defined with respect to the size of the toad itself), depending on whether the other animal is moving away from it or moving toward it. A large toad will sometimes respond to the approach of a small mammal such as a mouse or a shrew with a combative display, raising itself up "on point", puffing noisily, and attempting to repel the intruder by butting it with its head or flicking at it with its tongue.

In nature and in captivity, toads may engage in tongue-flicking duels followed by similarly aggressive displays directed at one another if they find themselves in competition for the same source of food. Tongue-flicking in this case is clearly an act of aggression or an attempt to assert dominance rather than displaced predatory behavior. When confronted with a potential predator such as a colubrid, a viper, or a stork, a toad will blow itself up and then raise itself up on tiptoe while remaining motionless and emitting puffing or panting sounds. Inflating its body in this case may have the practical purpose of preventing a colubrid from swallowing it, and a toad will sometimes succeed in driving off a viper by butting it with its head.

The venom in its parotid glands and the warty pustules on its back act as a strong deterrent for any mammalian predators that have already had some experience with toads; a dog that has bitten a toad will continue to salivate profusely for several hours and will be unlikely to show much future interest in toads. The toad's principal predators include raptors and wading birds, crows, colubrids, and hedgehogs; an otter will sometimes surprise a group of mating toads and make a feast of them.

The predatory behavior of congeneric species (cf. AGUA TOAD) as well as other terrestrial members of the family Bufonidae is quite comparable to that of *Bufo bufo,* taking into account that there is a substantial size differential among them (and thus a corresponding difference in the size of the prey that they are capable of subduing). Observations of Fowler's toad *(Bufo fowleri)* of North America conducted by Heusser (1958) suggest that the sense of smell plays some part in prey location in this species, but experiments conducted by Martof (1962) failed to confirm this in the case of the natterjack *(Bufo calamita)* of Western Europe. J.L.

COMPASS JELLYFISH *Chrysaora hysoscella* 121

121
Sub.: Cnidaria
Cl.: Scyphozoa
O.: Semaeostomae
F.: Pelagiidae

This species is found all along the Atlantic coast from Norway to Liberia as well as in the Baltic and the Mediterranean, and is one of the commoner large jellyfish encountered along the seacoasts of Europe in the summer. The translucent canopy (which lacks the external membrane called the velum) may be as much as 1.5 m in diameter; pigmentation is variable but generally comprises 16 V-shaped brown markings arranged symmetrically around the midpoint of the canopy. There are 24 highly extensible tentacles around the lower fringe of the canopy as well as 32 small protuberances called rhopalia, in which the sensory organs (statocysts and, in some other species, the eyespots) are located; the name "compass jellyfish" suggests an analogy between the rhopalia and the 32 points of a mariner's compass. The oral cavity is surrounded by a prehensile structure with 4 armlike prolongations called the manubrium.

The sessile polyp stage of the organism is called a scyphistome, which may live for several years and reproduces asexually by a process of lateral segmentation (strobilization); the apical tip of the scyphistome divides into segments, each of which develops into a small medusan (free-swimming jellyfish) only 2 mm in diameter. The ephyra, as this first stage in the medusal cycle is called, originally lacks tentacles, manubrium, and sex organs; by the end of 3 months, the medusan will have acquired all these appendages and attained a diameter of 30 cm. The medusans are protandrous hermaphrodites (which means that the male gonads develop first), incubating their larvae in the gastric cavity; after the larvae are released into the water, they attach themselves to the substrate, develop into sessile polyps, and the cycle begins again.

▶ The scyphistome feeds itself by straining out small vagile organisms from the current of water that is funneled into the oral cavity by the motion of its tentacles and by ciliary action. The full-grown medusan preys actively on other pelagic cnidarians (hydromedusans, siphonophores, ctenarians), polychaetes and arrowworms *(Sagitta)*, and fish hatchlings. While seeking prey, the medusan vertically extends the arms of the manubrium and horizontally extends its tentacles to form a lethal net that may be several times the diameter of the canopy itself.

Prey animals are immoblized by the stinging nematocysts (the venomous springlike filaments that are the characteristic weapon of the cnidarians) at the tips of the tentacles, whereupon the tentacles contract, bringing the prey animal within reach of the arms of the manubrium, by which the prey is immediately conveyed to the gastric cavity and digested.

◀ *Chrysaora hysoscella* provides food and shelter to at least 2 other species. An amphipod crustacean *(Hyperia galba)* actually lives inside the gastric cavity, and a small carangid fish *(Trachurus trachurus)* is secure from predators (of which the medusan itself has none) inside the ring of venomous tentacles that surround the edge of the canopy.

The common jellyfish (*Aurelia aurita*, family Ulmaridae) rarely exceeds 40 cm in diameter and is frequently encountered all along the European coastline. Though it can catch larger prey with its short tentacles and manubrium, it feeds primarily by collecting phytoplankton and zooplankton in a sort of ciliated trench that runs around the outer rim of the canopy, adjacent to the sites where the tentacles are attached. A centrifugal current is created by the motion of numerous flagellated ectodermic cells located on the upper and lower surfaces of the canopy, and the plankton that are sluiced down into this trench eventually come to rest in one of 8 recessed collection points, from which they are periodically removed by the arms of the manubrium.

The nematocysts of *Aurelia aurita* cause only minor irritation in humans, and in fact a jellyfish poultice is sometimes applied as a topical remedy for rheumatism and neuralgia at certain seaside curative resorts in Western Europe. Other varieties of jellyfish, notably the Portuguese man of war *(Physalia)*, have a more sinister reputation, and the venom of the tropical sea wasps (Cubomedusae), which are no larger in diameter than an orange, has been known to cause death in humans within minutes. R.P.

Condylostoma magnus 122

122
Sub.: Protozoa
Cl.: Ciliophora
S.cl.: Spirotrichia
O.: Heterotrichida

This is a species of heterotrich (suborder Heterotricha), a free-swimming ciliate protozoan that inhabits the intertidal zone of temperate seas and is generally found just above (or below) the surface of the sand. This single-celled organism is about 250 μm long and 50 μm wide; its form is elongated, somewhat flattened in front, rounded in the rear, and the entire surface of the cell is covered with longitudinal rows of cilia arranged in pairs. The rim of the buccal cavity, or peristome, is equipped with 2 or 3 rows of cilia on one side and a fringe composed of little tufts of specialized filaments called *membranelles* on the other.

▶ *Condylostoma magnum* feeds primarily on bacteria, microscopic algae, and other protozoa (ciliates and flagellates). The cilia that are attached to the surface of this minuscule predator are used exclusively for the purpose of locomotion, but the rhythmic pulses of the cilia that surround the buccal cavity creates a sort of two-stroke cycle that draws floating organic particles and microscopic creatures into the interior of the cell and then expels those particles or microorganisms that cannot be absorbed. Other cilia serve as a sort of filter, like a whale's baleen in minature, that traps these nutrients, which are then encysted in an alimentary vacuole formed by a projection of the naked cellular membrane that lines the rear of the buccal cavity and finally absorbed into the cell.

◀ We know of no single organism, protozoan or metazoan, that preys specifically on *Condylostoma magnum,* though there are of course an enormous variety of marine metazoans that feed indiscriminately on planktonic organisms of this kind.

There are 10 different species of *Condylostoma;* they differ somewhat in the shape and appearance of their cells, but all are to be found in salty or brackish water. Such freshwater genera as *Stentor* and *Spirostomum,* roughly similar in their basic organization, are among the largest of all protozoans; *Stentor* figures prominently in the diet of certain species of rotifers, notably *Notommata pseudocerberus.* J.G.

123
Sub.: Mollusca
Cl.: Gastropoda
S.cl.: Prosobranchia
O.: Neogastropoda
F.: Conidae

CONE SHELL *Conus sp.* 123

Conus is a genus of tropical sea snails that comprises about 500 species, most of them found in Indo-Pacific waters, characterized by the biconical structure of their shells. The shell itself is noted for its beauty and frequently for its rarity—a single specimen of the glory-of-the-seas cone *(C. gloriamaris),* the rarest of all seashells, may fetch a price of several hundred thousand dollars from a collector—and the snail for the efficacy of its venom, which is used both offensively and defensively and in the case of the larger species has sometimes proved fatal to human beings. Twenty-seven such fatalities have been attested in the literature since 1705.

From an evolutionary standpoint, the cones are regarded as the most highly developed gastropods. They range in size from 3 cm in the case of *C. mediterraneus,* an innocuous greenish sea snail found in European waters, to 22 cm in the case of the leopard cone, *C. leopardus.* The sexes are generally distinct, though internal fertilization sometimes occurs as well (i.e. some individuals are hermaphroditic). In either case, the eggs are encased in a gelatinous substance and deposited in a sheltered spot by the female, often in conjunction with other females of the same species.

The larva emerges about 10 days later and is initially free-swimming—the technical term is *veligerous,* because it is equipped with a veil-like organ of locomotion called the velum—though it begins to crawl along the bottom after a week or two. At first, the shell develops at a rate of several millimeters per month, though this rate of growth quickly levels off to about 1 mm per month, which suggests that an individual cone shell may live for as long as several decades. Cones are found among coral formations or on a sandy substrate (depending on the species) in the intertidal or subtidal zones. During the day, they conceal themselves in a crevice in a coral boulder or bury themselves in the sand and only become active at night.

▶ The cone shell has evolved an elaborate hydraulic and mechanical system that enables it to inject its venom into fish and other marine organisms considerably more mobile than itself. The venom is produced in a long tubular duct and transferred to the oral cavity, where it is stored, by the contractions of a muscular bulb. Many other mollusks "chew" their food by grinding it against a chitinous belt studded with sharp teeth, called the radula, located on the floor of the oral cavity; these teeth are continually being worn away and then replaced by an adjacent organ called the radular sac. In the cone shell, however, the teeth produced by the radular sac are hollow, with a channel running down the middle and an aperture at either end, and are not attached to a belt of chitin but simply deposited in the oral cavity, where the channel inside the teeth fills up with the venom that is stored there.

The cone shell's venom delivery system consists of an elastic muscular proboscis with an opening at its tip through which these pointed teeth can be extruded. If the first of these poison darts misses its mark, it will be discarded and another tooth is aspirated into the chamber so the cone will immediately be ready for another shot. The cone shell's venom contains several neurotoxins that have an immediately disruptive effect on the nervous and digestive systems of the victim, one of which will eventually cause the victim's heart to stop—in 10 minutes in the case of a fish, possibly within 2 hours in the case of a human being. The numerous cone shell species feed on fish, worms, or mollusks, and the chemical composition of the venom, the design of the poison dart, and other details of the venom delivery system differ accordingly. In habitats shared by several species, each species is fairly limited in its choice of prey (stenophagous) so that interspecific competition is avoided, though isolated cone shell species may be more eclectic in their feeding habits.

The geographer's cone *(Conus geographus)* and other piscivorous species produce the deadliest venom but are also the most restricted in their range. They are active at night while the blennies and gobies on which they prey are resting close to the bottom. The predatory behavior of one such species, *C. striatus*, has been observed in an aquarium: The cone shell buries itself in the sand, leaving only a siphon projecting from the surface, one of the functions of which is to maintain a continuous flow of water past an olfactory organ called the *osphradium*. When the presence of a fish has been detected, the proboscis of the cone shell is fully extended so that it attains a length equal to that of the shell itself and changes orientation, like the tip of a pointer, to track the movements of the prey.

Once contact has been made, the cone shell extrudes its poison dart from the tip of the proboscis and injects the fish with its venom. At the same time, the "muzzle" of the proboscis contracts around the base of the dart, and the paralyzed victim is drawn closer to the cone shell's mouth by means of a partial retraction of the proboscis. The prey is often as large as the cone shell itself and must be partially disgested in the oral cavity before it can be conveyed into the less readily extensible portions of the digestive tube. The complete process of digestion sometimes takes 24 hours or so, during which the proboscis is fully retracted inside the shell.

Molluscivorous cones prey on a large variety of marine gastropod species (order Opisthobranchia and Streptoneura), including other cones in the case of *C. textile*, the textile cone, and *C. marmoreus*, the marbled cone. Though relatively slow-moving, their prey are not entirely immobile, so that conchs of the family Strombidae, for example, will react violently to the approach of a cone. The process of predation is similar to that described above, except that instead of retracting the proboscis to draw the impaled and helpless victim toward its mouth, the cone shell moves forward to grapple with its prey.

Opisthobranchs with internal shells are swallowed whole and the shell is later regurgitated. In the case of other gastropod species with external shells, the cone presses its mouth over the opening of the shell and the shell opens, apparently of its own accord, after about an hour (though more probably because an infusion of *Conus* venom has brought about the relaxation of the columellar muscle that enables the gastropod to retract into its shell). The edible parts of the gastropod are swallowed whole, and digestion begins to take place in the oral cavity. The cone shell will only use its venom against cephalopod mollusks for the purpose of self-defense.

The vermivorous cone species, such as *C. mediterraneus*, are the most numerous. They feed primarily on polychaete (Neridae, Eunicidae) and hemichordate marine worms, and in this case, the proboscis, still clutching its poison dart, retracts entirely inside the shell as soon as the worm has been swallowed, and digestion begins to take place in the stomach.

◀ The eggs of the cone shell are devoured by wrasses and gobies, the veligerous larvae by numerous marine predators, and the larvae during their later crawling phase by brittle stars (ophiura), all of which constitute important natural checks on the cone shell population. Cone shells may be attacked in an aquarium by crabs or starfish, but apart from the molluscivorous cone species mentioned above, the gastropod *Cymatium nicobaricum* (order Streptoneura), which is able to insinuate its own proboscis into the crack of the cone shell and feed on its flesh, is the only predator they are known to encounter in nature; it has also been suggested that broken cone shells that are sometimes found cast up on the beach may have been cracked by the teeth of parrotfish or moray eels.

Two other families of marine gastropods, though otherwise quite different in their morphology, have evolved a similar apparatus for injecting their prey with venom; these are the Terebridae, or auger shells, with 150 species, and the Turridae, turrid shells, with 1200 species. M.M.

124
Sub.: Vertebrata
Cl.: Osteichthyes
S.cl.: Teleostei
O.: Anguilliformes
F.: Congridae

CONGER EEL *Conger conger* 124

The conger is a very large eel that may attain a length of 3 m and a weight of 65 kg. The average size of a 7– or 8–year-old specimen is between 1 and 2 m, the female being slightly largely than a male of the same age. Maximum longevity is probably 15 years; like other eels, the conger only mates once in its lifetime, generally between the ages of 5 and 15, and then dies shortly afterward. It is found along the Atlantic coast of Europe and Africa from southern Norway down to Senegal as well as throughout the Mediterranean and in certain estuaries. It lives in cracks and crevices in the rocks offshore down to a depth of 100 m; it migrates into deeper waters to spawn, though the depth at which this occurs is unknown.

▶ The conger eel is a nocturnal hunter with a well developed sense of smell. It patrols along vertical rock faces and the rocky bottom in search of its prey and feeds on all sorts of benthic (bottom-dwelling) organisms, including octopuses, shrimp, crabs, rock lobsters, lobsters, flounders, gobies, wrasses, and gadid fishes (cod, haddock, etc.) as well as gastropods, lamellibranchs, and even young seabirds, which it seizes by the legs as soon as they alight on the water and drags down below the surface. The free-swimming leptocephali (larvae) of the conger eel feed on plankton during their initial migration to the coast, where their diet is very similar to that of the adult.

◀ On occasion, a conger eel will attempt to take up residence in a crevice that is already occupied by a lobster or an octopus, and at such close quarters, the eel may not always prevail over its entrenched adversary. The leptocephali are preyed on by many kinds of fish and pelagic crustaceans as well as baleen whales.

Conger eels are occasionally caught with dragnets (on which they are also capable of inflicting serious damage) but more frequently with standing lines. The conger has to be carefully handled even when out of the water, since its jaws are easily strong enough to shear off a finger joint. The annual catch of the European conger eel fishery is currently about 12,000 metric tons, about half of which is taken by French vessels. The American conger, *C. oceanicus*, is less important economically but is still sought after by sports fishermen as a powerful and aggressive game fish.

J.-M. R.

125
Sub.: Vertebrata
Cl.: Aves
O.: Passeriformes
F.: Icteridae

COWBIRD *Molothrus ater* 125

The cowbird is a New World blackbird (icterid) that lives in meadows and pastures in the temperate regions of North America and is sometimes called the eastern cowbird to distinguish it from other species found in Mexico and Central America. Like the Old World cuckoo, the cowbird is entirely parasitic, laying its eggs in the nests of 185 different species of birds, including tyrants, vireos, and warblers (Sylviidae). Adult cowbirds are gregarious, flocking at dusk like starlings and roosting together in reedbeds.

▶ Before the destruction of the bison herds of the Great Plains, the cowbird fed primarily on ticks and other parasites retrieved from the hides and shaggy pelts of these creatures. Since then, they have entered into a similar symbiotic arrangement with domestic cattle, much appreciated by the latter, ridding them of ticks, larvae, and other parasitic arthropods that lay their eggs or burrow beneath the skin. The cowbird also feeds on fruits and seeds.

◀ The adults and nestlings may fall prey to a very wide variety of predators, including snakes, raptors, and carnivores. Cowbirds appear to welcome the competition of other icterids, since they often perch together and companionably search for parasites on the backs of the same animal.

The family Icteridae comprises 88 species altogether, including blackbirds and orioles, grackles, meadowlarks, caciques, and troupials. The troupials of Central and South America are primarily insectivorous; the grackles are omnivores, and their notched beaks and powerful maxillary muscles even allow them to crack nuts. The meadowlarks *(Strunella)* range from Canada to Brazil; their sharp, awl-shaped beaks are especially well suited for digging in the ground, turning up stones, and shredding rotten wood in search of insects. The caciques *(Cassica)* are best known for the complex structure of their nests, which are considerably elongated and shaped something like an airport windsock in order to protect the nestlings from marauding reptiles; the nest of Montezuma's cacique *(Gymnostinops montezuma)* may be as much as 2 m long. J.-F. T.

COYOTE *Canis latrans* 126

126
Sub.: Vertebrata
Cl.: Mammalia
O.: Carnivora
F.: Canidae

This doglike carnivore, long a familiar figure on the Western plains, has gradually increased its range since the colonization of North America by Europeans—as a result not only of widespread deforestation but also of the elimination of such formidable competitors such as wolves and lynxes—and is now found in many parts of the Midwest, southern Canada, and even New England. Maximum weight is about 25 kg with an overall body length of 130 cm (including 40 cm for the tail).

▶ The coyote has been known to prey on virtually every animal of the Americian prairie, from the grasshopper to the bison, though it feeds primarily on jackrabbits, followed (in apparent order of preference) by carrion, rodents, domestic animals, reptiles, birds, and insects. The coyote's fondness for carrion, into which poison can easily be introduced, almost proved to be its undoing a number of years ago, when coyote fur was very much in vogue; the commercial demand for this product has since largely subsided, though the coyote remains, of course, a frequent target of bounty hunters and aggrieved ranchers.

The coyote's diet and food-gathering methods vary with the seasons. It feeds on carrion primarily during the winter and fills out its diet of meat with summer fruit like a number of other carnivores. In spring, it may feed on the newborn calves of large herbivores and on wounded deer and other animals during the hunting season; by winter's end, the coyote will still be vigorous enough, as a rule, to bring down a half-starved elk or buck. Coyotes sometimes hunt in pairs, since cooperation is as essential in running down a jackrabbit as in penetrating the defenses of a porcupine. The coyote sometimes makes use of the services of an involuntary collaborator, the badger, in catching rabbits that the badger has flushed from their holes (and which the badger would be incapable of pursuing in any case). Coyotes have also learned to follow herds of cattle, as well as combine harvesters and other farm vehicles, in order to catch fleeing rodents, just as some cats have learned to do in Europe.

A coyote generally spends its life in the area in which it was born, and thus as it moves from one locality to the next, it has a very good idea of the sort of prey that is likely to be found there. (The coyote is an excellent swimmer, for example, and very fond of the flesh of ducks and other waterfowl). It catches rodents and other small animals by pinning them to the ground with its paws, as a cat does a mouse, before it bites them; it catches rabbits, as noted earlier, by running them down (in conjunction with a "tag team" partner) until the quarry is exhausted, then killing it with a bite to the back of the neck. To bring down a larger herbivore, such as a pronghorn or a buck, it seizes the animal by the throat and slashes the jugular vein after its hunting partner succeeds in pulling the animal off its feet.

◀ Apart from man, the coyote has a number of other formidable enemies, even including the golden eagle, which attacks it frequently. Other large carnivores, notably the lynx, the bear, and the wolf, will not tolerate competitors of their own or another species on their territories and will consequently attack the coyote on sight, as will the more aggressive herd animals on some occasions. The coyote's main defense against these competitors is its speed and agility, though in areas where coyotes have been hunted with packs of dogs, there have been reports of them rolling in the grass in order to stain their coats with plant juices and thus acquire a certain measure of protective coloration. P.A.

127
Sub.: Vertebrata
Cl.: Mammalia
O.: Pinnipedia
F.: Phocidae

CRAB-EATER SEAL *Lobodon carcinophagus* 127

The sexes are almost identical in size and weight—2.5–2.6 m and 225 kg, on the average. The seasonal migrations of the crab-eater seal follow the advance and retreat of the Antarctic ice shelf and do not extend much beyond that, so that the crab-eater is relatively sedentary in comparison to other seals. The pups already weigh 20 kg at birth; they are weaned at the age of 1 month, and attain their full size at 5 months. Maximum longevity is at least 30 years, and the current population is estimated at 15 million—or about half the total pinniped (seal, sea lion, walrus) population of the world.

▶ To begin with, the crab-eater seal does not and cannot eat crabs; its teeth are not strong enough the crack a crab's shell. It filters seawater through its closed jaws, somewhat in the manner of a baleen whale, and strains out small planktonic crustaceans, collectively known as krill and consisting primarily in this case of the species *Euphausia superba*. The crab-eater's upper molars have up to 5 cusps with which the lower teeth interlock very neatly when its jaws are closed, enabling it to bail out the seawater while retaining the krill.

◀ The killer whale and especially the leopard seal are the crab-eater seal's chief predators; almost all the adults bear scars inflicted by leopard seals. The leopard seal is also a competitor, in a small way, since it feeds on krill only when the supply is particularly abundant. More serious competitors, the great whales, have largely been exterminated in recent years, but humans (notably the Japanese) have recently begun to develop a commercial krill fishery, so the crab-eater will apparently not be allowed to reap the full benefit of this enormous ecological upset. M.P.

128
Sub.: Vertebrata
Cl.: Aves
O.: Falconiformes
F.: Accipitridae

CROWNED EAGLE *Stephanoaetus coronatus* 128

This spectacular bird of prey measures between 81 and 99 cm from beak to tail, weighs from 3 to 5 pounds, and is equipped with powerful talons and short, rounded wings; its cry is loud and raucous. Its average lifespan may be in the vicinity of 10 years: A single nesting site in Africa was occupied by 3 successive females and 2 males over a period of 20 years; one of the females lived to the age of 13 or 14, having reached maturity at the age of 3 or 4. The crowned eagle makes its nest in the tops of trees and inhabits the primary and gallery forests (the trees that grow along a watercourse in the savannah) of tropical Africa.

▶ Though it can overwhelm prey animals of 10 kg and more, its normal prey are small mammals weighing between 0.5 and 5 kg—young forest antelopes (duikers), monkeys, and rock hyraxes; it sometimes catches hornbills as well. It monitors the activities of antelopes and monkeys on the forest floor from its concealed perch in the treetops, then suddenly plunges down and pins the antelope to the ground or seizes it in its talons, taking care to avoid a kick from the animal's hind feet, and waits for the death of its victim. Monkeys, with their four hands and sharp teeth, are not so easily dispatched, but the eagle generally manages to stun them with the violence of its attack or by knocking them against the ground. If the crowned eagle kills a prey animal that is too big for it to carry off, it dismembers its prey on the spot and conceals the leftover pieces in the foliage.

◀ Because of its size and strength, the crowned eagle has no natural predators, or competitors for that matter. It avoids intraspecific rivalry by diligently patroling its own territory at a high altitude, emitting its piercing cry to warn other eagle pairs that it has a proprietary interest in that particular patch of forest. It can rarely be seen from the ground and thus is not often hunted by man, but deforestation currently poses a serious threat to its habitat in many areas.

There are several other species of tropical eagles with similar hunting strategies. The African martial eagle *(Polematus bellicosus)* also hunts hyraxes and antelopes but on the open savannah rather than in the forest. The harpy eagle, the most powerful of all eagles, lives in the Amazonian rainforest; another species is found on New Guinea. The rare monkey-eating eagle

(Spilornis holospilus), one of the most impressive of eagle species, now finds its domain severely threatened by logging and land-clearing in the Philippines. The hawk eagles (genus *Spizaetus)* of Asia and South America are a group of forest raptors intermediate between hawks and eagles.

J.-F. T.

CROWN-OF-THORNS STARFISH
Acanthaster planci 129

129
Sub.: Echinoderma
Cl.: Asteroidea
O.: Spinulosa
F.: Acanthasteridae

Found on coral reefs in the Indo-Pacific region, from the Red Sea and the east coast of Africa as far west as Hawaii and the Tuamotu Archipelago, this starfish is active at night and remains in concealment during the day. Local populations vary greatly in size and density from isolated individuals and small colonies to enormous hordes of many thousands; on one occasion, as many as 20,000 crown-of-thorns starfish were removed from the reef surrounding the small Japanese island of Miyako Jima, one of the Ryukyus.

After a planktonic larval stage lasting for 3 or 4 weeks (at a temperature of 27–29° C; it cannot survive below 25° C), the juvenile crown-of-thorns has 5 arms—between 9 and 21 additional arms will develop later—and measures about 0.3–0.5 mm in diameter. The maximum diameter of the adult crown-of-thorns is about 60 cm, and the spines, or "thorns," themselves are between 15 and 45 mm long. Average lifespan is about 3 years.

▶ The crown-of-thorns starfish has attracted considerable publicity, even notoriety, in recent years because it feeds on coral polyps and because it is subject to sudden and inexplicable population booms, so that it can quickly depopulate a coral reef. Its feeding technique is quite distinctive: It is able to extrude its stomach, which it applies, in the manner of a vacuum cleaner nozzle, to the surface of the coral; the multiple arms are helpful in spreading and unfurling the folds in its stomach. The crown-of-thorns secretes digestive enzymes that initially paralyze the coral polyp's stinging nematocysts (which are otherwise effective in repelling the onslaught of the starfish's tube feet) and eventually dissolve not only the soft body of the polyp but also its calcareous skeleton, and thus the coral reef itself. An adult crown-of-thorns starfish can eat its way through 6–12 square meters of coral in a single year; it only took two and a half years for these starfish to destroy 90% of the coral formations along a 38 k stretch of coastline on the isalnd of Guam.

The planktonic crown-of-thorns larva feeds on microscopic particles (unicellular algae less than 60 μ long) that it filters out of the water; the juvenile crown-of-thorns feeds on red algae that are found on coral reefs. The adult crown-of-thorns shows a clear preference for stony reef-building corals (madrepores), including *Acropora, Pocillopora,* and *Porites,* and particularly for massive formations of brain coral, though it will also attack staghorns and related branching corals—less frequently alcyonarians—on the open seabed as well as on the reef. The crown-of-thorns is a serious hazard for divers in tropical waters as well as for coral polyps; its thorns can inflict a painful wound, often with localized swelling, lymphangitis (inflammation of the lymph glands), and even temporary paralysis of the limb.

◀ There are many fish that feed avidly on the eggs of the crown-of-thorns, and the madrepore *Pocillipora damicornis* is an important predator of the planktonic larvae (as well as one of the favorite targets of the adult starfish). The main natural enemy of the crown-of-thorns, however, is a large marine gastropod called the giant triton *(Charonia tritonis)*; it was observed in the course of one study that 15 of these creatures had devoured 125 starfish over a period of 3 months. A reef fish, *Cheilinus undulatus,* also feeds on the flesh of the crown-of-thorns, though without inflicting any permanent damage on its victims, since an arm that has been nibbled off by a fish or other predator can quickly be regenerated.

Given the regenerative and even more impressive reproductive capabilties of the crown-of-thorns, it is not surprising that all attempts to limit the depredations of this species by introducing predators into the habitat have been unsuccessful. There are currently a number of other programs underway, sponsored both by national governments and international organizations, to eradicate the crown-of-thorns in areas where the wholesale destruction of coral reefs (with momentous and often unforeseeable consequences for the ecology of the region) is threatened.

The crown-of-thorns has only one close relative, *Acanthaster brevispinus,* which is found in deep coastal sediment as well as in the vicinity of coral reefs from the Philippines down to Queensland. Much more restricted in its range than the crown-of-thorns, *A. brevispinus* is so similar to *A. planci* in other respects that some authorities regard them as different varieties of the same species. A.G.

130
Sub.: Vertebrata
Cl.: Aves
O.: Cuculiformes

CUCKOO *Cuculus canorus* 130

This is the common Eurasian cuckoo, which lays its eggs in the nests of virtually every species of insectivorous passerine bird that shares its range as well as a number of others (including the pheasant), some 300 species in all. The cuckoo is a true parasite, insofar as it does not build a nest, incubate its eggs, or feed its nestlings; it is found in a great variety of different habitats between sea level and the snowline in the mountains, including forests and wooded groves, scrublands, and grasslands.

The cuckoo is about 34 cm long with a wingspan of 59–61 cm and weighs between 90 and 142 g. Its lifespan is probably fairly long, if only to compensate for its virtually effortless but extremely prodigal means of rearing its young. The cuckoo winters in southern and central Africa and returns to Europe at the end of March. It is polygamous, and the male defends a fairly extensive territory in which several females will deposit their eggs in the nests of other birds. The adults set off on their return flight in July, and the young cuckoos follow in August.

▶ The female cuckoo feeds on the eggs of the species it parasitizes, and the ejection of the nestlings of these species by the young cuckoo might also be considered a form of predation. More routinely, the cuckoo is an insectivore, feeding especially on hairy caterpillars disdained by other birds (notably the processionary and nun moth caterpillars) as well as the adult insects, myriapods, earthworms, and slugs. It hunts for caterpillars on the bark of pinetrees or along the ground, and carefully squeezes out the intestines before swallowing them; it can afford to be less fastidious about the caterpillars' stinging filaments, which adhere to the mucous lining of the cuckoo's stomach in feltlike tufts until an entire patch of squamous cells sloughs off and is regurgitated, hairs and all, by the cuckoo.

◀ It is estimated that only a single young cuckoo sets off on its fall migration for every 20 eggs that are deposited in other birds' nests in the spring; very frequently the nests are abandoned and the eggs or young cuckoos devoured by carnivores or birds of prey. The male gives its famous territorial call to fend off competitors of its own species, though the female is often set upon and treated rather roughly by birds whose nests it attempts to parasitize. The cuckoo's curious resemblance to a bird of prey (the precise adaptive value of which, if any, is unknown) has given rise to the legend that it can change itself into a sparrowhawk at will—for which reason it has reportedly been persecuted in some rural communities. J.-F. T.

131
Sub.: Arthropoda
Cl.: Crustacea
S.cl.: Coperpoda
O.: Cyclopoida
F.: Cyclopidae

CYCLOPS *Cyclops bicuspidatus thomasi* 131

Known as *Diacyclops thomasi* to European taxonomists, this is a species of copepod, a minute freshwater crustacean that is only about 0.07–0.09 mm long, the female being slightly larger than the male. The most striking feature of the genus *Cyclops,* collectively known along with several other copepod genera as water fleas, is the prominent simple eye dot. This particular species constitutes a substantial part of the plankton population in ponds and lakes on the North American prairies and in the mountains up to an altitude of 2000 m.

The females lay their eggs in May, and by July most of the larvae will have attained the fourth of their five devlopmental stages. They sometimes spend the hottest part of the summer buried in the mud in a state of suspended animation that is known as *diapause;* in early fall, the larvae will reemerge and complete their development. The nauplia and copepodids (larvae in the earlier stages of development) are found in open water whereas the stage IV and V larvae and the adults tend to congregate on the bottom and along the shore. Cyclops tends to be more active at night, though this pattern is occasionally reversed.

▶ Prior to stage IV, the copepod larvae feed exclusively on plant material. The older larvae and the adults are carnivorous, however, and often will attack creatures larger than themselves (though these may not be entirely consumed). The eye dot does not play a role in prey detection; cyclops locates its prey by means of chemoreception and by tracking the vibrations it makes in the water. Cyclops is a voracious predator, pouncing on the prey animal—very frequently a smaller member of its own species—and seizing it with its maxillulae (third pair of mouthparts), then often grasping it with the second pair of mouthparts (maxillae) and its leglike appendages (maxillipeds) and conveying the animal toward its open mouth. The creature's body is torn to pieces by the mandibles and the resulting fragments are thrust further back into the esophagus.

In a laboratory experiment in which *Cyclops bicuspidatus thomasi* was present in about the same concentration as one might expect to find it in nature (about 25–30 per liter of water), each individual cyclops consumed an average of either 1.8 nauplia or 0.7 copepodids of its own species or 1.6 calanid nauplia (stage I larvae of copepods of the family Calanidae) or as many as 6 rotifers (genus *Keratella)* when the concentration of the latter approximated 70 per liter. Clearly the appetites of this species constitute one of principal checks on its own population growth, and it was estimated that the older larvae and adults in a lake in British Columbia had managed to consume fully 31% of their own nauplia and 30.2% of the calanid nauplia that were originally present.

◀ In general, it is safe to assume that if two predatory copepod species are found in the same habit, they will attempt to prey on one another. Thus, notable invertebrate predators of *Cyclops bicuspidatus* include 2 other copepod species—the cyclopid *Tropocyclops prasinus* and the calanid *Diaptomus*—as well as the cladoceran (branchiopod crustacean) *Leptodora kindtii* and a genus of gnat larvae, *Chaoborus*. Copepods also make up a large part of the diet of many fish, especially the fingerlings; those thay prey specifically on *Cyclops bicuspidatus* include the yellow perch *(Perca flavescens)*, the bluegill *(Lepomis macrochirus)*, and the emerald shiner *(Notropis atherinoides)*.

The American smelt *(Osmerus mordax)*, a saltwater fish that has successfully been introduced into the Great Lakes region, feeds primarily on cyclops until it has grown to a length of 21 cm and is large enough to cope with more substantial prey. The alewife *(Pomolobus pseudoharengus)* is another saltwater species that has recently found its way into the Great Lakes region and has already acquired a taste for the younger larval forms of *Cyclops bicuspidatus* as well as other copepod species (notably *Diaptomus)* that congregate just below the surface of freshwater lakes; the bottom-dwelling and shore-hugging propensities of stage IV and V larvae and the adults would appear to exempt them from this predation.

The vast majority of species belonging to the family Cyclopidae are found in inland waters, including lakes, rivers, ponds, puddles, and underground streams. There are several cyclopid genera *(Microcyclops, Metacyclops*, some species of *Eucyclops)* that remain herbivorous after the initial larval stages, but most species eventually start to feed on zooplankton and somewhat larger aquatic organisms—notably tardigrads (mitelike arthropods), oligochaete worms, and fly larvae—as well as on algae and miscellaneous plant debris.

Other copepods may be found in salt water as well as fresh, are primarily free-swimming (though some species are parasitic), and constitute a sizable proportion of the planktonic biomass of the world's oceans. Insofar as they constitute an essential part of the diet of young fish, they can be said to be of considerable economic importance to man. F.L.-M.

Cyrtogonatopus 132

132
Sub.: Arthropoda
Cl.: Insecta
O.: Hymenoptera
Sup.f.: Bethyloidea
F.: Dryinidae

This genus of dryinid wasp is extremely common in fields and meadows in tropical and temperate regions; behavioral studies referred to below were conducted on the island of Guadeloupe. The female, wingless in most species and antlike in appearance, has a pair of chelae, or well-developed pincers like a lobster's claw, at the tips of its forelegs. These pincers are absent in the winged males, which have three pairs of undifferentiated walking legs. Some species are strictly parthenogenetic (exclusively female), others primarily so, in that males are relatively rare; in the latter case, females of a given lineage will produce either females (parthenogenetically) or males (by sexual reproduction), but never both.

▶ Like most other dryinid wasps, *Cyrtogonatopus* feeds exclusively on leafhoppers of the genus *Cicadella*. The latter process follows a familiar pattern—the female wasp alights on an adult leafhopper and temporarily paralyzes it with its stinglike ovipositor, then inserts an egg either wholly or partially under the abominal cuticle of the leafhopper, generally between two abdominal segments; some species deposit their eggs further up on the cuticle, at a point opposite the thorax or the head. Between 30 seconds and several minutes later, the leafhopper reawakens and resumes its normal activity, to all appearances unharmed. (However, if the wasp deposits its egg inside the body case of a leafhopper larva, the latter will be unable to molt its body case and complete its larval development).

When the larva hatches, it remains inside the leafhoppers's body shell, and at its initial molt, begins to surround itself with a tough protective covering (exuvium), which can be seen from the outside as a dark swelling beneath the aphid's shell. There are five larval stages, and at each molt, the larva secretes a new exuvial layer inside the previous one, and the latter is eventually ruptured as the larva grows in size. The head of the larva generally protrudes into the gut of the leafhopper whereas the rest of the body is coiled up inside an external vesicle with slitlike perforations (stigmata) that permit respiration.

By the end of the fourth larval stage, the larva will have consumed not only the contents of the gut but the leafhopper's internal organs as well, which rapidly results in its demise. At its fifth molt, the mature *Cytrogonatopus* bursts through the larval vesicle and the leafhopper's body shell and crawls off to find a suitable spot in which to undergo nymphosis. It spins a double cocoon, one layer inside the other, either in a shallow burrow in the ground or under a mat of dead leaves and dried grass.

In temperate regions, the developing dryinid larva sometimes goes into a state of suspended animation, or diapause, during the winter, but in Guadeloupe during the dry season, the leafhopper host simply contracts its range to the vicinity of streams and ponds and other moist areas, naturally taking its parasitic entourage along with it. During the rainy season, about 43 percent of the leafhopper population is parasitzed by *Cytrogonatopus;* the figure drops down to around 15 percent during the dry season. The host enables the parasite to survive the rigors of the dry season, though the numbers of both species are dramatically reduced, and when the rains begin again in spring and the grass returns to the meadows, the range and population of both species immediately returns to its former levels.

Female wasps raised in captivity have survived for 1–3 weeks and were able to parasitize as many as 40 leafhoppers during this period. The adult wasp was also obserevd to prey on leafhoppers of the genus *Cicadella* in the following manner: A female wasp that has been deprived of food for at least 24 hours begins by investigating the leafhopper with its antennae; at the first sign of movement, it seizes the prey with its chelae and, if found to be of convenient size and otherwise satisfactory, proceeds to pierce the leafhopper's body case with its mandibles and drain off its body fluids (hemolymph). The wasp will not attack a leafhopper that remains immobile, and not surprisingly, the hungrier the wasp, the more cursory its initial palpation of its potential victim. Conversely, the wasp will occasionally content itself with a less devastating, and thus nonfatal assault on a leafhopper; the leafhopper may subsequently be parasitized, however, and will thus eventually fall victim to the next generation of *Cyrtogonatopus.*

◀ The predators of this species, if any, are unknown. A.D.

D

DAMSELFLY *Calopteryx virgo* 133

There are numerous damselfly species, which look like frail, thin-bodied miniature dragonflies and comprise the suborder Zygoptera, which means "twin-winged" and refers to the fact that the damselfly's front and back pairs of wings are virtually identical. When the damselfly is at rest, its wings are folded vertically on either side of the abdomen; its head is transversely elongated and stalklike, and the eyes are thus rather far apart. The body of the damselfly larva, also elongated, terminates in three leaflike breathing tubes.

Three species of the genus *Calopteryx,* not always readily distinguishable, are the largest European damselflies; the wingspread of *Calopteryx virgo* is only a little less than 8 cm, with a total body length of 4.5 cm. Its wings are broad and rounded, very richly inervated, and, in the case of the males, adorned with patches of brilliant dark-blue pigment with emerald green highlights; the wings of the female are a uniform reddish color. The male's body is an deep, metallic blue-green, the female's a brilliant bright green, verging at times on bronze or golden.

There are 3 subspecies, ranging from north of the 60° parallel in Scandinavia to the Maghreb, all found in the vicinity of streams of clear, running water, whether subalpine mountain torrents or broad lowland rivers, but most frequently in clean, swiftly flowing upland streams with thick vegetation along their banks.

▶ As with all the European Odonata, the larva of *Calopteryx* is aquatic and otherwise bears little resemblance to the adult form (an instance of what is called incomplete metamorphosis, or hemimetabolism), though the larva lives on the bottom of the same sort of streams, of course, that are frequented by its parents. During the day, the larva conceals itself among the roots or beneath the overhanging banks at the edge of the stream and only goes out to hunt at night. Slow and deliberate in its movements, it appears to locate its prey primarily by touch, using its long antennae and its long, delicate forelegs to search for any signs of movement. Once the prey has been located, the larva aligns its body in that direction and extends its mask, or mandibular arm—a specialized extension of the lower jaw that acts like a pair of pincers—in order to seize its prey. The mask is then retracted, bringing the prey into contact with the larva's grinding mandibles; the larva lifts up its head, raises its antennae, and slowly consumes its victim before returning to the chase.

The adult *Calopteryx,* like most damselflies, is a home-loving creature that spends most of its time perched on a favorite leaf of a branch or plant that overhangs the stream. As soon as a small insect passes overhead, the damselfly executes a nearly vertical takeoff and pursuit, sometimes for more than a meter, intercepts its prey in midair by seizing it with its jaws and forelegs, then immediately returning to its perch to make a meal of it. The predatory mechanism of the adult damselfly is triggered by the same species whose larva are fed on by the larva of the damselfly; in the case of *Calopteryx,* this includes small, soft-bodied insects such as lacewings, mayflies, and horseflies.

◀ The larva of *Calopteryx* is a favorite food of the crayfish, many different species of fish *(Coregonus* and other members of the family Salmonidae), and even of the slightly larger larvae of the golden-ringed dragonfly *(Cordulegaster annulatus).* The adults are often caught on by insectivorous birds as well as the numerous spiders that spin their webs in the foliage along the riverbank.

Calopteryx splendens is found throughout much the same geographic range as *C.*

133
Sub.: Arthropoda
Cl.: Insecta
O.: Odonata
S.o.: Zygoptera
F.: Calopterygidae

virgo, though it tends to prefer the banks of wide lowland rivers to those of the swifter-flowing streams in the hills. Typically, the males of *C. splendens* display only a single narrow band of dark color on their wings, in contrast to the more extensive pigmentation of *C. virgo,* though the picture is somewhat complicated by the existence of several subspecies of each, so that only the males in which these patterns are the most pronounced can be readily differentiated in the field.

Calopteryx haemorrhoidalis is found along the Mediterranean coast. This species has a slightly thicker body than its congeners; its body colors are dark and smoky, verging on a metallic violet, and the males have a little slash of bright pink pigment on the underside of the abdomen—to which the species name makes fanciful allusion.

A. de R.

134
Sub.: Vertebrata
Cl.: Reptilia
O.: Squamata
S.o.: Ophidia
F.: Colubridae

DARK-GREEN SNAKE *Coluber viridiflavus* 134

This handsome relative of the North American blacksnake *(Coluber constrictor)* is green and yellow in color and may attain a length of 1.5 m or more. It is found in northern Spain, southern France, Italy, and Yugoslavia, as well as in Corsica and Sardinia and in a variety of habitats up to an altitude of 1500 m, including shrubs and grassy areas, stone walls, hedges, and all sorts of waste or uncultivated ground with good exposure to the sun. The dark-green snake is agile and aggressive, a good climber, and will not hesitate to swim across a pond or stream. It is an aglyphous species, meaning that it does not have poisonous fangs. Like other reptiles of the temperate zones, the dark-green snake goes into hibernation at around the end of September and does not emerge until the following March or April.

▶ The dark-green snake generally stalks a prey animal by sight (more rarely attacks from ambush), seizes it violently, and suffocates it by constriction. Ingestion is accomplished by the usual lateral chewing motions of the jaws, which, in conjunction with the snake's backward-pointing teeth, produce a sort of conveyor-belt effect. This species feeds primarily on small rodents and insectivores, notably voles of the genus *Pitmys, Arvicola,* and *Microtus,* field-mice of the genus *Apodemus* or *Sylvaemus,* the house mouse *(Mus musculus),* and shrews of the genus *Sorex, Neomys,* or *Crocidura.* The dark-green snake also feeds on lizards of every kind, birds, frogs, and, less frequently, on other snakes, though the proportion of its total diet represented by these different creatures varies considerably from one local habitat to the next. Young dark-green snakes, not large enough to come to grips with even the smallest mammals, feed primarily on lizards, bird's eggs, and insects.

◀ As with other terrestrial colubrids of Europe, the natural enemies of the dark-green snake consist essentially of diurnal raptors such as the short-toed eagle or the common buzzard, wild boar and free-ranging domestic hogs, geese, turkeys, and even chickens as well as human beings, by whom the dark-green snake is sometimes destroyed on sight though never (as with several viper and colubrid species of Western Europe) hunted for food. The dark-green snake is quick to defend itself against an attack, raising the entire forepart of its body off the ground, striking and biting vigorously and repeatedly if the aggressor persists in its approaches. If an attempt is made to grasp or restrain it, it lashes out violently with its tail as well (for which reason it is colloquially known in French-speaking countries as *fouet,* or "whip," just as another North American congener, *Coluber flagellum,* is commonly known as the coachwhip snake). Unlike the Aesculapian snake, for example, the dark-green snake never truly becomes accustomed to being handled in captivity.

In Western Europe, an immature dark-green snake might conceivably be confused with the young of the ringed snake, or European water snake *(Natrix natrix),* especially if seen swimming on the surface of the water. Two congeneric species that are very similar to the dark-green snake in appearance, *Coluber gemonensis* and *C. jugularis,* are both found on the eastern shores of the Adriatic.

J.C.

135
Sub.: Vertebrata
Cl.: Mammalia
O.: Chiroptera
F.: Vespertilionidae

DAUBENTON'S BAT *Myotis daubentoni* 135

Daubenton's bat, also known as the water bat, measures 4.1–5.1 cm in length (head and body) plus an additional 3–3.9 cm for the tail. It ranges throughout almost all of Europe and parts of Central Asia, and during the warmer months, it takes shelter in hollow trees or in the darker recesses of buildings during the day. It spends the winter in caves, cellars, and quarries, and is often called the water bat because it is often seen flying along the edges of ponds and streams. Its flight is sometimes slow and wavering, skimming just over the surface of the water, though it may also be seen flying in a rapid, zigzag course along the bank. Average life expectancy is about 7 or 8 years, and some individuals have lived to be 18 or older.

▶ Like other bats, Daubenton's bat tracks its prey by echolocation. It catches horseflies and deerflies (Tabanidae) in flight and also feeds on cladocerans (water fleas), ostracods, and other small freshwater crustaceans. This species is exceptional in that it also catches fish while fluttering slowly over the surface of the water (the first of the two characteristic flight patterns mentioned above). *M. daubentoni's* fishing abilities were deduced by the zoologist Brosset from the shape of the well-developed hind feet and from the presence of a powerful hind claw that pushes up the interfemoral membrane (the web of skin that connects the bat's hind feet) to keep it from interfering with these predatory activities; the same features are also found in the fishing bats of Asia and tropical America.

Four out of 7 samples of guano that were subsequently examined did contain fish scales, though it seems likely that *M. daubentoni* preys primarily on fish that are left floating on the surface after being disabled by fungal infections, since traces of these fungi were also identified in the fecal samples. In captivity, *M. daubentoni* will eat 7–10 fish, each measuring 20 cm, in a single day, and in nature, this species sometimes hunts in swarms of up to 200 individuals.

There are about a hundred *Myotis* species altogether, collectively known as little brown bats. A number of them are in the habit of catching insects over water, but *M. daubentoni* is the only one that catches fish. Natterer's bat *(M. nattereri)* feeds on gnats and midges, which it sometimes catches on the leaves of plants, and the echolocation system of Bechstein's bat *(M. bechsteini)* is particularly well adapted to the capture of stationary insects on foliage or the bark of trees. J.-J. B.

DEATH ADDER *Acanthopis antarcticus* 136

136
Sub.: Vertebrata
Cl.: Reptilia
O.: Squamata
S.o.: Ophidia
F.: Elapidae

Acanthopis antarcticus typically measures between 40 and 70 cm, sometimes attaining a meter in length, and though it is a member of the cobra family (Elapidae), it has the short, stout body and triangular head that is more characteristic of the vipers (hence, "death adder"). It is found in dry, sandy terrain on New Guinea and Australia, though it is replaced by its congener *A. pyrrhus* in the central desert regions of Australia.

▶ As with the other elapids, the dentition of this species is of the proteroglyphic type, which means that its venomous fangs are located in the front of its upper jaw, which is restricted in its mobility compared to that of the vipers. The death adder feeds on lizards, birds, small mammals, and occasionally on other snakes; the juveniles feed primarily on lizards.

It is active at dusk and at night, and hunts from ambush; it lies partially buried in the sand and when it senses the approach of a potential prey animal, it protrudes the tip of its tail above the surface and begins to wiggle it vigorously in imitation of the movements of an insect larva. When the prey animal approaches, the death adder lunges out to its full length and strikes, and does not withdraw its fangs until its venom has taken effect. The death adder's venom contains both neurotoxins and hemolytic agents, which destroy blood cells.

◀ Eagles, kites, harriers, falcons, crested hawks *(Aviceda)*, and kookaburras *(Dacelo)* are the chief predators of the death adder. Large numbers of Australian snakes are also destroyed by wild (feral) pigs, death adders perhaps most frequently of all since they will not take flight when attacked. The juveniles may even be attacked by centipedes and other large arthropods.

When approached, the death adder gives a warning before it strikes by puffing up its body to make itself look larger. The neurotoxins in its venom paralyze the muscles responsible for respiration, and the bite of the death adder is fatal to humans in about 50 percent of all cases when antitoxins are not promptly administered. It ranks among the most dangerous of the world's snakes. Large numbers of grazing sheep are also killed by death adders in Australia.

Apart from those already mentioned, the death adder exhibits a number of other striking resemblances to the vipers, which occupy the same ecological niche on the savannahs and arid steppes of Africa and Eurasia. Like them, the death adder is normally sluggish in its movements but quick to strike, it puffs up its body as an intimidatory gesture, and even uses the tip of its tail as a lure by imitating the movements of an insect larva (cf. HORNED VIPER). On the other hand, the death adder has virtually no competitors among the other Australian snakes. D.H.

137
Sub.: Vertebrata
Cl.: Osteichthyes
S.cl.: Teleostei
O.: Lophiiformes
F.: Ceraticidae

DEEP-SEA ANGLER *Cryptosaras couesi* 137

Unlike its relative the common anglerfish, or goosefish *(Lophius piscatorius;* see ANGLERFISH), which is a costal bottom-dweller, *Cryptosaras* is a bathypelagic fish that can operate quite comfortably at depths in excess of 1000 m. The female *Cryptosaras* is slightly smaller than *Lophius,* measuring up to 40 cm, and displays the disproportionately large head and enormous slotted mouth that is also characteristic of the Lophiidae. The parasitic male *Cryptosaras,* which only measures 10 mm when full-grown, attaches itself to the surface of the female's body at an early age; its superfluous organs are rapidly incorporated into the tissues of the female, until there is virtually nothing left of it but a sac containing the testicles.

▶ Deep-sea anglers feed on a variety of bathypelagic fish (including lanternfish, hatchetfish, and lampreys) as well as on squid, shrimps, and a number of other crustaceans that have adapted to life in the ocean depths. *Cryptosaras*'s feeding behavior has never been observed directly, but it appears to be analogous to that of the Lophiidae that live along the continental shelf, with one important modification: *Cryptosaras* has become a specialist in "jacklighting" (fishing by artificial light), though in this case the light is supplied by biolumniscent bacteria that inhabit the glowing bulb that is attached to the distal end of the anglerfish's lure.

The base of the lure is anchored in a bony socket, permitting it to rotate freely, and one of the vanes of the dorsal fins has evolved into a long flexible stalk (ilicium), so that *Cryptosaras* is able to cast its lure for a considerable distance and then slowly withdraw it toward its cavernous mouth once it has caught the attention of a prey animal. *Cryptosaras* is also equipped with numerous needle-sharp teeth and, as with many bathypelagic predators, its elastic stomach can expand to accommodate a very large fish, thus considerably augmenting the stock of prey animals available to it in those sparsely populated waters.

The light of the deep-sea angler's glowing lure varies from orange or yellow to yellow-green or bluish-green, according to the species, and is thought to be intermittent rather than continuous, thus produc- a strobe effect that would not only be more attractive to squid and shrimps and other potential prey animals but would also assist *Cryptosaras,* which is drably colored and thus virtually invisible at a depth of 1000 m, in maintaining its incognito. The eyes of deep-sea anglers are very small and probably barely functional, though the pronounced vibrations made in the water by a prey animal attempting to strike the lure should provide the waiting predator with ample warning of its approach.

Cryptosaras's ungainly form and poorly developed fins suggest that it hovers motionless in the water until the prey animal has approached very close to its mouth indeed, then opens it up abruptly and engulfs the quarry in an irresistible inrush of water. Thus, *Cryptosaras's* predatory strategy is comparable to that of *Antennarius chironectes* (q.v.), which lives in shallow water and attracts its prey by the physical oscillations of its nonluminous lure.

◀ Young deep-sea anglers may occasionally be devoured by their seniors, and the adult fish undoubtedly fall prey to sperm whales, giant squid, and certain sharks that occasionally descend into the depths in search of prey.

The family Ceraticidae comprises about a hundred species of deep-sea anglers, the largest of which, *Ceratias holboelli,* is about a meter long and can weigh up to 20 kg. F.M.

Dendrocoelum lacetum 138

This relatively large freshwater planarian measures about 30 X 5 mm and weighs between 7 and 25 mg. The forepart of its body consists of rounded cephalic lobes and a ventrally mounted adhesive disk; the entire body is white, with a wavy or scalloped fringe along the edges. This species is found in lakes and watercourses throughout Europe, and is not very susceptible to variations in temperature or current; it is also highly resistant to certain forms of organic pollution. At some time between December and August, *Dendrocoelum lacetum* attaches its eggcase (2–4 mm in diameter) to the bottom, and on the order of 10 offspring will emerge in 2–5 week, provided the temperature remains between 4 and 15°C. Life expectancy is 1 or 2 years, and the adults die shortly after mating.

▶ The planarian's basic predatory technique is described in the entry for *Dugesia polychroa*.

D. lacetum can exist only in an environment in which the freshwater isopods on which it feeds preferentially are present, so that these small crustaceans consitute a sort of ecological trump card that enables *D. lacetum* to compete successfully with other triclad (planarian) species. *D. lacetum* is also more rapid in its movements than other planarians, and its cephalic disk is especially well developed, so that it is capable of ingesting relatively large aquatic creatures, including arthropods as well as gammarids (amphipod crustaceans, "water fleas"). It also feeds on oligochaete worms, but is unable to sustain itself on a diet of these creatures alone. Unlike most planarians, *D. lacetum* is not particularly fond of snails.

◀ There are a number of aquatic creatures that feed on planarians in general—including several species of annelids (Hirudinea, which are also equipped with adhesive cephalic organs), caddis fly (Tricoptera) and stone fly larvae (Plecoptera), newts and salamanders, and various small fish—though there are none so highly specialized that they feed exclusively on *D. lacetum* or any other paludicolan (freshwater planarian). *D. lacetum* and the other planarians thus do not play a very important role in the aquatic foodchain, and unlike the Dugesiidae (cf. *Dugesia polychroa), D. lacetum* is not even particularly relished by its congeners or its own species.

The family Dendrocoelidae comprises 21 genera, or about 150 species altogether. The type genus *Dendrocoelum* includes 70 of these species, most of which are found in caves and underground streams; with the exception of a single North African species, the remainder are found in Europe and feed for the most part on crustaceans *(Stenasellus, Niphargus)* and oligochaetes (Haplotaxidae).

Bdellocephala punctata occurs sporadically in Europe (north of the Mediterranean Basin); like *D. lacetum,* this species is equipped with a well developed adhesive disk and feeds on mobile and relatively large prey animals, including isopods and gammarids. The body of the larval form is coated with an adhesive mucus with which it ensnares much smaller prey animals, including chironomid larvae and *Daphnia.* There are about ten congeneric species, all found in Asia.

The genus *Sphalloplana* consists of 8 primarily cave-dwelling species, one of which is found in Asia, the remainder in North America. *S. mohri,* found in subterranean streams in Texas, feeds on amphipods *(Stygonectes)* and various arthropods, and is most notable because it may have as many as 50 retractile pharynxes with which to ingest its prey. N.G.

138
Sub.: Plathelmintha
Cl.: Turbellaria
O.: Tricladea
S.o.: Paludicolae
F.: Dendrocoelidae

DESERT IGUANA *Dipsosaurus dorsalis* 139

139
Sub.: Vertebrata
Cl.: Reptilia
O.: Squamata
S.o.: Sauria
F.: Iguanidae

The desert iguana's total length is 23–40 cm, but the distance from snout to cloacal vent, excluding its tail, rarely exceeds 15 cm. It is agile, slender, fast-moving, and holds its body high off the ground when it runs. It is whitish-beige in color with brown bands and reticulated patterns; its muzzle is short, its body cylindrical, and it has a rudimentary, barely visible dorsal crest. This species is found in sandy or rocky terrain or among brush or bushes growing in clay soil throughout many of the desert regions of North America: in Nevada, western Arizona, southern California, Sonora, and Baja California (though not in the deserts of Chihuahua).

It scoops out a burrow in the ground or sometimes occupies the abandoned burrow of a rodent to protect itself from predators and the intense heat of the sun. The desert iguana ranges over a territory of about 1500 sq m on the average.

The desert iguana is active during the day; it emerges from its burrow in the late morning and sits motionless for a short time, warming itself in the sun. If the sun gets too hot at midday, it will cool off under a bush or retire to its burrow again. The normal body temperature of this markedly heliophilic species is 41°C, the highest of any reptile's, though the desert iguana and the sympatric collared lizard *(Crotaphytus collaris)* have the lowest metabolic rate of any New World lizard. The desert iguana's mating season is April and May; the female produces a single clutch of 3–8 eggs every year, and the eggs normally hatch some time between April and September. Even in the desert, this species is inactive during the colder months, beginning in October. Longevity is somewhere between 10 and 15 years.

▶ The iguana feeds primarily on the flowers, buds, and foliage of plants, which typically comprise between 80 and 95 percent of its stomach contents; the remainder of its diet consists of flies, ants, beetles, and other insects, which it hunts by sight. The diet of the juvenile iguanas contains a higher proportion of insects. The iguana also makes use of the olfactory organs located on the tip of its tongue to select its plant food, which it grazes lightly with its tongue before closing its jaws around a flowerhead or a clump of foliage and tugging it loose. The smaller insects that it picks off the ground are swallowed whole; the larger ones are chewed 2 or 3 times before being swallowed.

There is some seasonal variation in the desert iguana's diet as well as its level of activity. In the Mohave and Sonoran deserts, the flowers of a species of creosote bush *(Larrea divaricata)* are a staple of the desert iguana in April and May; the iguana turns to other food in the summer, when it appears to be less active. It was observed that as much as 41 percent of the adult iguana's diet consisted of insects during the summer rains in Baja California (68 percent in the case of the juvenile iguanas).

◀ Though the desert iguana is primarily active during the hottest part of the day, when many potential predators and competitors are obliged to take shelter from sun, there are a number of creatures that prey on it systematically, including the Western diamondback *(Crotalus atrox)*, whipsnakes *(Masticophis)*, bullsnakes *(Pituophis)*, buzzards and falcons, and the roadrunner *(Geococcyx californianus)*. Predation is a significant cause of mortality, particularly for the juvenile iguanas. Other birds and small carnivores, such as the shrike, the kit fox, and the collared lizard *(Crotaphylus)* may occasionally prey on the young iguanas as well.

Dipsosaurus dorsalis is the only representative of its genus, though three separate subspecies are recognized. Though the desert iguana is sympatric with dozens of other species of lizard, interspecific competition does not seem to be particularly intense. One reason for this is that the iguana browses on flowers and foliage, for the most part, whereas the smaller lizards are all insectivorous. The only exception is a chuckawalla, *Sauromalus obesus*, which is herbivorous and somewhat smaller than the desert iguana (maximum length 21.8 cm); however, this species lives on bare rock rather than the sort of stony or gravelly soil frequented by the iguana, so that the two are not competing for the same food source. C.G.

DESERT MONITOR *Varanus griseus* 140

140
Sub.: Vertebrata
Cl.: Reptilia
O.: Squamata
S.o.: Sauria
F.: Varanidae

This is the largest lizard found in the Sahara, and the genus and family names *(Varanus*, Varanidae) of the thirty—odd species of monitor lizard are derived from the Arabs' name for this one, which is *waran*. The adults sometimes measure over 1 m in length, occasionally as much as 1. 3 m. The desert monitor's body is cylindrical, and its tail tapers to a point and is somewhat longer than its trunk; this species is incapable of detaching its tail by autotomy. The desert monitor is sand-colored, as a rule, with more or less regular patterns of yellow or brownish stripes along its back; its underbody is white. Longevity in nature is uncertain, though probably at least ten years.

▶ The desert monitor is carnivorous and

extremely adaptable in its choice of prey. In a study conducted in the high dunes of the the northwestern Sahara (Great Western Erg), 28.6 percent of captures consisted of common skinks, 20.6 percent of beetles (tenebrionids and scarabs), 15.9 percent of lacertids, and 15.8 percent of rodents (gerbils, jirds, jerboas). Very similar figures (23 percent, 14.3 percent, 16 percent) were obtained in a similar study of the feeding habits of monitors living along the banks of a wadi, or intermittent desert watercourse (Wadi Saoura), though the species represented in each category were not the same in both cases.

The Wadi Saoura monitors fed largely on sand rats (*Psammomys obesus*) and lacertids (*Acanthodactylus scutellatus, A. boskianus*); the Barbary skink (*Sphenops boulengeri*), instead of the common skink, accounted for only 1.8 percent of the captures. Birds (wheatears, whitethroats, Dupont's lark [*Chersophilus duponti*] and related species), which often stop to drink at the temporary pools and waterholes in the wadis along their migration routes, made up a substantially larger proportion of the diet of the Wadi Saoura lizards (10.7 percent versus 3. 2 percent). Another staple article of diet for the Wadi Saoura lizards was the scorpion *Androctonus amoreuxi*, which is quite prevalent in the rodent burrows and galleries along the wadi, in which these lizards normally seek their prey.

Spiders, agamids, and snakes scarcely figured at all in the diet of the subject populations, though the desert monitor is known to attack sand vipers and horned vipers with impunity and appears to be entirely unharmed by their venom. The desert monitor is described by Pianka as a "widely foraging" predator, and it is likely to cover a distance of as much as 2–5 km in a single day, inspecting every tuft of grass and every rodent burrow that it comes across as well as digging up a great many of the smaller burrows made by arthropods and lacertids. It generally catches a great many nocturnal rodents, scorpions, and solpugids (sun spiders), which were all well represented among the stomach contents of the individuals examined.

Burrowing insects such as cockroaches and *Zophosis* (a tenebrionid beetle) as well as tunnelers such as the gecko *Stenodactylus* (q.v.), spiders, and jerboas that take the precaution of stopping up or concealing the entrances to their burrows are unlikely to fall afoul of the marauding desert monitor. There are a number of other species that are very common in the habitat whose absence from the inventory of prey animals is less readily explained, including *Scarites* (large carabid beetles, a staple item in the diet of the common skink) and *Anthia* (another carabid genus). In general, however, the desert monitor is a highly opportunistic predator and the relative frequency with which various species are preyed on by different populations tends to reflect their relative abundance in different habitats rather than any particular preference.

Desert monitors generally remain in their burrows for 2 or 3 days after they have had a successful day's hunting, and are capable of ingesting very large prey animals. A monitor weighing 585 g is reported to have swallowed a jerboa weighing 165 g (28 percent of its own body weight) and a monitor weighing 1100 g to have swallowed a spiny-tailed lizard (*Uromastix*) weighing 300 g (27 percent of its body weight). Herbivorous reptiles tend to have a sluggish metabolism and can live off a sort of vegetable compost heap maintained in the coecum or their reserves of body fat for almost a year; conversely, if the desert monitor goes for much longer than 2 months without feeding, it is unlikely to have the strength to track down even the smallest of prey animals. Rather than risk this possibility, all desert monitors go through a winter latency period that lasts up to 5 months (mid-October to mid-March), and about 80 percent are more or less inactive during the height of the summer as well (July-August), so that growth, reproduction, and all other vital processes are compressed into the relatively short period from April to mid-July.

◀ There are a few predators that may sometimes feed on the juveniles of this species, including the zorilla, the fennec, the sand fox, and the jackal; the African wild cat and the sand cat might also be added to this list, though they are not particularly prevalent in the Sahara, and the lanner falcon, the short-toed eagle, and the long-legged buzzard are also known to prey on desert monitors on occasion.

The adults are always very wary and alert, even when they are hidden away at the bottom of their burrows, and their sharp teeth and claws are sufficiently formidable to deter less persistent predators, possibly including humans.

When cornered, the monitor will try to make itself seem somewhat more formidable than it actually is, by rearing up or raising itself up on all fours, puffing itself out, and emitting a crescendo of panting noises that sounds very much like a old-fashioned steam locomotive pulling out of a station. If these intimidatory maneuvers are not successful, the monitor will suddenly lunge forward and lash at the intruder with its tail or charge straight towards him, jaws agape. Monitors of both sexes are not very tolerant of their conspecifics, and if they become too numerous in a confined area, frequent and extremely violent clashes are likely to result. Instances of cannibalism have also been reported on several occasions.

In the Sahara, desert monitors are hunted for food as well as for the alleged medicinal properties of their flesh. In the Great Erg, a nomad family of 6 or 7 persons may catch and eat as many as 50 monitors during a single season (May—September); the entire family, from the Bedouin patriarch down to the children, takes part in the hunt, which can be a very labor-intensive activity in view of the fact that population density among desert monitors does not exceed 2–5 per sq km.

The desert monitor's flesh is reputed to contain "forty medicines," and in the western Sahara, its fat is often used as an ointment to relieve muscular aches and rheumatic stiffness. A remedy for a stubborn cold or other bronchial complaint may be obtained, say the Bedouin, by boiling a desert monitor (with the hide left on) in a pot with garlic, corianders, fennel, and green peppers. Because the desert monitor feeds on vipers and scorpions and appears to enjoy a total immunity to their venom, its flesh is also regarded to be a very effective antitoxin, and the desert nomads often wear a dried monitor's head around their necks as a charm against snakebite. The monitor's hide is crudely tanned and fashioned into a number of more utilitarian articles, including handbags, tobacco pouches, and violin cases, which are offered for sale in every North African souk.

Unlike the Bedouins, the Tuaregs not only refrain from hunting or otherwise molesting the desert monitor, which they call the *arata*, but also try to prevent others from doing so, since the ancestor of the monitors is said to have been a Tuareg nobleman and the lizard is accordingly afforded the respect due to a paternal uncle of the Tuareg nation. The numerous legends surrounding this ancestral figure have gradually percolated into the mythology of other Saharan tribes as well, so that other tribesmen who are not ethnic Tuaregs have also taken the species under their protection.

The genus *Varanus* comprises 32 species altogether, collectively ranging throughout much of Africa, Arabia, Southeast Asia, Indonesia, New Guinea, and Australia (see KOMODO DRAGON, NILE MONITOR). The giant monitor *(V. giganteus)*, which is actually the third largest of these species, is found in central Australia and is thus also a desert dweller. Gould's monitor *(V. gouldii)* is more widely distributed throughout Australia, though it is also found in sandy terrain, and the Cape monitor *(V. exanthematicus)* is found on the dry savannahs of southern Africa. R.V.

DESERT NIGHT LIZARD *Xantusia vigilis* 141

141
Sub.: Vertebrata
Cl.: Reptilia
O.: Squamata
S.o.: Sauria
F.: Xantusidae

This small nocturnal lizard measures no more than 13 cm in length, and its body is olive green with black spots. Its pupils are vertical, and it has no eyelids. *Xantusia vigilis* is found in the arid regions of the American Southwest (southern Nevada, southwestern Utah, northwestern Arizona) as well as southern and Baja California, and an isolated subspecies, *X. v. extorris*, has also been discovered in the Bolsón de Mapimí in northern Mexico. *X. vigilis* matures late (2.5 years in the case of the males, 3.5 years in the case of the females) and has the lowest reproductive rate of any lizard. It gives birth to live young, and is similar in certain other respects to the common viviparous lizard of Europe. It is nocturnal and avoids sunlight; it generally inhabits deep crevices and recesses in the rock and the hollow stems of yuccas and agaves, and is found on buttes and mesas in arid and semiarid terrain.

▶ *X. vigilis* lives on insects, which are extremely abundant in the deep crevices in the rock; it hunts by sight, and feeds primarily on termites as well as on ants, flies, and small beetles. It also hunts for aphids on the stems on yuccas and dead agave plants, and has often been found in the burrows of pack rats *(Neotoma)*, where there is also an abundant supply of insects and their larvae. Its dentition is similar to that of herbivorous species, and traces of vegetable material (seeds, buds, leaves) have also been identified among its stomach contents.

◀ *Xantusia*'s secretive mode of life has protected it not only from predators and the vagaries of the desert climate but also from the prying eyes of zoologists, and little else is known of its habits. Its chief predators probably include the granite spiny lizard *(Scleoprus orcutti)* in California, the crevice spiny lizard *(S. poinsetti)* in northern Mexico, the night snake *(Hypsiglena torquata)*, the whipsnakes *(Masticophis)*, and the western patch-nosed snake *(Salvadora hexalepis)*. The barn owl and the great horned owl probably feed occasionally on *X. vigilis* as well. A high percentage of the adults have lost their tails by autotomy, though probably as the result of intraspecific combats rather than predator attacks; the tail is used to store fat reserves and is very easily detached.

Four of the 6 *X. vigilis* subspecies are found in the United States, as is the granite night lizard *(X. henshawi)*. The family Xanthusiidae, consisting of just a handful of species, are quite similar to the gekkos in many respects, though they are considered to be more closely related to the skinks and are thus usually included in the suborder Scincomorpha. C.G.

DEVIL'S COACHHORSE *Ocypus olens* 142

142
Cl.: Arthropoda
S.cl.: Insecta
O.: Coleoptera
F.: Staphylinidae

This species, the largest European rove beetle (staphylinid), measures 32–43 mm in length, is a uniform black in color, and is found in a variety of habitats, including forests. It has recently been collected in Southern California as well, where it prefers to live in cultivated areas.

▶ *Ocypus olens* feeds primarily on snails, sometimes by attacking the snail's foot through the opening of the shell, sometimes by snapping off the tip of the shell to feed on the adjacent internal organs. The larva only goes through 2 or 3 snails every 3 weeks, but the adult is somewhat less abstemious: A female *O. olens* was observed to eat 20 ordinary garden snails *(Helix aspersa)* is 22 days, so that at least this particular individual was in the habit of eating almost its own weight in garden snails every day. Once its hunger has been satisfied, however, *O. olens* becomes completely indifferent to the presence of other snails, so that this "devil's coachhorse" (a name that is also applied to other large black staphylinids in Britain) is hardly a wanton destroyer of prey animals.

O. olens seems to encounter its prey, which also includes insect larvae, more or less at random, guided at short range by touch and chemoreception. This species has been shown to be a significant biological control on *Helix aspersa* in localties where both species occur naturally. Since the latter is a common pest in gardens and croplands, various investigators have suggested that agricultural regions plagued by *H. aspersa* might benefit from the planned introduction of *O. olens*, though these suggestions have yet to be acted upon in any sort of systematic way.

◀ Birds are probably the chief predator of this large conspicuous beetle. When attacked, *O. olens* lifts up the tips of its abdomen and directs a stream of foul-smelling liquid in the direction of the aggressor.

The largest rove beetles are found in Africa. *Staphylinus stercorarius* feeds on small insects found in pats of dung, and *S. marinus* forages for insects and other small organisms in the tidal wrack along the shoreline. G.B.

DHOLE *Cuon alpinus* 143

143
Sub.: Vertebrata
C.: Mammalia
O.: Carnivora
F.: Canidae

The dhole, or Malay hunting dog, is similar in appearance to the domestic dog; it has a maximum body length of about 1.1 m, its tail measures 40–50 cm on the average, and it weighs between 15 and 20 kg. It varies in color, but is generally brownish-red with brownish-black strips on the front part of the forelegs. Several subspecies that live in the high mountains have very long hair on their backs.

Populations of the genus *Cuon* are still found in Siberia, India and Indochina, and on Java and Sumatra, though many local subspecies have become extinct in recent times. The dhole is very tolerant of climatic extremes, and is sometimes found at altitudes of 2000 m during the winter; packs have been known to pursue a prey animal up to an altitude of 4000 m. The dhole is rarely encountered by man, but those few observations that have been made indicate that it lives in packs of 3–15 individuals and that several packs will occasionally band together during the hunt.

▶ The dhole is found throughout a vast geographic range and preys on a great variety of other animals; it hunts in packs, like the wolf or the Cape hunting dog, and is thus capable of running down such large and powerful ruminants as the musk ox and the reindeer in Siberia, wild goats of various species, sambur, and wild boar in the mountainous regions of India and Southeast Asia. Like many other canids it feeds on vegetables as well as meat and is known to have a preference for the stems of wild rhubarb, though it prudently avoids the leaves, which are toxic.

The short-legged dhole makes up in endurance for what it lacks in speed, and a dhole pack is capable of pursuing its prey for several hours—though this is partly due to the fact that dholes hunt in relays in order to give the worn-out members of the pack an opportunity to rest while the others keep up the pursuit. Typically, one of the dholes will attempt to overtake a fleeing prey animal and cut off its escape route while the others fan out to encircle it. Then the pack tries to knock the animal off its feet by nipping at its legs; the dhole's teeth and claws are not strong enough to kill a large animal outright, so the pack continues to worry the fallen prey until they succeed in disembowling it. The carcass is immediately divided up among the members of the pack.

◀ The dhole is a superpredator, inasmuch as it has no natural enemies other than man. It is now protected in India, though it has been persecuted as a destructive competitor since prehistoric times. P.A.

144
Sub.: Pseudocoelomata
Cl.: Rotifera
O.: Monogonontes
F.: Dicranophoridae

Dicranophorus forcipatus 144

This benthic rotifer, which measures about 0.6 mm in length, is found on the leaves and stems of aquatic vegetation or on the bottom of ponds and lakes throughout the world. The body of a rotifer, or roundworm, only contains about a thousand cells (a number that never varies, in any case, throughout the animal's lifetime), and it moves along the bottom or across the surface of a clump of algae or an aquatic plant by means of the rotary motion of the fringe of cilia that surrounds its mouth. Rotifers belonging to the order Monogonota ("single-sexed") reproduce by parthenogenesis, as the name implies. After a brief period of growth lasting only 1 or 2 days, *Dicranophorus forcipatus* deposits a clutch of about 20 eggs on the bottom; individual lifespan is about 2 weeks at a temperature of 20°C.

▶ The body of a rotifer is transparent, and its internal organs can be clearly distinguished through the microscope, including the digestive, muscular, nervous, and reproductive system; it lacks a circulatory system. The digestive tube consists of a muscular pharynx called the mastax, the esophagus, the stomach, flanked by two digestive glands, and the intestine, which terminates at the cloaca (present in this species, though not in all rotifers).

The mastax functions as a sort of crop, since it contains two hard pincerlike projections which are attached at the base to the muscular wall of the mastax. These pincers (which give the species its name, *D. forcipatus)* are extremely tough and durable, and can also be extruded from the rotifer's mouth, which makes it a formidable predator indeed. When *D. forcipatus* has seized a prey animal with these pincers, they are retracted into the interior of the mastax and the prey animal's body is dragged along with them, to be further subjected to the grinding action of the pincers before it is sped on its way down the esophagus. *D. forcipatus* feeds on a variety of tiny marine organisms, though it prefers other rotifers (Bdelloidea, *Prolaes*) and preys with impunity on all but the largest species.

◀ Though we have no definite information on this point, it seems likely that the predators of *Dicranophorus forcipatus* would include Turbellaria (a class of flatworms) as well as the larvae of newts and salamanders and fish hatchlings.

Other *Dicranophorus* species are similar in size, morphology, and feeding habits. Rotifers of the genus *Encentrum,* which belong to the same family, are only about 0.2 mm long; the mastax, though similar in design, is less capacious, and these species can only feed on very small organisms (e.g. *Encentrum lupus,* which preys on ciliate protozoans). R.P.

DINGO *Canis dingo* or *lupus* 145

145
Sub.: Vertebrata
Cl.: Mammalia
O.: Carnivora
F.: Canidae

The dingo is not a true wild canid but rather a descendant of domestic dogs that have reverted to the wild. As far as its dentition and skeletal features are concerned, it has all the characteristics of the domestic dog and is highly variable in size and morphology; color varies from brown and reddish-brown to black or even spotted. Longevity is about 10–14 years. Dingos rarely bark but they sometimes growl or whine like dogs or howl like wolves; young dingos can be tamed quite easily.

The dingo was introduced to Australia by the first human settlers in prehistoric times and, in the absence of competition from other large predators, quickly established itself throughout the island continent. Some authorities maintain that dingos migrate seasonally, spending the summer on the eastern coast and the winter in the west. Like many other canids, the dingo scent-marks the boundaries of its territory with urine.

▶ Before the arrival of the British, the dingo's principal prey was the kangaroo; when the latter was largely exterminated to make room for the stockman's sheep, the dingo began to prey primarily on these, as well as on the rabbit, the introduction of which during the latter half of the nineteenth century was to have a disastrous effect on the ecology of the Australian grasslands. The dingo hunts in a pack like wolves and pursues its prey to the point of exhaustion, whereupon (in the case of the kangaroo, for example) the pack surrounds the quarry and harasses it with feint attacks while awaiting an opportunity to overpower it, knocking off its feet, killing and immediately dismembering it.

The kangaroo is a formidable opponent, however, and even if it is unable to make its escape, the dingo pack may have some difficulty in subduing it. It faces off against its pursuers, and if a dingo ventures too closely, the kangaroo may be be able seize it with its short front paws, then rear back on the base of its tail and disembowel it with the strong, sharp claws on its hind feet. The kangaroo has also been known to seize hold of its adversary and drag it out into the depths of a river or pool in order to drown it.

The social structure and food-sharing behavior of the dingo pack have not been systematically studied by Australian naturalists, and the dingo is held in fairly low esteem by nonaboriginal Australians, though tame dingo puppies are still trained by the Aborigines as hunting dogs; the behavior of the tame dingo is not unlike that of any other bred of domestic dog.

◀ Hunters can still collect a bounty on the dingo in some Australian states, whereas they are protected in others, depending on whether the dingo is locally regarded as a sheep-killer or the only effective natural check on the rabbit population. Rabbits have contributed enormousuly to the overgrazing of marginal Australian rangeland, and thus to the process of desertification—in some areas to a truly castrophic extent.

The dingo seems most closely related to the domestic dogs of Indonesia; the fact that dingos do not bark was once regarded as evidence that the dingo was more closely related to the wolf than the domestic dog, but this pont is somewhat mooted by the fact that the domestic dogs found throughout a large part of Africa, most notably the basenji, are also unable to bark. P.A.

DOMINICA BULLFROG *Leptodactylus fallax* 146

146
Sub.: Vertebrata
Cl.: Amphibia
O.: Anura
F.: Leptodactylidae

In spite of appearances, this large terrestrial "bullfrog" is actually a toad, though it lacks the characteristic warty skin of toads belonging to the genus *Bufo.* It is yellowish-brown in color with a few black spots or larger salmon-colored patches on its flanks; the females (112–149 mm, 123–400 g) are slightly larger and quite a bit heavier than the males (111–143 mm, 110–247 g). This species is still found on the Leeward Islands of Dominica and Montserrat. In the French-based creole of these islands, *Leptodactylus fallax* is simply known as *crapaud,* "toad," but its common name among English-speaking islanders, *mountain chicken,* suggests why it was hunted to extinction on Martinique and St. Kitts during the 19th century. On Dominica, it is currently confined to the valleys and steep slopes on the western (Caribbean) side of the island, more often in woodlands, brush, and on cultivated hillsides than in dense forests. It spends the daylight hours in its burrow and goes out to hunt at night; the male's short barking cry is sometimes heard at nightfall on a rainy evening.

▶ The Dominica bullfrog feeds primarily on crickets, grasshoppers, dung beetles, daddy longlegs (phalangids), myriapods, snails, and land crabs, though it is a voracious and opportunistic predator than will feed on virtually any creature that it is big enough to overpower, possibly including bats, highly venomous centipedes, and even a young (30 cm) boa constrictor. It occasionally preys on a related species, *Eleutherodactylus martinicensis,* a small leptodactylid tree toad. In other words, its diet is more varied than that of the family Bufonidae and quite similar to that of the larger Ranidae, the true bullfrogs. It hunts by sight, detecting prey by its movements and pursuing it with rapid bounds of over a meter in length. Rather than patiently stalking insects and other creatures and catching them with its elastic tongue in the manner of other toads, it overwhelms them by springing upon them abruptly and engulfing them in its very wide mouth.

◀ There are few vertebrates on Dominica, and apart from man, the West Indian subspecies of the common opossum *(Didelphis marsupialis insularis)* and the boa constrictor are probably the adult Dominica bullfrog's only natural enemies; the young are probably preyed on by colubrids (and definitely by larger members of their own species). The adult attempts to repel aggressors by standing up on tiptoe and inflating its body cavity by swallowing air, but will quickly flee—by means of a series of rapid and powerful leaps—if this maneuver is not immediately effective. As noted earlier, the succulent flesh of the "mountain chicken"—conventionally described as white, tender, and delicate—is the main reason that this species has become extinct throughout part of its range and is increasingly rare on Dominica in recent years. Its taste was extolled by the famous missionary Père Labat in an account of his West Indian travels published in 1722, and *"le crapaud des Antilles"* was accordingly given an entry in Diderot and d'Alembert's *Encyclopédie.*

Leptodactylus pentadactylus is widely distributed in Central and South America; other mainland species include *L. knudseni* and *L. labyrinthicus,* all 3 of which are slightly larger than *L. fallax* but similar in their diet and predatory technique. All of these large leptodactylids can be raised in captivity and are commonly fed on white mice, like boa constrictors. J.L.

DRILL *Thais lapillus* 147

147
Sub.: Mollusca
Cl.: Gastropoda
S.cl.: Prosobranchia
O.: Neogastropoda
F.: Muricidae

This carnivorous gastropod, variously known to Britons as the dog whelk, dog winkle, or horse whelk, is very common in the intertidal surge zone and is often found in the crevices of wave-battered rocks, particularly those occupied by mussels and fucus (brown seaweed, sea wrack); *T. lapillus* is also commonly found in mussel beds on both sides of the Atlantic. The shell is thick, relatively smooth, and measures between 2 and 3 cm in height. The exterior of the shell is generally white, yellow, brown, or purplish-pink, sometimes with stripes of various colors, and the purplish-violet or pinkish stains that are often visible on the interior (occasionally on the exterior as well) are produced by the secretions of the hypobranchial glands, colorless in themselves but which turn bright purple when oxidized. The famous "Tyrian purple" of antiquity was a dye manufactured from the glandular secretions of sea snails of the genus *Murex,* which belong to the same family.

Like most prosobranchs, the drill can fully retract itself inside its shell, securing the entrance with a movable hatch (operculum). The soft parts of the body consist of the visceral mass, which always remains inside the shell, and the cephalopodium; the muscular foot is less well developed than in the Naticidae (the moonshells; cf. NECKLACE SNAIL). The head is provided with a medially located trunk, which contains the mollusk's rasplike "tongue," the radula; the accessory boring organ, located along the anterior third of the foot, is normally retracted when not in use. The sexes are differentiated, and fertilization takes place internally; the spawning season lasts from February to April. Each female produces between 10 and 40 egg capsules (each containing 200– 600 eggs), which are attached to a rock, shell, strand of seaweed, or other support; the spawn (which in the case of mollusks refers to the hatchlings rather than the eggs) are morphologically similar to the adults (i.e. they do not undergo a larval phase).

▶ The drill preys exclusively on living mollusks, including acorn shells and limpets, though primarily on mussels. It feeds on acorn shells by pushing apart the valves of their shells, and it can occasionally gain access to soft parts of a limpet by pulling it up off its stalk, though it is usually necessary for it to bore a hole in its shell, as it invariably does when preying on mussels. The drill begins by firmly attaching its foot to one side of the mussel's shell, then contracts the forepart of the foot (propodium), opening up an airspace in the center into which the trunk can be inserted. Next, the radula begins to drill through the tough outer layer of the mussel's shell (the periostracum), and after about 15 minutes, the accessory boring organ, which secretes a corrosive substance that assists the purely mechanical drilling action of the radula, is applied to the same spot for about 30–40 minutes. This process continues in alternate stages until the drill is able to insert its trunk through the borehole and feed directly on the mussel's flesh.

Carnivorous snails are known from fossil evidence to have perfected this drilling technique as early as the Upper Cretaceous (roughly 100 million years ago), and the function of the radula has been understood since early antiquity. The role of the accessory boring organ was first discovered in the Naticidae (in which is located directly below the trunk and is visible without dissection), though it was realized only as recently as 1942–45 that the Muricidae are equipped with a similar organ, this time located in the interior of the foot, that serves the identical fucntion. Both, in fact, are essential to the success of the drill's predatory technique, since individual specimens of *Thais lapillus* from which either the trunk or the accessory boring organ has been removed are incapable of drilling the shells of other mollusks, at least until the missing organ has been regenerated.

◀ There are a number of predators that are capable of dealing with the drill's thick shell, including crabs, the spotted eagle ray *(Myliobatis aquila)*, and several shorebirds. Like moonshells, drills occasionally prey on one another, and like many other sea snails, they serve as intermediate hosts for a variety of parasitic flukes (trematodes), which generally results in the atrophy or destruction of the snail's reproductive organs (so-called parasitic castration). Even under normal conditions, the percentage of eggs that fail to develop or are washed up on the beach is probably very great; the female drill often attaches its eggs capsule to the byssus of a mussel (the beardlike fringe by which it attaches itself to a rock), with the result that the spawn are likely to be siphoned in by the mussel and devoured, which seems only poetic justice.

The Muricidae are a widely distributed family, all of which feed on other mollusks in more or less the manner described above. *Thais haemastoma*, a somewhat larger species, also feeds on mussels and is found in Mediterranean. The common name "oyster drill" is applied to several species that cause extensive damage to commercial oyster beds and that tend to be smaller than either *Thais* or *Murex*. *Oceanebra erinaceus* is found in the North Pacific, the North Atlantic, and the Mediterranean; *Urosalpinx cinerosa* and *Eupleura caudata*, two predominantly American species, are both notorious for their voracity, and in North American waters, the common name "drill" often refers to *Urosalpinx* rather than *Thais lapillus*.

The *Murex* species found in the Mediterranean are considerably larger than *Thais lapillus*. The shells of *M. brandaris* and *M.*

trunculus, for example, can measure up to 10 cm in height; the latter species is notable as the chief source of Tyrian purple in ancient times. These larger muricids prey on other mollusks not only by drilling their shells but also by more direct strong-arm tactics—wedging the valves of the victim's shell open by inserting their own siphons or the sharp frontal edge (peristome) of their own shells into the aperture. There are a number of very large tropical species, mainly found on coral reefs, whose shells can measure as much as 20 or 30 cm in height, including *M. ramosus* and *M. palmarosae* of the Indo-Pacific and *M. fulvescens* of the Gulf of Mexico. M.C.

148
Sub.: Vertebrata
Cl.: Mammalia
O.: Monotremata
F.: Ornithorhynchidae

DUCK-BILLED PLATYPUS
Ornithorhynchus anatinus 148

The most celebrated of the monotremes, or egg-laying mammals, was a subject of zoological controversy for several decades after the first preserved specimens were brought back from Australia in 1798. These were initially suspected of having been concocted by a playful taxidermist out of scraps from several different creatures, and it was not until 1824 that the duck-billed platypus was finally accorded its rightful niche in the Linnean system.

The platypus weighs between 0.5 and 2 kg and has a fairly long, flattened tail and thick fur. Its broad bill is softer and more flexible than the bill of a duck, and the male is equipped with venomous spurs on its hind feet; its venom is actually quite dangerous to humans if it is injected through a scratch or a break in the skin. The platypus is found in eastern Australia and on Tasmania from sea level to an altitude of 1500 m. It is semiaquatic and always digs its burrow on the bank of a stream or in close proximity to a watercourse of some sort; the platypus's burrow is always provided with several entrance tunnels. Because the platypus spends most of its time either underground or underwater, it is only in recent years that it has been observed closely, either in captivity or in the wild, and a great deal still remains to be learned about its behavior.

▶ The platypus finds its food by foraging on the bottom of a stream or pond with its sensitive bill; its eyesight is not very good, and in any case its eyes and ears are both covered up by a protective flap of skin while it is swimming underwater. It prefers to hunt in the early morning or the evening and feeds on worms, insect larvae, crabs, and fish; the adult platypus has no teeth and accordingly swallows gravel from the streambed to assist in grinding up its food. Larger prey animals are brought up to the bank before being ingested in this manner. In captivity, the daily ration of a platypus weighing 1. 5 kg might include as many as 540 earthworms, 20–30 crayfish, 200 mealworms *(Tenebrio)*, 2 small frogs, and 2 hen's eggs.

◀ The platypus is a quick and agile swimmer and has little to fear from predators; the fox and several large snakes appear to be the only creatures that it is afraid of, other than man. The platypus has also enjoyed the protection of Australian law since the beginning of this century; at present, population levels in the wild appear to be fairly stable, though as a semiaquatic species, the platypus is quite vulnerable to various forms of environmental pollution. P.A.

149
Sub.: Plathelminthes
Cl.: Turbellariés
O.: Triclades
S.o.: Paludicoles
F.: Dugesiidae

Dugesia polychroa 149

This common European planarian measures up to 22×4 mm; it is uniformly brown in color, sometimes spotted, and the forepart of its body takes the form of a flattened arrowhead with enlarged lateral barbs. It prefers a warm, quiet habitat, and is thus frequently found under rocks and (occasionaly) amid the vegetation on the bottom of ponds, marshes, lakes, and fairly slow-moving watercourses. This species normally reproduces by a process known as pseudogamy, which is to say that the spermatozoid merely activates the development of the ovule without contributing any genetic material of its own. In the spring, *Dugesia polychroa* deposits a series of small cocoons (1.5 mm in diameter), each attached to substrate by a tiny stalk, from which an average of 4 small planarians will emerge in 2 or 3 weeks' time, assuming a fairly constant temperature of 10–23°C. In regions where the temperature never departs from this range, as many as three generations may appear in a single year.

▶ The prominent eyespots of a planarian are mainly useful for distinguishing light and dark, and thus are not of much use in distinguishing prey animals from other features of the environment. Like other planarians, *D. polychroa* detects its prey with the aid of sensitive tactile receptors that are primarily located on the forepart of its body; planarians also use the lateral projections of the cephalic "arrowhead" (known as auricles) to immobilize the prey, assisted in many species by a more or less well developed adhesive organ, though this is lacking in the Dugesiidae.

Next, the planarian's protractile pharynx, which can be deformed into a variety of different shapes as circumstances dictate, is inserted into the body of the prey animal; the tissues are broken down by the proteolytic enzymes secreted by the salivary glands, and the nutrients are ingested by means of the peristaltic contractions of the muscular walls of the pharynx. Planarians can go for as long as 6–14 months without feeding, and even under normal conditions, the hypogeal species (those that live in underground streams and other subterranean habitats) feed no more often than once or twice a month; the epigeal (aboveground) species feed no more than once or twice a week.

D. polychroa, like most planarians, is highly specific in its choice of prey animals. It feeds preferentially on freshwater snails of various species *(Lymnaea, Planorbis),* and, like *Dendrocoelum lacetum* (q.v.) with its freshwater isopods, these creatures constitute a sort of ecological trump card, without which *D. polychroa* may not be able to make other planarians do its bidding. This species seems to be particularly attracted by snails that have been wounded and are leaking body fluids, and in addition to the standard and more direct predatory technique described above, *D. polychroa* is also capable of immobilizing an intact prey animal by entrapping it in an envelope of mucous secretions.

◀ *Dugesia lugubris,* otherwise almost identical to *D. polychroa*, is of interest because it preys on other planarian species *(Polycelis nigra, P. tenuis)* as well as on other members of its own species. Curiously enough, in fact, in planarian species that are not sexually differentiated, homophagia (cannibalism) may be the only means by which genetic material can be exchanged between individuals. Homophagia has also been observed in other members of the family Dugesiidae, namely *Dugesia gonocephala, D. tigrina,* and *Cura formani.*

Apart from these cannibalistic congeners, however, *Dugesia polychroa* need only concern itself with the relative handful of predators that feed on all planarians: hirudineans (aquatic annelids), certain insect larvae (dragonflies, caddis flies, stone flies), newts and salamanders, as well as a number of small fish. It is thought that the other potential predators may be deterred by the refractive filaments (rhabdites) that are found in the planarian's epidermis, though the precise function that they serve is still unclear.

The family Dugesiidae comprises 9 genera and 105 species of freshwater planarians, collectively found in both tropical and temperate habitats and on every continent. *Dugesia tigrina* and *D. dorotocephala* feed on mosquito larvae *(Culex, Aedes),* and have already been successfully employed as agents of biological pest control in North America; they may be in even greater demand in the future as the problem of pesticide-resistant lineages of malarial mosquitoes becomes increasingly critical. *D. dorotocephala* is the celebrity of the family, as the result of a famous series of conditioning experiments demonstrating that *dorotocephala* was capable of learning behavior—and, even more remarkably,that individual planarians could be successfully conditioned by feeding them RNA obtained from previous graduates of the program. These experiments have not been successfully repeated with any other species of planarian.

The Planariidae comprise 7 genera and 80 species, which are widely distributed throughout the Holaractic regions (the northern parts of Eurasia and North America) as well as much of East Asia. *Polycelis nigra* and *P. tenuis,* mentioned earlier as atypical prey of *Dugesia lugubris,* both feed primarily on oligochaetes (Lumbricidae,

Nadidae) as well as on mayfly and caddis fly larvae as well as gastropods and the occasional crustacean. Like *D. polychroa* and *D. lacetum,* they hold an ecological edge: a reserve of food (in this case furnished by the oligochaetes) that they share with few other planarians and that enables them to compete successfully. *P. torva,* like*D. polychroa,* feeds primarily on freshwater snails and to a lesser extent on tubifexes and isopods. N.G.

150
Sub.: Mollusca
Cl.: Cephalopoda
O.: Decapoda - Sepioidea
F.: Sepiolidae

DWARF SEPIA *Sepiola robusta* 150

This small, short-bodied Mediterranean squid measures no more than 5 cm and is found in coastal waters; its dark-brown coloration readily distinguishes it from the other bobtailed squids (Sepiolidae), which are invariably reddish-orange. The hatchlings are only a few millimeters long (dorsal length of mantle = 2.5 mm) but are otherwise similar to the adults in their behavior and appearance; they become sexually mature in 6 months, by which time they will have attained a length of 15–20 mm. The eggs are protected by a gelatinous coating, and the female attaches them, one by one, to a resistant surface, including the shells of tunicates, strands of *Posidonia,* or any one of the numerous plastic artifacts that currently abound in the coastal waters of the Mediterranean. This species has no fixed breeding season, so the eggs and the juveniles in various stages of development are likely to be encountered at any time of the year.

▶ The dwarf sepia stays buried in the sand during the day and sets out in search of prey at nightfall. It hunts by sight, tracking its prey by focusing its two eyes on it, catching it with its tentacles, quickly reeling it in, and slicing it into pieces with its beak before swallowing it. The tentacles are attached to the interior of the circlet of 8 arms and kept neatly rolled up when not in use, unlike those of the cuttlefish, which are retracted into a pouch

This species feeds on very small shrimp and shrimplike crustaceans, including mysidaceans, Palaemonidae, and probably Crangonidae *(Philocheras),* possibly young gobies and other very small fish, though it will not attack even the smallest crabs. Like the cuttlefish, the bobtailed squid sometimes catch their prey directly by pouncing on it and seizing it with their arms, though more frequently they intercept it at long range it by projecting their tentacles.

◀ Bass, groupers, and cuttlefish are the principal predators of the bobtailed squids, which can defend themselves by ejecting a cloud of sepia ink. This particular species is too small to be of much interest to fishermen, though it is sometimes kept in saltwater aquariums in Europe.

The subfamilies Sepiolinae (to which the genus *Sepiola* is assigned), Rossiinae, and Heteroteuthinae together comprise the family Sepiolidae. *Sepiola atlantica* is quite common in the coastal waters of the Eastern Atlantic, and 6 other species are found in the Mediterranean: *S. affinis, S. intermedia, S. rondeleti, S. ligulata, S. aurantiaca,* and *S. steenstrupiana.* S.v.B.

151
Sub.: Vertebrata
Cl.: Aves
O.: Strigiformes
F.: Strigidae

EAGLE OWL *Bubo bubo* 151

This large raptor measures 62–72 cm, with a wingspread of 150–188 cm, and weighs from 2 to 3 kg. It becomes sexually mature at the age of 2 or 3; record longevity for a captive specimen is 68 years. In Europe, it is generally found in rocky, wooded terrain and in mountainous regions, and it builds its nest on the ground or in an eyrie abandoned by another raptor. This species is found throughout Europe and Northern Asia (from Scandinavia to the Pacific and the South China Sea) but in India and North Africa as well, though not subsaharan Africa or in the Arabian Desert.

▶ The eagle owl is nocturnal and rarely takes to the air by day, though it is perfectly capable of doing so if alarmed or disturbed. It generally spends its days concealed among the rocks or perched on a tree or will occasionally take shelter underneath a bush on the ground. It goes out in search of prey at nightfall. It glides effortlessly and makes use of thermal updrafts to climb to a very high altitude and fly for long distances; it also skims along the ground in search of prey, though it sometime stakes prey that is perched on a branch or on a cliffside ledge as well as on the ground.

Even in Europe, the diet of the eagle owl varies greatly from one locality to the next; it feeds on mammals (small rodents, hedgehogs, small mustelids) as well as birds (ducks, grouse, coot, other raptors, cormorants, seagulls), amphibians, fish, and insects. It kills its prey with its talons, then decapitates the larger creatures; it cracks their bones with its beak and generally saves the head for another occasion. It skins the hedgehogs and plucks birds as well, usually in the vicinity of its eyrie. The owlets are fed on prey that is brought back the eyrie and minced up by the female, but in just a few weeks they are able to hunt on their own.

◀ When confronted with an intruder or an aggressor, the eagle owl hunches forward and spreads its wings, rocks back and forth from one leg to the other, puffs loudly and clacks its mandibles; this usually proves to be an effective deterrent. It has no specific predators, but it is sometimes mobbed or harassed by crows and certain diurnal raptors (the kestrel, for instance) if it shows itself in daylight, though it generally will not sustain any permanent injury. Formerly, a stuffed or a captive eagle owl was set out by hunters as a decoy to attract hawks and crows, a practice that contributed greatly to the marked decline of this species in Western Europe.

Eagle owls were relentlessly trapped and shot, and their nests destroyed, for many years because they prey on gamebirds; a professional bounty hunter in the Hérault (southwestern France) boasted of having killed about 30 of these birds in the years before the Second World War. Since 1930, for example, the eagle owl has disappeared completely from the subalpine ranges in the west of France (the Vosges and Juras as well as Burgundy) and now remains only in the eastern half of the Massif Central and in the Pyrenées.

The increasing popularity of rock-climbing has resulted in greatly increased human activity in the vicinity of its nesting sites, and the appearance of powerlines and cables for chairlifts, cablecars, etc. in the forests and valleys still frequented by the eagle owl has also resulted in many fatalities in recent years. Finally, the eagle owl is not a prolific breeder (a mating pair rears only a single chick every year, on the average), and its eggs sometimes fail to hatch when the nesting sites are disturbed by amateur photographers and other incautious observers. Fortunately, programs for raising eagle owls in captivity and later releasing them in the wild have been successfully carried out in Sweden and Germany.

In addition to the great horned owl *(Bubo virginianus)* of North America (the same vernacular name is sometimes used for the Eurasian eagle owl), there are 10 other eagle owl species, 6 of which are found in subsaharan Africa and 4 in Asia and Indo-Malaysia. They live, for the most part, in dense forests and feed on various kind of prey (including frogs and reptiles as well as birds and mammals). The akun *(Bubo leucostichus)*, found in West Africa and the Congo Basin, feeds almost exclusively on insects. Other European owls that are called by variants of same vernacular name as the Eurasian eagle owl *(hibou* in French, *Uhu* in German) belong to different genera *(Asio* or *Otus)*. J.-P. R.

EASTERN SPINY SOFTSHELL *Trionyx spiniferus* 152

152
Sub.: Vertebrata
Cl.: Reptilia
O.: Chelonia
S.o.: Cryptodira
F.: Trionychidae

The carapace of the males measures 15–20 cm, that of the females 20–45 cm, and these North American aquatic turtles can weigh up to 15 kg altogether. The brownish-green shell is rounded and very flat, which enables *T. spiniferus* to dig itself into a mudbank very quickly by rotating its saucerlike shell first in one direction and then in the other. The spines that give this species its name are located along the anterior edge of the shell, behind the head, and the carapace of the numerous species of soft-shelled turtles is said to be "soft" only because, unlike that of the snapping turtle, it is not protected by a layer of overlapping scales.

T. spiniferus' legs are highly mobile, with deep webbing between its three claws (*Trionyx* = "three-clawed"), so that this species is an especially strong swimmer. Its neck is very long; its head is also elongated, and its nostrils are located at the tip of its trunklike snout, which makes it possible for it to breathe air directly from the surface while keeping its feet firmly planted on the bottom. In addition, the mucuous membranes in the interior of its mouth are well provided with capillaries that can absorb dissolved oxygen directly.

T. spiniferus is found in fresh water of various depths, though it generally prefers a muddy or sandy bottom in a sheltered spot or with a thick cover of aquatic vegetation. It is found in swift-flowing rivers as well as in marshes and lakes, and it is often seen sunning itself along the bank or on a hummock or floating log not far from its underwater refuge, though it is sometimes active at night as well. The range of this species extends from northern Mexico just barely up into southeastern Canada; seven regional subspecies are recognized, not including a distinct but closely related species, the Florida softshell (*T. ferox*). In summer, the female deposits its 10–30 spherical eggs in a nest scooped out to a depth of 10–25 cm; the hatchlings measure between 25 and 40 mm in length. They will be sexually mature by the time they have attained a length of about 15 or 25 cm, in the case of the males and the females respectively. Longevity of this species exceeds 50 years.

▶ *T. spiniferus* is a carnivore, feeding almost entirely on crayfish and insects (each constituting almost 50 percent of its diet by volume) with the remaining small percentage composed of mollusks, worms, amphibians and their larvae, and fish. *T. spiniferus* swims or crawls along the bottom, rooting under rocks and in accumulations of vegetable debris; it locates its prey initially by touch, then sniffs it to determine whether it is actually edible and catches it by abruptly darting out its long neck like a snake. Larger prey animals may be gripped with the front claws to assist their passage down the gullet of *T. spiniferus*.

◀ The nests of soft-shelled turtles are frequently raided by raccoons and skunks. Snakes, other turtles, fish, and wading birds feed on the hatchlings, and alligators and humans on the adults. From a gastronomic standpoint, the eastern spiny softshell is rated very highly among the North American aquatic turtles and is widely consumed throughout its range. *T. spiniferus* is highly susceptible to water pollution, and though its preference for crayfish makes it competitive with certain freshwater food fish, it does not appear to be persecuted or controlled for this reason.

Soft-shelled turtles tend to rely on flight or protective coloration to defend themselves rather than simply taking refuge inside their shells. When these alternatives are unavailable, however, *T. spiniferus* is noted for the vehemence with which it resists being handled or picked up, its irascible temperament being ably assisted in this case by its sturdy legs, sharp claws, long neck, and flat, slippery shell.

The family Trionchyidae is further classified into two very distinct subfamilies; the defensive capabilities of the Cyclaorbinae of Africa and India are enhanced by movable panels on the undersides of their shells, and this subfamily tends to be less aggressive than the Trionchyidae, found in North America, Africa, and South and East Asia. This subfamily comprises 18 species (the number of genera is still not well settled) measuring between 25 and 150 cm in length and essentially similar to *Trionyx spiniferus* in their morphology and behavior; some of these species feed on mollusks, and consequently the cutting edges of their beaks are adapted for crushing and grinding rather than for tearing and biting.

R.Bo. E

EDIBLE FROG *Rana esculenta* 153

153
Sub.: Vertebrata
Cl.: Amphibia
O.: Anura
F.: Ranidae

This species normally measures between 60 and 90 mm, though it may attain a length of 120 mm. Considered by taxonomists to be a natural hybrid between two other species, the laughing frog, or marsh frog (*Rana rididunda*) and the pool frog (*Rana lessonae*), the European edible frog is found in marshes, pools, and along the shores of lakes. It prefers sunny, open spaces (as long as cover in the form of tall grass or other vegetation is available); the edible frog ranges throughout most of Europe and is found up to an altitude of 1100 m in the Alps.

▶ Like other anurans, the edible frog will only take prey that is moving—in this case, hydrometrids (relatives of the water strider), flies and mosquitoes, bees and wasps, chrysomelid beetles, mayflies, butterflies and moths. It also feeds on smaller frogs (including those of its own species), tritons, and small fish. It differs from its terrestrial cousin, the toad, in that it hunts from ambush rather than actively seeking its prey and is more likely to use its superior jumping skills, rather than patient stalking, in order to close with its victims. It will conceal itself in the vegetation along the shore, for example, and leap out at flying insects, including bees that come down to drink or gather nectar from the blooms of aquatic plants. Young frogs that have just emerged from the tadpole stage feed on springtails, small beetles, flies and mosquitoes, and ants.

◀ Watersnakes of the genus *Natrix* consume great numbers of frogs and tadpoles. Martin and Rollinat (1894) list almost a dozen mammals that feed on the edible frog—hedgehog, badger, common European shrew, water shrew, water vole, water rat, weasel, polecat, mink, and even, when other food is scarce, fox and wolf; along with its two parent species, it is frequently consumed by man as well. Avian predators include the honey buzzard, black kite, heron, stork, and spoonbill; tritons, the larvae of various aquatic insects, and diving beetles all feed on the tadpoles, and pike and chub prey on both the tadpoles and the adult frogs.

Another common European frog, *Rana temporaria*, is more clearly terrestrial in its habits than the edible frog and lives in humid meadows during the spring and summer. It feeds on snails, beetles, spiders, flies, mosquitoes, and true bugs (Hemiptera); its diet contains a smaller proportion of flying insects than the edible frog's and a smaller proportion of ants than the toad's. The juveniles feed on tiny flies and wasps (1–2 mm), springtails, aphids, and mites.

The laughing frog *(Rana ridibunda)* is quite a bit larger than the edible frog; it normally measures about 130 mm, and certain females of this species found in Eastern Europe have attained a length of 170 mm and may weigh as much as 300 g. It feeds on smaller frogs and young colubrids, small fish, small birds, and even small mammals. J.L.

EELGRASS PRAWN *Processa edulis* 154

154
Sub.: Arthropoda
Cl.: Crustacea
O.: Decapoda
S.o.: Caridea
F.: Processidae

This relatively large shrimp is found along most of the coasts of Europe, from the North Sea to the Mediterranean and the Black Sea and usually at a depth between 0 and 15 m in submerged salt meadows of phanerogamous (seed-setting) plants or in undersea beds of eelgrass—*Zostera marina* in the North Sea, the English Channel, and the Atlantic, *Posidonia oceanica* in the Mediterranean. It may also be found, though more rarely, under rocks or in beds of seaweed. It buries itself in the sand, headfirst, at first light of day and emerges only after dark.

▶ The planktonic larva of this species feeds on smaller organisms in the coastal zooplankton; unlike other species of shrimp that feed at least partially on seaweed and sessile organisms, the juvenile and adult eelgrass prawn prey almost entirely on such free-swimming marine creatures as polychaetes, small crustaceans (amphipods, isopods, Mysidacea, ostracods), and occasionally on small mollusks, fish hatchlings, and fish eggs.

The eelgrass prawn begins to hunt at nightfall, pursuing its prey through the stalks of eelgrass or foraging along the bottom. It comes out to feed every night and is particularly active during the first few hours of darkness, except when the temperature drops quite low in northern waters and while molting its shell. The molt occurs every 16–25 days, depending on the size and sexual maturity of the individual and the temperature of the water.

The shrimp locates its prey by vibrations in the water or by chemoreception and seizes it with its maxillipeds and its front legs, which, as with many decapods (shrimps, crabs, lobsters) have asymmetrically developed into a powerful pincer and a delicate claw. These are somewhat reminiscent (in certain genera of the family Processidae) of a knife and fork (e.g. *Processa* and *Nikoides;* the claws of *Ambidexter,* as the name implies, are symmetrical). Smaller, soft-bodied prey animals may be swallowed whole; larger or harder-bodied prey are dismembered with the claws and third maxillipeds. The eelgrass prawn also ingests a small quantity of sand which acts as an abrasive, something like the stones in a bird's crop, and facilitates the process of digestion.

◀ During its free-floating larval stage the eelgrass prawn may be devoured by any number of coastal marine organisms. The predators of the adult shrimp consist mainly of fish (wrasses, Gobidae, Cottidae, rays, scorpion fish, and small sharks) and cephalopods *(Sepiola* and other varieties of squid). Since these predators hunt by sight, the prawn's primary defense is to conceal itself in the sand during daylight; the molt is a critical period during which the prawn is not only soft-bodied and vulnerable but also finds it difficult to burrow into the sand.

Normally, however, the eelgrass prawn only sheds its shell after dark and by the morning the new shell will have hardened sufficiently so that the prawn is able to bury itself, at least superficially, in the sand once again. In addition, the eelgrass prawn is not readily visible against the sandy bottom; the juvenile prawn are practically transparent, and the ovaries and eggs cases are greenish in color, which affords the females some measure of protection in the beds of eelgrass and seaweed that they frequent.

In the aquarium, this species occasionally exhibits a tendency toward cannibalism, though it is not particularly pronounced. Four different *Processa* species are sometimes found in the same habitats in the Mediterranean, but we have no information as to whether they are specialized or competitive with respect to prey selection. *Processa edulis* is the only eelgrass prawn found in northern waters, where it may compete with prawns of the genus *Palaemon* or sand shrimps of the genus *Crangon* or *Philocheras.* As its Latin name *edulis* implies, this is an edible species and was formerly sold in fish markets in the south of France and elsewhere in the Mediterranean, though the total annual catch was quite small. These prawns can be caught in substantial numbers by dragging the eelgrass beds at night, but this practice has been prohibited since about 1960 in order to conserve this fragile marine habitat. This species cannot be caught in the baited traps, similar to crab pots, used for other crustaceans, since it will feed only on live prey.

The family Processidae consists of 55 species found in warm or temperate coastal waters, generally on a pebbly, sandy, or muddy substrate or in beds of seaweed or wrack grass; several species that live in deeper waters are active by day and conceal themselves in the sand at night. As far as humans are concerned, these shrimp are an important intermediate link in the marine food chain between the meiofauna and smaller macrofauna of coastal waters and the larger creatures, primarily fish and crustaceans, that are eaten by man.

P.N.

EGYPTIAN COBRA *Naja haje* 155

155
Sub.: Vertebrata
Cl.: Reptilia
O.: Squamata
S.o.: Ophidia
F.: Elapidae

As a rule, this long, svelte African cobra is uniformly brown, gray, or yellow in color, depending on the region; in North Africa, e.g., the skin is darker, almost black, and local populations in some areas are marked with alternating light and dark bands. The Egyptian cobra measures between 1 and 2 meters long, possibly 3 at the maximum; its weight may exceed 3 kg in exceptional instances. Longevity in excess of 30 years has occasionally been claimed for Egyptian cobras that were captured in adulthood, but it seems unlikely that more than a very few individuals could survive to this age in nature.

The Egyptian cobra is so called because of its close associations with the religion of ancient Egypt (in this context is is generally referred to as the "asp" or "uraeus"). In fact, this species is found all over Africa except along the coastal plain of the Maghreb, in the equatorial rainforests, and in the deepest deserts, though it has colonized certain oases and the beds of intermittent watercourses (wadis) in the Sahara where there is sparse vegetation; it has also been reported up to a height of 2700 m on mountain plateaus.

In the tropics, the Egyptian cobra is generally more active during the rainy season than during the dry season; in the subtropical regions in the northern or southern parts of its range, it hibernates during the winter—in southern Morocco, for example, starting around the end of September and the beginning of November and emerging sometime between March and May. At first, the cobra leaves its lair in the morning to warm itself in direct sunlight, then later reverts to its normal nocturnal pattern and may spend several days away from its lair on long journeys in search of prey (and more particularly in search of a mate).

▶ The dentition of the Egyptian cobra is of the type called proteroglyphic, which is to say that it has two hooked, elongated fangs located at the front of the upper jaw, each of which has a groove running down one side, the edges of which have fused to form a sort of tube. This tube has its anterior outlet at the base of the little hook at the point of the tooth and is connected by a duct to the poison glands behind the eye where the cobra's venom is produced.

The Egyptian cobra feeds on small mammals (sometimes as big as rabbits in the case of the larger individuals), birds and their eggs, reptiles (including other members of its own species), frogs, toads, and sometimes on fish. There are instances on record of an adult Egyptian cobra that had swallowed 2 puff adders, each 1 m long, and of a cobra measuring 2.2 m that had swallowed a 1.06 m monitor lizard. Cobras are eclectic in their feeding habits and will be content with whatever prey is most abundant in the area; some populations feed primarily on toads, for example, and those that live in the vicinity of human settlements tend to feed primarily on rodents.

The Egyptian cobra prefers to hunt at night, starting at dusk, though young cobras are sometimes active during the day; a cobra will venture fairly far afield in search of prey, though it always returns to the same lair each day—an abandoned burrow, a hollow tree, a pile of stones, a rocky crevice or termite mound. The cobra's vision is not very acute, and it seems to be able to track its prey visually only so long as the prey remains in motion. On the other hand, it can readily follow the faint molecular trail left by an animal; minute scented particles are conveyed by the tongue to an olfactory sensor known as Jacobson's organ.

This method of tracking is useful in leading the cobra to a bird or rodent's nest. Nestlings are swallowed immediately, whereas more substantial prey, such as frogs and toads, are injected with venom before being swallowed whole. Still larger and more agile prey, lizards and adult rodents, for example, are tracked visually by the cobra, which rapidly pursues its quarry in the manner of a blacksnake or other hunting snake of the genus *Coluber* until it is able to clamp down its jaws (sometimes after several unsuccessful approaches) on some part of the prey animal's body. If the prey can still defend itself with its teeth, the cobra will release it in order to get a better grip, but if the prey remains motionless, the cobra loses track of it immediately and will have to recover its bearings with the aid of its olfactory homing system.

Once a smaller prey animal has been killed or merely immobilized by the cobra's venom, the cobra will open its jaws and begin to swallow it, starting (in the case of a smaller prey animal) wherever it first happened to sink in its fangs or (in the case of a larger one) with the head. In the case of prey animals that are not very susceptible to cobra venom, notably other cobras, the aggressor sinks its fangs into the victim and then gradually works its way toward the head by means of alternate lateral motions of its jaws. Digestion takes from 3 to 8 hours, depending on the size of the victim and the external temperature; nevertheless, most of the cobras captured in nature have empty stomachs, which suggests that they normally feed at widely spaced intervals.

◀ Apart from man, the Egyptian cobra's more prominent predators are large diurnal birds of prey like the bateleur and the short-toed eagle, the secretary bird, and several species of falcons, eagles, and buzzards. With the exception of the secretary bird, these raptors all throw themselves on the cobra from above, seizing it in their talons and dazing it with repeated blows of their wings, finally killed it by crushing its head with their beaks. Though they have no particular immunity from the cobra's venom, the thickness of their plumage does afford them with some protection. There are also a number of other snakes, members of the Elapidae and Colubridae families, that prey routinely on the Egyptian cobra.

This species is occasionally but less systematically preyed on by nocturnal raptors, hornbills, wild and domestic pigs, storks and other wading birds, and monitor lizards. Cobras are sometimes caught by crocodiles or even large fish while attempting to swim across rivers and other bodies of water, and a number of smaller creatures, including monitor lizards and various rodents, feed on the cobra's eggs. The mongoose, a relative of the civets and genets (family Viverridae), is commonly portrayed as the cobra's deadliest enemy, though it should be regarded rather as an occasional predator. The mongoose is not highly susceptible to the cobra's venom, but it may succumb to a well placed bite, contrary to the legend that it is the mongoose that always emerges triumphant from these encounters. It is true, however, that the flexibility of the mongoose's ability and the quickness of its reflexes enable it to get the better of the cobra in a large percentage of cases.

When threatened or cornered, the cobra adopts its well known defensive stance—raising the anterior third of its body off the ground and by spreading and flattening out its flexible cervical ribs (which normally point backward) to create a broad "hood" of skin just below its head. Then, if the aggressor is undeterred by this menacing posture, the cobra will generally feint several times before it actually strikes, and it seems that it will do so only after the aggressor has approached within a certain distance. Once it has struck with its fangs, the cobra maintains its grip for several seconds, "grinding" its fangs in the wound to hasten the penetration of the venom; the neurotoxins in the cobra's venom take effect quickly and can cause death in humans by paralyzing the muscles responsible for respiration. The Egyptian cobra is an unaggressive creature, however, and has been blamed for relatively few fatalities among human beings; in many cases it appears that the victim made the fatal error of attempting to catch the cobra after blocking off all possible escape routes.

Several other cobra species are found in Africa besides *Naja haje*. The black-and-white cobra *(Naja melanoleuca)* is semiaquatic and almost exclusively a forest-dweller; it frequently feeds on fish and amphibians. The better-known black-necked, or spitting, cobra *(Naja nigricollis)* has mastered the startling defensive tactic of projecting its venom at the eyes of an aggressor in the hope of temporarily blinding it so the cobra can make its escape. Its venom is said to produce a painful burning sensation in the eyes and can indeed cause temporary blindness even in human beings.

The spitting cobra, the South African spitting snake, or ringhals *(Haemachates haemachatus),* and certain local races of the Indian spectacled cobra (*Naja naja*) can all project their venom over at least a short distance; the venom channel that runs down inside the fangs is L-shaped, so the venom can be expelled by means of hydraulic pressure from a rounded aperture that is located on the front surface of the cobra's fangs. Since the spitting cobra's awareness of comparative anatomy is necessarily limited, it can be induced to discharge its venom at any shiny reflective surface.

The king cobra *(Ophiophagus hannah)* of Southern Asia is the largest venomous snake in the world, with the individual record for the species currently standing at 5.58 m. The king cobra feeds almost entirely on other snakes, including several venomous species (cobras and kraits) as well as nonvenomous rat snakes of the genus *Ptyas.* In general, the cobra is undoubtedly the most fascinating of all snakes and plays a very prominent role in the mythology and folklore of Africa and India. It is perhaps even better known for being fascinated by man, though it must also be pretty well known by now that it is not the sound but rather the rhythmic motion of the snake charmer's flute—since snakes are totally deaf—as well as his cautious avoidance of the cobra's defensive perimeter that enables the snake charmer to charm the snake.

D.H.

ELECTRIC CATFISH *Malapterus electricus* 156

156
Sub.: Vertebrata
Cl.: Osteichthyes
S.cl.: Teleostei
O.: Siluriformes
F.: Malapteruridae

The electric catfish typically measures 60 cm, and a meter-long specimen can weigh as much as 20 or 25 kg. This species is found throughout the Nile and Congo basins; it spends its days concealed along the bottom or in a cavity in the bank, and searches for food at night.

▶ Like most catfish, *Malapterus* is eclectic in its feeding habits, preying on worms, fish, frogs, and a variety of small creatures that are found along the river bottom. In general, the various species of electric fish can produce either (1) a continuous but low-intensity field for purposes of navigation and prey detection (*Eigemannia* [see KNIFE FISH], *Gymnotus*, *Mormyrops*, Nile fish) or (2) an intermittent high-intensity electric discharge that is used as an offensive or defensive weapon (*Malapterus*). The electric eel has separate sets of electric organs that are used for all these purposes, though some authorities prefer to reserve the term "electric organs" for those of the second type.

Malapterus's electric organs were formerly thought to be of glandular origin, though it now appears, as with the other species mentioned, that they consist of plates of highly specialized muscle tissue that are located toward the base of the tail (the positive pole) and along the upper part of the body toward the head (the negative pole). A single discharge of *Malapterus*'s electric organs lasts for about 2 milliseconds and can generate a current of up to 350 volts, and are thus more than three times as potent as those of the electric eel.

A young *Malapterus* of no more than 5 cm in length can already produce a current of up to 30 volts—and of up to 150 volts by the time it has attained a length of 30 cm. Like other catfish, *Malapterus* presumably finds its prey with the aid of the sensitive olfactory receptors located on its barbels. Vibrations in the water can also be detected by the lateral line, and sound waves that are picked up by the inner ear are relayed via the ossicles and ligaments of the mullerian apparatus to the swim bladder, which acts as a resonating chamber, so the electric catfish undoubtedly has a very keen sense of hearing.

◀ The electric catfish is protected from predators by its truculent, aggressive temperament as well as by its formidable batteries; it is locally consumed by man, but is not of any great commercial importance. *Malapterus* was known to the ancient Egyptians and was even used by Arab physicians of the eleventh century in a pioneeringversionofelectrotherapy.

F.M.

ELECTRIC EEL *Electrophorus electricus* 157

157
Sub.: Vertebrata
Cl.: Osteichthyes
S.cl.: Teleostei
O.: Cypriniformes
F.: Electrophoridae

The electric eel probably lives between 10 and 20 years and attains a maximum size of 2.4 m. It is found only in northern South America, in the basins of the Orinoco and Madeira rivers and in the Guiana Highlands; it is usually found amid dense aquatic vegetation in shallow, oxygen-poor water that is often thick with alluvial deposits in the rainy season. The electric eel is not a true eel (Anguilliformes) but a cyprinoid fish (Cypriniformes), a distant relative of the carp, minnow, and goldfish. A closely related family of South American freshwater fish called the Gymnotidae has the same elongated, snakelike body but lacks the electric organs used by the electric eel to defend itself and to catch its prey; in some classification systems, these two groups are consolidated into a single family. In both, the abdominal cavity is not very extensive, the anus is very far forward on the ventral surface, and the remaining four fifths of the fish's body length is comprised of the highly specialized organs used to generate electricity (in the case of the Electrophoridae) and muscle tissue.

▶ Young electric eels live on tiny crustaceans (daphnia, copepods), insect larvae, and small annelid worms, then on larger crustaceans and adult aquatic insects as they grow larger. The adult electric eel catches frogs, small fish (families Cichlidae and Characidae), and immature members of the family Gymnotidae. After it has stunned its prey with a powerful electric discharge, it seizes it with a transverse motion of its head and immediately swallows it with an audible gulping sound.

Electrophorus has 3 separate electric organs of 2 different types, all of which are involved in predation. The principal organ is located farther forward and toward the underside of the body and essentially consists of a storage battery containing between 5000 and 6000 electric cells and capable of generating up to 550 volts—with enough current to kill a small fish on contact or to administer a memorable shock to a human or other large vertebrate (whence one of *Electrophorus*'s local names, *tremblador*). The discharges of the principal electric organ are not continuous but consist of a series of 3 to 6 very brief flashes of .002 second's duration separated by intervals of .005 second. A young electric eel only 7 to 10 cm long can already generate 100 volts, and the capacity of the principal organ continues to increase throughout the fish's lifetime.

The dorsal posterior electric organ, known as Sach's organ, generates a more or less continuous sequence of low-intensity impulses each lasting about .002 second and not exceeding 100 volts and serves essentially the same purpose as a radar "fish-finder" on a commercial fishing boat, since the electric eel has very poor vision, spends most of its life in muddy water, and is primarily active at night. Any perturbation in the electric field that is generated by Sach's organ is immediately detected by an array of sensory receptors, most of which are located on its head. When at rest, it emits a small number of these range-finding impulses per minute, about 30 per minute while swimming, and 50 per minute after its sensory organs have detected a disturbance in the field, which may be a potential predator or prey animal, or merely an underwater obstacle to be avoided.

In spite of appearances, the electric eel does not propel itself through the water by means of the undulating, eel-like motions of its caudal fin. Instead it swims by moving its anal fin, which runs along the underside of its body for a considerable fraction of its length. This is an important diference, since caudal propulsion would cause turbulence that would make it difficult for its electric detection system to operate smoothly. The Nile fish (q.v.), an electric fish not closely related to *Electrophorus*, has evolved a very similar style of locomotion.

◀ The third electric organ, Hunter's organ, is located roughly midway between the other two and is partly of use in immobilizing prey animals, partly in discouraging potential predators. Because of its nocturnal habits, the electric eel does not compete directly with the numerous predatory cichlids and characids that are found in profusion in the same waters. Young electric eels, more vulnerable than their elders, remain under their parents' protection, generally swarming around the head of an adult, until they have attained a length of about 15 cm. The electric eel sometimes plays a role in the religious and tribal rites of the human inhabitants of the region but has more to fear from certain inhabitants of more distant regions, since specimens are frequently collected for the purpose of scientific research of one kind or another.

Gymnotus carapo, the knife fish, is the electric eel's closest relative among the Gymnotidae. It grows to a length of about 50 cm, and its diet is very similar to that of *Electrophorus*, though it lacks electric organs with which to stun its prey.

F.M.

EMPEROR DRAGONFLY *Anax imperator* 158

158
Sub.: Arthropoda
Cl.: Insecta
O.: Odonata
F.: Aeshnidae

Since *anax* means "lord and master," *Anax imperator*, "lord and master emperor," seems slightly redundant, though perhaps no more so than most imperial titles. *Imperator* is the largest of the European dragonflies, about 8 cm long with a wingspread of over 10 cm. The male is a truly gorgeous creature with enormous, lustrous blue-green compound eyes, a luminous green thorax, and a brilliant sky-blue abdomen ornamented with a black dorsal band. The female's coloration is slightly subdued. Its wings are tinted yellowish-brown; the abdomen is a much less saturated bluish-green and encircled by a reddish median band.

Imperator, like most other odonata, favors standing water, from tiny pools to quite large ponds. The male is an indefatigable flier, performing its impressively frenetic aerobatic feats at a height of one or two meters above the surface of the water. The male is also highly territorial, defending its little corner of the pond against all rival claimants, these disputes being generally resolved in favor of the original proprietor in a swooping aerial dogfight of no more than a few seconds' duration. The female flies closer to the surface of the water, is less conspicuous in its movements, and alights frequently amid the vegetation along the shoreline to lay its eggs. *Anax imperator* is found throughout Western and Central Europe below 55° N as well as in Africa, the Near East, and elsewhere in Asia.

◀ The *imperator* larva may reach a length of 6 cm and is equalled only by the larva of the diving beetle for its ferocity and mobility in the invertebrate freshwater predator division. Apart from the "mandibular arm" used by most dragonfly larvae to seize their prey, the *imperator* larva can also make use of its so-called anal pyramid—a kind of rear-mounted battering ram formed of sclerotized plates surrounding the rectum—and the extreme flexibility of its abdomen either to impale a prey animal or to repel a potential predator. (*Sclerotized* refers to the hardened body armor of insects and other arthropods.)

Imperator larvae feed on the larvae of mosquitoes, chironomids, caddis flies, and dobsonflies (hellgrammites); they also consume large numbers of tadpoles and, as is characteristic of the odonata, smaller members of their own species. They are sometimes said to be a prime cause of mortality among fish hatchlings and young fry, and while it is true that a dragonfly larva may occasionally kill a very small fish, this does not happen frequently enough to be a matter of serious concern to anyone but the parties involved. In any case, these occasional depredations are handsomely compensated for by the destruction of very large numbers of diptera larvae.

The adult *imperator* feeds on many of the same species—mosquitoes, flies, gadflies, dobsonflies—as well as damselflies and fairly large dragonflies (notably *Libellula quadrimaculata*). It has also been known to catch moths as large as the oak bombyx and butterflies as large as *Silenus*, with a wingspread of about 7 cm. The adult *imperator* normally seizes and consumes its prey on the wing, though on those rare occasions when it does catch a large butterfly, it has to alight in order to make a meal of it. (It eats only the muscular parts and the viscera, detaching and discarding the appendages and sclerotized body case.) *Imperator* is also said to attack bees (though this seems very unlikely, since a bee's sting is known to be fatal to even the largest dragonfly) and to seize tadpoles and small fish swimming on the surface of the water (which is also possible but not very probable). Once again, the adult *imperator*'s occasional (and probably legendary) excesses as a predator seem scarcely to detract from the value of its services as a regular consumer of flies, mosquitoes, and other insect pests.

Closely related species include the Neopolitan dragonfly *(Anax parthenops)* of Southern Europe, which is slightly smaller than *imperator* ; it has a purplish-brown thorax and a greenish-brown abdomen with only the first few segments colored the same brilliant blue as *imperator's*. The female's colors are similar but more subdued, and the wings of both sexes have large yellowish-brown patches. *Hemianax epphipiger* is an Indo-African species that also may breed occasionally on the northern shores of the Mediterranean and is similar in appearance to *imperator*, except for the earth-and sand-colored pigment patches more appropriate to a semidesert enviroment. Only the male has the saddle-shaped dorsal patch of brilliant cerulean blue at the base of its abdomen (hence its unmelodious Latin name, *epphipiger*, "saddle-bearer"). Its hunting and other behavior seems comparable to that of *Anax imperator*.

A. de R.

Endaphis perifidus 159

159
Sub.: Arthropoda
Cl.: Insecta
S.cl.: Diptera
O.: Nematocera
S.o.: Bibioniformi
F.: Cecidomyiidae

This is an endoparasitic gall midge, a tiny fly about 2 mm long; the female's abdomen is bright red, with brownish appendages and a pair of transparent wings. In spite of its relatively conspicuous coloration, this species was unknown to science until the end of the last century; the type specimen was collected in Lorraine, in Eastern France.

▶ The parasitic host for the larvae of this species is an aphid, *Drepanosiphum platanoides*, that lives on the European sycamore maple *(Acer pseudoplatanus;* not the same tree as the North American sycamore). The female gall midge lays its eggs one at a time, attaching them to the ventral surface of the aphid's body shield. The newly hatched larva pokes its way through the flexible membrane between two abdominal segments and lodges in the aphid's abdominal cavity, where it gradually devours its host in typical endoparasitic fashion. After a couple of days, the larva can clearly be seen from the outside through the transparent green chiton of the aphid's body shield.

After this stage of its development is complete, the larva makes its exit through the aphid's anal orifice. The larva curls itself up, then abruptly extends its body, a maneuver (also found in the more numerous cecidomyiid genus *Contarina)* that sends it tumbling to the ground. It wriggles down through the soil particles with the aid of a specialized organ called the sternal spatula and spins its cocoon. The life cycle of *Endaphis perfidus* may be repeated several times in a single year; this species has no known parasites or predators.

A related cecidomyid, *Pseudendaphis maculans,* parasitizes an aphid species that is found on orange trees, *Toxoptera aurantii* Boy, in much the same manner. R.C.

160
Sub.: Vertebrata
Cl.: Mammalia
O.: Carnivora
F.: Mustelidae

ERMINE *Mustela erminea* 160

Along with several other European weasels, the ermine is also referred to as a "stoat," especially when it is still in its dark summer coat. The ermine is between 22 and 31 cm long (including the tail), and normally weighs between 125 and 350 g (sometimes as as much as 445 g); the female is smaller than the male. The ermine's summer coat is brown or beige on the upper parts of its body, white on the belly and along the insides of its legs; the tail—as many who have never seen the living animal are aware—is tipped with black, over about a third of its length. In the fall, the ermine sheds its summer coat and, at least in Northern Europe, assumes its white winter coat, though the tip of its tail is always black. Some populations (such as in western France) do not shed their coats at all, and many others, depending on local climatic conditions, remain fixed at various transitional stages throughout the winter.

The ermine ranges throughout most of Europe, with the exception of Italy, southern Spain, and the Mediterranean islands, and up to an altitude of 3000 m in the Alps. It is found in various habitats, including humid areas and those frequented by man—farm buildings and even human dwellings on occasion. It builds its nest in a crevice in the rocks, in a stone wall, a hollow in the ground, or in the abandoned burrow of a rodent or the tunnel of a mole (which it enlarges to suit its own requirements). The ermine is active both by day and at night; it swims and climbs very well, though not as well perhaps as the beech marten or the pine marten. Longevity is about 10 years.

▶ The ermine preys primarily on rodents, including snow mice *(Microtus nivalis)* in mountainous regions, rats, fieldmice, housemice, and voles in the lowlands, and young rabbits wherever it comes across them. It also preys on insects, snails, frogs, and lizards; it is quick and agile enough to catch small birds, and it sometimes feeds on eggs and fruit as well. The ermine depends partially on its keen sense of smell and hearing to locate its prey, though it appears to hunt primarily by sight. It often stops and sits up on its haunches for a better view of the terrain; though its body is very long and thin, it has very short legs and the slightest elevation of the ground, even a furrow in a plowed field, may be enough to block its view when it is trotting along on all fours.

Since its main strategy as a predator is to run down rodents in their tunnels and burrows, however, this peculiar conformation generally turns to its advantage once the prey has been located and pursuit is underway. In winters, voles (snow mice) and other rodents can protect themselves from raptors and other predators by tunneling through the snow, but here as well, the ermine can easily follow. The ermine has even been reported on several occasions to have attacked a full-grown hare—possibly one that was suffering from some physical impairment, since the European hare ordinarily outweighs the ermine by 10 or 20 times (3–6.5 k for the hare versus 300 g for the ermine).

Like other weasels, the ermine is sometimes said to "bleed" its prey and drink its blood; thus, it is asserted, the ermine needs to make many more kills than other predators because it does not eat the flesh of its prey. This assertion is actually based on a misinterpretation of two different phenomena. First, there is the ermine's habit, which it shares with various other mustelids, of maintaining caches of prey animals scattered about its territory. Thus, when the opportunity presents itself, it will generally make more kills that it can actually consume on the spot; the ermine returns to these food caches whenever it gets hungry and food is not readily available or when the day's hunting has been bad.

Second, the ermine usually dispatches a prey animal with a bite to the head or neck. At the moment of capture, it becomes tremendously excited, and often remains crouching over its prey for as much as several minutes, its jaws clamped firmly in place, and for some time after the animal is already dead; it may also lick the blood that flows from the wound. It is easy to see how this behavior might have given rise to the belief that the ermine actually drains the blood from its victims.

◀ Most of the other common mammalian predators of Europe are considerably larger than the ermine and will not hesitate to attack it, including the fox, lynx, wild cat, and feral cat, as well as various hawks and owls. For the least several years, the ermine has been legally protected in certain countries (it is partially protected in France), though it is still sought for its fur and still sometimes destroyed as a "harmful" predator; large numbers of ermines are also killed on the roads every year. F.T.

Erpeton tentaculatum 161

161
Sub.: Vertebrata
Cl.: Reptilia
O.: Squamata
S.o.: Ophidia
F.: Colubridae

The body of this remarkable-looking snake is slim, rigid, and covered with rough scales; its head is flat and elongated, quite distinct from its body, and on either side of the muzzle, there is a soft, fleshy appendage (hence, *tentaculatum)* that is flexible and covered with scales. Its body is brownish for the most part with lighter longitudinal bands running along its flanks; it never exceeds 1 m in length. *Erpeton* is found is lakes, rivers and streams and well as brackish and saline estuaries and lagoons throughout Indochina. It is strictly aquatic and lacks the ventral plates necessary for locomotion on land; *Ervteon* is ovoviviparous, and its young are born in the water.

▶ *Erpeton* normally conceals itself in aquatic vegetation and lunges out at fish that venture within a certain distance, though it does not pursue or hunt them actively. It feeds exclusively on fish, and it was once thought that its whiskerlike tentacles were used as bait like the tongue of the snapping turtle or the lure of the anglerfish. This does not appear to be the case, however; the tentacles are richly inervated and are used as feelers, to detect the movements of prey or even the prey itself in muddy or cloudy water. *Erpeton* has also been reported to follow the migrations of shoals of fish.

◀ Large carnivorous fish and, in certain localities, crocodiles, perhaps including the marine crocodile, are probably the most frequent predators of this species. *Erpeton* is only partially in competition with other reptiles; the Javanese file snake (q.v.) feeds on larger prey, and the only other member of the family Acrochordidae, *Acrochordus* (syn. *Chersydrus) granulatus,* is found exlcusively in saline costal waters and in brackish estuaries and lagoons.

The other members of the subfamily Homalopsidae, in which *Erpeteon tentaculatum* is included, are less strictly aquatic in their habits; some feed on frogs as well as fish, and others seem to prefer crabs, which are highly susceptible to their venom. D.H.

Eupelmus urozonus 162

162
Sub.: Arthropoda
Cl.: Insecta
O.: Hymenoptera
S.o.: Chalcidoidea
F.: Eupelmidae

This species of chalcid wasp is common in various European habitats. The adult is only 3–4 mm long, the male somewhat smaller than the female; its body is a deep metallic green, verging on black, and the central section of its terebra (augerlike ovipositor) is yellow.

▶ The male of this species feeds on nectar, sap, and other sugary substances; the female preys on the larvae of the same species (Coleoptera, Diptera, Lepidoptera, Hymenoptera, Homoptera) on which it nourishes its own ectoparasitic larvae. The female deposits its egg, which is white, elongated, and prolonged by a thin stalk, or peduncle, which is normally folded back over the chorion, or protective envelope for the egg, next to the body of an insect larva that it has paralyzed by injecting it with venom. The larva passes through five developmental stages, characterized by changes in the size and shape of its mandibles, among other things, then spins its cocoon beside the empty body case of its parasitic host.

The feeding technique of the adult female wasp, which is not always observed in connection with egg laying, is quite remarkable. It secretes a strand of mucus from the tip of its ovipositor, which quickly solidifies into a hollow tube; this tube can be used to pierce a plant stalk or other vegetable material in which the larva is encased, enabling the wasp to drain the larva of its hemolymph by applying suction with its mouth parts.

◀ The adult wasps are preyed on by birds, spiders, and other insectivores; the larvae, generally concealed inside a plant stalk or in some other secluded spot, have little to fear from these, though they are sometimes parasitized by larvae of their own or other chalcid species that are lucky enough to have hatched out a little bit sooner in the same vicinity.

The genus *Eupelmus* and the family Eupelimidae, both very well represented, comprise a large group of ectoparasitic wasps, found all over the world and preying on an enormous variety of other insects. Y.A.

163
Sub.: Vertebrata
Cl.: Aves
O.: Ciconiiformes
F.: Ardeidae

EURASIAN BITTERN *Botaurus stellaris* 163

The Eurasian bittern is a medium-sized Old World heron with brown striped plumage and a compact body; it is about 70 to 80 cm long with a wingspread of 125 to 135 cm and weighs between 1000 and 1300 g. A banded bittern is known to have survived for over 10 years, probably closer to a maximal than an average age. The bittern is found in marshy regions overgrown with reeds or bulrushes *(Phragmites, Typhus, Scirpes)* where the water is shallow; it builds its nest in areas where the reeds and other aquatic vegetation are completely above the tidal mark.

The bittern is a shy and solitary bird, rarely seen outside the shelter of the dense reedbeds where it makes its home. Its surprisingly bovine cry, usually described as a sort of hollow bellowing sound, has always been regarded as its most distinctive trait. The genus name *Botaurus* suggests *bos, taurus,* the Latin words for "cow" and "bull"; in France, the bittern is called the "marsh ox," in Britain the "booming bittern," and its cousin *B. lentignosus* is known by such names as "stakedriver" or "thunder pumper" in certain parts of North America. Some European bitterns make their way south to North Africa and even beyond the Sahara (a process that seems too erratic and capricious to be described as migration), and mortality is very high for those bitterns who choose to remain behind during the winter months.

▶ The bittern feeds on a variety of aquatic organisms, both vertebrate and invertebrate; it is especially fond of fish of the family Cyprinidae, and it also catches amphibians, insects, worms, mollusks, and even voles (in winter, when the marsh is frozen over). It also occasionally catches nestlings of other species and young watersnakes. It hunts during the day and at dusk, wading slowly through the shallow water and among the reeds in typical heron fashion, remaining motionless and vigilant for long periods at a time, then suddenly darting forward and impaling its prey with its long sharp beak.

Its variegated brownish plumage helps it to remain undetected during its hunting vigils, and its protuberant eyes allow it to survey a very large portion of its immediate environment without betraying itself with any sudden movements of its head. It swallows prey animals whole, often while still alive, though it sometimes takes the precaution of crushing the larger ones—frogs and eels, for example—between its mandibles. Since this is bound to be a very messy procedure, the bittern, like many other herons, grooms itself periodically by dusting the plumage of its neck and bib with a powdery substance produced by the down on its flanks.

◀ There are few animals that prey on the bittern, except in winter, when sick or enfeebled individuals may easily be captured by a variety of predators. The bittern has been intermittently persecuted by fishermen in many areas as a "harmful" competitor and is now a protected species in Europe. Nevertheless, the rapid shrinkage of its habitat in recent years due to the draining of marshes, the destruction of reedbeds and other wastelands, has produced a dramatic reduction in the bittern population (in France, estimated at 40 percent over the last 15 years with only about 300 pairs remaining).

There are 12 bittern species altogether, with the continental varieties of *Botaurus* (Eurasian, North American, South American, Australian) basically comprising a single cosmopolitan superspecies. Bitterns of the genus *Ixobrynchus*, notably the Eurasian little bittern *(I. minutus)* and the North American least bittern *(I. exilis)* and dwarf bittern *(I. sturmii)*, are only about half the size of *Botaurus stellaris*. J.-F. T.

EURASIAN DIPPER *Cinclus cinclus* 164

164
Sub.: Vertebrata
Cl.: Aves
O.: Passeriformes
F.: Cinclidae

The dippers, or water ouzels (from an old English name for the blackbird or thrush), are an unusual thrush genus that have learned to find their prey in icy mountain torrents. The Eurasian species is found in certain mountainous regions of North Africa, Western Europe (e.g., the Pyrenees and the Alps), Scandinavia, European Russia, and the northern Himalayas. It is about 18 cm long with a wingspread of 29 or 30 cm and weighs between 50 and 70 g. It nests on the ground, usually beneath the shelter of a rocky overhang; the nest itself closely resembles that of the common European wren, *Troglodytes troglodytes*. The Eurasian dipper is both solitary and sedentary, though it may be obliged to make a short migration if its stream freezes over in winter.

▶ The dipper is the only passerine bird that has adapted to life in the water, which is where it spends much of its time. It swims on or below the surface and walks along the bottom, and will sometimes its way upstream against a strong current with its head bent down, propelling itself forward by beating its wings beneath the surface, so that it is often said to be able to "fly underwater." It moves frequently from one rocky perch to another, skimming above the surface of the water, and while stationary its body appears to be racked by continuous nervous tremors. The dipper feeds on beetles and caddis flies and their larvae as well as small crustaceans (gammarids) and mollusks, sometimes on worms and leeches as well; it catches all of these prey animals in the water by seizing them with its beak.

◀ The Eurasian dipper has no natural enemies or competitors, though it is sometimes persecuted by fishermen who believe, quite erroneously, that it feeds on fish eggs.

Apart from *Cinclus cinclus*, there are 3 other dipper species, one of which is found in Eastern Asia and 2 in the Americas, notably *Cinclus mexicanus*, the North American dipper, found primarily on the western side of the Continental Divide between Alaska and Panama. J.-P. R.

EURASIAN PYGMY OWL *Glaucidium passerinum* 165

165
Sub.: Vertebrata
Cl.: Aves
O.: Strigiformes
F.: Strigidae

This tiny owl lives in mature deciduous-coniferous forests of Northern Europe and Asia, its range extending in a broad (600–1000 km wide) belt from Scandinavia and the Alps to the Pacific. The body of the Eurasian pygmy owl is only 15–17 cm long with a wingspread of 33 cm; it weighs between 50 and 77 g.

▶ Though it is active during the daylight hours, the pygmy owl prefers to hunt at dawn or dusk and is rarely observed by humans; it builds its nests in holes drilled by woodpeckers in a treetrunk and may spend the entire day on its perch in the treetops or in the shadowy interior of a thick pine tree. Its diet consists of roughly equal portions of small birds (wrens and tits) and mammals (rodents, moles and shrews), though it sometimes preys on slightly larger birds such as the thrush and the great spotted woodpecker *(Dendrocopos major)*. It stuns its prey by plummeting down on it from a considerable height and knocking it to the ground, and if the prey is not devoured on the spot, it may be stored in a woodpecker hole and eaten later on.

◀ As with other nocturnal species, the young leave the nest before they have learned to fly and are consequently sometimes preyed on by martens. The adult has no particular predators; when alarmed or disturbed, it drops down off its perch and flies off on an evasive zigzag course to a neighboring tree. Apart from the reduction of local forest habitats, man and his activities appear to have had little effect on this species.

The genus *Glaucidium* comprises 12 different species, found in all parts of the world except Australia and Oceania. *Glaucidium minutissimum,* which lives in the forests and on the pampas of South America, is only about 12–14 cm long.

J.-P. R.

166
Sub.: Vertebrata
C.: Mammalia
O.: Carnivora
F.: Canidae

EURASIAN WOLF *Canis lupus lupus* 166

The best known of the wild canids, as least as far as our folklore and fairy tales are concerned, was once prevalent throughout Eurasia and North America. The European gray wolf, the largest carnivore in Western Europe, has virtually been exterminated, with only occasional sightings or trappings to remind us of its existence. The North American wolf is now largely confined to the northernmost parts of its original range (Canada, northern Plains states, and especially Alaska). The Eurasian wolf is between 1 and 1.4 m long and weighs up to 75 kg in the case of the male; its pelt varies in color, generally brown with a sprinkling of black and white hairs, and the hair on the underbody is always very light in color. Average life expectancy in nature is about 10 years.

The numerous *C. lupus* subspecies found in Europe, Asia, and North America—some authorities recognize more than 25—may be distinguished by size and coloration (the red wolf of the southern United States, however, is usually considered a separate species, *Canis rufus*). The wolf has adapted itself to a wide variety of climatic conditions, and it appears that the southern extension of its range was originally limited only by the deserts of North Africa, Central Asia, and the American Southwest. Even today, though persecuted to the verge of extinction, *Canis lupus* has proved adaptable enough to maintain itself under the most adverse conditions in most of these different biotopes.

▶ The wolf pack relies primarily on endurance rather than speed, pursuing its prey until it is too exhausted to run any further. The pack's hunting territory varies in size, depending on the local abundance of game; a pack may have to cover a distance of 50 km or so before it encounters a game herd, and the chase may continue for several days. The pack never consists of more than about 20 individuals; wolves mark their territory with their urine, and a neighboring pack will sometimes follow this scent trail in order to form a temporary alliance, which disbands as soon as the hunt is over.

Wolves hunt in order to sustain the needs of the pack as a whole, so that individual wolves are capable of fasting for several days, though after a period of scarcity or a run of bad luck, a wolf is also capable of ingesting as much as 10 kg of meat in a single day. Where possible, the wolf pack preys on large ungulates (elk, reindeer, caribou, various species of deer), which an individual wolf would not be strong enough to bring down by itself. The gray wolf serves a selective function in eliminating the weak or sick members of the game herd. In winter, the wolf is sometimes obliged to feed on carrion, much smaller mammals, and even vegetable material.

The pursuit of a game herd is an organized process, carried out under the direction of a dominant individual (sometimes a female, though more frequently a male), in which each member of the pack plays a particular role. While one or two females stay by the den to look after the cubs, the rest of the pack follows the game herd in silence, keeping its distance until the herd stops to rest. When the herd catches the scent of the wolves and takes to its heels, the pack tries to isolate and surround any members of the herd that are running slowly or with difficulty; the wolves will generally try to knock the quarry off its feet and dispatch it with their powerful jaws—which are strong enough to crack the thighbone of a large cervid with a single bite—most often by severing the spinal cord.

The carcass is immediately parcelled out among the members of the pack. The pack leader tries to fend off the other wolves until it has chosen the choicest portions (the haunch and loin, the liver and heart) and eaten its fill; this prerogative is one that may be disputed by the other members of the pack, and violent squabbles frequently break out as a result. Certain pack members have the responsibility of bringing back food for the cubs and the females that have stayed behind to guard them; these individuals will regurgitate part of their share of the carcass when they return to the den.

When the pack has brought down a very large animal, it may remain at the site of the kill until the carcass has been completely stripped, though a cache of surplus meat will sometimes be left behind, either buried in the ground or concealed beneath a layer of dry leaves. The cubs do not take part in the hunt during their first year, though their games and mock combats play an important part in inculcating the rules by which the life of the pack is governed; the wolf is sexually mature by the age of 3. Older individuals that have lost their teeth and have become infested with parasites are sometimes expelled from the pack; no longer capable of hunting for themselves, they quickly succumb to starvation.

◀ The European wolf, the only European carnivore large enough to bring down a deer, has no competitors other than man. In prehistoric times, the wolf's social organization may have predisposed it to enter into an alliance with man, and at one time, man was similarly disposed to regard the wolf as an ally rather than a competitor. Much more recently, the systematic extermination of the wolf seems to have ceased in certain countries after their governments finally became aware of the vital role played by this species in controlling the population of game herds.

The Australian dingo (q.v.), though believed to be descended from domesticated dogs that reverted to the wild state, is noticeably similar to the wolf in its habits.

P.A.

EUROPEAN BEE-EATER *Merops apiaster* 167

167
Sub.: Vertebrata
Cl.: Aves
O.: Coraciiformes
F.: Meropidae

This Old World bird is 27 cm long, with a wingspread of 38–43 cm, and weighs between 56 and 65 g. It is found in dry, flat, open areas provided there are shrubs and branches on which to perch and a watercourse nearby; it nests in small colonies, in adjoining burrows in a bank of sand or clay at the edge of a road or a river. This species ranges from Southern Europe and the Balkans down through Africa and across southern Russia and Central Asia to the foothills of the Himalayas.

▶ The bee-eater eats not only bees but also a wide variety of other insects including wasps, bumblebees, dragonflies, butterflies, locusts, grasshoppers, and others; its catches its prey in flight, with its beak, and sometimes ingests it in flight as well, though more frequently after returning to its perch. It is a very active hunter, sometimes gliding through the air with its wings outstretched, sometimes flying with short, choppy strokes.

◀ The Montpellier viper and the ocellated lizard may sometimes feed on the nestlings. All the adult members of a bee-eater colony will attempt to mob any intruder or potential predator that approaches its nesting sites. The latter are sometimes destroyed by highway construction and canalization projects, but man has little direct effect on the bee-eater—though like many other birds that are brightly colored or otherwise striking in their appearance (kingfishers, roller, hoopoe, various raptors), the bee-eater is greatly in demand as a taxidermic specimen.

There are 21 other species of bee-eaters, found all across the subtropical regions of the Old World (including Indo-Malaysia and Australasia). All are insectivorous and are otherwise quite similar to the European species.

J.-P. R.

EUROPEAN BLUE DARNER *Aeshna cyanea* 168

168
Sub.: Arthropoda
Cl.: Insecta
O.: Odonata
S.o.: Anisoptera
F.: Aeshnidae

The blue darner, or southern aeschna, is the most conspicuous of large European dragonflies. All of the odonata are commonly referred to as dragonflies, though the term is sometimes reserved for the thick-bodied, strong-flying species, while the smaller, more delicate-bodied species are referred to as damselflies *(demoiselles, Jungfer).* In any case, the blue darner is the one most frequently observed by strollers in the parks and woodlands of Western Europe. The blue darner, like all the aeschnas, may be distinguished from the other odonata by its narrow, elongated cylindrical abdomen and by its relatively narrow wings. The male's body is spotted with blue, brown, and green pigments, the female's with brown and green, an effective camouflage against a background of vegetation. *Aeshna cyanea* is found in North Africa and the Mediterranean basin, all over Europe, and from Asia Minor up into Siberia.

▶ The mandibles of *Aeshna cyanea* are of the grinding or crushing type, and like all the odonata, this species is predatory throughout every stage of its development. The young adults leave the banks of the ponds and streams where they were born to live the life of nomadic and efficient hunters in the nearby fields and clearings, then return to the banks again when it is time for them to mate.

In its larval form, *Aeschna cyanea* is long and torpedo-shaped (fusiform) and lives amid the vegetation in stagnant bodies of water, sometimes on the mat of decaying vegetable matter that is found at the bottom of ponds, slow-moving streams, etc. After the eggs are hatched, the larvae feed on such minuscule prey as rotifers and copepods or the larvae of other aquatic insects (mayflies, chironomids). Various experiments have shown that it is the motion of the prey that triggers the capture reflex on the part of the dragonfly larva; it hunts exclusively by sight. Soon the larvae are able to handle more substantial victims, including other larvae of their own or related species (such behavior is known as homophagia), small freshwater mollusks, water fleas (amphipods), small fish, tadpoles, efts, and the immature forms of other amphibians. *Aeshna cyanea* can also catch prey that swims on the surface or is poised as much as one or two centimeters above the water; this is done by raising its head out of the water and rapidly extending its "mask," an armlike appendage equipped with pincers that is attached to the "chin" of the larva and is accordingly known to entomologists as the "mandibular arm."

The amount of food that is required by the larva, and thus its rate of growth (the length of time between molts), is largely determined by the temperature, but it is interesting to note that even at a given temperature certain indvidual larvae seemed to be "programmed" to grow at a faster rate (and thus require a larger intake) than others. The broad range of these individual variations provides *Aeshna cyanea* with considerable adaptive versatility as a species.

The adults hunt on the wing, like swallows; their diet consists of "aerial plankton"—primarily tiny flies and midges as well as some of the smaller beetles, lacewings, hymenoptera, and moths (Microlepidoptera). *Aeshna cyanea* remains active until nightfall, consuming enormous numbers of mosquitoes at dusk, and can accordingly be considered a species that is beneficial to man; the popular idea that the "darning needle" has a poisonous sting in its tail is entirely mythical, as is the idea that the larvae are destructive to food and game fish (the reverse is more likely to be the case).

◀ In the earlier stages of its life cycle, *Aeshna cyanea* is likely to be its own worst enemy, since homophagia is a prime cause of mortality among the smaller larvae. Later, the larva's most formidable enemies are a group of predatory water bugs (true bugs of the order Hemiptera, especially the Notonectidae, or backswimmers, and the Nepidae, or water scorpions) that can immobilize their prey almost immediately with an injection of posionous saliva, as well as the larval and adult forms of various diving beetles (Dytiscidae). The aeschna larva breathes by taking in dissolved oxygen (and expelling water and carbon dioxide) through its rectum, and at the approach of one of these predators, it may be lucky enough to propel itself out of reach by violently expelling a jet of water through this unconventional breathing apparatus.

Eight other species of the genus *Aeshna* are represented in Europe as well as 3 species belonging to closely related genera *(Boyeria, Brachytron, Anacyaeshna).* The smaller species *(A. affinis* Van der Linden, *A. mixta* Müller, *Brachytron hafniense* Müller) are found at lower altitudes and tend to exhibit various forms of ecological specialization (in comparison with *A. cyanea),* taking to the air very early in the case of the spring darner *(Brachytron),* very late in the case of *A. affinis* Van der Linden. Three species are more prevalent in northern Europe (where the larvae flourish in the acidic water of the peat bogs) and are only present, if at all, in mountainous regions farther south: *A. subarctica, A. coerulea, A.*

virdis. Two other species prefer a somewhat more northerly habitat (or a higher altitude) than *A. cyanea:* the greater darner *(A. grandis)* and the bulrush darner *(A. juncea).* The habitat of the 3 remaining species lies to the south of that of *A. cyanea: Boyeria irene,* which prolongs its hunting flights until nightfall or even after dark, *A. affinis,* and *Anacyaeshna isoceles.*

A. de. R.

EUROPEAN CRAYFISH *Astacus astacus* 169

This species is generally between 10 and 15 cm long, though it may attain a length of 21 cm and weigh as much as 135 g if it succeeds in reaching the age of 20 or 25. The mating season is October and November; the eggs are laid immediately afterward, and it is a fully formed little crayfish, 10–15 mm long, that emerges 6 months later, since the crayfish does not go through a larval stage; it will reach maturity at about the age of 4. *Astacus astacus* is found all over Europe; it prefers calm, tepid fresh water with a muddy bottom and is often found in holes in the bank of a stream or pond.

▶ The crayfish prefers to hunt for food in the dark or in dim light, and thus it is active not only at dusk and after dark but also when the sky is overcast or in turbid water. The young crayfish feed on infusoria, medina worms, nematodes, daphnia, *Cypris,* and other tiny aquatic creatures. The adults, like other large crustaceans, are quite eclectic in their tastes; they seem to prefer freshwater mussels and pulmonate snails, which pose no problem for a slow-moving predator like the crayfish. They also feed on a variety of other creatures that are incapable of rapid flight or of defending themselves vigorously: leeches, gammarids, water fleas, daphnia, insect larvae (caddis flies, craneflies, chironomids, libellulids), and adult aquatic insects (water scorpions, backswimmers), as well as tadpoles and frogs.

The larger crayfish are even capable of catching fish, particularly the relatively slow-moving bottom-dwellers like the tench and the loach and occasionally sick or injured fish of other species. Normally, the crayfish is easily evaded by the fish that populate the clear, open waters of a stream or pond, though it can sometimes catch water rats while swimming on the surface and pull them under to drown them. In short, the crayfish is a fairly indiscriminate feeder with an enormous appetite; it uses its large, strong claws both to capture and kill its prey, with its walking legs helping both to hold the prey in place and to tear it into pieces. The crayfish ingests its food with the aid of its maxillipeds and other mouth parts; it will eat carrion on occasion, but only when live prey is not immediately available.

169
Sub.: Arthropoda
Cl.: Crustacea
O.: Decapoda
S.o.: Macrura
F.: Astacidae

◀ The predators of the European crayfish include most of the large vertebrates that are found in the vicinity of slow-moving streams. The eel, a no less voracious and much more agile predator, is only one of many large freshwater fish that feed on crayfish; herons and other wading birds will eat even the smallest individuals, and it seems only fitting that the crayfish is very vulnerable to attack by the water rat when it is on (or under) the riverbank. Otters are also very fond of crayfish.

In France, crayfish can be legally caught only with a baited hook or with a special crayfish net called a *balance,* and then only in season; European crayfish populations were reduced severely toward the end of the 19th century by a fungal infection known as the "crab plague," and most species have been continuously in decline ever since. Overfishing, water pollution, poaching, and the use of less orthodox fishing methods (including the use of a baited bundle of brushwood, lowered into the water as a kind of homemade lobster pot) have also contributed to the disappearance of *Astacus astacus* from many European rivers, and a large proportion of the crayfish consumed by man are raised in captivity.

Other European species include the mud crayfish *(Astacus leptodactylus),* the torrent crayfish *(Austropotamobius torrentium),* and the "white-footed" crayfish *(Austropotamobius pallipes),* all of which are voracious carnivores, very much in the manner of *Astacus astacus.* North American crayfish (genus *Cambarus* syn. *Orconectes),* on the other hand, are almost exclusively vegetarian, though it has been discovered that during the summer months certain varieties of *Cambarus limosus,* if their more accustomed food is not available, will feed on the zebra mussel *(Dreissena polymorph).* As a result, this species has been introduced into Europe in order to keep hydroelectric equipment from being fouled by river mussels; in Poland, for example, it was determined that an average-sized *C. limosus* (9 cm) could account for as many as 100 young mussels in a day. *Cambarus limosus,* which is immune to the crab plague, is often raised on European crayfish farms. P.N.

EUROPEAN EEL *Anguilla anguilla* 170

170
Sub.: Vertebrata
Cl.: Osteichthyes
S.cl.: Teleostei
O.: Anguilliformes
F.: Anguillidae

The male of this species measures about 50 cm, the female about twice that, with a maximum weight of 3.5 kg. Migratory males live for 7 or 8 years, migratory females for 8 to 10. Eels in aquariums that are prevented from returning to their spawning grounds may live for an additional 20 to 40 years, sometimes attaining a length of 1.5 to 3 m and a weight of 4 to 6 kg.

The eel is said to be a catadromous ("down-traveling") fish, which means that after it has grown to maturity in fresh water, it swims downriver and returns to the open sea to reproduce. The European and North American eel are classed as different species, but both return to spawn in the Sargasso Sea in the middle of the Atlantic Ocean, at a depth of 30 to 200 m in the case of *Anguilla anguilla.* Both male and female die shortly after spawning.

After they emerge from the egg, eel larvae are transparent, almost 2–dimensional little creatures known as leptocephali ("thinheads"), or ghostfish. They spend the next 3 years negotiating the 4000 km of open water that separates the Sargasso Sea from the coast of Europe and begin to swim upriver in the winter or spring; by this time, though still transparent, they will have attained a length of 6–7 cm and are now referred to as elvers. During the next stage in their life cycle, which begins during the following summer, they live in the mud or under rocks on the river bottom and their appearance is considerably changed: the eyes are small, the snout broad and blunt, and the belly turns golden-yellow as the skin takes on pigment (hence, mature freshwater eels that are not ready to embark on their return migration are sometimes referred to as yellow eels).

After 3 more years in the case of the males, 5 to 8 in the case of the females, they will have undergone one final transformation: the eyes grow larger, the head becomes pointed, and the belly takes on a silvery color. It is at about this time that the eel begins its final voyage to its spawning grounds in the Atlantic, the precise details of which remain mysterious.

The European eel is found in fresh or brackish water from Southern Europe to Scandinavia and Iceland. The eels that are found in the Mediterranean basin and the rivers that flow into the Black Sea are considered to be members of the same species, but it is not known whether they also begin and end their lives in the Sargasso Sea.

▶ The eel is a rather sluggish and unaggressive predator, theoretically capable of attaining a speed of 4.5 kph, though 2 kph is its usual limit. In fresh waters, eels locate their prey primarily by smell and by touch; yellow eels consume substantial quantities of mollusks, insect larvae, crustaceans, minnows, sticklebacks, tadpoles, and adult frogs. In brackish water, they feed on saltwater species like the goby, marine worms, crabs, sandfleas, and other crustaceans. Silver eels do not appear to eat anything at all, since the digestive tract starts to atrophy, suggesting that they live exclusively off their body fat until they have reached the spawning grounds.

◀ In salt water, the tiny leptocephali are eaten by herrings, mackerel, and other pelagic (deep-sea) predators, the elvers in fresh water by pike and pike perch *(Lucioperca sandra).* Adult eels are sometimes devoured by the sheatfish, *Siluris glanis,* the largest of all freshwater fish, as well as ospreys and other fish-eating eagles, the European bald buzzard, otters, and badgers.

The European eel fishery also accounts for an annual catch of about 10,000 metric tons, about half of which are taken during their migration through the narrow straits that separate the Baltic from the North Sea. Yellow eels are caught in seines or purse nets (the latter are known in this case as eelpots) or with a special two-pronged eel spear; mature silver eels are caught with the help of purse nets or pens erected in shallow water that are known as "crawls" (from the Dutch *kraal*). Eels can be smoked or salted, and the "jellied" elvers are a traditional delicacy in Southern Europe as well as in Great Britain. In the Far East, elvers are trapped and raised to maturity on commercial fish farms. The annual worldwide output of these aquaculture facilities is about 35,000 metric tons, with Japan accounting for about 15,000 tons and Taiwan for 12,000.

Apart from the fact that at the beginning of its larval migration it catches the easterly rather than westerly currents at the edge of the Sargasso, the American eel (*Anguilla rostrata*) is very similar to the European species in all important respects.

J.-M. R.

EUROPEAN GENET *Genetta genetta* 171

171
Sub.: Vertebrata
Cl.: Mammalia
O.: Carnivora
F.: Viverridae

This catlike relative of the mongoose and the civet has a tail (40–50 cm) that is almost as long as the rest of its body (50–60 cm); it is slight and gracile and only weighs 1.2–2 kg, comparatively little for its size. Its pelt is beige or grayish, sometimes with black stripes. The genet, originally a native of subsaharan Africa, is now found in a small part of North Africa and on the Arabian pensinsula as well as in Southern and Western Europe. It is easily tamed and was a popular domestic pet in late antiquity and the early Middle Ages, and was apparently introduced into much of its present range as a result of human migrations. The genet was brough to Spain by Moorish invaders in the 9th century, and has since proved to be more successful in its conquest of Western Europe, though on a relatively modest scale; a few individuals have recently been observed, for example, in the forests of Belgium and Germany. The genet is not particularly specialized and has adapted well to various terrains; it hunts by night and spends the day curled up in a hollow tree, a crevice in the rock, an abandoned crow's nest or rabbit's burrow, most often in the vicinity of running water.

▶ The genet has excellent night vision, and can perceive immobile forms and objects in almost total darkness; the keenness of its hearing is reinforced by its large external ears (auricular pavilions), and if its sense of smell is inferior to that of the canids, it still surpasses that of the felids or its closer relatives in the family Viverridae. Its long whiskers and the sensitive pads on the soles of its feet enable it to creep securely from branch to branch, balancing with its long tail, and to run silently along the ground.

The genet spends its entire life in the same territory, and its predatory behavior is quickly codified into an immutable routine as it familiarizes itself with the unseen obstacles and other topographical features of its domain; it follows exactly the same itinerary every night. Field experiments revealed that the genet begins with a sort of slow-motion reconnaissance of an unfamiliar environment (a phenomenon that it would probably never encounter in nature). However, it quickly began to move about more naturally and confidently after it had apparently imprinted a detailed visual image of its new territory on its memory—continuing, for example, to "leap over" certain obstacles in the dark, even after these had been removed by the experimenters.

The genet brings itself within range of its prey by stalking like a cat—flattening out its body and crawling slowly along the ground—then throwing itself on its prey and immobilizing it by sinking its canines into the creature's neck. It is an excellent climber, and will sometimes jump as much as a meter to pluck a roosting bird from a branch before falling back to the ground with its prey. It feeds primarily on rodents, however, as well as insects, amphibians, and reptiles—or virtually any edible creature that it encounters in the course of its nightly perambulations and supplements its diet of animal prey with figs, grapes, apples, pears, and other fruit during the summer. In winter, it primarily catches birds in the trees, and in summer, it hunts for rodents in the fields and may occasionally raid a rabbit hutch or a chicken coop as well.

◀ The genet shares its habitat with a number of larger nocturnal predators that it is well advised to avoid—eagles and other raptors, foxes, lynxes, and wild cats. The fox can be evaded by climbing trees, but stealth and concealment is the genet's best defense against the others; at night, the grayish color and fuzzy outline of the genet's pelt makes it extremely difficult to detect. The genet will turn and fight when cornered, growling and snapping and fluffing up its fur to make it seem more formidable than it really is. This seems to be an effective deterrent against many potential predators, and without these various means of defense, the genet would probably have long since been exterminated in Europe. It is still sometimes persecuted by man in retaliation for its attacks on rabbits and poultry, though these are undoubtedly more than than compensated for by the genet's services as an agent of rodent control. P.A.

172
Sub.: Arthropoda
Cl.: Crustacea
O.: Decapoda
S.o.: Macrura
F.: Nephropidae

EUROPEAN LOBSTER *Homarus gammarus* 172

A full-grown male lobster may attain a length of 60 cm and weigh as much as 12 kg; the females are somewhat smaller, and the largest individuals are thought to be at least 50 years old. The European lobster is found all along the Atlantic coast, as far south as Morocco, and throughout the Mediterranean; it is common in rocky areas out to a depth of about 60 m. The lobster is nocturnal and takes shelter during the day in a crevice or a cavity in the rock; it crawls into its burrow tail first so its claws will be available to secure the entrance against intruders.

▶ The planktonic larvae feed on coastal zooplankton; the juveniles and the adults specialize in sedentary or slow-moving vagile prey, including mollusks and bivalves, crustaceans (isopods and decapods), polychaetes, hydrozoans, and echinoderms. The lobster is almost exclusively carnivorous and, unlike other large crustaceans, it eats very little seaweed. It seeks its prey by clumping along the bottom, gradually correcting its course in accordance with a series of olfactory cues; it is also aware of motion, so the sense of sight apparently plays some role in prey detection.

The lobster seizes its prey with its forefeet and its two large claws, then maneuvers it into place with its anterior walking legs (equipped with miniature pincers) and its outer mouthparts. Smaller shells can be dealt with by its grinding mouthparts, but the larger ones are broken up first by grasping them in the smaller claw and cracking with the larger. Unlike the spiny lobster, which swallows everything indiscriminately, the lobster ingests only the edible portions of its victims and daintily spits out the shell fragments and other debris.

◀ The adult lobster is well protected in its rockbound burrow, as well as by its powerful claws and thick carapace. It is most vulnerable when it molts its shell, but this occurs only about once a year, at least in the case of the larger individuals. The octopus, the conger, and the moray eel, which share its fondness for caves and crevices, are the European lobster's principal predators, along with the rock crabs *Cancer pagarus* and *Galathea,* which are somewhat similar in their habits and are also direct competitors for the lobster's food supply. Bottom-feeding fish like the cod and the angler will occasionally prey on juvenile lobsters.

Lobsters sometimes fight with one another for the possession of a burrow, and cannibalism is fairly prevalent; it is interesting to note, however, that during the breeding season the female releases a pheromone into the water that inhibits these cannibalistic tendencies. The extensive European lobster fishery is conducted partly as a fishing and partly as a farming operation—particularly with respect to the methods by which stocks that have been depleted by overfishing are replenished in certain areas.

The American lobster *(Homarus americanus)* may weigh as much as 16 kg and has basically the same diet as the European species; the Cape lobster *(Homarus capensis),* smaller and less common than either of these varieties, is found off the Cape of Good Hope. The Norway lobster, or langoustine *(Nephrops norvegicus),* is a member of the same family. It is found along the Atlantic coasts of Europe, from Iceland to Morocco, and is present, though not prevalent, in the Mediterranean as well. It burrows into the muddy bottom at depths between 50 and 500m and hunts by day as well as by night. 78 percent of its diet consists of vagile organisms such as crustaceans (benthic or bathypelagic shrimp, rock crabs and other crustaceans, as well as euphausids, opossum shrimp [Mysicdacea], and copepods), polychaetes (Aphroditidae, Phyllocodicae, Opheliidae, Sabellidae, Ampharetidae), and brittle stars. There are a number of bottom-feeding fish and large benthic crustaceans that feed on the Norway lobster, and its succulent flesh is much sought after by man as an ingredient in paella, bouillabaisse, and other seafood dishes. P.N.

EUROPEAN MINK *Mustela lutreola* 173

173
Sub.: Vertebrata
Cl.: Mammalia
O.: Carnivora
F.: Mustelidae

The adult's head and body typically measure 36 cm and the tail an additional 18 cm; average weight is about 1.5 kg, the males being slightly larger than the females. The pelt of the wild mink is chestnut-colored with a white border around the mouth, its legs are relatively short, and its feet partially webbed. It is found along the banks of watercourses, from the smallest streams to the broadest lakes and drivers, provided there is sufficient vegetation to screen the entrance to its den.

The range of the European mink has been steadily contracting over the past decades, and it must be considered one of the most seriously threatened European carnivores. At present it is found mainly in the southwest of France and in European Russia, ranging as far east as Siberia, and has already disappeared from most of the intervening territories (excluding parts of Finland and the Balkans). The European mink generally hunts at dusk, though it is also sometimes seen in daylight. The female bears a litter of 2–6 young every spring; the kittens will be sexually mature by the winter of their first year. Longevity in captivity is on the order of ten years.

▶ The European mink feeds on fish, amphibians, small mammals, mollusks, crabs, and other invertebrates, which it catches either in the water or along the bank. It is a good swimmer and diver, and it catches fish by corraling them in a tangle of submerged tree roots during the day or by creeping up on them stealthily while they are resting at night. The mink kills its prey with a bite to the head or the neck and carries it back to its den, where it stores the uneaten surplus in an underground larder. A Soviet naturalist reports that he found one or two dead frogs with deep bite wounds on their necks in the entrance to a mink's burrow every day, and a cache of 20 loach found buried about 50 cm under the roots of tree on a Siberian riverbank was also presumed to have been left there by a mink; a similar cache contained a pike and 3 good-sized orfes (*Leciscus idus*, an Eastern European relative of the chub and the dace). The mink has keen eyesight and good color vision (or at any rate, it has been shown that it can distinguish perfectly between red and green, though it has some difficulty with blue and yellow); the acuity of its other senses does not appear to have been tested.

◀ No long-term observations have been made of this species in the wild, and we have no definite information about its predators or competitors. Medium-sized carnivores such as the fox and the beech marten can presumably be included among the former; the predators of the kittens, though presumably more numerous, are unknown. If a European mink catches sight of an enemy while it is the water, it dives beneath the surface and bobs back up in a diferent spot some time later, though it cannot remain submerged for as long as an otter. Its den has an underwater entrance, but if this escape route is blocked, it will go to cover on the bank and may climb up into a bush or thicket on shore and often appears reluctant to back down to the ground again. Minks that are chased by dogs on land have been known to take to the trees as well.

The polecat, which feeds on frogs, and the otter, which prefers fish, are both potential competitors of the European mink, though the North American mink (*M. vison*) is probably a more serious competitor. American ranch minks were first brought to France in 1926 and are now believed to be at least as numerous in the wild in many parts of Europe. A recent study conducted in Sweden has shown that choice of prey and hunting behavior is essentially the same for both species.

The European mink is rarely hunted for its pelt, which is not of very good quality; poorly dressed skins were sometimes surreptitiously offered for sale at country fairs in former days (since this species is fully protected in Western Europe), but it seems likely that loss of habitat (water pollution, the removal of scrub vegetation from the banks of rivers and lakes) rather than this sporadic poaching must be held responsible for the dramatic decline of this species in recent years. M.-C. S.G.

174
Sub.: Vertebrata
Cl.: Aves
O.: Caprimulgiformes
F.: Caprimulgidae

EUROPEAN NIGHTJAR *Caprimulgus europeus* 174

The European nightjar, or goatsucker, is found in dry, sunny, open spaces (lawns, meadows, and forest clearings) in Western Europe, North Africa, and Asia; it spends the fall and winter in subsaharan Africa, arriving in Western Europe in April or May and flying south again in August. The nightjar is about 24 cm long with a wingspan of 55 cm and weighs about 80 g. Maximum longevity in the wild is probably 4 years.

▶ The nightjar is active at dusk and after dark; as with owls and other nocturnal birds of prey, its soft plumage muffles the sound of its wingbeats. Gliding or hovering, swooping and banking sharply, then plummeting abruptly to the ground, the nightjar is a breathtaking and unpredictable aviator. During the day, its plumage, which is speckled and earth-colored, also enables it to conceal itself quite effectively simply by lying motionless on the ground. The nightjar feeds exclusively on insects, which it catches in the air and on the ground. Contrary to popular belief, however (as reflected in the French name for this species, *engoulevent,* or "wind-gobbler"), the nightjar does not keep its mouth open to catch insects in flight; it opens its short beak only as long as is necessary to snatch a dragonfly, a moth, or a cranefly out of the air. It also hunts for crickets, grasshoppers, and dung beetles on the ground and often stations itself at the edge of a road or track to ambush flying insects.

The nightjar's common name in several European languages is the equivalent of the Latin genus name, *Caprimulges,* "goatsucker" (in English, this term is collectively applied to all the members of the family Caprimulgidae), and the peculiar idea that the goatsucker feeds on goat's milk seems to have been originated, or at least made respectable, by Aristotle. It was perhaps originally prompted by the bird's extremely wide mouth and the fact that it is often observed in meadows and pastures, and thus could easily give the impression that it was primarily interested in the goats themselves rather than in the flies and other insects that they attract—and in fact was surreptitiously making off with the goat's milk during the night when no one was looking. Rather than on goat's milk, then, the nestlings are fed on regurgitated insects from their parents' crops.

◀ The goshawk sometimes ambushes the goatsucker when it takes to the air at nightfall, and the female on the nest or the male or female at rest on the ground, no matter how carefully camouflaged, are occasionally caught by foxes. The nestlings are sometimes devoured by predators as well, though a sitting bird will attempt to decoy an intruder away from the nest by darting off while flapping its wings and noisily fluttering its feathers. The goatsucker is not directly affected by man, though its habitat may be affected by changes in land use (by the "improvement" of moors and meadowlands, for example), by the use of pesticides, and by automobile traffic, which takes a heavy toll of many nocturnal birds.

The Caprimulgiformes are essentialy quite similar in their habits and apearance, though they are classified into 3 different families and 70 species altogether. The Caprimulgidae, or goatsuckers, are found all over the world, except in Polynesia. Several species in tropical Africa (the standard-winged nightjar, *Macrodipteryx longipennis,* and the pennant-winged nightjar, *Semeiophorus vexillarius)* and America have developed very long tailfeathers or primaries, which enables them to perform spectacular aerial maneuvers. The whippoorwill *(Caprimulgus vociferus),* though heard more often than seen, is undoubtedly the best known member of the family in North America, though the white-throated poor-will *(Phalaenoptilus nuttallii)* is also notable as the first bird species in which hibernation was ever observed.

The frogmouths (Podargidae, *Podarges)* are a one-genus family of rather sedentary forest dwellers, found in Southern China, Indonesia and Malaysia, New Guinea, and Australia. The bizarre South American oilbird, or guacharo, is a family unto itself (Steatornithidae). Found from the Guianas to Peru and on the island of Trinidad, it resembles a bird of prey in appearance, with its hooked beak and flattened head. It will fly as far as 40 km in search of fruit, especially that of the palm tree (though it spits out the seeds), and in the darkened caves in which it builds its nest, it navigates by echolocation, like a bat; as this suggests, the oilbird is an agile flier, remarkably quick at making course corrections. The oilbird's nests are routinely robbed by the local Indians just as the fledglings are about to take to the air, and the rendered fat of these plump little creatures is said to make an excellent cooking oil. J.P.R.

EUROPEAN NUTHATCH *Sitta europea* 175

175
Sub.: Vertebrata
Cl.: Aves
O.: Passeriformes
F.: Sittidae

This Eurasian songbird measures 13–13.5 cm in length, with a wingspread of 26–27 cm, and weighs between 20 and 25 g. It builds its nest in a woodpecker hole or a natural cavity in a hollow tree and is found in lightly wooded areas (including parks and gardens) in the mountains as well as on lowland plains, though it avoids coniferous forests; the adults are sedentary, the juveniles are fairly nomadic (though not migratory). The range of this species extends in a broad corridor from Spain and Western Europe across Central Europe and Siberia to the shores of the North Pacific.

Like other nuthatches, *Sitta europea* makes its nesting cavity inaccessible to larger birds by reducing the diameter of the entrance hole, which it does by sticking little balls of mud or clay around the edges, then pressing them flat by using its beak as a trowel—which is why the European nuthatch is known as *torchepot* ("pot-scrubber") in French-speaking countries and *Kleiber* ("dauber") in German-speaking ones. Like the woodpeckers and tree creepers *(Certhia)*, the nuthatches are adept at walking up and down treetrunks—moving with their heads pointed forward, with equal facility in either direction—and along the undersides of branches. Since their feet are highly adapted for gripping, they are not very comfortable on the ground. Flights are generally limited to short hops between trees, and when the European nuthatch is obliged to fly across a patch of open country, its flight describes a shallow wave pattern, like that of the woodpeckers.

▶ The European nuthatch feeds on a variety of insects and their larvae, which it snaps up with its beak as it encounters them on treetrunks or on branches. It also feeds on spiders and snails as well as on acorns, beechnuts, hazelnuts, and other nuts and seeds, in the latter case especially in the winter when insect prey is scare.

◀ This species has no particular predators or competitors and is essentially unaffected by human beings and their activities.

There are almost 20 nuthatch species worldwide, of which those found in Australia and New Guinea are assigned to the genus *Neositta*, the remainder to the genus *Sitta*. Most are arboreal, though the rock nuthatch *(S. neumeyer)*, which ranges from the Balkans to Central Asia, and several other species have learned to clamber around as effortlessly on cliffsides and rock faces as do their congeners on treetrunks and branches.

This group also includes the last two species of Mediterranean birds to be identified, namely the Corsican nuthatch *(S. whiteheadi)*, confined to a few ancient mountain forests on the island, and the Kabyle nuthatch *(S. ledanti)*, a relict species found on a single mountain plateau in central Algeria; these two species were not discovered until 1884 and 1975, respectively. There are 3, possibly 4 *Sitta* species in North America, including the white-breasted nuthatch *(S. carolinensis)* and the red-breastednuthatch*(S.canadensis)*.

J.-P. R.

EUROPEAN POND TORTOISE *Emys orbicularis* 176

176
Sub.: Vertebrata
Cl.: Reptilia
O.: Chelonia
S.o.: Cryptodira
F.: Emydidae

This is a palustrine tortoise, which is to say that it inhabits marshy areas, including ponds, drainage ditches, bogs, swamps, or essentially any enclosed body of stagnant water where fish are abundant and that is bordered by reeds or rushes, remote from any human activity, and convenient to a patch of sandy soil in which the female can bury its eggs. The upper shell (carapace) of both sexes measures 16–18 cm; according to Rollinat (1934), the weight of the European pond tortoise may vary between 700 and 1000 g. Longevity is thought to exceed 70 years. This species is found in Central Europe and the Mediterranean basin (i.e., from northern Germany to North Africa and in France roughly south of a line extending from the mouth of the Loire to Lake Geneva) as well as in Western Asia.

▶ The European pond tortoise is a voracious and indiscriminate feeder, and will attack virtually all aquatic invertebrates as well as the smaller vertebrates, including crustaceans, mollusks, insects and their larvae, tadpoles, particularly slow-moving or disabled fish. After a heavy rain, it will also crawl up on land to catch earthworms, slugs, and snails. Rollinat reports that while the pond tortoise is incapable of catching larger birds, it occasionally catches ducklings and moorhen chicks by dragging them beneath the surface and drowning them.

Larger prey animals are invariably dragged back into the water before being eaten; the tortoise uses its sharp beak (and if necessary, its clawed front feet) to divide the prey into more manageable segments. It is curious to note that while it sometimes catches adult amphibians and picks them apart with its beak, it refuses to eat their flesh, presumably on account of the acrid venom contained in the cutaneous glands of these creatures. Although exclusively carnivorous in principle, the European pond tortoise sometimes inadvertently crops the leaves of aquatic plants that have fish or frog eggs adhering to them. It will also feed on fresh carrion, though only if the animal is question is rather small and quite recently deceased.

The pond tortoise emerges from hibernation in March but does not begin to feed again until the weather gets a little milder (having sustained in the interim a minimal weight loss of only 10–15 g). It eats a great deal from May to July (grounded junebugs and other coleoptera are a favorite prey on warm summer evenings), then begins to lose its appetite during the height of summer and stops hunting altogether in August or September. By then it will have gotten sufficiently plump to live off its fat reserves until the nest spring. Young pond tortoises feed primarily on tadpoles, small insects, and earthworms.

Rollinat states that the European pond tortoise is a highly competitive predator. For example, if one starts back into the water with a prey animal in its beak and it sees the heads of other tortoises popping up above the surface, it does an immediate about-face and returns to land. If one of its rivals is able to seize hold of the prey before it can leave the water, there may be a confrontation in which one or more of the combattants is bitten about the head and perhaps eventually a tug-of-war that results in both parties swimming off in opposite directions clutching a substantial fragment of the original prey.

◀ In early life, when its shell is still quite soft, the European pond tortoise is subject to predation by crows, magpies, and a variety of wading and web-footed birds as well as foxes, otters, and mustelids. Rollinat also mentions an instance in which a newly hatched tortoise fell prey to the powerful pincers of a large rove beetle *(Staphilinus)*. The European pond tortoise has been gathered or raised for food by man since prehistoric times; perhaps because of its great longevity, its shell was also employed in the funerary rites of several different cultures, and more recently the living animal has been recruited for pharmacological and other experimental purposes (Parent, 1981). The shrinkage of its habitat, however, as a result of drainage, land clearance, and pollution undoubtedly poses a more serious threat to the future of this species. Plans are currently afoot to reintroduce the European pond tortoise over part of its former range in France.

The family Emydidae, comprising 27 genera and 76 species altogether, is represented in all parts of the world except the polar regions. The semiaquatic Eurasian wood tortoise *Mauremys caspica*, subspecies of which are found in the Caucasus and Central Asia, North Africa, Turkey, and the Balkans, is competitive with the European pond tortoise in many areas.

J.F.

177
Sub.: Vertebrata
Cl.: Aves
O.: Falconiformes
F.: Accipitridae

EUROPEAN SPARROW HAWK
Accipter nisus **177**

This Old World hawk measures between 31 and 38 cm from beak to tail with a wingspan of 60–80 cm; the female (210–280 g) is quite a bit heavier than the male

(110–170 g). It is found throughout Europe and Asia in regions where woodlands (ideally, a fairly dense deciduous forest interspersed with stands of conifers) alternate with cultivated fields and meadows, hedgerows, brush and shrubs; only the northern populations are migratory. The mortality of the nestlings and young birds is very high (as much as 60–80 percent during the first year of life); banded birds have been known to reach the age of 11, but the average life expectancy is probably no more than a few years.

▶ This shy, solitary bird of prey with its short wings and long tail is well adapted for maneuvering through dense cover. Analysis of stomach contents revealed that out of several tens of thousands of prey animals consumed by sparrow hawks in Europe, 126 bird species could be identified, though of these, only 20 different species accounted for 80 percent of the total. The smaller male feeds primarily on sparrows, buntings, finches, and larks whereas the female is capable of bringing down slightly larger prey—blackbirds, thrushes, starlings, jays, and magpies. Small mammals (fieldmice, voles, moles, and shrews) accounted for less than 3 percent of the total.

The sparrow hawk may attack from cover, weaving nimbly through the leaves and branches of a tree in order to intercept its quarry in the open air; it sometimes patrols along the tops of hedges, "hedgehopping" frequently from side to side. It also may strike its prey on the ground and can readily retrieve a small bird or mammal from tall grass or other vegetation with its long slender talons armed with tapering claws. It kills the prey by means of spasmodic contractions of its talons, then plucks it and flies back to cover to finish its meal. The fledglings will attack small passerine birds as soon as they are able to fly on their own, but will continue to return to the nest for handouts from their parents until they are fully capable of fending for themselves.

◀ The hunting territory of a pair of European sparrow hawks is rarely less than 10 sq km, and when not engaged in rearing their young, the male and female are solitary. The differential in size between the male and female sparrow hawk and the male and female goshawk enables all four to occupy different ecological niches in the same habitat; competition between the sexes and the two species is very limited. Other raptors that feed on the same passerine small birds as the sparrow hawk (e.g., the falcon, the merlin, and the hobby) prefer to hunt in open country.

In the past, the sparrow hawk has been avidly persecuted by human predators, in the mistaken belief that it is destructive to songbirds and small game, whereas all systematic studies have shown that its presence has virtually no effect on the passerine population in a given locality. Since the sparrow hawk consumes no mammalian prey larger than a mole or a fieldmouse (and usually small ones at that), it can be regarded as totally innocent of the second count of this indictment. The sparrow hawk is now a protected species, and at least some local populations are beginning to recover from the devastating effects of this persecution and from the infiltration of toxic chemicals (chlorinated hydrocarbons, PCBs) into the environment over the past few decades. The clearing of hedges and woodlots in rural areas, on the other hand, has contributed to the disappearance of the sparrow hawk by depriving it of both its habitat and its prey.

The accipitrine hawks, very similar to one another in their habits and appearance, comprise about 50 different species worldwide. In North America, "sparrow hawk" is a the common name of *Falco sparvarius,* which is more closely related to the kestrel, and the sharp-shinned hawk is the North American equivalent of the European sparrow hawk. The goshawk *(Accipter gentilis),* also found in European forests and woodlands, is basically a slightly larger version of the sparrow hawk.

J.-F.T.

EUROPEAN TARANTULA *Lycosa narboensis* 178

178
Sub.: Arthropoda
Cl.: Arachnida
O.: Araneida
F.: Lycosidae

This large wolf spider measures about 2 cm in length and, along with the other members of its family, is referred to as a European tarantula in order to distinguish it from the tarantulas of tropical and subtropical America, a suborder (Orthognatha) of large hairy long-legged primitive spiders that can measure over 25 cm in diameter. *Lycosa narboensis* is found in heath, garrigue, and similar terrain in southwest France; it digs a burrow about 20–40 cm deep in the grass or in bare earth and can survive at altitudes of 700–800 m. Though not especially well disposed toward others its own kind, this species sometimes lives in small colonies, the distance between the individual burrows being determined by the age and size of the individual and the local abundance of prey.

L. narboensis generally reaches sexual maturity in the early summer. After mating, the male abandons its burrow and leads a wandering life for the month or so that remains to it, though the females often survive for 2 or 3 seasons. Their 200 or 300 eggs are encased in a silken cocoon that is attached to the walls of the burrow with silken strands, and during the day, the cocoon is hauled up to the mouth of the burrow to be incubated in the sunlight. The hatchlings climb up on their mother's back, where they remain until their first molt.

▶ *L. narboensis* generally hunts by night and rarely ventures more than 2 or 3 m from its burrow. Insect prey is generally quite abundant in the thick, low-lying vegetation in which this species makes its home, especially betwen spring and autumn, and *L. narboensis* feeds primarily on butterflies and caterpillars as well as locusts and their larvae. The early-stage pulli (larvae) can be sustained in the terrarium on a diet of fruit flies *(Drosophila)* and blowflies *(Calliphora)*; the older larvae and the the adults can be fed on locusts and their larvae.

Ordinarily rather slow-moving, *L. narboensis* can still make a rapid dash toward its quarry as soon as it catches sight of it; it then seizes the quarry with its forelegs and picks it up and devours it with its scissorlike mouthparts, or cheliceres. As long as fresh water is available, *L. narboensis* can go without food for several weeks, though cannibalism is quite frequent among the younger residents of neighboring burrows, and the competition for food and territory can be quite severe.

◀ At all stages of its life cycle, *L. narboensis* is a favorite quarry of the pompilid wasp, which drags the paralyzed spider off to its burrow to feed its larvae. The cocoons are sometimes parasitized by haplostomate flies and chalcid wasps, which may result in the destruction of the eggs.

A number of congeners, which are also referred to as tarantulas, are found in Italy and Spain. As late as the 17th century, the bite of these spiders was once thought to be responsible for a febrile disorder called tarantism, which could only be cured if the patient was allowed literally to dance until he dropped from exhaustion (and which is said to be the origin of the rather frenetic Sicilian folkdance called the tarantella). However, since the venom of these spiders has been shown to have no profound neurological effect on humans other than a painful swelling around the area of the bite, it has been suggested by modern scholars that this celebrated "dancing plague" may have been devised as a sort of medical fiction that made it possible for certain pagan festivals to continue to be celebrated throughout the Middle Ages without arousing the wrath of the Inquisition. The words *tarantula, tarantism,* and *tarantella* are all derived from the name of the city of Taranto, in southern Italy, where tarantism was said to be particularly prevalent. J.-F. C.

EUROPEAN WILD CAT *Felis silvestris* 179

179
Sub.: Vertebrata
Cl.: Mammalia
O.: Carnivora
F.: Felidae

The European wild cat looks very much like a slightly larger, shaggier, and fiercer version of the ordinary housecat, which is now thought to be descended from a related North African wild cat, the Kaffir cat *(Felis caffra* syn. *libyca). Silvestris* may attain a length of 80 cm; its tail is thicker toward the tip than at its base and measures an additional 30 cm. Its yellowish coat is thick, and it has black markings around its eyes and a black stripe running along its back.

Silvestris ranges throughout Europe as far east as the Urals and is well suited to a cold continental climate. It is found in wooded areas where ground cover is plentiful and in the adjoining clearings; it generally avoids heath or moorland, though at least some of the wild cats of Scotland seem perforce to have adapted to this terrain, and is particularly fond of thorny brushlands and the edges of forested highland plateaus as long as there is a stream, spring, or other running water in the vicinity.

The wild cat marks its territory in the same way as its domestic cousin, by shredding the bark of trees with its claws, thereby depositing a scented secretion of the sweat glands between its toes, or rubbing against a treetrunk and marking the site with a rather more pungent secretion of the scent glands situated around the anus. The wild cat makes its den in an abandoned fox's earth or badger's sett, underneath a treestump, in a disused quarry site, or at the base of a rocky ledge or of a slope covered with briars.

▶ The European wild cat feeds primarily on the voles (including water voles) and other rodents that it catches at the edge of forest clearings or by the banks of springs or streams. It also occasionally catches rabbits, birds, and large insects; an analysis of stomach contents suggests that rodents constitute about 92% of the diet of the average wild cat. Population studies in the Herbeumont region of southeastern Belgium have shown that wild cats are present in relatively large numbers in the vicinity of trout spawning grounds, which suggests that *silvestris* knows how to fish in shallow streams. The supposition that it also catches muskrats (originally introduced from North America and now quite common in Western Europe) has yet to be confirmed by direct observation. The wild cat normally hunts around the middle of the day and employs the usual feline strategy of stalking, pouncing, and dispatching its prey with a swift bite to the neck.

◀ The European lynx is a ferocious competitor of the wild cat, and a chance encounter between them is certain to end either in the death or the hurried flight of the smaller cat. Weasels also sometimes prey on the kittens, which do not reach maturity until the age of 3. The wild cat has frequently been persecuted by rural sportsmen in many parts of Europe as a supposed destroyer of rabbits and other small game, though in fact in areas (as in France) where it enjoys the protection of the law, it has been shown to be very useful in mitigating the effects of the "boom and crash" population cycles of voles and other rodents, which often have unfortunate consequences for agriculture. The European wild cat is frequently and understandably confused with the larger specimens of feral cat, which is simply a domestic cat *(Felis cattus)* that has reverted to the wild.

P.A.

EYED ELECTRIC RAY *Torpedo torpedo* syn. *ocellata* 180

180
Sub.: Vertebrata
Cl.: Chondrichthyes
S.cl.: Hypotremes
O.: Torpediniformes
F.: Torpedinidae

This species, so called because of the spotted markings on its back, generally measures about 60 cm in length; it is a sedentary cartilaginous fish, found along the continental shelf in the Eastern Atlantic, from Angola to the Bay of Biscay, as well as in the Mediterranean. The members of the family Torpedinidae are among the very few marine fish with differentiated electric organs; the others include the stargazers *(Astrocopus)* and 4 species of electric skate *(Raja)*. In the electric rays, there are two such organs, composed of modified muscle tissue and occupying a considerable amount of space inside the "disk" directly behind the head. These organs are so bulky, in fact, that the electric ray can scarcely use its pectoral fins for swimming; like other rays, it spends much of its time buried in the sand or mud in shallow water, though it sometimes is found out to a depth of several hundred meters.

▶ Each of the ray's symmetrical electric organs consists, in effect, of a voltaic pile containing as many as 500,000 separate electric plates arranged in adjoining columns. With this apparatus, the electric ray can generate a potential difference of 45 volts and a current of 4 5 amperes (which flows from the ventral side of the disk to the dorsal side), with occasional spikes of 60–80 volts; successive discharges gradually lessen in intensity, and a brief respite is required for the batteries to be recharged. Though the voltage is considerably less than that produced by the electric organs of the ELECTRIC EEL or the ELECTRIC CATFISH, the resistance of the medium (salt water) is quite a bit less in this case, with the result that the amperage—hence the ultimate impact on the prey animal—is fully comparable.

The eyed electric ray is nocturnal and feeds on a wide variety of small fish as well as mollusks and crustaceans. It catches its prey by flopping down on top of it and trapping it beneath its pectoral fins, then stunning it with a jolt of electricity. As with most rays, its mouth is ventrally located, so the prey animal need not be released from this trap before it its ingested. Prey animals of substantial size have been recovered from the stomachs of eyed electric rays, including eels weighing 1 kg and salmon weighing 1–2 kg; a larger congener, the black electric ray *(T. nobiliana)*, habitually feeds on bony fish of substantial size (mullet, plaice, flounder) as well as on dogfish and other small sharks.

◀ The eyed electric ray also uses its electric organs to ward off predator attacks, though these are likely to be inadvertent since this species has few natural enemies apart from the moray and the conger eel, which are sometimes able to circumvent its defenses by seizing it by the tail. The eyed electric ray is not fished commercially, though it is frequently collected for laboratory study; scientific experimentation involving this species can be said to have begun with the Romans, who treated migraine, gout, and a variety of other ailments with what, after all, was virtually the only sort of electrotherapy at their disposal.

The black electric ray, mentioned earlier, measures up to 1.8 m, can weigh 90 kg, and generates an electric discharge of 90 volts; though not as common as the eyed electric ray, it plays an important part in the benthic foodchain. The marbled electric ray *(T. marmorata)* is also found in the same waters and is intermediate in size between the two, measuring about 1 m.

F.M.

F

181
Sub.: Vertebrata
Cl.: Mammalia
O.: Carnivora
F.: Canidae

FENNEC *Fennecus zerda* 181

This small desert fox is perhaps most remarkable for its enormous ears, which can measure as much as 15 cm from base to tip, though the length of its body rarely exceeds 35–40 cm (not counting the tail, which contributes an additional 15–30 cm). The fennec's coat is yellowish-tan heightened with brown around the mask; its thick fur insulates it from climatic extremes, and its ears, well supplied with blood vessels, act as a sort of radiator, enabling it to regulate its body temperature by dissipating excess heat. The fennec ranges throughout the desert regions of North Africa and the Arabian peninsula; it digs a burrow for itself in the sand, generally consisting of a large underground chamber housing as many as 10 individuals, who seem to get along quite comfortably without displaying the least sign of aggression or irritability. Longevity (in captivity) is about10 years.

▶ In ascending order of preference, the fennec feeds on grasshoppers, gerbils, skinks, and birds. In semidesert regions, it includes some plant food in its diet as well and is accustomed to going for long periods without drinking water. It can survive almost indefinitely on the moisture contained in the bodies of its prey (a gerbil, for example, consists of about 80 percent water, and beetles and grasshoppers are also rather juicy, as insects go), though of course it will drink water from a pool or a wadi when circumstances permit.

The fennec spends the hottest part of the day in its burrow and patrols the dunes by night, locating its prey with the aid of its sense of smell and exquisitely sensitive hearing. The fennec's middle ear, directly outside the eardrum, has been modified into a sort of resonating chamber or, in effect, an amplifier. It can dig down into the sand very quickly to locate the source of any suggestive noise, and if the burrowing rodent or reptile evades its first approach, the fennec will often retrieve its supper by pinning it to the sand with a single agile bound.

◀ The fennec is sometimes hunted by man, for sport as much as for the pot, but it is not specifically preyed on by any other creature.

Along with the fennec, the caama *(Vulpes chama)* of southern Africa, Rueppell's sand fox *(V. rueppelli)* of the Sahara, and the kit fox *(V. macrotis)* of the American Southwest all display similar adaptations to desert life and are collectively known as sand foxes. P.A.

FISHING CAT *Felis viverrinus* 182

182
Sub.: Vertebrata
Cl.: Mammalia
O.: Carnivora
F.: Felidae

The fishing cat of India, Southeast Asia, Java, and Sumatra resembles the ocelot in size (about 70–80 cm plus another 30 cm for the tail) and coloration. Its coat is darker on its back than on its flanks with its black ocelotlike markings merging into stripes on its head and neck. The fishing cat is found along the banks of watercourses in a variety of different habitats, including marshlands, brushlands, and forests.

▶ The fishing cat is not as specialized in its dietary preferences as its name implies. It feeds on small mammals that burrow in or come down to drink on the banks of rivers, as well as on birds, fish, frogs, mollusks, and even insects; some of its mammalian prey may be nearly as large as itself. It is a solitary hunter for the most part and makes its den in a hole in the ground or a crevice in the rocks near a riverbank. It marks its territory by scratching the bark of trees. It hunts at dusk, stalking and seizing its prey in its powerful claws, pinning it to the ground, and dispatching it with its teeth in the manner of most hunting cats. The fishing cat is an agile climber and, more remarkably, a very good swimmer. It is not clear whether it ever takes to the water in pursuit of prey; it catches fish in much the same way as a bear—stationing itself on the bank, knocking the fish out of water with a quick lateral swipe of its paw, and devouring it on the spot.

◀ The fishing cat may occasionally come to grief as a result of an accidental encounter with a more powerful hunting cat of its own or some other species, though the destruction of its habitat by man poses a much greater threat to its survival.

There are a number of other small wild cats of India and Southeast Asia, including the flat-headed cat *(Felis planiceps)* of Sumatra and Borneo, which sometimes catches frogs and fish (according to some authorities) but is primarily vegetarian in its habits and appears to prefer the taste of sweet potatoes to fish or game. The leopard cat *(Felis bengalensis)* of India, the golden cat *(Felis temmincki)* of China and Southeast Asia, and the marbled cat *(Felis marmorata)* of Malaysia are all thought to feed primarily on birds and small mammals. The fishing cat, flat-headed cat (also known as the rusty tiger cat), and leopard cat are placed by some taxonomists in a separate genus, *Prionailurus.* P.A.

FLYING DRAGON *Draco volans* 183

183
Sub.: Vertebrata
Cl.: Reptilia
O.: Squamata
S.o.: Sauria
F.: Agamidae

This lizard is about 20 cm long; its tail comprises about three fifths of its body length. The dragon's ribs are greatly elongated, and its "wings" consist of two membranes stretched between the ribs and normally kept folded up against the sides of its body. When deployed for flight, their bright pigmentation (generally red or yellow) is revealed, in contrast to the utilitarian drabness of the rest of the dragon's body.

The flying dragon is found in the Philippines, Malaysia, and Indonesia (as far as the Moluccas); it lives in tropical rainfo rests or, where these are no longer in existence, in groves of tall trees in populated areas, notably rubber plantations and the like. It prefers the open, sunny areas about midway up the trunk of a tall tree where there is comparatively little foliage to interfere with its flight from branch to branch or from trunk to trunk. Average longevity is probably no more than a couple of years.

▶ The flying dragon is insectivorous, feeding primarily on the ants that it encounters as it climbs up and down the treetrunks; when the possiblities of one little patch of bark have been exhausted, it spreads its "wings" and glides to a new feeding ground, either on a different branch of the same tree, to a spot several meters further down the same treetrunk, or to a different tree altogether. The dragon is capable of short flights of 5 or 10 m, sometimes more, with a corresponding loss in altitude of 3 or 4 m. By changing the attitude of its "wings," the dragon controls the angle of its descent, which varies between about 20 and 30° below the horizontal, and is able to cushion its landing by leveling off a bit just before the moment of impact. The flying dragon is very active during in the morning hours, somewhat less so during the afternoon, and at dusk it climbs up to seek refuge in the canopy of the treetops; when it rains, it becomes completely immobile.

◀ While the flying dragon may occasionally use its remarkable gliding skills to escape from predators, its best defense is simply to flatten itself against the treetrunk, keeping its gaudy wing patches well concealed. Diurnal raptors, monkeys, snakes (notably of the genus *Chrysopelea),* and larger lizards appear to be the dragon's principal predators. The flying dragon exploits a rather different ecological niche from other lizards, insofar as it is diurnal, feeds primarily on ants, and seeks its prey on the surface of treetrunks at a considerable distance from the ground. Flying dragons are fiercely competitive, however, with other members of their own species, and any encounter is likely to result in a threatening confrontation in which flapping wing membranes and bulging throat pouches are called into play.

Several other *Draco* species are found in Southeast Asia, as well as *Draco dissumieri* in India. D.H.

184
Sub.: Vertebrata
Cl.: Osteichthyes
S.cl.: Teleostei
O.: Beloniformes
F.: Exocetidae

FLYING FISH *Exocetus volitans* 184

This common flying fish species measures 18–25 cm and is found in tropical and equatorial regions where the water temperature is approximately 23°C, including parts of the Central Atlantic, Pacific, Indian Oceans, and the southern Mediterranean. It prefers the open ocean to coastal waters and generally travels in very large schools, often intermingled with shoals of sardines.

▶ The young flying fish feed on copepods and other zooplankton, the adults on shrimps, small fish, and small planktonic crustaceans. There is nothing distinctive about the flying fish's hunting behavior except for the speed with which it pursues its prey, sometimes as high as 50–56 kph; it hunts by sight and usually preys on shoals of small fish.

◀ The flying fish's spectacular ability to escape from predators by skimming along through the air, several meters above the surface of the water, is certainly its most noteworthy characteristic. When the flying fish is pursued, it angles upward through the water until its head breaks through the surface; the lower fluke of the caudal fin, longer than the upper, vibrates back and forth very rapidly (up to 50 pulses per second), until the fish's body is lifted right out of the water like the hull of a speedboat.

The flying fish extends its highly developed pectoral fins and begins to skim over the water at a height of 2 or 3 m. In 2 or 3 seconds, it can cover a distance of up to 50 meters, and with a favorable wind, it may be able to glide for as much as 100 or 200 meters at a height of 5 or 6 meters above the surface, which is how flying fish sometimes come to be stranded quite far inland or picked up on the decks of ships. The flying fish actually glides rather than flies, since it pectoral fins are kept in a fixed position; it does not "flap its wings," in other words. Many of the flying fish's predators, which include swordfish, dolphins (both Coryphaenidae, the fish, and Delphinidae, the mammal), escolars (Gempylidae), and sharks, are able to execute long flying leaps and skim over the surface of the water, but none of them can match a performance of this kind—except for the frigate bird and other large seabirds that sometimes prey on *Exocetus* as well. The flesh of the flying fish is said to make very good eating, but no flying fish species supports a commercial fishery.

Of the several species of flying fish that make up the family Exocetidae, *Cypsilurus heterus* is also found in the Atlantic and the Mediterranean, though it prefers slightly colder water than *Exocetus;* it is also slightly larger, about 40 cm long. *Cypsilurus* resembles a biplane, since its pelvic fins, greatly elongated, as well as its pectoral fins are used as an airfoil. F.M.

185
Sub.: Vertebrata
Cl.: Reptilia
O.: Squamata
S.o.: Ophidia
F.: Colubridae

FLYING SNAKE *Chrysopela ornata* 185

This handsome arboreal colubrid typically measures about 1 m in length, sometimes 1.4 m or more; its slender body is green with black spots and covered with smooth scales. It generally lives in isolated tall trees (coconut palms) in forest clearings, but is also found in towns and villages, though never in dense forests; the juveniles generally remain on the ground, especially during the rainy season. This particular species ranges from Sri Lanka and the southern tip of India through Assam, Burma, Thailand, Vietnam, Cambodia, Hainan, Taiwan, and southeastern China to the Indo-Malaysian archipelago and the Philippines. Longevity is on the order of 10 years.

▶ *Chrysopela's* dentition, like that of many arboreal colubrids, is of the opisthoglyphic ("back-channeled") type, which means that its two large poison fangs are located toward the rear of its mouth; these are grooved to permit a not very copious flow of venom. It feeds on arboreal lizards—notably the so-called bloodsucker lizards *(Calotes)*, flying dragons *(Draco)*, geckos, skinks, and agamids—as well as terrestrial creatures, including frogs, agamids (e.g. *Sitana)*, and other lizards. Inventories of stomach contents typically include a large number of lizard's tails, presumably detached by autotomy and thus permitting their owners to remain at liberty. Depending on the region and the age of the individual, *Chrysopela* also preys occasionally on bats, young squirrels, and other small mammals as well as on birds, snakes, and insects.

Chrysopela is strictly diurnal and detects its prey by sight; it is primarly attracted by moving prey animals. It approaches its prospective victim slowly, then abruptly extends the forepart of its body, seizes the prey in its jaws, then works its jaws up and down to permit the fangs to penetrate the prey animal's body, which is promptly swallowed whole. (As noted elsewhere, the venom of opisthoglyphic snakes often serves an important predigestive function as well).

◀ *Chrysopela's* ability to climb tall, smooth treetrunks has provided it with access to an ecological niche unoccupied by other arboreal snakes, which generally climbing a tree by crawling from branch to branch rather than by slithering up a bare treetrunk. The corrugated pattern of rough scales on the underside of *Chrysopela's* tail and belly provide a purchase on the slightest irregularity in the bark, and once the resources of a particular treetop have been exhausted, *Chrysopela* flattens and depresses its underbelly into the shape of a trough or gutter, which provides maximum drag (and even a certain amount of directional control) as it launches itself into the air and glides heavily to the ground. *Chrysopela* can manage a vertical descent of 60 m, which also takes it some distance from the base of the tree, without sustaining any injury.

Chrysopela ornata and its congener *C. paradisi* have little to fear from reptilian competitors in the treetops, though they may sometimes be attacked by primates and birds of prey. The arboreal tokay *(Gecko gecko)* preys on the juveniles, and occasionally on the adults as well, and of course the flying snake is vulnerable to attack by a great many predators when it is on the ground, including raptors, jungle and domestic fowl, storks, large lizards, and other snakes. D.H.

FRESHWATER HYDRA *Hydra oligactis* 186

186
Sub.: Cnidaria
Cl.: Hydrozoa
O.: Hydrida
F.: Hydridae

This small hydrozoan is found in rivers and lakes throughout Europe, Northern Asia, the Americas, and Australia. Its vertical body column is about 3 cm long; its 4 to 6 tentacles may be 2 to 4 times that length. *Hydra oligactis* reproduces sexually as well as asexually, with new individuals developing from buds on the body column in the latter case. The sexes are differentiated, and the sexual organs also take the form of small knoblike protuberances on the body column; the spermatozoids are released into the water and impelled to unite with the ovocytes by the action of chemical attractants released by the latter. The newly hatched larva, known as a planula, comes to rest on the bottom, where it eventually surrounds itself with a spine-studded chitinous shell.

▶ We do not know what the hydra normally feeds on in nature, but in the aquarium it appears to flourish on a diet of small crustaceans *(Daphnia, Artemia)*, which it captures with the aid of its nematocytes, springlike filaments that constitute the offensive and defensive weaponry of the cnidarians (the name of the phylum is derived from the Greek word for stinging nettle). In the hydra, there are two types of nematocyte, both of which are arrayed in batteries along its tentacles. One type injects the prey with venom—with sufficient impact to penetrate the shells of small crustaceans—and the other enables the hydra to grasp the prey by adhesion and convey it to the oral cavity (a process known as prehension).

After the prey has been grasped by the tentacles and more or less immoblized, there is a latency period of several seconds before the tentacles move into "loading" position and the oral cavity opens. These nervous impulses are triggered by substances released into the water by the injured prey animal (in particular, a tripeptide called glutathion). Digestive enzymes called proteases begin to break down the tissues of the prey animal in the adjacent gastric cavity, and after about half an hour, the resulting nutrient broth is absorbed (phagocytized) and the process of digestion completed by specialized cells located in the internal body wall (endoderm).

The physiology of the hydra has been the object of extensive laboratory study, in which hydras of a number of different species *(Hydra, Chlorohydra)* have been involved. Several of these species are tinted green, due to the presence of symbiotic unicellular algae (zoochlorellae) in their tissues; a similar relationship exists between yellow-green algae and certain stony corals (scleractinarians), which also belong to the phylum Cnidaria. In each of these cases, the alga exchanges photosynthetic byproducts for nutrients metabolized by the host.

The hydra is a polyp, but unlike other cnidarian polyps (hydrozoans, coral animals), it can move about quite freely—either by swimming or by inverting itself and stalking along the bottom on its tentacles. In the Great Lakes, hydras sometimes congregate in huge planktonic mats several square kilometers in area and achieving densities of 5000 individuals per cubic meter. There are other freshwater cnidarians, including numerous species of jellyfish belonging to the order Limnohydrina, in which the alteration of polyp and medusan stages that is typical of this phylum also occurs. One species, *Craspedacusta sowerbii,* discovered in a London fountain in 1880, has since been identified in rivers and streams throughout Europe, Asia, and the Americas. The polyp is only a few millimeters long and lacks tentacles; the body of the medusan is translucent and measures between 1 and 2 cm in diameter. M.v.B.

187
Sub.: Vertebrata
Cl.: Reptilia
O.: Squamata
S.o.: Sauria
F.: Iguanidae

FRINGE-FOOTED SAND LIZARD
Uma exsul 187

The 4 species of sand lizard found in the vicinity of the U.S.–Mexican border are all well equipped for survival in arid, sandy terrain. Somewhat more is known of the habits of the 3 species that are found in the United States—*Uma inornata, notata,* and *scoparla*—than of the fourth species, *Uma exsul,* which is found exclusively in the desert basin called the Bolsón de Mapimi in northern Mexico.

This species may attain a length of 18.5 cm, though the distance from snout to vent (exclusive of the tail) is no more than 9 cm. Dense populations may be found among sand dunes characterized by a certain type of scrub vegetation and where numerous burrows dug by rodents are frequently occupied by *Uma exsul.* Two allopatric subspecies are recognized, *Uma exsul exsul* and *Uma exsul paraphygas,* "allopatric" meaning that they are found in separate regions (in this case, of the Chihuahua desert).

This small iguanid can run very quickly, sometimes at speeds exceeding 35 kph. Of the four species mentioned earlier, *Uma exsul* is the least narrowly specialized, morphologically and ecologically, though it has developed a number of anatomical adaptations to life in the sand dunes: Its nasal passages and ear canals can be sealed tight to prevent grains of sand from getting in; its overlapping eyelids and spoon-shaped jaws serve a similar function; the fringe of scales around the toes on its back feet that gives the species its name are of use into burrowing into sandy soil. The fringe-toed lizard is active when the soil temperature remains between 26°—41°C, which is at the lower range of temperatures actually encountered in the region (17–80°C).

▶ *Uma exsul* feeds almost exclusively on small insects, beetles and ants for the most part, which it normally catches by scratching violently at an anthill or some other spot when it suspects its prey may be lurking below the surface. In spring, it occasionally feeds on leaves and buds.

◀ The principal predators of the fringe-toed sand lizard include the leopard lizard *(Crotaphytus wislesenii),* rattlesnakes and colubrids, the shrike, and the kestrel. When threatened or alarmed, *Uma* has two characteristic intimidation tactics—it flattens its body obliquely to display the black spot on its underbody, or on occasion it may lean forward on its front legs, as if it were about to stand on its head, and curl its tail up over its arched back. In order to escape from predators as well as to avoid overheating it sometimes buries itself in the sand (to a depth of 4 cm), but more frequently seeks refuge in a burrow dug by rodents.

Uma exsul emerges from its burrow at a later hour of the morning than the 3 congeneric species mentioned earlier. It is often found in conjunction with the western whiptail *(Cnemidophorus tigris)* and the iguanid *Uta stansburiana.* Its potential rivals, the great and lesser earless lizards *(Cophosaurus* and *Holbrookia),* which have the same dietary preferences as *Uma exsul,* are absent from the high dunes. C.G.

FUNNEL-WEB SPIDER *Agelena labyrinthica* 188

Agelena labyrinthica is a large-bodied spider, about 2 cm long, that is especially common in the south of France. It is found in all sorts of open terrain—meadows, abandoned fields, and the arid brushlands that are called *garrigues*—where its spins a web in the form of a slightly concave and more or less horizontal silken carpet sloping into a funnel-shaped orifice and a hollow tube that leads down through the grass stems toward the ground. The female deposits her eggs in cocoons at the end of the summer; the eggs hatch out in spring, and the young spiders immediately set off to find suitable spots in which to weave tunnel webs of their own.

▶ *A. labyrinthica* feeds on virtually all flying or jumping insects that frequent the upper reaches of the grass "forest." The spider also weaves an irregular netting of sticky strands directly above the web, the purpose of which is to catch insects in mid-flight or mid-leap and to propel them downward into the web itself (and also to prevent them from jumping back out again once they have been caught). As usual, the vibrations of the web or the netting caused by the insect's struggles to free itself also alerts the spider to its presence. The venom of *A. labyrinthica* is not very potent, so it uses its crablike pincers (chelicerae) to subdue and dismember its prey.

◀ Hunting and parasitic wasps prey on both *A. labyrinthica* and its eggs, but the funnel and the tube at the bottom of its web provide the adult spider at least with a convenient escape hatch.

In Africa, there are several related species (among them *A. consociata)* that catch and consume their prey communally and whose conjoined webs may attain a volume of several cubic meters. Their behavior toward one another seems to be completely free of the aggressively cannibalistic tendencies usually displayed by spiders toward members of their own kind. Other genera of the family Agelenidae, notably *Tegenaria,* prefer a more dimly lit environment—cellars, attics, caves, or the base of a tree in the forest—but otherwise go about their predatory business in much the same way as *A. labyrinthica.* J.-F. C.

188
Sub.: Arthropoda
Cl.: Arachnida
O.: Araneida
F.: Agelenidae

G

GANGETIC DOLPHIN *Platanista gangetica* 189

This freshwater dolphin measures between 1.5 and 2.5 m in length and weighs between 50 and 80 kg, with substantial variations in size and weight being apparent among the three main populations of this species, which correspond to the three great rivers of the Indian subcontinent—the Ganges, the Indus, and the Brahamaputra—and their tributaries. The body of the Gangetic dolphin is massive and some-

189
Sub.: Vertebrata
Cl.: Mammalia
O.: Cetacea
F.: Platanistidae

what less streamlined than that of marine dolphins. It is found in muddy or turbid water, and as a result its eyes are almost completely degenerate, and it finds its way by echolocation.

▶ The Gangetic dolphin feeds on silurids (Eurasian catfish) up to 35 cm in length as well as crayfish and mollusks. It depends primarily on echolocation (hence, on its acute sense of hearing) to catch a fish or other mobile prey animal, gauging its speed and relative location by means of the ultrasonic echoes reflected back by the fish's body. The jaws of the Gangetic dolphin, though equipped with numerous needle-sharp teeth, are extremely long, slim, and fragile (like those of its potential competitor, the gavial) and thus well adapted to the task of seizing and restraining an elusive prey animal; they are not used for chewing, since the prey is swallowed whole.

◀ Though possibly a competitor, the gavial, which is now almost extinct in the wild in any case, is not a predator of the Gangetic dolphin.

The family Platanistidae comprises a homogenous group of small, slender-beaked freshwater dolphins, all more or less sightless, which are found in the larger rivers of Asia and South America; only the La Plata dolphin *(Stenodelphis blainvillei)* is found in coastal waters. V. de B.

190
Sub.: Vertebrata
Cl.: Aves
O.: Pelecaniformes
F.: Sulidae

GANNET *Morus bassanus* 190

The Latin name of this large North Atlantic seabird means "Bassan's booby," and its black-rimmed eyes and upturned mouth may indeed seem to give it a sort of deranged expression. The gannet is 78–85 cm long with a wingspread about 180 cm and weighs about 3.5 kg. Record longevity for this species is 23 years, but the average life expectancy for an adult is probably about 16 years, and only about 20 percent of the hatchlings survive to maturity. During the comparatively brief periods that the gannet spends ashore, it congregates in densely populated rookeries, or nesting colonies, located on sheer cliff faces and rocky offshore islets all along the shores of the North Atlantic, from Newfoundland to Norway; the most southerly gannet colony is on Sept-Iles, off the coast of Brittany, which currently contains about 4500 nesting pairs. In autumn, these colonies are deserted, since the yearling birds fly down to spend the winter on the coast of Africa and the older juveniles and the adults head further north.

▶ The gannet spends most of its life on the open sea, propelling itself along with vigorous wingbeats, then soaring effortlesly for many minutes, sometimes at a height of 20 m, sometimes just barely cresting the waves. It is also a very strong swimmer, though its movements on land are cumbersome and it comes ashore only to mate and rear its young. The gannet feeds on fish that assemble in shoals, like herring, mackerel, and sand launces *(Ammodytes)* as well as cephalopods. It dives for its food, either from the surface or from a height of up to 40 m—which can be a particularly impressive sight.

When the gannet has located a shoal of fish, it dives abruptly, sometimes vertically, its wings half-closed intially, then tucked back against its body immediately prior to impact. It seizes its prey in its beak on its way back up to the surface, using its webbed feet to propel it through the water. The bony helmet of the cranium contains air sacs that act as shock absorbers, cushioning the gannet's brain against the repeated impact of these vertiginous dives.

◀ The gannet's nests are often pillaged and the chicks devoured by great black-backed gulls and herring gulls when the parents' attention is engaged elsewhere. Skuas and jaegers also prey on the gannet in a less decisive fashion—harassing the adult birds with feint attacks and forcing them to disgorge the crops full of fish that they are bringing home to their nestlings. Gannets and their eggs have frequently been harvested by fishermen and islanders of the North Atlantic to be used as food or fish bait. In the 18th century, for example, the systematic harvesting of the island's gannet and fulmar colonies, along with the gathering of seaweed and Irish moss, was the principal occupation of the inhabitants of St. Kilda in the Hebrides, and historians of the region sometimes speak of an entire West Highland or Shetland culture that was based on the exploitation of the gannet.

There are 9 different species of gannets and boobies, collectively comprising the family Sulidae. The Peruvian booby, or piquero *(Sula variegata),* is one of the principal guano-producing species along the western coasts of South America and was thus of critical economic importance in the days before the discovery of the nitrogen-fixing process (for making fertilizer, munitions, etc.). The piquero feeds on the anchovies and flying fish that are (or rather, were) particularly abundant in the Humboldt Current and deposits its guano on the cliffs and offshore rocks on which it nests, in conjunction with the guanay cormorant *(Phalacrocorax bougainvillei),* in colonies containing several hundred thousand individuals. (Overfishing in the coastal waters of Peru and Ecuador has also resulted in a drastic reduction of the seabird population over the last decade or two).

J.-P. R.

GARDEN DORMOUSE *Eliomys quercinus* 191

This Old World rodent looks very much like a small rat: pointed muzzle, long ears, short legs, and long tail. The garden dormouse, or lerot, can be readily distinguished by the masklike bars of darker fur along the sides of its muzzle and around its eyes as well as by its furry tail. The adult garden dormouse is about 130 mm long, plus an additional 120 mm for the tail (if intact), and weighs up to 65 g. This species ranges from Mauretania to Finland (excluding the rest of Scandinavia and the British Isles) and is found primarily in sunlit, rocky terrain in the Mediterranean Basin, sometimes occurring at altitudes of up to 3800 m (as in the High Atlas of Morocco) and in the vicinity of alpine cottages and hikers' and climbers' huts at somewhat lesser altitudes.

The garden dormouse is a true hibernator, which confers an obvious adaptive advantage at these higher altitudes; toward the end of October, it finds a suitable spot for itself inside a hollow tree, in a cavity in a rock, or even inside a building and lapses into a state of hypothermia (its body temperature only slightly greater than that of the surrounding air); at the same time, its cardiac and respiratory rates are greatly decreased and its food intake, of course, is reduced to zero until awakens again at the beginning of April. During the warmer months, the garden dormouse is nocturnal, usually becoming active at dusk. In the spring, the female bears a litter of 4–8 young; the young dormice develop slowly, and life expectancy for this species may exceed 3 years.

▶ The sleepy dormouse is not usually thought of as a predator, and dentition, though typical for a rodent, seems ill suited to the task, lacking both canine teeth with which to grip its prey and carnassial teeth (specially adapted molars and premolars) with which to rend and shear tear its flesh. Nevertheless, analysis of stomach contents reveals that the European garden dormouse is a genuine omnivore, subsisting primarily on animal prey between the months of June and October, i.e. during most of its active life. This prey includes insects and insect larvae, earthworms, snails, young mammals, birds and nestlings, small amphibians, and young reptiles—a diet that is supplemented with vegetable food, including orchard and wild fruit, notably wild strawberries and other berries of all sorts.

191
Sub.: Vertebrata
Cl.: Mammalia
O.: Rodentia
F.: Gliridae

In captivity, the garden dormouse will live quite happily on grasshoppers and crickets, and it is capable of dispatching a small (20 g) fieldmouse in a matter of seconds with a bite to the back of the neck. The garden dormouse has also been reported (in captivity) to turn cannibal on occasion, devouring the flesh and entrails of other dormice but leaving the skin behind, often turned inside out like a glove. Its eardrums are relatively large, enclosing a spacious resonating chamber, so that it has remarkably keen hearing and can even detect the sounds made by a mealworm crawling slowly across the floor of its cage. On the other hand, its sense of smell is not very well developed, nor is its vision particularly acute; it has sometimes been observed to fix a potential prey animal with its large, protruding eyes but not to attack as long as the prey remains motionless. The garden dormouse is light and agile enough to climb trees and rob the nests of birds, small enough to run freely along the galleries dug by other rodents. It kills its victims by biting them with its strong chisel-like incisors; a single bite to the back of the neck is generally fatal to a mammal the size of mouse.

◀ Martens, owls, and vipers are the principal predators of the garden dormouse. The beech marten shares a predilection for haylofts, house foundations, and outbuildings with the garden dormouse; the pine marten is more likely to be encountered in the forest. The eagle owl and the tawny owl both devour large numbers of dormice, the screech owl relatively few. The dormouse may be quick and agile enough to escape from a terrestrial carnivore by dashing up a treetrunk, and it may find a safe haven from the pine marten in a narrow crevice in the rock and from the eagle owl by making its nest for the winter among the rafters of a building. The garden dormouse has a very high resistance to the venom of European vipers; the minimum lethal dosage of the venom of the European asp is 192 mg per 1000 g of body weight in the case of the garden dormouse, as compared with 31 mg in the case of the hedgehog (which is considered to enjoy a *relative* immunity to the viper's venom) and only 1.8 mg in the case of the guinea pig.

As a final means of defense, the garden dormouse is able to slough off the sleeve of skin and fur that covers its caudal vertebrae if a predator tries to seize it by the tail. There is no bleeding; the exposed vertebrae become necrosed and drop off a short time later. The dormouse is unable to regenerate the missing vertebrae, of course, so this defensive gambit (known as caudal autotomy) will been unavailable in future if a significant number of vertebrae have been lost the first time out. Most garden dormice appear to make use of this capability at one time or another, and a full-grown adult is rarely recovered from the wild with its tail intact.

The garden dormouse's principal competitor is the black rat, which lives in barns and farm buildings as well as in the wild in the Mediterranean Basin. Both species occupy the same niche and share the same dietary preferences, so that a house or barn that has been colonized by garden dormouse will rarely be occupied by rats and vice versa. Garden dormice that take up residence in human dwellings are often assumed to *be* rats and are poisoned or trapped accordingly. They also do considerable damage to orchards (due to their habit of taking a single bite from a fruit and moving on to the next) as well as to bedding and provisions stored in climbers' huts and mountain chalets; climbers and hikers, who have occasion to meet the dormouse face to face and are often amused by its effrontery, are generally more tolerant of these exactions than householders or nurserymen.

Other European dormice (family Gliridae) include the edible, or fat, dormouse (*Glis glis),* so called because it was once raised by the Romans as a gourmet delicacy, the forest dormouse *(Dryomys nitedula),* and the common dormouse *(Muscardinus avellarius).* All three are nocturnal hibernators and omnivores, though a considerably smaller portion of their diet consists of animal food, especially in the case of the edible dormouse. The tree dormouse and the common dormouse are also less likely to be found in the vicinity of manmade structures. M.C.-S.G.

GIANT AFRICAN OTTER SHREW
Potamogale velox **192**

192
Sub.: Vertebrata
Cl.: Mammalia
O.: Insectivora
F.: Tenrecidae

This West African carnivore is neither an otter nor a shrew, though like the shrews and its closer kin the tenrecs, it is included in the order Insectivora, and through the process of so-called evolutionary convergence, it has adopted the appearance as well as the feeding habits of a small otter. The otter shrew's body is as supple and streamlined as an otter's (though its feet are not webbed), and it uses its flattened tail as rudder; its nostrils are clamped shut by a cartilaginous septum while it is swimming underwater, and it can remained submerged for several minutes. It is found along the banks of watercourses in the tropical forests of West Africa, from Nigeria to Zaire, and digs a burrow for itself in the riverbank with an underwater entrance.

▶ The otter shrew feeds on freshwater crabs and other crustaceans as well as fish, amphibians, snails and other mollusks. Its whiskers can detect vibrations made by prey animals underwater, and it appears to detect its prey primarily by sight and hearing when it is on the surface. It catches crabs by turning them over on their backs, then ripping open the plastron (bottom shell) with its teeth.

◀ The predators of the giant African otter shrew, if any, are unknown. Its fur has been exploited commercially in the past, and it is still being trapped by local hunters.

A congener, the small African otter shrew *(Micropotamogale lamottei)*, was discovered in 1954 by French zoologists H. Heim de Balsac and M. Lamotte. It is found in shallow streams and marshy upland terrain in Guinea, Liberia, and the Ivory Coast and is less markedly aquatic in its habits than its much larger cousin, though it also feeds primarily on crabs. The Ruwenzori otter shrew *(Micropotamogale ruwenzorii)* was later discovered in the interior of Africa; it is only slightly larger than *M. lamottei*.

J.-J. B.

GIANT ANTEATER *Myrmecophaga tridactyla* 193

193
Sub.: Vertebrata
Cl.: Mammalia
O.: Edentata
F.: Myrmecophagidae

The giant anteater, or ant bear, which would certainly be difficult to mistake for any other creature, measures about 2 m in length, tail included, and can weigh up to 35 kg. Its coat is short and thick along its muzzle, then prolonged into a luxuriant mane extending over its back and tail, and it has patches of black fur on its chest and shoulders. The other animals assigned to this order of "toothless" (edentate) mammals, namely the armadillos and sloths, have rudimentary peglike teeth, but the 3 extant species of anteater are genuinely toothless. The giant anteater is exclusively terrestrial and does not confine itself to a particular territory; it is found in forests and savannahs virtually throughout South America except in the highest and most sparsely vegetated regions of the Andes.

▶ The giant anteater is active by night as well as by day, and it seeks out anthills and termite mounds primarily with the aid of very keen sense of smell. It has no difficulty in opening up a breach in the wall of a termite mound with its long, thick claws; next, it introduces its muzzle into the hole and attempts to locate the termites' principal living quarters at the center of the mound. It ingests the termites with the aid of its extremely long tongue; when the tongue is withdrawn into the mouth, the termites still adhering to it are crushed against the anteater's palate.

As with the CHINESE PANGOLIN, the walls of the anteater's stomach are extremely muscular, and are sometimes assisted by the grinding action of a handful of pebbles, which serve the same function as the grit in a chicken's crop. The giant anteater is capable of ingesting as many as 30,000 ants or termites in a single day. The anteater also preys on columns of foraging ants outside the nest, though it avoids those species that are protected by warriors with powerful jaws or that secrete acrid chemicals; the giant anteater also feeds occasionally on earthworms, insect larvae, and berries.

◀ When cornered, the giant anteater defends itself with powerful swipes of its claws, which measure up to 10 cm on its front paws and seem to have a deterrent (and perhaps occasionally a fatal) effect on the puma and the jaguar, the two most formidable predators of the region. The giant anteater is a skillful swimmer and can also conceal itself quite effectively in open country by squeezing itself down into a shallow trench in the ground.

The tamandua, or collared anteater *(Tamandua tetradactyla)*, and the silky anteater *(Cyclopes didactylus)* are both considerably smaller than the giant anteater; the tamandua is partially arboreal and the silky anteater exclusively so. Both are very similar to the giant anteater, however, in their feeding habits and in their geographic range.

P.A.

GIANT GROUPER *Promicrops lanceolatus* 194

194
Sub.: Vertebrata
Cl.: Osteichthyes
S.cl.: Teleostei
O.: Perciformes
F.: Serranidae

This large marine fish has been known to attain a length of 3.7 m and to weigh as much as 450 kg; it lives in a crevice between underwater rocks or underneath a coral outcropping, a retreat from which a previous tenant, a moray or an octopus, may have to be forcibly evicted (and in the case of the moray, it is often the grouper that gets the worst of it). The juveniles live closer to the surface and are thus more brightly pigmented (yellow and black) than the adults, which are blackish-gray with a few white spots; very large adult groupers are sometimes found in deeper water, on the edge of the continental shelf.

The giant grouper is a solitary species, and, like many of the serranids, it is hermaphroditic. Specifically, groupers are egg-laying females when they reach sexual maturity at the age of 5–7 and remain so for another 2 to 3 years, at which point they transform themselves into sperm-producing males after a transitional period during which they may display the characteristics of both sexes (or neither).

▶ The giant grouper feeds on both pelagic and benthic fish species, including sharks, as well as crustaceans, young octopuses, and even on carrion and sewage; when food is scarce, the grouper sometimes swims up into an estuary with the tide in order to feed on the refuse of human settlements. On a coral reef, it hunts from ambush, remaining concealed in its crevice until a prey animal is quite close by (within about 5 or 10 body lengths). Then the grouper lunges out of its burrow and seizes the prey animal while executing a lateral motion that propels it back into its burrow again. Its swallows its prey whole, which is the main reason why a large grouper (1.4 m) chooses to prey on fish that are only about 20 or 30 cm long. It is guided entirely by its sense of sight in detecting and intercepting its prey.

If the grouper makes its home in a sunken wreck, a submarine canyon, or some other spot where fish are less prevalent, it actively seeks its prey, swimming slowly over the bottom, propelling itself with gentle strokes of its caudal and pectoral fins in order to dampen the vibrations it creates as it moves through the water. In turn, the grouper will only attack a moving prey animal, though once it has gotten a fix on the prey, it may still attack it even if the prey remains stationary, which suggests that the giant grouper possesses some sort of visual memory.

While the grouper patrols along the bottom in this manner, it tips its head downward at a shallow angle in order to increase its field of vision as well as to diminish the size of the blind spot created by its own great bulk. Its eyes are able to move independently of one another, and though the giant grouper does not appear to have true binocular vision and depth perception, it is able to judge relative distances by making slight lateral alterations in its course (so that it appears to be swaying back and forth) in order to fix the prey with one eye and then the other.

When it is almost in contact, about 10–20 cm away, it opens its mouth and begins to swallow water; the prey is washed into the grouper's mouth by the current, and the grouper's teeth close on the prey while the water it has swallowed is sluiced out through the grouper's earholes. The grouper has about a hundred teeth, which can move freely in their sockets, tilting back toward the gullet to admit the prey animal, then tilting forward again and impaling the prey to keep it from escaping. Even if the prey is too large for the grouper to swallow, it will still have to undergo this ordeal by impalement before it can be released.

◀ Morays and triggerfish sometimes prey on young groupers, but the adults are generally quite safe in their crevices, although a contest of the type described above for the possession of a desirable crevice may end with the victor (often the moray) devouring the vanquished. The hatchlings, only 3 mm long at the outset, are extremely vulnerable to predation by many pelagic species, but sharks, other groupers, and humans are the only potential predators that a full-grown adult is likely to encounter as it searches for food along the bottom.

The traditional fishing techniques employed by Pacific islanders have had little effect on this species, but the same qualities that have made it an object of respect and even dread on the part of the islanders—namely, the grouper's intense curiosity and enormous size—have also made it a tempting target for vacationers and "sports" fishermen equipped with spearguns and scuba gear. These activities have had a statistically significant effect not only on total populations but on recorded maximum sizes of individuals in many areas; apart from being difficult to miss, groupers are also fairly easy to track down, because of the grouper's habit of returning to the same crevice for several years in a row. Spearfishermen who have been disappointed in their quest for a giant grouper have also begun to take a significant toll of the smaller species, which do not exceed 30 or 40 cm.

The red grouper *(Epinephelus flavocaerulus)* is typical of the smaller groupers of the Indo-Pacific region, and the Warsaw grouper *(Epinephelus nigritus),* also noted for its curiosity and its enormous size, is often encountered by reef divers. Sharks, morays, *Lutjans* (relatives of the snappers), and *Pterois* (which includes the venomous lionfish) are typical predators of the smaller species. M.D.

GIANT WATER BUG *Belostoma boscii* 195

195
Sub.: Arthropoda
Cl.: Insecta
O.: Heteroptera
S.o.: Hydrocorisa
F.: Belostomatidae
G.: Belostoma

The Belostomidae is a family of aquatic insects related to the Nepidae but lacking the elongated breathing tube that gives the latter their common name, water scorpions. They are found exclusively in the tropics, in the Caribbean and northern South America. It has a flattened oval body about 30 or 35 mm long and 15 mm wide and is brown with greenish highlights. It lives in ponds and sluggish streams that are more or less overgrown with aquatic vegetation, and it only leaves the water when its native pond or watercourse has dried up or during its short twilight nuptial flight.

Copulation takes place in the water and lasts from 1 to 3 hours. Shortly after that, the female deposits its clutch of about 180 to 140 eggs in a snug receptacle on the male's back, which totally or partially covers the elytra (wing covers) and prevents the male from flying for the next 5–8 days, until the eggs have hatched. The male brushes them regularly with its hind legs during this period; the eggs will not hatch if any of them are accidentally detached from the male's back. The lifespan of the adult giant water bug is highly variable, but specimens in aquariums have survived for as long as 7 months.

▶ The larvae feed for the most part on tiny planorbid snails *(Biomphalaria glabrata)* about 2 to 23 mm in diameter, supplemented by another freshwater snail species *(Physa marmorata),* various larvae, fish hatchlings, and tadpoles; they also feed on one another, which is an important point to keep in mind if one is thinking of trying to raise *Belostoma boscii* in an aquarium. Larvae and adults both pick up snails with their forelegs and turn them around until the opening to the shell is accessible, whereupon they inject a substance that promotes external digestion and ingest the liqufied soft tissues of the snail, leaving nothing but the empty shell behind. Tadpoles and small fish are attacked from ambush, quickly immobilized, and ingested in the same manner. Larvae and adult giant water bugs raised in aquariums consume 2 to 4 snails daily; in nature, the water bug performs an important service to humans, since its principal prey, the planorbid snail *Biomphalaria glabrata,* is an intermediate host for the organism that causes schistosomiasis, a serious parasitic disease endemic to several Caribbean islands.

◀ The smaller larvae are eaten by fish, tadpoles, and adult amphibians (perhaps destined to be devoured in turn when the surviving larvae have grown a little larger). The adults are eaten by large fish and aquatic birds, as well as toads during their nuptial excursions into the air; they are also attracted to light and thus readily picked off by insectivorous birds during these twilight forays.

Belostoma gigantea, the largest of the family, may attain a length of 6 or 7 cm and is capable of delivering a painful bite. Apart from the Notonectidae (backswimmers), members of 2 other families of aquatic insects, Corixidae and Naucoridae, are found in the same habitat and look something like miniature versions of the Belostomidae. A.D.

GILA MONSTER *Helodema suspectum* 196

196
Sub.: Vertebrata
Cl.: Reptilia
O.: Squamata
S.o.: Sauria
F.: Helodermatidae

This large multicolored lizard may attain a length of 60 cm (with its tail accounting for about one fourth of the total) and sometimes weighs more than a kilogram. Its small, round scales are pink, yellow, and black, with a texture that is strikingly similar to that of an old-fashioned beadwork purse. Longevity in captivity sometimes exceeds 20 years, and it seems highly probable that some individuals may survive much longer than this in the wild. The Gila monster is found in Arizona (which is where the Gila River is located), the southern tip of Nevada, the southwest corner of Utah and New Mexico, the extreme southeast of California, as well as in the Mexican state of Sonora. The Gila monster is found in rocky subdesert valleys and plains with limited vegetation (cactuses, thornbushes, sagebrush), though it is fond of water and will immerse itself whenever the circumstances permit.

▶ The Gila monster is nocturnal and spends much of its time in its burrow or in a crevice in a rock, an indentation in the soil, etc. It is comparatively active in April, which may be its mating season, and from July to August; its pattern of seasonal activity seems to depend on rainfall and on the availability of the eggs and nestlings on which it feeds. The Gila monster and its congener, the Mexican beaded lizard *(Helodema horridum)*, are the only venomous lizards in existence, though they do not have venomous fangs; the venom glands are located behind the lower lip, and the lizard's teeth are notched to allow the venom to flow into the wound.

The Gila monster feeds primarily on the eggs of birds and reptiles, nestlings, and young rodents, which it detects primarily with the aid of its sense of smell (localized in Jacobson's organ), perhaps secondarily by sight. It cracks eggshells with its teeth and laps up their contents with its tongue. Its jaws are very powerful, but it is too slow-moving to catch any creature that is not virtually helpless; its venom is used exclusively for defense and plays no part in predation. There are many predators that occasionally feed on eggs and nestlings, but the Gila monster is the only inhabitant of this region to do so almost exclusively; during those frequent periods when food is unavailable, and particularly during periods of drought, it almost never leaves its burrow and lives off the accumlated fat reserves that are stored in its tail.

◀ The coyote and certain raptors are believed to be the only natural enemies of the Gila monster. As with many venomous species, its gaudy coloration probably serves as a warning to predators. If it is attacked, the Gila monster bites down and locks its jaws like a bulldog, attempting to worry and lacerate its adversary's flesh so that its venomous saliva will be absorbed into the wound.

The Mexican beaded lizard *(Helodema horridum)* is found in wooded areas in western Mexico; its feeding habits are probably similar to those of the Gila monster. Both of these venomous lizards are too sluggish and unaggressive to pose much of a danger to man; fatalities are extremely rare. D.H.

GILTHEAD *Sparus aurata* 197

197
Sub.: Vertebrata
Cl.: Osteichthyes
S.cl.: Teleostei
Sup.o.: Acanthopterygii
O.: Perciformes
F.: Sparidae

The gilthead typically measures 25–50 cm, though larger individuals may attain a length of 70 cm and a weight of about 4 kg. It is found along rocky shorelines and in beds of undersea grass (posidonia) at a depth of 5–30 meters along the coasts of the Mediterranean, the Adriatic, the Black Sea (rarely), and the Atlantic from Portugal to the British Isles (rarely).

▶ The dentition of the gilthead is highly specialized: the teeth in the anterior part of the jaws are conical and pointed and perform the same function as canines; behind these, the jaws are lined with several rows of flattened teeth, similar to molars, that allow the gilthead to crack the shells of the mollusks and crustaceans that comprise the bulk of its diet. It patrols the bottom, either singly or in pairs, in search of this prey; its vision is very keen, and its eyes quite mobile as well.

◀ The adult gilthead has few natural enemies, among them a few larger species that inhabit the same costal biotopes, notably the Mediterranean grouper *Epinephelus gigas*. The gilthead is easily startled and generally attempts to evade pursuit by taking refuge in an undersea cave or a bed of wrack grass. Its flesh is greatly esteemed by man, and it is sometimes taken with trawls and longlines, though the total catch is of no great economic importance. Since 1972, giltheads have also been raised commercially in saltwater lagoons along the French and Italian coasts; the young fish are trapped in open water, then transferred to the lagoons—which serve the function of fattening pens—and harvested when they reach "portion" size (about 300 g), a little less than 2 years later. Annual production for the Italian fishery exceeds 1000 metric tons.

A related species, the royal gilthead *(Crysophrys major)*, has been raised commercially in Japan for many years; annual production currently exceeds 8000 metric tons. The gunner *(Pagellus centrodontus)* and the black bream *(Spondyliosoma cantharus)* are very similar to the gilthead in appearance, and the gunner is often sold in French fishmarkets under the name *dorade commune* ("common gilthead"). J.-M. R.

GLOSSY IBIS *Plegadis falcinellus* 198

198
Sub.: Vertebrata
Cl.: Aves
O.: Ciconiiformes
F.: Threskiornithidae

The glossy ibis is between 55 and 65 cm long, with a wingspread of 80–90 cm, and weighs between 500 and 600 g. A banded specimen survived to the age of 19 years and 10 months, but this is probably not typical of the species as a whole. It is found in warm, shallow water, both salt and fresh, in lagoons and river deltas, estuaries and floodplains, rice fields, and the rims of marshes. It nests in willow groves and reedbeds, usually in association with herons, cormorants, spoonbills, and other gregarious birds. The ibis is migratory and seminomadic, at least to the extent that it often changes the location of its colonies in response to fluctuations in the water level on its feeding grounds. It continues to nest in Hungary and the Balkans and along the coasts of Spain, Italy, and Asia Minor and is now also found on the west side of the Atlantic, in Florida and the Caribbean as well as the Middle Atlantic region and, in summer, as far north as the coast of Maine.

▶ The glossy ibis feds primarily on aquatic insects and their larvae, worms, leeches, snails (planorbids, *Lymnea*), crustaceans, tadpoles and young frogs, and reptiles. It detects its prey by probing in shallow water with its long curved beak, investigating mudflats, flooded fields, sedimentary deposits, and similar terrain.

◀ The nestlings in the colony are well protected by the proximity of other nesting birds; the adult birds are only rarely preyed on by raptors or carnivores, and interspecies competition is scarcely a problem for the ibis, which collectively and efficiently exploits the adundant food resources of its habitat. The glossy ibis is not hunted or otherwise persecuted by man; European ibis colonies have been considerably reduced by the use of pesticides and the destruction of habitat, though the populations recently established in North America have been steadily expanding their range.

The ibises make up a homogeneous group of 23 different species, a number of which are currently endangered. The ibis often figures as a motif in the art of ancient Egyot and traditional Japan, but only 9 specimens of the Japanese ibis *(Nipponia nippon)* were still in existence as of 1975 and the sacred ibis *(Threskiornis aethiopica)*, deified by the ancient Egyptians, has recently become extinct in Egypt, though it is still found elsewhere in Africa. The bald ibis *(Geronticus eremita)* was found throughout Southern Europe during the Middle Ages but is now confined to a handful of nesting sites along the coasts of Turkey and Morocco, and the red ibis *(Eudocimus ruber)*, formerly prevalent along the Caribbean coast of South America, has reportedly become endangered because its brilliant scarlet plumage is used in the manufacture of artificial flowers. A number of other species are found only in damp tropical forests, notably the green ibis *(Mesembrinibis cayennensis)* of South America. J.-F. T.

GLUTINOUS CAECILIAN *Ichthyophis glutinosus* 199

199
Sub.: Vertebrata
Cl.: Amphibia
O.: Gymnophiona (Apoda)
F.: Ichthyophiidae

A caecilian is a legless, frequently sightless burrowing amphibian with a smooth, segmented body that looks very much like a large earthworm. This particular species is about 40 cm long and lives on the island of Sri Lanka (Ceylon). The female lays its eggs in a burrow under the rocks or in vegetable debris on the banks of a small stream; the juvenile form, as with virtually all amphibians, is aquatic. The adult leads a largely subterranean existence and not a great deal is known of its habits.

▶ Tanner (1971) was able to observe the glutinous caecilian's solution to the problem outlined above. The caecilian seizes hold of one end of an earthworm, then rapidly snaps its jaws open and shut 2 or 3 times until the earthworm is in position to be swallowed lengthwise. At this point, the earthworm is fully extended and straining to escape—and would undoubtedly do so if the caecilian does not keep its jaws clamped tightly shut. However, if the caecilian spins around several times in place with the earthworm still in tow, the earthworm gets hopelessly entangled in itself and untenses its body segments so it is no longer capable of rapid motion. The caecilian can begin to swallow its prey, though it may find it necessary to repeat this pacifying maneuver several times.

The adult caecilian probably feeds on termites as well as earthworms, the juvenile form (once it has left its original home in the water) on insect larvae as well. Some caecilian species never leave the water; the predatory behavior of *Chtonerpeteon indistinctum,* a member of the family Typhlonectidae that is found in Brazil, is probably typical of these aquatic caecilians: It flails the front part of its body from side to side—somewhat comparable to the reconnoitering motions of blind man's cane—until it encounters a suitable prey animal, which it seizes in its jaws and drags deeper into the water, then executes a series of rapid motions, repositioning the animal in its jaws until it is more or less aligned with the caecilian's alimentary tract and thus ready to be swallowed. *Typhlonectes compressicaudus,* a related species of the Guianas and the Amazon Basin, has discovered that the bodies of fish caught in fishermen's gill nets quickly become distended, stretching and exposing the skin between the scales, which they can easily penetrate with their small teeth. J.L.

200
Sub.: Annelida
Cl.: Polychaeta
S.cl.: Polychaeta errantia
F.: Glyceridae

Glycera convoluta 200

The adult of this species of polychaete worm is 6–10 cm long; its body is rounded, about 3–4 mm wide, and tapering toward the rear. It is bright red or shell pink in color, due to the presence of liquid globules in the digestive tract that are rich in hemoglobin. *Glycera* is found in the intertidal zone, just above the low-tide mark; it tunnels through the muddy bottom beneath beds of wrack grass as well as through loose, gravelly sand at a depth of about 10 cm. When disturbed, it rolls itself up in a tight spiral, hence *convoluta.* This species is found along the shores of the Atlantic, the English Channel, and the Mediterranean (including brackish lagoons like the Étang de Berre).

The larva passes first through an active, free-swimming phase, during which it feeds on phytoplankton; its morphology is of the type known as tracopohorus. After its first metamorphosis, it embarks on its bottom-dwelling, or benthic, phase; the trunk with its 4 pincerlike jaws (see below) develop when the larva has attained a length of 1 mm, at which point it is capable of feeding on on animal prey.

▶ *Glycera convoluta* preys primarily on small crustaceans, and its predatory technique is similar to that of the ant lion: It lives in a network of shallow tunnels, including a number of vertical shafts terminating in an enlarged funnel-shaped pit at the surface. The prostomium (anterior part of the sandworm's body, including the brain) protrudes into this funnel from below and is equipped with receptors that are sensitive to vibrations, particularly those created by the struggles of a small crustacean that has blundered over the edge of the pit. The anterior end of the sandworm's digestive tube takes the form of an elastic trunk that is abruptly projected upward in order to envelop the prey and bring it into contact with the sandworm's four powerful jaws.

Each of these jaws consists of a curved pincer with an elastic lateral flap on each side; a canal runs down the outside of the pincer from a venom gland situated at its base. The venom is released into the prey through a series of perforations in the outer wall of the canal; it contains proteinaceous toxins, which appear to be particularly effective against small crustaceans, as well as proteases, enzymes responsible for the initial phases of the metabolism of animal protein. The pincers themselves are very sharp, consisting as they do of sclerotized proteins reinforced with molecules of iron and copper and other metals; their task is not only to immobilize the prey but also to pierce its shell so that the venom and predigestive enzymes can begin to take effect.

The sandworm swallows its prey immediately after the venom has been injected; a related species, *Glycera alba,* only has to eat about every 5 days, and the same is presumably true of *Glycera convoluta.* In aquariums, the latter has also been observed to feed on other polychaete worms, including members of its own species.

◀ Fish, large crustaceans, and seabirds are all known to prey on *Glycera convuluta,* particularly during the spawning season, when sandworms are often encountered on the surface of the water.

The giant sandworm *(Glycera gigantea),*

20–35 cm long, is found in the same biotopes as *G. convoluta,* though not as commonly. The most common species on the Atlantic coasts of Canada and the United States is *Glycera dibranchiata,* which is commercially exploited on a modest scale for use as live bait, especially by amateur salmon fishermen. Fossilized jaws of glycerids (scolecodonts) as much as 1 cm long have been recovered from Jurassic and Cretaceous strata, suggesting that these creatures may have attained a length of 80 cm, perhaps as much as 1 m. Until recently, there was some controversy in scientific circles as to whether the glycerids were genuine predators or merely detritovores, feeding on inanimate organic debris—a controversy that has since been resolved by direct observations of the predatory mechanisms described above.
C.M.

GOLDEN EAGLE *Aquila chrysaetos* 201

201
Sub.: Vertebrata
Cl.: Aves
O.: Falconiformes
F.: Accipitridae

The wingspread of both sexes varies between 1.90 and 2.27 m, though the female is somewhat longer and heavier than the male: 90–95 cm vs. 80–87 cm, 3.6–6.6 kg vs. 2.9–4.4 kg. Record longevity for the golden eagle in captivity is 46 years; a wild eagle in the Alps is known to have lived for more than 25 years. The golden eagle, sometimes called the imperial eagle in Europe, is sexually mature at the age of 4; its adult plumage appears the following year, but only 25% of young eagles survive to maturity.

Originally, the golden eagle probably flourished in all kinds of rough, open country from the seashore to the highest mountain, including the subdesert regions of North Africa and Mexico, the Canadian and Scandinavian tundra, the alpine massifs of Europe and North America; only dense wooded terrain is inhospitable to this species. In Western Europe, golden eagles build their aeries at altitudes of 200 to 2000 m, generally on a cliff face, rarely in a tree; their hunting grounds may include any kind of terrain with low or sparse vegetation: grasslands, brushlands, alpine meadows, and mountain crests.

▶ Young eagles may be rather venturesome, but when they reach maturity, they become creatures of very regular habits. The male-female bond is a close one, as with most large birds of prey, and the pair generally maintains a stately imperial progress between their several aeries and rest stops along a well defined intinerary. Legend has credited the golden eagle with the ability to carry off whole sheep, even 7– or 8–year-old children; stories have also circulated of attacks on full-grown humans.

In fact, though the golden eagle is theoretically capable of bringing down larger animals, extensive studies of its hunting behavior have shown that it will eat virtually anything from a fieldmouse to a fawn, including birds, snakes, and carrion. It prefers prey animals of between 1.5 and 5 kg, which in the alpine highlands of Europe would include marmots, hares, and capercailzie (a large European grouse, *Tetrao urogallus),* in the Mediterranean coastal regions, rabbits, hares, and red partridge, and in Scotland, varying hares and grouse. In other areas where the supply of plump rodents and game birds is less plentiful, the golden eagle may be obliged to restrict its diet to the smaller rodents (gundis, or comb rats, and gerbils of the genus *Meriones* on the subdesert steppes of North Africa) or to be much less selective in its choice of prey (voles, squirrels, adders and grass snakes, young chamois, and small carniviorous mammals in the Pyrenees).

Like most birds of prey, the golden eagle is an opportunistic hunter, which means that it prefers to go after easy prey—weak, sickly, or particularly incautious individuals. In patrolling its hunting grounds, it flies at low altitude, following ridgelines and the edges of clearings, darting out suddenly from behind a rocky outcrop or across a ravine to maximize the element of surprise. When the prey is sighted, the eagle tucks back its wingtips against its tailfeathers (so that its silhouette resembles an inverted heart shape) and plummets down in a high-speed vertical dive, pinning the prey against the ground or impaling it with its powerful talons.

Rarely, golden eagles will take their prey on the wing; for example, they have been observed attacking migratory cranes at an altitude of over 6000 m in the Himalayas. They have also learned the trick of breaking turtles' shells by dropping them from a great height (another eagle legend—the Greek orator Aeschines is said to have been killed in this manner when an eagle mistook his bald head for a rock).

The adult golden eagle normally requires about 230 g of meat daily, though it is capable of going without food altogether for several days. During the first few weeks of life, eaglets take scraps of meat from their mothers' beaks from which the hide or feathers have been removed. Before the fledglings are ready to leave the nest, their parents will be bringing entire carcases of small prey animals back to the aerie, and after leaving the nest, the eaglets will continue to accompany their parents on their hunting flights for 3 more months before setting out on their own.

◀ Due to their considerable adaptability with respect to diet and terrain, and the vast territories patrolled by a pair of golden eagles (70–400 km^2), the species does not face serious competition from other predators; in the Mediterranean region, only Bonelli's eagle *(Hieraetus fasciatus)* can be regarded as a competitor. This is something of a moot point, however, since because of its reputation as a stock-killer (or even an abductor of small children), systematic attempts have been made by mankind to exterminate the golden eagle, to destroy its eggs, its young, and its aeries with guns, traps, fire, and poison, in virtually all populated areas.

Before the species came under the protection of the federal government, for example, Texas ranchers organized airborne drives with light planes; one pilot is said to have brought down over 8,000 eagles. Today the golden eagle is protected throughout its range, but only about 2000 nesting pairs are thought to remain in all of Europe (excluding the Soviet Union). The beneficial effects of protection have probably been more than offset by an increased human presence (tourism, hiking, mountaineering) in its hunting and nesting grounds, and intensive hunting has also greatly reduced the numbers of the small mammals and game birds on which it preys and thus effectively exiled it from much of its former range.

The Linnaean system recognizes 2 genera of "true eagles," *Aquila* (10 species) and *Hieraetus* (5 species), though there are many varieties of large, primarily tropical raptors that seem to qualify as eagles from the standpoint of both appearance and behavior—*Haliaetus* (including the American bald eagle, Steller's sea eagle, and other fish-eating species), *Circaetus* (Old World snake-eating eagles), *Polematus* (including the African martial eagle), *Harpia* (the monkey-eating harpy eagles of South America and New Guinea), and others.

GOLDEN GROUND BEETLE *Carabus auratus* 202

202
Sub.: Arthropoda
Cl.: Insecta
S.cl.: Coleopteroidea
O.: Coleoptera
F.: Carabidae

This is by far the most common of numerous ground beetle (carabid) species on French soil; it is found in many different habitats depending on the region, with a general predilection for lawns, meadows, and gardens. The adult varies from 20 to 27 mm in size; the carapace is usually greenish in color with golden highlights. Certain local subspecies, limited in their range to Mont Ventoux, the Monts de Lure, and other isolated highland regions in the South of France, are black or bluish in color and greatly prized by collectors for their rarity. The elytra (wing covers) have three distinct ribs, more or less elevated or pronounced in the various subspecies.

▶ Like most other carabids, the golden ground beetle is flightless and seeks its prey exclusively on the surface of the soil. The adults generally hunt at night, but in the spring, their mating season when they are especially active, they may often be seen swarming through the "grass forest" in daylight, most often after a heavy rain. At other times of the year they are considerably less conspicuous. The adult golden ground beetle, like other carabids, is said to be *polyphagous,* which is say that while its tastes are somewhat more restricted than those of an omnivore, it feeds on a wide variety of prey—in this case, insects and their larvae as well as earthworms.

The larva lives almost entirely underground. Its body is tubular, elongated, and equipped with two long bristles on its tail; in these and other respects it resembles the adult form of the genus *Campodea,* thought to be a sort of evolutionary prototype of the carabids and a number of other insect families. For this reason, the body of the golden ground beetle larva is said to be *campodeiform;* with its small jaws and limited mobility, the larva is best suited to the task of dispatching earthworms, insect pupae, and other vulnerable prey. Like the adult beetle, the larva injects its digestive enzymes into the tissues of its prey, thus digesting its food "preorally" before it actually swallows it. The larva undergoes two preliminary moults before the nymphal molt, after which the adult beetle emerges from the underground chamber that the larva has dug for itself in the subsoil.

◀ During its annual population peaks after the spring mating season, the golden ground beetle is preyed on extensively by birds and small insectivorous mammals such as hedgehogs and shrews. Like many other carabids (including the bombardier beetle), the golden ground beetle is able to defend itself against these predators by expelling a corrosive, nauseating liquid from its rectum. During the last twenty years or so, the species has largely been exterminated in cultivated areas by the intensive use of insecticides and chemical fertilizers. The activities of collectors and amateur naturalists in quest of the rare local subspecies in the vicinity of Mont Ventoux have only served to increase their rarity—thus also decreasing their overall diversity of color and form, as well as doing a surprising amount of damage to the habitat and the other fauna of the subsoil.

J.-C.M.

GOLDEN JACKAL *Canis aureus* 203

203
Sub.: Vertebrata
Cl.: Mammalia
O.: Carnivora
F.: Canidae

From a human perspective (and perhaps because of its close kinship to the domestic dog), the jackal appears to be one of the friendlier and more appealing of the large African carnivores. There are 3 different species, the golden jackal, the side-striped jackal *(Canis adjustus)*, and the black- or saddle-backed jackal *(Canis mesomelas)*; all are quite similar in their morphology, differing only in the pigmentation of their coats and other external details. The jackal's body is from 90 to 100 cm long, terminating in a long bushy tail like a fox's that measures about 35 cm in the case of the black-backed jackal.

These 3 species are found in parts of Africa, Asia, and even in Eastern Europe; the Serengeti Plain of Tanzania is the only place in the world where all 3 are sympatric (i.e. their geographic ranges overlap). Like other canids, an individual jackal occupies a fairly extensive territory, in this case about 3.5 or 4 km^2. The jackal scent-marks the boundaries of its territory with its urine, and usually an aggressive display short of actual physical combat is sufficient to deter an intruder; naturally the jackal will only attempt to defend its territory against other jackals and not against a larger predator such as a lion.

▶ Since antiquity, the jackal has been disdained by humans as a scavanger and an eater of dead flesh. In fact, the jackal is an omnivore and a highly opportunistic predator that will eat virtually anything that comes its way—small mammals, the placentas and newborn calves of gazelles or other ruminants during calving season, older and enfeebled ruminants of moderate size during the remainder of the year, birds, eggs, and even windfall fruit, as well as carrion. The jackal hunts alone or in male-female pairs, though observations have shown that this second strategy is by far the more successful (80% of chases resulting in kills when jackals hunted in pairs as opposed to 17% when jackals hunted alone—though the difference is partially to be explained by the fact that the quarry is likely to be smaller and more elusive in the latter case).

When jackals hunt in pairs, the quarry is usually a young gazelle; one of the jackals acts as a decoy, provoking the mother gazelle into charging it and leaving its calf unattended. The other jackal, which has thus far remained concealed in the tall grass, emerges from cover and carries off the prey. Solitary hunting jackals will attempt to catch birds and insects and dig rodents out of their burrows, as well as feeding on the remains of kills abandoned by lions or other predators; alerted by the presence of vultures, as many as 15 or 20 jackals may be attracted to a carcass, though jackals do not hunt or assemble in packs on other occasions. The young remain with their parents, learning to hunt, until they are about a year old, and are usually capable of fending for themselves by the age of 8 months.

◀ The jackal may occasionally be taken from ambush by a leopard, but it is quick and agile enough to elude the other large

predators of the African plains. In parts of Asia and Eastern Europe where the jackal's diet has come to include domestic livestock, it is frequently persecuted by man.
P.A.

204
Sub.: Arthropoda
Cl.: Insecta
O.: Odonata
S.o.: Anisoptera
F.: Cordulegasteridae

GOLDEN-RINGED DRAGONFLY
Cordulegaster annulatus 204

The golden-ringed dragonfly is one of the largest members of the family Libellulidae that is found in Europe and is easily recognizable by the bright yellow rings around its black body and its green eyes that touch only at a single point (and are not partially fused together, as with most dragonflies). The posterior set of wings is the larger of the two (as is the case with all the anisopetrans) and is held out flat while the dragonfly is at rest and never held up vertically or folded back to cover the abdomen. Wingspan is close to 9.5 cm and the overall body length is about 8 cm (plus the ovipositor in the case of the female). The principal European subspecies, *C. annulatus annulatus*, ranges from England and the Netherlands through Central Europe, Asia Minor, and North Africa.

C. annulatus annulatus is found mainly along the banks of streams where the current is swift and the vegetation abundant. During the daylight hours, the males are constantly skimming over the surface of the water at a height of 30–50 cm and in a hesitant zigzag pattern that undoubtedly has some territorial significance; the female flies in a more decisive manner, at a slightly higher altitude, and may venture inland for some distance. The male occasionally interrupts its reconnaissance flights and rests while dangling vertically from the underside of a leaf.

▶ *Annulatus* seizes and devours its prey on the wing, even though the pursuit may require it to deviate from its unaccustomed flightpath and even leave its territory unguarded for a few brief moments. This prey consists of small flying insects (flies, craneflies and mosquitoes, microlepidopetra, caddis flies); unlike such such species as *Anax imperator, Cordulegaster annulatus* will not attack large butterflies or other more substantial prey. The larvae live on the bottoms of streams frequented by the adults; as with all the odonata (dragonflies), the larva is equipped with a fearsome predacious mechanism called the "mask" or "mandibular arm" (the latter an especially apposite term that was coined by René Antoine Réaumur in his *Mémoires pour servir à l'histoire des insectes* in 1742).

In the case of *annulatus*, the mask is a visorlike appendage that covers the top of the head (and the mouth when not in use) and terminates in a set of spatulate pincers with sawtoothed medial edges, perhaps slightly reminicent of the bucket of a steam shovel. The larva buries itself in the mud and hunts its prey from ambush, seizing a passing insect with the interlocking pincers of its mandibular arm so the prey can be brought within reach of the grinding mandibles on either side of its mouth.

The *annulatus* larva is powerful, vigorous, and quite voracious and feeds on the larvae of mayflies, hellgrammites, and mosquitoes that frequent the same biotype, as well as *Calopteryx virgo;* the *annnulatus* larva can go for many days without food, though in such cases, it eventually becomes impatient and sets out on a more active search for prey; an *annulatus* larva that had had nothing to eat for 3 weeks was observed to consume no fewer than 24 chironomid larvae in 8 minutes. By the time it is ready for its nymphal moult, the *annulatus* larva will have attained a length of more than 5 cm, making it the largest invertebrate (with the possible exception of the crayfish) in this particular environment.

◀ Like other anisopterans, the *annulatus* larva is equipped with an assemblage of sharp-pointed sclerotized wedges called the *anal pyramid,* the primary role of which is to assist in respiration but which may also be of some use in repelling smaller predators. The larva of this species may be devoured by fish of the salmon family, watersnakes (notably the ringed snake and the viperine snake, *Natrix maura),* and perhaps some birds and mammals as well. The adult may be caught by birds and by spiders that spin their webs in the riverbank vegetation.

The 4 subspecies of *Cordulegaster annulatus* differ somewhat in the size and arrangement of the yellow rings on the abdomen as well as in their geographical distribution; 2 related alpine species, *C. princeps* of Europe and Asia Minor and *C.*

bidentatus of the Atlas Mountains of Morocco, are quite similar in appearance and, as far as we know, in their predatory behavior. G.B.

GOOSANDER *Mergus merganser* 205

205
Sub.: Vertebrata
Cl.: Aves
O.: Anseriformes
F.: Anatidae

This species of diving duck, also known as the common merganser, is found throughout the Northern Hemisphere, though the North American subspecies *(Mergus merganser americanus)* is generally referred to as the American merganser rather than goosander. It measures between 57 and 75 cm, with a wingspread about 95 cm, and weighs between 1000 and 1900 cm. It prefers the cold, clear waters of rivers and lakes, wherever fish are abundant, and builds its nest in a hollow tree.

▶ Diurnal and sociable, even on its fishing grounds, the goosander feeds on small fish (generally about 10 cm long though sometimes as long as 30 cm) of whatever species is locally abundant, as well as various species of Salmonidae and Cyprinidae, eels, aquatic insects and their larvae, mollusks, and frogs. The goosander's hooked beak is 45–60 mm long and provided with sharp sawtooth serrations. It locates its prey visually—ducking its head below the surface to reconnoiter—then dives down to a depth of 2 or 3 m, chases and catches its fish, and brings it back up to the surface, holding it crosswise in its sawtoothed beak, and swallows it headfirst. The average duration of the dive is 50–120 seconds; several mergansers will often fish together, which probably increases their efficiency as predators.

◀ Newly hatched goosander chicks are often obliged to undertake a fairly long trek from the nest to the edge of the water, in the course which they may be attacked by crows and ravens, herring gulls, skuas and jaegers. The healthy adult goosander has no natural enemies, and though the red-breasted merganser *(Mergus serrator)*, the great crested grebe *(Podiceps cristatus)*, and various species of divers *(Gavia)* use essentially the same fishing technique, this does not appear to give rise to competitive pressures. Formerly much persecuted as harmful competitors by fishermen, mergansers are now known to play an important role in culling fish stocks and in cleansing the rivers; the goosander is now totally protected in France. The supply of suitable nesting sites may be threatened by deforestation and land development, but this could be remedied by the erection of nesting boxes.

The Mergini are a tribe of diving ducks that comprise 14 species classified into 5 genera, of which 6 species are generally regarded as mergansers; this includes the the Brazilian merganser *(Mergus octosetaceus)*, which is extremely rare and until recently was thought to be extinct, and the Auckland Island merganser *(Mergus australis)* of New Zealand, which has probably been extinct since the end of the nineteenth century. The smew *(Mergus albellus)* is the smallest of the mergansers and is found exclusively in the Old World.

J.-F. T.

GRAY HERON *Ardea cinerea* 206

206
Sub.: Vertebrata
Cl.: Aves
O.: Ciconiiformes
F.: Ardeidae

The gray heron, or common European heron, is 90 cm long, with a wingspread of 175–190 cm, and weighs between 1200 and 1900 g. Maximum longevity is about 25 years, though mortality may be as high as 79 percent during the first year of life. It is not overly particular in its choice of habitats; though it normally requires clear, shallow water in which to hunt for fish (marshes, reservoirs, irrigation ditches, rice fields), it is sometimes found in dry terrain (plowed fields) during the winter. It also frequents rocky shorelines and coastal mudflats, and nests at altitudes up to 1000 m and in a variety of localities—tall trees, willow groves, reedbeds, as well as on the ground. The gray heron spends many hours of the day resting in an open space or on its perch in a dead or leafless tree. Gray herons are sociable and may congregate in nesting colonies of several hundreds pairs. This species is found in Europe, Central Asia, and southern Africa. Some populations are nomadic, others migratory; the latter spend the winter in tropical Africa.

▶ The grey heron is a skilled fisherman, specializing in riverine species between 12 and 16 cm long (and sometimes up to 25 cm), including carp, rudd, roach, pike, bleak, eels, and perch. These may be supplemented by a variety of insect prey as well as worms, mollusks, frogs, colubrids, nestlings of other species (notably the dabchick), and small mammals (voles, moles, water rats, and even weasels). The heron's daily ration is about 330 g, and it has three important weapons as a predator: its sharp stiletto beak, its long reptilian neck, and its extraordinary patience. It stalks through the water very slowly, and after an almost interminable vigil, it darts out its neck and impales its victims with its beak. This maneuver is accomplished with considerable brutality, and though it will swallow smaller fish on the spot, it sometimes dispatches larger prey by clubbing it against the bank.

◀ The gray heron is not systematically preyed on by any other creature, though individual gray herons may occasionally be taken by a white-tailed eagle, a goshawk, or an eagle owl, rarely by a falcon. Its range overlaps with that of the purple heron *(Ardea purpurea)*, a tireless migrator that seeks its prey in pools in the midst of reedbeds and generally in shallower water, since its legs are shorter than those of the gray heron. The smaller night heron *(Nycticorax nycticorax)*, also found in Southern Europe, only becomes active at dusk.

The gray heron was systematically, and shamefully, massacred in many areas of Europe during the first half of this century as a threat to commercial fisheries. In France, where it is now a protected species, it is only just coming back up to strength (3500 nesting pairs, mostly the marshes of the east and west, with a few colonies in the north and the Camargue). The gray heron is undoubtedly guilty of pillaging commercial fish farms and hatcheries, but these depredations can more effectively be dealt with by simple preventive—rather than retaliatory—measures. The survival of the gray heron may still be threatened by high concentrations of toxic substances in the aquatic foodchain, including heavy metals, lead and mercury, and chlorinated hydrocarbons (DDT, PCB, etc.).

The heron family consists of 24 genera, 69 species altogether. The marsh-dwelling goliath heron *(Ardea goliath)* of Africa is the largest of these; the familiar great blue and great white herons, now recognized as two different varieties of the same species *(Ardea herodias)*, are the largest North American herons. The African black heron *(Melanophayx ardesiaca)* has a highly individualized fishing techique: It stands motionless in deep water and spreads its wings, creating a shady patch on the surface to which fish are attracted and are better seen by the elimination of distracting reflections on the water as well.

J.-F. T.

GRAY PHALAROPE *Phalaropus fulicarius* 207

207
Sub.: Vertebrata
Cl.: Aves
O.: Charadriiformes
F.: Phalaropodidae

The gray phalarope measures 27 cm, with a wingspread of about 37 cm, and weighs between 36 and 60 g. It nests on the shores of lakes and marshes in the North American tundra and spends the rest of its life on the open ocean; with its webbed feet and dense, water-repellent feathers, the gray phalarope is well adapted to life in cold northern waters. It is only an accidental guest in Western Europe. The female is larger than the male, and apart from the inevitable fact that it is the female that lays the eggs, the usual division of labor is completely reversed. It is also the female that wears the nuptial plumage and performs the courtship displays while (or rather, before) the male builds the nest, hatches the eggs, and finds food for the nestlings.

▶ In fresh water, the gray phalarope feeds on flies, mosquitoes, midges and their larvae, mollusks, small crustaceans, and worms, which it snatches off the surface of the water with a quick, darting motion of its head and beak. In salt water, it forages on floating rafts of seaweed and has even been observed picking the whale lice (a kind of crustacean) off the backs of whales. It is often seen whirling around in circles, both in fresh water or on the open sea; the purpose of this maneuver is to create a kind of whirlpool effect that sends submerged mosquito larvae or planktonic crustaceans bobbing up to the surface. Along the shoreline, the phalarope probes the mud with its elongated beak, like a snipe. The chicks share the same diet as the adults and learn to hunt for food by following their father on his rounds.

◀ The adult gray phalarope has no particular predators (with the possible exception of the gyrfalcon), though the eggs and nestlings are occasionally devoured by gulls, skuas, and Arctic foxes; with its peculiarly specialized feeding habits, it has no competitors at all. The gray phalarope is totally unafraid of humans (with whom it has admittedly had very little contact thus far), and some phalaropes will even allow themselves to be touched by a human being. The gray phalarope's migration routes have shifted northward in recent years, apparently in response to global climatic changes.

Three species of phalarope are found in the Arctic regions; in each case, the female can be distinguished by its colorful nuptial plumage during the brief period that it spends on land. J.-F. T.

GRAY SEAL *Halichoe rus grypus* 208

208
Sub.: Vertebrata
Cl.: Mammalia
O.: Pinnipedia
F.: Phocidae

The males may attain a length of 3m and weigh as much as 300 kg; the females generally do not exceed 2.5 m and 250 kg, and the average size is closer to 2.5 m for the males, 1.8 m for the females. The gray seal is found in the temperate and subarctic waters of the North Atlantic, more specifically along the rocky coasts of Iceland and the Faeroes (3000 individuals), Norway (1000 individuals), and Great Britain (35,000 individuals); smaller colonies are found along the North Sea and the English Channel. It generally frequents the base of cliffs and other places along the shore that are unapproachable from the landward side, and its rookeries are generally founded on inaccessible offshore islets. Longevity is on the order of 30 years.

▶ The diet of the gray seal varies according to the locality and the season; in general, it feeds on fish (cod, salmon, herring, flatfish) as well as squid, cuttlefish, and octopus, and occasionally on crustaceans. In clear water, the gray seal hunts by sight, but it is adept at finding prey in muddy or turbid water that is virtually opaque. Even though it keeps its nostrils tightly closed while underwater, it is thought that the senstive tip of its nose is equipped with additional chemoreceptors that can provide at least an approximate fix on a prey animal's location. The gray seal's whiskers are richly inervated, and when these make contact with a fish or other prey animal in murky water, the gray seal snaps its head abruptly and catches the prey in its jaws.

Adult gray seals can dive to a depth of 150 m, remaining submerged for up to 20 minutes, and they can eat as much as 5–8 kg of fish in a single day. They are accustomed to fasting for weeks on end, since the males are reluctant to leave their harems unguarded during the mating season, and the females are unable to hunt for themselves until their pups are weaned. Gray seals also eat comparatively little while they are shedding their winter coats; after the breeding season, when the blubber layers of both sexes are considerably depleted, they return to the sea and begin to hunt actively.

◀ The killer whale is the only marine predator of the gray seal. This species has never been of any great economic importance to man; the skin of the adults is occasionally made into leather, but, luckily for it, the ragged pelt of the young gray seal has little commercial value. The gray seal is protected throughout the North Atlantic region, though with certain significant reservations: Fishermen are permitted to shoot at seals that venture too close to their nets, and the populations of gray seal colonies in Scotland and Canada may be artificially controlled in order to avoid conflict with local fishing interests. It admittedly robs nets, though the gray seal's partisans contend that the dwindling catches are often the result of overfishing. The gray seal harbors the adult form of a parasitic nematode *(Porrocaecum decipiens)* that infests codfish and other gadids, making them unfit for human consumption. D.R.

GREAT BARRACUDA *Sphyraeana barracuda* 209

209
Sub.: Vertebrata
Cl.: Osteichthyes
S.cl.: Teleostei
O.: Perciformes
F.: Sphyraenidae

The adult barracuda attains an average length of between 1 and 1.5 m and a weight that varies between 10 and 30 kg; the record specimen measured 1.65 m and weighed 47 kg. Average lifespan is unknown. The great barracuda is a saltwater fish that is found throughout the tropics (with the exception of the Eastern Pacific) and along the Eastern Seaboard as far north as 40°; it may live in the open ocean but is also frequently encountered in the vicinity of coral reefs.

▶ The great barracuda is a voracious hunter, active during the day, that preys on virtually all species of benthic and pelagic fish as well as young seabirds. Its torpedo-shaped body enables it to approach its prey slowly and silently, often borne on the current; it detects its prey primarily by sight, though its sense of smell may also come into play at short range (e.g., it is possible that barracuda, like sharks, are drawn by the smell of blood). When it gets within striking distance, it propels itself forward with a violent motion of its caudal fin—sometimes attaining a speed of 44 kph, over very short distances—and seizes its prey in its very well developed front teeth. If the prey is too large to be ingested all at once, the barracuda snaps it in two during its intial onslaught, then turns around and collects the pieces.

Older and larger individuals are solitary, but young barracuda (smaller than 60 cm, more or less) hunt in shoals, pursuing their quarry persistently until it succumbs to exhaustion. The great barracuda is unquestionably dangerous to man; over 30 attacks on human beings have been reported, most of which have occurred in troubled water or as a result of some sort of provocation on the part of the victim. One authority (Fourmanoir, 1976) has assigned the responsibility for these incidents to a different barracuda species, *Sphyraena jello.*

◀ Young barracuda are sometimes preyed on by tuna, but the adult great barracuda has no natural enemies save for the two great superpredators of the deep, the shark and man. The flesh of the barracuda is consumed by man in some African countries, though in the Caribbean (including Florida waters) and the Pacific, the barracuda has frequently been found to be a carrier of one or more of the still unidentified toxins that have caused numerous cases of a serious disorder called ciguatera. The ciguatera toxin is not produced by the barracuda itself (or in any of the other predatory sea creatures in which it occurs) but is thought to be present in blue-green algae (Myxophyceae) and to be passed along the successive stages of the food chain; it has been determined, for example, that only the flesh of some barracuda in certain areas and at certain seasons is actually toxic to human beings. However, since such determinations can only be made after the fact, it has been thought advisable, notably in the French Antilles, to prohibit the sale of *Sphyraena barracuda* as a food fish, though this species is still especially prized by deep-sea fishermen for its fighting qualties.

There are 18 *Sphyraena* species in all, several of which are similar in their habits and appearance to the great barracuda. *S. sphyraena* lives in European coastal waters and may attain a length of 1.5 m. *S. querrie* is slightly larger (1.7 m), *S. beekeri* (1.1 m) slightly smaller than the great barracuda; the nocturnal *S. jello,* mentioned earlier, is larger still. Smaller species such as *S. forsteri* (80 cm) of the Pacific and *S. picudilla* (55 cm) of the Caribbean continue to live in shoals throughout their lives.

J.-M. R.

GREAT CORMORANT *Phalacocorax carbo* 210

210
Sub.: Vertebrata
Cl.: Aves
O.: Pelecaniformes
F.: Phalacrocoracidae

The great, or common, cormorant is the largest of the cormorants, with a body length of 0.8–1 m, a wingspan of 1.3–1.6 m, and a weight that varies between 1.7 and 2.8 kg. This species is sexually mature by the age of 3 or 5, and one banded individual survived to the age of 19, though mortality is around 70% during the first year of life. It is found in a variety of habitats on both sides of the North Atlantic, throughout Europe, Asia, Australia, and parts of Africa—flat coastal plains or escarpments, estuaries and watercourses, marshes, and manmade reservoirs; the great cormorant's most basic requirement is a broad expanse of water, devoid of floating vegetation, in which to hunt for fish. It builds its nest in trees or on sheltered ledges of a cliff face if these are available, as well as occasionally on the ground or in a reedbed; great cormorant populations in the north or south temperate zones may assemble in substantial troops along the shores of any large body of water, fresh or salt, where there is a suitable place (dead or leafless trees, sea cliffs) for them to roost at night.

▶ In salt water, the cormorant feeds primarily on flounders and other flatfish as well as sardines, cod, whiting, herrings, and sprats. In fresh water, it hunts for eels, perch, tench, and various members of the salmon family. In exceptional cases, it may feed on the chicks of other aquatic birds (ducks, sheldrakes, moorhens, turnstones) as well as on frogs and other amphibians. It typically consumes about 750 g of fish a day and regurgitates a pellet consisting of the scales, bones, and other inedible material.

The cormorant locates its prey by immersing its eyes and head in the water, then rocks back, wings folded, to position itself for its dive. In the water, its elongated, spindle-shaped body makes an excellent hydrofoil; alternate paddling motions of its feet help to sustain the momentum of the dive, which usually lasts between 20 and 45 seconds (though sometimes as long as 71). The cormorant may dive to a depth of 10 meters, though 3 or 4 is much more usual; it brings its prey back to the surface, tosses it into the air, then catches it and swallows it headfirst. The nestlings are fed on a fishy pap contained in the pharyngeal pouch of their parents.

◀ The great cormorant is sufficiently robust to deter most predators, with the occasional exception of the peregrine falcon and the white-tailed sea eagle *(Haliaetus albicilla)*; seagulls may succeed in snatching a fish from the cormorant while it fumbles with it on the surface. The genus *Phalacocorax* comprises about 30 species altogether. The crested cormorant *(Phalacocorax aristotelis)*, also found in European waters, is slightly smaller than the great cormorant.

The cormorant's hunting prowess has been both resented and exploited by man. In the first case, the cormorant has been perceived as a competitor and consequently persecuted by many fishing communities throughout the world, though it now appears that the size and species of fish consumed by it are of no real commercial value. In the second case, the great cormorant and a another smaller relative, the Japanese cormorant *(Ph. capillatus)*, have traditionally been trained to catch fish for their human masters in the Orient. Finally, such guano-producing species as the guanay *(Ph. bougainvillei)*, the white-backed cormorant *(Ph. atriceps)*, and the royal cormorant *(Ph. albiventer)*have often provided the chief export of the islands and coastal regions inhabited by them in western South America or in various parts of the Pacific.

J.-F. T.

GREAT CRESTED GREBE *Podiceps cristatus* 211

211
Sub.: Vertebrata
Cl.: Aves
O.: Podicipediformes
F.: Podicipedidae

The great crested grebe measures 47–56 cm, with a wingspread of 73–86 cm, and weighs between 700 and 1160 g. It becomes sexually mature in its second year of life; record longevity in the wild is 10 years. In spring, it frequents large freshwater ponds and lakes; in winter, it seeks saline estuaries and bays as well as large inland reservoirs—provided in each case that there are reedbeds along the shoreline. It is rarely found at altitudes in excess of 300 m. The great crested grebe currently ranges from the British Isles to Manchuria and is found throughout much of Europe and Central Asia as well as southern Africa and parts of Australia; it is curently expanding its range and population in Western Europe. A pair of great crested grebes will defend a territory that varies in size from 300 sq m to as much as 1 sq km (depending on the abundance of fish and other prey); intruders are warned off with an elaborate ritual in which both sexes participate. Though great crested grebes are less gregarious when food is scarce, they sometimes assemble in colonies on large, well-stocked lakes.

▶ This bird's legs are set very far back on its body, which is useful for swimming and diving, less so for locomotion on land; the great crested grebe may set foot in the shallows while it builds its floating nest and rears its young, but it never takes to the air except during its seasonal migrations. The adult bird consumes about 150–200 g of fish every day: bleak, roach, rudd, trout, eel, perch, bream, chub, and pike (up to 20 cm long). It also feeds on aquatic insects and their larvae, small crustaceans and mollusks, frogs, seeds, and bits of vegetation; the young grebes feed primarily on insects during their first few months of life.

The adult grebe sometimes catches insects and other small creatures on the surface, but it dives for fish and most of its other prey. It hunts and forages at a depth of 2–4 m, though sometimes as much as 30 m, and usually for about 30 seconds at a time. The grebe is able to reduce its buoyancy so that it can glide through the water "decks awash," making it easier to approach its prey or to escape from predators, and it increases its speed while swimming underwater by beating its wings as well as paddling with its feet.

◀ The adult great crested crebe is elusive and inconspicuous in its habits—it dives down to the bottom when threatened or alarmed—and consequently has little to fear from predators. The eggs and nestlings are sometimes devoured by buzzards, crows, and pike, and the adult grebes carry their chicks on their backs to protect them from pike and other underwater predators. As noted earlier, the norms of intraspecific aggression and territoriality are somewhat relaxed in open waters where food is abundant; smaller species such as the black-necked grebe and the dabchick feed on much smaller prey animals and are not competitive with the great crested grebe, though the latter may be exposed to competition from the common cormorant *(Phalacrocorax carbo)* while in its saltwater winter quarters.

Pursued by plume hunters during the last century and, until recently, persecuted by fish farmers as a harmful predator (though the fish that it feeds on rarely exceed 13 cm and are of no real commercial value), the great crested grebe is still very rare in certain areas. In Great Britain, for example, only 42 mating pairs were left in 1980, though the species has since become relatively common there. The eutrophication of Lake Leman and other large lakes on the Continent has attracted thousands of great crested grebes, which may stay on through the winter after nesting in the spring. J.-F. T.

212
Sub.: Arthropoda
Cl.: Insecta
S.cl.: Pterygota
O.: Coleoptera
S.o.: Adephaga
F.: Dytiscidae

GREAT DIVING BEETLE *Dytiscus marginalis* 212

The adult great diving beetle, or great water beetle, like the more notorious larval form (known as the "water devil" or "water tiger"), lives in stagnant bodies of water and sluggish streams—especially in ponds and drainage ditches that are overgrown with aquatic plants (elodea, Myriophyllum, Ceratophyllum). This species ranges throughout most of the the Northern Hemisphere, including much of Europe and Asia, Siberia, Japan, and North America. The adult's body is oval, flattened dorso-ventrally (i.e., top and bottom), and about 3.5 cm long, greenish or olive-green on top with a yellow border (hence, *marginalis)* and uniformly brownish on the bottom. The head of the great diving beetle is embedded directly in the thorax and terminates in two filiform antennae; its third pair of legs is flattened like the blade of an oar and fringed with long feathery bristles, providing a helpful "power stroke" as the beetle sculls through the water but dragging uselessly behind it as it clambers about on land. The female's elytra are usually fluted; the male's are always smooth. The adult lives for 2–4 years.

The body of the larva is greatly elongated, whitish or yellowish in color. Its head is broad and flat and provided with two strong, hooklike mandibles; its swimming legs are long and ciliated, and there are two lateral extensions at the tip of the abdomen, covered with water-repellent bristles and pierced with two slitlike stigmata that permit the larva to breathe while hovering just below the surface.

▶ The adult diving beetle is quite voracious and seeks its prey primarily during the daylight hours among the water weeds and in the mud at the bottom of its pool or stream; it prefers living prey—primarily mayfly, dragonfly, and chironomid larvae as well as planorbids and lymnaeids (freshwater snails), frogs and tadpoles, tritons, and fish hatchlings—though it occasionally feeds on carrion as well. The prey is minced up into very small pieces by the clashing mandibles and maxillae before being ingested by the beetle; in captivity, an adult *Dytiscus* eats about half a gram of food every day.

The diet of the *Dytiscus* larva, the "water devil," is much the same as that of the adult, though it shows a particular preference for tadpoles. It also hunts by day, stationing itself near the surface and plummeting down to seize its prey with its mandibles when opportunity permits. The devil's mouth is quite tiny, but its hooklike mandibles are hollow, like a cobra's fangs, enabling it to inject its prey first with a toxic liquid that kills it and subsequently with digestive fluids that liquefy its tissues, reducing them to a sort of broth than can readily be ingested, leaving only the empty skin of the tadpole behind.

◀ Herons, ducks, and other aquatic birds take a considerable toll of this species, particularly the softer-bodied larvae, which are generally to be found near the surface. Pike, perch, turtles, and crayfish also feed on giant diving beetles, and water devils sometimes fall prey to dragonfly larvae and, in the case of the very small ones, to slightly larger members of their own species, apparently an important cause of *Dytiscus* mortality during the first few weeks of life. Diving beetles of the genus *Dytiscus* and the related genus *Cybister,* as well as giant water bugs of the genus *Belostoma,* are eaten by man in China and in several other regions of the globe.

The family Dytisicidae comprises more than 3000 species worldwide, of which nearly 200 are found in Western Europe; these vary in size from 1 or 2 mm to several cm, and the size of their prey (all are carnivorous) varies accordingly, but all feed on larger or smaller individuals of the same species as *Dytiscus marginalis.*

GREAT EARLESS LIZARD
Cophosaurus texanus 213

This small iguanid is found in the arid plains and rocky uplands of the southwestern United States and northern Mexico. "Great" in this case is a relative term; maximal length is only 18.5 cm, though it still exceeds that of the lesser earless lizard *(Holbrookia).* It is distinguished by two transverse black patches highlighted by the light background color of its flanks as well as by transverse black bands on the underside of its tail, and when excited or alarmed, it shows off its stripes by curving its tail forward over its back.

"Earless" means that this lizard has no visible external eardrum, though it still may perceive sounds through the vibrations in its skull. This species is generally found in arid terrain; population density in the bajada of New Mexico ("bajada" refers to the sloping alluvial foothills of a mountain range or other formation) was found to vary between 3 and 20 individuals per hectare in accordance with fluctuations in annual rainfall. Reproduction takes place sometime between April and August, also depending on seasonal rainfall and the relative abundance of insect prey. The great earless lizard reaches sexual maturity very quickly; mortality during the first year of life is about 70 percent, and maximal longevity is on the order of 4–5 years.

This species is more active on hot, overcast days than when the sun is shining, though, unlike *Cnemidophorus,* its circadian cycle is continuous and it does not retire to its burrow at midday. It is quick and furtive in its movements, frequently darting from one rock to the next and then pausing for a moment as if to reconnoiter the territory ahead. It prefers high temperatures (38.5°C is the average temperature at which it is active), though some individuals emerge from their burrows during the first hour after sunrise and do not return until after nightfall.

▶ The great earless lizard hunts from ambush in open country; it pursues a "wait and see" strategy, perched motionless on top of a rock and preying opportunistically on crawling and flying insects, in more or less the following proportions: Isoptera, 28.5 percent; Lepidoptera (including caterpillars), 18.1 percent; Hymenoptera, 17.6 percent; Coleoptera, 8.6.

213
Sub.: Vertebrata
Cl.: Reptilia
O.: Squamata
S.o.: Sauria
F.: Iguanidae

◀ The great earless lizard and the western whiptail (q.v.), which is found in the same habitat, are both threatened by the same predators, including rattlesnakes and collared lizards, roadrunners, shrikes, kestrels, and sparrowhawks. Like the sand lizard *Uma exsul,* the great earless lizard will sometimes attempt to intimidate an aggressor by displaying the dark-colored stripes on the underside of its tail.

Cophosaurus texanus is the only representative of its genus, though 3 distinct subspecies are recognized. The lesser earless lizard *(Holbrookia maculata)* occupies a similar niche in the same ecosystem. In the Bolsón de Mapimi, *Holbrookia* is found only on the playa (floor of the basin) whereas *Cophosaurus* inhabits the sloping bajada that forms the walls of the basin, so that competition between the two species is minimized.

On the bajada, the great earless lizard

is most frequently found in conjunction with two whiptail species, *Cnemidophorus tigris* and *scalaris.* Though their dietary preferences and circadian cycles are very similar, their feeding techniques are sufficiently different that direct competition is avoided here as well—while the earless lizard pursues its policy of "watchful waiting" on a rock in open terrain the whiptail searches actively for insects by digging around the roots of bushes. C.G.

214
Sub.: Vertebrata
Cl.: Mammalia
O.: Chiroptera
F.: Rhinolophidae

GREATER HORSESHOE BAT *Rhinolophus ferrumequinum* 214

This species measures between 5.6 and 6.9 cm (exclusive of the the tail, which measures 3–4.3 cm), with a wingspread of 33–36 cm. The name "horseshoe bat" refers to the shape of the prominent nasal leaf, which is characteristic of the family Rhinolophidae. The greater horseshoe bat is widespread throughout Western, Southern, and Central Europe, and during the summer is generally found in attics, cellars, and abandoned buildings; during the winter it retreats into a cave or a cellar, either singly or in groups. Longevity is from 15 to 23 years.

▶ The greater horseshoe bat emerges at dusk, but does not begin to hunt until nightfall and continues to do so until sunrise, flying at an altitude of only.5–3 m. The sonar echolocation system with which bats navigate and pursue their prey in the dark is particularly well developed in this species, since the flexible nasal leaf acts as a sort of radar dish, focusing the ultrasonic vibrations (80–100,000 hz) that are produced in the horseshoe bat's nose (rather than in the mouth, as is the case with most other bats) on a moving target ahead of it. The horseshoe bat's ears can also move quite freely, which enables it to take advantage of the Doppler shift (the decrease in frequency as an object moves away from the observer) to track the relative speed and trajectory of flying insects with remarkable precision.

In summer, the greater horseshoe bat feeds primarily on noctua moths and scarabs in roughly equal measure, and primarily on dung beetles (Geotrupes) during the cooler months; inventories of stomach contents have also included a certain number of terrestrial prey animals (including cave spiders and wingless ground beetles), so this species presumably does not always catch its prey in flight. It catches large insects only while clinging to a perch of some sort and never in midair. The greater horseshoe bat feeds primarily on the soft tissues of arthopods, and its digestive juices contain a large proportion of chitinase, an enzyme that helps to break down the chitinous body parts; this process is only partially successful, however, since the guano of this species contains undigested fragments of the carapaces of beetles and the wing scales of noctua moths.

The lesser horseshoe bat *(Rhinolophus hipposideros)* is also widely distributed across Europe and feeds on smaller moths, grasshoppers, and other insects. It catches its prey by skimming over the tops of herbaceous plants, and, like its larger relative, will only ingest an insect when it is clinging from a branch, underneath which the legs, wings, and other inedible portions of the insect will later be found. J.-J. B.

215
Sub.: Vertebrata
Cl.: Osteichthyes
S.cl.: Teleostei
O.: Perciformes
F.: Trachinidae

GREATER WEEVER *Trachinus draco* 215

This marine fish typically measures between 20 and 30 cm and may attain a length of as much as 50 or 60 cm. It is found along muddy or sandy substrates in shallow coastal waters (5–80 m), though it is more frequently encountered at depths of up to 150 m during the winter. This species is common in the Eastern Atlantic from Norway to the south of Morocco as well as in the Mediterranean and the Black Sea. Aquarium specimens have been known to survive for longer than 20 years.

▶ During the day, the great weever stays buried in the sand with only its eyes showing, though it is quick to attack a fish or other creature that ventures near it. It hunts in small schools by night, tracking the vibrations produced by other fish by means of its lateral line detectors. The weever feeds voraciously on dragonets, gobies, and other small bottom-dwelling species as well as on shrimp and mollusks (or virtually any other small marine organism that it encounters).

◀ The greater weever generally avoids trouble by burying itself in the sand, but as soon as it senses that it has been found out, it rushes out to confront its tormenter with its venomous spines tilted forward. The spines on its gills and the spiny rays of its first dorsal fin serve as the outlets for subcutaneous venom glands, and the weever's venom can be very dangerous to humans, particularly those with a preexisting cardiac condition. The weever rarely (if ever) emerges from its hiding place to attack a human, but it is frequently stepped on by swimmers; the sensation is excruciatingly painful and is often accompanied by an edematous swelling and surface irritation. If left untreated, the pain may persist for up to twenty-four hours, and the wound frequently becomes infected.

The weever's venom destroys red blood cells, which may also result in cardiovascular and metabolic problems, and may have a neurotoxic effect as well. Since its venom is said to have a mildly euphorient effect when ingested in small doses, the greater weever is a poular food fish in Southern Europe; it is caught in trawls and stationary pens (crawls), but does not support a substantial fishery.

The lesser weever *(Trachinus vipera)* is often found in shallow water and thus represents an even greater danger for swimmers. This species measures only about 20 cm in length; the spotted weever *(T. araneus)*, which is only found in the Mediterranean, and the streaked weever *(T. radiatus)* are about the same size as the greater weever. J.-M. R.

GREAT GREEN BUSH CRICKET *Tettigonia viridissima* 216

216
Sub.: Arthropoda
Cl.: Insecta
S.cl.: Orthoptera
O.: Ensifera
F.: Tettigoniidae

This Old World katydid (family Tettigoniidae) is easily recognized by its large size (28–42 mm) and vivid green color. Its back is flecked with brownish spots, and its wings are twice as long as its body, so that the female's ovipositor, called a "saber" by country people, scarcely extends beyond the tips of the elytra. *T. viridissima's* long soaring flights (up to 300 m) are among the most spectacular sights of a summer afternoon in the countryside. The male perches in a large bush at dusk and begins its tireless stridulating song (which evolved as a sexual attractant) at nightfall and as long as the temperature stays above 12°C, continues to chirp until dusk. The larva is a vivid dark green in color with a wide reddish-brown stripe running the entire length of its body and is otherwise similar to the adult in its morphology.

▶ The adult *T. viridissima* is a voracious predator, feeding on flies, locusts, butterflies, and caterpillars for the most part. It hunts by sight, stationing itself near the tip of a twig or branch and pouncing on any living creature that approaches within a fairly short distance. It seizes the prey with its forelegs and holds it firmly in place with the adhesive pads on its tarsi, then quickly disables it by biting through the central ganglia in the head and thorax. *T. viridissima* also feeds on a variety of nutritious vegetable material including fruits, the blossoms of peas and other legumes, seed capsules, and green heads of grain on the stalk.

The larva feeds primarily on grass seeds and other plant material, more rarely on insect prey, which it catches in the "grass forest" of a meadow rather than among the foliage of a bush. It will survive in a terrarium on a diet of lettuce exclusively.

◀ A number of solitary sand wasps (notably the Languedocian sand wasp, *Sphex occitanus*, as well as *Sphex maxillosus* and several *Tachyspex* species) stock their burrows with paralyzed but still living grasshoppers of this species, which are devoured piecemeal by the developing wasp larvae. R.C.

GREAT SKUA *Stercorarius* or *Catharacta skua* 217

217
Sub.: Vertebrata
Cl.: Aves
O.: Charadriiformes
F.: Stercorariidae

The great skua spends much of its time on the open ocean, and is found throughout the North and South Atlantic. Its nesting colonies are located on Iceland, the Faeroes, and Scotland, as well as on the Falklands, the southern tip of South America, and the coast of Antarctica; the nesting sites, generally on grassy coastal terrain, are sparsely populated, and the great skua will vigorously defend its nest against intruders. This species measures about 60 cm, with a wingspread of 130 cm, and weighs between 500 and 600 g.

▶ The great skua generally flies slowly over the open water, though it is capable of maneuvering very rapidly while engaged in aerial combat with another seabird (see below). It feeds partially on surface-dwelling fish that it catches for itself and partially on fish that have been caught by other seabirds—including herring gulls and black-backed gulls, gannets and boobies, tern, and cormorants—that are returning to their own nesting colonies, their crops heavily laden. It is not uncommon to observe a skua pursuing a gannet or a cormorant in flight, nipping its victim's wing or tail in its beak, forcing it to land and, finally, to disgorge the contents of its crop; the regurgitated fish is gobbled by the skua for the benefit of its own nestlings. The great skua also feeds on the eggs and nestlings of kittiwakes, puffins, and other gregarious seabirds. In the Antarctic, organized raiding parties of skuas snatch isolated chicks from the communal nurseries of penguin colonies and even steal the eggs from underneath sitting birds; in this latter case, one skua acts as a "stall," harrassing and distracting the penguin, while another makes off with the egg.

◀ The great skua has no predators or competitors, apart from other members of its own species. During the 19th century, the eggs and plump skua chicks were sometimes harvested from the nesting colonies of northern Scotland and the island of Foula (Shetlands) by the local inhabitants.

In some classification systems (particularly those in use in North America), the great skua is assigned to a separate genus *(Catharacta)* and the closely related species of the genus *Stercorarius* are referred to as jaegers rather than skuas. In others, the great skua is also assigned to *Stercorarius*, and all of the members of are accordingly referred to as skuas. The pomarine jaeger or skua *(S. pomarinus)* nests in northern Canada, northern Russia and Siberia, and in Greenland, but may be found in more temperate regions during the spring and summer. The parasitic jaeger or skua *(S. parasiticus)* nests in the same regions as well as in northern Scotland (including the Shetlands and the Orkneys), the Faeroes, and Scandinavia. The long-tailed jaeger or skua *(S. longicaudis)*, notable for its two very long, pennantlike medial tailfeathers, nests in Scandinavia, Lapland, Siberia, northern Canada, and Greenland, but is occasionally seen in Western Europe or the more temperate regions of North America. These 3 species are smaller than the great skua; all 3 are parasitic; they may prey on birds (swallows, small sandpipers, buntings) and even on lemmings and other small mammals. J.-P. R.

218
Sub.: Vertebrata
Cl.: Aves
O.: Cuculiformes
F.: Cuculidae

GREAT SPOTTED CUCKOO *Clamator glandarius* 218

This bird is between 39 and 45 cm long with a wingspread of 60 cm; it weighs between 140 and 220 g. Like other cuckoos, the great spotted cuckoo has evolved a parasitic reproductive strategy, and the cuckoo's eggs usually mimic those of the host species (in this case usually the magpie). Several female cuckoos may lay their eggs in the same magpie's nest, though the young cuckoos of this species make no attempt to eject their foster siblings from the nest; thus, mixed broods of magpies and cuckoos are not uncommon. To assist in this deception, the juvenile plumage of the young cuckoo is entirely black, which increases its resemblance to a young magpie.

The great spotted cuckoo is found in scrublands and pine groves in Southern Europe and on the savannahs and arid plains of subsaharan Africa. In Spain, the great spotted cuckoo is parasitic on azure-winged magpies *(Cyanopica cyana)* and crows; in Africa, it lays its eggs in the nests of bulbuls (family Pycnonotidae) and butcherbirds. European populations may also spend the fall and winter in Africa, setting out on their migratory flights across the Sahara in July.

▶ Adult cuckoos are nomadic, though they will linger briefly in certain areas where there is a substantial food supply as well as during the courtship and "nesting" periods. The adults feed almost exclusively on processionary caterpillars, which they catch while walking or hopping along the ground; the young cuckoos are fed on insects and their larvae and on snails.

◀ The great spotted cuckoo has no particular predators or competitors. In the south of France, the practice of paying a bounty for the destruction of magpies' nests—the magpie is considered a destructive species by hunters—inevitably takes its toll of a great many eggs and nestlings.

Four *Clamator* species are found in Africa

and Asia; *Clamator coromandus* nests in southern Africa and winters in equatorial Africa. J.-F. T.

GREEN CRAB *Carcinus maenas* 219

219
Sub.: Arthropoda
Cl.: Crustacea
O.: Decapoda
S.o.: Brachyura
F.: Portunidae

This species of portunid (swimming) crab is found on both sides of the North Atlantic as well as in the Mediterranean, where it is sometimes considered to be a separate species, *C. aestuari.* It is common in the intertidal zone and in shallow water where the bottom is rocky or sandy. The maximum length of the adult's shell is about 5.5 cm, its breadth about 7.2 cm.

▶ The planktonic larvae of the green crab feed on zooplankton and in the home aquarium can be raised on a diet of *Artemia* nauplii (first-stage larvae of a genus of brine shrimp). Juvenile and adult crabs are primarily carnivorous, though they are not as quick or as active as the blue crab and the creatures they prey on tend to be slower-moving: fish, shrimp, sea snails, bivalves, annelids, polychaetes, and echinoderms. The green crab occasionally feeds on inert organic material, such as fish spawn and carrion, as well. It locates its prey either by sight or with the aid of olfactory organs that are located near the tips of its walking legs.

The green crab opens the shells of mollusks with a combination of leverage and persistence. It picks up the shell with its anterior walking legs and holds it up to its mouth with its third set of maxillipeds, then uses the larger of its claws to crack the shell and detaches portions of the mollusk's fleshy body with the smaller and more delicate claw, as well as its maxillipeds, and conveys them to its mouth. It can sometimes break open the shell or crack the columella (central axis) of a gastropod by main force, but in the case of a more substantial bivalve, it may be obliged to work away at the edges of the shell with its larger claw in the hope of tiring out the muscles that keep the shell clamped shut. Normally the green crab will only attempt to pry open the shell of a young or medium-sized mollusk and even then will only persist in its efforts for a minute or two before moving off in search of easier prey.

◀ At low tide, the green crab is frequently exposed to attack by seagulls and other predators of the intertidal flats. The coloration of its shell may have evolved in response to predator attacks on the juvenile green crab, which is consumed by squid, octopus, fish, and other crustaceans; some species of shrimp are large enough to prey on the adults as well. The green crab's main defense against these predators is to conceal itself under rocks or in beds of seaweed, though it will occasionally attempt to escape by scuttling across the sand. There is no specialized fishery that commercially exploits the green crab, though it is frequently brought up in fishermen's nets in Europe (its claw-snapping, irascible demeanor on these occasions is responsible for its common French name, *la crabe enragée)* and is sometimes eaten in chowders, *zuppa de pesce,* and other seafood dishes.

Apart from the blue crab and the green crab, there are several other portunid species whose flesh is esteemed by man. *Liocarcinus puber,* known as the "currycomb crab" *(étrille),* is an edible species found in the Eastern Atlantic and the Mediterranean. It feeds on sea snails, bivalves, and even on sea hares (a large sea slug of the genus*Tethys);* its shell may attain a length of 7.8 cm and a width of 10.6 cm. *Scylla serrata,* an estuarine swimming crab of the Indo-Pacific region, also feeds on sea snails and bivalves as well as on other crabs (including hermit crabs). In many parts of its range, it is hunted by man both as a competitor—since it occasionally preys on oyster beds—and as a locally important source of food. P.N.

GREEN LIZARD *Lacerta viridis* 220

220
Sub.: Vertebrata
Cl.: Reptilia
O.: Squamata
S.o.: Sauria
F.: Lacertidae

The green lizard is very common in France (south of a line between Normandy and Alsace), and its range extends eastward through southern Germany, Italy, and the Balkans as far as southwestern Russia. Its maximum body length is 30 cm; its back is a delicate green color, generally somewhat opalescent and covered with fine black speckles in the case of the male. The females generally have 2 (occasionally 4) transverse whitish dorso-lateral stripes and occasionally a sprinkling of black spots as well. In both sexes, the color of the underbody varies from yellowish green to pure yellow, and the males have bluish throat patches during the mating season.

This species is generally found in small patches of scrub or brush (rocky slopes overgrown with briars, the edge of a woodlot, or a rocky outcrop in a forest clearing thickly covered with shrubs and bushes) in the midst of sunny, exposed terrain, so that the lizard can take advantage of the slightest ray of sun and still have a convenient refuge in case of danger. Very quick and agile, especially when its body has been warmed up by the sun, the green lizard climbs trees and leaps fearlessly from branch to branch, only rarely taking to the water.

The green lizard hibernates from about mid-October to mid-March. Mating takes place in April and May, and is preceded by a round of violent inter- and intrasexual combat. The female lays its eggs in a hole that it has dug for this burrow in friable soil; the young lizards hatch out from mid-August to mid-September. The green lizard attains sexual maturity at the age of 3. Average longevity in nature is on the order of 4 or 5 years, so that a maximal life expectancy of 10–12 years is quite conceivable.

▶ Like most of its congeners, *Lacerta viridis* is carnivorous, feeding primarily on prey that is mobile and highly active, which it detects exclusively by sight. As soon as it catches sight of a prey animal, generally a terrestrial arthropod, it rushes toward it and tries to catch it in its jaws as best it can. If it misses its hold and the victim seems about to escape, it repeates the operation until the prey has been successfully impaled on its needle-sharp teeth, then partially masticates before swallowing it with with several powerful motions of its jaws while shaking its head shortly and sharply from side to side. On occasion, a prey animal that is too large to be swallowed or that struggles too vigorously will be allowed to escape.

The favorite prey of the green lizard includes many different kinds of insects and their larvae (cockroaches, grasshoppers, aphids and other homopterans, wasps, bees, and others), woodlice and other terrestrial crustaceans, myriapods (Polydesmida, Lithobiidae, *Glomeris*), and spiders. It also occasionally feeds on snails and earthworms and less frequently on vegetable material (grape seeds and cherry pits); it has also been known to prey on smaller lizards *(Lacerta muralis, vivipara)* and even the young of its own species, though this seems only to happen rarely. Adult green lizards have been maintained successfully in captivity on a diet of newborn mice, though this species has never been known to feed on mammalian prey in nature.

◀ The predators of the green lizard include corvids, diurnal raptors, and even domestic fowl, as well as a great many reptiles and several small mammals. The smooth snake and other *Coronella* species as well as young European vipers feed almost exclusively on lizards of various sorts, including *Lacerta viridis,* and the dark-green snake, the Aesculapian snake, the Montpellier snake, and the ladder-backed snake all prey on the green lizard in substantial quantities. Shrews *(Neomys, Sorex,* and *Crocidura)* sometimes feed on very young green lizards, and the weasel *(Mustela nivelis)* and other mustelids, as well as the domestic cat, may also prey on the adults.

The green lizard itself is rather aggressive, and its attempts to defend itself with its strong jaws and sharp teeth may serve as a deterrent to many potential predators. If it is handled or restrained in any way, it begins to struggle violently, and if it is able to get its teeth into its attacker, it will hang on grimly until it is released—at which point it will let go immediately and try to make good its escape. Like most lizards, *Lacerta viridis* can voluntarily detach its tail by autotomy when the latter is seized by a predator.

There are a number of other lacertids of roughly the same size and coloration that might be confused with *Lacerta viridis* in various parts of its range (and elsewhere): certain males *(Lacerta agilis)* in France as well as juvenile ocellated lizards *(L. lepida)* in the south of France. In northern Spain, the green lizard is replaced by *L. schreiberi,* which it also closely resembles. In the Balkans, Greece, and Turkey, the Balkan green lizard *(L. trilineata)* can be distinguished from *L. viridis* only by a close comparative examination of two individuals. J.C.

GREEN WOODPECKER *Picus virdis* 221

221
Sub.: Vertebrata
Cl.: Aves
O.: Piciformes
F.: Picidae

This common Eurasian woodpecker typically measures 30 cm, with a wingspread of 50 cm, and weighs between 150 and 210 g. It makes use of its well-known talent by drilling out a nesting cavity in a treetrunk as well as in searching for prey. It is found on flat or wooded (but not densely forested) terrain, e.g. in woodlots and hedgerows, at the edge of a forest, particularly if there meadows nearby, and in stands of larch or other trees in mountainous regions. Its range extends from Western Europe (including southern Sweden but excluding Finland) into Central Asia and as far east as Iran and India.

▶ The green woodpecker searches for insects on treetrunks and rotten stumps as well as on the ground, moving along in a series of short hops, and apparently with with equal ease, on a vertical or a horizontal surface. It pecks energetically with its bill to remove strips of bark from a treetrunk or to dislodge xylophagous (wood-eating) insects (beetles, borers, caterpillars) from their burrows in soft or rotten wood. On the ground, it forages for worms, snails, and insects; it hunts for other insects more or less at random, but will deliberately seek out an anthill, digging it up with its beak and ensnaring the ants and their larvae with its long protractile tongue, which, like that of most ant-eating creatures, is coated with sticky saliva.

◀ The green woodpecker has no specific predators, though it may occasionally fall prey to a raptor or mustelid. If is attacked or alarmed while on the ground or in open country, it will simply take to the air. If an intruder appears on the scene while it is perched on a treetrunk, it will immediately try to conceal itself by flattening out against the trunk, remaining motionless for a short time and then hopping rapidly up the trunk so it can still keep the intruder under surveillance while remaining out of harm's way.

Several other Eurasian woodpeckers have similar feeding habits—notably the black woodpecker *(Dryocopus martius)* and the black-naped green woodpecker *(Picus canus)*—but are not, strictly speaking, competitors of the green woodpecker. This species is not seriously affected by humans, though many of the routine activities of the tree surgeon or forester (removal of dead trees or branches) are highly undesirable from the woodpecker's point of view.

Woodpeckers, comprising 200 species in all, are found everywhere on the globe with the exception of Madagascar, Australasia, and Oceania. Many species are primarily terrestrial, such as the wrynecks of Eurasia *(Jynx torquill)* and tropical America *(J. pectoralis)*, or primarily arboreal, such as the Eurasian black-naped green woodpecker, mentioned earlier, or the North American pileated woodpecker *(Dryocopus pileatus)*. Others, including the green woodpecker and the black woodpecker, are less specialized in their feeding strategy. A number of different species are omnivorous, notably the Eurasian greater spotted woodpecker *(Dendrocopus major)* and the yellow-shafted and red-shafted flicker *(Colaptus auratus, C. cafer)* of North America, which feed on seeds, berries, and fruit, insects and even small vertebrates. The sapsuckers of North America *(Sphyrapicus)* so are so called because they drill through the bark of birches or maples to make sugar taps and feed on the sweet sap that flows out. The ivory-billed woodpecker *(Campehilus principalis)* of the southeastern United States may already have surrendered its title as the largest extant woodpecker (as well as the rarest of North American birds), since an authenticated sighting of this species has not been reported for many years. J.-P. R.

GRENADIER *Coryphaeanoides rupestris* 222

222
Sub.: Vertebrata
Cl.: Osteichthyes
S.cl.: Teleostei
O.: Gadiformes
F.: Macrouridae

This bottom-dwelling fish is also called the "rattail" or the "rattail grenadier," because of the fancied resemblance between its long pointed tail and the long pointed queue formerly worn by human grenadiers. It attains a maximum length of about 1 m and is found along the continental shelf and on the abyssal plains of the Northeastern Atlantic; it may venture a short distance into open water, but never ascends very far toward the surface. The male grenadier, like other members of the family Macrouridae, has a set of special muscles that cause the walls of its swim bladder to vibrate and produce a kind of drumming sound; the grenadier also has a keen sense of hearing, but the precise function of this mode of communication (if that is indeed what it is) is still uncertain.

▶ The grenadier was formerly believed to be a detritovore, feeding on organic particles in the sediment on the bottom; it does forage along the muddy bottom with its snout (which is more highly developed than in related species such as *Coelorhynchus, Nezumia,* and *Macrourus,* in which the mouth has migrated to a ventral location), but exclusively in search of live prey. It feeds on a fairly wide variety of free-living vagile organisms, including polychaetes, squids and other cephalopods, gastropods and lamellibranchs (though rarely), and (primarily) on shrimps, euphausids, amphipods, and other crustaceans as well as on bony fish of the family Myctophidae and other bathypelagic species that feed on plankton. The young grenadiers feed on similar prey, though correspondingly reduced in size and quantity.

In spite of the impracticability of observing its feeding habits in nature, a number of reasonable assumptions can be made from what we know of the grenadier's anatomy and morphology. Thus, the grenadier feeds along the bottom, for the most part, with its head downward and its body inclined at a fairly acute angle (less so in the case of other Macrouridae whose mouths are ventrally located) so it can more efficiently turn up the muddy sediment in which it finds its prey. As noted earlier, the grenadier will often venture a few meters into open water in quest of free-swimming prey (fish, squid, crustaceans), as evidenced by the large number of these fish that are taken in semipelagic trawls, which are dragged along at comparable depths above the ocean floor.

The grenadier most probably locates its prey with the aid of the sensory receptors along its lateral line, which detect vibrations in the water; the "rattail" elongation of the fish's body increases the number and thus the effectiveness of these organs. The grenadier also has disproportionately large eyes, and the optic regions of the brain are equally well-developed; the retina consists almost entirely of rods—a configuration that is associated with night vision in terrestrial creatures. Even in the total darkness of the abyssal plain, the grenadier might very well use use its eyes to hunt for squid, euphausid crustaceans, and other bioluminescent organisms. The grenadier's olfactory organs are also well developed, and its sense of smell probably also plays a role in prey detection.

◀ The principal predators of the grenadier and the other Macrouridae include selachians (rays and sharks) as well as large bony fish of the family Gadidae (cod and hake); the latter are often found along the bottom, though it is probably when the grenadier ventures into open water that it is most subject to predation. The flesh of the grenadier is very delicate and has long been appreciated by the Eskimos; the grenadier and other Macrouridae have been harvested on a substantial scale by Soviet trawlers for the past two decades or so. In Western countries (apart from Northern Canada and Greenland), the Macrouridae only enter the human foodchain as an article of diet of the larger Gadidae and other commercially exploited species.

F.M.

GULPER *Saccopharynx flagellum* 223

223
Sub.: Vertebrata
Cl.: Osteichthyes
S.cl.: Teleostei
O.: Anguilliformes
F.: Saccopharyngidae

The gulper, or gulper eel, is the common name of several species of slow-moving bathypelagic eels that are normally found at depths of 500–3000 m, though they sometimes venture quite close to the surface. The body (unlike the head and jaws) is conventionally eel-like, very elongated (1.8 m) with a tapering, pointed tail and a rudimentary caudal fin. Longevity is unknown.

▶ The gulper's mouth is enormous, and the hinges of its jaws are set quite far back behind its head; the jaws are about 4 or 5 times the length of the rest of the skull and equipped with numerous long, pointed teeth. The open mouth of the gulper (its customary position) is reminiscent of that of a cobra or some other large snake as it prepares to engorge its prey; the gulper's elastic and already capacious stomach can also be greatly expanded if necessary. The gulper feeds primarily on other bathypelagic fish (lanternfish, bristlemouths [*Cy-*

clothones]), which may be as large as itself or even larger; thus, a small gulper, 15 cm long, was found to have recently swallowed a gadid 22 cm long.

Because of the inaccessibility of the gulper's habitat, its feeding habits have not been observed directly, but it is thought that the gulper simply swims through the water with its mouth agape, serving the function of a trawl net. Smaller prey are immediately engulfed and conveyed directly into the gulper's stomach; larger prey are probably ingested slowly, by aligning the prey with its gaping gullet and alternately protruding the upper and lower jaw (rather than with alternate lateral motions as in the case of the cobra and other large snakes) and successively "annexing" the larger fish a few centimeters at a time.

The gulper's eyes are very tiny, and it probably locates its prey primarily with the aid of the sensory organs located along the lateral line; the extreme elongation of the gulper's body increases the number and effectiveness of these organs—in much the same way that nocturnal creatures on land have developed very large eyes. The gulper can be considered a predator of the second rank, as it were, since it feeds on other species that are themselves predators. Other abyssal fish have evolved similar mechanisms for devouring very large prey, apparently as a means of increasing the available food supply in these cold, infertile, and sparsely populated waters.

Another member of the same family, *Eurypharynx pelecanoides*, is found at even greater depths—from 1500 to below 3000 m—and its anatomical peculiarities are correspondingly exaggerated: the length of the jaws is from 7 to 9 times that of the skull, and the hinge is accordingly set back quite a bit beyond the rear of the skull. *Eurypharynx* is only 60 cm long, and it is thought that its pharyngeal mucosa act like the whalebone strainers of baleen whales, filtering out small crustaceans and other zooplankton from the water. The stomach of this species is not expandible, since its prey is considerably smaller than itself.

The giganturids (family Giganturidae), only 20 cm long and found at depths between 1000 and 4000 m, are also capable of feeding on fish much larger than themselves. Their eyes are situated at the ends of long "telescopic" stalks, though the purpose of this modification is unknown. Less ambiguously, its mouth is very large and equipped with numerous sharp, needlelike (acerate) teeth; these can even be folded back against the lining of the mouth in order to accommodate the passage of an exceptionally large prey animal. A 7 cm giganturid has been found with a 12 cm sea viper *(Chauliodis sloani)* in its stomach, almost as favorable an intake ratio as that enjoyed by the gulper eel.

F.M.

Gymnosphaera albida 224

224
Sub.: Protozoa
Sup.cl.: Actinopoda
Cl.: Heliozoa
O.: Centrohelida

This free-living spherical protozoan, between 0.07 and 0.10 mm in diameter, is found in the shallows (0–10 m) along the coasts of the Western Mediterranean and Northern Atlantic, in clear, richly oxygenated waters where an abundant supply of organic nutrients is also to be found. Like other heliozoans, *Gymnosphaera* consists of a central cellular body and an array of thin, wandlike stalks of cytoplasm (axopodia) extending outward in all directions. Thus, it can rest the tips of its axopodia on a hard substrate (usually sponge or rock, including rocks encrusted with algae or bryozoans or the secretions of tubeworms), though not on muddy sediment, and it can even roll along the bottom as well as float through the water, though usually not far above the level of the bottom. When the amount of dissolved oxygen and organic nutrients drops below a certain level, *Gymnosphaera* can retract its axopodia and encyst itself until conditions become more favorable. This species reproduces by fission, generally in the spring.

▶ Like other Actinopoda, *Gymnosphaera* is not so much a predator as a passive collector of nutrients, including some living organisms, that adhere to the mucous surfaces of its axopodia. It feeds avidly on dinoflagellates (unicellular mobile algae, about 1–10 μm in diameter) and will also consume micellar aggregates, bacteria, yeast cells, diatoms, and organic debris that is rich in microorganisms. Certain organisms are immediately rejected, such as the dinoflagellate *Oxyrrhis marina*, which provokes an immediate "gag reflex" on the part of *Gymnosphaera*, possibly because the algae releases a toxic secretion or because of some sort of chemical incompatability between their respective cell membranes. *Gymnosphaera* is continually taking in nutrients; only the encysted cells do not feed.

The cellular mechanisms of ingestion are similar to those encountered in many other protozoa: initial contact between the two cell membranes, followed by the formation of an alimentary vesicle in the interior of the "predator," followed in turn by the formation of a digestive vacuole. In this case, when another organism comes into contact with the axopodia, this touches off a flurry of intense activity in the underlying cytoplasm; organites called exosomes, which are involved in the secretion of immobilizing toxins and adhesive mucuous through the cell membrane, quickly maneuver into position.

Next, the cell membrane surges up on either side of the prey animal and engulfs it; when the membrane has completely enclosed the prey, it forms a sort of bubble called an alimentary vesicle, which is detached from the cell membrane and conveyed by the cytoplasmic stream into the interior of the cellular body, and thereafter referred to as a digestive vacuole. Organites assemble around the edges of the vacuole, releasing enzymes called lysosomes that begin the process of digestion and absorption.

◀ *Gymnosphaera albida* is undoubtedly preyed on by other heliozoans from time to time; some helizoan species are involved in symbotic partnerships with algae of the genus *Chlorella* and others are occasionally parasitized by cilates and rotifers, but *Gymnosphaera albida* is not among them.

The order Centrohelida includes about a dozen species of marine heliozoans, most of which are free-living and are found in the same sort of habitat as *Gymnosphaera albida*. A few species are permanently attached to the substrate by means of an inert or contractile stalk (peduncle) and feed on organisms swimming just above the level of the bottom.

Most heliozoans, like *Gymnosphaera albida*, feed on very tiny marine organisms and on organic particles, but there are a few—such as *Actinophrys* and *Actinosphaerium* (order Actinophryida), *Hedraiophrys* and *Actinocoryne* (order Centrohelida)—that prey on relatively large creatures, some of which can swim quite rapidly, including ciliata, nauplian larvae of certain crustaceans, larvae of sedentary polychaetes, nematodes, and others. The difference in feeding habits can be explained by the fact that the cellular membrane is rather excitable in this latter group, less so in the case of *Gymnosphaera*, and the organism's response to any stimulus (whether or not produced by contact with a suitable prey animal) is very prompt, whereas it appears to be relatively sluggish in the case of *Gymnosphaera* and the more typical heliozoans.

C. F.-C.

H

225
Sub.: Vertebrata
Cl.: Mammalia
O.: Insectivora
F.: Solenodontidae

HAITIAN SOLENODON *Solenodon paradoxus* 225

This species actually ranges throughout Hispaniola and is not confined to the Haitian side of the island. Its body measures about 35 cm and its tail an additional 15 cm, thus providing an excellent example of what is called insular gigantism—the tendency of relatively small island creatures to grow larger than their relatives (in this case, shrews, hedgehogs, and other insectivores) that remain on the mainland. The extinct dodo of Mauritius (a member of pigeon family) is probably the most celebrated instance of this phenomenon, and insular dwarfism has also been observed in larger animals, including cervids and extinct elephants and hippos. The solenodon's long pointed snout is covered with very fine whiskers (vibrissae), and this species is also noted for its habit of emitting high-frequency clicking sounds, possibly as a form of echolocation. A captive specimen is reported to have survived to the age of over 11.

▶ The solenodon feeds mainly on insects and other small invertebrates, and finds its food by probing in the ground or in rotten wood or other vegetable debris with its snout; its sensitive whiskers thus play a primary role in prey detection. The solenodon catches an insect its its jaws, picks it apart with its front paws, then gathers up the pieces and eats them while sitting back on its haunches and supporting its weight with its tail. Its lower incisors are grooved to enable it to inject its toxic saliva into a prey animal, and it is also capable of subduing larger, surface-dwelling creatures such as reptiles and domestic fowl.

◀ There are only two *Solenodon* species, both of which are sometimes known by the Taino Indian name almiqui. The Haitian solenodon and the Cuban solenodon *(S. cubanus)* developed in the absence of larger carnivores and were apparently much more prevalent before the introduction of dogs and cats by the Spaniards. The disastrous introduction of the mongoose into the sugar islands during the 19th century, ostensibly as a control on reptiles in the canefields, brought both species very close to the brink of extinction; the Haitian solenodon is now believed to be the rarer of the two. P.A.

Halicarcinus planatus 226

Halicarcinus platanus is a species of marine crab found in subantarctic waters at depths ranging from 0 to 60 m, notably off the coasts of the Kerguelen Islands in the Indian Ocean, New Zealand, and Patagonia. It is locally quite abundant amid the floats and fronds of the Pacific giant kelp *(Macrocystis)*—the adhesive suckers that anchor the kelp to the substrate are particularly hospitable to the juvenile crabs and breeding female—and in mussel beds off the Kerguelen Islands, where the population sometimes attains a density of 160 crabs per square meter.

▶ The planktonic larva feeds on cold-water zooplankton. The adult forages actively among the kelp and under rocks, though in these chilly waters the selection of prey animals is limited mainly to marine worms (annelids, free-swimming [vagile] polychaetes), sea urchins, and copepods. The crab seizes its prey in its claws and, if necessary, shreds it into pieces with its maxillipeds before ingesting it.

◀ *Halicarcinus platanus's* most notable characteristic is the extreme softness of its shell, the calcium content of which is only 9%, versus 25–35% in other species. Thus, its vulnerability as well as its extreme availability in certain areas makes it attractive to a number of different predators, and it provides an essential link in the foodchain of this relatively sparsely populated habitat. Such bottom-feeding fish as rays, *Notothenia* (the Antarctic blenny), and *Chaenicthys* (the crocodile icefish) are known to consume substantial quantities of these crabs; invertebrates such as starfish and sea urchins have also been observed to prey on *Halicarcinus,* as has the kelp gull *(Larus dominicanus)* and occasionally the skua and the rockhopper penguin as well. The crab's only effective means of defense is to conceal itself under a rock or in a thick bed of kelp; the coloration of its shell helps it to blend in very convincingly with the rocky seabed, which protects it from attacks by seabirds if not from other predators.

The family Hymenosomatidae comprises 10 different genera, all of which are represented in the coastal waters of Australia and New Zealand. Some species are found in the northern Indian Ocean; others in the waters off Japan and elsewhere in the Pacific, and a few are found in brackish or fresh water; all the members of this family have soft shells with a relatively low calcium content. Though much remains to be learned of the exact nature of their relationships among themselves and with other crabs, they have been provisionally assigned to the superfamily Oxyrhyncha (spider crabs and their relatives). "Soft-shelled crab," by the way, is a term that is often encountered in a culinary rather than a zoological context; it refers to an edible crab, more often a blue crab, that has recently moulted and whose shell has still not entirely hardened.

B.R. de F.

226
Sub.: Arthropoda
Cl.: Crustacea
O.: Decapoda
S.o.: Brachyura
F.: Hymenosomatidae

227
Sub.: Arthropoda
Cl.: Insecta
O.: Strepsiptera
F.: Halictophagidae

Halictophagis lopesi 227

This tiny insect belongs to the order Strepsiptera, the so-called twisted-winged parasites, which in some classification systems is treated as a superfamily of the order Coleoptera rather than as a separate order. The female is a wingless, legless degenerate parasite that remains embedded in the abdominal wall of its leafhopper host for its entire lifetime, and the visible part of its body consists of a rudimentary cephalothorax (about 0.5 mm long) that appears to be nothing more than a grayish or dark-brown swelling on the intersegmental membrane of the host.

The winged, free-living male is brown, with two fan-shaped (flabellate) antennae and two large compound eyes, laterally mounted and with large facets (10 of which are visible in the dorsal view). The first pair of wings is greatly atrophied (hence "twisted-winged"); the second pair is fully functional, so that the male *Halictophagis* is capable of flight. The first segment of the abdomen is embedded in the pterothorax, and the abdomen is also brown in color; the male's body is 2.5 mm long.

Because of the peculiar morphology of the female, the mating habits of this species are unconventional: In addition to the visible cephalothorax, referred to above, the remainder of the mature female's body consists essentially of an abdomen swollen with thousands of eggs, and the male may opt to insert its sexual organ either through the genital pores or though the mouth of the female in order to fertilize them. In either case, the fertilized eggs develop into triongulid larvae no longer than a few microns.

▶ The strepsipteran larvae station themselves on grass stems and when a leafhopper of the appropriate species alights nearby, the larvae make their way into the abdominal cavity of their prospective hosts by burrowing through the intersegmental membrane, where the cuticle is relatively thin. The females are entirely concealed inside the abdominal cavity until they reach sexual maturity, at which point the cephalothorax of the parasite finally begins to protrude through the host's abdominal wall. The males undergo nymphosis while still inside the host's abdominal cavity. The tip of the nymphal cocoon likewise protrudes through the abdominal wall, providing an aperture through which the adult male strepsipteran eventually emerges. The leafhopper generally survives this ordeal, though it is frequently unable to reproduce, since its genital organs will have either been devoured by the parasite or atrophied as a result of its presence inside the abdominal cavity (so-called parasitic castration).

Four strepsipteran species were studied in the fields and grassy meadows of Guadeloupe. *Halicophagis lopesi,* described above, is parasitic on *Hortensia similis,* a small green leafhopper that is very common in cultivated fields and fallow land on the island. In the tropics, the life cycle of the host (and thus of the parasite) is not interrupted during the winter as it would be in the temperate zone, so that both these species may be observed year round, though they are most common in moist areas during the rainy season. Seasonal fluctuations in the strepsipteran population are directly proportional to those of the host population.

◀ *Halictophagis lopesi* has no specific predators, though its survival may be threatened by other environmental factors. The most serious of these is drought, which has both a direct and an indirectly adverse effect on the development of the triongulid larvae; when the pith of the grassy shoots on which the leafhopper feeds begins to dry out, the number of available parasitic hosts is reduced accordingly. It is possible for as many as 5 strepsipterans to coexist in the abdominal cavity of the same host (multiparasitism), as was observed in the case of one leafhopper of a different species, *Sogatella kolophon meridiana,* but a leafhopper that has been parasitized by more than 2 strepsipterans will probably not be able to sustain both itself and its parasitic brood for very long. The strepsipterans also face competition from the semi-ectoparasitic larvae of *Cyrtogonatopus* (q.v.) and other dryinid wasps, and when a leafhopper has been parasitized by both species, it is unlikely that any of the parasite larvae will reach maturity.

Two closely related strepsipterans found in the same habitat, *Strenocranophilus quadratus* and *Elenchus tenuicornis,* are parasites of *Saccharosydne saccharivora,* another species of leafhopper, and *Sogatella kolophon meridiana*, mentioned earlier. A.D.

HARP SEAL *Pagophilus groenlandicus* 228

228
Sub.: Vertebrata
Cl.: Mammalia
O.: Pinnipedia
F.: Phocidae

Adults of both sexes measure between 1.8 and 2 m and weigh between 100 and 150 kg; the thickness of the blubber layer, and thus the weight of the individual harp seal, varies greatly with the season. Like other seals, harp seals assemble in coastal rookeries during the breeding season, but unlike other seals, they appear to be monogamous; no other form of social organization has been identified. The 3 principal breeding grounds are located in the White Sea, around the Arctic island of Jan Mayen, and along the coast of Newfoundland; during the summer, these separate populations are dispersed throughout a large part of the North Atlantic above 70° N. lat., from Hudson's Bay to Spitzbergen. The harp seal (called *Sattelrobbe,* or 'saddle seal' in German) is named for the (roughly) harp-shaped marking on its back.

▶ Young harp seals feed primarily on euphausids (shrimplike crustaceans); the adults feed on euphasids and, perhaps more frequently, on shrimp. Free-swimming and benthic crustaceans are ingested by suction (or in more or less the same way that we eat oysters); capelin, herring, cod, and other fish are swallowed head-first. The harp seal has to consume at least 1.5 percent of its body weight every day, and, given its ability to descend to a depth of 200 m and remain submerged for almost 30 minutes, it has substantial food resources at its disposal.

◀ Harp seal pups are ocasionally caught by polar bears while they are still on shore; the killer whale and the Greenland shark *(Somniosous microcephalus)* are the only marine predators of the adult. Two species of finback whale, the common rorqual *(Balenoptera physalis)* and the lesser rorqual *(Balenoptera acutorostrata)* also consume substantial quantities of capelin and thus can be regarded as direct competitors of the harp seal. Much as with the crab-eater seal, however, the near-extinction of these formidable competitors has partially been offset by the rapid development of a commercial capelin fishery in the North Atalantic.

As is well known, the continuing massacre of "whitecoats" (seal pups) by human hunters on the ice fields off the coast of Labrador and elsewhere produced an enormous international outcry during the 1970s and 80s. A partial ban on the clubbing of seal pups was finally imposed by the Canadian government, but environmentalists have continued to object that current legal restrictions on seal hunting are not only far too lenient but also commonly ignored or evaded by the hunters (mostly natives of Labrador). Environmental groups are currently concentrating on a campaign of political protest and direct physical interference with the spring seal hunt, as in the past, as well as intensively promoting a consumer boycott of sealskin coats and other fur products. M.P.

Hastigerina pelagica 229

229
Sub.: Protozoa
Sup.cl.: Rhizopoda
Cl.: Foraminifera
F.: Globigerinidae

This is a large planktonic foraminiferan, very prevalent in tropical and subtropical waters, where it sometimes represents as much as 20 percent of the planktonic biomass. The cellular body (endoplasm) is surrounded by a calcareous shell, 0.8–1 mm in diameter, that contains 6 or 7 intercommunicating planospiral chambers, of which the outermost is the largest. The external surface of the shell is covered with long radiating spines intermingled with microscopic pores; a cytoplasmic envelope (ectoplasm) exudes from these pores and from the outer opening of the shell chambers, which covers almost the entire surface of the shell and is in a state of constant, bidirectional motion. The outer surface of the ectoplasm may send out a network of branching prolongations called rhizopods (see below), which will increase the total diameter of the protozoan's body to as much as 2 cm.

▶ *Hastigerina* is exclusively carnivorous and displays a preference for multicellular organisms sych as coelenterates, marine worms, copepods, and tunicates. Anderson and his colleagues have reported that the maximum survival rate for this species was achieved in the laboratory when food was offered every 6 days. The rhizopods play an essential role in predation: as soon as a prey animal makes contact with the surface of the ectoplasm, a branching network of rhizopods erupts out of the cytoplasm all around it; the prey is quickly engulfed, then conveyed toward the external opening of the shell chambers by the cytoplasmic stream, and finally absorbed into the endoplasm. Cross-sectional sampling of the endoplasm by electron microscopy reveals extensive traces of the foraminiferan's previous meals, including more or less competely digested fragments of muscle tissue.

◀ Pelagic foraminiferans provide nourishment for an immense variety of creatures that extract plankton from seawater by means of a filtration system of one kind or another, from tunicates to baleen whales.

We should mention that other planktonic foramaniferans are phytophages (plant-eaters), including *Globigerina bulloides* and *Globorotaliella truncatuninoides*, omnivores, including *Globigerinella aequilateraliis, Globorotaliella menardii, Globigerinoides ruber* and *Pulleniatina obliquiloculata*, as well as one additional carnivore *(Globigerinoides sacculiferer).* C. F.-C.

230
Sub.: Vertebrata
Cl.: Reptilia
O.: Chelonia
S.o.: Cryptodira
F.: Chelonidae

HAWKSBILL TURTLE *Eretmochelys imbricata* 230

The carapace of this relatively small sea turtle measures slightly less than 85 cm, though the shell of a hawksbill observed in the West Indies measured more than a meter, and the weight of the entire animal was 126 kg. The hawksbill is generally encountered along coral reefs or along sandy or rocky substrates covered with marine vegetation or a thick layer of sediment. The two very similar subspecies range throughout the warmer waters of the Atlantic, Pacific, and Indian Oceans, and the female comes ashore to lays its eggs on isolated beaches along the coasts of St. Vincent, St. Lucia, and the Greater Antilles, the Comoros, Oman, the Maldives, Sri Lanka, Australia, and Fiji. At least one individual has survived beyond the age of 15 in captivity.

▶ The hawksbill is omnivorous, feeding on an assortment of marine invertebrates (including both pelagic and benthic species) as well as on seaweed and other marine plants. An inventory of the stomach contents of a hawksbill recovered from the subtropical waters of the Eastern Atlantic included the remains of sea snails (violet snails [*Janthina*], periwinkles and whelks, and *Amychina)*, bioluminous cuttlefish, crustaceans, echinoderms, sponges, and especially on sea anemones, by-the-wind-sailors *(Vellela,*a genus of large pelagic jellyfish), and other cnidarians. Bryozoans *(Amthia, Steganoporella)*, barnacles and oysters (including *Crassostea gigas)* also figured among the prey of hawksbills examined. Earlier investigators had linked this diet of soft-bodied marine invertebrates with the observation that a high percentage of hawksbills appear to be infested with intestinal worms.

The hawksbill is so called because the slightly serrated horny sheaths on both its upper and lower jaws are more developed than in the other Cheloniidae (marine turtles), thus forming a powerful beak with which the hawksbill can browse on seaweed and crop off morsels of larger prey animals. The means by which this species detects its prey in not definitely known, though some authors have suggested that certain prey animals may be detected by smell; we may suppose that a hawksbill pushing its way slowly through a bed of wrack grass hunts primarily by sight, flushing out its prey by stirring up the bottom with its flippers. Even less is known about the diet and feeding habits of the juveniles, which are generally encountered far out to sea.

◀ Marine turtle hatchlings are notoriously vulnerable to attack by shorebirds and raptors during their initial migration from the nesting site at the head of the beach to the shoreline. In the Comoros, for example, they are attacked by black kites during the day and probably also by gray herons and screech owls at night, and on the beaches of Australia, they make easy prey for swift terns and silver gulls. At sea, the adults are sometimes preyed on by tiger sharks and killer whales. The heads of 4 hawksbills were discovered in the stomach of a tiger shark *(Galeoceras)* caught off the coast of Senegal.

The green turtle *(Chelonia mydas)* and the hawksbill are the two species of marine turtle most extensively exploited by man. The green turtle is hunted primarily for its flesh, and the sharp scales on the hawksbill's carapace are the source of the highest grades of commercial tortoiseshell. The market for natural tortoiseshell was almost moribund for many years, but has been reactivated during the past few decades; though it has largely been replaced by plastic in the manufacture of combs, eyeglass frames, and other utilitarian objects, the blond carey (shell scales) of the hawksbill is the only variety that is thick enough to be fused into blocks and fashioned into a variety of art objects, curios, and "retro" fashion accessories.

There are five different methods by which hawksbills are caught on Madagascar: The females are ambushed on the nesting beaches, and turtles of both sexes are caught with nets, harpoons, or grappling hooks, or, most ingeniously, with remoras (pilotfish with sucker disks on top of their heads that will attach themselves to any large marine creature) tied to the end of a line. The scales are sometimes removed from the living turtle on the beach, and while it is claimed that this operation is not necessarily fatal, it seems more accurate to state that it is not immediately fatal.

There are six other species of marine turtles (Cheloniidae), assigned to the genera *Chelonia*, *Lepidochelys*, and *Caretta* and found throughout the warm seas of the world; the LEATHERBACK TURTLE has been assigned to a separate family of its own, Dermochelyidae. Green turtles migrate for thousands of kilometers between their nesting beaches and the submarine meadows of wrack grass on which they feed. In captivity, the immature *Chelonia* , which raised by the hundreds in fattening pens and provided with an ample supply of protein-rich feed cakes, display profoundly aggressive (if not cannibalistic) tendencies, often inflicting severe bite wounds on the heads and flippers of their penmates. J.F.

HEDGEHOG *Erinaceus europeus* 231

This small Old World mammal is between 22 and 27.5 cm long, plus an additional 2–4 cm for the tail; its height as the shoulder is 12–15 cm, and it weighs between 750 and 1200 g. It is brownish-gray with small but still visible external ears, and the top part of its body is studded with sharp acerate spines between 2 and 3 cm long and about a millimeter in diameter. It reaches sexual maturity in its twelfth month of life and lives for about 8 or 10 years altogether. The hedgehog is found throughout Western Europe from the Mediterranean coast up to 63° N and ranges to the east as far as Hungary and the Balkans, Poland, Finland, and the Baltic states. It is most prevalent in woodlands with dense undergrowth or in heavy brush, though it is also found in gardens, orchards, and hedgerows, and rarely in forests of tall trees or resinous conifers; it is sometimes encountered up to an altitude of 1200 or 1500 m.

The hedgehog is nocturnal and takes shelter under a bush or in a woodpile, a haystack, or a bed of leaves during the day; it does not dig its own burrow. It goes out to hunt at dusk and returns before sunrise. When the temperature drops below 10°C in winter, it falls into a kind of hypothermic torpor, from which it occasionally emerges for a brief excursion if the temperature rises or if it is disturbed in some way, but soon drops off to sleep again.

▶ The hedgehog feeds on insects, earthworms, slugs, snails as well as small frogs, lizards, and snakes and will eat the eggs of ground-nesting birds—provided the shells are thin enough for its teeth to penetrate; they slide harmlessly off the shell on a hen's egg, for example. It also feeds on fallen fruit (apples, grapes, pears) and acorns. The hedgehog's sense of smell is very keen, and though not completely immune, it has a very high resistance to certain toxins—including cantharadin, secreted by certain beetles, and the venom of the common viper. The hedgehog is capable of dispatching the viper in a bristling, broadside onslaught, as described by Hainard: "It bunches up the spines on its forehead while pulling them down on its blind side and deploys them against the viper, awaiting a propitious moment in which to seize the snake, now worn out and wounded by these attacks, and crush its head or its backbone."

◀ The polecat and other members of the weasel family, the fox, and the domestic dog will all prey on the hedgehog, though the owl is perhaps its most redoubtable predator. The eyries of these birds are often littered with hedgehog spines and even whole skins; hedgehog spines are

231
Sub.: Vertebrata
Cl.: Mammalia
O.: Insectivora
F.: Erinaceidae

very often found in the pellets regurgitated by the eagle owl. Tens of thousands of hedgehogs are also killed on the roads every year, since the hedgehog still reposes too much confidence in its protective spines and defends itself against an oncoming car in the same way that it would against any other large, threatening creature—by rolling up in a ball in the middle of the road instead of scurrying out of the way.

In addition, the hedges and brushwood in which the hedgehog makes its home are increasingly giving way to one-crop agriculture—a much less suitable habitat from the hedgehog's point of view. Though legally protected in France since 1977, it is still sometimes destroyed by hunters because it allegedly robs the nests of pheasants and partridges. Though notoriously hospitable to fleas, ticks, mites, and other parasites, the hedgehog is still worthy of the esteem of farmers and gardeners as a destroyers of harmful insects; in rural French cottages and Gypsy caravans it was once considered a great delicacy as well.

The Algerian hedgehog *(Erinaceus algirus)* is lighter in color and the spine on its forehead are "parted" into two lateral tufts; this species is found in North Africa and along a thin strip of the Mediterranean coast of Spain and has been reported in southeastern France as well. Other species found in the desert regions of North Africa are assigned to the genus *Paraechinus*.

F.T.

232
Sub.: Vertebrata
Cl.: Reptilia
O.: Squamata
S.o.: Sauria
F.: Gekkonidae

Hemitheconyx caudicinctus 232

This gecko has small but muscular legs, a large head, a short, stubby tail (containing fat reserves), and conveys an overall impression of solidity; the conical shape of the tip of the tail appears to mimic the shape of the snout. This species differs from most other geckos in two respects: its eyelids are movable rather than fixed, and it lacks adhesive pads on its fingertips. Its body is light beige with wide, dark-brown transverse stripes, and its overall length rarely exceeds 20 cm (of which its tail accounts for about a third). This species ranges from Senegal to Nigeria, extending far to the north (Mali, Burkina Faso) in gallery forests along the banks of rivers; well camouflaged and inconspicuous in its movements, it has rarely attracted the attention of naturalists.

▶ The diet of *Hemitheconyx caudicinctus* seems to consist entirely of insects and other arthropods and their larvae. Inactive during the day (a captive specimen chose to spend the day sleeping under a slab of bark; this species does not dig its own burrow), it sets out at dusk on a slow, "systematic" investigation of its territory. It uses its tongue as both a tactile and olfactory organ, repeatedly flicking it out to "taste" the air, and paws through the vegetable litter on the ground in search of insect larvae. It attempts to approach its prey very slowly and seize it in its jaws, but it is very clumsy and is likely to be eluded by all but the slowest-moving creatures. It may attempt pursuit, but only for a short distance and then seems to lose interest in its quarry altogether. It crushes the larger insects or those with hard shells in its teeth before swallowing them.

◀ We have no specific information on the predators of this species, though there are a great many birds, mammals, and snakes that seek their prey amid the litter and organic debris on the forest floor. The gecko's first line of defense would be its protective coloration, which obscures the outline of its body and makes it somewhat difficult to see against the background of the forest floor. It is probably significant that the shape of the tip of its tail appears to mimic that of its head, since *Hemitheconyx* would able to shed its tail by autotomy and escape from a predator that had mistakenly seized it by the tail.

The other species that comprise the subfamily Eublepharinae are desert– rather than forest-dwellers: *Coleonyx variegatus* is a small gecko that is found in Mexico and the American Southwest; *Eublepharis macularis* is larger and found in the deserts of Western Asia.

D.H.

233
Sub.: Arthropoda
Cl.: Crustacea
O.: Decapoda
S.o.: Anomura
F.: Paguridae

HERMIT CRAB *Pagarus prideauxi* 233

The hermit crab's cephalothorax is about 15 to 19 mm long; the gelatinous abdomen is coiled up and relatively small. This species is found down to a depth of 400 m throughout the eastern Atlantic, from the Cape Verde Islands up to Norway, and the Mediterranean.

▶ The planktonic larvae feed on other

plankton; young and adult hermit crabs search out a wide variety of creatures that inhabit the sandy or muddy subtrate—small lamellibranchs (genera *Venus* and *Cultellus)*, echinoderms *(Echinocyamus, Echinocardium*, brittle stars), amphipods, shrimp, other hermit crabs, barnacles, free-swimming and sedentary polychaeates—as well as foraminifera, diatoms, and sometimes seaweed. The hermit walks slowly over the bottom, sifting through the sediment with its claws and its third set of mouth parts (maxillipeds, located directly behind the mandibles and maxillae), by means of which food particles and prey animals are conveyed to its mouth. Hermit crabs of the family Paguridae use their right claws to break the shells of snails or bivalves; other hermit crab families are "left-handed" or ambidextrous.

◀ The hermit's limited mobility and soft abdomen make it very attractive to a large number of benthic predators, including such bottom-feeding fish as rays, gurnards, Gadidae (cod, haddock, pollack, etc.), dogfish, and sharks, as well as crabs (of the genus *Calappa* or the family Portunidae), lobsters, octopuses, and cuttlefish. The hermit's first line of defense is of course to take refuge in an empty mollusk shell, and hermits sometimes battle among themselves to establish possession of an especially roomy or sturdy specimen. *Pagarus prideauxi's* symbiotic association with the anemone *Adamsia palliata*, whose stinging tentacles provide this particular hermit species with an additional measure of protection, has already been discussed (see COMMENSAL ANEMONE). As a last resort, a hermit that is not provided with an anemone can easily bury itself in the sand or mud with only its eyestalks protruding above the substrate.

Other hermit crab species are roughly similar to *Pagarus prideauxi* in their feeding habits. Some, like *Clibanarius erythropus*, a small hermit crab of the European tidal zone, frequently found in rocky areas, tend to feed more on organic wastes and debris *(detrivorous* is the technical term). The Atlantic hermit *Pagarus bernhardus* specializes in mollusks and other creatures that attach themselves to rocks: barnacles, very small oysters, and tube worms of the family Serpulidae. Hermits of the family Galathidae follow a similar regime to that of the Paguridae, concentrating on the smaller crustaceans (amphipods, copepods and their eggs) and other tiny marine organisms (diatoms, dinoflagellates); the feeding habits of the Galathidae that live in the ocean depths, well below the plankton layer, are unknown. Hermit crabs are rarely eaten by man, with the partial exception in recent years of one relatively large species, *Galathea strigosa;* commercial crabmeat is primarily obtained from Northern Pacific spider crabs of the family Lithodidae (see KING CRAB). P.N.

HERRING *Clupea harengua* 234

234
Sub.: Vertebrata
Cl.: Osteichthyes
S.cl.: Teleostei
O.: Clupeiformes
F.: Clupeidae

This familiar species is common in the North Atlantic, Barents Sea, White Sea, and the North Sea, and is often found in the English Channel, though rarely in the Bay of Biscay. It is about 40 cm long and generally lives for about 9 years, though sometimes for as long as 18 or 19, or even 25. There are several geographic races of herring, distinguished by the rapidity of growth, the size of the adults in spawning season, and the route and timing of their migrations between their coastal spawning grounds and deeper waters where plankton are seasonally abundant. The herring is a typical pelagic fish, never descending below a depth of 250 m, and in general, it prefers cold water of low salinity; it is also extremely gregarious, congregating in gigantic shoals, called *Heringsberge*, or "herring mountains," that may contain several thousand tons of fish.

▶ In the spawning season, the fertilized roe sinks down to the continental shelf, and the hatchlings, attracted by the light, swim up toward the surface to feed on algae and planktonic crustacean larvae. As they grow larger, the sprats (a term that is somewhat imprecisely used to refer to young herrings in general as well as the species *Clupea sprattus* in particular) begin to feed exclusively on animal prey—zooplankton (copepods), shrimp, pteropods ("sea butterflies," or free-swimming gastropods), and ammodyte (sand launce) larvae.

Adult herrings feed during the day, for the most part, moving slowly through the water (at a rate that has been calculated at 0.2–5 kph) and filtering out zooplankton and other marine organisms with the aid of the long, sharp spines, called gill rakers,

that are attached to the bony arch (branchial arch) that supports the gills. This feeding strategy enables the herring shoal to extract large quantities of nutrients from an even more substantial volume of water.
◀ The herring's sharp gill rakers, apart from supplying it with its food, make it both difficult and disagreeable to ingest, and they may also afford it a last-minute escape from the jaws of a potential predator. More experienced predators are probably inclined to avoid it for the same reason. The anterior part of the herring's swim bladder, which is in direct contact with the outer wall of the auditory capsule, has been modified into a sort of resonator that enables it to detect the sound of a predator swimming toward it through the water. Cod, mackerel, and many varieties of seabird feed on herrings, and the shoal's only collective means of defense is precipitous retreat. Human consumption of the Atlantic herring is currently reckoned at about 2 or 3 million metric tons per annum.

The sprat *(Clupea sprattus),* sometimes assigned to a separate genus *(Sprattus),* is found along the European coasts of the Mediterrananean and the North Atlantic. It also feeds primarily on zooplankton (especially copepods) and is preyed on by many larger species of fish. The European pilchard *(Sardina pilchardus)* and the anchovy *(Engraulis encrasicholus)* are both found in great numbers off the Atlantic coasts of Britain, France, and Spain, as well as in the Mediterranean; the pilchard feeds on planktonic crustaceans, which it ingests orally, one at a time, instead of by gill filtration. Unlike the herring, it prefers warm, salty water. In commercial parlance, the term "sardine," like "sprat," refers primarily to the size of the fish rather than the species and is thus often applied to young (or small) clupeids of commercial size other than *Sardina pilchardus*.

F.M.

235
Sub.: Vertebrata
Cl.: Aves
O.: Charadriiformes
F.: Laridae

HERRING GULL *Larus argentatus* 235

This familiar seabird, now common throughout the Northern Hemisphere, measures between 60 and 65 cm, with a wingspread of 128–150 cm, and weighs between 0.8 and 1.3 kg. It is found along rocky or muddy shorelines, in coastal marshes, and in the vicinity of croplands and cultivated fields, landfills, and garbage dumps, as well as along the shores of rivers and inland lakes and other large bodies of fresh water. Record longevity is 49 years in captivity, 26 years in the wild, but only about 10 percent of all herring gulls survive beyond the first year of life.
▶ Gregarious, omnivorous, and active both by day and night, the herring gull is also sturdy, resourceful, and notoriously successful in exploiting the nutritional possibilities of human refuse whether found in dumps and landfills, around wharves and docksides, sewer outlets, or floating in the wake of freighters and fishing boats. The herring gull also hunts on damp or freshly turned soil for earthworms, grasshopper, cockchafers, and small rodents, feeds on carrion that has been washed up on shore, dispatches sick or injured animals, and snatches fish from the beaks of terns and other seabirds on their fishing grounds. It also feeds on the eggs and nestlings of other gulls as well as terns, puffins, shearwaters, and sea ducks.

The herring gull is not an accomplished diver; it prefers to patrol the tidal flats in search of crabs, marine worms, stranded fish, and mollusks, which its retrieves by kneading in the soft mud with its feet. It cracks open the shells of bivalves by dropping them from a height of several meters onto a flat rock, a tar road, or the roof of a building—behavior that places it among the small fraternity of tool-using birds, which also includes the crow, which cracks nuts, the lammergeier vulture, which splits marrowbones, and the golden and imperial eagles, which crack open the shells of turtles—all using more or less the same technique as the herring gull. The young gulls are fed by regurgitation; a red spot on the lower mandible triggers an instinctive feeding response on the part of the parents.
◀ Gulls nest in very large colonies, and few predators other than foxes will approach their nesting sites; on the other hand, a large percentage of the eggs and nestlings are destroyed by other gulls, and adult gulls are occasionally taken in flight by gyrfalcons and peregrines. At the beginning of this century, the herring gull was much less common than it is today, and some European nesting grounds were legally protected. Since then, both the her-

ring gull and the great black-backed gull have taken advantage of this protected status and, more importantly, of the abundant refuse of twentieth-century civilization to expand their range and greatly increase their numbers on both sides of the Atlantic. (In Brittany, for example, the herring gull population grew from 6000 or 7000 pairs in 1955 to 25,000 pairs in 1970.) A number of less resilient shorebird species, including terns and sandpipers, are now threatened with extinction as a result of the encroachments of the herring gull, and in several countries large numbers of nests and adult birds are now being routinely destroyed, in the absence of natural checks on this runaway population boom.

The North American herring gull is treated as a subspecies of *Larus argentatus* in some classifications, a separate species *(L. dominicanus)* in others; some 15 subspecies, found in various regions of the Northern Hemisphere, are recognized altogether. As noted, the great black-backed gull *(L. marinus)* is also increasingly common on both sides of the Atlantic; a close relative, the lesser black-backed gull *(L. fuscus)*, is somewhat smaller and found only along the coasts of Europe. The mew, or European gull *(L. canus)*, is intermediate in size between the larger transatlantic species and such smaller European gulls *(Mowen, mouettes)* as the kittiwake, black-headed gull, and little gull. Several European species are endangered or very rare, including Audouin's gull *(L. audouinii)*, only 2000 pairs of which are still nesting on islets in the Balearics or off the Corsican coast, and the slender-billed gull gull *(L. genei)*, also found along the Mediterranean coast. The rather mysterious gray gull *(L. modestus)* is found in the salt deserts in the interior of Chile, the driest region on earth, but little is known of its habits.

J.-F.T.

HONEY BUZZARD *Pernis apivorus* 236

236
Sub.: Vertebrata
Cl.: Aves
O.: Falconiformes
F.: Accipidridae

The honey buzzard is an Old World hawk, about 52–60 cm long, with a wingspread of 135–150 cm, and weighing about 600–950 g. One banded specimen is known to have lived for 28 years. It makes its summer home in Europe, particularly in moist, sparsely wooded areas or brushlands where bees are relatively abundant or in mountain forests up to a height of 1500 m; it migrates in large flocks to spend the winter in the equatorial forests of Africa, crossing mountains ranges and "fording" the Mediterranean at the Straits of Messina, the Bosphorus, and Gibraltar. The fledglings set off with their parents on their southbound migration shortly after leaving the nest; very plump at the time of departure, the honey buzzard lives entirely off its body fat during its long migratory flight.

▶ The honey buzzard is not a bird of prey in the usual sense, since it feeds primarily on insects, especially on the larvae of wasps and honeybees, which it digs out of their underground nests, normally capturing a number of adult hymenopterans in the process, prudently snipping off their stingers with its beak before it swallows them. It also eats beetles, grasshoppers, moths and butterflies, caterpillars, insect larvae, as well as lizards, snakes, frogs, baby birds, fruits, and nuts; in its African winter home, it lives primarily on hymenopterans and termites in about equal measure. It catches insects in flight and digs up wasps' nests with its talons, which are less curved than those of other hawks; the thick, scaly plumage around its beak and eyes helps protect it from being stung.

◀ The honey buzzard is so specialized in its feeding habits that it faces competition only from other members of its own species. A pair of honey buzzards (of which there are still thought to be several tens of thousands in Western Europe) presides over very extensive territory. The eggs and nestlings may occasionally be eaten by martens or wild cats; in spite of its innocuous insectivorous regime, the honey buzzard is difficult to distinguish from other hawks and thus is in some danger of being persecuted by man as a "destructive" predator. A considerable toll of this species is also taken by hunters in Sicily and Malta during its spectacular migratory passage across the Mediterranean.

J.-F.T.

HOOPOE *Upupa epops* 237

237
Sub.: Vertebrata
Cl.: Aves
O.: Coraciiformes
F.: Upupidae

This Old World bird measures 25 or 26 cm, with a wingspread of 44–47 cm, and weighs between 51 and 80g. It found in sun-drenched open country that is dotted with old trees and ranges throughout much of Eurasia (excluding the British Isles, northwestern Spain, northern Germany as well as the Middle East, and the Himalayas) and parts of Africa (excluding the Sahara and the equatorial rain forests). It arrives in Western Europe at the end of March and sets off again for southern Africa in August. The hoopoe is a solitary bird that nests in a cavity in a hollow tree (a hole made by a woodpecker, for example) or in a stone wall. Its flight is wavelike and fluttering, its wingbeats soft; it is quite at home on the ground, bobbing its head back and forth as it walks along.

▶ The hoopoe feeds on a variety of insects and their larvae (beetles, grasshoppers and crickets, butterflies and moths, etc.) as well as spiders, mollusks, and earthworms, which it sometimes finds by probing in pats of cowdung; it catches its prey in its beak and kills it by beating it against the ground, and sometimes tosses it up in the air and catches it in its mouth while holding its beak wide open. The adults take turns feeding the nestlings, holding insects out to them in the tips of their beaks.

◀ The hoopoe is not systematically preyed on by any other species, though migrating birds are sometimes taken by Eleonora's falcon. When the hoopoe is anxious or alarmed, it opens and closes its brushlike crest like a fan, and sometimes when it catches sight of a bird of prey, it assumes a defensive stance, spreading its wings and fanning out its tail, lifting its head and pointing its beak in the air. The young hoopoes and the females are also capable of secreting a foul-smelling liquid from the anal gland, which is assumed to be a deterrent against predators. The hoopoe may compete with the starling for nesting sites.

The hoopoe's striking appearance and curious behavior invested it with mythical associations for the ancient Greeks and Romans, likewise for the inhabitants of the Sahara, who observe a taboo against hunting it, though for the modern Greeks living on Crete and the Peloponnesus, it is just an ordinary gamebird. In Morocco, on the other hand, the brains of the hoopoe are thought to have medical or magical properties and are accordingly very much in demand; in Europe, its appeal is purely visual, and it furnishes a favorite demonsration piece for the taxidermist.

The family Upupidae includes one other species, found only in tropical Africa, and 2 other genera *(Phoeniculus, Rhinopomastus)* that are collectively known as wood hoopoes, arboreal birds that are also found in Central Africa. J.-P. R.

HORNED VIPER *Cerastes cerastes* 238

238
Sub.: Vertebrata
Cl.: Reptilia
O.: Squamata
S.o.: Ophidia
F.: Viperidae

This species typically measures between 60 and 70 cm, though it may attain a length of up to 90 cm. The horned viper is common in the Sahara and is found in deserts (except in regions of shifting sand) and arid steppes from Mauretania to Iraq, ranging southward as far as the southern edge of the Sahel. The horned viper can be readily identified by the two sharp excrescences over its eyes, measuring about 5–6 mm, which may be reduced to less prominent pyramidal bumps in older individuals. The horned viper's body is covered with markedly carinate ("keeled," i.e. ridged or wedge-shaped) scales and is sand-colored with brownish or blue-violet spots.

▶ As with the other true vipers, the horned viper's dentition is of the solenoglyphic type: It injects its venom through a channel in each of its two hollow fangs that serves as the outlet for a venom gland (actually a highly modified salivary gland) in the upper jaw. In the spring and autumn, when the heat of the sun is less intense, the horned viper stays buried in the sand with only the tip of its snout protruding near the entrance to its burrow or underneath a bush and waits for its prey to come.

During the summer, as long as the temperature is at least 25° C, it actively seeks its prey at dusk and on into the night (generally from around 6 to 11 p.m.), sometimes covering several kilometers in the course of a single night's hunting and generally examining every tuft of grass and poking its snout into every burrow it comes across. The horned viper's marauding strategy is largely conditioned on the scarcity of prey animals in a given locality, and it does not remain within the confines of a well-defined hunting territory for several years as some European vipers do.

The juveniles feed largely on lizards *(Acanthodactylus*, agamids), young rodents stolen from the nest, *Galeodus* (a solpugid, or sun spider), grasshoppers, tenebrionid beetles, and other arthropods. The adult supplements this basic menu with a variety of larger prey animals, including gerbils, gerboas, jirds (a relative of the foregoing), and sand rats *(Psammophys)*. The horned viper also attempts to catch birds by using the tip of its tail as bait, wiggling it occasionally to simulate the spasmodic movements of an insect larva, though the effectiveness of this method seems to be limited. Horned vipers have repeatedly been observed curled up in the sand surrounded by a wide circle of bird tracks, with a virgin expanse of sand lying between them. Nevertheless, the horned viper is able to catch house buntings, redstarts, wheatears, warblers, and even large woodchat shrikes on occasion.

Horned vipers also feed on a more nourishing and less elusive quarry, the African spiny-tailed lizard *(Uromastix acanthurinus)*, a species that is slow-moving and unaggressive and whose only drawback, from the viewpoint of the horned viper, is that it is almost too large for the latter to ingest. For example, the stomach of a horned viper measuring 65 cm and weighing 217 g was found to contain a spiny-tailed lizard measuring 24 cm and weighing 83 g (or 38 percent of the body weight of the predator!).

The predatory reflex of the horned viper is triggered by the movements of a prey animal as it passes close by the viper's lair or place of ambush. The initial perception is followed by a brief pursuit, and in some cases the prey is released shortly after it has been injected with venom (cf. ASP VIPER) and then quickly, when the venom has already done its work; smaller prey animals may be devoured on the spot, starting as usual with the head.

The process of digestion may last for several days or several weeks, depending not only on the size of the prey animal but on the external temperature as well. Thus, in April it took 6–7 days for a horned viper to digest a great gerbil *(Grebillus gerbillus)*, during which time the temperature ranged between 14 and 30°, whereas the undigested remains of a mourning wheatear *(OEnanthe lugens halophila)* swallowed on February 15 were not excreted until 22 days later, and the viper remained inactive, concealed under a rock, throughout this entire period.

The horned viper also refrains from feeding during the winter latency period, which lasts for several months; the snakes are still sluggish at the end of this period, in March, and though the females begin to feed quite actively, the males may not break their fast for another two months; they begin to hunt again before the end of the mating season, which generally lasts from about May 15 to mid-June. The females continue to feed during the mating season—and even during copulation, since the sexes typically remain coupled from 36 hours up to as long as 4 days—up until the beginning of the winter latency period.

◀ The horned viper's habit of burying itself in the sand serves to conceal it from potential predators as well as prey animals and also helps to regulate its body temperature. It coils itself up on the sand and begins to move its tail from side to side, pushing the sand away on either side to form a trench, the walls of which eventually collapse, leaving the viper's tail buried in the sand; these lateral movements are gradually picked up by the foreparts of the viper's body, until only the head is left protruding above the surface. Finally, the viper ducks its head forward, then draws it back, and with a single quick, oblique motion, jabs it into the sand and disppears from sight.

When there is no time for it to conceal itself in this manner, the horned viper will attempt to warn or intimidate an aggressor by emitting a peculiarly disquieting sort of a rustling or rattling sound through its nostrils, puffing up its body and scraping its scales together as it uncoils. Both the loftiest and the humblest of the viper's natural enemies, namely the eagle (in this case, the short-toed eagle) and the hedgehog are both found in the Sahara, and the desert monitor (q.v.) also preys on the horned viper, clamping its jaws down just behind its head and shaking it violently to break its back. The monitor's long sharp teeth penetrate deep into the viper's flesh, so that it sometimes succumbs to internal hemorrhaging. Since the viper's venom appears to have no affect on the desert monitor, these contests almost invariably end with the viper being swallowed by the monitor.

Several other members of the family Viperidae are also found in the arid steppes and desert regions of North Africa; all of them have large heads with a clear demarcation of the neckline, strongly carinate (ridged or wedge-shaped scale) scales, and the same lateral ("sidewinding") method of locomotion. This adaptation, by virtue of evolutionary convergence, is shared with two unrelated desert-dwellers that are accustomed to traveling across soft sand, namely the Namib viper *(Bitis peringueyi)* of southwestern Africa and the sidewinder *(Crotalus cerastes)* of North America. The sand viper *(Cerastes vipera),* unlike the horned viper, is found in the high dunes; it is a light yellow-orange in color, though the tip of the tail is occasionally black. It is smaller than the horned viper (40 cm), it lacks horns, and its eyes are located on top and toward the edges (rather than on the sides) of its flattened head. Its hemolytic venom seems to be identical in its effects to that of the horned viper.

The saw-scaled viper *(Echis carinatus),* found in subdesert terrain in Africa and Western Asia, is similar to both the horned viper and the sand viper in its habits. It also lacks horns, and its sand-colored skin is a deeper yellow than that of the sand viper and marked with a pattern of darker lozenges with a lighter-colored spot in the center of each one; the underbody is pinkish-white and speckled with very distinctive brown dots. R.V.

239
Sub.: Annelida
Cl.: Kirudinea
O.: Gnathobdellae
F.: Pharyngobdellae

HORSELEECH, DOG LEECH
Hamaeopis sanguisuga and *Erpobdella octoculata* 239

The great majority of leeches are free-swimming annelids found in fresh or brackish water that attach themselves as temporary external parasites on other creatures (invertebrates as well as vertebrates), though a few species are marine or terrestrial. Leeches are found in every region of the globe, and their parasitic activities sometimes result in the death of smaller organisms; the bloodsucking freshwater species constitute a serious public-health problem in Southeast Asia, Sri Lanka, and other tropical regions. The anticoagulants in the saliva of the medicinal leech *(Hirudo medicinalis)* are still employed in certain specialized surgical procedures, though on a considerably more restricted scale than in premodern medical practise; the English word leech word comes from the Old English lecce, 'doctor,' 'bloodletter,' though some scholars maintain that the word was first applied to the hirudinean and only later to its human counterparts.

The horseleech *(Haemopis sanguisuga;* German *Pferdegel)* is found in Europe as far north as Norway, in North America, North Africa, and Asia and normally lives in stagnant water, though it is occasionally encountered on land as well. The dog leech *(Erpobdella octoculata)* lives under rocks in streambeds in essentially the same regions and has been reported as far north as Archangelsk. Both species are predatory rather than parasitic, though because of structural diference in the musuclar pharynx (see below), they are assigned to different suborders—Gnathobdellae and Phayngobdellae, respectively. The longevity of both species is two years.

Most leeches have both a posterior and an anterior adhesive disk, the latter of which extends outward from the creature's mouth. Leeches are hermaphroditic and have well-defined genital organs; the glands of the clitellum, a clearly differentiated section of the epidermis in the middle of the body, secrete a cocoon to receive the eggs and in which the embryos develop directly into their adult form without undergoing an intermediate larval stage.

▶ Though it belongs to the same family as *Hirudo medicinalis* (and despite the bloodthirsty connotations of the species name), *Haemopis sanguisuga* feeds on earthworms, young snails, insect larvae, the eggs of frogs, snails, and fish as well as tadpoles and small fish. It feeds most actively in the spring following its winter dormancy, and is capable of going for long periods without feeding at all. Its wide mouth opens directly onto an extrudable muscular pharynx that contains two sparse rows of teeth and is well equipped for chewing, and instead of subsisting on the body fluids of its prey, like most leeches, the horseleech devours its prey in its entirety.

Erpobdella octoculata feeds on freshwater

mollusks, aquatic insects, chironomid larvae, and occasionally on bits of carrion; most of these are creatures it simply encounters at random as it crawls along the streambed, though it can catch a chironomid larva only if it approaches it from behind. The movements of the toothless pharynx are controlled by longitudinal bands of muscle (rather than annular, as is the case with *H. sanguisuga)*, and it absorbs the prey by means of strong peristaltic contractions. It only consumes 1 or 2 chironomids, for example, in the course of a single day, and it may take from 7 to 26 days to digest its meal. *E. octoculata* begins to stir as the daylight wanes and begins to hunt more actively at night, it is somewhat less active during the winter, but will continue feeding as long the water temperature remains above 2°C. Its predatory activity remains on a more or less constant level at temperatures between 7 and 12°C.

◀ Hemipterans (true bugs), dragonfly larvae, and other aquatic insects, as well as aquatic birds and fish, all feed on leeches from time to time.

As noted above, the medicinal leech *(Hirudo medicinalis)* is a parasitic species assigned to the suborder Gnathobdellae; it feeds on the blood of its victims by biting a triangular hole in the skin and secreting its anticoagulant saliva into the wound. Species assigned to the suborder Rhynchobdellae are equipped with an extrudable trunk through which they can suck out the bodily fluids of their victims, though not necessarily with fatal results in the case of the larger host species. *Glossiphonia* specilaizes in mollusks, *Hemiclepsis* in fish, *Theromyzon* in aquatic birds, and such genera as *Glossiphonia* and *Helobdella*, which devour the bodies of their victims outright after they have drained out the fluids, can reasonably be considered predators rather than parasites. C.M.

HORSE MACKEREL *Trachurus trachurus* 240

240
Sub.: Vertebrata
Cl.: Osteichthyes
S.cl.: Teleostei
O.: Perciformes
F.: Carangidae

The horse mackerel, also known as the scad, is very common in the Eastern Atlantic all the way from North Cape (the tip of Norway) to the Cape of Good Hope; the Mediterranean subspecies is sometimes classified as a separate species. It is a gregarious deep-sea (pelagic) fish, very similar to the mackerel in its habits and appearance. It grows to a length of about 25 cm by the age of 4 and attains a maximum length of 50 cm. It is found in coastal waters and in winter may retreat to a depth of 100 to 500 m; younger individuals are found in large numbers in shallow coastal waters where the bottom is sandy.

▶ The hatchlings feed initially on diatoms (microscopic plants) but later turn to a diet of zooplankton (copepods, crustacean larvae, fish eggs) when they have grown a little larger. The adults will also feed on zooplankton (pteropod mollusks and crustaceans), though they prefer small clupeid fish (herrings, sprats, pilchards, anchovies), ammodytes (sand launces), and the smaller members of the codfish family (Gadidae). Squids are also sometimes found among their stomach contents. The horse mackerel hunts in schools, like mackerel, and members of these two species are often found in close association.

◀ Dolphins, tuna, and dogfish are among the principal predators of both species, and their habit of assembling in schools appears to be the horse mackerel's best defense against predation. Young horse mackerel in shallow water will often seek protection from predators under the canopy of a stinging jellyfish of several different species, notably *Aurelia aurita* or *Cyanea capillata*, or even of a nonstinging jellyfish like *Rhizostoma octopus*. The young horse mackerel rarely ventures far from the safety of this protective canopy and will retreat inside the jellyfish's defensive perimeter at the slightest sign of danger.

There is no commercial market for the horse mackerel in Northern Europe, where its flesh is used for bait or reserved for the consumption of local fishermen, whereas in Portugal and West Africa, it is commonly smoked or eaten fresh and sometimes ground up into fishmeal. Schools of horse mackerel are often attracted by shining a light on the water, a ruse to which they are very susceptible; they are ordinarily caught with surface nets or dragnets or with a line, like mackerel. On the Pacific coast of North America, the name *horse mackerel* was once frequently applied to the bluefin tuna—a usage that has largely been eradicated by the efforts of publicists for the commercial tuna fishery. F.M.

241
Sub.: Vertebrata
Cl.: Aves
O.: Gruiformes
F.: Otididae

HOUBARA BUSTARD *Chlamydotis undulatus* 241

The houbara bustard measures between 55 and 65 cm, with a wingspread of 135–170 cm, and weighs between 1100 and 2500 g; the males are somewhat larger than the females. Captive specimens become sexually mature at the age of 2, though this is not necessarily true in nature. Like all bustards, this species is found in open terrain—in this case, the subdesert steppes of Africa and Asia, particularly in areas with rocky or gravelly soil and only occasional tufts of vegetation. The houbara's drab, earth-colored plumage fits in very well with this backdrop, and this species is largely terrestrial in its habits, though some populations are migratory and others make susbtantial forays into the desert to feed on plants and insects after a sudden downpour. The male houbara bustard is noted for its bravura courtship technique; houbara bustards live in pairs during the mating season and travel in small flocks during the rest of the year.

▶ The houbara bustard patrols its hunting ground with lengthy strides, snatching up insects and other small creatures with its beak. It preys on crickets, locusts, cockroaches, beetles and their larvae (mealworms, *Buprestis*, chrysomelids), ants and termites, butterflies and moths, as well as spiders, snails, snakes, and lizards; it also feeds on a variety of vegetable material. The nestlings are fed on insects during their first few weeks of life; the adults can live on dew and will often go for lengthy periods without drinking groundwater.

◀ Bustard chicks are occasionally taken by the saker falcon or the peregrine, though the adults have learned to flatten out against the ground, making themselves more or less invisible from the air. The adults can also to defend themselves quite ably with their wings and beaks, even blinding their attackers with a sticky gob of guano if circumstances permit. A female bustard with chicks is reported to have attacked a fox and driven it off with a flurry of wings, and, more surprisngly, a flock of bustards to have attacked a falcon that was harrying a calf.

The Bedouins have traditionally hunted the houbara bustard with falcons, though in contemporary Arabia, it has become the custom to hunt bustards with Jeeps and automatic weapons (!). Perhaps overly confident of its protective coloration, the houbara bustard will allow a vehicle to approach within a very few meters before it attempts to run away, and even though the houbara bustard has been clocked at up to 40 kph on the straightaway, it is obviously no match for a Jeep. Irrigation and the reclamation of semidesert terrain for agriculture has somewhat reduced the habitat available to this species; intensive hunting (particularly of the type described) threatens its extinction throughout much of its range in the next few years.

The bustard family is a homogeneous group of 22 species, 15 of which are found in Africa; most of them, like the houbara, are currently threatened by overhunting and habitat reduction. Two species are found in Europe, though both are becoming increasingly rare: The greater bustard *(Otis tarda)* has not been reported in France since around the turn of the century; it is an enormous bird, the males sometimes exceeding 15 kg. The lesser bustard *(Otis tetrax)* was common until fairly recently in the low-lying, cereal-growing regions of Western Europe, though its numbers have recently been reduced by the effects of monoculture as well as chemical fertilizer and pesticide residues. J.-P. T.

242
Sub.: Vertebrata
Cl.: Aves
O.: Trochiliformes

HUMMINGBIRD *Trochilidae sp.* 242

The Trochilidae, or hummingbirds, are among the most homogeneous and most easily characterized of all bird families. There are about 320 different species, all native to the New World and found in virtually all possible habitats—tropical forest, desert, subarctic tundra—betweeen Alaska and the Straits of Magellan. Though related to the sparrows and other songbirds (Passeres), the hummingbirds have their own characteristic morphology, essentially a series of adaptations that arose in conjunction with their unique feeding habits. It was long assumed that hummingbirds fed exclusively on nectar, but in fact the nitrogen-rich nutrients they derive from insect prey are essential in maintaining their energy-intensive, hovering and darting mode of flight.

The bee hummingbird *(Mellisuga minima)* of the Caribbean is about 5 or 6 cm long, and thus the smallest of all birds, but even the largest member of the family, the giant hummingbird *(Patagona gigas)* of the Andes, is only about 20 cm long and weighs barely 20 g. It has been known to survive (in captivity) for as long as 8 or 9 years, which seems like something of an accomplishment for such a delicate creature with such a rapid metabolism. The majority of hummingbird species are found in tropical and subtropical America, and they naturally show a preference for any region where flowers and insects are abundant all year round. Thus, the greatest concentration of species is found in the foothills of the Andes, and the depths of the Amazon rainforest provide a comparatively unfavorable environment. Some hummingbird species that live in colder regions are migratory; others undergo a sort of hibernal dormancy during which their body temperature drops as low as 18°C.

▶ The hummingbird is highly territorial, and the brilliantly colored plumage of the males is employed for purposes of sexual display. Specialized adaptations of the heart and lungs enable it to increase its oxygen consumption if necessary, but, as noted earlier, the hummingbird depends on a rich, highly concentrated diet to meet the considerable energy demands of its specialized modes of feeding and locomotion. Part of this diet consists of nectar and other floral secretions as well as the sap of plants and trees (in the latter case, where the bark has been pierced or damaged in some way), the remainder consists of insects—mosquitoes, aphids, blackflies (family Simuliidae), and spiders as well as some chitonous insects such as beetles.

The rapidity of the hummingbird's wingbeats enables it to climb, dive, bank, and, most importantly, to hover motionless outside a flower while collecting nectar and insect prey. The hummingbird eats spiders along with the creatures imprisoned in their webs and catches insects on the wing—somewhat in the manner of the Old World flycatchers—setting off in pursuit from its unconcealed perch, to which it returns a few moments later with its prey.

The young of Viellot's hummingbird *(Hylocharis leucotis)*, to choose one example, are fed exclusively on spiders until they are 7 days old, when they are big enough to accept larger and less readily digestible arthropods, a progression that culminates in insects with tough chitinous shells. Hummingbirds in temperate regions feed on insects and spiders during the winter and on plant sap during the warmer months. To assist in extracting nectar and dislodging insect prey from the interior of a flower, the hummingbird's tongue is protractile, and its beak, though invariably long and thin, has undergone a host of adaptive specializations, the character of which can often be deduced from the common name of the species, e.g. the sword-billed hummingbird *(Ensifera ensifera)* or the sicklebill *(Eutoxeres)*.

◀ Each hummingbird species defends its territory with a display of daredevil aerobatics interspersed with threatening vocalizations. On these occasions, a hummingbird will not hesitate to attack a larger bird, even a hawk or other raptor, and because of the swiftness and agility of the hummingbird's flight, it has little to fear from these or other predators at any time. During the 19th century, hummingbirds were occasionally caught with limetwigs, stuffed, and mounted on pins, brooches, and ladies' hats as a kind of readymade costume jewelry. Fortunately this practice has long since been discontinued, and no permanent harm seems to have come to any hummingbird species as a result of it.

J.-F. T.

HUMPBACK WHALE *Megaptera novaeangliae* 243

243
Sub.: Vertebrata
Cl.: Mammifera
O.: Cetacea
F.: Balenopteridae

The humpback whale, found in all the oceans of the world, may attain a length of 16 m and weigh as much as 70 metric tons, the females being slightly larger than the males. Longevity is in excess of 20 years. The humpback's migrations tend to follow the coastline, and its feeding habits and other behavior have been more extensively observed than those of other ceteceans. Considerable interest has been focused on the "songs" and other complex vocalizations by means of which the humpback appears to be able to communicate across vast oceanic distances.

▶ Like the blue whale, the humpback feeds on herring, capelin, and other small fish, as well as small crustaceans; the basic process—that of opening its mouth very wide and swallowing large volumes of water—is simple enough, but has been subject to various tactical modifications in order to increase its efficiency. Thus, when the humpback prepares its initial assault on a shoal of herring in the North Atlantic, it approaches from below, generally swimming upward at an angle of 20–40°, in ever decreasing circles to create a funnel of bubbles, opening its mouth and increasing its speed as it rises, then snapping its mouth shut as soon as it breaks the surface.

A large part of the humpback's body often heaves up above the surface, and the distension of its gullet and the water streaming out of the corners of its mouth can be seen quite clearly before the whale drops back into the water, accompanied by a tremedous upsurge of foam and spray. After the humpback disappears beneath the surface once again, it literally lies low—at a depth of 5 or 6 m—for as long several minutes before it attempts another pass at the herring shoal.

The humpback uses various strategies to corral the herring shoal into an even more compact mass—by swimming in circles around it, compressing it against the surface of the water by approaching the shoal from below, or swimming slowly toward the surface in a decreasing spiral while expelling bubbles of air from its blowhole. This maneuver gives rise to a more or less conical curtain of bubbles, referred to as a "bubble net," and the fish, alarmed by this unusual phenomenon, tend to congregate in the center of the cylinder where they can more efficiently be devoured by the humpback. The humpback appears to be a solitary hunter, and cooperative "herding" behavior has never been observed.

◀ The humpback is smaller and slower than the blue whale, and thus perhaps more likely to fall victim to a pack of killer whales, which prefer to prey on the calves in either case. The humpback was one of the principal objects of the summer whaling season in the cold waters of the North Pacific, North Atlantic, and Antarctic Oceans as well as during the spring calving season in warmer waters and even during the intervening coastal migrations. By the beginning of this century, the humpback population of the North Atlantic was already severely depleted; commercial whaling operations in Antarctic waters were also concentrated on the humpback, resulting in a even more precipitous decline. The taking of humpbacks was finally forbidden in the North Atlantic in 1955, in the Antarctic in 1964, and in the Pacific in 1966; since then, a slight recovery of the humpback population has been reported in Antarctic waters. The North Atlantic population is currently estimated at about 1000. D.R.

Hygrobia tarda 244

244
Sub.: Arthropoda
Cl.: Insecta
O.: Coleoptera
F.: Hygrobiidae

This aquatic beetle is widespread throughout the central part of Western Europe; both the larvae and adults are found along the muddy bottom of stagnant ponds and pools, and always in conjunction with tubifex worms, on which the larva of this species feeds exclusively. *Hygrobia* is thick-bodied and highly convex in shape, and measures about 8.5–10 mm long. Its upper body is brownish with rust-colored spots or patterns, which are generally rendered less conspicuous by a thick coating of mud. The lower surface of the elytra (in both sexes) is provided with a vibratory mechanism that permits *Hygrobia* to stridulate (to whirr or buzz loudly like a cricket or cicada) when it is frightened or upset—as it usually becomes if it is picked up with two fingers, for example, in which case the sound can be heard very distinctly. The hind legs (all 3 pairs of legs in the larva) are slightly flattened and fringed with filaments that assist in swimming, and the first few segments of the tarsi on the male's forelegs are somewhat dilated and provided with a feltlike "heel" (perhaps more closely resembling a strip of Velcro) that has adhesive properties.

The body of the larva is yellowish and elongated; its abdomen is tipped with 3 long projections, and it may attain a length of 18 mm by the time its development is complete. Life expectancy of the adult is 1 to 3 years.

▶ The adult generally remains buried head downward in the sediment, with only the tip of the abdomen protruding; occasionally it bobs back up to the surface to replenish its air supply and then quickly conceals itself in the mud again. It makes its way along the muddy bottom in search of insect larvae (chironomids, dragonflies, caddis flies) and oligochaetes (tubifex worms) especially; these prey are minced up very quickly by mandibles and maxilla, and immediately ingested. It primarily hunts for food by day.

As noted, the larva feeds exclusively on tubifexes. Its tracheobranchial breathing apparatus relieves it of the necessity of returning to the surface, and it conceals itself on the bottom and attacks its prey from ambush. When a tubifex shows itself at the entrance to its tunnel, the *Hygrobia* larva throws itself toward it and seizes it with the claws at the tips of its forelegs. The larva will sometimes crawl down the tunnel itself if preliminary investigation with its palps and antennae reveal it to be occupied; the worm is devoured on the spot after being reduced to more manageable pieces by the larva's mouthparts.

◀ Little else is known about this species; it is probably attacked by the same predators as the great diving beetle, *Dytiscus marginalis* (q.v.).

The family Hygrobiidae consists of only 4 different species worldwide; 2 of the others are found in Australia, the third in China, and their habits are probably quite similar to those of *H. tarda*. J.-L. D.

I

ICHNEUMON *Herpestes ichneumon* 245

245
Sub.: Vertebrata
Cl.: Mammalia
O.: Carnivora
F.: Viverridae

This species of mongoose ranges widely throughout Africa and is also found in Italy, Yugoslavia, and the Iberian peninsula. It is basically similar to other mongoose species (and numerous other viverrids) in form, with very thick fur, a bushy, black-tipped tail, and nonretractible claws. It is also called the Egyptian mongoose, since the ichneumon was one of many animal divinities revered by the ancient Egyptians, and hunting mongooses were often portrayed on the walls of Egyptian temples.

The ichneumon is found in thickly wooded areas, brushwood, reedbeds along the shores of ponds and lakes, and anywhere where there is dense vegetation to provide a refuge for it. The juveniles stay with their parents until they reach sexual maturity, and a mongoose family always travels in single file, each one with its nose right under the tail of its predecessor. Generally speaking, the ichneumon does not scent-mark or defend its territorial boundaries.

▶ The ichneumon is omnivorous, feeding on worms and insects, reptiles (lizards and snakes, including venomous species), birds and their eggs, fieldmice and other small mammals, as well as summer fruit. (In general, the dietary adaptability of the mongoose has enabled it to flourish in the various regions of the New World and the Pacific to which it has been introduced by man.) It was once thought that the ichneumon hunts at night, but observations recently made in Spain by F.R. de la Fuente have shown that it is active from dawn to dusk, with a peak of activity at around the middle of the day.

The ichneumon sometimes hunts in groups, patrolling its territory in the hope of flushing a rodent or other small creature, though if the quarry appears to be capable of outrunning its pursuers, the ichneumon quickly gives up the chase. When a group of ichneumons comes upon a rabbit warren, each one of them heads down a separate burrow so that rabbits will be unable to escape. When it hunts on its own, the ichneumon is more methodical, moving silently and stopping from time to time to sniff the air and to make a visual reconnaissance of the surrounding territory. The ichneumon cannot run very fast, but it is extremely agile and stealthy, and excels at stalking its prey and attacking from ambush; it generally dispatches a prey animal with a single bite to the neck.

Like other mongooses, the ichneumon has three important adavantages in its legendary duels with the cobra and other venomous snakes. Although it is not entirely immune, it is only one sixth as susceptible to the snake's venom as is a rabbit, for example, and its thick coat supplies further protection against the cobra's fangs. However, the ichneumon's principal advantage lies in its agility; it allows the snake to tire itself by striking repeatedly and ineffectively, then, when it senses that the snake's reflexes are sufficiently blunted, the ichneumon waits for it to strike once more, then seizes the base of the cobra's skull in its jaws, pins it to the ground, and quickly bites its head off.

◀ The ichneumon has few predators but is nevertheless prepared to meet an attack with a sequence of intimidating or bewildering displays; it begins by bristling its fur, hitching up its hindquarters, and pointing its muzzle at the intruder, probably in an attempt to appear larger and more threatening than its really is. If this proves ineffective, the ichneumon wraps itself up in its bushy tail, and an adversary that comes away from its initial attack with nothing but a mouthful of fur generally decides to have nothing more to do with the Egyptian mongoose. P.A.

246
Sub.: Vertebrata
Cl.: Reptilia
O.: Crocodilia
F.: Gavialidae

INDIAN GAVIAL *Gavialis gangeticus* 246

The gavial, or gharial, is a freshwater crocodilian, measuring 4–5 m (possibly as much as 7 m) and formerly quite common in streams and rivers in India and Burma, now virtually extinct in the wild.

▶ The gavial's snout is thin and tapering, four times as long as it is wide and terminating in a knoblike swelling that readily distinguishes it from the crocodile. Its jaws are lined with numerous fine, sharp, backward-pointing teeth, all identical, that enable it to seize hold of even the slimiest fish. Once it has caught its fish, it shifts it into position by champing its jaws abruptly and proceeds to devour it headfirst. The gavial's elongated snout provides it with broad field of maneuver and allows it to catch virtually any fish or other aquatic that ventures too close with a quick lateral motion of its jaws.

The young gavials feed on tadpoles, young watersnakes and fry, and aquatic insects. They locate their prey by smell, and are more energetic hunters than their elders, actively searching for prey amid submerged roots or in cavities along the bank. The adult gavial sometimes catches small mammals on the bank as they come down to drink, but usually remains submerged as it lies in wait for its prey. In former times, when gavials were much more numerous in the Ganges and other major rivers, they performed an essential sanitary service to the human inhabitants of the river basin by disposing of corpses that had been set adrift in accordance with Hindu custom.

The feeding behavior of the African slender-snouted crocodile *(Crocodylus cataphractus)* is identical to that of the gavial. It is locally suspected of robbing fishnets with its long pointed snout, and though it is not hunted for its hide like the Nile crocodile, it is also regarded, perhaps with some justification, as dangerous to man and is frequently destroyed in mass crocodile roundups. The marine crocodile *(Crocodylus porosus)*, which ranges from Australia to the Philippines, is the world's largest crocodilian. It pursues a similar hunting strategy in brackish creeks and inlets and coastal estuaries; apart from man, its principal predator is the seagull, which devours substantial numbers of its newly hatched young. Man also numbers, at least potentially, among its prey. M.D.

J

JAGUAR *Panthera onca* 247

247
Sub.: Vertebrata
Cl.: Mammalia
O.: Carnivora
F.: Felidae

This large New World carnivore may attain a length of up to 2.3 m, perhaps weighing as much (according to some authorities) as 180 kg, though there is considerable morphological variation among individuals. The jaguar's coat is tawny-yellow overlaid with a regular pattern of black spots, though due to a genetic anomaly, the "spots" may be continuous, covering the entire pelt in certain individuals. The jaguar is highly adaptable and is found in virtually all the terrain types of Latin America—with the exception of the high mountains, which it cedes to its only potential competitor, the puma. It lives in rain forests and deciduous forests, brushlands and spiny scrub, pampas, and the banks of rivers and lesser watercourses. Melanic (all-black) individuals are apparently more common in in the jungle than in deciduous forests or brushlands in temperate regions.

Younger jaguars are comparatively sedentary, ranging through a territory not much larger than 2.5 5 km on each side; older males that have been expelled from their original territories by their younger rivals, and other solitary males, exhibit territorial behavior only during the rut, when they will tolerate the presence of females, of course, but not that of other males.

▶ In regions that have been sparsely settled by man, such as the Amazon Basin, the jaguar prefers to hunt by day, though it has become nocturnal in more heavily populated regions. The jaguar is a solitary hunter; since it has virtually no competitors and prey is abundant throughout much of its range, there has been no reason for it to develop any cooperative hunting behavior. The frequency with which it hunts depends on the size of the prey, which varies considerably. Jaguars feed on deer, peccaries, capybaras and pacas, birds, reptiles (caymans, boas, turtles), and fish (in substantial quantities) as well as on wild fruit. Older jaguars sometimes acquire the habit of poaching stock, and in the rain forest, jaguars prey on sloths, monkeys, and other medium-sized mammals that are locally abundant. In general, individual jaguars will often specialize in game of a particular species, though they will always be amenable to a change if the chosen prey is no longer available.

The name *jaguar* comes from a Tupi or Guaraní Indian word meaning "it kills with a bound," which is an accurate description of the jaguar's hunting strategy: It stretches out along a branch overhanging a trail that leads down to the banks of a river or a pool of rainwater and waits for animals, singly or in groups, to come down to drink. When the quarry is in position, the jaguar drops from the branch, and the animal's back is often broken by the impact. In less convenient terrain, it stalks its prey from cover, working its way slowly toward its quarry and being careful to remain downwind; when it comes within range, it leaps out at the prey animal, claws foremost. If the animal is not killed outright or if it puts up a vigorous defense, the jaguar will frequently break off the attack. In other circumstances, however, even if the quarry is not killed by the shock of the initial attack, the jaguar will finish it off with a bite to the neck or (in the case of larger animals) by suffocation.

The jaguar is also reported to slap its paw or dip its tail into the water in order to attract fish, which it snatches up with a backhand motion of its paw. This technique is also practiced by the Amazon Indians, who claim to have learned it by imitating the jaguar. Some authorities insist that while the jaguar has no objection to getting its feet or its tail wet, its own fishing technique is much less subtle—it simply scoops the fish out of the water, somewhat in the manner of a bear.

The jaguar indisputably does attack caymans when they crawl onto the shore and are virtually defenseless, and it catches turtles by flipping them over on their backs. The jaguar always carries its kill off to a sheltered spot before it begins to feed; it is remarkably strong and can carry the 200 kg carcass of a deer in its mouth without apparent effort. It generally eats about half of a large kill and leaves the rest to the scavengers that abound in the tropical forest.

◀ A number of other members of the cat family are found in the Amazon rain forest, but most of these are arboreal and feed on rodents, small birds, and other prey disdained by the lordly jaguar. A rough equilibrium between the predator population and the available food supply, which is generally abundant, is maintained by the intraspecies rivalry between the younger and older jaguars, as a result of which the older jaguars are eventually displaced from the communal hunting grounds.

Though jaguar attacks on humans have been very rare, the jaguar has been regarded since the Conquest as too large and dangerous a predator to coexist with man; by adopting a nocturnal mode of life, it has been able to maintain its numbers even in populated areas. After many decades of relative tranquility, this species now finds now finds itself threatened in the last of its territorial strongholds, the Amazon Basin, as the result of widespread deforestation and the flooding of low-lying terrain in connection with hydroelectric projects. P.A.

248
Sub.: Vertebrata
Cl.: Reptilia
O.: Squamata
S.o.: Ophidia
F.: Acrochordidae

JAVANESE FILE SNAKE *Achrochordus javanicus* 248

The Javanese file snake, also called the elephant's trunk snake, is a watersnake with a broad, flat head and a heavy body clad in a wrinkled, brownish granular skin; its small eyes and circular nostrils are both mounted well up on top of the head. Adults range in size from 1.3 to 1.6 m, though some of the females may exceed 2 m; the tail, which is not flattened as in many watersnakes, represents 14 to16% of the watersnake's total length. A good swimmer, though rather sluggish by nature, *Achrocordus* spends most of its time in deep water and lacks the specialized ventral plates that would enable it to get around efficiently on dry land. It is ovoviparous (egg-laying), and generally found in fresh or brackish water and rarely along the seacoast except in marshy areas or in mangrove swamps, from Malaysia to south Australia. (Smith, 1943)

▶ The diet of the Javanese file snake consists essentially, if not exclusively, of fish, which it locates on the muddy bottom, among the water weeds, or in underwater crevices and burrows in the bank by means of a chemical sensor known as Jacobson's organ. It shares this technique with all snakes that have adapted to a primarily aquatic existence; in *Acrochordus,* the normal organs of smell, the olfactory epithelia, are clearly atrophied and probably vestigial. It does not have venomous fangs, but it does have powerful jaws and numerous sharp, backward-pointing teeth, which enable it to grasp its prey securely; then, by means of alternating lateral movements of its jaws, it gradually works its way toward the head of its victim, which it swallows alive.

Achrocordus is one of only two species of snake that have been reported not to be exclusively carnivorous; this supposition, however, is based on Hornsted's discovery (1787) of "undigested fruit" in the interior of a dissected specimen, and Bergman (1958) has suggested that the fruit in question was more likely to have been unfertilized eggs, since 5 well-developed embryos were also found in the same vicinity. Tirant (1885) asserts that the members of this species that inhabit fresh water live on amphibians as well as fish, but does not furnish other particulars; this assertion, though much more plausible on its face than Hornsted's, has yet to be confirmed by other evidence.

◀ Nonvenomous, heavy-bodied, and not notably alert, *Achrocordus* is probably preyed upon by the normal aquatic predators of the region, notably the crocodile and, in the case of the young at least, carnivorous fish. *Achrocordus* is also consumed by man and, particularly in Indonesia, also prized for its skin, which is covered with fine, overlapping scales, easily cured, and especially durable; we have no information that would suggest that these activities constitute a threat to the survival of the species.

The only other member of the family, *Acrochordus granulatus,* has a thinner body and rarely exceeds a meter in length; it lives in estuaries, coastal areas, and a few large rivers from the Gulf of Bengal to the southern Philippines and northern Australia. Its diet also consists primarily of fish, possibly supplemented with an occasional crab; since a few individuals have been found several kilometers out to sea, the shark as well as the crocodile can probably be included among its predators.
H.S.G.

JELLYFISH *Phialidium hemisphericum* 249

249
Sub.: Cnidaria
Cl.: Hydrozoa
S.cl.: Thecata
O.: Conica
F.: Campanulariidae

This is probably one of the commonest hydrozoans of the tropical and temperate oceans; it is found in coastal waters and will attach itself to the surface of any submerged object. In its polyp stage, it looks like a little bush, a few centimeters across, with long branches radiating out from a central trunk, called the stolon. The individual polyps, known as hydranths, are found at the tips of the branches. The entire organism (or colony) is protected by a slotted sheath into which the polyps can withdraw if danger threatens. The individual polyps are essentially separate organisms, though differentiated into two different types: the feeding polyps, capable of injecting venom, and the reproductive polyps, known as blastozoids, fingerlike stalks attached directly to the stolon and surrounded by a protective covering (gonotheca) inside which buds appear and develop into free-swimming medusans.

In the medusan stage, the upper part of the organism takes the form of a flattened, hemispherical canopy, shaped something like the crystal of a watch and fairly soft in consistency (hence, "jellyfish"). Suspended from this canopy is a structure called the manubrium, which contains the mouth of the organism, surrounded by protruding "lips" equipped with stinging filaments called nematocysts. The medusan's stomach is located in the thickest part of the canopy; a circular marginal canal runs around the inner rim of the canopy, connecting with the stomach by means of four radial canals. The outer rim of the canopy is fringed with tentacles, between 16 and 32 in number, which are also equipped with nematocysts. Finally, the medusan has a kind of a simple inertial guidance system—a collection of organites called statocysts, each consisting of a minute grain of sand balanced on the flagellum of a sensory cell that enables the organism to keep track of its orientation in the water.

▶ In its medusan stage, this species is noted for its voracity. A small jellyfish, barely 1 cm across, will consume as many as 8 crustacean larvae *(Artemia* nauplia) in sequence—a meal that leaves it stomach so distended that it fills up the entire canopy—and will then be ready for more about 4 hours later. The medusan's tentacles can be elongated to about 4 or 5 times the diameter of the canopy, forming a barrier through which floating zooplankton must pass. Any creature that comes in contact with one of the tentacles will be paralyzed by the venomous nematocysts; the discharge of a nematocyst also transmits a nervous impulse through the canopy, as a result of which the manubrium is extended to bring the mouth into contact with the prey animal while the curtain of tentacles draws back so as not to interfere with this procedure.

Phialadium hemisphericum is exclusively carnivorous; analyses of stomach contents have revealed that it feeds on very small herrings, wrasses, and pilchards, fish eggs, arrowworms *(Sagitta bipunctata),* crab larvae (zoids), copepods *(Temora longirostris, Calanus finmarchicus, Acartis clausi),* other medusans *(Obelia, Sarsia),* polychaete larvae, appendicularians *(Oikopleura dioica),* and siphonophores *(Muggiae atlantica).*

◀ Its tentacles protect it from predators, but they are powerless to prevent a variety of amphipod crustaceans (hyperiids) from taking up residence in the space between the inner wall of the canopy and the gastric cavity, or even in the gastric cavity itself, where they are sure to find abundant nourishment.

Another thecatan, *Obelia dichomata,* is found in the same habitat; in this species, which is only 3 mm across, the canopy is flattened into a disk; it is also a very fast swimmer, and all of three factors probably combine to protect it from internal parasites. J.G.

K

250
Cl.: Vertebrata
O.: Mammalia
F.: Otariidae

KERGUELEN FUR SEAL *Arctocephalus gazella* 250

The Kerguelen fur seal already weighs between 5 and 7 kg at birth and measures 60–66 cm; the adult males measure between 1.7 and 2 m and weigh between 125 and 200 kg, the females 1.1–1.5 m and 25–50 kg. This species is found in sub-Antarctic waters (i.e. north of about 65° S lat.), particularly on the islands of South Georgia and Kerguelen. It was intensively hunted during the 19th century and at one time was thought to be extinct; the population on South Georgia now numbers as many as 350,000, though the species only began to reestablish itself on Kerguelen as recently as 1954.

The Kerguelen fur seal assembles in large colonies during the mating season, and the male actively defends a territory occupied by his harem of 5–15 females. The males are sexually mature by the age of 6 or 7, the females by the age of 3 or 4; maximum recorded longevity is 13 in the case of the males and 23 in the case of the females. The colonies disband at the end of the breeding season, but the Kerguelen fur seal's subsequent migrations, if any, have yet to be observed.

▶ This species feeds primarily on tiny marine crustaceans called krill *(Euphausia superba* for the most part); the juveniles consume larger quantities of fish than the adults, though the remains of seabirds and the beaks of cephalopods have occasionally been identified among the stomach contents of the latter. The Kerguelen fur seal can remain submerged for about a quarter of an hour and seek prey at depths of up to 75 m. As with other pinnipeds, the molars are tipped with pointed cusps and thus well adapted to gripping prey animals but of little use in chewing.

◀ The leopard seal sometimes feeds on young fur seals, though this predation does not appear to be a significant check on population levels. Killer whales also feed on Kerguelen fur seals and have more than once been observed toying with their victims like a cat with a mouse, tossing them in the air and then catching them in their jaws. In effect, the rapid recovery of the Kerguelen fur seal colonies on South Georgia and Kerguelen can be regarded as an incidental benefit of whaling operations in the Antarctic zone, which have reduced the great whale population to such an extent that it has been estimated that an additional 153 million metric tons of krill have thus been made available each year to the Kerguelen fur seal.

The name fur seal is commonly applied to two different genera, *Arctocephalus* and *Callorhinus* (see NORTHERN FUR SEAL). J.C.

251
Sub.: Vertebrata
Cl.: Mammalia
O.: Cetacea
F.: Delphinidae

KILLER WHALE *Orcinus orca* 251

The largest member of the dolphin family is massive and thick-bodied, though elegant and streamlined in profile. The males attain a length of 9.5 m, the females 7 m, and the weight of both sexes varies between 7 and 9 metric tons. The killer whale, also called the orca, is a cosmopolitan species, found in substantial numbers in all the oceans of the world from the Arctic to the Antarctic, though it seems to be most abundant in the cold and temperate waters of the Atlantic and Indian Oceans. The killer whale lives in packs of several dozen individuals, including adult males, females, and calves; concentrations of several hundred individuals have been reported on occasion.

▶ The killer whale swims very rapidly and can remain submerged for lengthy periods; it is also an intelligent, versatile predator, which shares the same basic diet as other dolphins—including squid, octopus, cod, herring, tuna, mackerel, and other fish—and feeds on a variety of other creatures as well, including seabirds, penguins, seals, dolphins, and even baleen whales (the killer whale is said to tear out the whale's tongue and leave the rest of the carcass). Killer whales locate their prey either by sight or by means of echolocation, which allows them to monitor their underwater environment in great detail; a pack of killer whales exhibits complex co-operative hunting behavior in pursuing seals and sea lions, penguins, or other gregarious prey animals of substantial size.

The pack begins by encircling and then harassing the quarry, trying to drown them or butting them with their snouts, even picking them up and tossing them further out to sea when they attempt to swim toward the shore, until finally, at a signal from one of the older males, the pack moves in for the kill. Killer whales have occasionally been observed clambering up on a piece of floating pack ice on which seals have taken refuge in an effort to overturn it and dump their quarry back into the water; they will routinely come very close inshore, and occasionally up on the beach, in pursuit of a seal or a penguin.

The prey animal is torn into several pieces and then bolted down in great chunks, since the simple conical structure of the killer whale's teeth does not permit it to chew its food very thoroughly. The remains of as many as 13 dolphins and 14 penguins have been retrieved from the digestive tract of a single killer whale; these were probably ingested over the course of several days. A daily ration of 200 kg of fish is thought to be adequate for a captive adult male. Killer-whale attacks on humans have occasionally been attested, but the authenticity of such reports remains in doubt.

◀ Young killer whales may sometimes be attacked by sharks, but the killer whale has no other known predator apart from man. V. de B.

KING CRAB *Maia squinado* 252

252
Sub.: Arthropoda
Cl.: Crustacea
O.: Decapoda
S.o.: Brachyura
F.: Majidae

The shell of the king crab, or spiny spider, frequently measures 13 to 16 cm across, sometimes as much as 20 cm in the case of the male, which makes it one of the largest of crab species. Longevity is difficult to determine, since unlike most other crustaceans, the king crab stops molting its shell after it has reached sexual maturity. It ranges all along the Atlantic coast from the south of England to the Gulf of Guinea and throughout the Mediterranean; in the course of its seasonal migrations, it may be found on a rocky or sandy bottom or on seaweed, but it avoids deep sediment. The king crab spends the fall and winter in very deep water (below 300 m), then moves up to the shoreline by spring while the female incubates its eggs. When the larvae hatch out in May or June, the females immediately return to the depths while the males remain in shallow water for a little bit longer.

▶ The planktonic (free-floating) larva of the king crab feeds on zooplankton; the adult is an unspecialized benthic (bottom-dwelling) predator that (most probably) locates its prey by sight or by means of chemical stimuli. It catches its prey with its pincers and feeds primarily on sessile or slow-moving organisms of the sort that are quite abundant in its various habitats: echinoderms (sea urchins, brittle stars, and common starfish), mollusks (bivalves and sea snails), polychaeate worms, and certain crustaceans, including other crab species (brachyura), hermit crabs (anomura), and isopods. It also eats seaweed.

◀ The common octopus *(O. vulgaris)* is the king crab's most frequent predator, though lobsters may sometimes feed on the smaller individuals. The king crab vigorously defends itself with its pincers, which are quite powerful in the case of the male, and seems to possess a simple social organization, with anywhere from a dozen to several thousand individuals congregating in a "heap," with the females and the young on the inside, the males on the outside, and the strongest males staked out along the perimeter. Individuals also defend themselves against predation by camouflaging their shells with bits of seaweed and other debris, or by burying themselves in the sand. Though normally very sluggish in its movements, the king crab is capable of a brief burst of speed when about to be seized by a predator.

The flesh of the king crab (called *grande araignée de mer* on French menus) is quite delicious, and the annual catch of the European fishery, predominantly in France and along the Mediterranean coast, was estimated at 2000 metric tons in 1968. King crabs are caught with mask and speargun in shallow water, with special nets and baited pots in deeper coastal water, and are also brought up in dragnets on the open sea.

Two related species in the Mediterranean are either too small *(Maia verrucosa)* or too rare *(M. goltziana)* to be exploited commercially. The biggest king crabs are found in the Pacific; *Loxorhyncus grandis* is big enough to prey on *Octopus vulgaris* in the waters off California. The snow crab, *Chironectes,* of the North Pacific is also the object of an important fishery in the coastal waters of Alaska and Japan; its own diet includes bivalves, hermit crabs *(Pagurus),* shrimp, acorn barnacles (Balanidae), amphipods, as well as polychaetes, sea urchins, and snails.

Some of the smaller spider crabs are much more specialized in their feeding habits. Hydrozoans (jellyfish, siphonophores, and the like) comprise 80% of the total intake of some species of the genus *Pisa,* for example; *Libinia* feeds exclusively on jellyfish, and *Macropodia* on seaweed and other crustaceans exclusively. *Inachus* has a rather more extensive diet, including lamellibranchs, snails, hydrozoans, free-swimming polychaetes, crustaceans, brittle stars, and other small marine organisms; similarly, *Hyas* feeds on mysidaceans (small shrimplike crustaceans), sedentary polychaetes, starfish and brittle stars, etc.

P.S.

253
Sub.: Vertebrata
C.: Aves
O.: Sphenisciformes
F.: Spheniscidae

KING PENGUIN *Aptenodytes patagonica* 253

The king penguin is between 85 and 95 cm long and weighs between 13 and 20 kg. There are small colonies on the Falklands and on Staten Island, near Cape Horn, but the principal concentrations of this species are found further south, on the islands of South Georgia, Kerguelen, Macquarie, and Marion.

Like all penguins, the king penguin is remarkably well suited to life in the water, and in very cold water at that. Its body is streamlined, its feet are attached at the rear, like a rudder, and its wings have been modified into flippers. Its plumage is very thick, and it is abundantly provided with subcutaneous fat and other thermoregulatory amenities. The king penguin moves through the water at a speed of about 10 kph, swimming underwater (at a depth of 6–12 m) and on the surface at alternate intervals, like a dolphin. The king penguin is flightless, and its movements on land are slow and awkward; it forms vast nesting colonies on the shore or on the edge of an ice flow. The female lays a single egg, and shortly after the chicks hatch out, they are herded together into nurseries, over which the adults mount guard.

▶ The king penguin dives down to catch fish and cephalopods in its sharp beak, sometimes catching prey up to 90 cm long. It can swallow its prey underwater and can store a whole sequence of prey animals in its crop; the chicks are fed partly on the fish and squid they retrieve from their parents' crops, partly on nutrients that are secreted from the stomachs of the adults.

◀ The leopard seal is the principal predator of both the adult and the juvenile king penguin. When a swimming penguin catches sight of a leopard seal, it slaps the surface of the water with its flippers and makes for the edge of the iceflow while squawking loudly. This behavior is imitated by all the other penguins in the area, so that the leopard seal, bemused by the choice of so many potential targets amid this churning mass of flippers, feet, and bubbles, may very well end up with none.

The king penguin may also be caught by a killer whale while swimming in open water, and eggs and chicks that are not properly guarded by the adults are often devoured by skuas, southern giant petrels *(Macronectes giganteus),* and sheathbills *(Chionis).* The king penguin has no competitors, strictly speaking, except for the members of its own species; this competition manifests itself primarily when the penguins come ashore to select a suitable nesting site. The king penguin was exploited by man during the 18th and 19th centuries; its flesh was eaten (principally by seafarers on long voyages), its skin was used for clothing, and its rendered fat was used in the curing of leather.

The penguins comprise a single order and a single family, consisting of about 17 species (in some classifications, 18) altogether, which are widely distributed throughout the Southern Hemisphere, from Antarctica as far north as the Galapagos Islands. All penguins feed on marine organisms—fish, mollusks, and/or krill (small crustaceans). The emperor penguin *(Aptenodytes forsteri)* is the largest of these, capable of diving to a depth of 250 m or more in search of fish and remaining submerged for up to 9 minutes. The rockhoppers and macaroni penguins *(Eudyptes)* are quite small and readily identified by a sheaflike crest of soft plumage that is located above the eye. The genera *Eudyptes* and *Spheniscus,* including the Galapagos penguin *(Spheniscus mendiculus),* are found in the warmer waters off the coasts of South Africa and South America.

J.-P. R.

KING'S FRILLED LIZARD
Chlamydosaurus kingi 254

254
Sub.: Vertebrata
Cl.: Reptilia
O.: Squamata
S.o.: Sauria
F.: Agamidae

The frilled lizard is between 45 and 90 cm long, though its tail may account for about two thirds of this length. It owes both its popular and its scientific names (*chlamys* was the name of a togalike garment worn by the ancient Greeks) to the broad cutaneous flap that runs around both sides of its neck and just behind the angle of its lower jaw; the frill is normally folded back against the neck and can be spread open, when required, by flexible cartilaginous "ribs," actually long spinelike projections of the hyoid bone. Coloration varies from one population to the next, ranging from dark gray to reddish brown; the frill is flecked with black, white, pink, and brown. This particular species is found in grasslands and wooded regions in the northern part of Australia (from southeast Queensland across to Western Australia) and on New Guinea.

▶ King's frilled lizard is primarily insectivorous, feeding on ants, beetles, and grasshoppers, though it occasionally catches smaller lizards and small mammals; in certain areas, it has a reputation as a henhouse robber. It is strictly diurnal and hunts from ambush, posting itself on a favorite treetrunk to await the aproach of prey, which it detects entirely by sight and then jumps down to the ground to devour. It laps up ants and smaller insects with its tongue, seizes larger arthropods with tough chitinous shells in its mouth and crushes them with its jaws before swallowing them. It sometimes patrols the surrounding area in search of food, but scurries back up a treetrunk at the slightest hint of danger.

◀ Pythons and other snakes, raptors and a number of other birds, will all feed avidly on the frilled lizard. The lizard has two basic lines of defense: to make itself as inconspicuous as possible, or as terrifyingly conspicuous—those defeuses are used against aerial and terrestrial predators respectively. When it sees a raptor's shadow on the ground, it flattens itself against its treetrunk to avoid casting a shadow of any kind; its drab coloration (when the colored spots on the frill are tucked out of sight) enables it to blend in perfectly with the surface of its chosen treetrunk. If it is caught out in the open, it makes for the nearest tree, running along on its hind legs with its tail held aloft like a balance pole.

When cornered, it puts on a spectacularly intimidating display—opening its mouth up wide and hissing noisily while stretching out its spiny frill to its fullest extent. It may also rear up on its hind legs, keeping its balance by shifting its weight from one leg to the other as it whips its tail violently back and forth, finally lunging forward and attempting to bite its attacker. Apart from serving as an effective defense against predators, some of this behavior is also incorporated into the frilled lizard's mating display.

D.H.

KISSING BUG *Reduvius personatus* 255

255
Sub.: Arthropoda
Cl.: Insecta
O.: Hemiptera
F.: Reduviidae

This predacious bug is the largest member of the assassin bug family (Reduviidae) to found in Europe, sometimes attaining a length of 20 mm. Its body is blackish-brown and sprinkled with very fine hairs, its legs are reddish-brown, and its wings are well developed. It is nocturnal and attracted by light. It produces a faint but distinctly audible clicking sound by rubbing the tip of its beak against the stridulating pads on its forelegs. The female lays its eggs, which are smooth, oval, orange in color, and about 1 mm long, in June or July, and the larvae develop very slowly.

▶ *R. personatus* is often found in human dwellings, particularly in attics, outbuildings, and storage rooms, where it feeds on virtually every insect with which its shares its habitat, including flies, deathwatch beetles (Anobiidae), larder beetles (Dermestidae), mealworms (Pyralididae), and clothes moths (Tineidae), as well as house spiders. The kissing bug also feeds on bedbugs, and occasionally attempts to feed on sleeping humans while engaged in this activity; *R. personatus's* beak is short, curved, and painfully blunt. The larva preys on the same species of insects as the adult.

Reduvius hunts by night, creeping up very slowly on its prey, then seizing it with its forelegs and stabbing it with its beak. The prey is paralyzed instantly by an injection of venomous saliva, and *Reduvius* continues to feed until the soft tissues of the prey are entirely liquefied, which may take some time.

The larva also hunts by night, though it relies primarily on camouflage rather than stealth. Its body is coated with a viscous film to which dust, dirt, and other debris adheres as it crawls along the ground or across the floor of an attic storeroom. It appears to use the tarsi of its posterior set of legs to attach this camouflage material to its body, since it is incapable of doing so when these segments have been artificially removed. The common name "masked hunter" is sometimes applied to the larva in particular, and the species name *Reduvius personatus* also refers to this behavior, since *personatus* means 'masked' or 'concealed' and *reduvius,* the Latin word for hangnail, also refers to anything that has been shed or discarded. The *Reduvius* larva also stalks its prey very slowly, moving toward it in a series of abrupt, seemingly random progressions.

◀ As far as we know, this species has neither predators nor competitors.

R.C.

KIWI *Apteryx australis* 256

256
Sub.: Vertebrata
Cl.: Aves
O.: Apterygiformes
F.: Apterygidae

The 3 species of the genus *Apteryx* are found exclusively on New Zealand, and *Apteryx australis,* the common or brown kiwi, is found only in damp forests of a kauri pine with a dense undergrowth of ferns. It weighs between 1.3 and 4 kg and measures about 50 cm; its wingspread is also about 50 cm. Since its wings are rudimentary, the kiwi is unable to fly.

▶ The kiwi is a solitary, nocturnal hunter; it spends the day concealed in the undergrowth and sets out at dusk in search of food. During the rainy season, it feeds on earthworms, insects, and insect larvae, which its digs out of the ground with its bill. The kiwi's sense of smell is extremely well developed, to the extent that it can sniff out an earthworm buried under 10 cm of topsoil. During the dry season, it feeds on fallen fruit and large quantities of foliage. The adults do not provide food their nestlings, since the newly hatched kiwis are capable of foraging for themselves within 24–48 hours.

◀ The kiwi's natural enemies are all mammals that have been introduced to the islands during the last millennium, including the pigs and dogs (introduced by the Maoris), feral cats, stoats, and ferrets (introduced by the British). Both the Maoris and early British settlers hunted the kiwi for food; the kiwi's long, cylindrical feathers were woven into the robes worn by the Maori chieftains. Today, the kiwi—as befits the national emblem of New Zealand—enjoys the strict protection of the law, though only after having been considerably reduced in its range and numbers by the depredations of these imported predators and the effects of deforestation (land clearance for agriculture and stock grazing).

The great gray kiwi *(Apteryx haasti)* and the lesser gray kiwi *(Apteryx oweni)* are very similar to the common kiwi in their habits. Collectively, the kiwis are anatomically very different from other flightless birds, and some authors have suggested that they may be a sort of dwarf variety of the extinct giant moa, one of the largest birds that ever existed (more than 3 m tall); various species were still extant on New Zealand in the 16th—19th centuries.

J.-F. R.

KNIFE FISH *Eigemannia virescens* 257

257
Sub.: Vertebrata
Cl.: Osteichthyes
S.cl.: Teleostei
O.: Cypriniformes
F.: Rhamphichthyidae

This South American electric fish measures between 30 and 35 cm, and, as with many electric fish, the knife fish's abdominal cavity is greatly reduced and its tail is disproportionately long. It is found through the Guianas, the Amazon Basin, and in several shallow South American rivers that carry a large burden of sediment (Rio de la Plata, Paraguay, Parana). It is active at dusk or at night, and spend its days concealed amid the stalks of aquatic plants or under a mat of water hyacinths or other floating vegetation.

▶ Like young electric eels, young knife fish feed primarily on insect larvae, annelids, and crustaceans, though unlike the adult electric eel (and other related species such as *Gymnotus carapo,* which is also frequently referred to as the knife fish), the feeding habits of the knife fish are restricted by the small size of its mouth, so that it primarily preys on insects and medium-sized crustaceans. The knife fish's electric organs, analogous to Sach's organ in the electric eel, consist of a number of tubular coils of specialized muscle tissue that are embedded in the musculature of its flanks (extending to the tip of its tail) and ventral region (directly in front of the abdominal cavity).

These organs produce an almost continuous but very weak electrical discharge (250–600 impulses per second, of less than 1 volt); the electric field that is thus created is used for navigation as well as prey detection, since submerged obstacles and other features of the environment as well as other aquatic creatures will all be registered by the fish's electrosensory organs as disturbances in the field. Since the electric field would also be disturbed by vigorous locomotion on the part of the knife fish, it propels itself through the water by means of gentle undulations of its anal fin. As with the electric eel and the Nile fish, the caudal fin is absent altogether and the tail, which is reduced to a sort of stalk like the tail of a file, is not involved in locomotion. The knife fish can swim forward or backward with equal facility by reversing the motion of its anal fin, which makes it very easy for it to maneuver among the roots and stalks of aquatic plants.

◀ The knife fish does not occupy a very lofty position in the aquatic foodchain, since most of the creatures it feeds on are small herbivores or detritivores, and there are quite a number of bigger creatures (including several species of very large catfish) that feed on it. The knife fish's electric organs serve no specifically defensive function (apart from warning of the approach of potential predators), though the substantial fraction of its body length that lies astern of its greatly compressed abdominal cavity can be regenerated—several times if necessary—if it is lost to a piranha or other predator, an ability that it probably shares with other members of its family that are constructed along similar lines. Larger members of the family *(Sternopygus, Rhamphichthys)* that attain a length of 1 or 2 m are locally consumed by man, and the knife fish *Gymnotus carapo* has become a popular aquarium fish.

The numerous species belonging to the family Rhamphichthyidae are primarily dintinguished by the length of the tail-stalk and the form and size of the mouth. There is a related family of electric fish, the Apteronotidae, which have a small caudal fin at the extremity of the tail-stalk and are capable of generating as many as 1000 electrical impulses per second; in fact, these species are more readily differentiated by the frequency and other characteristics of their electrical discharges than by their morphology. It is safe to assume that there are a great many other varieties of electric fish in the Amazon Basin that are still unknown to science. J.-F. M.

KOMODO DRAGON *Varanus komodoensis* 258

258
Sub.: Vertebrata
Cl.: Reptilia
O.: Squamata
S.o.: Sauria
F.: Varanidae

In spite of its very conspicuous size and commanding appearance, this largest of extant lizards remained unknown to science until the beginning of this century and was not formally described until 1912. This species, also known as the Komodo monitor, is found on the western side of the Indonesian island of Flores, which lies a little to the south of Wallace's Line (separating the Oriental and Australasian biogeographic regions), as well as on the neighboring islands of Komodo and Rintja, each of which measures only 20 km by 30, and the even smaller island of Padar, which is no more than 100 sq km in area. Even apart from the presence of these giant lizards, the terrain on the smaller islands is not very hospitable to human settlement, so that encounters between the *buaja darat* (''land crocodile'') and the human inhabitants of the region were not widely reported until the late 19th century.

The maximum length attained by this species has sometimes been exaggerated. Reports of individuals measuring 7–8 m in length have never been confirmed, and the largest officially recorded specimen was a male that measured 3.05 m, though the largest females never exceed 2 m in length. The Komodo dragon is much more heavily built than the other monitors (a male measuring 3 m can easily weigh as much as 135 kg), its muzzle rounded and less tapering, and its skin more coarsegrained in appearance. Both sexes are dark in color, the males generally brick-or chestnut-colored, the females greenish-black with a sprinking of small yellow spots running down the neck from the base of the skull.

The Komodo dragon is most frequently found on dry savannah interspersed with small groves of trees or gallery forests, particularly along dry watercourses, and is also occasionally found in the mountainous regions of the interior (up to an altitude of 500 m) and the coastal mangrove swamps. Komodo dragons have survived for over 15 years in captivity, and since the adults continue to grow very slowly throughout their lives, it has been estimated that the very largest dragons are probably more than 50 years old.

▶ The Komodo dragon appears to be dependent to a large extent on the Timor hog deer *(Cervus rusa timorensis),* which is most abundant on those parts of the islands where the largest concentration of dragons are found (and is correspondingly absent from the neighboring islands that are not inhabited by the Komodo dragon). The presence of these deer may be regarded as a necessary but not sufficient ecological condition, however, since the dragon also feeds on a number of other mammalian species, including wild boar, water buffalo, wild horses, and monkeys; the juveniles subsist primarily on shrews and rats.

The Komodo dragon also feeds occasionally on the eggs and young of megapodes and other ground-nesting birds as well as on the eggs of sea turtles dug up on the beach, and it seems likely that the dragons that live in the coastal mangrove swamps feed on fish, mollusks, or small crustaceans when warm-blooded prey is unavailable. It sometimes ambushes dogs, goats, and other domestic animals on the outskirts of villages and steals fishermen's catches that have been set out to dry; attacks on humans have occasionally (though rarely) been reported as well.

Like the other monitor lizards, the Komodo dragon locates its prey in essentially the same way that snakes do: it follows a scent trail by picking up tiny particles from the ground with its long forked tongue and transferring them to the olfactory receptors in the lining of Jacobson's organ. The dragon's sense of smell is very keen, and they are routinely enticed out of the underbrush with the malodorous carcass of a hog deer for the benefit of tourists and scientific observers. Because of this fondness for putrid meat, some authorities have concluded that the Komodo dragon feeds exclusively on carrion, though it is in fact quite capable of dispatching a living prey animal. It conceals itself in ambush in the tall grass of the savannah and frequently lashes the quarry violently with its tail before attempting to seize it in its jaws. As far as the larger game animals are concerned, the Komodo dragon preys primarily on the very young or the old and infirm.

The larger dragons have a well defined hunting territory, which they generally patrol along the same paths every day. Dragons frequently feed in groups, however, and as many as 8 of them have been observed around the carcass of a water buffalo, wild boar, or other large animal. (A certain order of precedence seems to be observed as they approach a carcass that has been set out for them, which implies they recognize some form of dominance hierarchy.) It is not known how often the Komodo dragon customarily feeds; like other large reptiles, it is probably capable of lengthy fasts, whereas on the other hand, a dragon measuring 2.5 m has been known to dispose of as much as 20 kg of meat at a single feeding.

◀ The very young dragons may fall prey to raptors, snakes, and the larger members of their own species, but the adult Komodo dragon has virtually no natural enemies. Shortly after the species was discovered, it was intensively hunted for its hide, but it soon became apparent that the numerous bony plates in its its skin made it unsuitable for tanning. This species has been legally protected since 1925, and the greatest danger that currently faces the 5000–6000 remaining dragons is that of famine and overpopulation, since the deer, boar, and other large mammals on which they feed are rapidly being exterminated by human hunters on all four islands.

The two-banded monitor *(Varanus salvator)* is only slightly smaller than the Komodo dragon, since it may attain a length of almost 3 m; this semiaquatic monitor is found along the banks of watercourses and in coastal areas, and ranges from Bangladesh to Indonesia and the Philippines. It is also found on the island of Flores, and since it preys on small rodents, it may be regarded as a direct competitor of the juvenile Komodo dragon. It is more agile in its movements than the dragon and has been widely hunted for its hide, which is much more suitable for commercial purposes than that of the Komodo dragon though still considered to be of poorer quality than crocodile skin.

R.V.

L

LAMPERN *Lampetra fluviatillis* 259

259
Sub.: Vertebrata
Cl.: Cyclostomata
O.: Petromyzonoidae
F.: Petromyzonidae

The range of this common species of European river lamprey extends from the Atlantic coast as far upstream as the shallow, swift-flowing headwaters frequented by trout and graylings. In spite of their eellike appearance, the lampreys are not true fish (Osteichthyes) but are assigned to the class Cyclostomata (or the superclass Agnatha) on the basis of a number of fundamental structural differences—among them the fact that the lamprey has no jaws. Instead, the lamprey's mouth consists of a perforated suction disk; its tongue takes the form of a toothed piston or plunger, which functions in much the same way as the blade of an electric food processor or a carpenter's router. The suction disk attaches itself to the body of the prey animal by creating a vacuum seal; then the piston detaches small bits of tissue and draws them into the lamprey's alimentary canal on the upstroke.

As with other lampreys, the lampern's development is quite different from that of fish or other vertebrates. Its eggs are laid in fresh water and hatch within 10–20 days. The eyeless larva, called an ammocete, buries itself in the mud, taking in oxygen through its skin and filtering out microplankton and organic debris from the water. The ammocete stage lasts for 3 or 4 years, depending on the relative abundance of these nutrients; then, prompted by the secretion of a pituitary hormone, the ammocete undergoes a total metamorphosis, anatomical as well as physiological. The eyes, fins, and mouthparts now appear for the first time, and the immature lampern sets off on its journey downstream to the sea, which also marks the beginning of its career as a predator. When the mature lampern, now between 40 and 50 cm long, has accumulated sufficient fat reserves to sustain it on its voyage upstream—usually after a year or two—it stops feeding altogether and its teeth and intestines begin to atrophy. The lampern makes its way upstream by night, moving from areas of higher to lower salinity. The female lampern seeks out a sunlit spot and undergoes a final round of structural modifications before it is ready to deposit its eggs (10–20,000 in number) in holes scooped out in the streambed, then eventually succumbs to exhaustion or starvation.

▶ The lampern preys on a number of different species of benthic and pelagic fish, including salmon, smelt, herring, cod, plaice, and others. Its probably locates its prey visually and conducts an active pursuit, though this behavior has been observed only in the aquarium and never in open water. The lampern attaches itself to its prey by means of its suction disk, then literally begins to bore its way into the fish's side, abrading deeper into the muscle tissue (and ingesting the fragments thus detached) in spite of prey's desperate struggles to escape. Even if the prey does succeed in breaking loose, the lampern is generally able to reattach itself to some other part of the fish's body; this process frequently (though not necessarily) results in the death of the prey animal.

◀ Sharks, tuna, cetecean mammals, and other large marine predators—against which it has no means of defense other than rapid flight—feed on the lampern in the open seas. Bullheads, catfish *(Siluris)*, and other bottom-foragers feed on the ammocete larvae in fresh water. The lampern is fished in considerable quantities during its upstream migration, particularly in Eastern and Southwest Europe, where a single night's work is likely to bring in up to 400 kg. The various species of European river lampreys are currently in decline, however, since the ammocete larvae are highly susceptible to water pollution.

The life cycle of the European sea lamprey *(Petromyzon marinus marinus)* is almost identical to that of the lampern, except that the marine phase accounts for a somewhat larger proportion of its lifespan. This species is about 1 m long, and the scars left by its suction disks have been found on the hides of cetecean mammals, which gives an indication of its aggressiveness as a predator.

A second subspecies, the American sea lamprey *(Petromyzon marinus dorsalis)*, migrated to the Great Lakes region via the inland waterway during the 1940s and now spends its entire life in fresh water. In this case, a comparison of the annual yield of the Lake Michigan trout fishery before and after the sea lamprey's arrival in inland water provides us with even more impressive evidence of its aggressiveness and voracity—viz., 3000 metric tons in 1944 as against 16 *kilograms* in 1955. None of the various attempts to contain this ecological scourge (involving "selective" poisons, weirs and lock gates, traps, etc.) has been totally successful, though the Great Lakes lamprey plague has been somewhat overshadowed in recent years by the even more extensive effects of chemical pollution.

The lampreys' closest relatives and the most primitive of all living vertebrates are the hagfishes (family Myxinidae). Unlike the lamprey, the hagfish develops continuously from the egg to the adult form without undergoing a metamorphosis, though the sexes are incompletely differentiated in the hagfish, which is regarded as a reversion to a primitive hermaphroditic state. A representative species is the Atlantic hagfish *(Myxine glutinosa)*, which lives in populous colonies half-buried in mud on the ocean bottom. The hagfish locates its prey with the aid of barbels around its mouth (which is arranged very much the lamprey's) that serve both a tactile and an olfactory function, and it has an olfactory cavity in its nostrils that is sensitive to the presence of decaying flesh over distances varying between 60 cm and 1 m.

The hagfish feeds on live prey—including marine worms, sessile or disabled mollusks and other slow-moving marine organisms, including fish that have been injured, caught in a net or on a line, or otherwise incapacitated—as well as on carrion. After it has secured a piece of flesh with its suction disk, it sometimes knots itself into a figure eight and loops its tail around a projecting rock or other object, which provides additional torque and thus assists it in twisting loose its prize from the carcass. M.D.

LANGUEDOCIAN SCORPION *Buthus occitanus* 260

260
Sub.: Arthropoda
Cl.: Arachnida
O.: Scorpionida
F.: Buthidae

This dust-colored Mediterranean scorpion, sometimes known as the common yellow scorpion, may attain a length of 6 cm. It is found in Spain and southern France (ranging westward from the Var and northward along the Rhône valley), particularly in sparsely vegetated areas that are exposed to direct sunlight at midday, where it excavates a small burrow for itself under a rock. Both sexes, which are not greatly differentiated, take part in an elaborate mating dance in which the partners touch pincers and pretend to sting one another. Scorpions are ovoviparous, and the newborn scorpions ride around on their mother's back until their first molt, or for 2–8 days. Scorpions generally live for several years.

Buthus occitanus is basically similar to other scorpions in appearance, though its venom is considerably less potent than that of subtropical species like *Centruoides noxius* and its congeners (said to cause about 2000 fatalities in Mexico every year) and *Androtocnus australis* of North Africa. *Buthus's* sting can still be very painful, all the same, though the aftereffects are more pronounced in children than in adults. Outside its native region, *Buthus occitanus* is best known as one of the species studied intensively by the Provençal entomologist and nature writer Jean-Henri Fabre at around the turn of the century.

▶ The scorpion preys on virtually every small invertebrate that it encounters, including spiders, harvestmen, and many species of insects. It eats irregularly and may go without food for over a year without suffering any ill effects (36 months appears to be the record, in the case of a female that was given water but no food). It also refrains from eating during the days preceding and following a molt and (in the case of the female) while it is carrying its offspring on its back. The appetite of *B. occitanus* quickens considerably in the spring, and in April, according to Fabre, it indulges in "scandalous orgies of gluttony," even indulging in cannibalism on occasion.

B. occitanus' four pairs of eyes seem to play only an auxiliary role in prey detection (since blanking them out with a coat of varnish does not impair its effectiveness as a predator in the laboratory), and it is alerted to the approach of a prey animal primarily by the sensory hairs (trichobothria) on its maxillipeds (the second of its four pairs of legs), which resonate sympathetically with the vibrations created by insects walking along the ground. *B. occitanus* turns immediately toward the source of these vibrations, rearing up on its walking legs with its pincers held open and its stinger curved upward.

It uses its pincers to seize the prey and to crush the carapace of insects (though in the case of certain hard-shelled ground beetles, it will abandon the attack at this point), and if the prey continues to struggle, it stretches its abdomen up over the forepart of its body to bring its stinger into contact with any point on the body of the prey. After it has injected its venom, it remains in this position until the activity of the prey animal begins to diminish in intensity, then drags the prey off to a sheltered spot to eat it.

Grasping the prey animal in one of its pincers, the scorpion minces up the soft parts of its body by means of the stabbing, scissorlike action of its sharp mouthparts (cheliceres), then aspirates the resulting fragments along with the body fluids of the prey. The cheliceres stop moving as soon as these edible portions of the prey animal have been eaten, and the carapace and other chitinous remnants are generally abandoned. The feeding technique of the scorpion may thus be seen as a prototype of the process of external digestion, which is most highly developed in the insects. *B. occitanus* feeds only on living prey, and the hunting behavior of the immature scorpions is similar to that of the adults. Some species found on more humid areas, notably the African *Pandinus imperator*, will feed on carrion as well.

◀ *B. occitanus* has no systematic predators or parasites that we know of. Scorpions may sometimes engage in combat with centipedes (scolopendromorphs) and spiders, possibly to their detriment on occasion, and lizards, corvids, owls, and domestic fowl may also feed on scorpions, though rather rarely.

Panuroctonus mesaensis, found in desert regions of North America, can detect vibrations made in the sand by insects at a distance of up 50 cm. G.B.

LANTERN FISH *Myctophum punctatum* 261

261
Sub.: Vertebrata
Cl.: Osteichthyes
S.cl.: Teleostei
O.: Salmoniformes
F.: Myctophidae

This bathypelagic fish measures about 10 cm and is found throughout in the Mediterranean and the Eastern Atlantic (from Norway to Mauritania). During the day, it descends to a depth of up to 750 m, then swims back up at night, typically up to a depth of about 250 m, sometimes venturing fairly close to the surface on a dark moonless night. The lantern fish has no swim bladder, which makes it easier for it to equalize internal and external pressure and thus to carry out these daily vertical migrations, and it lights its way through the sunless depths with its numerous bioluminescent organs, called photophores. The cold light emitted by the photophores varies in wavelength from one species to the next (the common name lantern fish is applied to all the members of the family Myctophidae), generally ranging from yellow to blue or green. The intermittent flashing of the some of the photophores is thought serve as a territorial marker, by which the lantern fish proclaims its presence and attempts to banish rivals from its particular patch of ocean.

▶ The lantern fish is thought to feed primarily at night, when large planktonic organisms (copepods, euphausids, amphipods, pteropods, squid, and various kinds of larvae) are particularly attracted by the light of the photophores; the lantern fish is very quick in its movements and can snap its head rapidly to one side in order to catch these curiosity-seekers in its jaws. The larvae of the lantern fish, only 6 mm long at first, begin by feeding on surface plankton and gradually work their way down to a depth of about 100 m, by which time they will have attained a length of 1.8–2 cm and have assumed their adult form.

◀ The lantern fish's daily migrations expose it to a broad cross-section of potential predators, including tuna, dolphins *(Coryphaena)*, and sea lions, when it feeds near the surface at night. Lantern fish feeding on banks of euphausid crustaceans (krill) may occasionally be swallowed by baleen whales, which have been known to make away with whole flocks of penguins in similar circumstances. Hake, as well as a large number of bathypelagic species that are less well known, prey extensively on the lantern fish during the day. The more than 150 species of lantern fish comprise a substantial biomass on which a great many species, including several of primary economic importance to man, are dependent for their survival.

The largest member of the family Myctophidae is *Lamparyctus crocodilus*, which measures between 25 and 30 cm and is commonly found in both the Mediterranean and the Atlantic. The members of the family Sternoptchyidae, the hatchet fish (so called because of their extreme lateral compression), are somewhat similar to the lantern fish in their mode of life and the position they occupy in the pelagic foodchain. The silvery-white skin of the hatchet fish is studded with numerous photophores, and the hatchet fish also feed on small herbivorous organisms, as well as furnishing an important source of food for such large surface predators as the Thunnidae in the Bay of Biscay and elsewhere. F.M.

LAPWING *Vanellus vanellus* 262

262
Sub.: Vertebrata
Cl.: Aves
O.: Charadriiformes
F.: Charadriidae

This common Eurasian plover measures 30 cm in length, with a wingspread of 70 cm, and weighs between 170 and 265 g. The lapwing prefers humid, open country and in winter is often found in flooded fields and bottomlands as well as hedged or sown fields, and particularly in pastures that have been close cropped by cattle. The lapwing nests on the ground at the edge of a marsh or lake, sometimes in a cultivated field near fresh water, a peat-bog, or in a polder directly along the seacoast, and ranges throughout most of Europe and as far eastward as Manchuria. In 1964, there were 40,000 pairs of lapwings nesting in France, mostly in the northern half of the country; this permanent population is supplemented by large numbers of winter migrants as soon as the cold weather begins. Mortality is very high during the first year of life, and longevity is probably on the order of several years.

▶ The lapwing is an adaptable forager that probes for insects and other small invertebrates in soft, moist earth, in the muddy bottom of a pool, shallow stream, or ditch, or along costal mudflats. It is omnivorous, though about 90 percent of its diet consists of animal prey, primarily of earthworms and sandworms, insects and their larvae, and mollusks. The lapwing hunts for food by night as well as by day, and the nestlings begin to follow their parents and peck for food almost as soon as they are hatched.

◀ As with the other plovers, the speckled eggs of the lapwing are almost invisible on a shingle beach or a patch of gravel. The nestlings flatten themselves against the ground and remain completely motionless at the slightest sign of trouble; both eggs and nestlings are frequently devoured by crows, magpies, harriers, gulls, and foxes. The adults of the colony try to protect the nesting sites by mobbing all intruders (including humans), their attempts to harry and peck at them accompanied by a chorus of plaintive cries, and the bereaved parents can sometimes make good their losses by hatching out several clutches of eggs in a single season. The adults, though acrobatic fliers, are sometimes taken by the peregrine falcon as well.

The lapwing is a sociable bird with no direct competitors and is often found in association with the black-headed gull and the golden plover; during the mating season, however, the male lapwing will defend its chosen nesting site against its rivals. There is a French saying to the effect that "he who has never eaten a lapwing [*un vanneau*] has never had a tasty morsel [*bon morceau*]." True or not, the lapwing is at least prolific enough to have preserved its numbers in spite of intensive egg collecting, particularly in Northern Europe, as well as the netting of migratory flocks. The first of these practises is now severely restricted in the Nordic countries, the second almost universally prohibited. The lapwing does require that it be undisturbed in its nesting sites, however, with the result that it now tends to nest in cultivated fields and pasturelands rather than in marshes and coastal areas that are actually more likely to be frequented by humans during the spring hunting season.

The plover family (Charadridae) comprises 66 species, of which 25 are classified as lapwings: primarily Old World terrestrial birds found in wet, open country and active by night as well as by day. Some species are distinguished by their brightly colored wattles, caruncles, and other facial ornamentation *(V. aliceps, V. senegalensis,* and *Lobibyx miles),* others by the small metacarpal spurs on the undersides of their wings *(V. armatus, V. spinosus).*

There are also 38 species of plovers *(Charadrius, Pluvialis, Squatarola,* etc.), fast-moving terrestrial birds found on sandbars, gravel banks, and beaches on every continent. The ringed plover *(Charadrius alexandrinus)* is found along the coasts of Western Europe, the little ringed plover *(C. dubius)* on gravel banks and islets in large rivers. The dotterel *(Eudromias morinellus)* was formerly a common migratory visitor in Britain and Western Europe, but was evidently too trusting and unwary to coexist successfully with humans and is now rarely seen. The ruddy turnstone *(Arenaria intrepres)* is so called because it searches for small marine organisms by turning over pebbles or foraging in the tidal wrack along the seashore.

J.-F. T.

LARGE MOUSE-EARED BAT *Myotis myotis* 263

263
Sub.: Vertebrata
Cl.: Mammalia
O.: Chiroptera
F.: Vespertilionidae

In fact, *Myotis myotis* is one of the largest Eurasian bat species; its head and body measure 6.5–8 cm, its tail 4.8–6 cm, and its wingspread may be up to 37–43 cm. The large mouse-eared bat is found from Portugal in south to Sweden in the north and ranges eastward into Asia; it has been reported at altitudes of 2000 m in the Pyreenes. It sometimes makes journeys of up to 100 km in search of seasonal lodgings, which may include attics, the rafters of barns and other buildings, wells, the undersides of bridges, and similar enclosed spaces as well as natural caves, where it spends the summer as well as the winter. This species is quite gregarious; substantial numbers of females often congregate in the spring, and in the winter a hibernating colony may number up to 4500 individuals. Large mouse-eared bats nestle together very closely (and as far as possible from the entrance to the cave) but do not wrap themselves up in their membraneous wings as many bats do; they are inactive for practically the entire winter.

▶ The members of a colony of large mouse-eared bats go out to hunt in small squadrons. The first of these generally leaves the cave at about 20 minutes after sunset, the others follow at 5 minute intervals. The first squadron to leave the cave will also be the first to return, generally at about 11 p.m., and the remaining squadrons will continue to return at 5 minute intervals until about 1 a.m. The bats fly slowly while hunting, generally in a straight- line course and at an altitude of 5–8 m, along streets and over forest clearings and other open areas.

The large mouse-eared bat feeds on large

beetles (e.g. junebugs and scarabs exceeding 18 mm in length) as well as moths, diptera, caddis flies, and other insects. It detects and tracks its prey by echolocation, vigorously propelling a column of air through its larynx and altering the size and shape of the passageway (at a point adjacent to the glottis) to produce a sequence of extremely brief ultrasonic squeaks, each about 1–5 milliseconds in duration. One bat specialist has likened this tireless nocturnal marauder to "an efficient guided missile that taps the energy stored in the target to fuel its flight."

If the large mouse-eared bat cannot find a convenient open space on which to alight, it will search for an overhanging support it can cling to while it devours its prey. It teeth are well adpated to the task of snagging insects on the wing and crushing their shells; a large mouse-eared bat can catch about 6–8 insects per minute and may ingest half of its own weight in the course of a single night's hunting.

◀ The large mouse-eared bat has few natural enemies; it clings to the roof of its cave (or the equivalent manmade structure) during its long winter sleep, out of the reach of small carnivores. Though it normally flies a straight course, it is skilled at evading attacks by owls, which occasionally feed on it. It is certainly to be regretted that these bats are still sometimes destroyed as the result of human ignorance or superstition.

The little brown bat *(Myotis leuctifulgus)* is a common North American species.

J.-J. B.

264
Sub.: Vertebrata
Cl.: Osteichthyes
S.cl.: Teleostei
O.: Perciformes
F.: Centrarchidae

LARGEMOUTH BASS *Micropterus salmoides* 264

More properly known as the largemouth black bass, this freshwater fish usually attains a length of about 40 cm and a weight of 2 kg; one record-breaking specimen weighed in at 8 kg, with a length of 70 cm. The adult largemouth may be found in the deeper waters of a pond or lake or in a pool in a stream with a sluggish current; the young fingerlings live among the vegetation along the shore. This species is a North American native (east of the Rockies exclusively) that was widely introduced into Western Europe during the previous century.

▶ As with many other species, the fingerlings feed on tiny freshwater crustaceans until their mouths have grown large enough to accommodate more substantial prey—in this case, tadpoles, worms, snails, and eventually crayfish. The adult largemouth occupies much the same lofty position on the freshwater food chain as the pike and the perch, with which it competes for much the same sort of prey: primarily other fish and their eggs as well as tadpoles, dragonfly larvae, crayfish, and even (in the case of the very largest pike and bass) water rats, bullfrogs, and aquatic birds. The largemouth is voracious and very aggressive; it conceals itself in the vegetation and detects its prey by sight and by vibrations; if the prey is not caught instantly, the largemouth is capable of sustaining a reasonably hot pursuit for a short time (at a speed of from 0.2 to 2 kph).

It sometimes catches birds by leaping up out of the water and dragging its quarry back down into the depths, but in general the largemouth is an opportunistic predator that feeds on sick, injured, or otherwise enfeebled prey animals. For that reason, it is sometimes stocked as a natural means of quality control on commercial fish farms for the purpose of eliminating unsaleable competitors (such as the ruff, *Acerina cernua,* a freshwater perch often found in European carp ponds).

◀ The eggs and hatchlings, both highly vulnerable to predation, are zealously guarded by the adults. The adults themselves are sometimes caught by fish-eating birds and mammals and are appreciated by sports fishermen both for their tenacious fighting qualities and the succulence of their flesh.

F.M.

LEATHERBACK TURTLE *Dermochelys coriacea* 265

The leathery shell of the largest sea turtle typically measures 1.67 m in length (maximum 1.92 m) and 0.92 m (maximum 1.2 m) in breadth. A female leatherback caught off the English coast is said to have weighed 1016 kg, though this figure is undoubtedly exaggerated; Pritchard (1969) estimates that a female measuring 1.8 m probably weighs on the order of 600 kg. This species has been reported in all the warm and temperate waters of the world and is generally met with far out to sea. Is principal nesting beaches are located along the coasts of southern Africa, Malaysia, Mexico, French Guiana, and Trinidad.

The female comes ashore 6 or 7 times at intervals of about 10 days to lay its eggs, which in the interim are maintained at an optimal temperature of 7°C by means of an internal thermoregulatory system. The adults can breathe dissolved oxygen while swimming underwater, though it is still not clear how the hatchlings manage to breathe during the several days that it undoubtedly takes them to struggle their way up through the thick layer of sand that covers the nest. Leatherbacks do not generally survive very long in captivity, and their life expectancy in nature is unknown, though it seems likely that, along with some of the larger land tortoises, they are among the longest lived of all the vertebrates.

Authorities hesitated for some time in deciding whether this comparatively little-known species is actually carnivorous, herbivorous, or omnivorous. The stomach and intestinal contents of the first specimens to be examined included a miscellaneous assortment of animal and vegetable material, and it seems likely that the leatherback, though essentially carnivorous, ingests a substantial quantity of seaweed and other marine plants along with its animal prey. More recent studies of the stomach contents of beached leatherbacks suggest that this species feeds on mackerel, horse mackerel *(Trachurus)*, and other fish, crustaceans, mollusks, sea urchins and other echinoderms, and especially on by-the-wind-sailors *(Vellela)*, Portuguese men-of-war *(Physalia)*, and other large pelagic jellyfish, tunicates, and sponges.

The most detailed study of the feeding habits of the leatherback was conducted by Duron (1975) in the Pertuis Charentais, a coastal district of southwest France opposite the islands of Réand Oléron in the Bay of Biscay, where large concentrations of the pelagic jellyfish *Rhizostoma pulmo* (q.v.) are periodically reported. Loggerheads of both sexes also congregate in these waters between July and September, and a single turtle may be observed to eat as many as 50 jellyfish during its 5–hour period of daily activity. *Rhizostoma's* nutritional value is extremely slight, but, as Duron has pointed out, the leatherback's metabolism requires a considerable amount of water to flush out its system (to correct its electrolyte balance), whereas a reptile's need for protein is not all that great.

Thus, a typical daily intake of 50 jellyfish supplies the leatherback with roughly 200 liters of water and 8–10 kg of protein, which appears to be sufficient for its purposes. Even so, the leatherback also has a pair of lachrimal ducts, known as salt glands, that help to prevent a dangerous, possibly lethal level of sodium and potassium from building up inside its body, since it is unable to eliminate enough of these elements in its urine as mammals do; the viscous discharge from these glands can generally be seen running down from the corners of the eyes of female leatherbacks on the nesting beaches.

In order to catch a submerged jellyfish, the leatherback rocks forward in the water so that its hind flippers and supercaudal spur (the pointed extension of the carapace that covers the base of the tail) are tilted up in the air. It snaps up the jellyfish very quickly (keeping its eyes closed to avoid the stinging tentacles), then lifts its head out of the water, first to swallow its prey and then to take a breath of air. The leatherback is not a very discriminating browser, with the result that—given the current state of the coastal waters of the Mediterranean and Eastern Atlantic especially—it frequently runs the risk of choking on a floating plastic bag or some other artifact of man that it has mistaken for a jellyfish. The diet of the hatchlings and the immature leatherbacks is still unknown, though we were sucessful for a time in maintaining the hatchlings on a diet of cooked oysters and chopped squid in the St. Malo aquarium.

◀ The tiger shark and the killer whale are thought to be the only marine predators of the adult leatherback turtle, and many leatherbacks bear scars on their leathery shells and even have missing flippers (presumably) as a result of these attacks. It has also been pointed out that a

265
Sub.: Vertebrata
Cl.: Reptilia
O.: Chelonia
S.o.: Cryptodira
F.: Dermochelyidae

leatherback is far more likely to have lost one of its right flippers than its left, suggesting that the leatherback generally tries to evade pursuit by veering off in that direction. The shells of leatherbacks are occasionally creased by bullets and harpoons as well, though this species is generally only hooked or netted accidentally in temperate waters. Its flesh, though frequently toxic, is still consumed locally in tropical regions, and its abundant fat has been put to a number of different uses (notably the manufacture of marine varnish).

On the nesting beaches, the females are not only vulnerable to attack by jaguars and humans, but they may also find their way barred by various inanimate obstacles, including dead trees and other debris deposited on the beach by storms, steep banks of clay, and coastal marshes. The eggs are buried at a depth of about 80 cm and, in French Guiana, for example, are often dug up by the ghost crab *Ocypode quadrata*, crab-eating raccoons, coatis, and humans, though among the coastal Kalina Indians a taboo against eating either the flesh or the eggs of the leatherback is observed by some clans.

The hatchlings are sometimes intercepted by the great black hawk *(Buteogallus urubutinga)* and the night heron *(Nycticorax nycticorax)* as well as the Indians' dogs, raccoons, and coatis during their brief migration down to the shoreline. They are sometimes collected by the Kalinas themselves to be used as shark bait, and even those that manage to avoid all these perils still frequently fall prey to sharks and catfish in the shallows.

The leatherback's closest relatives are the HAWKSBILL and the other marine tortoises belonging to the family Cheloniidae. J.F.

266
Sub.: Vertebrata
Cl.: Mammalia
O.: Pinnipedia
F.: Phocidae

LEOPARD SEAL *Hydrurga leptonyx* 266

The leopard seal can readily be distinguished from other pinnipeds that live in Antarctic waters by its elongated body, snakelike head, and long, finlike flippers. The adult male weighs between 200 and 455 kg and measures between 2.5 and 3.5 m; the adult female weighs between 225 and 591 kg and measures between 2.4 and 3.4 m. Average weight is 324 kg for the males, 376 kg for the females.The leopard seal is generally a solitary creature and is more or less uniformly distributed along the fringes of floating pack ice that surround the Antarctic ice shelf, and during the summer months, it may venture as far north as the sub-Antarctic islands of South Georgia. This species is currently thought to number about 500,000 individuals altogether; longevity may exceed 26 years

▶ The leopard seal is an opportunistic hunter, though it generally seeks its prey in the open water rather than on the ice and relies primarily on the swiftness of its underwater pursuit. In Antarctic waters, it preys on the Adelie penguin *(Pygoscelis adeliae)*, and leopard seals have often been observed trying to tip over a floating ice cake on which a penguin has taken refuge. The pups ordinarily feed on krill; young crab-eater seals also make up a substantial part of the adults' diet. In sub-Antarctic waters, the adult leopard seal preys on seabirds, notably rockhoppers *(Eudyptes chrysocome)* and macaroni penguins *(Eudyptes chrysolophus)*, a regime it occasionally varies by catching young elephant seals and trying to drown them by holding their heads under water. In contrast to that of the crab-eater seal, the dentition of the leopard seal is well adapted to a carnivorous diet, and in fact the leopard seal is the only pinniped that feeds almost exclusively on warm-blooded creatures.

◀ The leopard seal is no longer hunted by man, which leaves the killer whale as its only natural enemy. Crab-eater seals and penguins, which also feed on krill, may be regarded as direct competitors of the leopard seal pups; the consumption of krill by baleen whales presumably cuts into the food resources available to the these two species, and consequently to the adult leopard seal, which feeds on them. M.P.

LEOPARD TOAD *Bufo regularis* 267

This species is found in tropical and subtropical Africa from the Sahara to the Cape of Good Hope, particularly in the vicinity of human dwellings, cultivated soil, and grazing grounds for stock; in urban areas, it is often found on recently watered lawns and lighted verandahs and the like, where it hunts for insects that are attracted by the light. The size of the adult varies between 45 and 100 mm.

▶ The leopard toad comes out to hunt at nightfall, like *Bufo bufo* in Europe and North America and *Bufo marinus* in tropical America; it feeds on ants, beetles, true bugs, crickets, sowbugs, spiders, coack-roaches, and flies (in descending order of frequency). Some individuals are especially partial to termites; the stomachs of 8 specimens collected by this writer in Senegal contained between 60 and 560 termites. The predatory behavior of the leopard toad is analagous to that of *Bufo bufo*, described above (see COMMON TOAD).

◀ Snakes, small mammals, and particularly birds figure among the predators of the leopard toad. Precise confirmation is lacking in these first two cases, but this writer once saw a heron gobble all the tadpoles in a small pond in Senegal. Adopting a more strictly quantitative approach, G. and M.Y. Morel (1962) have calculated that a particular nesting colony of 2,000 pairs of cattle egrets in northern Senegal would consume 500 kg of invertebrates per day, including 1,200,000 insects, as well as 150 kg of batrachians (*Bufo regularis, Bufo pentoni,* and various Ranidae), which would amount to 16,000 toads and frogs.

The smaller African toads (less than 70 mm) feed primarily on ants rather than beetles and other large insects; in this category, according to Inger and Marx (1961), may be included *Bufo funereus* (average size 47 mm), *Bufo ushoranus* (23 mm), and *Bufo melanopleura* (21 mm). The same might be said of a small bufonid, *Nectophrynoides occidentalis,* which is found in the Nimba Mountains in Guinea, measures between 10 and 26 mm, and feeds almost exclusively on ants (Lamotte, 1959). J.L.

267
Sub.: Vertebrata
Cl.: Amphibia
O.: Anura
F.: Bufonidae

Leposoma guianense 268

The genus *Leposoma* is assigned to the Teiidae, a family of New World lizards of which the tejus are perhaps the most prominent representatives. *Leposoma guianense* is fairly typical of the smaller members of the family, with a maximum length of only 39 mm from snout to vent; the tail, when present, may represent as much as 64 percent of total body length. Its upper body is brown and its flanks are dark with black spots, which enables it to move inconspicuously through the liter of fallen leaves in which it spends most of its time; the underbody of the males is salmon-pink, that of the females dirty white.

Leposoma guianense is widely distributed throughout the forests in the eastern part of the Guiana plateau, including parts of Guyana, Suriname, French Guiana, and the Brazilian state of Amapa. It is diurnal and occasionally makes itself visible in order to inspect the upper surface of the forest litter. It is already active at sunrise, and its periods of activity appear to be independent of weather conditions—it is undeterred by a light rain and is not attracted to the warmth of a patch of sunlight. In exceptional cases, several individuals may observed within an a few meters of one another, but in general (and bearing in mind the difficulties of taking an accurate census of such a timid and inconspicuous creature) the population density appears to be considerably lower.

▶ Analysis of stomach contents reveals that *Leposoma guianense* feeds on insect larvae and very small adult arthropods that are found in the uppermost layers of the soil and forest litter, in more or less the following order of preference: springtails (collembolans), particularly the Entobryomorpha, followed by spiders, insect larvae, adult Diptera, Homoptera, crustaceans, and acarians (mites). Though acarians are by far the most abundant constituents of the soil biomass (as much as 70 percent), *Leposoma* nevertheless exhibits a very strong preference for the springtails, perhaps because its predatory instincts are triggered by the abrupt, spring-like leaps executed by these tiny creatures. It is also possible that *Leposoma's* pursuit of the springtail is triggered by certain olfactory cues which are lacking in the case of the spiders, flies, and other prey that are equally likely to be encountered in the same environment.

◀ It is interesting to note that though *Leposoma guianense* immediately tries to conceal itself when a larger creature approaches, it does not simply bury itself in the litter but is careful to provide itself with a peephole (a crevice in the ground

268
Sub.: Vertebrata
Cl.: Reptilia
O.: Squamata
S.o.: Sauria
F.: Teiidae

or a small hole in a leaf) through which it can continue to observe its surroundings.

In spite of its leaflike protective coloration, *Leposoma* is sometimes devoured, along with other small lizards, by the motmot *(Momotus momota)* and other carnivorous birds that seek their prey in the underbrush, as well as by vine snakes, notably *Oxybelis argentus* and *aeneus*. These snakes station themselves on a low-lying branch with the slender forepart of their bodies dangling down like a thin liana, their heads touching the ground. *Oxybelis* has relatively large eyes and a narrow snout, so that its field of vision is largely directed forward; in detecting prey approaching from other quarters, it appears to make use of its olfactory organs (tongue and Jacobsen's organ), since while it is posted in ambush, it periodically flicks out its tongue and maintains it in that position for a moment. When a small lizard or other creature draws within range, it abruptly extends the forepart of its body, seizes the prey in its mouth, and subdues it with an injection of paralyzing venom from short fangs located in the rear of its mouth.

Arthrosaura reticulara and *Alopoglossus* are two other small lizards found in the same shadowy regions of the tropical forest floor; both are somewhat larger than *Leposoma* (50–70 mm), and springtails do not appear to figure in their diet. J.-P.G.

269
Sub.: Arthropoda
Cl.: Crustacea
O.: Cladocera
F.: Leptodoridae

Leptodora kindti 269

This cladoceran species is an important constituent of the plankton layer of large freshwater lakes. Its morphology is well adapted to a predatory mode of life and is thus substantially different from that of herbivorous cladocerans such as *Daphnia*, the common water flea. Its body is long and slender, offering little resistance to the water, and it swims faster than other species. It measures about 5–6 mm at maturity, though it may eventually attain a length of up to 15 mm. Its appendages are well suited to prehension (seizing and grasping prey) but are not provided with branchial lamellae (gill-like respiratory organs), so that *Leptodora* can only exist in richly oxygenated water.

At the edge of the carapace is a dorsal pouch (marsupium) in which between 1 and 13 eggs are deposited at the time that *Leptodora* molts its shell. The embryos complete their development in 12 days (at 20° C), at which time they are released into the water at the time of the following molt. Though these embryos are parthenogenetically conceived, some of them develop into males, though the factors (whether environmental or endogenous) that are responsible for this transformation are not well understood. At the same time, certain females produce a small number of haploid eggs, which are fertilized and then sequestered in a part of the carapace known as the eppiphium ("saddle"). These fertilized eggs, which are dark in color and well provided with vitelline nutrients, will remain in a state of arrested development and will not hatch unless external conditions are favorable, thus ensuring the survival of local populations through brief periods of adversity. The females that emerge from these eppiphial eggs are exceptional among cladocerans, since they take the form of metanauplia (crustacean larvae with 7 sets of appendages); most other cladoceran (and all other *Leptodora)* hatchlings are morphologically identical to the adults. Maximal life expectancy is on the order of 1 1/2 months.

▶ Virtually every species of plankton has been identified among the stomach contents of *Leptodora*. In the laboratory as in nature, the juveniles have no difficulty in catching rotifers, though the adults exhibit a clear preference for herbivorous cladocerans rather than rotifers (too small?) or copepods (too fast?). Their prehensive organs are not well developed enough to seize prey larger than 1.5 mm. *Leptodora's* daily intake varies with the age and sex of the individual as well as with the amount of time that has elapsed since its last feeding and—since the success of *Leptodora's* hunting technique depends on random encounters with other planktonic organisms of the appropriate size—with the density of the plankton population. When this population exceeds a certain optimal level, *Leptodora* tends to become a messy and wasteful feeder, killing other organisms but only partially consuming them.

At a temperature of 15–20° C, an adult female catches an average of 15–30 plankton every day, and herbivorous cladocerans, such as *Daphnia,* have evolved a number of defensive tactics in response to these depredations. *Daphnia* occasionally increases the length of its transparent body capsule (a process known as cyclomorphosis) in response to water turbulence or an increase intemperature; it is now thought that *Daphnia* measuring 1–1.5 mm may also do this to push themselves over the edible limit as far as *Leptodora* and other tiny carnivores (Cyclopidae, gnat larvae of the family Chaoboridae) are concerned. Herbivorous plankton may also take refuge in the shallows, or in the deeper waters of a lake or pond where the water is too poorly oxygenated to provide a suitable habitat for *Leptodora.*

◀ *Leptodora* is consumed in large quantities by large plankton-eating fish. Though relatively large, much of its body is transparent, which may make it less vulnerable to predation by fish that detect their prey visually.

Leptodora kindti is the only representative of the genus *Leptodora* and the family Leptodoridae. The family Polyphemidae, in addition to several marine genera, includes two genera of freshwater plankton that are similar to *Leptodora* in their feeding habits. One of these, *Bytotrephes,* is rare and has not been thoroughly studied; it appears to have a preference for copepods in their earlier larval stages. The other is represented in Western Europe by a single species, *Polyphemus pediculus,* most notable for its single enormous compound eye, which occupies the entire cephalic region. *Polyphemus* is found in forest pools and along lakeshores that are overgrown with aquatic vegetation. Its ventral appendages are fringed with branchial lamellae and tipped with little claws; the globular dorsal pouch can accommodate as many as 25 embryos. *Polyphemus'* optimal daily intake consists of 8–10 cladocerans or as many as 100–200 rotifers, which of course are considerably smaller. R.P.

LEVANTINE VIPER *Vipera lebetina mauritanica* 270

This large viper typically measures between 80 and 130 cm, though it may attain a length of 150 or even 200 cm and a weight of 3 kg. The *mauritanica* subspecies ranges along the coast of northwest Africa, from Rio de Oro to Libya, and into the steppes, mountains, and subdesert regions of the interior. *V. l. mauritanica* is found only where there is some vegetation or other cover, including groves of eucalyptus and cork oak, plantations of *Sideroxylon* (a domesticated berry bush), grassy plains dotted with shrubs, rockslides and banks of scree, or hedges and stone walls dividing cultivated fields. It is also found in mountainous regions up to an altitude of 2000 m and in the beds of dry wadis (intermittent watercourses) in subdesert regions as long as there are a few thornbushes for cover.

▶ The adult *V. l. mauritanica* seems to feed exclusively on warm-blooded animals, including birds, gerbils, jirds (a relative of the gerbil), ground squirrels, and other small mammals up to the size of a rabbit; the very young vipers feed primarily on lizards and occasionally on baby mice. As with other vipers, the dentition of this species is solenoglyphic (that is, the venom channel is entirely enclosed inside the tooth), and the short upper jaw can be opened very wide so that the fangs, which are firmly fixed to the upper jaw, are pointed forward rather than downward when the viper is about to strike. The Levantine viper will not withdraw its fangs from the body of a bird or lizard, but a larger prey animal, especially one that defends itself vigorously, will be released until the venom has done its work and then will be immediately retrieved.

The females hunt most actively in the spring after emerging from hibernation, the males after the end of the mating season, in the late spring and summer. During the colder months, *V. l. mauritanica* is largely or completely sedentary; it conceals itself in ambush and hunts only by day. During the summer, it begins to seek its prey more actively, beginning at dusk and continuing into the night.

◀ Though it is often killed on sight by humans, the adult *V. l. mauritanica* has no other natural enemies. The juveniles are sometimes preyed on by the Egyptian cobra *(Naja haje)* or the Montpellier snake *(Malpolon monspessulanus)* in various parts of

270
Sub.: Vertebrata
Cl.: Reptilia
O.: Squamata
S.o.: Ophidia
F.: Viperidae

the range of this subspecies, though small mammals that normally feed on snakes (including hedgehogs and mongooses) seem to have little relish for the Levantine viper.

Lataste's viper *(V. latasti),* which preys primarily on lizards, is thus competitive with the juvenile *V. l. mauritanica* in some parts of North Africa, and the puff adder *(Bitis arietans),* an African species that ranges up to the southwestern tip of Morocco, also feeds on lizards, birds, and small mammals; *B. arietans* has considerably more tolerance for direct sunlight than the Levantine viper, however, and is not necessarily found in the same biotopes. The other large snakes of the region are less specialized in their tastes or have different circadian cycles.

The remaining 6 subspecies of *Vipera lebetina* are found in the Near East and Central Asia (ranging as far east as Afghanistan) as well as on Cyprus and several of the Cyclades. D.H.

271
Sub.: Vertebrata
Cl.: Aves
O.: Passeriformes
F.: Fringillidae

LINNET *Acanthis cannabina* 271

This Old World songbird is 13 cm long, with a wingspread of 24 cm, and weighs between 16 and 21 g. It is found in a variety of different biotopes—essentially wherever there is low-lying herbaceous ground cover in which it can find its food as well as shrubs and bushes in which it can build its nest. These include grasslands interspersed with thickets, chalk downs or stony ground overgrown with brushwood, fields of gorse, hedges around houses or gardens, small vineyards, etc. This species is found in virtually all of Europe (excluding some parts of the South of France and northern Scandinavia), Asia Minor, and parts of Central Asia.

The linnet is a sociable bird, nesting in loose-knit groups and gathering in flocks on the plains where it searches for food in winter. Most populations are sedentary, though some European linnets fly southwest to spend the winter in North Africa. Linnets often flock together with other fringilids, including finches, greenfinches, and tree sparrows.

▶ The linnet is primarily a vegetarian, though insects and other animal prey constitute an important supplement to its diet of plant foods during the summer. The linnet eats the seeds of a great many different plants—wild and cultivated Cruciferae (including cabbages and colza in the latter category), wild grasses growing in vineyards or fields of stubble, thistles and other Compositae, and Polygonaceae. The adults also feed on beetles, butterflies, and their larvae as well as spiders, and feed their young on a puree of seeds, caterpillars, and insects.

◀ The linnet has no known competitors. The linnet's nest, however, is not very well concealed and the eggs and nestlings are destroyed in substantial numbers by crows, magpies, and even buzzards. The adults are sometimes caught by kestrels and sparrowhawks, and migratory linnets (as well as finches and other songbirds) en route to North Africa are still frequently caught with spring nets and deadfalls along the southwest coast of France. Though the linnet, like other songbirds, is legally protected all over Europe, the practise of netting them illegally appears to be on the increase in country districts. J.-F. T.

LION *Panthera leo* 272

Africa's best-known carnivore has always had a special place in the human imagination; Bible stories and classical myth have told us how the greatest heroes of our own species have tested their strength and courage in combat against this king of beasts. The Nemean lion was deprived of its pelt by Hercules, for example. The lion, along with the other big cats (the tiger, jaguar, leopard, and the panther) is assigned to the genus *Panthera.* A full-grown adult measures between 2.2 and 2.9 m and may weigh as much as 275 kg; the male is larger than the female and can readily be distinguished by its shaggy mane, which extends down to the breastbone. At birth, the cubs are covered with dark spots, which will have almost completely disappeared by the age of 2 or 3, the same age at which the mane first appears in the males.

Two thousand years ago, the species flourished from Greece to the East Indies as well as on the continent of Africa. Today, however, only a few hundred African lions survive; the Indian subspecies, *Panthera leo goojratensis,* has retreated into

272
Sub.: Vertebrata
Cl.: Mammalia
O.: Carnivora
F.: Felidae

the game reserves in the Gir Forest, and other subspecies have vanished entirely, the Barbary lion, the Cape lion, and the Persian lion among them. The African lion lives on grasslands and open savannah; the pride is a stable group composed of 4 or 5 individuals as a rule, though prides on the Serengeti have been found to number more than 40.

The size of a pride's hunting territory varies with the abundance of game and may occasionally be as much as 260 sq km; limits of the territory are defined with the usual olfactory markers, deposited by the males on trees and other prominent features of the landscape. The mere presence of a lion—as signalled by these urinary clues, as well as by the conspicuous silhouette and thunderous roar of the adult male—is usually enough to deter other felids from poaching on its preserve. Providing for the safety and security of the pride is the only responsibility of the males; all other duties (notably those of providing food and rearing the cubs) are undertaken by the females.

▶ Whether nomadic or sedentary, the lion is a superpredator. The females prey on zebras and gnus during the rainy season; during the dry season, when the great migratory herds are far away, the lionesses turn their attention to the hartebeest, Thomson's gazelle, warthog, and Cape buffalo. They will try to pick out quarry that is heavy and easy to bring down (preferably the gnu, *Connochaetes taurinus*), but the lion is an opportunistic hunter and will attack any prey animal that it encounters. It keeps a close watch on other predators and attempts to make off with their kills; it will supplement its diet with small mammals, reptiles, and even carrion if the necessity arises. Cannibalism is a routine feature of leonine society—at least to the extent that the pride will feed on the bodies of their own dead as well as on cubs belonging to another pride.

The lion generally eats every 3 days, and 6–10 lbs of meat is the adult's minimum daily requirement, but lions are certainly capable of eating 4 or 5 times that much when game is particularly abundant, and are accustomed to eat twice as much during the gnu's seasonal migrations. Like other felids, the lion's endurance is limited, and it is incapable of sustained pursuit in the manner of the hyena or the Cape hunting dog. The hunting pack invariably operates as two separate groups; the first group charges the herd and disperses it, driving the herd toward the various spots where the members of the second groups have concealed themselves.

Some authorities claim that the males sometimes take part in the hunt, though this has been disputed; the claim has also been made that the lionesses in the first group, the "beaters" as it were, will approach the herd from upwind, making the quarry even more likely to panic when the lion scent is carried to them, and that the second group, which actually makes the kill, is careful to station itself downwind from the quarry and thus remain incognito for as long as possible. This dispute will undoubtedly be resolved fairly soon with the use of more sophisticated observational techniques.

After one of the lionesses in the second group has marked its victim, it attacks at maximum speed so that the shock of the impact will knock the antelope off its feet. Next, the lioness bears down on the fallen antelope with all its weight, seizes it by the throat in an attempt to constrict the airway and the major blood vessels, and maintains this pressure until the antelope dies of asphyxiation, which is generally rather quickly. By this time, the other members of the pride will probably have arrived on the scene, and the carcass is dragged off to a shady spot and speedily dismembered and devoured. Unlike canids, lions feasting on a kill extend no particular privileges to their cubs, who are obliged to seize such scraps as they can from the melee and are often violently rebuffed by the adults. This behavior helps to explain the fact that mortality among the cubs may be as high as 90 percent during the dry season, when game is scarce.

The cubs' games and mock combats teach them assert themselves and thus to win the respect of older members of the pride. By the age of 3 months, they are old enough to accompany the females into the hunting field, though strictly in the role of observers. They will actually begin to take part in the hunt when they are almost a year old and may be big enough to make their first kill shortly thereafter. At the age of 3, the young males are old enough to compete for the favors of the lionesses and are expelled from the pride by the dominant male.

◀ As the largest of African carnivores, the lion has little to fear from other animals, excluding man. A hyena pack (though not an individual hyena) will dispute the posession of a kill with a lion, though it is often a question of the hyenas reclaiming their lawful prey; the cubs may occasionally be attacked by leopards, which are in the habit of preying on young carnivores.

P.A.

273
Sub.: Vertebrata
Cl.: Osteichthyes
S.cl.: Teleostei
O.: Scorpaeniformes
F.: Scorpaenidae

LION-FISH *Pterois antennata* 273

This species measures between 20 and 25 cm and rarely exceeds 30 cm in length. Like all scorpion fishes of the genus *Pterois,* which are collectively known as lion-fish, *P. antennata* lives on coral reefs in the South Pacific. It swims slowly, either singly or in pairs, along the edges of a reef or in the open water, and is sometimes seen swimming in a vertical attitude and even head downward. Longevity in captivity is 6–7 years; longevity in nature is unknown.

▶ The species name *antennata* is something of an understatement, since the lion-fish's spiny fins are elaborated into brightly colored feathery lures with which it attracts and subsequently entraps crustaceans and small fish, herding them against a wall of coral by slowly shifting the position of its fins, then snapping them up with an avidity that seems surprising in a creature that is otherwise so languid in its movements. The lion-fish also hunts from ambush, stationing itself along the edge of a wall of coral or at the bottom of a canyon and remaining motionless until a small prey animal approaches within striking distance, then abruptly opening its mouth and gulping down the prey animal, along with the water that it was peacefully immersed in a moment or two earlier.

◀ As with all the scorpion fishes, the lion-fish's dorsal, anal, and pelvic fins are ribbed with long, sharp vanes, each of which has a T-shaped slot running along its length that is filled with venom glands. The lion-fish is unaggressive toward other fish, though when it is approached by a larger creature, it assumes a defensive posture, swiveling as many as its spines as possible in the direction of the intruder. Not surprisingly, the lion-fish itself has no specific predators, though the armored hide and tough, parrotlike beak of the trigger-fish permit it to gnaw off the spines of other scorpion fish without suffering any ill effects.

It was formerly believed that the lion-fish's venom could cause fatalities in humans, but its effects, as painful and unpleasant as they might be (intense pain lasting for several hours accompanied by an edematous swelling and loss of sensation), are strictly transitory; the same holds true for the other three lion-fish species: *P. radiata, eunalata,* and *vobitans.* The Pacific islanders' traditional remedy is to cauterize the wound, since the lion-fish's venom is deactivated by intense heat. The flesh of the lion-fish is locally appreciated as a delicacy, but the not very extensive commercial traffic in this species is conducted almost exclusively for the benefit of medical research laboratories and tropical-fish fanciers.

J.-M. R.

274
Sub.: Arthropoda
Cl.: Insecta
S.cl.: Pterygota
O.: Hymenoptera
F.: Sphegidae

Liris nigra 274

The males of this species of delicate, black-bodied digger wasp are about 20 mm long; they appear over the course of the summer and, after they have mated with the females, die out during the fall. The females go into hibernation, lay their eggs during the summer, and die with the approach of another fall.

▶ The males are predatory during their brief lifespan, but the females probably account for more substantial quantities of insect prey over a longer period. It is not certain whether the females begin to feed before their second season of life, when they emerge from hibernation, though they sometimes exhibit typical predatory behavior during their first season.

Both the adults and the larvae feed on crickets, and the predatory strategy of *Liris nigra* has been very thoroughly described by A. Steiner (1957, 58): The wasp probably locates its prey by sight, in the course of an aerial reconnaissance of its territory. It approaches the cricket from behind and attempts to clamber up on its back; the cricket tries to dislodge the wasp by kicking vigorously with its powerful third pair of legs, its jumping legs. If the wasp has taken hold of only one of its legs, as happens frequently, it may be detached by autotomy, enabling the cricket to escape.

In those cases where the cricket fails to dislodge the wasp immediately, the wasp next tries to position the tip of its abdomen (where the stinger and venom gland are located) over one of the soft, flexible membranes that covers the joint between the cricket's legs and its body. This is where the neural ganglia that control the motions of the cricket's legs are located. The tip of *L. nigra's* abdomen is extremely sensitive, and the wasp generally has no difficulty in inserting its stinger into the ganglion. Once the ganglia are blocked by *L. nigra's* venom, the adjoining muscles are unable to contract, since the ganglia are no longer capable of transmitting instructions relayed through the central nervous system. Where possible, the wasp begins by paralyzing the cricket's jumping legs.

Four venom injections are generally enough to render the cricket totally helpless, though as many as 6 may be necessary if the wasp fails to hit the ganglion with its usual pinpoint accuracy. In the case of an extremely large cricket, where the wasp not only has further to travel, in effect, from one injection site to the next, the chances are also greater that its aim will be thrown off by its victim's vigorous bucking and kicking. It has been noted that not all wasps have attained the same high standards of marksmanship, and individual performance also seems to vary from one occasion to the next.

Sometimes, after the wasp has been dislodged in the course of an unsuccessful attack from the rear, it may attempt a broadside onslaught, in which case the first injection of venom will be administered from in front of rather than behind the joint where the jumpings legs are attached to the body. Similarly, if the cricket manages to escape during the first attack and takes refuge under a rock, the tip of the wasp's abdomen is sufficiently sensitive that it may still be able to find its way to the target while the wasp is clinging to the sides of the rock, since the conventional tactic of clambering up on the cricket's back is now clearly impossible. After the cricket is immobilized, it may be devoured straightaway by the adult wasp, or the female may deposit an egg in the same spot, so that the cricket can eventually be devoured by the larva that emerges from it.

◀ When *Liris nigra* is seized by an insectivorous bird and is not killed outright, it will attempt to use its venomous stinger as best it can to compel the bird to release it. The predatory behavior of other *Liris* species has not been studied in such detail as with *L. nigra;* presumably they each tend to feed on a different cricket species, though the basic strategy is probably very much as described above. G.B.

Lithobius forficatus 275

275
Sub.: Arthropoda
Cl.: Chilopoda
O.: Lithobiomorpha
F.: Lithobiidae

This common centipede is a typical representative of the order Lithiobiomorpha. Its body length variies between 18 and 35 mm, and its armored cuticle contains about 78 mg of living tissue. *Lithobius forficatus* undergoes several larval stages and takes 4 or 5 years to attain its adult form. It is by far the commonest centipede in France and Western Europe, and is also found in parts of the Mediterranean Basin and at least as far east as the Caucasus and the Urals as well as along the western shores of the Atlantic, including Newfoundland. This species lives in leafmold, under rocks, and in the topsoil in virtually all terrestrial ecosystems up to an altitude of 2000 m.

▶ *Lithobius* is an unspecialized, opportunistic hunter that feeds on almost any prey of the appropriate size, provided the latter is not protected by a thick carapace or chemical repellents (such as the millipede *Diplopodes).* It seems to prefer to feed on isopods (woodlice and other small terrestrial crustaceans), mites, springtails (collembolans), aphids, earwigs, small spiders, and the larvae of many different arthropods; *Lithobius* also preys on other centipedes, even those of its own species, though cannibalism appears to be more commonly reported in terrariums and other artificial habitats than in nature.

Lithobius forficatus also feeds on carrion to a small extent (if the bodies of recently deceased isopods can be so called), though it is not really one of those necrophagous species that performs the principal work of disposing of dead arthropods and other tiny inhabitants of the forest litter and the topsoil layer. Leafmold, humus, and other organic debris, on the other hand, do furnish a substantial part of *Lithobius'* diet during the winter.

Lithobius actively seeks its prey in the spaces between dead leaves in the forest litter and in crevices in the ground. The method by which it locates its prey is not well understood; perhaps by chemoreception or by detecting the vibrations made by prey animals. Its eyes are rudimentary, consisting of juxtaposed eyespots (ocelli), and its vision is not very keen, though it may be capable of detecting animal movements by this means. *Lithobius* cannot be said to engage in predatory behavior, and for the most part, contact is made with a prey animal more or less at random, and the prey is immediately seized in *Lithobius'* curved clawlike pincers.

These appendages, technically known as forcipules, developed from the walking legs of the first body segment. The centipede injects its venom into the prey through an aperture at the tip of each of its pincers. *(Lithobius'* venom, though lethal to a creature the size of a woodlouse, is not harmful to man.) After the prey has been immobilized in this fashion, it is torn to pieces by the centipede's mandibles and swallowed immediately.

◀ Though *Lithobius* may be vulnerable to predation on the part of larger centipedes and many other arthropods both during its earlier larval stages and during the molt, when it has neither its tough cuticle plates or its sharp pincers to protect it, the adult centipede has few natural enemies other than large spiders, carabid beetles, and ants. Because *Lithobius* is generally concealed in the upper layers of topsoil or leafmold, it has little to fear from larger creatures, though it may occasionally fall prey to frogs, salamanders, and lizards (notably the green lizard). It is sometimes caught by birds, though rarely, and by such insectivorous mammals as the mole, the shrew, and the hedgehog, against which it has no effective defense other than flight.

About 40 Lithiobiomorpha species are found in Western Europe, including a number of larger centipedes that may be found in the same habitat with *Lithobius forficatus* and share the same food resources; these include *L. variegatus,* found in the forests of Britain, and *L. pilicornis,* which replaces *forficatus* at higher altitudes, though the 2 species do coexist in upland valleys and on high plateaus.

J.-J. G.

LITTLE EGRET *Egretta garzetta* 276

276
Sub.: Vertebrata
Cl.: Aves
O.: Ciconiiformes
F.: Ardeidae

This elegant little heron, once prized by milliners and plume hunters for its long, curved scapular (shoulder) plumes, measures between 50 and 65 cm from beak to tail, from 90 to110 cm from wingtip to wingtip, and weighs between 400 and 600 grams. It becomes sexually mature toward the end of the first year of life or (more frequently) during the second. Mortality among nestlings is very high.

The little egret lives in many different parts of the Old World. In 1974, there were 23 colonies in France (south of the Loire), or a total of 1815 nesting pairs, the vast majority of which (1700) inhabited a marshy region in Provence called the Camargue; by 1981, the number of colonies had risen to 29, or 2253 nests (1441, or 64%, of which were in the Camargue). The little egret winters along the Mediterranean coastline, sometimes venturing as far south as equatorial Africa, and prefers the shallow, relatively warm, whether fresh or brackish, waters of coastal marshes and estuaries, lakes, irigation ditches, and ricefields—as long as there is a stand of trees nearby suitable for a large nesting colony. The female lays an average of 4 eggs every year; both parents take turns on the nest.

▶ The little egret feeds on a variety of small aquatic creatures (none exceeeding 15 cm), both vertebrate and invertebrate. In the Camargue, about half its diet consists of fish, notably Cyprinidae (minnows and shiners) and eels. Other prey includes frogs, toads, newts, efts and tadpoles, and occasionally reptiles, worms, mollusks (planorbids, or freshwater snails), crustaceans, insects and their larvae (diving beetles, dragonflies, water scorpions, backswimmers, crickets, locusts).

The little egret actively stalks its prey, wading with long strides through the shallow waters of its hunting grounds, sometimes approaching with extreme stealth, or remaining motionless, assuming the familiar hunting stance of the heron family with its head tucked back like a cobra's, ready to dart out suddenly and impale anything edible that comes its way. The egret also hunts on the wing, snapping up its prey from the surface of the water, or forages in the mud along the bottom in search of fish, amphibians, or insect larvae.

◀ The egret flies slowly and, with its brilliant white plumage, is quite conspicuous but seems to have few natural enemies as an adult. The eggs and the nestlings are very vulnerable, and both parents are obliged to stand guard over the nest. The egret's habit of living in large, raucous colonies discourages predators to some extent; crows and gulls as well as other members of the heron family (Ardeidae) may attempt to raid the nests. Adult egrets are reportedly taken by falcons and ospreys on occasion.

Since food is quite abundant in the sort of terrain favored by the little egret, there is little direct competition with other species. Its closest rivals in the Camargue, the crab-eater (another egret, *Ardeola ralliodes)* and the night heron *(Nycticorax),* feed on different-sized prey animals at different times of the day and use a different hunting technique. Competition during the first few months of life, however, is very intense, with the sturdier nestlings often being responsible, directly or indirectly, for the elimination of their weaker siblings.

The beauty and delicacy of the egret's nuptial plumes, known as "aigrettes" to the millinery trade, brought about the wholesale extinction of many colonies—and almost of entire species in North America—at the hands of plume hunters, particularly during the Edwardian era, when the aigrette was at the height of its vogue. Since the 30s, the advent of conservationist legislation and more utilitarian fashion trends has enabled the egret to at least partially replenish its former numbers in many areas.

The heron family comprises 69 species worldwide, primarily tropical and almost exclusively aquatic. The crab-eater, mentioned above, is another all-white egret frequently found in Southern Europe; a third is the great white egret *(Egretta alba),* known as the common egret in America, though it is rather rare in the Old World. The African cattle egret *(Ardeola ibis),* noted for its commensal relationships with various grazing animals, has recently established itself in the southeastern United States.

LITTLE OWL *Athene noctua* 277

277
Sub.: Vertebrata
Cl.: Aves
O.: Strigiformes
F.: Strigidae

The little owl is a widespread Eurasian species ranging all the way from Brittany to the Pacific and from as far north as Scotland and Denmark to the Sahara. Its body is from 22 to 27 cm long with a wingspread of 57–61cm; it weighs between 124 and 198 g, and its lifespan is not thought to exceed a couple of years. It can live in all sorts of open terrain, including parklands, gardens, nurseries, etc., as well as steppes and deserts.

▶ The little owl prefers to hunt at dawn or dusk but, unlike other nocturnal raptors, it is sometimes active by day as well. Its wavering flight recalls that of the European green woodpecker, and it sometimes travels along the ground in search of earthworms, which it plucks out with its beak in the manner of a starling or a blackbird. The little owl preys on small mammals, particularly rodents, frogs, small birds, and (in arid regions, primarily) on insects. It catches insects in flight or stations itself on a convenient perch to await the approach of larger prey, which it pins to the ground with its talons and its beak. The parents take turns feeding the nestlings, which leave their nest cavity at the age of 4 to 5 weeks, before they have even learned to fly.

◀ During this first, terrestrial phase of its career, the little owl sometimes falls prey to marauding martens, foxes, or even stray cats and dogs. The adult is not subject to direct predation, though its numbers have undoubtedly declined in Western Europe due to the widespread use of insecticides, the disappearance of suitable nesting sites (notably dead trees and hedgerows), and increased automobile traffic—to which both the little owl and the barn owl *(Tyto alba)* have frequently fallen victim. The little owl will freely make use of the nesting boxes put up for it by farmers and gardeners, and is not particulary shy. When it is perched on a fencepost or a rooftree or at the entrance to its nest, it will allow a solitary human to approach fairly closely without budging from its perch; then, suddenly alarmed, it executes a series of nervous, ticlike bowing motions—very characteristic of this species—before taking to the air or retreating into its nesting cavity.

The genus *Athene* comprises only 3 species. *Athene brama,* whose habits are quite similar to those of the little owl, is found in India and the Middle East; *Athene blewiti* lives in the forests of Northern India and the Himalayas. The insectivorous burrowing owls of the New World (genus *Speotyto)* are almost entirely terrestrial. *Speotyto cunicularia* is a denizen of the arid grasslands (prairies or pampas) of North and South America; colonies of these burrowing owls build their nests or take up permanent residence in burrows abandoned by rabbits, prarie dogs, or other rodents. The elf owl *(Micrathene whitneyi)* is fairly common in the Southwestern United States and frequently builds its nest in the holes made by woodpeckers in the trunk of the giant saguaro; the elf owl and the South American pygmy owl *(Glaucidium minutissimum)* are the smallest of owl species.

J.-P. R.

278
Sub.: Vertebrata
Cl.: Osteichthyes
S.cl.: Teleostei
O.: Scorpaeniformes
F.: Cottidae

LONG-SPINED BULLHEAD *Taurulus bubalis* 278

Generally speaking, the members of the family Cottidae are known as sculpins in North America, bullheads in Great Britain (a term that has been reapplied to a genus of catfish in North America). Since *Taurulus bubalis* is found primarily along the Atlantic coast of Europe, from the Bay of Biscay to the Norwegian Sea (and to some extent in the Baltic and Mediterranean as well), we will refer to it as the long-spined bullhead. The male attains an overall length of 8–12 cm, the female as much as 25 cm; individual "long-spined bullheads" of 33 or even 40 cm referred to in the literature were almost certainly members of another sculpin species, of which there are about 30 altogether. Both sexes attain maturity and maximum size by the age of 2; the longevity of this particular species is unknown.

The long-spined bullhead is found in shallow coastal waters, particularly in areas where the bottom is rocky and overgrown with seaweed (though in exceptional cases, it may range to a depth of 100 m), and it is frequently found stranded in rock pools at low tide. The larvae are initially free-swimming (pelagic), though they settle on the bottom as soon as they have reached a length of 1.3 or 1.4 cm.

▶ The long-spined bullhead is a voracious predator with a very wide mouth that enables it to ingest a nearby fish or other prey animal (sometimes half the size of itself) simply by gulping down large quantities of water, so that the prey is in effect drawn in with the tide. The bullhead hunts from ambush and feeds primarily on small fish such as gobies as well as young fry and immature members of larger species. It will give chase if necessary, but it is not a very active hunter and also uses this technique to ingest slower-moving sea creatures (crustaceans as well as mollusks and brittle stars) and floating organic material (the eggs of fish and mollusks).

◀ The long-spined bullhead has two defenses against predation. The first is protective coloration; its skin is brownish-gray with black spots, which makes it more difficult to detect against a background of seaweed and submerged rocks. Secondly, the gill covers (opercula) of the bullhead, like those of other scorpaenid fishes, are equipped with long sharp spines, which make it not only much more formidable in appearance (it looks about twice as big as it really is, at least when viewed from the front) but also very difficult to swallow. A larger fish may occasionally make the attempt, but the bullhead is generally expelled unharmed as soon as its sharp gill spines make contact with the lining of the would-be predator's mouth. On the other hand, the anemone, which is sightless and able to digest even the most refractory morsels at its leisure, is not deterred by either of these tactics and feeds frequently on the long-spined bullhead.

This fish is not exploited commercially, though a few of the larger European sculpin species are used to make fish meal. Traditionally the North American sculpins, notably the sea raven, have been among the most conspicuous "trash fish" as far as New England fisheries were concerned, and the word *sculpin* is still used by old-time New Englanders to describe a worthless or superfluous object or person; interestingly, *bullhead* is also a term of mild abuse in some dialect areas. The miller's-thumb *(Cottus gobrio)* is a common European freshwater sculpin, perhaps so called because of its prevalence in medieveal millponds. J.-M. R.

M

MABUYA *Mabuya buettneri* 279

279
Sub.: Vertebrata
Cl.: Reptilia
O.: Squamata
S.o.: Sauria
F.: Scincidae

The male of this species of African skink measures between 60 and 70 mm from snout to vent and weighs an average of 6 g; the female is somewhat larger, measuring 75–90 mm and weighing 7–8 g. This species is notable for the length of its tail, fully three times that of the rest of its body. Average life expectancy is about 11 months; maximum longevity, as reported both in nature and in captivity, is 18 months.

Mabuya buettneri is prevalent throughout the savannahs of West and Central Africa; in the Ivory Coast, where it was studied by Barbault (1974), the life cycle of this species coincides very closely with the alternation of dry and rainy seasons on the savannah, where the vegetation is burned off each year by grassfires around the middle of the dry season in February: Thus, the eggs are laid in November and December and the last adults of the previous generation die out in January, but the eggs do not hatch until March and April.

▶ Like most small and medium-sized lizards, the mabuya is an insectivore, feeding on the spiders and locusts, mantids, and other insects that live in the grassy vegetation of the savannah. Attracted by the movements of an insect, it lunges and swallows it in a single gulp. Ants and termites are very abundant in the same biotope, but even in captivity, *M. buettneri* cannot be induced to eat them. Young captive specimens might make a few initial attempts to eat the ants that are offered them but immediately spit them out and refuse all subsequent offers of the same kind. In nature, the mabuya consumes about 6–10 percent of its own weight in insects every day.

◀ *M. buettneri* is preyed on by many different creatures, and this is a significant cause of mortality in this species. Apart from a number of "generalist" predators—small mammals, birds, and snakes that may occasionally feed on lizards of various sorts—there are quite a number of predators that prey specifically (though not exclusively) on *M. buettneri,* notably the lizard buzzard *(Kaupifalco monogrammicus)* and a half-dozen snake species, including *Psammophis sibilans* and *elegans* (sand snakes), *Lycophidion irroratum* and *semicinctus* (wolf snakes), and *Meizodon coronatus* and *regularis. M. buettneri's* only defensive strategy in the face of this considerable predatory pressure is confine itself strictly to the thick carpet of grassy vegetation and never to venture out onto open terrain. Its survival as a species is also greatly assisted by its rapid maturation (6–7 months) and its extreme fecundity (the average female produces 2 successive clutches of 8 or 9 eggs).

Two congeneric species are also found on the West African savannah. *Mabuya maculibris* is a common terrestrial skink that may also be found in scrub forest or in groves of Palmyra palms *(Borassus);* it is about the same size as *M. buettneri* and its diet is much the same, except for a higher proportion of hemipterans (true bugs) and a lower proportion of mantids. *Mabuya perroteti* is much larger than its congeners (155 mm) and is strictly terrestrial in its habits; it also feeds on larger prey than the other species, primarily beetles and cockroaches as well as some grasshopper and locusts. R. Ba.

280
Sub.: Vertebrata
Cl.: Reptilia
O.: Squamata
S.o.: Squamata
F.: Gekkonidae

MADAGASCAR GECKO *Phelsuma madagascariensis* 280

The species that make up the genus *Phelsuma* form a distinct group within the family Gekkonidae. Even the largest species, found on Madagascar and in the Seychelles, is less than 25 cm long, and unlike most geckos, *Phelsuma* is brightly pigmented, green with red spots. Like the chameleons and the anoles, though not to the same extent, it can vary the intensity of these pigments in order "blend in" with its background. The Madagascar gecko is arboreal, living among the foliage of banana plants and coconut palms, as well as among the rafters of human dwellings.

▶ *Phelsuma* is active both by day and at dusk. It goes out in search of insect prey, which it detects by sight, and sometimes attacks from ambush. It also feeds on flowers and fruits, which it prudently tastes beforehand by flicking it with its tongue before swallowing it.

◀ All the *Phelsuma* species are fiercely territorial, and the males are particularly intolerant of intrusions by members of their own species; they will fight until one of them breaks off the engagement and retreats, sometimes mortally wounded. Snakes, birds, and other lizards are all known to prey on the Madagascar gecko. As noted, its intense pigmentation probably serves some protective purpose, and if this disguise is insufficient, it can often escape in one bound—it is an excellent jumper—and scurry off to shelter or conceal itself behind a branch or treetrunk. One species found on the Seychelles, which also lives in the treetops, hurls itself confidently into space when it sees the outline of a kestrel against the sky and plummets to the ground while controlling its speed and direction, more or less, with the movements of its tail.

Phelsuma abotti, also found on the Seychelles, is in the habit of hitching rides on the back of the Seychelles giant tortoise *(Testudo gigantea).* The gecko feeds on insects that are attracted by the tortoise's droppings and may seek refuge when alarmed by darting inside one of the "armholes" of the tortoise's shell.

Geckos of the genus *Ptychozoon,* found in Southeast Asia, are able to parachute, rather than plummet, from the treetops in a less precipitious fashion. They have developed a cutaneous fringe that runs all the way around the body (including webbing between the toes), so that when they spread their legs out flat and launch themselves into space, they can glide smoothly to the ground—once again, using the fringed tail as a rudder—and land softly on the jungle floor some distance from the base of the tree from which they departed. The wide cutaneous fringe around the body is found, in rudimentary form, in other arboreal geckos *(Uroplatus,* for example) and is thought to have developed as a means of obscuring the outlines of the gecko's body when flattened out against a treetrunk by eliminating cast shadows.

D.H.

281
Sub.: Vertebrata
Cl.: Aves
O.: Pelecaniformes
F.: Fregatidae

MAGNIFICENT FRIGATE BIRD *Fregata magnificens* 281

This very large New World seabird nests on the wooded or rocky shores of islands in tropical or subtropical waters, along coral reefs, and in mangrove swamps. It is about 95–110 cm long with a wingspread of 215–245 cm (among the largest of any living bird's) and weighs between 1.1 and 1.6 kg. The female is slightly larger than the male; the adult plumage appears during the fourth year of life, but the age at which the frigate bird attains sexual maturity is unknown. Average longevity is also unknown but probably fairly high. The frigate bird spends much of its time soaring over the warm seas and coastal regions at very high altitudes, and is a strong enough flier to ride out a hurricane, though it never swims or even alights on the water. It is active during the day and sociable, even gregarious, during those periods that its spends in its dormitories and nesting areas ashore.

▶ The frigate bird feeds on flying fish, gray mullet, herrings and sardines, mackerel, menhaden, and other pelagic fish, as well as squid and cuttlefish, jellyfish, macroplankton, and young sea turtles. When it sights a potential prey animal, often from a very high altitude, it goes into a vertical or steeply pitched dive and intercepts the prey on the surface of the water or on the shore, seizing it with a swift lateral motion of its long hooked beak. It catches flying fish either just below the surface or in the air, and like the jaeger and the skua, it will sometimes harass a more industrious bird—usually a booby or noddy tern—by circling and making feint attacks until the victim disgorges its catch of fresh fish, which the frigate bird summarily devours. The frigate will also feed on carrion and dead fish, and since it is not at all shy of humans, sometimes clusters around fishing boats, wharves, and even slaughterhouses to feed on offal and other organic refuse.

◀ The parents watch over their single chick very carefully to keep it safe from predators, including other members of the frigate colony. The adults themselves have little to fear from predators or from competing species, for that matter, since food resources are abundant and the frigate is unrivalled as a hunter and aviator in the biotopes that it frequents. As noted, the frigate's relations with man are familiar and on essentially the same footing as the seagull's.

There are 5 species of frigate bird, all belonging to the genus *Fregata*.

J. F.-T.

MANED WOLF *Chrysocyon brachyurus* 282

282
Sub.: Vertebrata
C.: Mammalia
O.: Carnivora
F.: Canidae

This South American carnivore is about the same size as the Eurasian or North American wolf, though otherwise rather foxlike in appearance, with its pointed muzzle and long pointed ears. It is about 1 m in length, with its tail accounting for an additional 40 cm, and its weight varies between 20 and 25 kg. Its pelt is reddish-brown, with black feet and a white patch at the throat, and its back is covered by a thick black mane. Its long legs constitute a striking adaptation to life in the tall grass, and it is found exclusively on the savannahs of southern Brazil (Mato Grosso), eastern Bolivia, and eastern Paraguay (Gran Chaco).

▶ The maned wolf feeds on the cavy *(Cavia aperea tschudii*, the wild ancestor of the guinea pig) and other rodents, birds, frogs, lizards, and even on grasshoppers. Unlike most other canids, the maned wolf does not run down its prey in packs; it is a solitary hunter (the male will not even tolerate the presence on its hunting territory of its mate or its cubs, outside the mating season and after the cubs are weaned) roaming through the tall grass at night, sniffing along the ground, listening closely for the noises made by a bird or a small mammal.

Once the quarry has been located, the maned wolf stops in its tracks before it pounces; the prey animal rarely escapes, and the maned wolf dispatches it by holding it down with its front paws and biting deeply into its vital organs. A bird or other small creature will be snapped up on the spot, feathers and all; a larger prey animal will be carried off to a sheltered spot, where the maned wolf devours it while sitting up alertly on its haunches.

◀ The adult maned wolf has no natural enemies or competitors, other than man. Though protected by law, its range has been considerably reduced in recent years as large tracts of Brazilian grasslands have been opened to stock-raising and cultivation; the maned wolf is too wary to remain for very long in the vicinity of human settlements.

P.A.

MANTIS SHRIMP *Squilla mantis* 283

283
Sub.: Arthropoda
Cl.: Crustacea
O.: Stomatopoda
F.: Squillidae

The mantis shrimp is a medium-sized stomatopod crustacean (the true shrimp are decapods); the female is about 19.5 cm long whereas the male never exceeds 17 cm. This species is found at a depth of 5–50 m in the Atlantic or the Mediterranean; it digs its own burrow in the mud or muddy sand of the bottom. A single female sometimes lays more than 50,000 eggs, which hatch out some 7–9 weeks later.

▶ The planktonic larva of the mantis shrimp, referred to as an alima, is a very attractive little creature that feeds on zooplankton. The full-grown mantis shrimp is armed with two formidable claws (the second of its five pairs of maxillipeds), the distal segment of which is equipped with six toothlike projections that can be partially sheathed in a notch in the second segment, like one of the attachments on a Swiss army knife. The mantis can immediately unsheath these razor-sharp claws in order to seize or slash at an approaching prey animal. It preys on a variety of small crustaceans that live in the muddy bottom (alpheid shrimp, crabs and hermit crabs) as well as annelids, small fish, and other soft-bodied organisms; the mantis also seeks out and devours small thin-shelled bivalves.

The mantis shrimp prefers to hunt at dusk or at night, detecting its prey both by sight and with the aid of chemoreceptors located on its antennules, which are effective over a greater distance. Once the prey has been captured, the mantis shrimp carries it back to its burrow, where it is minced up into little pieces by the mantis' mouthparts before being ingested. The female incubates its eggs in its mouth, so it is unable to eat anything at all during this period.

◀ The mantis shrimp's slashing claws appear to serve as an effective deterrent against attack; other stomatopods, which are very numerous in certain biotopes, are its principal predators. Cannibalism appears to exist among mantis shrimp, but it does not play a very important role, and territorial disputes between individuals are rarely fatal, ending with the discomfiture of the weaker party. In the south of France, the mantis shrimp (where it is known by the dialect name *galère)* sometimes serves an an ingredient in bouillabaisse or other seafood dishes, though in general the stomatopods are not consumed by man in any great quantities.

About 300 stomatopod species are found in the warm and temperate seas of the world (see, for example, the SMASHER SHRIMP). Most of them are rarely seen, living in burrows and crevices in the sea bottom, and little is known of their habits. *Squilla empusa,* found of the shores of the southeastern United States, is an important predator of true shrimp living in shallow water, and *Harpiosquilla harpax,* found in the Indian Ocean, is noted for its ability to catch fish as long as 11 cm and to devour them in less than 4 minutes. P.N.

MARINE NEMATODES *Adenophora* 284

284
Sub.: Nemathelminthes
Cl.: Nematodea
S.cl.: Adenophora

Nematode worms are generally thought of as parasites, and a great many species are parasitic not only on vertebrates but on invertebrates and plants as well. The subclass Adenophora consists of about 450 genera and 4000 known species of free-living, predominantly marine nematodes that are found on all types of substrates and throughout the shallow seas of the world. In banks of sediment along the shoreline, population densities range from several thousands to many tens of millions per sq m.

This biomass typically amounts to several grams per sq m, and adenophorans almost invariably constitute the dominant group (50–95 percent) of the meiofauna (organisms measuring <2 mm in length) that live among the grains of sand or in the moisture in the sedimentary layer on the bottom. Because their rate of biological productivity (their ability to convert nutrients into their own tissues) is so high, the annual increase in this biomass may be as much as five times that of the benthic macrofauna (organisms measuring >2 mm in length).

▶ The role played by these free-living nematodes in the ecology of the benthic region is not all that well understood. They are found on the surface of submarine rock faces, on clumps of seaweed, and most frequently in the sand and sediment on the bottom. As many as 100–200 different species may be present in a single liter of sand, and the diet and feeding habits even of those species that share the same biotope may be fairly diverse. Based on a study of the teeth and mouth capsules of these adenophorans, four different types of feeding behavior were identified.

The group designated 1A consists of nematodes with small mouth capsules that feed selectively on discrete particles (particularly certain species of bacteria) in the sedimentary layer. The 1B group browses on the thin film of sediment that colllects on the surface of sand grains; the 2A group feeds by nonselectively ingesting larger quantities of sediment, and the 2B group are all predators that feed actively on living prey or omnivores that also feed on large sedimentary particles. The relative proportions of these groups in a given sampling vary in response not only to the organic richness of the habitat but also to the size and shape of the individual grains of sand.

Predatory nematodes with well developed teeth may attack from ambush or actively pursue their prey among the grains of sand. The prey may be swallowed whole (Sphaerolainidae) or minced up (Enoplidae, Halichoanolamidae) before it is swallowed, and since only prey animals with nonperishable body parts can be reliably identified through an analysis of stomach contents, the presence of copulatory spicules may suggest that they feed primarily on other nematodes, but the extent to which these species also feed on soft-bodied protozoans has not been ascertained. Other species feed by selectively siphoning water from the surface of rocks and from in between the particles of sediment on the bottom and straining out the bacteria, and some species even practice a kind of bacterial husbandry by secreting mucus that serves as an excellent culture medium.

◀ It has been suggested that the meiofauna, and the free-living nematodes in particular, constitute a kind of ecological dead end as far as the benthic foodchain is concerned. In spite of their almost universal prevalance on the seabed, there is still some disagreement among specialists as to what extent they constitute a source of food for other marine organisms. As far as the micro-and meiofauna are concerned, *Protohydra* (a marine hydrozoan), numerous species of turbellarians (a class of primitive flatworm), and immature ploychaetes all forage actively for nematodes in friable substrates.

Among the macrofauna of the seabed, holothurians (sea cucumbers), sipunculids (a phylum of unsegmented worms), and polychaetes *(Nereis diversicolor, Abarenicola, Nephthys, Pontoporeia)* all feed by straining out relatively large quantities of sediment and as a result are likely to ingest substantial numbers of nematodes as well, since between several hundred to several tens of thousands are likely to be contained in every liter of sediment. Nevertheless, nematodes never comprise more than 20 percent of the stomach contents of these species.

Other organisms that feed selectively on suspended particles (such as Terebellidae, a family of polychaetes, and the bivalve *Scrobicularius)* rarely consume free-living nematodes, which are largely confined to the denser layers of sediment. Mysidaceans and the common shrimp *(Crangon crangon),* which have similar feeding habits, do appear to feed on nematodes in nature, and common shrimp can be maintained in the aquarium on an exclusive diet of nematodes, though they will not grow as large as those that are provided with a more balanced diet.

The larvae of flatfish and gobidae also feed actively on the benthic meiofauna, and free-living terrestrial nematodes have also been successfully used as fish food in the aquarium. *Nemachromadora* is a predatory nematode that, depending on the height of the tide, sometimes assemble in concentrations of several hundred per sq m on the intertidal mudflats of the Bay of Morlaix in Brittany, where they are the dominant species of the meiofauna; as many 250 of these nematodes have been recovered from the stomach of a small goby *(Pomatochristus minutus),* which does not exceed 8 cm in length.

Shorebirds and other fish such as the mullet that feed less selectively on sediment also consume an unknown but probably undoubtedly quantity of nematodes. Earlier observations may have prompted the conclusion that the vast numbers of free-living nematodes found on the seabed are, in effect, underutilized as a food source, though to some extent these findings may merely reflect the difficulties involved in identifying the fragmentary remains of soft-bodied nematodes in the digestive tubes of their invertebrate predators. G.B.

MASON WASP *Odynerus parietum* 285

285
Sub.: Arthropoda
S.: Insecta
O.: Hymenoptera
F.: Vespoidea

Odynerus parietum is one of a number solitary wasp species, found in both the Old and New World, that are commonly known as mason wasps. It is average-sized for a wasp, about 9–12 mm long, with a bullet shaped abdomen (blunt-ended in front and pointed in back) circled with yellow stripes; the rest of its body is black with spots of deeper black. *Odynerus* builds its nest of dried mud in a cavity in a wall; the nest consists of several adjoining cells and is surmounted by a chimneylike vestibule. Sixty *Odynerus* species are found in Europe (300 worldwide), of which *Odynerus parietum* is both the most common and the least particular in its choice of nesting site, provided that the cavity in question is of the appropriate depth and diameter.

▶ The adult wasp feeds on the nectar of a number of different plants, especially herbs of the parsley family (Umbelliferae) and figwort *(Schrophularia)*. The female begins to construct its nest in June; it begins by spitting a drop of liquid on the ground, probably a mixture of water and saliva. It forms a ball of mud with its mandibles and kneads it until it becomes pliable, then carries it back to its nesting site and slowly builds up the walls of the egg cell, leaving only a circular entrance hatch for each cell. Before the cell is stocked with provisions, the female attaches a single egg, about 3 mm long, to the ceiling by means of a short silken filament so that the egg hangs suspended just above the floor of the cell.

The genus name *Odynerus* is derived from a Greek word meaning "painful," but of course one of the principal functions of *Odynerus*'s stinger is to anesthetize the caterpillars on which the wasp larva will feed. The female wasp seizes the caterpillars with its forelegs and inserts its stinger into the central ganglia to prevent them from struggling; 8–12 caterpillars are deposited in each cell before the entrance is sealed up. The egg hatches out 4 days later; the larva is yellowish in color and initially finds itself heanging head downward from the ceiling of the egg cell, its body still enclosed inside the egg case.

Without moving from this position, it extends the forepart of its body and begins to grope along the floor in search of a caterpillar; only after it has attached its mouthparts to the body of one of the caterpillars (which are still totally immobilized but very much alive) does it allow itself to drop down to the floor. The larva feeds on the bodily fluids of the caterpillars for the next three weeks or so, then begins to spin its cocoon, which is made of brownish silk and adheres to the inner walls of the egg cell. The larva remains in its cocoon until the following May, when it completes its transformation into a nymph and then an adult wasp.

◀ On occasion, a cuckoo wasp *(Chrysis ignita, Chrysis micans)* or a sarcophagid *(Amobia odyneri)* succeeds in depositing its egg in the mason wasp's nest before the egg chamber is sealed off, with the result that the mason wasp larva is devoured and its store of caterpillars appropriated by the larva of the parasite. R.C.

MATAMATA *Chelus fimbriatus* 286

286
Sub.: Vertebrata
Cl.: Reptilia
O.: Chelonia
S.o.: Pleurodira
Fam.: Chelyidae

The matamata turtle is found in swamps and sluggish streams in the Amazon Basin, ranging as far as the Mato Grosso to the south and Colombia, Venezuela, the Guianas, and the Brazilian state of Amapa to the north and northeast. Its carapace (upper shell) is 32 cm long and 27 cm wide, on the average, with a recorded maximum length of 53 cm and a recorded maximum width of 36 cm. The female lays its eggs in a thick layer of vegetable debris on the bank. Longevity of this species in nature is unknown.

▶ This species is sufficiently rare and its habitat sufficiently inaccessible that its feeding habits in nature have never been observed. In captivity, it prefers to feed on small fish and amphibian larvae and will also accept bits of red meat; it most probably preys on fish that are sick, injured, or otherwise impaired in their movements, as well as on other small aquatic creatures. The matamata hunts by sight and probably by sensing vibrations in the water. In an aquarium, it hunts from ambush, with the plastron (lower shell) resting on the bottom; the carapace, covered with bumps and indentations and encrusted with algae, is very difficult to distinguish from a flat mossy rock.

While this vigil continues, the matamata slowly stretches out its neck until its tubular beak protrudes above the surface. When it sees a fish approaching, the matamata suddenly opens its jaws very wide, forming a sort of funnel into which the helpless fish is drawn by aspiration as the matamata starts swallowing water in great gulps, then immediately expels the water along with any mud it might have swallowed along with it. The lower jaw is poorly developed, consisting of little more than a curved rod, so that the matatmata is unable to grind up its food between the two halves of its beak in the way many turtles do.

The naturalist Broyer observed in 1862 that the matamata "is the dirty color of the mud where it lies in wait for its prey" and "it surpasses the serpent and the caiman in malevolence." Related species reflexively extend their necks in order to intercept an approaching prey animal, but this behavior is less frequently observed in the matamata. Gadow has suggested that the flattened reddish appendages on its neck may serve as lures to attract fish, but Bourlière maintains that this function is served instead by the grainy skin that covers the neck and parts of the head as well as the cloak of algae atop the carapace.

◀ The black caiman, which may attain a length of 5 m and is known to feed avidly on turtles of other species, is probably the chief predator of the matamata; young caimans probably also compete for some of the same food resources. The jaguar, the superpredator of the Guianas and the Amazon Basin, may also occasionally prey on the matamata as it searches for prey on marshy ground or comes down to drink or catch fish in a small watercourse.

This species is not exploited by man, except for occasional specimens that are collected for the benefit of terrarium enthusiasts—the matamata has acquired renown as an "exotic pet" on account of its bizarre appearance and feeding habits. This traffic is not numerically or economically significant, but since a rare species is involved, it might be advisable for it to be made subject to the provisions of the Washington Convention.

The family Chelyidae comprises 8 genera and 23 species. *Chelus, Hydromedusa, Phrynops,* and *Platemys* are found in South America, *Pseudoemydura, Chelodina, Elseya,* and *Emydura* in Australia and New Guinea. *Phrynops nasutus* ingests its prey by aspiration, like the matamata. The head of *Platemys platycephala* is triangular in shape, orangish-brown in color, and looks very much like a fallen leaf when the turtle submerges itself in the waters of a marsh with just its head protruding; this camouflage is probably of some value in evading predators as well as in attracting prey.

J.F.

MEDITERRANEAN SHEARWATER
Calonectris diomedea 287

287
Sub.: Vertebrata
Cl.: Aves
O.: Procellariformes
F.: Procellariidae

The Mediterranean shearwater measures 50 cm in length, with a wingspread of about 1 m, and weighs about 500 g. Shearwaters are good swimmers and divers, though clumsy in their movements on land. They live on the open sea except during the breeding season, when they build their nests on an island cliff or on the sandy shores of an offshore islet. This species is found throughout the Mediterranean, from Spain to Asia Minor, and is particularly numerous on Corsica and Malta. The individual record for longevity (29 years) is held by a related species, the Manx shearwater *(Puffinus puffinus).*

▶ The Mediterranean shearwater swoops and glides over the troughs and crests of the waves, occasionally alighting on the water to search for prey; it feeds on small fish and cephalopods that swim close to the surface, sometimes accompanying shoals of porpoises on their fishing expeditions, and occasionally feeding on the debris of fishing boats and other vessels.

◀ The nests, nestlings, and adult birds are preyed on by rats and feral cats, foxes, and other carnivores on the nesting sites, and the fledglings are sometimes taken by gulls when they first try out their wings. Thus, two of the Mediterranean shearwater's terrestrial predators are likely to have been introduced by man onto the small islands and other inaccessible areas it favors as a nesting site; individual nesting colonies have occasionally been disturbed by the development of tourism and other recreational activities along the seashore.

There are about 15 shearwater species altogether. The Manx shearwater *(P. puffinus)* is very common in the Eastern Atlantic, and a subspecies, the yelkouan shearwater *(P. puffinus yelkouan)* nests throughout the Western Mediterranean as well as in the Canaries. The sooty shearwater *(P. griseus),* which follows a clockwise migratory path throughout the North and South Atlantic and is also found in the Indian Ocean, sometimes appears along the coasts of Western Europe during the summer. Cory's shearwater *(Procellaria diomedea),* which nests in the Canaries, also makes a complete circuit of the Atlantic in the course of the year, skirting the coasts of Iceland as well as South America. The nesting sites of these last two species have been systematically exploited by man, especially in the North Atlantic; as with the fulmars and other gregarious seabirds, the bodies of the adults are (or were) salted down, the eggs consumed locally, and the plump bodies of the nestlings rendered to make lamp oil. The shearwaters of the Southern Hemisphere are similarly impressive in their migrations; the great shearwater *(P. gravis)*, for example, nests on the island of Tristan da Cunha from November to April, and sometimes makes its way up to Newfoundland or even Baffin Bay by August before turning south again. J.-P. R.

288
Sub.: Arthropoda
Cl.: Insecta
O.: Hymenoptera
S.o.: Aculeata
F.: Methocidae

Methoca ichneumonides 288

This red-and-black wasp measures between 1 and 1.5 cm; the female is wingless.

▶ Several studies have confirmed that *M. ichneumonides* is one of the few insect predators that dares to attack the larva of the tiger beetle (Cincindelae), itself a formidable predator, in its lair. Probably after reconnoitering the mouth of the beetle larva's tunnel to make sure it is occupied, the wasp circles the mouth of the tunnel, sometimes actually stepping on the larva's head as it searches for an opportunity to insert its abdomen into the mouth of the tunnel and inject its venom into the neck (or thorax) of the larva.

The larva occasionally strikes preemptively, however, thrusting up its head and attempting to seize the wasp with its mandibles. The larva's outsize pincers harmlessly encircle the wasp's slender pedicel ("waist") without crushing it or even especially hampering it in movements, and some observers have concluded that the wasp's initial approach is actually designed to provoke this reaction. The wasp wriggles in the beetle larva's grasp until it has positioned the tip of its abdomen over the larva's neck or thorax and then injects its venom.

Some observers have reported that the effect on the beetle larva is drastic and instantaneous; the larva releases the wasp immediately and slumps down to the bottom of its tunnel, totally paralyzed though not killed—and in such cases we may assume that the larva was stung directly in the neck. Others have observed the larva writhing and struggling violently for as long as 10 minutes, presumably because the wasp has stung it in the thorax and the venom has not taken effect as quickly.

In either case, the wasp waits until the larva no longer shows any signs of life before crawling slowly down to the bottom of the tunnel, where it stings the larva several more times and, as a further precaution, gnaws on the strip of tissue between the larva's legs, which is where the locomotor ganglia are situated. When all this has been accomplished, the wasp attaches its eggs to the body of the helpless larva and crawls back out of the tunnel, carefully sealing the entrance behind it. The wasp will sometimes choose to devour its victim on the spot rather than leaving it for its posterity, and, for reasons that remain mysterious, the wasp also sometimes makes its way into the larva's tunnel and stays there overnight, postponing its attack until the following day.

◀ Predators of this species may exist, but they have yet to be identified.

G.B.

289
Sub.: Vertebrata
Cl.: Osteichthyes
S.cl.: Teleostei
O.: Cypriniformes
F.: Characidae

MEXICAN BLIND CAVEFISH
Anoptichthys jordani 289

This species measures about 8 cm and exists only in a small number of limestone caves and underground streams in central Mexico, where the mineral content of the water is very high and the water temperature is a constant 26–27°C. The blind cavefish is descended from a genus of tropical characids *(Astyanax)*, and appears in some classifications as *Astyanax mexicanus;* it seems to have been isolated in its present habitat as the result of a series of severe earthquakes.

As a result, its eyes have completely atrophied, and the lateral line detectors, which are responsive to changes in water pressure as well as vibrations in the water, have greatly increased in senstivity. Since *Anoptichthys* has evolved in an environment from which predators are completely absent, it can afford to be more conspicuous in its movements than surface-dwelling fish. It navigates by echolocation: the wave patterns generated by the vigorous undulations of its fins and body bounce off the walls of the cave and other submerged obstacles and are reflected back, so that the relative location of these obstructions are registered by *Anoptichthys's* lateral line detectors.

▶ *Anoptichthys* continually forages along

the bottom in search of insects and small cave shrimp, which are quite abundant in this particular habitat. *Anoptichthys's* sense of smell and touch have also become quite acute, and once it has sniffed out the approximate location of a prey animal, it continues to explore the area until it actually brushes up against the prey, then makes an abrupt sideways lunge and catches it in its jaws.

◀ Because of its exotic appearance and energetic behavior (and in spite of its propensity for inflicting painful bites on humans), the Mexican blind cavefish has become a popular aquarium fish throughout the world.

About 30 species of fish live exclusively in caves, sinkholes, underground lakes, and similar lightless habitats, and have developed heightened senses of touch and smell (as well as hearing) to compensate for their loss of sight. To cite just two examples: *Typhlogarra widowzoni* lives in underground streams in Iraq, and *Caecobarbus geertsi* is a representative African cave fish. J.-Y. S.

MIMETIC ROBBER FLY *Hyperechia bomboides* 290

290
Cl.: Arthropoda
S.cl.: Insecta
S.o.: Diptera
F.: Asilidae

This large fly is deep violet in color, its thorax bristles with fine yellow hairs, and its black wings, when not in use, lie flat and partially overlapping, like a pair of scissors. Its legs are also very black and shaggy, terminating in 2 claws and 2 powerful suction pads. The sexes are very similar in appearance; *Hyperechia bomboides* is found in Senegal, West Africa. The name *mimetic robber fly* refers to the fact that *Hyperechia* also exhibits a strong resemblance to 2 related species of carpenter bees *(Xylocopa pubescens* and *olivacea)*, both the larva and adult of which are preyed on by the robber fly larva and adult, respectively. The ectoparasitic robber fly larva, which measures up to 30 mm when fully developed, has no clearly differentiated head or legs; its body is divided into 11 segments with a pronounced constriction between each segment.

▶ The adult robber fly stations itself on an elevated observation post and awaits its prey, swiveling its head around with great agility, somewhat in the manner of a praying mantis, for a panoramic view of its surroundings. It catches a female carpenter bee on the wing by seizing it between its forelegs, then pierces the tegument of its victim with its sclerotized rostrum (beak) and injects its poisonous saliva; the bee succumbs almost immediately and is drained of its body fluids by the robber fly.

It is possible that the robber fly's mimetic resemblance to the carpenter bee has developed as an aid to predation, but it seems more probable that the primary, perhaps the sole, purpose of this deception is to permit the female robber fly to deposit its eggs in the hollow plant stalks in which the carpenter bee constructs its larval cells. The female robber fly deposits its eggs either directly in or adjacent to the nest of the carpenter bee, and since the robber fly is particularly vulnerable to attack at this moment, it might easily be driven away by the female carpenter bee if not perceived as a member of its own species.

The robber fly larva is equipped with barbed mouth parts that slash the skin of the bee larva, whereupon the robber fly larva ingests the fluid that flows from the gash; a single robber fly larva may eventually devour a number of bee larvae, depending on their size, before its development is complete. The robber fly larva spins its nymphal cocoon in a separate part of the stalk that is not occupied by the bee, scraping away all but the thinnest layer of bark in one small area to provide itself with an escape hatch. The nymph has spines on its head that can easily tear through this fragile barrier while at the same time it bears down on the set of spiny bristles (spicules) on its abdomen and wriggles its body in order to free itself both from the cocoon and the nymphal body case and emerge as a fully formed adult.

There are a number of related robber fly species, all found in Africa. R.C.

MINUTE PIRATE BUG *Orius vicinus* 291

291
Sub.: Arthropoda
Cl.: Insecta
S.cl.: Pterygota
O.: Hemiptera
F.: Anthocoridae

Orius vicinus is a small bug, between 2.1 and 2.5 mm long, with an oval, slightly elongated body; the head and thorax are blackish, the hemelytra (anterior wings) are light brownish-yellow, sometimes darker at the tips. The antennae are fairly long, and the rostrum (beak) consists of three jointed segments; its three pairs of legs, golden-yellowish in color, are functional walking legs. Two or three generations of pirate bugs are born in single year, the last of which spends part of its adult stage in hibernation, generally underneath the bark of a tree. The larvae are morphologically similar to the adults and undergo 5 molts before their development is complete; the wings begin to appear, in rudimentary form, during the final stages.

Orius vicinus is found throughout most of Europe and Central Asia. It is quite common in Western Europe, and both the larvae and the adults may be found on the leaves and branches of plum tree and apple trees, though they also occur on other trees (linden, willow, oak) and ground-hugging plants such as goosefoot *(Chenopodium heracleum).*

▶ Both the adults and the larvae are active during the daylight hours, as long as the temperature remains above 15° C. The minute pirate bug is omnivorous, feeding on apple pollen and other vegetable material as well as on mites and insects, particularly aphids; laboratory studies have demonstrated that an exclusively vegetarian diet is not sufficient to ensure normal development in this species. The pirate bug is thus considered a highly desirable tenant by nurserymen, since a single individual has been shown to be capable of devouring as many as 300–600 fruit-tree red-spider mites *(Panonychus ulmi)* or a hundred aphids during the course of its larval development. Cannibalism is fairly prevalent when mites or aphids are unavailable.

Like most hemipterans (true bugs), the minute pirate bug ingests the non-chitinous body parts of its prey by the process of external digestion. As soon as it locates an aphid or other prey animal, it inserts its rostrum into the aphid's body; the aphid is almost immediately paralyzed by the pirate bug's poisonous saliva, and the liquefied contents of the aphid's body case will be completely consumed by the pirate bug during the next several minutes.

◀ The predators of the minute pirate bug include beetles and syrphus fly larvae as well as other hemipterans.

The family Anthocoridae comprises about 450 species worldwide; there are numerous *Orius* species, all of which are predatory, though some, like *Orius vicinus*, feed on vegetable material as well. J.-L. D.

MOLE CRICKET *Gryllotalpa gryllotalpa* 292

292
Sub.: Arthropoda
Cl.: Insecta
O.: Orthoptera
S.o.: Ensifera
F.: Grillotalpidae

This is the only European representative of the large family of mole crickets (Gryllotalpidae), comprising 45 species worldwide. *Gryllotalpa* is a burrower that lives in flowerbeds and garden plots where the soil is light, sandy, and slightly moist. It is about 50 mm long and its front legs have been transformed into spadelike appendages, somewhat reminiscent of a the highly developed forepaws of its namesake, the mole. The elytra are short and the wings rather long; the thorax is encased in a rounded shieldlike plate and covered with fine hairs.

The adults emerge in May or June, go into hibernation in the fall, and breed the following summer; the eggs hatch out 3 weeks later, toward the end of July. The adults live only for about a year, but the larvae develop slowly, requiring 2 or 2 1/2 years to complete their development; shortly after hatching, they band together in clusters, then go their separate ways after the first few weeks.

▶ The larval and adult *Gryllotallpa* feeds primarily on the larvae of click beetles, noctua moths, and other burrowing grubs as well as on adult beetles and earthworms, which it catches in its mandibles while tunneling through the soil. The mole cricket is considered an agricultural pest, though only in a minor way, since it damages the roots of plants when it excavates its shallow underground galleries.

◀ The mole and the hoopoe, which feeds large insects to its young, are the chief natural enemies of the mole cricket. In the southern part of its range, *Grylloptalpa* may also be parasitized by a species of sphecid wasp, *Larra anathema.* R.C.

MONK SEAL *Monachus monachus* 293

293
Sub.: Vertebrata
Cl.: Mammalia
O.: Pinnipedia
F.: Phocidae

The adult male typically measures 2.4 m and weighs 260 kg; the female measures 2.6 m and weighs 300 kg. The monk seal's coat is a smooth dark-brown (thus thought to resemble the robe of a Franciscan friar) with a yellowish patch on its underside. The pups are born throughout the year, though most frequently in summer, and already weigh between 17 and 24 kg at birth. The monk seal reaches sexual maturity at the age of 4; its average life expectancy is unknown. Since the monk seal is wary of humans and extremely rare, a great deal remains to be discovered about its habits and population dynamics. The monk seal has been sighted up to 30 km out to sea, but is largely sedentary and does not undertake any notable migrations. It is found along rocky coastlines, particularly limestone cliffs that are honeycombed with caves and submarine grottoes inaccessible to humans. It is suspected that in former times this species may have been less security-conscious in its choice of habitat.

At one time this species was found throughout the Black Sea and the Mediterranean, as well as along the Atlantic coast of North Africa, though today it has virtually disappeared from the coasts of France, Italy, Yugoslavia, Spain, Egypt, Israel, Rumania, and the Soviet Union (among others). Of 17 sites along the Corsican coast frequented by monk seals in 1955, all but 4 had been abandoned by 1973. In 1978, the total population was estimated at no more than 500–600 individuals, and the current population of the largest single colony, on Cap Blanc in Mauritania, is estimated at only 55–100.

▶ The monk seal is active during the day and seeks its prey in shallow water. Inventories of stomach contents suggest that it feeds primarily on eels, sardines, tuna, spiny lobsters, octopus, numerous flatfish, red mullet, yellow mullet, and bogue *(Boops boops)*, another member of the mullet family. The monk seal also browses on seaweed, but nothing else is known of its feeding habits in the wild. A captive female weighing 120 kg will eat up to 10 percent of body weight every day.

◀ The monk seal is the only pinniped found in the Mediterranean; it has no predators or competitors other than man, and enjoys legal protection throughout its range. Though a few individuals may still be killed wantonly or inadvertently (e.g., slashed by powerboat propellors), the principal threat to the survival of this species comes from the pollution of the Mediterranean and the disappearance of suitable breeding grounds, since the seals will abandon any site that is even occasionally frequented by human beings. There is even some concern that the Cap Blanc colony, which seems to represent the only hope for the long-term survival of this species in the wild, may be endangered in the future if commercial fishing is intensified off the coasts of Mauritania and the Spanish Sahara.

The West Indian monk seal became extinct during the early part of this century; the Hawaiian monk seal, reduced to a total population of 150 individuals by 1955, has been saved from a similar fate by the exercise of intelligent and intensive conservation measures; the current population is estimated at around 1350. M.P.

MONTAGU'S HARRIER *Circus pygargus* 294

294
Sub.: Vertebrata
Cl.: Aves
O.: Accipitriformes
F.: Accipitridae

This small European hawk is 40 to 42 cm long with a wingspread of 106 to 115 cm and weighs between 250 and 420 g; it lives in many different types of treeless terrain, as long as there is some sort of thick, low-lying vegetation—gorse, heather, briar, marsh, farmland overgrown with weeds, or wet meadow. It has adapted in some areas to similar manmade environments—fields of grain, alfalfa, and stands of young conifers. Montagu's harrier is found from Morocco to eastern Siberia during the spring and summer and spends the colder months in Central or (especially) East Africa. It reaches maturity at 2 or 3 and sometimes survives to the age of 16.

Montagu's harrier is an agile and graceful flier, a conspicuous aerial presence over its territory of several square kilometers around its nest, which is hidden amid dense vegetation on the ground. It sometimes lives in loose-knit colonies consisting of several breeding pairs that build their nests a few tens of meters from one another.

▶ This species is strong enough to remain airborne for long periods of time, enabling it to conduct its hunting sweeps over a vast expanse of territory. Its long legs and keen sense of hearing help it to locate and seize its prey in the tall grass and dense foliage; it feeds primarily on small mammals and ground-dwelling birds—shrews, mole rats (family Spalacidae), hamsters, larks and crested larks, and pipits. It occasionally raids the nests of other birds, especially larks and partridges; insects, especially grasshoppers and locusts, form an important part of its diet in winter. It occasionally eats earthworms, frogs, toads, and snakes and, very rarely, baby hares or rabbits. It captures about 5 to 10 prey animals a day, or as many as 20 to feed a brood of 5 nestlings. The weight of its individual victims varies from as little as 1.5 g to as much as 250.

The parents teach their young to hunt by flying above them and dropping a prey animal for them to catch at progressively greater distances; the transfer of freshly caught prey from the male to the female on the nest takes place in a similarly aerobatic manner.

◀ Adult harriers are zealous in the defense of their nestlings and will drive away any herons, seagulls, crows, or other raptors that appear in their nesting area; there is not much they can do about foxes and mustellids (members of the weasel family). The weakest member of a harrier brood is often devoured by its brothers and sisters if not by predators. In addition, harriers' nests in cultivated fields or pastures are frequently destroyed by tractors or trampled by domestic animals. Adults are sometimes, though rarely, taken by peregrine falcons or other raptors.

Montagu's harrier can coexist quite comfortably with its close relatives the hen harrier *(Circus cyaneus)*, sometimes called St. Martin's bird, and the marsh harrier *(C. aeruginosus)*. Its agility and the extraordinary keenness of its hearing permit it to specialize in prey animals that are too quick or too skillfully concealed for these other harrier species. The principal European "mousing hawks," the common buzzard and the kestrel, prefer to hunt in terrain that affords less concealment to their prey.

It is thought that the number of Montagu's harriers that return to Western Europe in the summer, now estimated at several thousands breeding pairs, has been reduced by about 50 percent in recent years. This is largely a result of the disappearance of swamps, marshes, and large tracts of wasteland, as well as its persecution as a "harmful" bird of prey since earliest times (as suggested by the name *hen harrier* and the word *harrier* itself, derived from the verb "to harrow," meaning "raid, rob, plunder," etc.—presumably the henroost in this case).

The marsh harrier is a larger bird, now almost extinct in Great Britain, that lives exclusively in marshy areas and reedbeds. The nonmigratory hen harrier is otherwise very similar to Montagu's harrier in appearance,habitat,andbehavior.

J.-F. T.

MONTPELLIER SNAKE *Malpolon monspessulanus* 295

295
Sub.: Vertebrata
Cl.: Reptilia
O.: Squamata
S.o.: Ophidia
F.: Colubridae

The Montpellier snake may attain a length of 2.4 m, and is thus one of the largest of European snakes. It is found at many different points along the Mediterranean coastline, including southwestern France and Asia Minor, up to an altitude of 2000 m. It is a terrestrial, primarily diurnal species that prefers rocky terrain or brushland well exposed to the sun. Like most other reptiles of the temperate zones, it spends the winter in hibernation, curled up inside the abandoned burrow of a rodent or other creature, an old stone wall or a pile of rocks, or the base of a rotten stump. The Montpellier snake is greenish or olive-green in color and may be fairly readily recognized by the 2 lateral ridges (each of which is called a *canthus rostralis)* running from below the eye to the snout. Its head is massive and elongated; its vision is quite acute, and its eyes are somewhat larger than those of other snakes.

▶ The Montpellier snake is one of 3 European colubrids equipped with venomous fangs. However, its dentition is of the opisthoglyphic ("back-channeled") type, which is to say that the fangs are mounted toward the rear of the upper jaw so the venom can be injected into a prey animal that has already been partially engorged; the venom appears to be sufficiently potent to cause death in several minutes in the lizards and other small creatures that it preys on.

The Montpellier snake pursues and brings down its prey in typical colubrid fashion. A smaller prey animal will be swallowed immediately and thus automatically injected with venom (which after all is only a sort of superactivated saliva) before it begins its passage down the snake's gullet. In the case of a larger prey animal, the snake may arrest the process at this stage and wait several minutes for the venom to take effect before it attempts to ingest the prey completely.

As noted, about two thirds of the diet of the Montepellier snake consists of lizards of various sizes, including the green lizard, the wall lizard, the sand lizard *(Lacerta agilis)*, and the jeweled lacerta *(L. lepida)*. Rats, fieldmice, voles, and other small mammals (up to the size of a young rabbit) and nestlings, particularly of ground-dwelling species, may provide an additional 28–30%. The Montepellier snake is also a bit of an ophidophage, which is to say that it sometimes preys on other snakes, including in this case other Montpellier snakes. The young of this species feed primarily on beetles and young lizards.

◀ The predators of the Montpellier snake include the short-toed eagle and several species of snake-eating buzzards and harriers as well as domestic pigs and wild boar. In the south of France, this species has traditionally been consumed by man under the name "hedge eel" *(anguille de haie)*—a practice that may not entirely have been discontinued. Apart from the sort of deliberate persecution suffered by all large snakes at the hands of human beings, the Montpellier snake is frequently killed by automobiles and by largescale disturbances of the habitat such as brushfires and forest fires.

When prevented from fleeing, the Montpellier snake adopts a threatening and rather cobralike posture, raising the forepart of its body off the ground while inflating and flattening its neck and hissing noisily. It is not at all reluctant to strike if any attempt is made to capture it, and though its venom has no noticeable effect on humans, its teeth are very sharp and the larger specimens especially can inflict a very painful bite.

The two other venomous colubrids of Southern Europe also have the opisthoglyphic dentition of the Montpellier snake. *Macroprotodon cucullatus* is found in the south of the Iberian peninsula and *Telescopus fallax* in the Balkans and along the shores of the Eastern Mediterranean. Both species are timid and do not generally venture out until dusk. J.C.

MORAY EEL *Muraena helerea* 296

296
Sub.: Vertebrata
Cl.: Osteichthyes
S.cl.: Teleostei
O.: Anguilliformes
F.: Muraenidae

The moray often exceeds 1 m and occasionally 1.3 m in length and weighs between 4 and 5 kg. Longevity in captivity may exceed 10 years. The moray is a benthic (bottom-dwelling) creature; more specifically, it lives in caverns and crevices in the rock about 50–80 m from the bottom. It is quite common in the Mediterranean and is also found (though with decreasing frequency at higher latitudes) along the Atlantic coast of Europe from Gibraltar to the Channel.

▶ Like most of the Anguilliformes, the moray is nocturnal, locating and tracking its prey by smell until it is close enough to get a clear view of it. Unlike most other carnivorous fish, the moray attacks with a slashing lateral motion of its jaws; with its wide mouth and its sharp, very numerous teeth, the moray can easily immobilize its prey by snapping its head forward prior to slicing it into smaller pieces with its teeth and ingesting it. The moray feeds on fish, large crustaceans, and even carrion on occasion.

◀ As mentioned elsewhere, the moray is sometimes forced to contest the ownership of its crevice with a previous occupant or an interloper, usually a lobster, an octopus, or a grouper. All of these combattants are about evenly matched, though the octopus is perhaps the most formidable opponent that this species is likely to encounter. The octopus's ink contains a substance that blocks the moray's olfactory receptors, making it more difficult for the moray to come to grips with it while both combattants are enveloped in billowing clouds of ink.

The moray's reputation for ferocity is somewhat exaggerated, though not entirely without foundation; it is not aggressive toward humans, though it may well bite if it is disturbed or attacked inside its lair. This may result in a very serious predicament for a diver, since the moray's jaws are very powerful; it hangs on as tenaciously as a bulldog once it has sunk its teeth into an adversary. In addition, it sometimes wraps its snakelike tail around an outcropping to prevent its being dislodged from its crevice, giving rise to the belief that the moray deliberately and maliciously attempts to catch hapless divers and drown them. The moray's bite can be very painful and the wound often becomes infected, though the moray's saliva is not venomous, as it sometimes alleged.

The moray was prized as a delicacy by the ancients and fattened for the table in special tanks. The emperors Tiberius and Nero were both said to have kept them as pets, and the moray's sinister reputation was enhanced by the story of a wealthy Roman named Pollio, who was reputed to have kept a tankful of truly enormous specimens that were fed on the flesh of slaves who were tossed into their tank as punishment for various infractions. Nowadays, the moray is not exploited commercially, though it often finds its way into fishermen's nets while attempting to feed on fish that have become entangled there and may do substantial damage in its efforts to free itself.

A great many muraenid species are found in tropical waters, including those belonging to the genera *Gymnothorax* and *Echidna.* Like the great barracuda (q.v.), jack *(Caranx),* and other large predatory fish (as well as some that merely browse on algae), the larger species may contain a poisonous substance called ciguetara, which is now thought to be an accumulation of toxins originally secreted by blue-green algae and greatly concentrated in the course of their passage up the foodchain. Ciguetara poisoning is sometimes fatal to humans.

Most of the Muraenidae are virtually identical to *Muraena helerea* in their habits, though *Gymnothorax pictus,* found in the Pacific, is diurnal and locates its prey by sight; it has even been reported to leave the water on occasion in pursuit of small crabs. J.-M. R.

297
Sub.: Vertebrata
Cl.: Osteichthyes
S.cl.: Teleostei
O.: Mormyriformes
F.: Mormyridae

Mormyrops boulengeri 297

This freshwater fish of tropical Africa, a distant relative of the Nile fish *(Gymnarchus;* q.v.), measures between 80 and 100 cm; its body is slender and somewhat laterally compressed, and the anal and dorsal fins are set well back on its body. It lives among the underwater vegetation in rivers and streams that are often swollen with floodwaters and heavily laden with sediment from flooded banks and marshes. Since it frequently finds itself in muddy water, *Mormyrops* has little use for its eyes, which are small and poorly developed. Instead, it orients itself with the aid of its lateral line and the electrical field that surrounds its body.

As with the electric eel, Nile fish, and other electric fish, this field is generated by specialized organs located in front of the caudal fin; *Mormyrops* has 4 such organs, each consisting of a battery of 150–200 electric "plates" (actually a modified form of muscle tissue). Underwater obstacles, approaching fish, and other creatures create disturbances in the field that are detected by sensory receptors, called mormyromasts, distributed all over the fish's body, which is kept exceptionally rigid at all times in order to maintain the symmetry and uniformity of the electric field.

▶ *Mormyrops's* electric organs generate brief, intermittent, and very weak electric impulses at a rate of about 50–100 per second; these are sufficient for purposes of navigation and prey detection but are not strong enough to stun or kill a prey animal. *Mormyrops* has a very wide mouth studded with sharp teeth and is a fairly strong swimmer, fast enough to pursue a smaller fish (never exceeding 20 cm in length), for which reasons its feeding habits are sometimes likened to those of the pike.

◀ *Mormyrops* is forewarned of the approach of predators by its electric detection system and, with its slender body, it can easily conceal itself in a clump of underwater weeds or a crevice between two stones. It is too large for most aquatic predators to take on, with the exception of the Nile fish, the crocodile, and the dogfish. The larger specimens are sometimes dried and eaten by local fisherfolk, but this by no means constitutes a commercially important fishery.

The family Mormyridae consists of over 100 species, classified into 12 different genera; these in turn can be broadly classified into 3 different categories, based on the shape of the snout and location of the mouth:

(1) Larger fish such as *Mormyrops* and

Mormyrus, the sacred fish of the Nile, which have wide mouths, terminally located, and a large number of teeth.

(2) Smaller fish such as *Marcusenius, Petrocephalus,* and *Stomatorhinus* which have shorter snouts, ventrally located mouths, and substantially fewer teeth.

(3) Smaller fish such as *Gnathonemus,* which have elongated trunklike snouts, small mouths with fewer teeth (located only in the front of the mouth). Sometimes the lower lip is prolonged into a single barbel, like one of the whiskers of a catfish, which is very thick but still flexible, well supplied with sensory organs (presumably playing a role in prey detection), and may measure as much as a tenth of total body length, as in the case of *Gnathonemus petersii,* which is 13 cm long and in which this distended lower lip—tentacle measures 13 mm.

The smaller species in categories (2) and (3) are sedentary bottom-feeders, foraging in the mud or in the crevices between rocks with their snouts or barbels; they feed on small invertebrates, which they ingest by aspiration, somewhat in the manner of the seahorses and their kin (Syngnathidae). The electric organs are invariably present, though the manner and intensity of their discharge varies with the species (and the sex of the individual, in some cases); the eyes of all the mormyrids are atrophied. F.M.

MUD POT WASP *Eumenes arbustorum* 298

298
Sub.: Arthropoda
Cl.: Insecta
S.cl.: Hymenoptera
O.: Apocrita (Petiolata)
S.o.: Aculeata
F.: Vespoidea
S.f.: Eumenidae

Eumenes means "graceful," and this solitary mason wasp has a slim, elongated body, half yellow and half black, and a petiolate abdomen with a raised band running around the middle; the abdomen of the female is 18–20 mm long, that of the male only 15 mm. The mud pot wasp is so called because it builds its nests with an earthen paste reinforced with tiny pebbles about the size of a peppercorn. The nest, about 2 cm in diameter, consists of several cells and is attached to the outer walls of a building or the surface of a boulder in a sunny spot. The mud pot wasp can often be seen approaching a spring or an outdoor tap to fill its crop with water, relatively large amounts of which are required for construction purposes, though it sometimes collects ready-mixed building materials in the form of liquid mud. This species is found primarily in Southern Europe.

▶ The adult wasp feeds from nectar-rich blossoms, such as those of the common rue and the larger Umbelliferae, and captures caterpillars only to feed its still unhatched larvae. The mud pot wasp is the only solitary wasp that lays an egg in an empty cell of its nest before the cell has been stocked with paralyzed but still living prey. Henri Fabre was the first to observe that just before it deposits its egg in the cell, the female mud pot wasp attaches a thin strand of mucus to the roof of the cell, which hardens rapidly in contact with the air. This mucus is secreted by glands adjacent to its sexual organs, and the other end of the strand is attached to the egg (as is also the case with the eggs of lacewings of the families Chrysopidae and Hemerobiidae), which emerges last of all and is thus suspended from the ceiling of the cell.

Fabre suggested that the purpose of this arrangement is protect the egg from the thrashing and wriggling of the caterpillars that will soon be occupying the floor of the cell and which, though partially paralyzed, are still able to move their legs and jaws to some extent. All the caterpillars deposited in a particular cell will all be of the same size and species, though, in general, the mud pot wasp feeds its larvae on butterfly and moth caterpillars of many different species (families Noctuidae, Lycaenidae, Pieridae). The female wasp places either 5 or 10 caterpillars in each cell, probably depending on whether the larva is male or female, since the female eventually grows to be twice the size of the male.

◀ The mud pot wasp only has two known predators, the cuckoo wasp *Stilbum calens* and the mutillid wasp *Dasylabris maura.* R.C.

MUDFISH *Amia calva* 299

299
Sub.: Vertebrata
Cl.: Osteichthyes
S.cl.: Holostei
O.: Amiiformes
F.: Amiidae

This North American freshwater fish ranges more or less between the Great Lakes and the Rio Grande. It attains an average size of about 50 cm in the male, 70 cm in the female, with a maximum length of around 90 cm. Maximum weight is around 3.6 kg. *Amia calva*, commonly called the mudfish, bowfin, or beaverfish, is primarily of interest to science as the sole survivor of 6 families of Mesozoic fish that constituted the order Amiiformes, widely distributed throughout the New and Old World during the age of the dinosaurs (Jurassic and Cretaceous periods, or about 80–160 million years ago). The life expectancy of an individual mudfish is about 24 years.

▶ A voracious predator, the mudfish feeds on worms, mollusks, aquatic insects, crayfish, other fish, and even amphibians. It actively seeks its prey and propels itself through the water with wavelike motions of its very long dorsal fin (hence its alternate common name, bowfin), swimming forward or backward with equal facility by changing the direction of these undulations. The mudfish stations itself in a clump of aquatic plants and waits for the approach of a prey animal, then emerges abruptly from ambush by means of a powerful stroke of its caudal fin (which ordinarily does not play as important a role in locomotion as it does with other species). It seizes its prey around the middle of the body, then turns it around so it can swallow it headfirst. The mudfish hunts primarily by sight and may also be able to detect vibrations in the water by means of its sensitive swimming bladder.

The hatchlings and young fry are born carnivores, feeding of course on much smaller aquatic prey. They remain under the supervision of the male mudfish, whom they ordinarily accompany in a tight formation, until they have attained a length of about 8–10 cm. They grow very quickly and measure about 15–20 cm by the end of their first year.

◀ In contrast to the young fry, the adult mudfish has only a few natural enemies—bears, bald eagles, and pumas. Its flesh is described as soggy and bad-tasting, but the mudfish is still greatly esteemed for its fighting qualities by Midwestern sports fishermen. F.M.

N

300
Sub.: Vertebrata
Cl.: Mammalia
O.: Cetacea
S.o.: Odontoceata
F.: Monodontidae

NARWHAL *Monodon monoceros* 300

The narwhal is found exclusively in the deeper waters of the Arctic Ocean and never ventures south of 70° N. lat. in the course of its seasonal migrations; its range is thus the most northerly of the ceteceans, though the narwhal's "horn" (actually the male narwhal's single 1.5–2 m long tusk) has occasionally found its way to more southerly latitudes, thus providing the ancients with convincing proof of the existence of the legendary unicorn. The narwhal measures 1.6 m at birth and weighs 80 kg; the adult male may attain a length of 4.7 m and a weight of 1600 kg, the female a length of 4 m and a weight of 900 kg. The narwhal is a gregarious species, living in pods of as many as 50 individuals; it has rarely been observed in the wild, and comparatively little is known of its habits.

▶ Our only information on the narwhal's feeding habits is based on observations made in Canadian Arctic waters, where the narwhal appears to feed primarily on squid and cuttlefish and on a variety of small to medium-sized fish, including polar cod, Greenland halibut, herring, and skates, as well as the occasional crustacean. The narwhal has no functional teeth, and its prey is grasped firmly in the animal's strong jaws before it is swallowed whole. The tusk of the male narwhal appears to be used exclusively for purposes of intraspecific combat and sexual display.

◀ The Greenland shark, killer whale, and the polar bear are the known or presumed animal predators of the narwhal; this species is also still extensively hunted by the Eskimos, who use its blubber for food and fuel, feed its flesh to their sled dogs, and sell the skulls (with tusks attached) of the males to dealers and collectors. The narwhal is still prevalent in Arctic waters but unlikely to maintain its present numbers if hunting continues at the current level of intensity.

The beluga, or white whale *(Delphinapterus leucas),* lacks the narwhal's distinctive tusk but otherwise very similar morphologically; it ranges further south than the narwhal and often frequents shoal waters and the mouths of large rivers.

V. de B.

NECKLACE SNAIL *Natica catena* 301

301
Sub.: Mollusca
Cl.: Gasteropoda
S.cl.: Prosobranchia
O.: Mesogastropoda
F.: Naticidae

This marine snail is common all along the Mediterranean and Atlantic coasts of Europe; it burrows in mud or sand below the shoreline and is often exposed by the tide in areas where the tidal coefficient is equal to 80 or more. Its shell is very similar in size and shape to that of the edible European land snail; it is about 4 cm in diameter, smooth, globular, and pinkish-beige in color with a scattering of brown spots along the identations formed by the whorls. There is also a central depression (umbilicus) at the origin of the whorls; a shell of this type is said to be "perforate." The semicircular aperture of the shell may be enitrely obscured by the ear-shaped operculum.

The only part of the snail's body that protrudes from the shell is an appendage called the cephalopodium (so called because it is formed by the conjunction of the head and the foot). The head is equipped with two small "horns," each of which has a basal eye, and in the center, a long trunklike mouth, which contains the radula and the so-called accessory boring organ (the functions of both of these organs will be described shortly). The foot is very well developed; the forepart, or propodium, can be pulled all the way back over the head and the front part of the shell when fully extended; the rear of the foot, or metapodium, can similarly be extended to cover the rear of the shell.

The foot is able to expand in this fashion by soaking up seawater, like a sponge, and the water is abruptly expelled when the snail pulls back inside its shell. The necklace snail uses the propodium to bury itself in the sand or mud (and is incapable of doing so when this organ is excised); the distinctive track that the snail leaves behind it in the sand is also made by the propodium, which, as we shall see, plays a role in prey detection and prehension as well.

▶ The necklace snail feeds primarily on burrowing lamellibranchs (members of a large class of mollusks that includes the clam and the oysters), especially *Nucula, Donax, Tellina, Mactre, Mya, Venus,* and *Tapes,* among others. The necklace snail detects its prey with the aid of sensory cells located on the propodium. The mollusk is quickly enveloped in the folds of the snail's extensible foot, and the snail begins to bore through the shell, initially with the radula alone. After a short while the radula is withdrawn and the bore hole is deepened and enlarged by the acidic secretions of the accessory boring organ over a period ranging from several minutes to almost an hour before the radula is inserted once again; in the case of an especially hard-shelled mollusk, it may take several days for the necklace snail to finish the task.

The completed bore hole varies between 1 and 4 mm in diameter, which is just large enough to allow the snail to insert its trunklike mouth into the shell and to detach the soft parts of the mollusk's body from the inner surface of the shell. This process had not been observed directly until recently, since the mollusk is entirely engulfed by the snail's extensible foot and both predator and prey are normally buried in the sand. The damage inflicted by necklace snails on a bed of edible mollusks may be considerable, as evidenced by the number of shells washed up on the shore that have been pierced by bore holes. Considerable numbers of necklace snail shells are also found in this condition, suggesting that cannibalism is not uncommmon in this species; young necklace snails are capable of boring out mollusk shells as soon as they emerge from the egg.

◀ There are numerous seabirds that are able to crack the shell of the necklace snail with their beaks, though this normally only occurs at very low water when the snail's burrows are exposed. The snail's eggs are often washed too far up the beach to be retrieved by the outgoing tide, which appears to constitute a significant environmental check on the proliferation of this species. Finally, necklace snails may also be parasitized by trematodes (a type of flatworm), and the rapid multiplication of these parasites inside the snail's body sometimes results in the parasitic castration of the host.

The Naticidae are carnivorous burrowing snails of worldwide distribution, all of which prey on other mollusks in more or less the manner described above. The moonshells of Atlantic coast of North America are somewhat larger than the European necklace snails—about the size of a human fist. *Natica nitida* is a very small European species that is often found in conjunction with *N. catena, N. josephina,* and *N. millepuncta* are found in the Mediterranean. M.C. and J.F.

302
Sub.: Vertebrata
Cl.: Osteichthyes
S.cl.: Teleostei
O.: Beloniformes
F.: Belonidae

NEEDLEFISH *Belone belone* 302

This European marine fish measures about 70 cm (maximum length, 90 cm) and weighs about a kilogram; it is somewhat similar in appearance to the tropical gar (q.v.) and related North American freshwater ganoids (thin, elongated body, dorsal and anal fins set well back toward the tail) and is also sometimes referred to as a "gar" or "garfish." It is found along the Atlantic coasts of Western Europe and Morocco as well as in the Mediterranean and the Black Sea. The Atlantic populations begin to migrate toward their spawning grounds in the North Atlantic in the spring and then return to the waters to the south of the British Isles in the fall; the migrations of the Mediterranean and Black Sea populations are naturally less extensive.

The needlefish attaches its eggs to the leaves of marine plants along the edge of the continental shelf, and the young fry remain in relatively shallow water for some time. Their jaws, initially very short, develop into a sort of long pointed beak; the lower jaw is always a little longer than the upper one. Average life expectancy is probably 3 or 4 years, though individual needlefish in the Black Sea sometimes survive to the age of 18.

▶ The young needlefish feed on copepods, and the adult, a voracious predator that resembles the nothern pike and the alligator gar in its appetites as well as its appearance, feeds on herring, sprats, sand launces, and sticklebacks, as well as on young mackerel and small crustaceans. The needlefish's elongated jaws make it easier for it to seize its prey—when a needlefish attacks a shoal of herring, for example, it picks its target with precision and generally strikes at the middle of the body—and its numerous pointed teeth also make it more difficult for the prey to escape before it has been swallowed whole by its captor. The needlefish's olfactory sacs are rudimentary, and it detects its prey primarily by sight, though it may be alerted to the proximity of a shoal of fish by the vibration-detecting sensors along its lateral line.

◀ The needlefish occupies a similar position to that of the mackerel in the marine foodchain, namely as a predator of the second rank (or third, if man is included); the dolphin and tuna are its principal predators, though, since it always swims near the surface, the needlefish is particularly adept at evasive action, leaping out of the water and changing its course in midair by flipping its tail to one side—a maneuver that has been perfected by the needlefish's relatives, the flying fish. The needlefish is sometimes caught with hook and line or in wooden fish weirs, but it is not an important food fish, since its flesh is rather dry and, due to the presence of a mineral called vivianite, $Fe_3 (PO_4)_2$-8 H_20, its skeleton is a deep bluish-green, which is admittedly not to every taste.

The saurie *(Scombresox saurus)*, another close relative of the needlefish, is about 40 cm long and, like the Atlantic mackerel, has small auxiliary fins, or pinnules, directly behind the dorsal and anal fins. It lives in schools in the eastern Atlantic (from the Canaries to the White Sea) and never comes close to shore, feeding on pelagic crustaceans and small clupeids (herrings and their kin) and being pursued in turn by dolphins and tuna. A school of sauries will occasionally be targeted by a flock of seabirds, with devastating results, but this species is not as important a component of the marine food chain as the needlefish. F.M.

303
Sub.: Vertebrata
Cl.: Aves
O.: Passeriformes
F.: Muscicapidae

NIGHTINGALE *Lucinia megarhynchos* 303

This Old World songbird measures about 16 cm in length, with a wingspread of 24 cm, and weighs between 21 and 24 g. It prefers open country: heath, garrigue, or other regions of dense, low-lying vegetation where the ground is covered by a thick layer of humus. It avoids coniferous woodlands and regions with a markedly oceanic climate (rainy in the spring and summer, i.e., as in Brittany, Normandy, and the coastal regions of Great Britain), but is found throughout most of Western Europe and in parts of Asia Minor as well. A tireless traveler, it typically covers distances of as much 500 km between dawn and dusk of a single day in the course of its yearly migrations.

Though long celebrated in European poetry, mythology, and folklore, this species is very difficult to observe in its natural surroundings. The nightingale is active by day and by night, but it also stays well concealed, close to the ground, in the midst of a thicket of a brush or shrubbery, and only reveals its presence when its melodious, liquid song (very inadequately represented in English folklore as "Jug, jug") breaks the silence of a summer evening.

▶ The nightingale forages for insects in the humus layer and among the lower branches of a thicket, feeding on most of the smaller species as well as their eggs and larvae, including butterflies and moths, beetles, ants' eggs, as well as spiders. During the summer it also feeds on berries.

◀ The nightingale's nest is very well concealed by low-lying herbaceous vegetation, but a large percentage of the nestlings still fall prey to snakes, mustelids, and other terrestrial predators. During their migrations, the adults are also vulnerable to predation; the nightingale figures among the five passerine species most frequently taken by Eleonora's falcon. The males are highly territorial and intolerant of trespassers of their own species, and since the nightingale is drab and inconspicuous in appearance, it has learned to defend its territory with a display of musical virtuosity rather than brightly colored plumage or some other physical attribute. The decline in local populations in recent years is most probably traceable to the spread of cereal monoculture (at the expense of countless hedgerows, heaths, and woodlots) as well as the gradual replacement of deciduous by coniferous species in the woodlands of Northern Europe.

The thrushes and their kin are sometime considered a separate taxonomic family (Turdidae) and sometimes merely a subfamily (Turdinae) of the very large family Muscicapidae (see the note at the end of the entry for the Old World Robin). In any case, this group includes quite a number of other accomplished singers, including the sole congener of the nightingale, the nightingale thrush *(L. luscinia)*, as well as the OLd World blackbird *(Turdus merula)*, or merl, and the song thrush, or throstle *(T. philomelos)*, all of which frequently heard in the parks and gardens of Western Europe. The hermit thrush *(Hylocichla guttatus)* of the Eastern United States and the related wood thrushes of the American West, some of the "nightingale-like thrushes" *(Cathurus)* of Latin America, and the forest robins *(Cossypha)* of tropical Africa, are also noted for their melodious song, though the popular cagebird known as the Japanese nightingale, or Japanese robin *(Leiothrix lutea)*, is actually a species of titmouse.

J.-F. T.

NILE CROCODILE *Crocodylus niloticus* 304

304
Sub.: Vertebrata
Cl.: Reptilia
O.: Crocodilia
F.: Crocodilidae

Already 25 cm long when it emerges from the egg, the Nile crocodile frequently attains a length of more than 7 meters when full grown; average lifespan is probably around 50 years, though many individuals may survive to the age of 70. This species differs relatively little from fossil ancestors that were alive on the earth several hundred million years ago, though in the interim the crocodile has become somewhat better adapted to the aquatic life. For example, the crocodile is the only cold-blooded animal in which the chambers of the heart are separated in a way that prevents the intermingling of venous and arterial blood, which makes the oxygenation of the blood much more efficient. During a deep dive, however, it may become necessary to reduce the differential in pressure across the system, and for this purpose, the crocodile is equipped with a sort of saftey valve, a small orifice called the Panizzae foramen.

The long internal passageways between its internal and external nostrils (nares) and the lobes of its lungs are also divided or partitioned to facilitate the exchange of oxygen and carbon dioxide, especially important for an air-breathing creature that spends long periods underwater. Its tail is flattened laterally and surmounted with a scaly crest that can significantly increase the body's surface area, which is useful in regulating body temperature. The flattened tail serves as a rudder while the animal swims underwater, as well as a kind of springboard that enables it to leap out of the water in order to seize its prey.

The crocodile has a transparent membrane that comes down over its eyes while allowing it to see quite clearly while swimming underwater; similarly, its ear canals can be tightly sealed with a flap of skin. Its nostrils are mounted well up on top of the muzzle and are closed by muscular constriction as soon as the animal dives below the surface. The crocodile's gullet can be closed off by the expansion of soft tissues at the rear of the mouth, a feature that enables it to drag a prey animal down into the water and drown it without swallowing water; the crocodile continues to breathe through its nostrils, which it maintains above the surface of the water. Finally, the crocodile is in the habit of swallowing rather large stones (gastroliths), which may serve a digestive function or may be used as ballast.

▶ The Nile crocodile is sexually mature at the age of 7 or 8, by which time it will have grown to a length of about 1.7 or 2 m. Baby crocodiles (less than a year old) feed primarily on insects, crustaceans, and mollusks, which account for about 70–80% of their total diet with insects comprising about 90% of this subtotal; the remaining 20–30% consists largely of frogs and other amphibians. Young crocodiles never venture very far from the water, though they occasionally catch their prey on exposed mudbanks or on land. The prey is located by sight or hearing, and a young crocodile will not hesitate to give chase; it closes with its prey by means of an abrupt forward impulse of its entire body, accompanied by a snap of its jaws. After a few perfunctory chewing motions, the prey is swallowed almost immediately.

Adult crocodiles feed on fish (about 50% of the total diet), other vertebrates (about 20%), and invertebrates (about 30%). When prompted by hunger, an adult crocodile will begin to move slowly toward the prey, one foot at a time. In the case of a fish or other aquatic creature, it draws almost close enough to touch the prey with its muzzle, then makes a sudden lateral movement of its muzzle, mouth open, and seizes the fish in its jaws. It clashes its jaws several times in order to turn the fish around so it can be swallowed headfirst; the tongue, which is fixed to the floor of the mouth for most of its length, is of no great use in these proceedings.

Older crocodiles tend to hunt from ambush, lying submerged in the water or at least partially so—frequently with only their eyes and nostrils protruding above the surface—and awaiting the approach of a suitable prey animal. When an antelope or some other creature comes down to the edge of the water to drink, the crocodile makes a sudden forward bound and seizes the animal by its muzzle or one of its legs. The crocodile raises itself up on all four feet and slowly retreats into the water, dragging the prey behind it and eventually drowning or smothering it.

Smaller creatures (rodents and small dogs) will be devoured on the spot. Larger prey, such as the antelope just mentioned, will be dragged into the crocodile's underwater burrow, or game larder, in the riverbank to steep until it is soft enough for the crocodile to eat it. The bluntness of its teeth and the articulation of its jaws prevent it from biting through hide and muscle that have not been treated in this manner, and even so, the crocodile is frequently obliged to sink its teeth into the animal's body and spin around in the water like a top in order to detach an edible morsel.

The crocodile makes use its sense of smell to home in on its burrow, and several individuals will often feed on the same kill. Large numbers of crocodiles can be seen swimming after the carcases of drowned animals (e.g., after migrating herd animals have attempted to cross a rain-swollen river), but the Nile crocodile is a solitary hunter and never displays any sort of cooperative behavior in the pursuit of prey. The diet of older crocodiles consists of 90% vertebrates (about two thirds mammals). The stomach of one Nile crocodile was found to contain over 50 identification tags, the dog collars to which they had originally been attached having long since been digested; assuming that only a very small proportion of the dogs in tropical Africa are outfitted with collars and identification tags, this may give some idea of the sort of environmental havoc that might be caused by even a single adult crocodile in the vicinity of a populated area.

It is the older crocodiles that are also responsible for attacks on humans that come down to the river to bathe, to wash clothes, or to get water. Fortunately, older individuals are also quite sedentary and rarely stray very far from the area in which they have established themselves, so that the precise location of these "crocodile holes" is generally well known. An older crocodile will make no attempt to pursue its intended prey once its initial asault has failed; it immediately retreats back into the water and submerges itself. In fact, if an older crocodile does attempt to run, the metabolic effects are frequently catastrophic; it almost immediately depletes its body's sugar reserves and lapses into a sort of hypoglycemic coma.

▶ The eggs of the Nile crocodile are frequently dug up and devoured by monitor lizards just as they are about to hatch, since the monitors are attracted by the high-pitched, birdlike cries emitted by the baby crocodiles that are about to break through the shell. The female crocodile stays near the eggs during the 90–day incubation period and often sleeps on top of the nest; the shade and moisture provided by the female's body (the water that trickles down off its flanks) are helpful in regulating the temperature of the nest. However, once the nest has been dug up by monitor lizards, the remaining eggs will be quickly gobbled up by marabout storks, hyenas, jackals, and other scavengers.

Young crocodiles are preyed on by ospreys and other raptors, watersnakes, softshelled turtles *(Tryonix)*, and older crocodiles. They defend themselves from these predators by taking refuge among the roots of trees that have been undercut by the current or in burrows that they have excavated in the bank (an enterprise in which several young crocodiles may be involved). Adult crocodiles are sometimes attacked by big cats or constrictors, though generally as the result of an unsuccessful initial assault on the part of the crocodile.

Older crocodiles have virtually no enemies other than man. Apart from being hunted for its hide, the Nile crocodile has been exterminated in many areas as a destructive predator or an undesirable competitor. As often happens in such cases, hitherto unsuspected benefits conferred by the presence of the Nile crocodile have occasionally become apparent in its absence. For example, the Nile crocodile was discovered to have played an essential role in controlling the spread of disease among *Tilapia* (a sort of large sunfish, an important freshwater food fish raised commercially in many parts of Africa). It reduced the number of other, even more destructive predators of *Tilapia* (otters, for the most part). These facts came to light only after the crococile had been removed from the scene and the harvest of *Tilapia* had begun to deteriorate both in quality and quantity.

Similarly, there was a marked increase in the incidence of canine rabies on the island of Madagascar as the crocodile population began to dwindle as the result of another such campaign. Previously, rabid dogs that developed the insatiable thirst that is characteristic of one phase of the disease had apparently been killed by crocodiles while they were lapping up water on the banks of streams—and before they had had the opportunity to transmit the disease to other animals. The Nile crocodile is now protected by international convention and is being raised in captivity, both for its hide and even with a view to its eventual reintroduction into certain habitats. On some of these "crocodile farms," the animals have been induced to breed in captivity; others are dependent on eggs that have been recovered from the wild. Interestingly, about 90% of the latter are successfully hatched and raised to maturity; in nature the figure would be closer to 5%. M.D.

NILE FISH *Gymnarchus niloticus* 305

305
Sub.: Vertebrata
Cl.: Osteichthyes
S.cl.: Teleostei
O.: Mormyriformes
F.: Gymnarchidae

This species attains a maximum length of about 1.5 m and lives in the muddy, murky waters of the great rivers of tropical Africa. It is eel-like in appearance as well, with an elongated body, a long dorsal fin, and no tailfins, and, as with other electric fish, the size of its abdominal cavity seems to have been considerably reduced in order to accommodate its electric organs.

▶ The Nile fish feeds primarily on fish, frogs, and young birds, which it swallows whole. Its mouth is very wide and lined with sharp teeth, and it lunges at its prey, somewhat in the manner of a pike. Otherwise, it propels itself through the water with decorous, wavelike motions of its dorsal fin, its body remaining motionless to avoid disturbing the electric field that is continuously generated by its electric organs. These consist of 4 separate bands of highly modified muscle tissue emitting 250–300 pulses per second and creating a weak electric field of several volts, and are its principal means of detecting prey, since the eyes and olfactory organs of the Nile fish are both poorly developed. The Nile fish's head corresponds to the positive pole, its tail to the negative pole; even the slightest perturbation in the intensity of the field, whether caused by an animate or an inanimate object (a snag or submerged rock) of different conductivity, is registered by a cluster of electrosensitive receptors located on the Nile fish's head, which are as much as 500,000 times more sensitive than the barbels, or "whiskers," on the head of a chub or a catfish.

◀ The Nile fish is able to use its generating capability to warn it of the imminent approach of predators, though not to defend itself or to stun its prey; it occupies an intermediate position in the foodchain, and its best defense against crocodiles and other large riverine predators liesin flight or in concealment in the aquatic vegetation. The Nile fish can shed its tail if it is seized by a predator and later regenerate it; its flesh is consumed by man in some localities, but it is not an important food fish.

The sacred fish (genus *Mormyrus)* of the Upper Nile—so venerated by the ancient Egyptians because it was thought to have eaten the flesh of the slain god Osiris—possesses an even more limited generating capacity than the Nile fish; the electric organs of *M. kannume* emit about 100 pulses per second, creating a low-intensity field of only about 2 volts. F.M.

306
Sub.: Vertebrata
Cl.: Reptilia
O.: Squamata
S.o.: Sauria
F.: Varanidae

NILE MONITOR *Varanus niloticus* 306

The Nile monitor may attain a length of up to 2 m and is brown or brownish-green in color with eye-shaped yellow speckles (ocelli) along its back and arrayed in transverse rows along its tail. It is semiaquatic in its habits, and is found along the banks of watercourses and on sandbars in sub-Saharan Africa. The Nile monitor is very wary and alert, and immediately dives into the water when disturbed rather than seeking the shelter of its burrow; it can remain submerged for up to an hour and only reappears on the surface when it is convinced that the danger has past. It is an excellent swimmer, propelling itself with eel-like lateral undulations while it keeps its feet pressed flat against its body and steers with its laterally flattened tail. It spends the night and the hottest part of the day in the hollow trunk of a tree or in a burrow along the bank, and like other monitor lizards, it is primarily active during the day.

▶ The female Nile monitor sometimes buries its eggs in a termite mound, thus providing an optimal incubation temperature for the eggs as well as an abundant food supply for the hatchlings. Both juvenile and adult monitors also feed on a great variety of other small invertebrates besides termites, their larvae, and their eggs. The Senegalese population studied subsisted largely on beetles (carabids, scarabs, and diving beetles), which comprised 46.8 percent of their total diet, as well as on crickets, grasshoppers, and locusts (25 percent), caterpillars (15 percent), millipedes (12.5 percent, though excluding julids, which are very common in the region but are well protected from predators by their alkaline secretions), cockroaches (6.2 percent), arachnids (6.2 percent), and snails (3.1 percent).

Forty percent of the specimens examined had also fed on vertebrates, including lungfish *(Protopterus annectens)*, frogs and toads, terrapins *(Pelusios subniger)*, lizards (agamas and mabuyas), young Cape monitors *(Varanus exanthematicus)*, weavers *(Ploceus)*, and small mammals. The Nile monitor is an inveterate robber of crocociles' nests, feeding on the hatchlings as well as the eggs, and sometimes coming into villages to steal hens' eggs and chicks; it also feeds on carrion.

The Nile monitor generally attacks from ambush, though after a monitor has dug a hole in the riverbank in the hope of intercepting the burrow of some small creature, it will subsequently return to the same spot, like a hunter checking his traps, since the hole may be used for a wallow for a frog, a toad, or even a lungfish. The predatory activity of the Nile monitor depends largely on rainfall, and in drier regions, it may either go into diapause during the summer months or subsist exclusively on vertebrates.

◀ The short-toed eagle (which primarily attacks the juveniles), the rock python, and the crocodile are the Nile monitor's chief natural enemies, and the Cape monitor frequently feeds on the eggs of its congener. The Nile monitor is widely hunted by humans, though, as with the DESERT MONITOR and the Tuaregs, it has been adopted as a totemic animal by certain tribes (such as the Serer of Senegal), who accordingly refrain from eating its flesh.

Apart from the two-banded monitor (described at the end of the entry for the KOMODO DRAGON), there are a number of other semiaquatic monitor lizards, or water monitors, including Merten's monitor *(Varanus mertensi)*, Mitchell's monitor *(V. mitchelli)*, and *V. semiremex*, all found along the northern coast of Australia; the latter also ranges eastward into Queensland. The Indian monitor *(V. indicus)*, so called because it is an inhabitant of the Indies rather than India, is found in Australia and New Guinea as well as in Indonesia. R.V.

NINE-BANDED ARMADILLO

Dasympus novemcinctus **307**

307
Sub.: Vertebrata
Cl.: Mammalia
O.: Edentata
F.: Dasypodidae

This species, which is the only armadillo found in North America, measures between 40 and 50 cm in length and weighs about 6 kg. The nine-banded armadillo is very powerfully built for its size, with muscular legs, thick, blunt claws well suited for digging, and a thick protective carapace consisting of 8 or (less frequently) 9 jointed bands. The tail is also sheathed in armor, and the underbody sparsely covered with hair. Appearances to the contrary, the armadillo's jointed shell is remarkably flexible, and it can roll itself up into an impregnable armored ball almost instantly.

A native of northern South America, this species has fairly rapidly extended its range up through Central America and Mexico; it first appeared in Texas about a hundred years ago, and has since made its way into a number of Southeastern and Midwestern states as well. The nine-banded armadillo digs its burrow in a marshy or wooded area, generally along the banks of a watercourse; the main burrow is lined with dead leaves and may be supplemented by several additional boltholes, which are always grouped quite close together.

▶ The nine-banded armadillo feeds exclusively on earthworms and insects, and its teeth are too soft to chew more substantial food. Its eyesight is very weak, and it detects its prey almost entirely by smell; it begins to forage at nightfall, moving along with its snout tilted toward the ground, and it can sniff out an earthworm at a depth of up to 15 cm and dig it up with truly astonishing speed. Armadillos require very little oxygen and can hold their breath for up to 5 minutes, which enables them to dig very rapidly and stir up tremendous amounts of dust without inhaling very much of it. On a cool day, the nine-banded armadillo will often emerge from its burrow to hunt before nightfall.

◀ The jaguar, a creature that outweighs the nine-banded armadillo by a factor of 10 or 20, appears to be the only predator with sufficient muscular strength to cope with an armadillo that has rolled itself up in a ball. The nine-banded armadillo's legbones are also provided with crests to which powerful muscles are anchored, thus enabling it to dig itself into the ground very quickly, which is its preferred method of escaping from predators when the terrain permits.

This species is sometimes hunted for food in South America, though in North America it has generally been encouraged as a destroyer of harmful insects and was deliberately introduced into Florida and other areas that it had not already colonized under its own power. Because of their poor eyesight and nocturnal habits, armadillos are very frequently run over on the roads, and because of its susceptibility to the bacillus that causes Hansen's disease (leprosy), the nine-banded armadillo in particular has often been conscripted for medical research.

There are about 20 armadillo species in South America, all basically similar in their habits to the nine-banded armadillo. There is also a six- and a three-banded species, and a considerable range in size among them all, with the giant armadillo weighing about 50 kg and the pink fairy armadillo weighing as little as 1 or 2 kg.

P.A.

NORTH AMERICAN OVENBIRD
Seiurus aurocapillus **308**

This North American warbler, about 13 cm long, nests in deciduous and mixed forests, primarily in areas where the underbrush is not too thick; it is so called because of its basket-shaped nest, reminiscent of an old-fashioned baker's oven. Like other warblers, the ovenbird picks its way through the underbrush; its call, which it gives almost continually, is one of the most familiar sounds of the American forest. It flies south to spend the winter in tropical and semitropical regions, from Central America up to Florida and Louisiana.

▶ The ovenbird actively searches for insects on the ground, in the tall grass, in bushes and brush, and among the foliage as well as underneath the bark of trees. It also feeds on berries.

◀ The American kestrel and Cooper's hawk are the ovenbird's principal predators. Several species of warblers (collectively known as water thrushes) are also assigned to the same genus, but though they are insectivorous and are found in essentially the same habitat, they are not directly competitive with the ovenbird, since they find their prey in slightly different locales—underneath particular kinds of trees, for example, or along the banks of streams.

The Parulidae, or American warblers, comprise 125 different species and are found throughout the New World, from Argentina to Alaska. Many of these species are only slightly differentiated and crossbreeding frequently occurs, notably between the blue-winged warbler *(Vermivora pinus)* and the golden-winged warbler *(Vermivora chrysoptera)*. The diet of the banana quits *(Coereba)* and other tropical honeycreepers is very similar to that of the hummingbirds: they catch spiders and insects and feed on nectar, sometimes by piercing the base of the corolla with their beaks if this proves to be an obstruction. (These birds are assigned to a separate family, Coerebidae, by some authorities).

J.- F.T.

308
Sub.: Vertebrata
Cl.: Mammalia
O.: Carnivora
F.: Procyonidae

NORTH AMERICAN RACCOON
Procyon lotor **309**

This familiar North American carnivore measures between 60 cm and 1 m in length and varies considerably in weight, from as little as 1.5 kg to as much as 22 kg. Its thick coat consists primarily of gray hairs interspersed with black and white; its tail (typically measuring about 40 cm) is ringed with alternate black or brown and light-gray stripes, and it has a very distinctive black mask around its eyes.

Zoologists distinguish up to 25 different species of *Procyon lotor*, which is found both in wooded areas and regions of low-lying vegetation (including semidesert regions) throughout most of North America, though always in close proximity to a pond, lake, or river. The North American raccoon was inadvertently introduced into the wild in Western Europe in about 1930; descendants of escapees from commercial fur farms have acclimated very well. Life expectancy in the wild (under optimal conditions) may be as much as 16 years; a captive specimen is reported to have survived to the age of 22.

▶ The North American raccoon is solitary and unaggressive; hunting territories may overlap, but the individuals involved prefer to avoid one another; if two raccoons set out after the same prey animal, the dispute is quickly resolved by a few intimidatory postures rather than actual combat. A raccoon rarely ventures more than 1 or 2 km from its burrow and exploits the resoruces of territory quite efficiently.

It preys (depending on the season) on insects, earthworms, crustaceans, mollusks, snakes, fish, frogs, young or small mammals (muskrats), and sometimes aquatic birds, and also browses extensively on fruit, leaves, and even bark. In an inventory of the stomach contents of 520 specimens, 52 percent was determined to be of animal origin, 48 percent consisted of vegetable matter. The raccoon builds up an impressive fat reserve for the winter, and will have lost about half of its original body weight by spring. Young raccoons begin to follow their mother into the hunting field at the age of 5 weeks or so and will set out on their own at the age of 4 months.

The raccoon primarily detects its prey with the aid of its senses of smell and touch, foraging with its delicate front paws along the banks of streams and other areas where the small creatures on which it feeds are likely to be found. When it comes upon a likely object, it turns it around in its front paws and sniffs it carefully before deciding whether to eat it or toss it away. This is the origin of the popular belief that the raccoon invariably "washes its food"; in fact, it never does so in the wild, though captive specimens, accustomed to fishing frogs and crayfish out of shallow water, will often engage in a kind of simulated hunting behavior—picking up small objects (edible or otherwise), dunking them in the water, and swirling them around as it pretends to catch them before putting them in its mouth.

◀ The puma and (in certain regions) the jaguar are the only predators than pose a serious danger to a full-grown adult raccoon, though the cubs may fall prey to lynxes, wolves, foxes, ocelots, martens, and various raptors. "Coon hunting" has become as ritualized an activity in the American South as foxhunting in England, and the North American raccoon is particularly adept at crossing its tracks, doubling back, and throwing the pack off the scent by cutting across running water. The raccoon is still widely exploited for its fur, which is primarily (though not exclusively) obtained from animals bred in captivity rather than trapped in the wild.

The South American crab-eating raccoon *(Procyon cancrivorus)* is the only congener that is distributed throughout a wide geographical area; three insular species *(P. maynardi, P. minor, P. pygmaeus)* are found off the coasts of Florida and Mexico and in the West Indies. P.A.

309
Sub.: Vertebrata
Cl.: Aves
O.: Passeriformes
F.: Parulidae

NORTHERN FUR SEAL *Callorhinus ursinus* 310

310
Sub.: Vertebrata
Cl.: Mammalia
O.: Pinnnipedia
F.: Otariidae

The male of this species already weighs 5.4 kg at birth; it weighs between 182 and 272 kg when full grown and attains an average length of 2.1 m. The female weighs 4.5 kg at birth, between 43 and 50 kg when full grown, and measures 1.4 m on the average. During the breeding season, the northern fur seal comes ashore on a number of small islands off the coasts of Siberia and Alaska, including the Kurile, Komandorskie, and Pribilof Islands, and Seal Island in the Sea of Okhotsk. After the pups are old enough enough to fend for themselves, groups of 2–7 individuals set off on on a lengthy coastwise migration that may take them as far south as 32° N. lat.; part of the Pribilof population spends the winter off the coast of California (notably on San Miguel Island) whereas the Siberian populations follow the Asian coastline.

The female is sexually mature by the age of 3, and though the males mature at the age of 4–5, they do not begin to mate until about the age of 9 (during this intervening period they are referred to as "young bachelors"). The pups are born from the end of June through the month of August and are weaned when they are 4 months old. Maximum longevity for this species is about 25 years.

▶ The northern fur seals feeds on herrings and squid for the most part, though the remains of other fish—anchovies, capelins, mackerel—are frequently identified among its stomach contents; it is capable of descending to a depth of 70 m in pursuit of such deepwater species as *Bathylagus collorbinus,* a variety of balcksmelt. This species begins to hunt at dusk and continues through the night, consuming about 10 percent of its own body weight per day—which suggests that the fur seal population of the Pribilof Islands, estimated at 1,300,000, would require a daily ration of 965,000 metric tons of fish.

◀ During the course of its seasonal migrations, and particularly while feeding on herring and salmon off the coast of British Columbia during the winter months, the northern fur seal may find itself competing for food sources with other pinnipeds, notably Steller's sea lion *(Eumetopius jubata)* and the harbor seal *(Phoca vitulina).* Steller's sea lion sometimes attacks juvenile fur seals, and more formidable predators of this species are thought to include sharks and killer whales, though more precise information on this point does not appear to be available.

During the late 19th century, commercial demand for the pelt of the northern fur seal, which was frequently offered to the public under the name "sea otter" or *"loutre de mer,"* was so great that the species was thought to be on the verge of extinction by 1911, when strict limits were established on the taking of the northern fur seal in the Pribilofs and other seal nurseries in the North Pacific. The annual quota, most of which is reserved for the Aleutian Islanders and other native peoples of the region, is currently fixed at 35,000; the total population is estimated at 1,765,000.

An additional 7000 fur seals are thought to be killed each year as the result of commercial fishing operations in this region—drowned as the result of becoming entangled in fishnets that have been deliberately or accidentally cast adrift on the open water (in the former case, they are referred to as drift nets) or in plastic bags and other casual debris of the fishing industry. It is anticipated that intensive "industrial" fishing operations in these waters will have a more seriously adverse effect on this species in the years to come. M.P.

311
Sub.: Vertebrata
Cl.: Osteichthyes
S.cl.: Teleostei
O.: Salmoniformes
F.: Esocidae

NORTHERN PIKE *Esox lucius* 311

The northern pike, found in both hemispheres, reaches a maximum length of 1.5 m and a maximum weight of 35 kg. Its lifespan is on the order of 10 years. The pike lives in lakes or pools in running water where the current is not especially strong, usually among the vegetation along the shore, in the region that is frequented by minnows, chubs, etc. (Cyprinidae). It is occasionally found in brackish water, even in the Baltic, which has a very low saline content.

▶ The pike is a predator of lengendary voracity, even when raised on commercial fish farms; it feeds exclusively on live prey throughout its life. It starts out with daphnia, cyclops, and other minute creatures, then moves on to insect larvae and small crustaceans (isopods), finally to young fry of other species (minnows, carp, sticklebacks, roach) as well as frogs and ducklings and the young of other aquatic birds.

The pike hunts from ambush, concealing itself in the water weeds, and can detect its prey by sight at a distance of 15 to 20 m; its eyes are placed very far up on its flat wedge-shaped skull, enabling it to look upward as well as forward at the same time. Its sense of smell also plays a role in prey detection, though the sensory receptors along the lateral line and in the numerous small pores on its head, which detect vibrations in the water, are undoubtedly of primary importance in this respect. The pike will attack only a moving target, which is why fishermen prefer to attract its attention with a metallic spoon lure—which is shaped like a shoehorn with a cluster of hooks at one end and is designed to shimmer and vibrate in a manner reminiscent of the scales of an injured fish.

The pike remains absolutely motionless until the prey draws within striking distance; though normally a sluggish swimmer (clocked at 300 meters per hour), it is capable of attaining quite respectable speeds of 10 to 15 kph at this critical moment. It tries to strike its prey from the side, as close to the head as possible, then slowly, without unclenching its jaws, maneuver the head of the prey animal into its mouth. The pike almost always swallows its prey headfirst without benefit of further mastication—to do otherwise would be highly inadvisable where the spiny fins of the perch or the spines of the stickleback are concerned. The pike is capable of ingesting prey up to half its own size, though sometimes at the risk of choking, since its backward-pointing teeth will not allow it to disgorge a fish that proves to be too big for it to handle once the process of ingestion is underway.

◀ The pike is one of the superpredators at the very top of the freshwater food chain, which does not of course prevent it from being eaten by aquatic birds like the common Old World heron *(Ardea cinerea)*. Cannibalism is a frequent cause of pike mortality, particularly in aquaculture facilities where sufficient live food is not made available. The pike is considerably more important as a commercial food fish in Europe than in North America, though it is greatly appreciated as a gamefish by sports fishermen on both continents; pike fishing is a pastime (or craft) that requires considerable technique and experience. Like the largemouth bass, the pike is often stocked in European carp ponds as a means of culling out the sick or the injured and eliminating competitive interlopers of other species.

There are 3 other pike species in North

America: the chain pickerel *(Esox niger)*, the red pickerel *(E. americanus)*, and the muskellunge *(E. masquinongy)*. Confusingly, the word *pickerel* is sometimes applied to a small or a young northern pike; the walleyed pike, redoubtable fighting fish of the northern U. S. and Canada, though similar to the northern pike in its habits and appearance, is actually a kind of perch. Another perch species, the pike perch, or zander *(Lucioperca luciperca)*, is the northern pike's principal competitor in European waters, and since its dietary requirements (from the standpoint of quantity) are somewhat less exacting, it has been successful in supplanting the pike in certain areas. F.M.

NORTHERN PILOT WHALE *Globicephala melaena* 312

The pilot whale, or blackfish, can either be regarded as a large dolphin (family Delphinidae) or a small toothed whale (suborder Odontoceti). The males are about 6 m long and weigh between 1200 and 1500 kg, the females somewhat smaller (5m, 800–900 kg); pilot whales are already 1.8–2 m long at birth. The pilot whale's body is relatively svelte and streamlined, but its head, due to the considerable size of the oilcase in its forehead, is massive and bulbous; the jaws and teeth seem disproportionately small in relation to its overall size and weight. The pilot whales, among the most widely distributed of the Delphinidae, are found in all the subpolar seas and oceans of the world, including shallow coastal waters and narrow straits and channels. The pilot whales are extremely gregarious, and pods consisting of several hundred individuals, "piloted" by a handful of larger males, may frequently be observed.

▶ The pilot whale feeds avidly on squid and cuttlefish and follows them in their seasonal migrations between deeper and shallow water. In the absence of these cephalopods, they may feed on codfish, mackerel, and halibut, but only reluctantly as it seems; some captive specimens have refused to accept fish if their first-choice food is not forthcoming. The pilot whale consumes about 4 percent of its weight every day, which represents between 40 and 50 kg in the case of males. This substantial intake is usually spread out over three daily meals of about 15 kg apiece; the process of digestion requires 7 or 8 hours.

Pilot whales also prefer to catch their prey in the most efficient and least laborious manner, which means that they will only feed on large shoals of cephalopods and will not attack isolated animals, which is why they are obliged to remain in close contact with shoals of squid and cuttlefish throughout the latter's migrations between the ocean depths and the shallows. Like other toothed whales, they use echolocation to locate and identify their prey; they rely very little on vision and accordingly sometimes hunt at night.

◀ Sharks and possibly killer whales are the main pelagic predators of the pilot whale. Pilot whales are still routinely massacred and devoured by the human inhabitants of the Faeroe Islands in the North Atlantic—where pilot–whale beachings and strandings are deliberately induced by local fishermen, in a sort of ritual feeding frenzy that has been repeatedly condemned by pro-wildlife organizations. Pilot–whale strandings and "drownings" have been reported in many coastal areas around the world, notably in Florida and New England; the ultimate causes of this are still unknown, though the immediate cause is quite simply that pilot whales continue to follow their leaders even after the latter have become disoriented (perhaps because their sonar navigation system is not very effective in muddy or shallow water) and begin to run aground.

A similar species, the Indian blackfish, or short-finned pilot whale *(Globicephala macrorhynca)*, is found only in tropical and temperate waters. V. de B.

312
Sub.: Vertebrata
Cl.: Mammalia
O.: Cetacea
F.: Delphinidae

NORTHERN TEJU *Tupinambis nigropunctatus* 313

313
Sub.: Vertebrata
Cl.: Reptilia
O.: Squamata
S.o.: Sauria
F.: Teiidae

The northern teju (or tegu), also called the jacura, was dubbed "the protector" by the first naturalists to study the fauna of the Amazon region. It is the largest terrestrial lizard of the Amazon forests and the Guianas, sometimes attaining a length of up to 30 cm from snout to vent. It digs its burrow in the middle of a thicket, often

located near the steep banks of a river, and continues to occupy this same burrow throughout its life; though these lizards are relatively common throughout their range, they are never found in dense concentrations, even when quite young. Little else is known of the habits of this species in the wild; tejus occasionally move about in small groups of 2 or 3, perhaps during a preliminary phase of the mating season.
▶ In the forest, the teju forages in the topsoil, somewhat in the manner of the AMEIVA, pausing frequently whenever a gap in the forest canopy allows the sunlight to spill down onto the ground. The teju feeds on large insects, crustaceans, large spiders, frogs and toads, birds and their eggs, small rodents and opossums, and has gained some notoriety for its habit of stealing chickens from the dooryards of settlers in forest regions. In captivity, the teju is greatly excited by the smell of meat; its knifelike dentition and the shearing force of its jaws enable it to feed on prey animals that are too large for it to swallow all at once, and it can also lap up liquid nourishment (eggs or beef broth in captivity) with its tongue.
◀ A teju can outrun a dog and is large enough to confront one face to face if compelled to, so that the puma and the jaguar are the only mammals large enough to prey on this species, though we have no evidence that they actually do so. The younger lizards are occasionally caught by medium-sized hawks *(Buteo)* that hunt from ambush at the edge of forest clearings.

The common name "teju" is locally applied to a variety of terrestrial lizards belonging to the family Teiidae (sometimes including the ameiva), though strictly speaking it should be reserved for the genus *Teius,* which is smaller than *Tupinambis* and currently restricted in its range to a roughly triangular area of southeastern Bolivia and northern Argentina. South of the Amazon, the northern teju *(T. nigropunctatus)* is replaced by the common teju *(T. teguixin),* which also ranges down into Argentina. The semiaquatic caiman lizard *(Dracaena guianensis)* lives along marshy estuaries and feeds on mollusks by cracking their shells with its teeth, whose rounded cusps are well suited to this purpose. The terrestrial members of this family are all heliophilic, live in burrows, and are notable for the rapidity of their flight—sometimes even resorting to bipedal locomotion in the case of the younger teiids; in these respects they can be regarded as occupying the same ecological niche as the monitor lizards of Africa and Asia.

J.-P. G.

NORWAY RAT *Rattus norvegicus* 314

314
Sub.: Vertebrata
C.: Mammalia
O.: Rodentia
F.: Muridae

This familiar species, also known as the brown rat, typically measures 23 cm altogether, with the tail measuring 18.5 cm alone, and weighs about 350 g. Its brownish fur is short and smooth; its legs are short, its feet are long, and its ears rather poorly developed. Its long tail is hairless and covered with scales. A native of the steppes of Central Asia, the Norway rat began to replace the European black rat during the late Middle Ages and is currently found in every region of the globe that is permanently occupied by man with the exception of the northernmost Arctic regions.

The Norway rat is commonly and contemptuously known as the "sewer rat," which not only gives an indication of the fear and distaste with which this species has traditionally been regarded by our own but also reflects the fact that *R. norvegicus* prefers a moist, subterranean habitat; it is a strong swimmer and diver, and has been known to swim across rivers and saltwater inlets. This species is highly gregarious and generally lives in packs comprising all the surviving members of a single lineage, which can recognize each other by their distinctive pheromones. The Norway rat has an elaborate social organization, and though it ranges far too widely in its search for food to be able to defend its hunting grounds, the pack will defend its home territory against encroachment by members of a strange pack.

The males are sexually mature at the age of 3 months, the females a little later, and each female is capable of producing at least 5 litters of 6–11 offspring each year. Thus, it is theoretically possible that a single pair of Norway rats would have left behind more than 800 descendants at the end of single year, assuming that conditions were absolutely optimal for their survival; in nature, of course, this is never the case, and typical life expectancy is no more than a few months.

▶ In general, the Norway rat prefers live prey when it can get it, and feeds not only on invertebrates but on small mammals, birds, fish, amphibians, and even reptiles. Rats will also devour young domestic animals (lambs, piglets, etc.) and occasionally attack the adults as well, and bites inflicted by them generally become infected, sometimes with fatal results. Zoo elephants have been reported to have died after having the soles of their feet gnawed by rats, and of course human infants have all too frequently been attacked in their cribs in rat-infested tenement buildings.

Rats will also feed on carrion, including carcasses in a slaughterhouse: in a single night, the bodies of 35 horses are reported to have disappeared from the premises of a European horse butcher in this fashion. The Norway rat has had a far more devastating ecological effect when accidentally introduced as the result of shipwreck, maritime commerce, or human settlement into the vicinity of seabird nesting colonies on remote islands, as a result of which entire species have been exterminated in very short order by the descendants of a single boatload of rats. Some rat packs feed exclusively on vegetable food (grain, legumes, fruit) stored in silos and warehouses, and in general, the Norway rat consumes between 10 and 40 g of food per day, depending on its size, and drinks between 17 and 35 ml of water.

The Norway rat's sense of smell plays a primary role not only in prey location but also in mating and all other intraspecific relations. Rats will rarely consume their prey or other booty on the spot but generally carry it back to their home territories, when substantial caches of food are maintained, and are frequently left uneaten and allowed to spoil. Rats kill their prey by biting them with their incisors, which are quite powerful in addition to being relatively long; they generally hold their food up their mouths with their front paws, a habit that is shared with some of the other Muridae. A fieldmouse, for example, will eat a mealworm while grasping it in its front paws and nibbling on it like a child eating a graham cracker; the bank vole, on the other hand, clamps the worm against the ground with one of its front paws and chews cautiously on the protruding end without attempting to pick it up in its paws.

◀ The Norway rat has a quite a number of natural enemies, though very few of them will actually attack a full- grown adult, which, when cornered, turns to face its adversary and then lunges toward it with its chisel-like incisors at the ready. Most dogs (except for a few breeds such as the Jack Russell terrier that have been specially bred to the task) and almost all domestic cats will hesitate to attack a full-grown Norway rat, though they frequently prey on the pups. With wild rabbits in short supply as a result of the myxomatosis epidemic, however, the red fox has become a major predator of the Norway rat in Europe, and, among the raptors, the eagle owl preys avidly on the adults as well; the eyries of some of the few remaining eagle owls in Central France are located near public dumping grounds and literally paved with the skulls of Norway rats. The screech owl and the tawny owl, on the other hand, also hesitate to attack the adults and only prey on the pups.

Cannibalism is fairly frequent among Norway rats, and in addition, there are "killer rats" that cannot tolerate the presence of members of another pack, so that when two such individuals belonging to different packs encounter one another, a fight to the death generally ensues. The Norway rat does not face serious competition from the other Muridae, since the black rat (commonly known in North America as the roof rat) normally prefers a warm, dry attic to a cold, damp cellar and relies to a lesser extent on animal prey. House mice are competitive with Norway rats to some extent in grain silos, warehouses, and similar facilities, but they too generally prefer a much drier habitat and are obviously no match for the Norway rat in size or strength.

The Norway rat is intensely persecuted by man as a destructive competitor (Hainard reports that as many as 16,000 rats were killed in one Parisian slaughterhouse in a month) as well as a source of infection. The Norway rat has also been implicated in the spread of typhus, hemorrhagic fever, and a number of other serious infectious diseases. Some epidemiologists have suggested that the pandemic outbreaks of the bubonic plague during the 14th century (the Black Death) occurred in Europe after infected black rats were introduced from Central Asia—and that the effects of subsequent outbreaks were considerably abated as the black rat was displaced or eradicated by the Norway rat, which prefers not to live in quite such intimate proximity to human beings. Laboratory white rats are specially bred albino strains of *R. norvegicus*, so that this species has once again become an inadvertent benefactor of humanity, sacrificed by the thousands every year in toxicity tests for drugs and consumer products, studies of the effects of hormones and other physiological processes, cancer research, and in numerous other commercial, medical, and scientific investigations.

There is considerable morphological variation within the genus *Rattus*, but apart from the black rat *(R. rattus)* and the Norway rat, the only other recognized species is the Pacific rat *(R. exulans)*, which is also essentially parasitic on man and does considerable damage in copra warehouses and coconut groves. M.-C. S.G.

315
Sub.: Pseudocoelomata
Cl.: Rotifera
O.: Monogontes
F.: Notommatidae

Notommata pseudocerberus 315

This free-swimming rotifer, or roundworm, measures 0.6 mm in length and is found in freshwater ponds and forest pools. *N. pseudocerberus* normally crawls along the leaves and stems of aquatic plants, strands of filamentous algae, or dead branches that have fallen into the water, using the rotary motion of the fringe of cilia that surrounds its mouth to propel itself forward. This species is also capable of swimming through the open water with the aid of a pair of ciliated finlike projections that are retracted into its body when not in use.

Like the other Monogonota (cf. *Dicranophorus forcipatus)*, this species is oviparous and parthenogenetic. *N. pseudocerberus* lives for about 2 weeks and, toward the end of this period, it deposits a cluster of about 30 eggs on the bottom. Though the precise mechanisms involved are not well understood, there is an alternate mode of reproduction which only occurs once in several generations: the females of this type produce small brown hard-shelled eggs that lie dormant on the bottom until the following spring, and from these eggs, females of the first type emerge to continue the cycle.

▶ *N. pseudocerberus* feeds primarily on ciliate protozoans of the genus *Stentor*, including both sessile species such as *S. polymorphus*, which attach themselves to the stem of a plant or some other support, and free-swimming ciliates such as *S. niger*. *N. pseudocerberus* has no special means of detecting prey at a distance, and sustains itself by means of random attacks on most *Stentor* species that it encounters as it propels itself through the water or crawls along the stems of aquatic plants. As with most rotifers, the muscular pharynx (mastax) contains a set of sawtoothed grinding plates, used by *N. pseudocerberus* to breach a hole in the cell wall of its protozoan victims, and, as in all the Notommatidae, the mastax also serves as a kind of hydraulic pump, which sucks the protoplasm that begins to flow from this wound into the pharynx and down the esophagus of the predator.

There are some *Stentor* species (notably *S. coeruleus)* that are attacked by *N. pseudocerberus* only when no other choice of prey presents itself, and even then, the rotifer quickly detaches itself after making one or two attempts to siphon off the cytoplasm of this protozoan, which it appears to find highly distasteful. At a temperature of 20°C, *N. pseudocerberus* typically consumes 4 or 5 protozoans per day. ◀ We know of no creature that preys specifically on *N. pseudocerberus*, though like other rotifers, this species is most probably subject to predation by fish hatchlings, newt and salamander larvae, flatworms (Turbellaria), and heliozoans.

Some *Notommata* species feed on algae rather than protozoans, and *N. glyphura* feeds on other rotifers that crawl along the bottom or on the leaves and stems of aquatic plants; it attacks the adults of soft-bodied vermiform species and swallows them whole, by aspiration, and devours the eggs of other rotifers by cutting a hole in the capsule and then inhaling the contents, in much the same way that *N. pseudocerberus* feeds on *Stentor*. Members of a related family, the Trichocercidae, feed on free-swimming rotifers in a similar manner and attach their eggs to the teguments of other rotifers, as well as to strands of filamentous algae. R.P.

NUMBAT *Myrmecobius fasciatus* 316

316
Sub.: Vertebrata
Cl.: Mammalia
O.: Marsupialia
F.: Myrmecobiidae

The numbat, or marsupial anteater, is the only representative of its family. It measures about 25 cm, plus an additional 15 cm for its long bushy tail, and in appearance is more reminiscent of a squirrel than of the anteaters of South America (to which it is not closely related); its fur is dark colored with 6–12 lighter-colored transverse stripes running across its back. The numbat lives in a grass-lined nest in a hollow tree and, unlike other marsupials, is active by day and sleeps soundly through the night; it is now uncommon outside wildlife sanctuaries, though it is still occasionally encountered in the forests of southern Australia.

▶ The numbat feeds almost exclusively on ants and termites; it spends its days progressing from one dead tree to the next and foraging for termites' nests (though it will not attack termite mounds on the ground). It shreds and detaches bits of wood with its powerful front claws and pauses frequently to explore the termite tunnels inside a piece of wood with its long pointed tongue, which, like that of the aardvark and the South American anteaters, is coated with viscous saliva.

◀ The dingo, introduced into Australia by the Aborigines several tens of thousands of years ago, and other domestic and wild carnivores (including the fox) introduced from Europe in more recent times are the main natural enemies of the numbat, against which it has no defense other than to scurry up the nearest tree-trunk. The practice of systematically burning off brushwood to clear land for grazing has further reduced the remnant populations that still exist outside wildlife sanctuaries. P.A.

NURSERY-WEB SPIDER *Pisaura mirabilis* 317

317
Sub.: Arthropoda
Cl.: Arachnida
O.: Araneida
F.: Pisauridae

This spider measures about 15 mm and is commonly found in open, grassy areas and among the lower branches of shrubs and bushes in Western and (especially) Southern Europe, including the British Isles. Larval development is complete by mid-June to mid-July, and mating occurs almost immediately; the female uses its cheliceres (clawlike pincers) and maxillipeds (grasping mouthparts) to transfer the eggs to a spherical cocoon of silk slung underneath its abdomen. When the eggs hatch out or shortly thereafter, the female spins another cocoon at the top of a herbaceous plant stalk or the tip of one of the lower branches of a bush, then lashes the surrounding leaves together with additional strands of silk to form the "nursery web" for which the species is named. The female watches over its brood of pulli (first-stage larvae) until they have completed their first larval molt, when they are ready to leave the nursery and fend for themselves.

▶ The nursery-web spider makes its way slowly among the grassy stalks and along the leaves of bushes, waiting for gnats, mosquitoes, and other small insects to alight within reach of its cheliceres. Its venom its not very potent, but a single bite is generally sufficient to pierce the body wall of an insect and kill it instantly.

◀ Hunting wasps, including *Sceleriphon* and the Pompilidae, are probably the most frequent predator of the nursery-web spider. There are other species of parasitic wasp that prey specifically on the embryos or the various larval stages, and as is generally the case with spiders, cannibalism is quite prevalent among local concentrations of two or three hundred individuals.

The numerous species that belong to the family Pisauridae are mainly found in humid areas or along the banks of watercourses; most of them spin a permanent web and specialize in a particular type of prey, as with the Agelinidae (see FUNNEL-WEB SPIDER). There are several very large species, known as the African fishing spiders, that are able to catch fish measuring several centimeters by belaying themselves to an overhanging branch with a rope of silk and pouncing on their quarry from directly overhead, sometimes from a height of a meter. J.-F. C.

O

318
Sub.: Vertebrata
Cl.: Aves
O.: Passeriformes
F.: Muscicapidae

OLD WORLD BLACKBIRD *Turdus merula* 318

The Old World blackbird, or merl, is 23–24 cm long, with a wingspread of 37–38 cm, and weighs between 75 and 120 g. It is found in woodlands and scrub vegetation up to an altitude of 1600 m, as well as in urban and suburban areas. In woodland and rural areas, it is naturally diffident and shuns all contact with humans, whereas in more populated areas it is quite tolerant of humans and will eagerly take advantage of the smallest patch of greenery. Its range, bounded by northwestern Africa and southern Scandinavia (including Finland), extends eastward in a broad belt as far as southeastern China and Mongolia. Some European populations (as in France) are relatively sedentary; others (as in Switzerland) include substantial numbers of migratory and nomadic birds. Average life expectancy is 2 years, and juvenile mortality is about 10 percent; the highest recorded longevity is 8 1/2 years.

▶ Old World blackbird feeds primarily on fruits and berries and on a considerable variety of animal prey (a selection that varies greatly with the season and locality, and with the nature of its habitat). It hops along the ground quite easily, feeding on earthworms that it digs out of the topsoil with its beak and picking through the litter of dead leaves with its beak and claws in search of insects and their larvae (beetles, ants, caterpillars), sowbugs, spiders, snails and slugs.

It also feeds on tadpoles and small frogs on occasion, and on the seacoast it can often be seen foraging below the tidal mark, inspecting rocks and clumps of seaweed for small crustaceans. The blackbird is a strong flier, threading its way skillfully through wooded or heavily overgrown terrain and setting a straight course and flying with strong, deep wingbeats when obliged to cover more substantial distances.

◀ The blackbird builds its nest close to the ground, and the male is energetic in defense of its nesting ground, pecking at other blackbirds with its beak if intimidatory displays are not sufficient to drive them away. The sparrowhawk, peregrine falcon, long-eared owl, tawny owl, jay, and magpie all prey on the adult birds or the nestlings, and in populated areas, a great many fledgling blackbirds are devoured by household cats.

During the last century, the blackbird was considered to be a "wild" bird, i.e. primarily a woodland and scrubland species, but it has since adjusted very comfortably to our encroachments on its former habitat, especially as far as suburban lawns and garden allotments are concerned. On the other hand, the medieval practice of shooting or netting blackbirds for the pot is still prevalent in certain areas around the Mediterranean, notably the south of France and Corsica.

The Old World blackbird is actually a member of the thrush family (Muscicapidae), which comprises more than 300 species worldwide, whereas the red-winged blackbird *(Agelaius phoenicus)* and other New World blackbirds belong to the family Icterinae. To compound the confusion, a congeneric species *(Turdus migratorius)* was thought to be a robin—and is still so called—because the first English settlers in North America were misled by the similarity of its plumage to that of the Old World robin *(Erithacus rubecula)*, which is not closely related. (The New World robin is more sensibly called the *merle* by French Canadians, originally the French word for blackbird.) The main difference in their habits is that the Old World blackbird rarely migrates, the New World robin almost invariably does. Another congeneric species, the ring blackbird, or ring ouzel *(Turdus torquatus)*, lives just below the treeline in the mountain forests of Europe, North Africa, and Asia Minor. J.-P. R.

OLD WORLD ROBIN *Erithacus rubecula* 319

319
Sub.: Vertebrata
Cl.: Aves
O.: Passeriformes
F.: Muscicapidae

The Old World robin, or redbreast, measures 13 cm in length, with a wingspread of 22 cm, and weighs between 12 and 19 g. It is found in cool, humid, wooded terrain, on the lowland plains or in the mountains; it seeks out hedges and dense brushwood, city parks, and suburban lawns and gardens, and is often found in the vicinity of human dwellings, especially in winter. It generally stays concealed in the interior of a hedge or thicket, where its presence can most easily be verified by its clear, liquid song.

The robin is currently one of the 10 most common species in France, and though most Western European populations are sedentary, night-flying migrants from Scandinavia may be found in southern France and elsewhere in the Mediterranean Basin during the winter. This species is more closely related to the New World bluebirds *(Sialia)* than to the New World robin *(Turdus migratorius),* which is a congener of the Old World blackbird.

▶ The robin forages for insects and other small invertebrates very early in the morning or at dusk and generally in the shadow of thick underbrush. It hops from branch to branch, frequently alighting to search through the leafmold and humus layer. Since it is not strong enough to do so on its own account, a robin will eagerly follow along in the wake of a gardener turning up the soil with a spade (possibly a rooting wild boar as well) in search of earthworms, sowbugs (myriapods), spiders, insect larvae, ground beetles, snails, and the like. The yellow gullets and the pale-edged beaks of the nestlings serve as visual cues, prompting their parents to stuff the target areas full of insects whenever possible. In the winter, the Old World robin also feeds on berries and will come to peck at suet set out on a bird feeder.

◀ The robin's nest is well concealed in a crevice or a recess in the underbrush, generally quite close to the ground, and the nestlings are protected by their cryptic black-speckled pigmentation, but only about 23 percent of their number will survive the first year of life; many are drowned in rainstorms, others fall prey to mustelids, housecats, and other terrestrial predators. A census of 58,000 birds (of 126 separate species) taken by sparrowhawks included 1,723 adult robins.

The Old World robin is aggressive in defense of its home territory (about 6–8000 sq m), during the winter as well as in the nesting season, and the male's red breast and melodious song both evolved as warning signals to deter trespassers; the male robin will even attack any bird with reddish plumage, a tuft of red feathers, or its own image in the mirror.

Although in Britain and Western Europe the robin remains a sentimental favorite with birdwatchers, wildlife illustrators, and the public in general, it is one of a number of songbird species that are still avidly consumed in rural areas throughout the Mediterranean Basin. In Provence robins are illegally caught by the thousands in snares baited with earthworms and flying ants, and in Italy, Spain, Cyprus, Malta, and North Africa, robins continue to be destroyed by the hundreds of thousands (500,000 each year on Malta alone) along the migratory flyways for the sake of the few grams of flesh that they contain.

The family Muscicapidae comprises more than a thousand species, or about one eighth of all living birds. The subfamily Turdinae (often considered a separate family, Turdidae), to which both the Old and New World robins belong, consists of 45 genera and 309 species. All are diurnal or crepuscular in their habits, and most have large eyes and actively seek for prey in open terrain; the song of many of these species is complex and melodious. The nestlings generally display some form of cryptic coloration, and sexual dimorphism is observed in many species; the subfamily Turdinae, now of cosmopolitan distribution, originated in the tropical and temperate regions of the Old World.

A congener of the Old World robin, the bluethroat *(Erithacus svecicus* or *Cyanosylvia svecica),* found in Northern Europe and Asia, sometimes nests along the coasts of Western Europe. The common bluebird *(Sialia sialis)* of the Eastern United States has a reddish breast and a blue back, whereas the plumage of the mountain bluebird *(S. currucoides)* of the Rockies is uniformly blue. J.-F. T.

320
Sub.: Vertebrata
Cl.: Amphibia
O.: Urodela
F.: Proteidae

OLM *Proteus anguinus* 320

This European cave-dwelling salamander is very frail, and even rather unhealthy-looking: Its body is dead white—with its blood-engorged external gills providing the only spot of color—and greatly elongated; it has no visible eyes, and its spindly legs terminate in toeless knobs. It has characteristics in common with certain extinct surface species of the Cretaceous Period, though it is currently restricted to the subterranean regions of the Slovenian Karst, a stark limestone plateau that is honeycombed with sinkholes and grottoes—a habitat whose limited food resource can support only a small population. This species was identified in 1662 (the name proposed by its discoverer was "the little dragon of the caverns"); "olm" is derived from its German name, *Grottenolm*, "cave salamander."

▶ *Proteus* is a capable swimmer, but it generally seeks its prey by creeping stealthily along the streambed, particularly in spots where the force of the current is deflected by blocks of stone that have tumbled down from the walls and ceiling of the cave. It will also hunt from ambush, stationing itself along the clay-lined banks where small crustaceans come to feed on the fringe of organic debris that is deposited by the stream; *Proteus* will often lie concealed for many hours without betraying itself by the slightest movement. Its favorite prey is an amphibious white cave shrimp *(Tithanetes)*, though it also feeds on a variety of other subterranean amphipods and isopods (more specifically, Asellidae [*Asellus*, *Parasellus*], Stenasellidae, Gammaridae [*Niphargus*, *Bogidiella*], and Atyidae) and more rarely on snails and worms, as well as on mosquito larvae and even tadpoles that have been carried down from the surface.

When captive specimens are presented with suitable prey, they show no reaction at all until several seconds or even minutes have elapsed (in the case of the juveniles and adults, respectively). *Proteus* makes its way toward the quarry by means of a series of successive approximations; then, when it is only a few centimeters away, it stops abruptly, lifts up its head, and often pushes itself up on its forefeet. It swivels its head from side to side as its samples the gusts of air from various directions, and finally points its muzzle directly at the prey animal. Unlike surface-dwelling proteids such as *Necturus* (a sighted, pigmented, and somewhat sturdier North American salamander), *Proteus* will feed only on living prey.

Studies have demonstrated that *Proteus* can detect the presence of a prey animal, living or dead, from some distance away by chemoreception alone—a technique of prey location that is very well suited to a terrain that is sparsely populated and very dificult to traverse, quite different from the muddy streambeds frequented by *Necturus*. However, *Proteus* still requires an additional cue before it will attack a prey animal, an almost imperceptible movement or some not-too-vigorous sign of life. If a prey animal that has hitherto remained motionless suddenly makes a last-minute bid for freedom, *Proteus* will not attempt to track it down. If the prey animal appears to be neither dead nor excessively lively, *Proteus* moves in for the kill. Tilting its head a little to one side, it aspirates the prey into its mouth, holding it secure with its delicate rows of teeth. (It sometimes examines dead animals and bits of organic debris by taking them into its mouth in the same way, but then invariably spits them out again.)

◀ Apart from a few parasites, *Proteus* has no natural enemies, and is in fact the most formidable predator in its subterranean habitat. Its eggs are hatched in a remote part of the cavern, at some distance from the streambeds where its other inhabitants are likely to be found, and during the first 4 months of life, the young salamanders are still feeding off their vitelline reserves. By the time their mouths have fully developed and they are ready to hunt for themselves, they are already too big to be devoured by a hungry amphipod or any other creature they are likely to encounter. On the surface, both the juveniles and the adults are totally helpless. J.-P. D.

OPOSSUM SHRIMP *Siriella armata* 321

Opossum shrimps are mysidaecans, members of an order of shrimplike crustaceans that are distinguished from true shrimps by the presence of whiplike appendages (exopodia) on their thoracic legs and by the organ of balance (statocyst) located on top of the caudal fins. The common name "opossum shrimp" alludes to the fact that the female incubates its eggs in a pouch atop the thorax, called a marsupium. The newly hatched young are morphologically similar to the adults and do not undergo an intermediate larval stage; the hatchlings are always released from the pouch at night.

Siriella armata is a relatively large mysidacean, measuring about 2 cm and weighing 22 mg. It lives on submerged rocks and in beds of seawed or wrack grass in clear, shallow water, from the Mediterranean to the North Sea. During the day, *Siriella* congregates in small formations consisting of several tens of individuals, all swimming in place and all oriented in the same direction (depending on the current and direction the light is coming from); in regions where *Siriella* is especially prevalent, a number of these formations will sometimes merge into a shoal numbering several hundred individuals.

▶ Both the juveniles and the adults feed on various sorts of coastal zooplankton. *Siriella* propels itself through the water by rotating the whiplike exopodia at the tips of its thoracic legs like so many little propellors. It can thus actively hunt for larger prey, which it catches by touching the tips of its thoracic legs to form a sort of basket directly below its mouth. When *Siriella* finds itself swimming through a dense concentration of copepods, it begins of "hover" in place, creating eddies with its twirling expodia and using its conjoined thoracic legs as a strainer to catch the copepods that are swept along by the tiny currents that are created. It uses the same technique to collect dead insects and other organic debris that it finds floating on the water. The prey animal is transferred to the mandibles, where it is chewed up into more manageable bits and ingested piecemeal.

Siriella feeds by day and detects its prey by sight; once the prey animal has been trapped in the basket formed by its thoracic legs, *Siriella* may select some particularly tempting morsel by chemoreception (by tasting it, in other words) and then reject the rest. Thus, it has been observed to trap a female copepod in this manner, to remove the egg sac and devour it, and then to release the copepod itself, unharmed. Similarly, in the laboratory, *Siriella* has been observed to pick out the highly pigmented eyespots of other individuals of its own species.

◀ *Siriella armata* is preyed by on by various fish, notably saupes *(Boops salpa)*, base *(Sargus)*, and combers *(Serranus cabrilla)* in the Mediterranean and codfish in the North Sea. When a shoal of opossum shrimp is approached by a potential predator, it immediately disperses; each individual shrimp makes an abrupt motion with its caudal fin (uropods and telson) that sends it skittering backward over the surface of the water, though as soon as the danger is past, the shoal reconsititutes itself immediately.

Siriella's body is virtually transparent when its chromatophores (pigment cells) are contracted, except for the eyespots, which always remain black. Otherwise, the line of chromatophores extending along *Siriella's* ventral surface makes it look something like a slender twig or stalk when its chromatophores are fully dilated. In its transparent phase, a shoal of these opossum shrimp looks very much a school of sand smelt *(Atherina hepsetus)* hatchlings; this may in fact be a form of protective mimicry, though the precise details (e.g. the identity of the predators that might be deterred by this resemblance) are still obscure.

The Mysidacea comprise some 780 species worldwide, most of which are marine, ranging in habitat from the deepest ocean trenches to the brackish waters of river mouths and estuaries; some are found in fresh water (notably the Great Lakes) and even in underground caverns and streams. Opossum shrimps and other mysidaceans are very common in temperate waters, though they often pass unnoticed because of their inconspicuous size and appearance. They are not usually taken by con-

321
Sub.: Anthropoda
Cl.: Crustacea
S.cl.: Malacostraca
O.: Mysidacea
F.: Mysidae

ventional fishing (or shrimping) operations, though two East Asian species, *Acanthomysis longirostris* and *Neomysis awatchensis*, are widely employed in Oriental cookery in the form of "shrimp paste."

Other common mysids furnish an essential link in the littoral foodchain, feeding on algae and organic particles as well as copepods and being eaten in their turn by fish. *Praunus flexuosus*, common in the English Channel, is an omnivorous mysid of this type. Salmon and trout are both extremely fond of *Mysis relicta*, a North American freshwater species that has been accordingly introduced into lakes and rivers in the United States, Canada, and Sweden with a view to increasing the yield of both these fish. Other mysids are no doubt equally suitable for use use as "fodder" on commercial fish farms and shrimp hatcheries. J. C.-R.

322
Sub.: Vertebrata
Cl.: Reptilia
O.: Squamata
S.o.: Ophidia
F.: Viperidae

ORSINI'S VIPER *Vipera ursinii* 322

Orsini's viper, also known as the field adder, is the smallest European viper; the adults never exceed 50 cm in length. Its coloration varies from light brown to gray, with a wavy dark brown or black band running down its back. Two subspecies are in fields and grasslands interspersed with bushes and other low-lying vegetation in Central and Eastern Europe, ranging eastward as far as the Caucasus; two others are found in meadows and fields of dwarf juniper in the Alps and in the mountains of Southern Europe, sometimes above 2000 m.

Vipera ursinii spends the colder months (from the end of September to the beginning of April) in an underground gallery dug by rodents or in a fairly deep cleft or recess in the soil. It is diurnal in its habits, and never ventures out on the surface until the soil temperature is as least 15°C. In July or August, the female gives birth to 3–12 viperlings (in accordance with the general rule among cold-blooded animals, the older the female, the larger the number of offspring).

▶ Unlike its larger congeners, Orsini's viper susbsists to a large extent on insects (particularly crickets, locusts, grassshoppers, and beetles), spiders, harvestmen, and other arthropods. The juveniles are too small to catch larger prey, and arthropods and lizards each furnish about 40 percent of the adult's diet; the remainder consists (in proportions that vary with the season and the habitat) of small mammals, slugs, and amphibians. This species could probably survive quite comfortably on an average intake of 10 g of animal protein every 5 or 6 days.

◀ Apart from the two greatest enemies of all venomous snakes, namely humans and birds of prey, the predators of the Western European subspecies include the Montpellier snake, the dark green snake, and possbily other species as well. The French population is largely confined to Mont Ventoux, the Monts de Lure, and a few other parts of the Maritime Alps; the first of these remnants has virtually disappeared in recent years, evidently as much the result of natural or artificial alterations in its habitat as of direct predation or persecution. Orsini's viper is fairly timid, and though it will attempt to strike if attacked, its venom is not dangerous to humans or other large mammals, including grazing sheep.

This species is made up of several isolated subspecies that became differentiated very early on, and in some respects (notably its feeding habits and the retention of certain other primitive characteristics) it appears to be less highly evolved than other vipers, even though the viper family is a very homogeneous group and *V. ursinii* differs very little from the smaller European vipers in the principal features of its morphology. J.C.

323
Sub.: Vertebrata
Cl.: Aves
O.: Passeriformes
F.: Emberizinae

ORTOLAN *Emberiza hortulana* 323

The ortolan is an Old World bunting (a kind of finch) with a plump body and a large beak, measuring about 14 cm with a wingspread of 24 to 27 cm and weighing between 19 and 27 g (increasing to as much as 40 or 50 g when the ortolan is fattened in captivity as a gourmet delicacy). Like most songbirds of the Northern Hemisphere, its lifespan rarely exceeds 2 or 3 years. Its name is derived from the Latin word for gardener, *hortulanus*, perhaps because it is frequently found in vineyards and cultivated fields in Southern Europe, especially on the terraced slopes of mountains up to the treeline, as well as in scrub and brushlands of the type called *causses* and *garrigues* in Southern France. The ortolan is the only migratory bunting; after the nesting season is over, it spends the month of August in fattening itself for its long journey to Ethiopia or Subsaharan Africa.

▶ The ortolan is a quiet, inconspicuous bird that eats seeds and grain, which it cracks with its powerful beak, though it feeds its nestlings on caterpillars, grasshoppers, and insect larvae that it finds in overgrown wasteland areas rather than cultivated fields.

◀ The ortolan is a ground-nesting bird, and though its nest is usually well concealed, it is sometimes pillaged by crows, magpies, jackdaws, and small carnivorous mammals. The adults are preyed on by sparrowhawks and, in the course of their migrations, by the Eleanora's falcon *(Falco eleanorae)*, a raptor of the Western Mediterranean islands that preys on migratory flocks before embarking on its own migration to Madagascar in the late fall. The ortolan has no competitors except from members of its own species; the male sings from its perch in a bush or vine to advertise the limits of its territory.

Since Roman times, the ortolan has been regarded as a supeme gastronomic delicacy, somewhat on a par with the truffle; Louis XVIII would never eat a partridge unless it had been stuffed with both. The gourmets of the Ancien Régime devised recipes for ortolans that had been fattened and drowned in cognac, among other refinements. Nowadays, perhaps 10,000 ortolans are netted illegally every year in southwestern France along with almost a million other songbirds of less distinguished pedigree—many of which at least acquire the posthumous distinction of being served up as "ortolans" in three-star restaurants. Also, in recent years, vineyards are less frequently found in association with cereal crops, thus depriving the ortolan of its favored nesting grounds.

Old World buntings comprise 3 different genera, 40 species in all. The yellowhammer *(Emberiza citrinella)*, the corn bunting *(E. calandra)*, and the cirl bunting *(E. cirlus)* are all found on the plains of Western Europe, along with the reed bunting *(E. schoeniclus)* in marshy areas. A number of colorful North American finches are also referred to as buntings, including the indigo, lazuli, and painted buntings; the latter are among the most colorful (or the gaudiest) of North American songbirds. The snow bunting *(Plectrophenax nivalis)* and the Lapland longspur *(Calcarius lapponicus)* are natives of North America and Eurasia, respectively, and are often found in the polar regions of both hemispheres.

J.-F.T.

OSTRICH *Struthio camelus* 324

324
Sub.: Vertebrata
Cl.: Aves
S.cl.: Ratites
O.: Struthioniformes
F.: Struthionidae

The ostrich, the largest of living birds, has a long naked neck, a small head with a short, broad beak, very long legs and two-toed feet somewhat reminsicent of wooden shoes. The male has black feathers on its body, white plumes on its wings and tail; the female is usually gray. Height often exceeds 2 m; weight varies between 70 and 150 kg. Maximum recorded longevity is 27 years.

The ostrich was formerly found in Syria, Arabia, and parts of Asia; nowadays its range is limited to the desert and subdesert regions of Africa, particularly south of the Equator. It was also introduced into several of the desert regions of Australia during the previous century. The ostrich avoids woodlands and prefers the savannah or other open country. During courtship, the males fight by kicking and pecking each other, sometimes in the presence of the female; the ultimate victor in these contests may accumulate a harem of 3 or 4 females. The female lays a clutch of about 20 eggs, each of which weighs 1.5 kg. The incubation period is 6 weeks.

▶ The ostrich is very well adapted to the migratory life on the savannah. Its wings, though equipped with long flexible plumes, are used only to provide stability while running and for purposes of aggressive or courtship display. When running, the ostrich can attain a maximum speed of 60–70 kph for brief periods, covering as much as 15 m in a single stride. In the dry season, ostriches travel in small bands, dispersing into still smaller family groups during the rainy season. The ostrich can go for several days without water but is not averse to total immersion in a waterhole.

It feeds primarily on leaves, fruit, and seeds, also occasionally on such small rodents and reptiles (lizards and tortoises) as it encounters in the course of its migrations. Contrary to onetime popular belief, the ostrich has a rather delicate digestion and is obliged to fill its crop with sand and bits of gravel, which assist in breaking down the fibers of ingested vegetable matter. There have been reports from South Africa, not always well authenticated but not inherently implausible, of ostriches that have been found with diamonds in their crops (though none of the stuff of legend, such as horseshoes or alarm clocks).

◀ The ostrich is strong and fast enough to escape from the larger carnivores of the African savannah, though the eggs and young chicks are sometimes eaten by jackals and by the Egyptian vulture *(Neophron percnopterus)*, which breaks the thick shell by bombarding it with stones. When approached by a predator, the adult ostrich attempts to discourage or disorient its pursuer by running in wide circles at top speed (rather than by burying its head in the sand, a legend that was probably inspired by the posture adopted by the ostrich while grazing).

The ostrich propulation was considerably reduced by plume-hunters, in spite of the fact that the ostrich's plumes can be readily and painlessly removed from the living animal (current demand is supplied by commercial ostrich farms in South Africa and elsewhere); the ostrich was exterminated in the Middle East primarily by local sportsmen. The ostrich is easily domesticated and also flourishes as a protected species on a number of South and East African wildlife preserves.

Related ratite (flightless) birds include the rhea *(Rhea americana)* of the pampas and high plateaus of South America, which lives in bands of 20 to 30 individuals and feeds on leaves, seeds, roots, worms, insects, snails, and lizards; the emu *(Dromaius novaehollandiae)*, found exclusively in Australia, prefers a dry woodland or open terrain and feeds on insects and caterpillars during the winter; it also eats plants and does some damage to crops during the summer. Several cassowary *(Casurarius)* species are found in Australia as well as New Guinea and neighboring islands. They all prefer dense, humid forests; the Australian species is exclusively vegetarian while the others also feed on several varieties of insects. J.-P. R.

OWL-FLY *Ascalaphus libelloides* 325

325
Sub.: Arthropoda
Cl.: Insecta
S.cl.: Nevropteroidae
O.: Planipennia (Neuroptera)
S.o.: Myrmeleonoidea
F.: Ascalaphidae

Purists might prefer to call it "the little dragonfly that looks like an owl"; in any case, this elegant insect, roughly the size and shape of a small dragonfly, also has enormous compound eyes and two long feathery antennae with a sort of knob at the end. It measures about 30 to 40 mm; the head and thorax are very bristly, and the body is black. The owl-fly is generally found patroling over grassy areas in the chalk hills of Southern Europe, e.g. in the region between the Maritime Alps and the Seine.

Owl-flies mate on a sunny summer day while both sexes are gliding gracefully through the air; the male seizes the female, rather brusquely, with the pair of pincers at the tip of its abdomen, and copulation takes place immediately. The larvae grow very slowly, taking two years to develop fully, at the end of which time they will have attained a length of 15 to 20 mm.

▶ The owl-fly larva has a quadrangular head, considerably recessed into the thorax and short, filiform (spindle-shaped) antennae; its rudimentary eyespots (stemmata) are mounted on stalks that protrude from the side of the head, which gives it a panoramic view of its immediate surroundings. *Ascalaphus* is a relative of the ant lion, but the larva, which hunts in broad daylight, does not construct a trap or a tunnel; instead it conceals itself beneath a few blades of grass or a heap of gravel and awaits the approach of its prey, which is usually another larva or a small crawling insect. Its legs are very short, its pace is sluggish, and the element of surprise is undoubtedly of the essence, so it further increases its chances of approaching its prey without being detected by outfitting itself with a bulky camouflage net consisting of scraps of lichen and vegetable debris.

Like those of the ant lion larva, the mouth parts of *Ascalaphus* consist of 2 separate maxilla set directly atop its 2 mandibles, all 4 of which are greatly elongated, very sharp and serrated, and, when closed, form

2 separate channels through which the liquefied tissues of captured insects can be absorbed. After the prey is caught, the larva begins to secrete its paralyzing saliva through the salivary ducts located between both sets of mouth parts; the saliva also contains a protelytic enzyme that expedites the process of external digestion.

On sunny summer days, the adult owl-fly hunts on the wing, remaining aloft for long periods, frequently gliding or "soaring" without moving its wings at a height of 2 or 3 meters above the ground. It feeds on flies and hymenoptera, and its large compound eyes allow it to catch sight of an insect as soon as it leaves the protective cover of the "grass forest." The owl-fly seizes its prey with its forelegs and mandibles, like a dragonfly, and tears it to pieces with its jaws.

◀ The adults are probably preyed on by diurnal insectivorous birds, and it has been suggested that they are parasitized by various ichneumon fly (e.g., braconid) and microhymenopteran species as well, though we have no definite information on either of these points. It is also not clear whether the elaborate camouflage adopted by the larva also plays a defensive role in protecting it from predators.

The family Ascalaphidae comprises 65 different genera, a total of about 300 species, most of them found in tropical or subtropical regions. R.C.

OYSTER CATCHER *Haematropus ostralegus* 326

The oyster catcher is about 41 cm long, with a wingspread of 79–82 cm, and weighs between 400 and 600 g. It attains sexual maturity at the age of 2 or 3, and a banded specimen is known to have survived to the age of 36, attesting to the hardiness of this species, which generally lives for about 10 years. The oyster catcher is found along sandy or rocky shorelines; it is active by day and night, noisy and restless, and its contrasting black-and-white plumage is readily visible on the sandbars and mudflats where it hunts for mollusks and other marine invertebrates. It is particularly gregarious outside the nesting season, when it sometimes congregates in very large flocks. *Haematropus ostralegus* is found in Europe and Asia, a susbspecies, *Haematropus ostralegus palliatus,* in both North and South America.

▶ The oyster catcher's attractive yellow-orange beak is versatile as well as ornamental. It uses the point of its beak to pry open or bore through the shells of bivalves (cockles and mussels for the most part—it does not actually feed on oysters). The flattened edge of its beak is useful is scraping limpets off the rocks, and it probes through the mud in search of worms *(Arenicola marina)* and crustaceans. It also uses its beak to investigate the tidal wrack, turning over clumps of seaweed that may have insects and small crustaceans clinging to them. Finally, it has been known to raid tern colonies and feed on the eggs and nestlings.

◀ The oyster catcher's eggs are perfectly concealed amid the sand and the shingle, but the nestlings are sometimes devoured by skuas and other large gulls. The oyster catcher has developed a highly specialized technique for harvesting bivalves of a certain size and thus has no direct competitors. Its gregarious instincts enable it to exploit the food resources of a particular beach with greater efficiency and also protect it from predators to some extent. As far as its relations with man are concerned, the oyster catcher is still regarded as a legitimate gamebird and is sometimes accused (erroneously) of causing damage to commercial mussel beds. Increased recreational use of coastal areas has deprived this species of some of its nesting sites as well. J.-F. T.

326
Sub.: Vertebrata
Cl.: Aves
O.: Charadriiformes
F.: Haematopodidae

P

327
Sub.: Mollusca
Cl.: Cephalopoda
O.: Octopoda – Incirrata
F.: Argonautidae

PAPER NAUTILUS *Argonauta argo* 327

The paper nautilus is an epipelagic cephalopod that lives in tropical and temperate waters all around the world. (*Epipelagic* means that it lives in the sunlit zone just below the surface of the ocean.) The female's body is partially protected by a streamlined calcite cap, or nacelle, which it begins to secrete very early on and which continues to grow throughout its lifetime. The paper nautilus in its nacelle certainly bears a family resemblance to the true nautilus (see NAUTILUS) in its shell, though the latter is composed of aragonite, which, like calcite, is a form of calcium carbonate, the crystalline structure of which has been rearranged in order to produce a harder and denser material. The chambered shell of the true nautilus is also a great deal more commodious than the thin nacelle of the paper nautilus, the primary purpose of which is provide protection for the eggs that the female deposits in its interior.

The male is quite a bit smaller than the female (15 cm) and, like many cephalopods, is equipped with a nuptial appendage called a hectocotylus, the tip of which is detachable and by means of which the male deposits its sperm in the female's mantle cavity or directly in the nacelle. The eggs of the paper nautilus are among the smallest of any cepahalopod's (0.8 mm), and the young nautilus is barely a milimeter long when it emerges from the egg. It spends the first few months of its life in the plankton layer and probably lives on tiny crustaceans, but nothing definite is known about its habits. The female probably becomes sexually mature at the age of 8 or 10 months, judging from the fact that the smallest nacelles that have been recovered measure about 4 or 5 cm; the nacelle may eventually achieve a size of about 10 to 15 cm.

▶ The feeding habits of the paper nautilus have been observed in aquariums but never in nature. The paper nautilus has two sets of tentacles, dorsal and ventral. The ventral tentacles are each flattened out into a broad membrane, and these are wrapped around the nacelle to create a kind of prey detector. As soon as a reasonable-sized marine organism comes into contact with the membranes, the nautilus reaches up with its ventral tentacles and drags it down to its beak. According to some observations, the paper nautilus feeds exclusively on fish; according to others, it is also very fond of small crustaceans.

◀ The paper nautilus is a slow swimmer, and the fragile nacelle does not afford very much protection from predators. The beaks of paper nautiluses are frequently found among the stomach contents of large pelagic fish, notably tuna, and marine ceteceans. Mimicry and camouflage are probably the nautilus's best defense, but recent observations of a still unclassified *Argonauta* species in the waters off Hong Kong have suggested another possibility. A group of 6 female nautiluses had linked themselves into a single "supernautilus" convoy by clinging to one another's nacelles with the suckers at the ends of their tentacles; all 6 were able to swim in perfect unison and undoubtedly presented a much more daunting target to a potential predator.

There are at least 6 species of *Argonauta,* of which 2 others, *A. hians* and *A. nodosa,* are found all over the world. S.v.B.

328
Sub.: Vertebrata
Cl.: Aves
O.: Passeriformes
F.: Emberizidae

PARADISE TANAGER *Tangara chilensis* 328

This small tropical bird measures about 11 cm in length and is found throughout the canopy forests of the Orinoco and Amazon basins, or from Venezuela and the

Guianas to southern Brazil; its species name notwithstanding, it does not range as far south as Chile. The paradise tanager is considered an indifferent singer, and is so called because its brilliantly colored plumage is surpassed among the birds of tropical America only by that of the hummingbirds.

▶ The paradise tanager supplements its normal diet of fruit and berries by searching for insects and spiders among the patches of lichen, moss, and epiphytic vegetation that cover the upper branches of the forest canopy. Birdwatchers sometimes succeed in coaxing it down to the ground by setting out fresh bananas on a feeding table.

◀ Even when partially concealed by foliage, the paradise tanager is still quite conspicuous and is probably preyed on by various accipitrine hawks.

The tanagers (Thraupinae) constitute a numerous subfamily of New World buntings, comprising 73 genera and 236 species altogether. Of the 4 species that occur in North America, the best known are the scarlet tanager *(Piranga olivacea)* and the summer tanager *(P. rubra)*, both of which catch insects on the wing, somewhat in the manner of the European flycatchers, and feed on the larvae of ground-nesting wasps. *Piranga* tanagers raise larger broods than their sedentary South American cousins and may spend the winter as far south as Bolivia and Peru. The gray-headed tanager *(Eucometis penicillata)* is an Amazonian species that seeks its prey along the forest floor and in forest clearings rather than in the canopy, feeding especially on insects that have been flushed out of their burrows by driver ants, anteaters, or domestic fowl. J.-F. T.

PARDEL LYNX *Lynx pardinus* 329

329
Sub.: Vertebrata
Cl.: Mammalia
O.: Carnivora
F.: Felidae

The lynx is the largest feline and one of the powerful carnivores in Western Europe. The pardel, or Spanish, lynx measures between 80 and 110 cm and weighs between 12 and 20 kg. It has a short tail, tufted ears and long side-whiskers, and its pelt is brownish-yellow with black spots. The range of the pardel lynx is steadily diminishing, and it is now found primarily on the Iberian peninsula and in northern Greece. It is a solitary hunter with an intimate knowledge of its compact hunting territory of 3–5 sq km, continually scent-marking trees, rocks, and underbrush with its urine and blazing the bark of trees with its claws.

▶ The pardel lynx exhibits a clear preference for mammalian prey, including rabbits, otters, various rodents, and the young (though rarely the adults) of such larger species as the wild boar, though it sometimes preys on waterfowl and other birds as well. In Spain, the lynx devotes most of its time to hunting rabbits. The lynx hunts from ambush; during the day, it stations itself on a elevated vantage point or in the underbrush—though because of its excellent protective coloration, it makes little effort to conceal itself—stretches out full length on the ground, and awaits the approach of a suitable prey animal.

The lynx is noted for the keenness of its vision, and Linderman has been able to establish that it can distinguish the outline of a roe deer at a distance of 500 m, a rabbit at 300 m, and a rat at 75 m—a claimed record for the animal world. At night, even though it may still able to distinguish moving shapes and forms, the pardel lynx depends on its sense of hearing to identify prey animals, so its predatory mechanisms will be triggered, for example, by the sounds made by a rodent in the underbrush, but not of a wild boar rooting in the topsoil. If the lynx hears the scream of a rabbit that has just been caught, which means that another predator has been poaching on its territory, it rushes out immediately to repel the intruder.

Once a prey animal has been sighted (and no matter how far away it is), the lynx immediately crouches down and creeps rapidly toward it, the thick pads on its feet enabling it to stalk its prey through the underbrush without making the least sound. When it is only about 4 m from the prey animal, which is still totally unaware of its approach, the lynx suddenly bounds toward it; if the prey animal is successful in escaping from this initial assault, the lynx will not attempt to pursue it for very long. If the lynx's initial assault is successful, it holds the prey animal down with its claws and dispatches it with a powerful bite to the back of the neck. The pardel lynx has also shown itself to be an excellent swimmer when it hunts for ducks

Two or three lynxes (though never more) will occasionally band together in order to harry a yearling buck or an older large herbivore that has become somewhat enfeebled; in these cases, the lynx tries to dig its claws into the animal's hide and worry its throat with its teeth until the animal dies of asphyxiation. But since the lynx needs to eat only about 1 kg of meat per day and never maintains a cache of surplus food, these cooperative attacks on larger prey animals are seldom attempted. The pardel lynx never eats a prey animal at the site of the kill, always removing it to a carely chosen, sheltered spot where it can eat without interrupting its continuous surveillance of its territory. When it has eaten its fill, it buries whatever remains of the carcass and covers the hole with leaves.

◀ The pardel lynx is a superpredator and its refusal to tolerate other carnivores (which might include the fox, European wild cat, beech marten, badger, and ichneumon) on its territory effectively serves to maintain the ecological balance of a given region by regulating the number of predators. *Lynx pardinus* is the most severely threatened of the lynxes. Though all European lynxes are protected, the pardel lynx continues to be persecuted as a henhouse raider and to be dispossessed by the destruction of the last great forests in the Mediterranean Basin, so that its populations have been declining virtuallly throughout its range.

The lynxes are no longer regarded as a separate genus by many authorities, and the pardel lynx, for example, may be identified as *Felis pardinus* in recent classifications. The European lynx, currently found in Scandinavia and Central Europe, is lighter in color and somewhat more robust in build than the pardel lynx; this species has recently been reintroduced into the Alps and the Jura ranges in order to control the number of foxes and thus limit the spread of rabies. The Canadian lynx and the bobcat (also called the red lynx) are both found in North America. The caracal, or Persian lynx, found in desert regions of Africa and the Middle East, has evolved a remarkable hunting technique—catching small birds in flight by executing spectacular leaps into the air. P.A.

Pareas carinatus 330

330
Sub.: Vertebrata
Cl.: Reptilia
O.: Squamata
S.o.: Ophidia
F.: Colubridae

This small, thin-bodied Old World colubrid is about 60 cm long. It has very large eyes, its head is short and broad, and its body is brownish. Basically arboreal in its habits, it usually found in dense foliage—a hedge, thicket, or among the lower branches of a tree—or occasionally curled up between the roots of a tree or concealed beneath a fallen log or a pile of forest litter. Its range extends southerly and easterly from Burma and South China as far as Java.

▶ This species feeds almost exclusively on snails and other pulmonate gastropods; it hunts by night, and it locates its prey by chemoreception. *Pareas's* lower jaw is much less flexible than that of other snakes and is equipped with two long curved fangs; it hooks these fangs into the snail's soft body and twists it out with a rotary motion of its head (thus, more in the manner of a corkscrew than a snail pick) while pressing the shell against the ground with its body.

Flight and concealment are *Pareas's* only defense against predators, which are undoubtedly numerous. Ophidophagous (snake-eating) snakes are particularly common in Southeast Asia; raptors and large terrestrial birds that seek their prey among the forest litter probably also feed on the adults, as well as frogs and toads on the juveniles.

Several other Southeast Asian species belonging to the genera *Aplopeltura* and *Pareas,* as well as the members of the family Dipsadniae in tropical America, occupy a similar ecological niche. D.H.

PEARLY NAUTILUS *Nautilus macromphalus* 331

331
Sub.: Mollusca
Cl.: Cephalopoda
O.: Nautiloiiae
F.: Nautilidae

A true living fossil, the pearly nautilus has not substantially altered its appearance since the Cambrian Era, 300 million years ago. It is the only cephalopod that has retained its external shell, which, like the cuttlebone of the cuttlefish, acts as a flotation chamber and allows the nautilus to propel itself almost effortlessly through the water. The nautilus lacks the ink sac that is found in other cephalopods. This particular species is found off the coasts of New Caledonia and the Loyalty Islands; it lives on the edges of coral reefs at a depth of 50–600 m.

The pearly nautilus is rarely encountered by man since it shuns the light and comes up only to feed in shallow water at night. In spite of numerous laboratory studies conducted in recent years (notably in Noumea, New Caledonia), the embryonic and juvenile development of the pearly nautilus, which normally occurs at depths in excess of 400 m, have not been observed thus far. The substantial size of the eggs suggests that the hatchlings are about 1.5–2 cm long when they emerge the egg. The pearly nautilus is sexually mature at the age of 3, by which time it will have attained a length of 16–18 cm and its shell will have up to 30 chambers. Longevity in nature is unknown, but indivudals specimens have been kept alive in aquariums for periods exceeding one year. Other aspects of the biology of this species have recently been clarified by aquarium studies conducted in Tokyo and Monaco, but a great many to the details still remain to be filled in.

▶ The upper and lower edges of the pearly nautilus's powerful beak are calcified, and its mouth is surrounded by some 40 tentacles, which are prehensile, retractable, and adhesive, though not equipped with suckers like those of an octopus. The nautilus feeds primarily on hermit crabs and other small crustaceans as well as rock lobsters and fish. In the aquarium, the nautilus will not attack any creature that is capable of rapid motion; it will accept bits of fish, chicken, or shrimp in lieu of living prey.

◀ The nautilus is protected by the tough mantle that covers its head, and it is able to retract its entire body into the outermost chamber of its shell when threatened. When cracks appear in the "living chamber" of the shell—presumably as a the result of intraspecific combat or of attacks by unknown predators in the depths of the ocean—the nautilus is able to repair them with a cementlike secretion of its mantle. Netted traps for pearly nautilus, something like lobster pots, are sometimes set out along the bottom; its flesh is sometimes eaten by Polynesians, though the principal object of this fishery is the nautilus' shell, which is exported or sold locally as curious and *objets de décor*.

Nautilus pompilius ranges from the Andaman to the Fiji Islands; a subspecies, *Nautilus pompilius stenomphalus,* is found off the coast of the Philippines and along the Great Barrier Reef, *N. scorbiculatus* off the coast of New Guinea; an entirely new species, *N. belauensis,* was recently discovered off the island of Palau in the Carolines. The pearly nautilus is somewhat similar to the extinct ammonites, a tribe of mollusks that lived in waters of medium depth during the Mesozoic; these creatures traveled in schools and most were likely formidable predators, particularly such species as the giant *Cameroceras,* which attained a length of up to 10 m.

S v. B.

PEL'S FISHING OWL *Scotopelia peli* 332

332
Sub.: Vertebrata
Cl.: Aves
O.: Strigiformes
F.: Strigidae

This species is found all across subsaharan Africa, from the Atlantic to the Indian Ocean. It lives in wooded areas in the vicinity of rivers, lakes, and swamps. It measures between 51 and 63 cm with a wingspread of 80–90 cm. Its wingbeats are deep in amplitude and virtually silent, like those of other owls; it is basically nocturnal but sometimes ventures abroad by late afternoon. It builds its nest in an abandoned eagle's aerie, but we have no particular information on the longevity or the reproductive cycle of this species.

▶ Pel's fishing owl is thought to feed primarily on fish and amphibians; its hunting behavior has not been observed directly, but is thought to resemble that of most other fishing owls, which skim over the surface of the water and seize in their victims in its talons while they are swimming just below the surface. The hooked talons of Pel's fishing owl are well adapted to this purpose; the tarsi and toes are featherless, and the inner surfaces of the toes are covered with abrasive scales that make it easy to grasp a wet slippery fish or frog. The fishing owl retains the prey in its talons until it is ready to feed, then tears it to pieces with its beak.

◀ This species appears to have little to fear from predators or competitors of any kind, including man. There are 2 other *Scotopelia* species in subsaharan Africa; neither is as widespread in its range as *Scotopelia peli,* and even less (which is to say, virtually nothing) is known of their habits. There are no New World fishing owls, but the 4 different species of the genus *Ketupa* are found over a very wide range in Asia, from the Kolyma Mountains of Eastern Siberia to the Malay peninsula and from the shores of the East China Sea to the Eastern Mediterranean. These Asian owls prey on small mammals, birds, reptiles, crustaceans, and insects as well as fish. Some species seek their prey by wading through the water as well a by patrollings on the wing, notably Blakiston's fishing owl *(Ketupa blakistoni),* which hunts for crayfish along the beds of shallow streams in Eastern Siberia and the interior of Manchuria (and is the only fishing owl that has feathers on its feet). J.-P. R.

333
Sub.: Onychophora
F.: Peripatidae

PERIPATUS *Peripatus acacioi* 333

Peripatus is the most famous representative of an obscure invertebrate phylum, Onychophora; it looks something like an enormous cutworm with numerous stubby, laterally mounted walking legs,which are known as lobopods. This particular species measures about 30 cm when in repose, more than twice that when fully extended. The males weigh about 170 mg, the females about 470 g, or 2.5 times as much as the males; the males have only 24–27 pairs of lobopods, the female 26–29 pairs. Peripatus's body is encased in a thin shell, about 1 μm thick, which is continually being renewed by exuviation (which is to say that a new shell is contiually being laid down beneath the old one, which is molted about every 3 weeks).

Peripatus mates only once during its lifetime. The spermatozoids are deposited by the male in a series of receptacles in the female's body (copulatory vesicles), and once impregnated, the female will continue to produce a "litter" of 4 live-born offspring every year. The young grow slowly (though the weight of females increases at a rate approximately 2.5 times that of the males), and life expectancy in captivity is about 3.5–4.5 years for the males and about 7–8 years for the females. This particular species is found in the countryside around Ouro Preto, Minas Gerais, Brazil, at about 1200 m above sea level.

Peripatus acacioi lives in cracks and crevices in the soil, an environment in which the temperature is constant at 18°C and the humidity at 100 percent. During the dry season, *P. acacioi* retreat en masse into the deeper crevices and huddle together conserving moisture until the rains begin again—at which point the group returns to the surface to go their separate ways. Other invertebrates found in the same biotope include terrestrial planarians *(Geoplana),* several species of harvestmen *(Gonyleptides),* cockroaches *(Parahormetica),* and scorpions *(Bothrirus)*

▶ In nature, *P. acacioi* feeds on small arthropods for the most part, including bristletails, springtails, and spiders, as well as baby scorpions that have tumbled off their mothers' backs, and preys opportunistically on dying or disabled crickets, cockroaches, or other large insects. Captive specimens in the laboratory became accustomed to a hearty meal of minced cockroach about every two weeks.

The liplike border of peripatus's mouth can be extended into a sort of suction cup that adheres to the prey while the tissues enclosed by the cup are partially liquefied by the proteolytic enzymes in peripatus's saliva and partially shredded by the shearing action of two pairs of so-called mandibular blades, working in alternation, which are also housed inside peripatus's protractile mouth. The digestive products are separated from the lining of the gut by a disposable membrane, which permits the passage of peripatus's digestive fluids in one direction and dissolved nutrients in the other; one of these membranes, along with its cargo of uric acid crystals and other indigestible waste products, is sloughed off and expelled through the anus every day.

Peripatus is also equipped with a pair of tentacles, located on either side of its mouth, through which it can project a whitish gluelike substance that solidifies on contact with the air, and in which a large insect can readily become entrapped. This chemical weapon is fairly accurate within a range of about 50 cm, and (though there is some disagreement on this point) appears to serve both an offensive and a defensive function. *P. acacioi* has been observed to discharge these jets at an aggressor before retreating, but if the adversary is completely immobilized by this parting shot, *P. acacioi* will probably return to make a meal of it shortly afterward.

◀ Numerous specimens (including onychophorans other than peripatus) have been recovered from the wild with lobopods missing and long scars on their body shells. In Australia and New Zealand, these mutiliations have been attributed to large terrestrial planarians. In Brazil, centipedes are known to prey on other peripatus species, and a North American herpetologist has recently identified a coral snake *(Micrurus hemprinchi)* that feeds primarily on onychophorans. We suspect that the numerous small scars observed on the body cases of *P. acacioi* were made by the pedipalps of scorpions *(Bothriurus)*, since *P. acacioi* will allow a number of other arthropods (spiders, harvestmen, cockroaches) quite literally to walk all over, but the approach of *Bothriurus* inevitably provokes a discharge of sticky glue from *P. acacioi*. (This of course does not preclude the possibility of attacks on *P. acacioi* by large planarians and centipedes, which are also found in the same bitotope.) Numerous instances of homophagia (cannibalism) have been reported in the literature on peripatus, but this has not been observed in the case of *P. acacioi*. The phylum Onychophora comprises about 80 species, divided into 2 main families. The family Peripatidae is distributed in a wide belt across the tropics (the eastern Himalayas, Malaya, Sumatra, the Caribbean and tropical America, Central Africa), and Peripatopsidae is found in Australasia, Melanesia (New Britain, New Guinea), Chile, Patagonia, and South Africa. Average length varies considerably among al the species—between 15 mm and 15 cm—and the number of lobopods may vary between 14 and 43 pairs. A few of the Australian species are oviparous (egg-laying); the others give birth to between 1 and 20 live young at a time.

Onychophorans can survive only in dark, moist areas, such as the underground crevices favored by *P. acaioi;* other species may be found beneath the bark or under the roots of trees, under rocks, in rotten logs, leafmold, or forest litter. They are exclusively carnivorous, feeding on small arthropods for the most part (and never on vegetable debris, as has sometimes been asserted). Fossils discovered in Cambrian schist confirm, as their appearance certainly suggests, that peripatus and its relatives have changed very little over the past 500 million years. It has been suggested that the onychophorans represent an intermediate stage between the annelids and the arthropods (and in some classifications, Onychophora is downgraded to a class of Arthropoda), but recent research tends to confirm that these curious creatures evolved quite independently of other main lineages. R.L.

PHANTOM GNAT *Chaoborus flavicans* 334

334
Sub.: Arthropoda
Cl.: Insecta
O.: Diptera
F.: Chaoboridae

The aquatic larva of this flying insect, a close (albeit non-stinging) relative of the mosquito, measures about 15 mm and is found in lakes, ponds, marshes, pools, and similar freshwater habitats. The larva's body is perfectly transparent and has a sort of flotation chamber at either end; it lives near the surface at night, then fills these floats with water and descends to the bottom at daybreak, sometimes burying itself in the sediment, and does not return to the surface until nightfall. It absorbs dissolved oxygen through its skin, so it does not need to come to the surface to breathe like many aquatic larvae. *Chaoborus* undergoes 4 brief larval stages before it pupates and emerges as an adult; the interval between successive generations may be as little as a few weeks in the summer or as much as several months in the winter, due to a period of larval diapause (equivalent to hibernation).

▶ The larva of the phantom gnat may remain immobile for long periods, and will jerk its body spasmodically in order to move in one direction or another. The larva becomes of aware of the proximity of the tiny planktonic crustaceans on which it feeds by detecting the vibrations they make in the water. First-stage larva prefer to feed on the tiny rotifers that are disdained by the larger fourth-stage larvae, though these in turn, when presented with a choice in the laboratory, clearly preferred the smaller and more manageable *Daphnia* to the larger ones.

The shape as well as the size of the potential victim is a factor in prey selection. *Chaoborus* prefers to attack small crustaceans whose bodies are elongated, rather than globular, and thus easier to catch hold of. Not surprisingly, rapid flight appears to be the most effective of the defenses evolved by prey populations in order to escape the selective depredations of the voracious but slow and erractically moving larva of *Chaoborus flavicans*.

◀ The older *Chaoborus* larvae sometimes prey on the younger and smaller larvae, though the potential for homophagia (and other forms of intraspecific competition) is limited, since different-stage larva, when present in the same habitat at the same time, are generally found at different depths. *Chaoborus* larvae are also a great favorite with fish hatchlings of many different species as well as some amphibians and adult aquatic insects. Since *Chaoborus* takes refuge in the poorly oxygenated waters along the bottom, or the layer of sediment on the bottom, during the daylight hours, it has little to fear from plankton-eating fish that track their prey by sight; the total transparency of its body is clearly a useful passive defense as well.

The adapative value of *Chaoborus flavicans's* daily descent to the bottom is illustrated by the counterexample of a related species, *C. obscuripes*, which remains near the surface during the day and is likewise never found in habitats where plankton-eating fish are present. R.P.

335
Sub.: Arthropoda
Cl.: Crustacea
O.: Amphipoda
S.o.: Hyperiidea
F.: Phronimidae

Phronima sedentaria 335

Phronima sedentaria is a pelagic amphipod crustacean, which only occurs accidentally near the shoreline. The adult female measures between 25 and 40 mm; the male is substantially smaller, measuring only 8.5–12 mm. This species is found throughout the oceans of the world out to a depth of 800 m, and, like many planktonic crustaceans, it remains in deeper water during the day and returns to the surface to feed at night. In the Mediterranean, *Phronima* is sexually mature at the age of about 2 months. This species has no common name in English, though it French it is sometimes called the *tonnelier de mer*, or "sea cooper," for reasons that will be explained shortly.

▶ The female *Phronima* finds shelter as well as food by preying on pelagic tunicates, primarily salps and pyrosomes (colonial tunicates whose fingerlike communal shells can measure up to 14 cm in some Mediterranean species). *Phronima* snips away the soft parts of the tunicate's body and devours them, then scrapes off the interior of the tunicate's barrel-shaped shell, which is open on both ends, and crawls inside it. *Phronima* may have to appropriate several successfully larger tunicate shells in the course of its life, and since the appearance of the original cellulose "tunic" is barely recognizable after it has been remodeled by *Phronima*. The identity of the tunicate species involved was not conclusively established until a series of precise biometric studies had been undertaken.

The tunicate's shell is thick and flexible, and *Phronima* clings to the inside of the barrel with 2 pairs of its forelegs and 2 pairs of its hindlegs while propelling itself through the water with oarlike swimming appendages, called pleopods. *Phronima* does not have to leave its shell in order to catch prey animals, all of which are slow-moving, relatively soft-bodied organisms of a type sometimes referred to as "gelatinous macroplankton," including siphonophores (pelagic cnidarians) and ctenophores (comb jellies, so called because their organs of locomotion consist of a series of comblike vibratory plates) as well as tunicates of the same species from which *Phronima* obtains its shell.

Observations of *Phronima's* feeding habits have been made in nature and, somewhat more extensively, in the laboratory, after specimens freshly caught with a plankton net were brought to the marine biological station at Villefranche, near Nice. *Phronima's* third and fourth pairs of forelegs protrude from one of the open ends of the barrel, and as it approaches a siphonophore, for example, its needle-sharp claws penetrate the prey animal's body. *Phronima* retracts its legs, drawing the siphonophore into the barrel, where it is transferred to *Phronima's* mouthparts by the first two pairs of forelegs, specifically modified prehensive organs called gnathopods. A siphonophore or small salp will be entirely consumed in several minutes, and nematocysts of the same siphonophore species that *Phronima* has been observed to attack in the aquarium have also been recovered from the stomach of specimens netted at sea.

◀ *Phronima's* borrowed tunic also provides protection for its numerous brood of larvae, which hatch inside the barrel and are supplied with food by their mother. There are a number of female insects and spiders that provide food and protection for their offspring, but *Phronima* provides the only comparable example among the crustaceans.

The large pair of pincers at the tip of *Phronima''s* fifth pair of legs appears to be used in fending off predators, since it is not involved in the capture of prey animals. *Phronima's* body is perfectly transparent except for any undigested food that might be contained in its stomach. The gelatinous coating of the tunicate's tunic is not greatly to the liking of most fish, though *Phronima* is sometimes found, barrel and all, in the stomachs of tuna and other large pelagic species. *Phronima's* appropriation of the tunicate's shell might incidentally be regarded as a form of protective mimicry.

The suborder Hyperiidea comprises more than 300 known species, all of which are to some degree parasitic on jellyfish and other soft-bodied macroplanktonic organisms. The term *parisitoid* is usually applied to an insect that develops inside the body of a host and gradually devours it from within; this also seems to describe the career of the hyperiid *Lestrogonus schizogonus,* which measures no more than 3–4 mm and spends its entire larval phase inside the body of a leptomedusan that measures 10–20 mm in diameter. The female *Lestrogonus* deposits its egg in the genital gland or on the manubrium of the jellyfish, and the newly hatched larva soon migrates to the underside of the canopy, where it begins to feed on planktonic organisms that have killed by the jellyfish's tentacles. Finally, when *Lestrogonus* has grown to full size, it can no longer sustain itself on the jellyfish's scraps and devours its host outright.

Hyperia galba, found in the coastal waters of the Atlantic and in the English Channel (and often found inside large beached jellyfish) measures between 10 and 24 mm and is mentioned several times in this book as a commensal (or parasite) of the larger scyphozoans (see COMPASS JELLYFISH, *Rhizostoma pulmo).* It often lives inside the gastric cavity and has been known to feed on the body (though more frequently on the stomach contents) of its host.

Phronima sedentaria constitutes an even less clearcut case, since it appropriates the shell of its host only after it has entirely devoured the soft tissues, in much the same way that the hermit crab makes off with the shell of a mollusk that it has just made a meal of. In spite of *Phronima's* highly evolved behavior patterns—with the empty shell of the tunicate serving as protection for the adult female as well as a sort of brood chamber for the larvae, which are provided with food by their mother—it must still be interpreted in light of the parasitoidal tendencies of the more typical hyperiids, whose behavior is only beginning to be understood as the result of direct underwater observations and a few instances in which both the crustacean and the host species have been successfully maintained in the aquarium. P.L.

Phryxe caudata 336

336
Sub.: Arthropoda
Cl.: Insecta
S.cl.: Pterygota
O.: Diptera
S.o.: Cyclorrhapha
F.: Tachinidae

Phryxe caudata is an endoparasitic tachinid fly that is very common in the Mediterranean region; its distribution naturally corresponds very closely to that of its primary parasitic host, the pine processionary moth *(Thaumetopoea pityocampa).* The adult tachninid measures between 5 and 10 mm; there is a more or less distinct reddish spot on both sides of the second and third abdominal segment, and the head, thorax, and abdomen are uniformly covered with thick bristles. Even specialists may have some difficulty in distinguishing *Phyrxe caudata* from other tachinid species by purely visual inspection.

▶ The female tachinid rarely strays very far from the nests of the pine processionary, which, like the more familiar ''tents'' of the gypsy moth, are made of of innumerable silken strands spun between adjoining branches. After mating, the female tachinid seeks out a likely caterpillar in its nest and approaches it with its legs fully extended and its abdomen tilted forward, thus bringing its long (1 cm) retractile ovipositor to the fore. Most of the caterpillar's body is covered with long venomous spines, but the parasite is careful to insert its ovipositor into one of the various unprotected regions—the intersegmental membranes, the base of one of the "false legs" on the underside of the abdomen, or the ventral surface of the abdomen.

Before it has exhausted its clutch of eggs, the female tachinid will have parasitized as many as 150 caterpillars, on the average, depositing a single egg on the body of each one. The maggot crawls out of the egg case (chorion) almost immediately and begins to burrow its way through the caterpillar's body wall with its long sharp beak; after a few minutes, it will be comfortably ensconced inside the body cavity. It spends its first two larval stages in the interior of the caterpillar; the third larval stage, during which the edible portions of the caterpillar's body are finally consumed by the maggot, is comparatively brief.

The pupal stage corresponds to the period of winter dormancy, with the pupa lying concealed among the silky filaments on the exterior of the caterpillar's nest. The adult finally makes its appearance in the late winter or early spring; copulation occurs almost immediately, and the cycle begins again. *Phryxe caudata* can be induced to deposit its eggs on caterpillars of other species in the laboratory, but in nature it exhibits an almost invariable preference for the pine processionary.

◀ The tachinid's parasitic strategy is frequently defeated by the presence of *hyper*-parasitic larvae in the body of the same caterpillar, frequently chalcid or pteromalid wasps *(Dibrachys, Habrocytes).* It sometimes happens that several tachinids will lay their eggs on the same caterpillar, though no more than two tachinid maggots can develop normally in the body of the same host; the presence of more than two parasites, in fact, is likely to prove fatal to the host, and thus to the parasites as well.

M.M.

Phytoseiulus persimilis 337

337
Sub.: Arthropoda
Cl.: Arachnida
S.o.: Gamasidae
F.: Phytoseiidae

This useful mite, a typical representative of the family Phytoseiidae, is normally found on the leaves and stems of dicotyledons (deciduous trees and herbaceous flowering plants). This species is native to the Mediterranean basin and Chile, and has since been introduced in substantial numbers into other agricultural regions with a comparable climate, as well as nurseries and greenhouses in colder regions. *Phytoseiulus* is active at temperatures of 20°C and above.

The body of the female phtyoseiid is ovoid, about 0.2 mm long, and weighs 25 μg; it is whitish in color and semitransparent, so that the brightly colored spider mites that it feeds on are sometimes clearly visible in its interior. Its mouthparts consist of 1) a pair of clawlike pincers (cheliceres); 2) simple pedipalps without terminal pincers, which serve as sensory organs *(Phytoseiulus* is eyeless, like virtually all gamasids) and are also useful in cleaning its other appendages; and 3) a long pointed beak (rostrum) pierced by a channel that gives access to the phytoseiid's mouth. It has four pairs of legs, the anterior pair serving as tactile and olfactory organs, the remaining three as walking legs.

▶ *Phytoseiulus* preys exclusively on red-spider mites of the genus *Tetranychus,* which feed on the sap of dicotyledonous plants. A female phytoseiid consumes an average of one *Tetranychus* egg per hour; it detects living prey by the simple process of waving its anterior pair of legs in the air until they come in contact with a mite that it considers edible. It pierces *Tetranychus's* body case with its pointed beak and injects its saliva; the soft tissues of its victim are liquefied by the process of external (also called pre-oral) digestion with the result that *Phytoseiulus* does not excrete solid feces. Even when *Tetranychus* is in short supply, *Phytoseiulus* may still be able to survive by drinking rainwater and occasionally resorting to cannibalism.

◀ *Phytoseiulus* is not very susceptible to parasites or virus and microbial infections, though in its original habitat it is probably preyed on by other small (unidentified) arthropods.

The Phytoseiidae are a cosmopolitan family *(A. largoensis,* for example, is found in bogs and on the tundra of the Canadian Arctic), comprising several hundred species altogether. A number of others besides *Phytoseiulus persimilis* are sometimes employed as agents of biological pest control, notably *Amblyseius potentillae,* an arboreal species that feeds on red-spider mites (also of the genus *Tetranychus)* that infest vineyards and fruit orchards. C.A.-H.

PIED BARBET *Tricholaema leucomelan* 338

338
Sub.: Vertebrata
Cl.: Aves
O.: Piciformes
F.: Capitonidae

About the size of a sparrow, the pied barbet has a thick beak and fairly large head surrounded by a collar, short legs with 2 toes, and remarkably bright plumage. It weighs between 29 and 33 g and is found in the tropical regions of the Old World, particularly on the savannahs of southern Africa. It excavates a home for itself in the soft, fleshy trunk of the giant African spurge *Euphorbia candelabra,* in a dead tree or stump, a termite mound, and sometimes in the ground. It prefers forest clearings or thinly forested terrain.

▶ The pied barbet feeds on fruit, whether wild (guavas, papayas, figs) or cultivated (tomatoes, bananas), as well as insects; the vibrissae, the bristly feathers that grow around its beak, are thought to keep insect prey from escaping, though they may have a protective function as well. It feeds its young exclusively on caterpillars and beetle larvae, and digs into termite mounds with its powerful beak and devours their inhabitants. It is also said, though not on the basis of very firm evidence, to have a taste for the eggs of other birds.

◀ The barbet can easily take refuge from predators by retreating into the foliage or into its nest, though the young barbet's rightful place in the nest is sometimes usurped by its scapegrace parasite cousin, the honey guide. (This bird gets its name from its habit of pointing out the location of a bee tree to human beings or other large animals in the hope of feasting on the honey, which it cannot readily obtain for itself). The double-banded barbet spontaneously enlarges its nest after it has hatched out one of these parasites, perhaps to reduce the chances that its own offspring will be ejected by the large and clamorous honey guide chick. The barbet's song is described as shrill and irritating—and it lacks the profit-sharing instincts of the honey guide—but in other respects it coexists quite comfortably with man.

The barbet family consists of 76 different species divided among 16 genera, more than half of which live in Africa. The tiny yellow-throated barbet *(Pogoniolus subsulphureus),* a cavedweller of Gabon and Uganda, is the smallest member of the family; the largest, the giant barbet *(Megalaima virens)* of Kashmir, southern China, and Vietnam, is about the size of a starling. J.-F. T.

PIED WAGTAIL *Motacilla alba* 339

339
Sub.: Vertebrata
Cl.: Aves
O.: Passeriformes
F.: Motacillidae

The pied, or white, wagtail lives in open terrain (cultivated fields, riverbanks, shortgrass meadows) throughout Europe and Asia; some spend the winter in temperate latitudes while others migrate to North or Subsaharan Africa. This species is about 20 cm long and weighs from 18 to 27 g; wagtails are gregarious birds that flock together at dusk to return to their nighttime rookery in the trees or a bed of reeds. They makes their nests on the ground in a wide variety of terrains and locations. During the breeding season, the males vigorously defend their territories from the encroachments of other wagtails.

▶ Well adapted to the pedestrian life, the pied wagtail spends most of its time on the ground, flicking its tail in a nervous, staccato rhythm as it walks along; it flies very quickly, in a sinuous rather than a straight-line trajectory. It usually catches its insect prey on the ground (occasionally in flight), often on the banks of a watercourse, or on the surface of the water; it feeds on flies and gnats, butterflies, and beetles, as well as on worms and seeds.

◀ The pied wagtail has no particular predators, though its nestlings are sometimes victimized by the parasitic activities of the gray cuckoo. In any case, the fledglings are ejected from the nest by their parents after a suitable period of time has elapsed in order to make room for a second clutch of eggs. Human beings appear to have little impact on this species, except in some conservative rural areas (notably in Occitania, or Southwestern France), where it is still shot for the pot or snared (illegally) in a fowler's net.

The family Motacillidae comprises some 70 species divided into 2 main groups. The first is the wagtails, well represented in Europe and Northern Asia, with a couple of species found in India, Africa, and on Madagascar as well. Apart from the pied wagtail, the yellow wagtail *(Motacilla flava)* and the gray wagtail *(M. cinerea)* are the most common in Western Europe.

The pipits (genus *Anthus)* make up the second group, about 40 species in all. They are well distributed around the globe, including the Antarctic islands of South Georgia, and found in flat, open areas such as tundra, Alpine meadows, and sandy beaches; the pipits generally prefer running along the ground to flying, with the exception of the tree pipit *(Anthus trivialis)*, which uses its perch both as recital stage and observation post. The rock pipit *(Anthus spinoletta)* of Western Europe is differentiated into 2 subspecies, the descendants of breeding populations that were separated by the great ice sheets of the Pleistocene. The rock pipit lives in the tundra and the Alpine meadows of Northern and Western Europe (the latter including Corsica and the Pyrenees as well as the Alps); it feeds on snails and insects. The maritime subspecies *(A. spinoletta maritimus)* is found along a narrow coastal strip that extends from France to Norway; it feeds on sand fleas and periwinkles, which it catches below the tidal mark.

J.-P. R.

340
Sub.: Vertebrata
Cl.: Mammalia
O.: Marsupialia
F.: Peramelidae

PIG-FOOTED BANDICOOT *Chaeropus ecaudatus* 340

The pig-footed bandicoot is a marsupial with a long tubular nose, pointed ears, long legs; its toes have grown together into a single "trotter," which is what gives the species its name. It is about 20 to 25 cm long; its tail measures 10 cm. (The Latin name *ecaudatus* refers to the fact that the type specimen had been deprived of its tail, which is by no means typical of the species as a whole.) The bandicoot family consists of 8 different genera, 13 species in all, which are found exclusively in Australia. The family exhibits considerable diversity; there are spiny bandicoots as well as bandicoots with soft, silky fur, so-called rabbit bandicoots. The pig-footed bandicoot inhabits the savannahs and semidesert regions of South Australia; it is strictly nocturnal and spends its days curled up in a grassy nest.

▶ All bandicoots hunt at night. The pig-footed bandicoot trots noiselessly through the outback in search of worms and insects; other species feed on roots. The rabbit bandicoot is equipped with claws rather than trotters, and it digs extensive underground galleries at a remarkable rate of speed, devouring mice and insect larvae that it encounters in the course of its excavations.

◀ Like most Australian marsupials, the bandicoot has few natural enemies except for those that have been introduced by man (the fox, the dingo) and man himself. In spite of the valuable services that the bandicoot renders to Australian agriculture by destroying insects and parasites, the pig-footed bandicoot has been hunted to the verge of extinction. P.A.

341
Sub.: Vertebrata
Cl.: Osteichthyes
S.cl.: Teleostei
O.: Perciformes
F.: Percidae

PIKE PERCH *Lucioperca lucioperca* 341

This Eurasian freshwater fish, also called the zander, typically measures between 35 and 55 cm and weighs about 1 kg, though the very largest specimens sometimes attain a length of 1.2 m and weigh as much as 18 kg. Long appreciated as a food fish in Central Europe, it originally ranged from Bavaria to western Siberia and has also been introduced into Western European rivers over the last few decades, including the Rhône (1940), the Seine (1966), and a number of other river systems during the period 1962–71.

The pike perch lives singly or in small schools, prefers open water to heavily vegetated pools and backwaters, and is somewhat less particular in its choice of spawning areas than the Pike (q.v.). Average longevity is 5–6 years, but individual lifespans as long as 14 and even 20 years have been reported.

▶ The pike perch hatchlings feed intially on freshwater zooplankton, then, at the age of 2 months, begin to move up the foodchain to prey on mosquito larvae and fry of other species. The young pike perch are voracious feeders and grow very rapidly, and before much longer are capable of feeding on schooling fish such as whitefish *(Coregonus),* roach, and bleak as well as perch. Though the pike and the pike perch are roughly the same size, the pike perch has a smaller mouth and generally preys on smaller fish.

The pike perch frequently hunts from ambush, hovering in a single spot by swimming against the current; unlike the pike, it will continue to pursue its quarry if its initial assault is unsuccesful. Its eyesight is excellent, and it sometimes actively seeks out its prey, searching carefully for the telltale movements made by smaller fish swimming along the bottom. It disables the prey with a snap of its powerful jaws, sometimes biting it clean in two (in which case, it swallows one of the fragments immediately, keeping a close watch to make sure that some other predator does not make off with the other one).

◀ The pike perch is at the summit of the freshwater foodchain, and its only potential predators are the pike (which tends to prefer a slightly different habitat, closer to the bank and more heavily vegetated) and other members of its own species. The pike perch *(sandre, Zander)* is greatly appreciated for its fighting qualities as a gamefish, and is an important food fish whose flesh is both good-tasting and nutritious.

F.M.

PINE MARTEN *Martes martes* 342

342
Sub.: Vertebrata
Cl.: Mammalia
O.: Carnivora
F.: Mustelidae

This Old World carnivore is about 55 cm long and weighs between 800 and 1800 g. Its body is long and slender, and its thick bushy tail measures about 30 cm. Its lustrous fur is dark in color, with lighter patches at the throat. The pine marten is represented by different subspecies in various parts of Europe and Central Asia, ranging as far east as Iran. It is found in dense forests up to an altitude of 2000 m and is common throughout Europe (average population density in Alsace, for example, is .76 per 100 hectares).

Each animal occupies a territory about 5 km across, which it patrols both by day and by night, always confining itself to the same trails, which, on the average, are about 9 km long altogether. A pine marten make several dens for itself in hollow trees or abandoned bird's nests and furnishes them with moss or dry leaves. The marten routinely scent-marks these trails with its urine or with special glandular secretions, which are extremely pungent.

▶ Pine martens in Sweden were observed to feed primarily on small rodents and squirrels, then, to a lesser extent, on birds and their eggs and on fruits. Other European martens are known to feed on wasps and other insects and on the young of various forest mammals; they are especially fond of a kind of wild fruit called the sorb apple. This diet is subject to some seasonal variation, and martens will also kill other animals—fish, snakes, and frogs, for example—that they do not eat, apparently, as some authorities have suggested, because they become possessed by a kind of killing frenzy once they begin to hunt.

At night, the pine marten uses its senses of smell and hearing to locate its prey, during the day, it relies on its considerable agility and sense of balance as well as on the keenness of its vision; it may sometimes be seen chasing a squirrel from tree to tree and from branch to branch, making prodigious leaps of almost 3 m. Smaller prey animals are killed instantaneously with a bite to the back of its neck; the pine marten dispatches its larger catches by slashing the jugular and crouching over with its teeth clamped down on the prey animal's neck. It removes its kill to a secluded spot before eating it, and the uneaten remnants are then transported to a special cache it maintains for that purpose, usually in a hollow tree.

◀ The pine marten has few natural enemies and avoids all encounters with other mustelids. Its anal scent glands are less developed than those of the skunk, and their contents cannot be discharged at an adversary. Nevertheless, the marten begins by turning its back if it is threatened or approached. If the aggressor is not deterred by the acrid aroma of its anal secretions, the pine marten makes for the nearest tree, where it can be quite confident of evading all pursuit. The pine marten is protected by law in Europe, though it continues to be sought for its fur and has already been exterminated in certain localities.

The sable *(Martes zibellina)*, whose fur is the most precious of all, had almost become extinct in the wild but, as a result of returning captive specimens to the forests of Siberia and European Russia, has subsequently reoccupied virtually all of its former range. The tayra *(Eira barbatus)* is quite similar to the pine marten in its morphology, dietary preferences, and predatory technique, though it is somewhat longer (60–70 cm, its tail accounting for an additional 40–50 cm) and heavier (4–5 kg). The tayra is the sole representative of its genus, and is found in wooded or heavily overgrown terrain from southern Mexico to Argentina. P.A.

343
Sub.: Vertebrata
Cl.: Mammalia
O.: Chiroptera
F.: Vespertilionidae
344
Sub.: Arthropoda
Cl.: Crustacea
O.: Decapoda
S.o.: Caridea
F.: Alpheidae

PIPISTREL *Pipistrellus pipistrellus* 343

The pipistrel is the smallest European bat, with a body length of 3.3–5.2 cm (excluding the tail, which adds an additional 2.6–3.3 cm) and a wingspread of 19 cm. It is also one the most common and is often found in the vicinity of human dwellings (even in urban areas). It is nonmigratory, and its summer and winter refuges are generally not very far from each other. In summer, it conceals itself during the daylight hours in a crack or a crevice in a wall, a hollow tree, even behind curtains and drapery. The pispistrel leaves its hiding place shortly after sunset, occasionally even before that. It flies at a height of at least 2–6 m from the ground.

▶ Like the vast majority of bats, the pipistrel feeds on insects, which it detects in flight by echolocation; because it tracks its prey very closely, making constant course corrections, its flight is very erratic, full of abrupt loops and sideslips and somewhat reminiscent of a butterfly's. The pipistrel feeds on gnats, midges, and other small dipterans, and during the summer a temporary colony of pipistrels will sometimes assemble in a marsh, on a patch of boggy ground, or in some other area where gnats or midges are swarming. When it encounters a larger insect in fight, instead of swallowing it immediately, the pispistrel sometimes snags it with one corner of its wing and wraps it up in its interfemoral membrane, which serves as a sort of game bag. The pipistrel eats the equivalent of its own weight in insects every night; it feeds while clinging to a branch or other perch, or while still on the wing in open country.

◀ The pipistrel's chief predator is the screech owl, which also feeds on the bodies of bats that have been killed by cars on the highway.

Kuhl's pipistrel *(Pipistrellus kuhli)* is found in Southern Europe and hunts at low altitude, skimming over the tops of stone walls and hedges. The serotine bat *(Eptesicus serotinus)* is a larger relative of the pipistrel; a powerful flier, it feeds primarily on cockchafers and other hard-shelled insects. The common noctule bat *(Nyctalus noctula)* is slightly larger than the serotine and flies at a very high altitude (up to 50 or even 100 m). It feeds on moths, cockchafers, and dungbeetles, and is often taken on the wing by owls and other birds of prey. J.-J. B.

PISTOL SHRIMP *Alpheus californensis* 344

Alpheus is a genus of small or middle-sized coastal shrimps (up to 4.5 or 5 cm) of warm or temperate waters, comprising many different species that are morphologically quite similar to one another and are collectively known as pistol shrimp or snapping shrimp. They usually live in pairs in small burrows in the sand. *A. californensis* is found along the Pacific coast of North America; its lifespan is at least 8 years, which is quite remarkable in a creature of this size.

▶ The pistol (or snapper) in question is the shrimp's enormous claw (chela), which may be either the left or the right, is at least as long as the shrimp's entire body, and can serve as either an offensive or defensive weapon. *Alpheus* stations itself outside its burrow, which serves as a refuge in case of attack, and awaits the approach of a prey animal of reasonable size. Smaller and slower creatures, segmented worms (annelids), for example, are seized with the lesser claw and dispatched with the larger one. With larger, more vigorous prey, such as shrimps or minnows, *Alpheus* approaches stealthily, picking its way carefully over the sand and pebbles, stuns its prey with a snap of its pistol claw, then seizes it and drags it back to its burrow. It shares its catch with its burrow-mate and tidily discards leftovers and inedible remnants outside the mouth of the burrow. *Alpheus* normally hunts by sight, but like other crustaceans, it also has a modest ability to sniff out its prey.

◀ During its first (planktonic) stage of life as a defenseless microscopic larva, *Alpheus* is relentlessly snapped up by a great variety of other zooplanktonic species. In adulthood, the burrow and the pistol both seem to serve as effective deterrents against smaller predators, though *Alpheus* species that live in tunnels in the muddy bottom may be devoured by dogfish and other bottom-feeding selachians (sharks and rays). No member of the family Alpheidae is ordinarily consumed by man.

Alpheus and its close relatives exhibit a number of behavioral idiosyncrasies, few of which have been the object of systematic study. We have already mentioned *Alpheus armatus* in connection with its commensal relationship with the anemone *Bartholomea* (see below, COMMENSAL ANEMONE). In the Red Sea, *Alpheus purpurilenticularis* lives in association with a fish called the goby, which gives a kind of semaphore signal to warn the shrimp of the approach of predators, flicking its tail rapidly from side to side while the shrimp's antennae are touching the goby's body.

A European cousin, *Synalpheus fitzmuelleri*, has devised a way of feeding on the flesh of the sea snail *Coralliophilia caribaea* without putting itself to the trouble of breaking the creature's shell, as crabs and lobsters do. It uses its larger claw to pry open the operculum (the horny plate that serves as a hatch when the snail is fully retracted into its shell), then reaches inside and snips off a few morsels of the sea snail's foot with its smaller claw. P.N.

PITTED GROUND BEETLE *Carabus variolosus* 345

This large semiaquatic ground beetle varies in size from about 23 to 32 mm; its shell is flat black, and the elytra are pitted with tiny depressions (*foveolae* is the technical term) that give the species its name. It is found throughout Central Europe, from Savoy and the Vosges and Northern Italy to the Carpathians in the east; the western subspecies is sometimes classed separately as *Carabus nodulosus* Creutzer. Though widely distributed in range, this species is not known to be common in any particular locality; it prefers sandy (siliceous) soil and wooded mountainous terrain, and is never found more than 5 or 6 m from running water. The adults spend their winter hibernation holed up in a burrow in the bank of a stream or spring, underneath the bark of a tree, or inside a rotten log.

▶ *Carabus variolosus* is a typical carabid as far as appearance is concerned, but its predatory strategy is quite atypical. Both the adult and the larva seek their prey on or below the surface of the water, propelling themselves along with a kind of alternate swimming motion of their legs, which are not terribly well adapted to locomotion in this medium. Consequently they prefer to hunt in very shallow water that is well provided with stepping stones or floating mats of peatmoss (sphagnum) and other aquatic mosses (*Fontialis, Philonotis, Bryum* . . .).

The adults are active by night as well as by day; on sunny days in the spring or summer, they station themselves on flat rocks in the streambed and will dive to a depth of about a meter in pursuit of their prey, which includes mollusks and the slower-moving aquatic insects as well as dead fish. The average dive lasts for about 4 minutes; some individuals have been known to stay submerged for as long as 20 minutes. The beetle travels with its own supply of oxygen, a bubble of air trapped beneath its elytra. The adult *Carabus variolosus* also seeks its prey on land, where its diet is considerably more varied as a function of its greatly increased mobility.

During its first stage of development, the larva is strictly terrestrial, feeding on insects, snails, and organic debris. During the second and third stages, it sculls along on the surface of the water, with its head and thorax just barely immersed, feeding on the snails and insect larvae that crawl along the surface of the underwater rocks and plants. Both the adult and larva are equipped with strong jaws with toothlike projections that enable them to grip and tear their prey.

◀ The larvae may occasionally be swallowed by fish; the nymphs, hibernating adults, and the larvae during the initial

345
Sub.: Arthropoda
Cl.: Insecta
S.cl.: Pterygota
O.: Coleoptera
F.: Carabidae

stage of their development on land may be devoured by ants and insectivorous mammals. For the most part, however, this species does not appear to be especially subject to predation.

The genus *Carabus* comprises about 500 species, only 5 of which have adopted the same semiaquatic strategy as *Carabus variolosus;* 4 of these are also found in Western Europe. M.M.

346
Sub.: Vertebrata
Cl.: Osteichthyes
S.cl.: Teleostei
O.: Pleuronectiformes
F.: Pleuronectidae

PLAICE *Pleuronectes platessa* 346

This species of flatfish (Pleuronectiformes) typically measures between 25 and 40 cm. The males generally live about 14 years, the females to about 16 years—by which time the males will normally will normally have attained a length of up to 50 cm and a weight of 2 kg, the females a length of up to 60 cm and a weight of 3.5 kg. These figures are subject to considerable variation, of course, and the largest specimen on record was 90 cm long, weighed 7 kg, and was probably 50 years old when caught.

Pleuronectes platessa is found in the Eastern Atlantic, from Gibraltar to the North Cape and beyond, as well as in the White Sea, the Black Sea, and the Mediterranean. Plaice reach sexual maturity between the ages of 3 and 12, depending not only on the sex of the individual but on the average temperature of the habitat. Males living in the North Sea mature at 3–4 years, females living in the North Sea at 6, males living in the Barents Sea (off the northern coast of European Russia) at 8–10 years, and females living in the Barents Sea at 11–12 years.

▶ The plaice, like all flatfish, is a bottom-dweller, normally found on a muddy or sandy substrates at depths of 50–100 m, though the juveniles remain much closer to shore (0–15 m). The plaice is basically a sedentary species, though with the approach of the winter, the adults swim out into deeper water (200–250 m); they also become quite active during the spawning season, sometimes covering up to 30 km in a single day. The pelagic larvae feed on other larvae, primarily those of polychaetes and mollusks; when they have attained a length of 1.3–1.7 cm, they will have assumed their adult form and begin to feed on polychaetes, thin-shelled bivalves, and small crustaceans that they catch along the bottom.

The adults detect their prey primarily by sight and secondarily by olfaction, since they occasionally seek for prey at night or at depths where the sunlight does not penetrate. The plaice gradually approaches its quarry by means of gentle oscillations of its dorsal and anal fins, and when it has arrived within striking distance, abruptly thrust itself forward with a snap of its caudal fin and swallows the prey immediately. As might be deduced from its appearance, the plaice is not a very strong swimmer; normal cruising speed is only about 0.1–0.5 kph, though it is capable of short bursts up to 3.2 kph.

◀ The plaice frequently buries itself in the sand, leaving only its bulbous eyes exposed or, when confronted with a less cooperative substrate, camouflages itself by dilating or contracting the chromatophores (pigmented cells) in the dermis. A visual image of the immediate area registers on the hypothalamus (the ventral portion of the diencephalon), which in turn stimulates the secretion of a pituitary hormone that controls the relative size of the individual pigment cells. In this way, the spots of pigment on the plaice's body are resolved into a pattern of lights and darks that correspond as closely as possible to the visual texture of the surrounding substrate—in (very roughly) the same way that a photographic image appears on a piece of film that is coated with photographic emulsion.

The plaice also changes its pigmentation pattern in response to stress or fright, and a sudden color change of this kind may prove useful in throwing off a potential predator's estimates of the size and trajectory of its quarry. In spite of its remarkable gift for mimicry, however, the plaice is preyed on by a variety of other benthic (bottom-dwelling) species, including members of the codfish family (ling, cusk), sharks and rays (cat shark, tope shark, flapper skate, thornback ray), and other flatfish (halibut).

The plaice fishery is one of the most important in Northern European waters. Plaice are caught with trawls and seine nets, and fish smaller than 25 cm cannot legally be taken in the North Sea. The annual catch currently amounts to about 120,000 metric tons, of which Danish and British vessels are each accountable for about a third. Though considerably augmented by the suspension of deepwater fishing during World War II, stocks have since been reduced dramatically by intensive fishing, and both the overall tonnage and the average size of the individual fish taken have been steadily diminishing in recent years. J.-M. R.

POLAR BEAR *Thalarctos maritimus* 347

347
Sub.: Vertebrata
Cl.: Mammalia
O.: Carnivora
F.: Ursidae

The polar bear weighs between 320 and 450 kg; the male may attain a length of 2.5 m, the female a length of 1.8–2.1 m. It inhabits the polar regions of Europe, Asia, and North America, though in fact it spends almost its entire life on the edge of the pack ice and rarely sets foot on dry land. Pregnant females spend the winter in hibernation (the temperature of the polar bear's den hovers around 0°C); the cubs are born in the middle of the winter, and though other polar bears may occasionally venture forth to search of food, females with newborn cubs subsist exclusively on their stored-up fat reserves and do not emerge from their dens until the end of March or even April.

▶ The polar bear feeds primarily on the ringed seal—on the pups during the spring and on the adults during their summer molt; when polar bears catch seals during the winter, they eat only the skin, the blubber, and intestines, and abandon the rest of the carcass. Polar bear begin to feed on newborn seal pups even before they are weaned; the females are very solicitous of their cubs' education, and the cubs take turns in accompanying their mother on hunting expeditions, learning to stalk a seal by imitation.

The polar bear begins by investigating the edges of glaciers, icebergs, and other natural shelters where seals are likely to be sleeping; it eyesight and sense of smell are both very acute, and it is able to identify a seal's den from a considerable distance. It approaches stealthily with its head lowered or even by crawling on its belly, immediately flattening itself against the ice if the seal wakes up and raises its head—even cradling its face in its paws or pushing a block of ice ahead of itself in order to conceal its conspicuous black nose. When the polar bear has arrived within striking distance, it dispatches its quarry with a single blow of its paw.

When a polar bear catches sight of a seal of the edge of an ice floe, it begins to backtrack, slipping soundlessly into the water, and sinking out of sight below the surface as it approaches closer, and finally hurtling up out of the water and cutting off the seal's retreat. A polar bear will sometimes station itself by a seal's breathing hole in the ice, waiting patiently until the seal comes up for air; ringed seals also dig dens for themselves and a nursery for their pups in the thick layer of snow that covers the pack ice. In this case, the breathing hole, which also serves as an entrance, is generally concealed by a layer of soft snow, though the polar bear's sense of smell is sufficiently keen that it can sniff out the entrance to a seal nursery through as much as a meter of snow. It scratches away the top layer of snow with its paws to expose the breathing hole, then presses down with all its weight around the edges of the hole, which causes the roof of the nursery to collapse and thus precipitates the polar bear in among its occupants, with fatal results for the latter.

In summer, the polar bear is omnivorous, feeding on seaweed, lichens, mosses, and berries in addition to its normal fare; it also hunts for reindeer, caribou, musk oxen, and turns over rocks to catch rodents. In shallow water, it catches salmon and other fish by flipping them onto the bank with its paw, and will sometimes feast on the carcass of a beached whale or other large cetecean.

◀ The adult polar bear has no predators, though a female with cubs will keep them on the pack ice, since they are likely to be attacked by wolves on dry land. When threatened, both the female and the cubs will begin to flee, though the female will often turn to confront their pursuer; male polar bears frequently rob the females of their kills and will even destroy their cubs if they are left unprotected. Male polar bears occasionally succumb to the wounds they sustain in combat with other males during the mating season.

The relations between polar bears and humans have customarily been distant but correct, the polar bear evincing a mild curiosity, perhaps, but no signs of either fear or aggression. The Canadian town of Churchill, in northern Manitoba on the shores of Hudson Bay, has recently gained some notoriety because it lies along the route of the polar bear's seasonal migrations. The bears began by feeding on garbage at the town dump, then more recently have gotten into the habit of invading houses in search of food and belligerently confronting their inhabitants; arrangements have been made for persistent offenders to be relocated to distant, uninhabited regions. J.J.B.

348
Sub.: Vertebrata
Cl.: Osteichthyes
S.cl.: Teleostei
O.: Tetraodontiformes
F.: Diodontidae

PORCUPINE FISH *Diodon hystrix* 348

The porcupine fish attains a maximum length of 80 cm—1 m and is found in tropical waters and occasionally in the Mediterranean, most frequently on coral reefs but also in coastal lagoons, tide pools, rock pools, and tidal estuaries. It is active primarily at dusk and avoids direct sunlight, which is injurious to its large, very senstive eyes. The porcupine fish lives in pairs during the spawning season but is otherwise solitary and territorial; the hydraulic pumping system that constitutes its primary defense against predators (see below) is also involved in one phase of the male's courtship display.

▶ The teeth of the porcupine fish have fused into two continuous ridges, forming a sort of beak and thus enabling it to prey on hard-shelled marine organisms such as mollusks and sea urchins and even to browse on coral, crunching the coral skeleton with its beak to get at the soft-bodied polyps. In fact, it is more or less obliged to feed on hard-shelled prey, since the two halves of its beak keeping growing continuously and (like the incisors of a rabbit) have to be ground down by constant use.

The eyes of the porcupine fish are large and can pivot independently in their sockets; it locates its prey primarily by sight, searching along the bottom, for example, for the faint ridges made by mollusks and other burrowing creatures. To dislodge a prey animal from its burrow, the porcupine fish fills its stomach with water and then, with the aid of powerful abdominal muscles, expels it in a concentrated stream, washing away the layer of sand above the prey and then snapping it up in its jaws. It chews its food with a continous grinding motion, then spits out the shell fragments and other inedible leftovers.

In order to catch a crab, it pounces on the quarry from above, then begins to swim around it in a tight circle, like a prizefighter waiting for an opening. Eventually the crab lets down its guard for a moment and the porcupine fish lunges forward and catches one of the crab's claws in its beak; when the crab has not only been disarmed but totally immobilized by having all of its legs snipped off in its manner, the porcupine fish moves in to crack the carapace with its powerful beak. It sometimes succeeds in tipping a crab or a sea urchin over on its back by directing a concentrated stream of water against the underside of its shell; then (in the case of a sea urchin), the porcupine fish starts enlarge the hole that covers urchin's mouth by nibbling around the edges, where there are comparatively few spines, until it is able to insert its beak into the urchin's shell.

◀ Young porcupine fish may sometimes fall prey to sharks, barracudas, groupers, and other large reef-dwelling predators, though the adults are more or less invulnerable to attack by any predator other than man. The porcupine fish's body is covered with short, sharp spines that are normally held flat against its body. These spines are erected when danger threatens and the muscles of the gills and stomach begin pumping in water, so the body of the porcupine fish is distended into a large prickly, inedible ellipsoid. The muscles of the esophagus and throat prevent the water from backing up, and the entire process is regulated by an instinctive shut-off mechanism, which comes into play when the skin is stretched completely taut.

The porcupine fish's spine—which has no ribs to impede its flexibility—is extremely supple, so that the fish can remain in this inflated position for several minutes; when the danger is passed, the pumping mechanism is reversed and the water is abruptly expelled through the mouth, gills, and anus. If the danger has not passed, however, the porcupine fish can reinflate itself as many as ten times in succession before it is overcome with fatigue. (The fish reacts to being removed from the water in a similar fashion, puffing itself up with air until it is almost competely spherical, though it may take several minutes for it to deflate itself).

When it is no longer capable of pumping itself up, it immediately attempts to flee, propelling itself rapidly through the water with its unpaired fins. It can make a tight 180° turn in mid-flight by agitating its caudal fin, which may be of use in throwing off its pursuer, and then seeks out a crevice or some similar refuge and securely wedges itself in place by pumping water through its mouth and gills. Normally, however, the porcupine fish tends to be curious rather than cautious, and sometimes gets into trouble on this account. Porcupine fish frequently bear the scars of an encounter with the sharp venomous spines of the lion-fish, for example, and the large staring eyes of the porcupine fish make an easy target for triggerfish and sharp-nosed puffers *(Canthigaster).* Once blinded by *Canthigaster's* beak or the triggerfish's ballistic missiles, it is quickly dispatched by these predators, although other potential predators, notably *Heniochus* and *Forcipiger* (two genera of butterfly fish), seem to mistake the black spots that cover the porcupine fish's body for its eyes, and thus rarely succeed in inflicting any serious damage.

The spines of several smaller species of porcupine fish *(Chilomycterus)* are kept permanently erect, though they protrude a little more prominently when the fish pumps itself full of water and its skin stretches taut. The related family of puffers, or globefish (Tetraodonidae), have smooth skins and are so called because the top and bottom halves of the beak are incompletely fused *(Tetraodon* = "four-toothed"), though they are comparable to the porcupine fish in their feeding habits; there are a number of freshwater species (e.g. *Tetraodon miurus* and *fahaka,* both found in the Nile). Dried puffers and porcupine fish, frequently varnished and paintedl, have been fashioned into a variety of curios and artifacts—including oil lamps in the case of the smooth-skinned puffers and, in the Pacific islands, war helmets and shields in the case of the porcupine fish.

Saltwater puffers of the genus *Tetraodon* secrete a toxin (tetrodotoxin, or tetrodoxin for short) that can cause serious illness or even death when consumed by humans. The toxin is localized in the ovaries and digestive organs, and even when these fish are carefully prepared by specially trained chefs (as is required by law in Japan), there is still a certain risk involved for the customer. This peculiar brand of culinary bravado has recently spread to North America, where there are said to a number of restaurants specializing in fugu, which has become the *nom de cuisine* of several East Asian species in English as well as Japanese. In captivity, puffers that have been alarmed or upset in some way have been known to discharge this toxin directly into the water, which is generally fatal to the other occupants of the aquarium. M.D.

PORTUGUESE MAN-OF-WAR
Physalia physalis **349**

349
Sub.: Cnidaria
Cl.: Hydrozoa
O.: Siphonophora
F.: Rhizophysaliae

Also known as caravels (in Britain) and bluebottles (in Australia), these large venomous jellyfish are actually complex colonial organisms, consisting of both polyps and medusans that have become differentiated in various ways in order to perform different anatomical functions. Dense flotillas of several hundred men-of-war are often observed in the tropical and subtropical regions of both the Atlantic and Pacific (notably along the coasts of Florida and eastern Australia).

The founder of the colony is a larval polyp that eventually develops into a large hollow floatation chamber surmounted by a comblike crest, the only part of the organism that is visible on the surface.The crest varies in color from light green to sky blue to pink or reddish violet, and the floatation chamber is kept inflated with a gas, manufactured by an internal gland, that assures the buoyancy of the man-of-war. The crest serves as the sail of the man-of-war and is kept tilted into the wind at a 45° angle by the internal pressure of the float and the contraction or relaxation of the muscles at its base.

The rest of the organism consists of various offshoots of the original polyp, which are suspended from the floatation chamber and have developed into new polyps of three different types: (1) the dactylozoids, so called because they are equipped with tentacles, which can sometimes measure up to 10 m when fully extended, (2) grapelike clusters called gastrozoids, which collectively serve as the organism's digestive system, and (3) the branching gonozoids, or reproductive organs, on which the rudimentary sessile medusans from which a new colony of polyps will develop first appear as small budlike protuberances.

▶ As with other cnidraians, the tentacles of the man-of-war are equipped with venomous stinging filaments (nematocysts), which are potent enough to catch fish, crustaceans, and other fairly large marine organisms. Once the prey animal has been paralyzed in this manner, it is moved into position by the tentacles, aspirated into the interior of one of the gastrozoids, then broken down by the digestive enzymes that are secreted by these specialized polyps, and finally absorbed into the organism.

◀ As is frequently the case with large cnidarians (primarily scyphozoans), there are at least two species of fish that live in association with the man-of-war, *Numeus gronovii* and *Schedophilus maculatus*. They treacherously feed on fragments of their host while they enjoy the protection of its tentacles; the man-of-war does not derive any apparent benefit from this relationship. Though certain sea turtles are known to feed avidly on the Portuguese man-of-war, it has no other predators.

The man-of-war's venom is known to cause reactions in humans, ranging in severity from localized stinging and burning accompanied by nausea to heart failure and sudden death. (It is worth noting, however, that a number of fatalities that have been attributed to the Portuguese man-of-war, particularly in Australia, were probably caused by cubomedusans, or sea wasps). These most effective antidote to the venom of the man-of-war is an antihistamine, which can be taken orally or applied as a topical ointment. Failing that, the welts can be washed with dilute ammonia, urine, picric acid, or isopropyl alcohol, or simply rubbed with sand. The purpose of all these remedies is to neutralize the hemolytic proteins in the venom as well as to dislodge any remaining nematocysts that might not yet have popped out of their capsules.

In some other siphonophore species, the polyps have become even more highly differentiated than in the Portuguese man-of-war. In *Porpita* and *Vellela*, the floatation chamber is disk-shaped (and surmounted by a veil or canopy in the case of *Vellela*). On the underside of the disk, the gonozoids are arrayed around a single large gastrozoid, from which a number of gastrodermic canals transport nutrients to the floatation disk and the other specialized polyps. The tentacles are attached directly to the outer rim of the disk, and the dactylozoids themselves have completely degenerated. R.P.

PRAWN *Palaemon serratus* 350

350
Sub.: Arthropoda
Cl.: Cructacea
O.: Decapoda
S.o.: Caridea
F.: Palaemonidae

A prawn is a large edible shrimp; this one is found along the Atlantic coast of Europe from Denmark to Mauritania and in the Mediterranean as well, particularly in areas where the bottom is covered with rocks and seaweed, out to a depth of 40 m. The females are slightly larger than the males and may weigh as much as 8 or 9 g and grow to a length of 7.5 cm in their first year; by the age of 17 months they will have moulted 33 times and perhaps attained a length of 9 cm. Average for the species is 5–7 cm, with a maximum of 10.5 cm. Typical longevity is probably 2 years, though 3 or 4 is not impossible. This species tends to be more active at night than during the day; prawns make short migrations along the coast or into deeper water in response to seasonal variations in temperature.

▶ The planktonic larva of this species feeds on coastal zooplankton, and the juvenile form on slightly larger marine organisms of the intertidal zone; juveniles of less than 3 cm in length feed primarily on ostracods where available. The adult can subsist on seaweed to some extent, but normally prefers to feed on smaller shrimp, young crabs, and other decapods, ostracods, bottom-dwelling copepods, amphipods, small gastropods and polychaetes, and foraminifera. The search for food in the intertidal zone takes place by day as well as by night, but in deep water, where many predators hunt by sight, the prawn is more likely to hunt for its food at night.

The prawn locates its prey first by smell, then by touching it with its antennae, and finally seizes it with its first two pairs of walking legs, both of which are equipped with pincers, or with its maxillipeds. Smaller, soft-bodied prey can be ingested directly with the aid of the inner mouthparts, the maxillulae; larger or denser prey are torn into smaller morsels by the outer sets of mouthparts, the mandibles. The prawn's stomach frequently contains little bits of inorganic debris—shell fragments, the spicules of sponges, bristles of worms and other creatures—which presumably serve as digestive aids, like the pebbles in a bird's crop. The various buccal palps, though elaborate in structure and design, play a relatively small part in preparing the prawn's food for digestion.

◀ The planktonic larvae are one component of the thin nutrient broth that makes up the coastal zooplankton. The juveniles that inhabit the intertidal zone frequently concentrate in tidal pools and shallow depressions where they may be vulnerable to attack by shorebirds. In deeper water, adult prawns are eaten by squid, large crustaceans, and a number of different fish that frequent rocky coastal areas, including wrasses, bass, eels, and cod and other Gadidae. The prawn tries to escape from these predators by concealing itself during the daylight hours in a hole or crack in the rocks, a fissure between two blocks of stone, or in a bed of seaweed.

Juvenile prawns are almost entirely transparent, and the adults rely heavily on protective coloration in certain environments. A prawn that has been placed on a bed of white sand needs about 20 minutes to adjust its body pigmentation to the color of the background; if it is placed on black sand or dark-colored seaweed, it takes about an hour to adjust to its new circumstances. A prawn might be able to escape with a few rapid flicks of its tail from an especially keen-sighted predator that is able to penetrate this disguise; like many crustaceans, it also has the ability to dislocate and shed some of its limbs (autotomy) in case of an emergency of this kind.

Prawns are caught in substantial numbers in baited traps, similar to crab or lobster pots, that are set along rocky coastlines; they are also collected by amateur fishermen at low water and generally at night, between boulders in tidal areas and in beds of seaweed. Fortunately, the prawn seems fairly resistent to oil slicks, oil spills, and other forms of hydrocarbon pollution, probably because it is accustomed to migrate into deeper water whenever surface conditions are unfavorable.

Prawns of the family Palaemonidae are found along rocky shorelines in all the warm or temperate seas of the world. One prawn genus, *Palaemonetes,* is found in water of low salinity and feeds on aquatic insects and other fauna characteristic of brackish estuaries and the like: nematodes, copepods, polychaetes, oligochaetes. Giant shrimps of the genus *Macrobrachium* are found in tropical waters, both salt and fresh; they frequently reach a length of 30 cm and are increasingly being raised on commercial fish farms in many areas. Unlike their smaller relatives, these giants are primarily vegetarians and scavengers (detritivores). P.N.

PRAYING MANTIS *Mantis religiosa* 351

351
Sub.: Arthropoda
Cl.: Insecta
O.: Dictyoptera
S.o.: Mantodea
F.: Mantidae

Originally found in South and Central Europe, this species was deliberately introduced into North America and has since become quite common. The praying mantis is diurnal and thermophilic, especially fond of patches of bright sunlight on the ground when searching for prey; it stations itself in a bush or on a stalk of tall grass, swiveling its head—which is particularly mobile—in the direction of any promising movement, including directly behind it.

When the eggs hatch out, the emerging larvae are wormlike in shape and apparently legless, since their appendages, like those of libellulid dragonflies, are originally enclosed in a sort of sheath or coverall that is discarded at the time of the first molt. The larvae begin to appear in June and grow very quickly; their weight doubles and their length increases by a factor of 1.26 during each of the five larval stages that precede the final molt, which lasts for 3 hours. The imagos (adults) are already in evidence by the middle of August and linger on until the first frost.

The adult praying mantis measures between 42 and 76 mm, depending on its sex; some adults are green, others brown, depending on the background color of the leaf, stalk, etc., that the mantis was resting on during its final molt. Since the praying mantis feeds indiscriminately on every insect that crosses its path, including those of its own species, it is obliged to lead a solitary existence. Copulation lasts for a long time, and it is thus not uncommon for the male to be devoured either directly afterwards or even during the act.

▶ When an insect approaches, the praying mantis lifts up the forepart of its body and extends its long prehensile forelegs in what appears to be an atttitude of prayer (called the "intimidatory posture" by entomologists). The mantis gets its name in a variety of European languages: *mante religieuse* in French, *prega-Diou* in Provençal, *louva Dios* in Catalan, and *Gottesanbeterin* in German; the word *mantis* itself comes from the Greek word for prophet. When the forelegs are upraised in this manner, two spotted patches on the underside of the femur are exposed, which may play some role in fixing the quarry's attention.

The mantis abruptly extends its forelegs, catching the prey animal in a sort of hammerlock with the second and third segments of the forelegs (the tibia and the femur), which are studded with a double row of sharp spines all along their length to make it impossible for the prey to escape. The larva begins its predatory career by feeding on aphids, then moves on to progressively larger insects during its rapid larval development; in Europe, the imago feeds primarily on large locusts and grasshoppers such as the wart biter *(Decticus).*

◀ The praying mantis is clearly most vulnerable to predation as a developing larva or even before it has emerged from the eggcase. The larvae may be attacked by a variety of ectoparasitic wasps, paralyzed by the wasp's venomous stinger, and dragged off to its burrow to provide a source of food for the wasp larvae. A number of these species are found in the South of France, the best known being those that were studied by J. H. Fabre: *Stizus distinguendus, Tachyspex julliani,* and *Tachsyspex costai.* The eggs of the praying mantis may also be parasitized by a chalcid wasp, *Podagrion pachymerum,* and the scelionid wasp *Mantibara manticida.*

There are 1800 mantis species wordwide, most of which are found in the tropics.

R.C.

Prospaltella berlesei 352

352
Sub.: Arthropoda
Cl.: Insecta
S.cl.: Hymenopteroidea
O.: Hymenoptera
F.: Aphelinidae

This tiny parasitic wasp measures only about 0.4–0.7 mm; the head of the adult is light brown, its legs and antennae are yellow, and the rest of its body is black. The unique parasitic host of the larva is the white peach scale *(Pseudaulascapis pentagona),* a common orchard pest that generally occurs in dense clusters on the branches of mulberry bushes and peach trees. The aphelinid wasp *Prospaltella berlesei* was originally found in the Far East (probably in Japan) and has since been introduced as an agent of biological pest control into many different regions, including the Mediterranean Basin and elsewhere in Europe, the southern United States, and Latin America. *Prospaltella* reproduces exclusively by parthenogenesis (there are no males); three to four generations reach maturity in a single year.

▶ *Prospaltella* gets around primarily by running or hopping along a branch rather than by flying. It can readily be observed on sunny days (assuming a mulberry or peach orchard infested with *Pseudaulascapis pentagona* is nearby) and is always found in the vicinity of the host insect; the imagos feed on nectar and honeydew, and are not carnivorous. The adult wasp scurries from one scale insect to the next, evaluating each one as a potential foster home for its brood by measuring off the scale insect's body with its quivering antennae; the size and capaciousness of the body shield is the only criterion for selection.

When a suitable host has been selected, the wasp crouches down on top of the peach scale's body shield, thrusts its ovipositor through the shield and the dorsal teguments, and inserts its egg directly into the hemolymph (corresponding to the bloodstream) of the prospective host. The entire process takes less than a minute, and when it is completed, the wasp stops and rests for a moment before setting off in search of another victim. The body of the wasp larva is oval and fairly elongated; it begins by extracting nutrients by osmosis directly from the hemolymph, then goes on to devour the internal organs of the host when the hemolymph is exhausted. It undergoes nymphosis while still inside the body of the host, and the emerging adult gnaws a circular hole through the tegument and body shield, and makes its way out into the open air.

◀ The competitors and predators of this species are one and the same, since the principal danger to the *Prospaltella* larva is that a faster-developing larva of a native species (e.g. *Aphytis proclia* in southern France) will hatch out inside the same scale insect and end up devouring *Prospaltella* along with the internal organs of the host. Either outcome will result in the destruction of the white peach scale, of course, but from from the standpoint of the nurserymen, it is more desirable for *Prospaltella* to survive, since this species preys exclusively on mulberry scales and is thus more effective as an agent of biological pest control. Also, *Prospaltella* is quite susceptible to chemical herbicides and pesticides, so that an orchard into which this species has been introduced cannot be too vigorously sprayed and dusted—since it is likely to be rapidly overrun with scale insects once the proportion of hosts to parasites rises above a certain level.

A related species, *Prospaltella perniciosi*, has also been introduced into fruit orchards in many diferent localities as a biological control on the San Jose scale *(Quadraspidiotus perniciosus)*. C.B.

PUMA *Puma concolor* 353

353
Sub.: Vertebrata
Cl.: Mammalia
O.: Carnivora
F.: Felidae

The puma, also known as the mountain lion, catamount, or "painter" (=panther), measures between 1 and 1.6 m, with its tail measuring an additional 80 cm. Its coat varies from brown to red but is most frequently a shade of brownish gray, and it is generally found in mountainous areas that are overgrown with brush, occasionally in forested lowlands. At one time, the puma ranged throughout most of North America and virtually all of South America. Today it is only found west of the Rockies (British Columbia to Baja California) in North America; it is still widely though sparsely distributed through South America, and it is considered to be endangered throughout its entire range.

The puma is a solitary hunter; the male occupies a hunting territory of 10–50 sq km, the female of 5–20 sq km, and both sexes claw the bark of trees and scent-mark boundary stones and other features of the terrain with their urine; they make their dens in a hollow tree or a crevice in the rock at the approximate center of this territory. Pumas generally avoid contact with one another and do not fight territorial skirmishes; though the territories of two males will never overlap, a male will sometimes tolerate the presence of a female on its hunting grounds.

▶ The puma preys on virtually every mammal that is found in North America, including mice, rabbits, jackrabbits, squirrels, wolf cubs, buffalo calves, sick or enfeebled deer and elk, as well as on certain birds; it is an important factor in the control of cervid populations. The puma's hunting tactics are similar to those of larger felids: It hunts at dawn and dusk, stations itself in ambush and awaits the arrival of a prey animal, stalks it by creeping cautiously along the ground, then pounces on its back (or simply throws itself on top of it in the case of a smaller animal). If the quarry escapes from its initial onslaught, the puma makes no attempt to pursue it; it kills its victims by biting them in the back of the neck and severing the spinal cord.

◀ The jaguar is a potential competitor of the puma, though the puma generally prefers to live at a higher altitude; in regions where these species are sympatric (notably the Amazon Basin), the puma carefully avoids any contact with the jaguar. The puma's only potential rival in North America is the wolf pack, which it will also attempt to avoid by taking refuge in a tree. Accounts of puma attacks on humans are regarded rather skeptically by naturalists, but pumas do undoubtedly prey on livestock from time to time, and the puma has been systematically persecuted (as well as hunted for sport) in both North and South America. P.A.

PYRENEAN DESMAN *Galemys pyrenaicus* 354

354
Sub.: Vertebrata
Cl.: Mammalia
O.: Insectivora
F.: Talpidae

This small insectivorous mammal is 11–13.5 cm long (with an additional 13–15.5 cm for the tail) and weighs between 50 and 80 g; its upper body is dark brown, its underparts are white, and it is perhaps most remarkable for its highly flexible proboscis, about 2 cm long. This species is found in Portugal and northern Spain as well as in the Pyrenees, usually in the vicinity of small streams (particularly mill-races and mountain torrents) at altitudes between 300 and 1200 m.

▶ Not very much is known of the habits of this small nocturnal hunter, which lives in holes and hollows in the riverbank, though it may occasionally dig a burrow for itself. The Pyrenean desman feeds on insects (chiefly Coleoptera) and their larvae and small crustaceans (amphipods). It can sometimes be observed as it seeks for prey along the banks of a stream and among the submerged rocks in the streambed, trotting along briskly for a while and then sitting back on its haunches, though it probably finds most of its prey underwater.

The desman can stay submerged for fairly long periods, poking along through the sand and lifting up pebbles with its proboscis, which is covered with fine bristles, or vibrissae, that are sensitive to the slightest vibration. Since the desman is obliged to keep its nostrils closed underwater and is practically sightless even on land, these are its only effective means of detecting prey while foraging along the streambed.

◀ The desman is in direct competition with the dipper, or water ouzel, and several other diving birds in the same habitat; it is preyed on by raptors and perhaps by carnivorous mammals and large fish as well. The Pyrenean desman is now seriously threatened by a number of greater or lesser alterations of its environment, all of human origin: by the "correction" of the course of mountain streams for the purpose of flood control; and by the use of rock salt to remove snow from highways, which produces highly corrosive runoff that has a very destructive effect on the various small creatures on which the desman feeds. The most serious to the desman's survival, however, is posed by the presence of increased concentrations of dissolved atmospheric pollutants in the waters of the mountains streams in which it seeks its prey, since the desman can only survive in the purest water.

The Muscovite desman *(Desmana moschata),* slightly larger than the Pyrenean desman, is found in European Russia; other species are found in Siberia, China, and Japan. J.-J. B.

R

355
Sub.: Vertebrata
Cl.: Mammalia
O.: Carnivora
F.: Canidae

RACCOON DOG *Nyctereutes procyonoides* 355

The raccoon dog is strikingly different in appearance from other canids; its muzzle is more rounded and its legs much shorter. It is typically about 50 or 60 cm long, excluding the tail, which measures another 15 cm or so. Its coat is brownish-yellow with black hairs around the muzzle. As noted earlier in connection with the European badger and its competitors, the raccoon dog is a native of Manchuria, Japan, and Siberia that was deliberately introduced into European Russia and has subsequently migrated across Eastern and Central Europe over the last 50 years. Descendants of these original escapees from commercial fur farms in the vicinity of Moscow have recently been encountered on the west bank of the Rhine. The raccoon dog is a solitary hunter that may occasionally assemble in small packs of 5 or 6 individuals.

It digs a den for itself in the ground, never more than 500–800 m from a source of fresh water, or it may find a less permanent refuge in a hollow tree, a crevice in the rocks, or under a bush. In Western Europe, the raccoon dog seems to prefer a humid woodland terrain interspersed with clearings and ponds or other bodies of water; an individual raccoon dog will occupy a small territory of about 1 or 2 km^2.

▶ The raccoon dog feeds on fruit, acorns, and other vegetable material as well as on insects, fish, small rodents, and occasionally on birds and their eggs; it seems especially fond of fish and, where circumstances permit, may eat as many as 10 in a single session. Analyses of stomach contents undertaken by Soviet scientists revealed that 64% of the individuals in the sample had most recently fed on vegetable material. About half the animal food (by bulk) consisted of insects, followed by rodents, fish, amphibians, and birds. The raccoon dog is nocturnal; it begins to hunt shortly after sunset and seeks its prey primarily by means of its well developed sense of smell.

◀ The keenness of its senses and its agility at escaping detection have enabled this curious animal to coexist with a number of other larger and more effective predators, notably the lynx, the fox, and the wolf. In the Soviet Union, the raccoon dog is still raised for its fur, particularly in the region directly to the southeast of Moscow. P.A.

RATEL *Mellivora capensis* 356

356
Sub.: Vertebrata
Cl.: Mammalia
O.: Carnivora
F.: Mustelidae

This little-known mustelid, somewhat similar in appearance to the badger, attains a length of 60–70 cm (not including its tail, which measures about 20 cm) and weighs no more than 11 kg. Its long, thick coat is dark on the underside and lighter on the upper body, where it is interpersed with occasional black hairs, and is dense enough to protect it from the fangs of snakes, the teeth of dogs, the quills of porcupines, and other perils. Like all mustelids, the ratel has well-developed front claws and is an expert burrower.

The ratel is found in a variety of habitats (forest, savannah, arid steppes and semi-desert wasteland, though excluding the Sahara) throughout Africa and ranges widely throughout Asia as well. During the dry season, it digs a burrow in a variety of terrain types and then moves to higher ground, occupying a crevice between two boulders or a hollow tree, during the rainy season. The male ratel is fiercely territorial and will drive away any other carnivore (including a lioness, according to one report) that ventures too close to the mouth of its burrow. Longevity in captivity sometimes exceeds 20 years.

◀ The ratel is not strictly nocturnal or diurnal and feeds on a variety of small creatures (rodents and other small mammals, birds, locusts, beetles, termites), though it seems to live primarily by raiding the hives of wild bees, often in symbiotic partnership with the HONEY GUIDE. The ratel spends hours on its own observing the flight of wild bees in order to learn the location of the hive, though its efforts are often considerably expedited by the aerial reconnaissances of the honey guide.

There are a number of different species (including the rhinoceros) that are alerted by the distinctive cry of the honey guide, a bird that feeds avidly on bee larvae but is incapable of digging up the hive by itself (the hives of African bees may buried up to a depth of 2m underground). The ratel follows the honey guide to the spot where the hive lies buried and quickly disinters it, feeding greedily on the honey, wax, nymphs, and adult bees; the thickness of its coat prevents it from being stung. The honey guide waits until the ratel has finished its meal and then swoops down to investigate the scattered remains of the hive.

◀ The ratel generally flees from an aggressor by quickly digging itself into the ground, though when cornered, it will not hesitate to attack a human being or other large mammal, relying not only on its jaws and powerful claws but also on the pungent secretions of its anal gland to drive off the intruder. The ratel's hide is very thick and elastic, which makes it difficult for carnivores to seize in their jaws.

P.A.

RAZOR-BILLED AUK *Alca torda* 357

357
Sub.: Vertebrata
Cl.: Aves
O.: Charadriiformes
F.: Alcidae

The razorbill measures between 39 and 43 cm, with a wingspread of 61–70 cm, and weighs between 545 and 920 g; it reaches sexual maturity at the age of 3 or 4. The Alcidae (auks, puffins, guillemots) are all marine birds, found in those latitudes of the North Atlantic and Pacific where polar and equatorial currents converge and where fish and marine invertebrates are particularly abundant.

The razor-billed auk spends most of its life on the open water and only comes ashore during the breeding season, when it assembles in small colonies (the largest of which contain no more a few thousand birds), often in association with puffins and guillemots. Nesting sites are found along the coasts of Newfoundland and Labrador, southwest Greenland, and northwestern Europe (from Brittany to the White Sea). European populations spend the winter in the North Sea and the Baltic as well as in the Mediterranean and off the Atlantic coast of Morocco.

▶ The razor-billed auk is an excellent swimmer and diver as well as a strong flier, though its flight is rather stiff and ungainly. It can dive to a depth of tens of meters and "flies" through the water by paddling vigorously with its wings half-outstretched; it sometimes seeks its prey on the surface, and normally feeds on small fish, annelids, crustaceans, cephalopods, and sometimes mollusks as well. The nestlings are fed exclusively on small fish, and an adult can carry as many as 6 small fish in a single trip, with the heads and tails lolling out on either side of its beak.

◀ In mixed breeding colonies in the North Atlantic, razorbills, puffins, and guillemots have slightly different topographical preferences, so there is no competition for nesting sites. In northern waters, ospreys and gyrfalcons prey routinely on razorbill nesting colonies, as did the peregrine falcon along the coasts of Brittany, though it is now extinct in those regions. The razor-billed auks of Normandy were exterminated by market hunters more than 50 years ago, and the fewer than 100 pairs remaining in Brittany are currently threatened by oil spills and "black tides" (slicks discharged from the fuel tanks or storage tanks of oceangoing vessels) in the Channel.

The great auk *(Pinguinnis impennis)*, a flightless species, was formerly quite common along the coasts of Iceland, the Faeroes, and Labrador; it was hunted to the brink of extinction by sailors and fishermen, and finally wiped out by egg hunters and specimen collectors toward the middle of the 19th century. The family Alcidae consists of 13 genera and 21 extant species altogether; 6 of these are found in the North Atlantic, the remaining 15 in the North Pacific (see ATLANTIC PUFFIN). Alcids of the genus *Uria* are especially numerous; in 1960, the combined population of two species, the Atlantic murre *(Uria aalge)* and Brunnich's guillemot *(U. lomvia)*, was estimated at 60 million.

J.-P. R.

358
Sub.: Vertebrata
Cl.: Aves
O.: Passeriformes
F.: Laniidae

RED-BACKED SHRIKE *Lanius collurio* 358

This Old World shrike measures 17.5 cm, with a wingspread of 27–29 cm, and weighs between 26 and 38 g. It is found in hedges, brush, and thickets adjoining open country. Its summer range extends from the Atlantic coasts of Europe (including southern England and Scandinavia) to the northern frontiers of China. The red-backed shrike arrives in Southern and Western Europe in May, and sets off in its migratory flight to sub-Saharan Africa in September or October.

▶ The shrike surveys its territory from the top of a bush or a fencepost and when it has sighted a prey animal, it glides quickly and noiselessly down to the ground, without flapping its wings, to intercept it. (During its migrations, it displays a curious side-slipping style of flight, and sometimes when its prey has gone to ground, it succeeds in hovering motionless in mid-air). The red-backed shrike feeds primarily on insects (notably bees and wasps) and a variety of other creatures, including spiders, snails, voles, shrews, small birds, lizards, and salamanders.

It uses its beak as well as its claws to catch its prey; the tip of its beak is hooked, like that of a bird of prey, which enables it to grasp the prey firmly while transporting it to a nearby thorn tree (hawthorn or sloe plum) or possibly a barbed-wire fence. There, the shrike immobilizies the prey by impaling it on a thorn (or a metal barb), then either devours it on the spot or returns to eat it later. Before attempting to swallow a bee or wasp, it picks it up in its beak and scrapes the abdomen against a fence rail or a branch until the stinger and the venom sac have been dislodged.

◀ The red-backed shrike aggressively defends its patch of brush or shrubbery against other members of its own species. It shares the Eurasian portion of its range with two very similar species, the woodchat shrike *(Lanius senator)* and the lesser gray shrike *(Lanius minor)*, whose diet includes a much larger proportion of insects and which are not strictly competitive with the red-backed shrike. As far as predators are concerned, the red-backed shrike and the woodchat shrike together comprise a substantial part of the diet of Eleonora's falcon (as much as 18 percent, according to one study, of all prey taken by this species during its breeding season).

The marked decline in all three European shrike species over the last decade may be due to climatic factors; in the case of *L. senator* and *L. collurio*, it is more likely because of chemical pesticides and herbicides, which are injurious to the birds themselves and fatal to the large insects and other small creatures on which they feed. Drought, desertification, and the loss of habitat (brushwood) in sub-Saharan Africa may also be implicated in the decline of all three species.

Shrikes are found primarily in the Old World, ranging from Europe to New Guinea; a few species are found in the Americas, notably in the American Southwest. The gray shrike *(Lanius excubitor)*, found in Western Europe and North America as well as North Africa and Asia, is a very aggressive predator; it often attacks songbirds (sparrows, finches, larks) on the wing, particularly when insects are in short supply, and it will not hesitate to confront a buzzard or even a goswhawk or a sparrowhawk. Louis XIII, an ardent falconer (and the author of a cantata celebrating the joys of blackbird hunting), had gray shrikes trained to accompany him into the hunting field, a practise that was continued into the 19th century in India and the Near East. J.-P. R.

RED-BILLED HORNBILL *Tocus erythrorynchus* 359

359
Sub.: Vertebrata
Cl.: Aves
O.: Coraciadiformes
F.: Bucerotidae

This smallest of African hornbills is about 50 cm long and weighs about 180 g. Its plumage is brown with white spots on its back and pure white underparts. Its bill is curved and rather large (though not remarkably so for a hornbill) and surmounted with the usual horny casque. It is found in the southern Sudan and Uganda, in sparsely wooded regions of thorny scrub, provided there are trees large enough for it to build its nest in. The red-billed hornbill hops from branch to branch in pairs or in small groups, emitting its loud, irritated-sounding cry. Like many hornbills, this species builds its nest in a hollow tree, and to protect the nest against predators, the female is walled up inside this cavity by the male, a task in which the female also participates. The male passes food in to the female and the nestlings through a small aperture in this wall, which is made primarily of mud; when the chicks are large enough to be fed in the same manner, the female breaks out of her confinement (a process that may take several hours) and the wall is built back up again until the chicks are ready to leave the nest.

▶ The red-billed hornbill catches swarming termites on the wing and hunts for locusts on the ground; it also feeds opportunistically on insects and other small creatures fleeing from brushfires and forages around the head of a column of army ants in much the same way. Its long curved bill is also suitable for breaking through the rock-hard mud of termite mounds, and this species also pillages the nests of other birds on occasion. The nesting period coincides with the rainy season, during which all food-gathering chores necessarily devolve upon the male. The hornbill also feeds on fruit and other vegetable material.

◀ The red-billed hornbill is normally very wary, especially during the nesting season. Its mania for security seems to pay off by virtually eliminating the chances that its eggs or chicks will be devoured by snakes or marauding baboons.

J.-F. T.

RED-BILLED TROPIC BIRD *Phaëthon aethereus* 360

360
Sub.: Vertebrata
Cl.: Aves
O.: Pelecaniformes
F.: Phaethontidae

This Indo-Pacific seabird measures about 1 m in length, has a wingspread of 100–110 cm, and weighs between 450 and 750 g. The tropic birds are so called simply because they are found in tropical waters, from the western Indian Ocean to the Eastern Pacific. The red-billed tropic bird is a solitary oceanic species that comes ashore only to build its nest on a coral reef or a mat of vegetation. It is especially graceful in its flight (the species name *aethereus* refers to the fluttering of its greatly elongated medial tailfeather) and often follows in the wake of ships at sea, though it does not feed on refuse.

▶ The red-billed tropic bird feeds on flying fish (typically measuring 10–25 cm), which it catches in the air as well in the water, and dives for shrimps and cuttlefish that are swimming just below the surface. The tropic bird skims over the waves until it has sighted its prey, then soars to a much higher altitude and begins its dive, sometimes swallowing its victim before it returns to the surface. It feeds its nestlings at first on partially digested fish and squid, though they already will be eating fish up to 30 cm long by the time they are ready to fend for themselves.

◀ Apart from rats, the tropic bird has few natural enemies on shore, though it staunchly refuses to abandon its nest at the approach of human predators, and its most effective defense in such a case is a well concealed nesting site. Congeners that nest on the Cape Verde Islands and along the Senegalese coast have suffered greatly from raids by fishermen in recent years. Even on a secluded coral atoll, the mortality among nestlings may be very high, largely from competition for nesting sites among birds of the same or different species (notably the yellow-billed tropic bird, *Phaëthon lepturus).* J.-F. T.

361
Sub.: Arthropoda
Cl.: Insecta
O.: Coleoptera
F.: Elateridae

RED CLICK BEETLE *Ampedus rufipennis* 361

This species measures between 10 and 14 mm, and can easily be recognized by its red elytra and black pyothorax, which is lightly covered with downy hairs, as well as by the male's sawtoothed antennae. The eggs are laid shortly after mating and hatch out about 15 days later; the larval stage lasts for 2 years, and nymphosis takes place in the fall, though the adults remain in their nymphal lodgings in a hollow oak or beech tree until the following April.

The adults are found in treetrunks and on sawed-up or fallen logs, and can sometimes be observed on live trees where the bark has swollen and burst spontaneously or on flowering hawthorn bushes. The larvae stay buried inside dead treetrunks or fallen logs, particularly where the wood has been broken down by the combined activities of bacteria, fungi, and insect larvae into a sort of powdery humus.

▶ The adults feed on pollen and nectar, both of which are abundant in May and June, and as on dead insects found along the ground or on the leaves of plants as well as on other organic debris. The larvae of *Ampedus* and a number of related species feed initially on particles of decayed wood and sawdust as well as on drops of moisture, the sloughed-off teguments of other insects, and similar humble fare; in autumn, the *Ampedus* larva stops feeding and remains in a state of total diapause throughout the winter.

After it awakens into its second year of life, *Ampedus* begins to feed on stag beetle larvae *(Lucanus, Dorcus, Platycerus,* and probably *Aesalus scarabeoides* as well). These larvae generally occupy a tiny cavity in a treetrunk, scarcely larger than themselves, and, though not entirely immobile, are rarely in a position to escape when attacked by *Ampedus. Ampedus* begins to carve up its victim with its short, powerful mandibles, and the stag beetle larva quickly succumbs to its wounds.

◀ The predators of the red click beetle, though they may exist, are unknown. R.C.

RED CORAL *Corallium rubrum* 362

The familiar red coral used in jewelry and other decorative artifacts since ancient times is the internal skeleton of a colonial polyp found in the western Mediterranean and eastern Atlantic. It is not to be confused with rose coral, a more deluxe article produced by several tropical species of the genus *Isophyllia,* "the value of which is considerable when infused with that carmine hue so attractive to behold that the Italians, with their fondness for figurative expressions, have designated it by the name *angel skin*" (Lacaze Duthiers, 1864).

Red coral is clasified as a variety of horny coral (as opposed, e.g., to stony coral, which is the type that is found in coral reefs) and may actually be pink or white, rigid, and irregularly branching, some-

362
Sub.: Cnidaria
Cl.: Anthozoa
O.: Gorgonacea
F.: Coralliidae

times confining its growth to the same plane. According to Weinberg (1976), the average height of a colony varies between 10 and 30 cm, though Pax and Müller (1962) have observed that some colonies may exceed 1 m in height and weigh more than 30 kg.

The red coral polyp is actually white in color and gives birth to live young—more precisely, a somewhat elongated larva measuring 3–5 mm emerges from the oral cavity of the adult. Colonies may be found in clear, shallow water where comparatively little light can penetrate (e.g., attached to the roof of an underwater grotto) or in deeper water "in the lee of a broad overhang or beneath the narrower eaves of an undersea cliff" and finally "on vertical surfaces and even relatively shallow slopes at depths of between between 135 and 165 m; at 200 m, colonies have been observed on horizontal surfaces" (Carpine and Grasshoff, 1975). In the last case, the authors suggest that these colonies consist of larvae rather than adults, which clearly have very little tolerance for light.

Corallium rubrum is found in the coastal waters of Spain, France, Corsica, Sardinia, Sicily, the Maghreb, the Canaries, and the Cape Verde Islands at depths ranging from 16 to 200 m, though the species was formerly present in shallower water, at a depth 7–8 m off the Algerian coast (Bounhiol, 1909).

▶ A coral colony is made up of two different kinds of polyps, autozooids and siphonozooids. The autozooids, normal polyps equipped with 8 feathery tentacles, are responsible for assimilation and reproduction whereas the siphonozooids, degenerate polyps consisting of little more than a minute contractile pore, are believed to play a role in regulating the water-filtration system of the colony. In virtually all alcyonarian species, "nutrition is provided by minute planktonic organisms that are ingested through the buccal orifices of the autozooids" (Tixier-Durivault, *Traitè de Zoologie).*

In several instances, the remains of brine shrimp and mollusk larvae have been identified in the gastric cavity of gorgonians, including one species, *Leptogorgia viminalis,* that had devoured a very large number of lamellibranch larvae, with a devastating effect on the local population (Bayer, 1956). Food-gathering involves 2 different organs: first, the stinging venomous cells at the tips of the tentacles (nematocysts) that strike out at all prey that passes within reach and, second, a narrow ciliated passageway (the siphonoglyph) that hastens the prey animal on its way toward the pharynx and the digestive cavity.

◀ In recent times, the Mediterranean coral fishery made use of a crude mechanical device that could be operated from the surface and was capable of detaching and retrieving coral colonies in shallow water—and only at the risk of considerable breakage and wastage of the product; this method is still practised by a group of Italian and Tunisian fishermen around the port of Tabarka. Red coral has largely been eradicated in shallow water and is currently collected by scuba divers at depths of up to 140 m (so that the risk, now that the quarry is all the more precious, has been transferred to the harvesters themselves). According to Carpine and Grasshoff (1975), red coral is still relatively common at greater depths, though the larger colonies are rapidly disappearing. It would seem advisable to subject the commercial exploitation of this species to strict regulation if it is not to be eradicated entirely in certain areas.

The information in the scientific literature on suspected marine predators of this or other coral species is fairly sketchy. Bayer (1956) mentions the presence of the zooanthid *Palythoa* (a zooanthid is a kind of anemonelike coelenterate) and polychaete worms of the genera *Harmothoe* and *Polynoe* on *Corallium* but refrains from speculating on the nature of their association. Thèodor (1967) was able to establish that the egg shell *Simnia spelta* feeds on the coenchyme layer (gelatinous connective tissue) of the Mediterranean polyp *Eunicella singularis* and that the adult *Simnia patula* can live for months by feeding on the tissues of *Alcyonium,* another alcyonian.

Less conclusively, Bouqet al. (1978) observed the ovulid *Simnia patula* among clusters of alcyonarians, probably *Alcyonium digitatum,* a species commonly known as dead man's fingers and prevalent in the English Channel and the Bay of Biscay. Ovulids, or egg shells, a family of mainly tropical sea snails, are especially noted for their ability to change the coloration of their mantles in order to deceive both predators and prey, and these authors make the general assertion that they feed on gorgonians or on alcyonarians, and often assume the color of their prey.

The genus *Corallium* contains 27 extant species, including corals that are found at depths down to 2400 m and in many of the world's tropical and subtropical seas, including the Atlantic (2 species in addition to *Corallium rubrum* plus an additional 3 species that are found in the waters off Madeira), the Indo-Pacific region, and the coastal waters of Japan (8 species), Hawaii, and the Philippines. M.-J. d'H.

363
Sub.: Vertebrata
Cl.: Mammalia
O.: Carnivora
F.: Canidae

RED FOX *Vulpes vulpes* 363

This familiar carnivore measures 35 cm at the shoulder; its head and body are between 58 and 77 cm long, its tail between 35 and 40 cm long, and it weighs between 5 and 10 kg. Its relatively short legs identify it as a member of the dog family (Canidae); its upper body is reddish-brown, and the throat, belly, insides of the legs, and the tip of the tail are white or sand colored; there are sometimes black patches on the belly, throat, and shoulders as well.

The red fox ranges throughout Europe and North America, as well as North Africa, most of Asia, and Australia. It is found in a variety of habitats (up to an altitude of about 2500 m in Western Europe), including the edges of forests and woodlands, small groves of trees, and cultivated fields interspersed with woodlots. The red fox digs a burrow, which serves primarily as a refuge for the pregnant female and the pups. Some foxes are fairly sedentary, others less so, and though the red fox is essentially nocturnal, it can sometimes be observed by day in areas not much frequented by humans. The red fox is sexually mature at the end of its first year of life, and generally lives to the age of 10 or 12 in the wild.

▶ In Western Europe, the red fox feeds on a variety of small mammals, ranging in size from voles to hares and including moles, fieldmice, rats, and hedgehogs, though its diet varies considerably with the region, the season, and primarily the circumstances. Thus, foxes have been known to feed opportunistically on ringdoves and small passerine birds (as well as domestic fowl of course), frogs and toads, snakes, insects, and earthworms. The author once discovered a group of scats in a mountainous area that consisted almost entirely of the wing cases of beetles and the legs of grasshoppers, though the droppings of the red fox more frequently have a feltlike appearance, on the account of the large proportion of rodent hair that they contain.

Since the number of wild rabbits in Western Europe has been greatly reduced by myxomatosis in recent years, foxes now tend to feed largely on voles, which they consume in substantial numbers (as many as 40 on a single session, in the case of one individual). The cubs are fed primarily on moles and small rodents. The red fox also feeds on apples, pears, berries, mulberries, and other fruit, and (as the evidence of its skats also discloses) appears to be particularly fond of cherries. Finally, apart from feeding on carrion, the red fox also routinely investigates garbage pits, landfills and dumping grounds of various sorts, refuse containers on campgrounds, etc.

The red fox's versatility as a predator has unfortunately been demonstrated in Australia, where it was introduced during the 19th century in order to combat the plague of rabbits, which had originally been introduced to provide food for settlers in the bush and had, in the course of a few decades, turned vast tracts of marginal grazing land into a lifeless desert. As it turned out, the foxes acclimated very well and quickly increased in numbers, but they preferred to prey on the indigenous marsupial fauna, which were easier to catch than the rabbits, and have thus contributed toward the extinction of several different species. On the other hand, the red fox will occasionally specialize in a particular prey animal that is locally very numerous; the relationship between the 8–12 year population cycles of the snowshoe hare and the red foxes that prey on them (notably on Isle Royale in Lake Superior) has been been intensively studied by ecologists for many years.

The red fox has little difficulty in locating its prey with the aid of its excellent vision, hearing, and sense of smell. R. Hainard furnished this description of a fox hunting fieldmice in a meadow like a cat: "It walks slowly, lifting up its feet very high and setting them down again carefully, with its muzzle tilted downward and its ears pricked up, then suddenly leaps up and falls on its prey in a single bound." And like cats, foxes are often seen playing with voles, fieldmice, and other small animals they have caught, tossing them in the air and catching them, pretending to let them run away and forcibly retrieving them at the last moment; while this behavior may seem distasteful to humans, it provides the fox with a necessary outlet for the aggressive instincts that are stirred up by the experience of chasing and catching the prey.

◀ The lynx and wolf (both of which which are almost extinct in Western Europe) are the chief predators of the red fox; the cubs may occasionally fall prey to eagle and the eagle-owl as well. The lynx can also be regarded as a successful competitor, since it has recently increased its range at the expense of the red fox in several widely separated territories, notably Canada, Scandinavia, and the Carpathians; according to M. Fernex, veteran trappers in northern Sweden report they can no longer make a living from the sale of fox pelts in regions that have been reoccupied by the lynx. The European wild cat and the feral cat are also competitors of the red fox in more southerly regions of Western Europe.

Apart from being trapped for its fur and hunted for sport, the red fox is frequently persecuted by man as an unwelcome competitor, since it preys not only on domestic fowl but also on rabbits and other game animals. Pheasants that have been raised in captivity and introduced into the wild just before the opening of the hunting season are particularly vulnerable to predation by foxes, since they have no experience in fending for themselves or in concealing themselves from predators—it is easy enough for humans to approach within 10 m of these birds without making any particular effort to "stalk" them.

Mange and rabies are also an important cause of mortality among foxes in Western Europe, and a variety of draconian measures have been adopted to prevent rabies from being communicated to domestic animals; these include intensive drives (battues) involving large numbers of hunters and beaters, poisoned bait and spring guns, as well as the use of poisonous gas (Zyklon and chloropicrin) to suffocate the females and the cubs in their dens. These methods are rarely effective, particularly insofar as any territory that is depopulated in this fashion will soon be resettled by foxes from adjoining areas, so that the net effect of these extermination campaigns is to increase the mobility of the red fox population throughout the surrounding region—which tends to promote rather than inhibit the spread of the disease.

For several years, the World Health Organization has recommended that domestic animals should be vaccinated against rabies, particularly since those rare cases in which humans are bitten by foxes (rabid or otherwise) can be treated on an emergency basis, whereas it is impossible to tell whether dogs, cats, or even livestock, might have come in contact with a rabid wild animal. In short, the disease can much more readily be controlled on this level.

The arctic fox (*Alopex lagopus;* q.v.) is quite similar to the red fox, though its ears are smaller, its legs longer, and its tail is uniform in color; the winter coat of some individuals is pure white. Other wild canids such as the wolf, the coyote, and the jackal are considerably larger than the foxes; their legs are proportionately longer, and their tails shorter. F.T.

RED PIRANHA *Serrasalmus nattereri* 364

364
Sub.: Vertebrata
Cl.: Osteichthyes
S.cl.: Teleostei
O.: Cypriniformes
F.: Characidae

S. nattereri is found in lakes, rivers, and smaller watercourses in Brazil and northern South America, at altitudes where the normal temperature range is 23–26°C. Though occasionally solitary, it more frequently congregates in schools of several hundred or even several thousand, and is one of four piranha species considered dangerous to man. The red piranha typically measures between 15 and 20 cm and weighs between 500 g and 1 kg, though some individuals may attain a length of 40 cm and a weight of 3–4 kg; average longevity is unknown.

▶ While it appears to be true that *most* piranhas feed exclusively on smaller fish, primarily of the characid family, and are thus harmless to humans or other mammals, most authorities would still confirm a modified version of the piranha feeding frenzy as depicted in films and popular legend—an incautious horse, cow, human, or other large animal steps into the waters of the Amazon and immediately disappears from sight—with the important qualification that piranha attacks of this kind occur only in certain localities and under fairly specific conditions. Even in the four piranha species that have been known to attack large mammals, aggressiveness seems to vary considerably among local populations, the size of the piranha school and the time of year being the main determining factor.

Thus, it appears to be true that *some* piranhas are attracted not only by the smell of blood but also by any kind of splashing or commotion on the riverbank, and are even capable of overwhelming a cayman and devouring it down to the bone (a much more challenging feat than a human or a cow). It is important to emphasize, however, that these attacks occur only occasionally, and more or less at random, so that the piranha cannot be said to prey systematically, or even habitually, on mammals.

The scholarly piranha controversy is probably based on a similar difference of opinion among the Indians of the Amazon. Watercourses said to frequented by piranhas are strictly avoided by many tribes (though it is not apparent whether the tribe in question is observing an empirical safety precaution or a religious taboo), whereas in other areas even the smallest children swim and bathe in the rivers without suffering any ill effects. Piranhas sometimes devour (or merely mutilate) larger fish that are caught in fishermen's nets, and though aggressive feeding frenzies have never been observed in groups of piranhas small enough to be maintained in an aquarium, occasional instances of cannibalism have been reported among piranhas in captivity, mainly involving fish that were sick, wounded, or old and enfeebled.

◀ Isolated piranhas may sometimes fall prey to electric eels, young anacondas, matamatas and other turtles, and aquatic birds, though schools of piranhas are unlikely to be molested by any other creature. The flesh of the piranha is greatly esteemed by man in certain regions, though it can only be caught with heavy-gauge wire fishing leader, since its razor-sharp teeth will shear through anything less substantial.

The 16 piranha species are assigned to four different genera—*Pygopristis, Pygocentrus,* and *Rooseveltiella,* as well as *Serrasalmus.* The largest piranha, *Serrasalmus* (in some classifications, *Pygocentrus) piraya,* may attain a length of 60 cm, and apart from the red piranha, several other species have also become popular with tropical-fish fanciers in recent years. J.-Y. S.

REDSTART *Phoenicurus phoenicurus* 365

365
Sub.: Vertebrata
Cl.: Aves
O.: Passeriformes
F.: Muscicapidae

This Eurasian songbird, a relative of the Old World robin, measures 13.5 cm in length, with a wingspread of 23 cm, and weighs between 12 and 17 g. The characteristic vibration of the male's handsome red-orange tail is accentuated by stress or emotion, and appears to serve as a recognition signal for this species. In Western Europe, the redstart is found in deciduous woodlands interspersed with clearings, including orchards, public parks, meadows bordered with hedgerows, and woodlots containing hollow trees, even lawns and gardens in urban areas, and it builds its nest in any sort of natural or artificial cavity. The male redstart energetically defends a territory of about 0.5 ha by singing its melodious song and displaying its vibrant colors. The redstart is diurnal in its habits, though it travels by night during its long migratory flight to sub-Saharan Africa, where it spends the winter in acacia thickets on the savannah.

▶ The redstart stations itself on an upper branch of a tree or on top on a wooden upright and swoops down on flying insects (primarily flies and mosquitoes, butterflies, bees and wasps) as they pass. It also hops lightly along the ground, pecking at snails and earthworms as well as terrestrial insects, or flies from branch to branch in search of spiders and caterpillars among the foliage of a bush or tree. The feeding reflex of the adult redstarts is triggered by the bright orange gullets and yellow-edged beaks of the nestlings.

◀ The redstart's eggs are sometimes stolen from the nest by two carnivorous rodents, the garden dormouse (q.v.) and the fat dormouse *(Glis glis)*. The young birds leave the nest before they can really fly, flattening themselves out on the ground and attempting to escape the notice of predators; they are assisted in this endeavor by the speckled fledgling plumage that is typical of the thrush family, though many fall victim to domestic cats in populated areas. The adults may be preyed on by the sparrowhawk as well as by Eleonora's falcon during their migratory flights. As with many migratory species, redstarts are still being netted and trapped in substantial numbers in the Mediterranean Basin, though the destruction of habitat is a far more serious problem for the redstart.

Species counts have been declining steadily since 1970, probably because the thorny brushwood in which the redstart makes its winter home in the Sahel has rapidly been disappearing from those regions as the result of drought, desertifcation, and human demands for fuelwood. In Europe as well, suitable nesting sites may also be in short supply, as the result of cereal monoculture (destruction of hedgerows and woodlots) and the replacement of deciduous by coniferous species in European woodlands. The increased use of insecticides is probably a factor in both these regions as well.

The black redstart *(Phoenicurus ochruros)* has been luckier in its choice of habitats, since it builds its nest on the sides of cliffs or sheer rock faces and the roofs of houses, and its chirping song may still occasionally be heard in the midst of European cities.

J.-F. T.

RED-TAILED PIPE SNAKE

Cylindrophis rufus **366**

This thick-bodied burrowing snake measures between 60 and 80 cm; it has a conical head (with no distinct separation between the head and the neck) and a short tail. The upper body is dark brown patterned with about 30 narrow beige trasnverse stripes that either meet or alternate along the midline of the body. The underbody is black, patterned with about 50 wider stripes, which are white along the body and orange or red along the underside of the tail, which is flattened dorsoventrally. This species is prevalent thoughout Indochina (below 20° N. lat.), especially in lowland forests with a thick layer of topsoil; the red-tailed pipe snake is generally found in the uppermost layer of the topsoil, in burrows excavated by other creatures, under tree stumps, etc.

▶ The pipe snake is not exclusively nocturnal in its habits, though it is encountered on the surface only at night. It feeds on other snakes, particularly on *Typhlops* (see BLIND SNAKES) and other fossorial (burrowing) species and a number of small, very abundant terrestrial species (notably *Oligodon taeniatus)* that live in burrows. The pipe snake's predatory behavior has never been observed in nature. In captivity, it approaches its potential prey quite matter-of-factly, and the prey seems equally unconcerned by its approach. The pipe snake leans its snout against the other snake's body to conduct a brief olfactory examination (though it does not necessarily protrude its tongue), then seizes the prey in its mouth and attempts to sever its spinal column with a violent grinding motion of its jaws.

The bodies of the two snakes will frequently become entwined as the prey struggles in the grip of the pipe snake, though the latter does not kill its prey by constriction and in fact relies primarily on the strength of its neck muscles, which are invariably well developed in burrowing snakes. After the prey's first violent reactions have begun to subside, the pipe snake works its jaws toward its victim's head and swallows it fairly rapidly. The pipe snake generally feeds on snakes much smaller than itself (Deuve, 1970); it is not known whether these tactics are effective against venomous species.

366
Sub.: Vertebrata
Cl.: Reptilia
O.: Squamata
S.o.: Ophidia
F.: Aniliidae

The pipe snake has been reported to feed occasionally on skinks and other burrowing lizards, on small mammals, amphibians, insect larvae, earthworms, and other invertebrates, though this finding suggests an unusually diversified regime for a fossorial snake and should probably be verified by further reasearch.

◀ The pipe snake probably runs afoul of venomous kraits *(Bungarus)* and small carnivorous mammals during its rare nocturnal excursions on the surface. It attempts to defend itself by tucking its head under one or more of its coils and bending back its broad flat tail to display the vivid colors on the underside, which presumably encourages an attacker to strike at its tail rather than its head, or else possibly to retreat from this alarming spectacle at once. Full-grown pipe snakes have few natural enemies underground, though the juveniles may fall prey to small burrowing cobras or *Calliophis*.

Several other *Cylindrophis* species are found in Southeast Asia, all very similar in their habits and appearance. The genus *Anomachilus* includes several other species found in Malaya and Sumatra; they rarely emerge from their burrows, and very little is known of their habits. The false coral snake *(Anilius scytale)* of tropical America attains a length of 90 cm and is somewhat similar in coloration to the venomous coral snakes of the New World; the false coral snake, like the pipe snakes, is assigned to the family Aniliidae and is another semifossorial species that tends to feed on other snakes. The unrelated milk snake of North America is also sometimes known as the false coral snake (see CALIFORNIA KINGSNAKE). H.S.G.

367
Sub.: Cnidaria
Cl.: Scyphozoa
O.: Rhizostomeae
F.: Rhizostomatidae

Rhizostoma pulmo 367

Along with the compass jellyfish *(Chrysaora hysoscella;* q.v.), *Rhizostoma pulmo* is one of the most common large jellyfish found along the seacoasts of Western Europe, as well as in the Mediterranean and the Black Sea. A third species, *Rhizostoma octopus*, found in the English Channel, the North Sea, and the coastal waters of the Atlantic, is regarded by some authorities as a subspecies or variant form of *R. pulmo*, which it greatly resembles. This species is known as the "great jellyfish" in several European languages; its North American congeners are less ceremoniously referred to as "football jellyfish."

The opalescent canopy of the great jellyfish has the consistency of cartilage and may attain a diameter of 90 cm. The female gonads are reddish in color; the male gonads are blue or violet; the small tonguelike projections, called rhopalia, that house the statocysts (organs of balance) are distributed around the rim of the canopy, but *Rhizostoma* (in its medusan phase) has no tentacles. The *Rhizostoma* larva, called a planula, attaches itself to the substrate and develops into a polyp 12 mm long, equipped with 32 tentacles; the base of the polyp is protected by a chitinous exoskeleton. This sessile polyp (scyphistome) phase of *Rhizostoma's* life cycle has been observed in the laboratory but never in nature.

▶ In this species, the 4 pairs of armlike projections of the manubrium have fused into a single complex structure with numerous folds and projections (known as epaulets). The surface of the manubrium is perforated with hundreds of small orifices, the mouths of canals leading inward and upward to the gastric cavity, each of which is surrounded by buds (papillae) that are liberally supplied with stinging nematocysts. *Rhizostoma* propels itself through the water by contracting its canopy. It feeds on a variety of marine organisms that come into contact with the nematocysts that cover the surface of the manubrium, including copepods, isopods, zoids and nauplia (crustacean larvae), polychaetes and arrowworms *(Sagitta)*, and small fish.

◀ The amphipod *Hyperia galba*, which feeds on the undigested contents of the gastric cavity of numerous scyphozoans, is very often found in association with *Rhizostoma*. There are several species of small fish *(Gadus merlangus, G. morhua, G. luscus)* that find a refuge under the canopy of the jellyfish, feeding on their fellow parasites, the amphipods, as well as on scraps overlooked by the jellyfish and on fragments of the body of their host.

The life cycle of other cnidarians apart from the scyphozoans includes a free-swimming pelagic phase, which assists in the geographic dispersal of individual organisms. The siphonophores, an order of hydrozoans that includes the Portuguese man of war, are colonial animals, comprising several medusans (in this case called zoids) and numerous polyps. The trachymedusans and narcomedusans consist entirely of free-swimming medusans, having dispensed with the polyp stage altogether. M.v.P.

RINGED SNAKE *Natrix natrix* 368

This Old World colubrid, also called the European water snake, is between 1.2 and

1.7 m long and is usually gray, brown, or olive-green with a slightly haphazard checkerboard pattern of black and white scales on its belly. The ring that gives this species its common name consists of several rows of orange-or creamy-yellow scales around the neck and is readily identifiable though not invariably present; some individuals are entirely black.

The ringed snaked is prolific (the female lays 10–15 eggs in July) and adaptable, found in a variety of different habitats in Europe, from Great Britain to Asia Minor, and North Africa. It is especially fond of humid areas—marshes, bogs, the banks of ponds, streams, or mountain torrents—as well as in wooded areas and in meadows along hedgerows, and can survive at an altitude of 2500 m in the Alps. It is a strong swimmer and can stay submerged for as long as half an hour. The ringed snake is aglyphous (i.e. its teeth are solid and have no venom channel) and is not at all dangerous to man or other large mammals.

▶ The ringed snake either hunts from ambush or actively pursues its prey along the ground or in the water. It seizes the prey by striking out with the forepart of its body, then grasps it in its jaws for several seconds. In the case of a larger animal, it may proceed to coil its body around it and suffocate it by constriction. Otherwise it begins to swallow the freshly killed prey animal immediately, working it slowly down the gullet with alternate lateral movements of its jaws in the manner of most snakes. (Apart from its well known ability to open its mouth extremely wide, a snake's trachea extends forward to the tip of its snout so it can keep breathing during the lengthy process of digestion.)

Between 75 and 80% of the ringed snake's diet consists of toads, frogs, salamanders, newts, and other amphibians, the remainder of shrews, voles, and other small mammals as well as fledglings and perhaps some insects as well. The ringed snake is one of the few predators that is not deterred by the cutaneous poison sacs of the common toad *(Bufo bufo)* or the spotted salamander *(Salamandra salamandra).*

◀ The ringed snake may be preyed on by diurnal raptors, wild boar, and even free-ranging domestic fowl. It has also been suggested that hedgehogs sometimes eat ringed snakes, presumably very young ones, though this has yet to be confirmed. This species is still quite common throughout its range, though some individuals may be destroyed (though never eaten, so far as we have been able to determine) by human beings. When confronted with a potential predator, it first attempts to flee, and if unable to do so, it hisses and inflates its neck in a threatening manner but rarely ever bites. It may also play dead by rolling over on its back and opening its mouth and remaining in this posture until the aggressor has lost interest in it.

As a last resort, the ringed snake is able to eject a foul-smelling liquid from its cloacal gland, a defensive capability that is shared by two congeneric species, the viperine snake *(Natrix maura)* and the checkered water snake *(Natrix tessellata).* The viperine snake *(Natrix maura),* is a semiaquatic nonvenomous snake (so called simply because it *looks* like a viper) that feeds on fish and is found on the Iberian peninsula, in southern France and Switzerland, Sardinia, and the Maghreb. *Natrix tessellata,* also semiaquatic and piscivorous, is found in Italy, Central Europe, and the Balkans. J.C.

368
Sub.: Vertebrata
Cl.: Reptilia
O.: Squamata
S.o.: Ophidia
F.: Colubridae

RING-NECKED SNAKE *Diadophis punctatus* 369

369
Sub.: Vertebrata
Cl.: Reptilia
O.: Squamata
S.o.: Ophidia
F.: Colubridae

This small North American colubrid generally measures between 25 and 40 cm, though it may attain a length of 80 cm in some areas; color is also highly variable among different populations. It ranges from Mexico to southern Canada and, though originally found in wooded areas, has come to occupy a variety of moist, shady terrains, including meadows and river bottoms, the banks of watercourses, abandoned fields and rangelands, plus such manmade habitats as gravel pits, the banks of canals and reservoirs, etc.

The ring-necked snake makes its way through the uppermost layer of the topsoil, though it is not a burrower and generally seeks its prey in galleries and tunnels dug by rodents, moles, insects, and other creatures; it is often encountered on mossy ground, or on a stump, plank, sheet of metal, or other flat surface. Ring-necked snakes hibernate, sometimes en masse, during the coldest months of the year (e.g. December—February in Kansas), though the hibernation period may be very brief in the more southerly parts of its range.

The males are the first to emerge in the spring, and the peak period of activity for this species is from April until June; the ring-necked snake retreats down into the subsoil during the hottest part of the summer, when it is rarely seen on the surface, then becomes active in September, particularly in damp weather. In many areas, a pattern of small-scale seasonal migrations has been observed, since this species tends to seek out sunlit ground during the spring mating season and moister, shadier ground during the summer. The ring-necked snake is diurnal, though fairly discreet in its habits while on the surface, and it tends to be active primarily at dusk or even at night during the height of summer.

▶ The ring-necked snake actively seeks its prey in burrows and galleries excavated by other creatures, and typically preys on earthworms, slugs, insects and their larvae, small salamanders, young frogs and toads, small skinks and other lizards, or other snakes (notably *Virginia, Sonora, Tantilla),* with the staple diet of local populations varying greatly across the range of this species. It detects its prey by chemoreception and catches it simply by clamping down its jaws on the prey animal until its venomous saliva has taken effect.

Its dentition is of the aglyphic ("unchanneled") type, which is to say that the saliva seeps gradually into the wound and is not injected through venom canals in the the fangs. Consequently, when the ringneck catches a relatively large prey animal, such as another snake (which may be up to 45 percent as large as itself by weight), it may have to hang on patiently for several hours until its venom has done its work and the prey is completely immobilized.

Digestion takes 3–7 days under normal climatic conditions, and an individual ring-necked snake is thought to make between 25 and 30 kills per year. The identity of the prey seems to be conditioned primarily by its prevalence in a given biotope and the ease with which it can be captured rather than by genuine preference, though some individuals will refuse to touch prey that is found quite palatable by ringnecks living in other regions. In the Northeast, for example, ring-necked snakes feed largely on salamanders, whereas in the arid regions of the Southwest, they tend to feed on other snakes, and one population in Kansas, which was the subject of a particularly intensive study, subsists almost entirely on earthworms.

◀ The copperhead *(Agkistrodon contortrix),* pygmy rattlesnake *(Sistrurus milarius),* coral snake *(Micrurus fulvius),* whipsnake *(Coluber constrictor),* kingsnake *(Lampropeltis spp.),* and, on occasion, various other snakes, along with hawks, owls, and corvids are the chief predators of the ring-necked snake. The young ringnecks may occasionally fall prey to bullfrogs and large toads, even large insects, and at certain times of the year, the diurnal rhythms of the ring-necked snake may correspond (with fatal results for the latter) to those of other predators that search for prey by foraging under rocks or in the forest litter; in this case skunks and badgers down to spiders, centipedes, ground beetles, and even certain fish. Hibernating ringnecks are sometimes rooted up by predators in the winter, when they are too sluggish to take flight.

Normally, when seized by a predator, the ring-necked snake attempts to make itself less appetizing by voiding the contents of its intestines, along with the foul-smelling secretions of its cloacal glands, a ruse that appears to be successful in many cases. When threatened or surprised, it often adopts a curious defensive posture, raising its tail off the ground in spiraling coils to display its bright red underbelly to the fullest effect. In populations where the contrast between the drab dorsal surface and yellow or orangish underside is less pronounced, the ring-necked snake merely lifts up its tail without coiling it.

Thus, the ring-necked snake has evolved both cryptic and aposematic (warning) pigmentation, on its dorsal and ventral surfaces respectively. It is thought that the purpose of the maneuver just described is to remind a bird of prey (or any other predator endowed with color vision) of the unpleasant odor these snakes give off when trifled with, without actually requiring the snake to void its bowels. Alternatively, it is possible that the gaudy pigment on the underside of the tail is designed to attract the predator's attention toward this non-vital portion of the ring-necked snake's anatomy.

The common name "ring-necked snake" is applied to several species of *Diadophis.* Other snakes have adopted a similar defensive posture including the RUBBER BOA *(Charina botttae)* of the western United States and the RED-TAILED PIPE SNAKE *(Cyclindrophis rufus)* of Southeast Asia.

D.H.

RIVER PERCH *Perca fluvialis* 370

370
Sub.: Vertebrata
Cl.: Osteichthyes
S.cl.: Teleostei
O.: Perciformes
F.: Percidae

The river perch is a rather sedentary fish that becomes less gregarious as it grows older. The young fry congregate in substantial schools, the 3–year-olds in smaller groups of about half a dozen; the older perch are completely solitary. The adult perch typically measures 25 cm and weighs up to 3.5 kg. Average longevity is 6–9 years, with a recorded maximum of 13 years. This species is found in lakes and ponds (provided the water is richly oxygenated, with at least 3 ml of dissolved oxygen to every liter of water) as well as in river and streams, even where the current is fairly strong. It ranges throughout the temperate regions of Eurasia and is found up to an altitude of 1000 m.

▶ After they have absorbed the vitelline sac, the perch hatchlings feed on *Daphnia,* copepods, and other tiny organisms that congregate in the sunlit regions near the surface of a lake or stream. At the age of 2 months, by which time they will have attained a length of 15–20 mm, the young perch either descend into deeper water or move in closer to the bank, and begin to feed on insect larvae, gammarids, crayfish, and roe. Young perch that feed exclusively on other fish (roach, bream, minnows, elvers) are said to grow faster than those that also feed on insect larvae and crustaceans.

The perch actively pursues its prey, rather than hunting from ambush like the pike, though it may have to conceal itself in aquatic vegetation to protect itself from predators while waiting for a suitable quarry to present itself; the vertical stripes along its sides are thought to mimic the stalks of aquatic plants. The perch appears to detect its prey by sight rather than by smell; it swims with its jaws agape and gives chase (attaining speeds of up to 5 kph) until it is close enough to clamp down its fine, sharp teeth on the tail of its quarry, then rolls the smaller fish around in its jaws until it is in position to be swallowed headfirst. Sometimes the quarry manages to wriggle free during this maneuver and gets away with nothing worse than a lacerated tail.

◀ Young perch try to decrease the chances of being singled out by their numerous predators (pike, pike-perch, eels, watersnakes, otters, herons, storks, ospreys, and gulls) by swimming in schools and by concealing themselves among the stalks of aquatic plants as the adults do. The dorsal and anal fins of the adult perch are stiffened with sharp spines, so that predators that are accustomed to swallowing their food whole may find themselves unable to do so without running the risk of strangulation or impalement.

Between 20 and 30,000 metric tons of river perch are caught in European waters every year, by commercial fishermen with nets, purse seines, and fish weirs as well as by sportsmen with hook and line. Attempts to restock rivers and lakes with perch have occasionally come to grief, since the fish create a local population surplus that tends to result in a shortage of food, followed by an outbreak of cannibalism among the young fry and a subsequent population crash. Even the fish that survive under these conditions generally reach sexual maturity and stop growing before they have reached average size, which is 15–20 cm.

The yellow perch *(Perca flavescens)* is found in the Great Lakes and along the Eastern Seaboard.

F.N.

RIVER TROUT *Salmo trutta fario* 371

371
Sub.: Vertebrata
Cl.: Osteichthyes
S.cl.: Teleostei
O.: Salmoniformes
F.: Salmonidae

This inland subspecies of the European brown trout measures about 30 cm in length and weighs between 1 and 1.5 kg. The river trout is a solitary creature that is only found in the chilly, richly oxygenated waters of upland lakes and streams and mountain torrents.

▶ Both the fingerlings and the adults feed on insect larvae, gammarid crustaceans, and small worms, though the adults also may feed on the fry of other species, and the very largest river trout feed on chub and minnows (as well as trout fingerlings). River trout seek their prey by turning over the gravel on the bottom or attack from ambush by concealing themselves in the shadow of a rock or behind a branch that protrudes down into the water. Trout can often be seen leaping out of the water as they feed on small flying insects just above the surface. They will attack only moving prey (though the movements of the prey animal need not be particularly vigorous) and, as fly fishermen are well aware, they are particularly attracted to yellow and red and other bright colors.

◀ The fingerlings may fall prey not only to the larger members of their own species but also to watersnakes, kingfishers, pike, perch, zander (pike-perch), otters, polecats, and domestic cats. The largest river trout occupy a place at the top of the aquatic foodchain, which is to say that they have no specific predators—though of course they may occasionally run afoul of superpredators such as ospreys, bears, and human fishermen. River trout are raised in hatcheries not only to meet the commercial demand for *truite bleue* but also for the purpose of restocking European lakes and streams whose original populations have been depleted by pollution.

Two larger subspecies of *Salmo trutta*, the European lake trout and the oceanic trout are found, respectively, in the larger inland lakes and the coastal waters of Europe. The well-known rainbow trout *(S. gairdneri)*, originally a native of East Asia and the American West, is found in warmer water and consequently at a lower altitude and has been widely introduced into European rivers and streams; its habits are similar to those of the European brown trout.

J.-Y. S.

372
Sub.: Arthropoda
Cl.: Crustacea
O.: Decapoda
S.o.: Brachyura
F.: Canceridae

ROCK CRAB *Cancer pagarus* 372

Cancer pagarus, one of the largest species of true crab, measures up to 20 cm in length and 30 cm across. It is found along rocky substrates in shallow water (out to a depth of about 90 m) all along the seacoasts of Europe. *C. pagarus* mates in the fall, though the females continue to incubate the eggs inside their bodies throughout the following year; the zoid larva that emerges from the egg will have attained the legal limit for commercial crabbing by the age of 3–5 years; maximum longevity of this species is on the order of 10 years.

▶ The planktonic zoids feed on small zooplankton, and the adults prey avidly on numerous mollusk species that are found along the offshore rocks, including scallops, oysters, mussels, and snails (particularly the Muricidae). *C. pagarus* hunts perhaps more frequently by night than by day, and accordingly detects its prey both by sight and by chemoreception, "tasting" the water with the sensory bristles that cover its third pair of maxillipeds. It cracks the shells of mollusks with its powerful pincers, breaking up flattened shells in the most efficient manner by placing its pincers against the thin edges of the shell and quickly rejecting those that are too tough for it to crack. Once the shell is broken, *C. pagarus* very carefully picks out the soft body of its victim and immediately spits out any shell fragments that it might have overlooked; this operation is performed with such delicacy that only the opercula of sea snails are recognizable when the stomach contents of this species are examined.

◀ The adult *C. pagarus* is large and truculent enough to fend off attacks by predators and consequently has few natural enemies, though, like all crustaceans, it is highly vulnerable when it has molted its shell and may sometimes fall prey to octopuses, lobsters, or rays and skates. The lobster also feeds on many of the same species of mollusks as *C. pagarus* and is thus a direct competitor. The juveniles, and occasionally the adults, of this species sometimes defend themselves by playing dead, so that *C. pagarus* is sometimes known as the *dormeur* ("sleeper") in French (more frequently as *tourteau*, which derives from the word for "twisted" and refers to its sideways method of locomotion). *C. pagarus* is also regarded by humans as a highly edible species, and several hundred thousand metric tons are harvested and consumed in Europe every year.

Cancer bellianus is found in deeper waters and is not nearly common enough to support a commercial fishery, though several congeners are taken in more or less substantial numbers in North American waters, including *C. magister*, *C. antennarius*, *C. productus*, *C. irroratus*, and others.

P.N.

ROCK PYTHON *Python sebae* 373

373
Sub.: Vertebrata
Cl.: Reptilia
O.: Squamata
S.o.: Ophidia
F.: Boidae

The rock python typically measures between 3–5 m and weighs between 15 and 30 kg, which makes it the largest African snake and one of the largest in the world. Some individuals attain a length of as much as 8 m (9.8 m according to one authority) and weigh up to 70 kg. Longevity is in excess of 50 years. The head of the rock python is clearly demarcated, and its body is gray with brownish spots; this species is found in various habitats (from arid savannah to dense tropical forest) throughout sub-Saharan Africa and at altitudes of up to 2600 m, though it prefers to live along the shores of lakes and streams, particularly in drier regions. Near the Equator, the rock python remains active all year round (discernibly less so during the dry season, however); in more temperate regions (South Africa), there is typically a 2–4 month winter latency period.

▶ The adult rock python will feed on any warm-blooded creature that it is capable of ingesting, including hares, porcupines, various rodents, monkeys, hyraxes, domestic dogs and pigs, small antelopes, guineafowl, cranes, ducks, geese, and other large birds. Newborn rock pythons can already catch rodents, and perhaps frogs and toads, for themselves, and before long the juveniles will be able to prey on small birds and mammals (rats, bush babies). Pythons are notorious, of course, for their ability to ingest very large prey animals, and though a creature as large (and especially as tall) as a human could hardly be considered a habitual prey of the python, it seems that the very largest specimens (> 8 m) may occasionally feed on human beings (though not on cows or other large four-footed mammals); there is a well authenticated account, for example, of a 13–year-old boy who was devoured by a rock python in Uganda in 1951.

The rock python becomes active at dusk and continues to hunt through the night; it is a good swimmer and climber, and during the day it takes refuge under a bush, in the branches of a tree, or along the bank of a lake or stream. During the dry season especially, it may immerse itself almost totally, with only its nostrils protruding above the surface of the water, preying on animals that come down to drink. It actively seeks its prey as well, searching out rodents and other creatures in their burrows at night, and on these occasions, it uses its forked tongue as a tactile organ to verify the location of a sleeping prey animal.

The python also detects its prey both by chemoreception (sampling molecules from the air with its tongue and conveying them to the olfactory cells inside Jacobson's organ) and thermoreception (as with other boids, nerve endings in the python's labial fossae, comparable to the pits of a rattlesnake or other pit viper, are sensitive to infrared radiation). Its vision is rather poor, perhaps only keen enough to distinguish a moving animal from its stationary background—though it is this visual cue that finally triggers the rock python's lightning-quick strike reflex.

The python abruptly thrusts its head forward, seizing the prey animal in its jaws and immediately surrounding it with several of its coils; the subsequent powerful contraction of the python's coils can kill a small animal almost instanteously by causing its major blood vessels to burst, whereas larger prey animals are killed by suffocation. When the prey animal's heart stops beating, the rock python slowly unwraps its coils and, beginning with the head, still more slowly ingests the prey, a process that may take 20 or 30 minutes. If the rock python is disturbed during this time, it will immediately disgorge the prey animal—which is lavishly lubricated with saliva but otherwise intact, since the python does not chew its food—before attempting to make its escape.

Adult rock pythons generally sustain themselves on a relatively small number of large prey animals; thus, a python measuring 4.8 m and weighing 44 kg was observed to feed on nothing but 5 small kobs (weighing about 25 kg) over a 9–month period. The actual process of digestion, fairly efficient in the juveniles, may take up to several weeks in the full-grown adults; rock pythons have been known to go without food for as long as a year without suffering any ill effects.

◀ Adult rock pythons have few natural enemies other than man. A leopard, for example, may have comparatively little difficulty in subduing a python that is less 3–4 m long, but once past that point, the struggle may be resolved in favor of the python. Warthogs and other members of the pig family (Suidae) prey fairly frequently on the smaller and medium-sized pythons, which may also fall prey to crocodiles while swimming across a river or lake. Young pythons are beset by all the predators that normally prey on smaller snakes, including eagles, vultures, otters, wild cats, mongooses, etc., and the python's eggs are sometimes devoured by monitor lizards.

The python defends itself by striking violently at any large creature that ventures within a distance equal to about a third of its total body length, but will attempt to protect itself by rolling up in its coils only if it is actually seized by a predator. As the only very large African snake, the adult python has no competitors. The python is intensely hunted not only for food but also for its skin, which is still highly prized as a deluxe article of decor. Habitat degradation (especially grassfires, brushfires, etc.) has also taken a heavy toll of this species in recent years.

The royal python *(P. regius)* of the West African savannahs rarely exceeds 1.5 m in length; when alarmed or attacked, this species rolls itself up in a compact ball with its head tucked away inside its coils. The Indian python *(P. molurus)* and the reticulate python *(P. reticulatus)* of Southeast Asia are similar in their habits to the rock python; the reticulate python, the world's longest snake, sometimes attains a length of 10 m and weighs more than 100 kg.

The boa constrictor *(Boa constrictor)* of tropical America is also similar in its predatory strategy, though it rarely exceeds a length of 4 m. The giant anaconda *(Eunectes murinus)* is a very large semiaquatic constrictor that is generally found along the banks of streams and is only slightly smaller (up to 9 m) and considerably heavier (150 kg) than the reticulate python. Its diet varies from one region to the next but essentially consists of fish, caimans and large lizards, birds, and even fairly large mammals.

The green python *(Chondropython viridis)* of New Guinea and the emerald boa *(Corallus canina)* of tropical America are both strictly arboreal; the emerald boa never exceeds 2 m, and its emerald-green skin is spotted with yellow. The black-headed python *(Aspidites spp.)* of Australia is unique among the Boidae in that it feeds primarily on other snakes. D.H.

ROOK *Corvus frugilegus* 374

374
Sub.: Vertebrata
Cl.: Aves
O.: Passeriformes
F.: Corvidae

Corvus frugilegus is the common field crow of the Palaearctic zone (the common North American crow is *Corvus brachyrhyncos)*. The rook hunts for food in cultivated fields and meadows and builds its nest in groves of tall trees and frequently in city parks and along tree-lined streets, since it is hardly too diffident to neglect the foraging opportunities provided by human settlements. Longevity in captivity has been known to exceed 20 years; average body length is about 43–45 cm with a wingspread of 90 cm; weight varies between 380 and 590 g.

The most notable characteristic of this species is its extreme gregariousness; rooks assemble in enormous flocks to forage in the fields and especially to roost in the trees at night. In spring, they nest together in clamorous rookeries—a term that has been borrowed to describe the crowded nesting or breeding grounds of other species or even a collection of human dwellings where the premises are comparably noisy, overpopulated, and insanitary. The rook is a powerful, even acrobatic flier. Populations that spend the summer in Russian or Northern Europe fly south to France or other more clement regions for the winter; the populations that nest in Southern or Central Europe are essentially sedentary, though they may move a few hundred kilometers further south in winter.

▶ About 60% of the rook's diet consists of seeds, fruit, and berries, the remainder primarily of insect larvae (cockchafer grubs and wireworms), earthworms, caterpillars, snails, and spiders. It also, though more rarely, preys on small rodents, nestlings, and the eggs of other birds and ekes out its diet with carrion and other organic debris during the winter. Rooks locate most of their prey by probing and turning over the soil with their beaks, which are bare of feathers at the base; the nestlings are fed on a regurgitated pulp of seeds, insects, and larvae.

◀ There are few raptors bold enough to launch an attack on an entire flock of rooks, though fledglings that fall from the nest are occasionally devoured by foxes, martens, and other carnivores. Rooks are frequently confused with crows—to their detriment, of course, since this is the reason they are persecuted by hunters as supposed destroyers of the eggs and fledglings of more desirable game birds; farmers, perhaps more justifiably, have also held them responsible for eating seed grain, particularly in freshly sown fields of winter wheat. In recent years, the practices of putting out poison for rooks and shooting them on the nest have abated somewhat and the rook has extended its range in some parts of Western Europe, notably into the south of France.

The common Eurasian crow *(Corvus corone)* can be distinguished from the rook by its solitary habits. It is an omnivore, and a large part of its diet consists of small mammals, eggs, chicks, and nestlings of various species, including birds of prey; two distinct subspecies or geographical races are recognized, the carrion crow and the hooded crow. The raven *(Corvus corax)* is found along rocky coastlines and in mountainous regions in both hemispheres. It is primarily a carrion-eater, though otherwise somewhat reminiscent of the larger raptors in its behavior, particularly its penchant for competitive aerobatic displays.

The jackdaw *(Corvus monedula)* is a gregarious Eurasian corvid that often nests on the roofs of builings, though the name *jackdaw* may also be applied to a family of North American birds (Icteridae) more conventionally known as grackles. Two smaller Eurasian corvids, the Alpine, or yellow billed, chough *(Pyrrhocorax graculus)* and the Cornish, or red-billed, chough *(P. pyrrhocorax)*, formerly common in Western Europe, are now rarely encountered except in mountainous regions. Altogether the family Corvidae comprises over 100 species, found in all parts of the world; many are noted for their intelligence and adaptability, and some are brightly colored (the jays and magpies are both).

J.-F. T.

RUBBER BOA *Charina bottae* 375

375
Sub.: Vertebrata
Cl.: Reptilia
O.: Squamata
S.o.: Ophidia
F.: Boidae

The rubber boa is a small snake (about 50 cm, never exceeding 80 cm) with smooth, rubbery scales and rudimentary eyes. Its short, conical tail is rather difficult to distinguish from its head, and it is also known as the "two-headed snake" for this reason and the "ball snake," for reasons that will be discussed shortly. The rubber boa is a uniform olive-brown, reddish-brown, or yellow-brown in color. It is native to the Pacific coast of North America, from California up to British Columbia (which makes it the northernmost of the boas in its range) and east as far as Montana, Wyoming, and Utah. It lives in mountainous regions up to a height of 3000 m and prefers the humid terrain of a coniferous forest or mountain meadow, where it is often found under rocks, stumps, and dead trees.

▶ The rubber boa explores the uppermost layers of topsoil and shallow tunnels dug by other animals in search of its prey, which consists primarily of lizards (genus *Sceloporus)* and skinks, garter snakes, and small mammals. The rubber boa locates its prey by touch and chemoreception, and like all the Boidae, kills its prey by constriction (cf. ROCK PYTHON, CALIFORNIA KINGSNAKE) or, in this case, by crushing it against the walls of its tunnel when it does not have enough room to uncoil. The rubber boa is a reclusive creature and little else is known of its habits; in winter, it undergoes a period of dormancy that lasts from 4 to 6 months, depending on the latitude.

◀ The natural enemies of the rubber boa include a fair number of small carnivorous mammals that seek their prey under rocks, rotting stumps, leafmold or other organic litter. When threatened, the boa adopts a remarkable defensive posture, coiling up into a ball with its head tucked away inside its coils and flourishing its tail and even pretending to "strike" at the aggressor like a rattlesnake. Adult boas with scar tissue on their tails are frequently encountered in nature, which suggests that this ruse is at least sometimes successful in inducing predators to concentrate their attentions on the boa's tail rather than its head, thus giving the boa a second chance to make its escape.

The rose boa *(Lichanura trivirgata)* of Mexico and Southern California is similar in its habits, though it is less of a burrower and feeds primarily on small mammals. The genus *Eryx,* collectively known as sand boas, comprises several species that burrow in parched or sandy soil and are found in the arid regions of the Old World from West Africa to India. D.H.

376
Sub.: Mollusca
Cl.: Gasteropoda
Inf.cl.: Pulmonata
O.: Stylomatophora
F.: Stenogyridae

RUMINA DECOLLATA 376

This odd-looking land snail has a sturdy, nearly cylindrical truncate shell, 2.5 to 4 mm long; in the case of a snail's shell, *truncate* specifically means that its pointed tip, still present in young individuals, is broken off in adulthood, a phenomenon alluded to by the Latin name *decollata,* "decapitated." *Rumina decollata* is found in most Mediterranean countries, particularly in regions of chalky soil and scrub vegetation, such as the *garrigues* in the South of France; subfossil specimens (as well as a single living one) have been recovered from the depths of the Sahara. Like most land snails, its daily period of activity begins at dusk; it moves very slowly, of course, with its shell inclined almost totally toward the horizontal.

▶ *Rumina decollata* is an extremely voracious species that feeds on vegetable material and on other land snails. Experiments have been conducted at the University of California at Riverside to determine whether the species might be suitable as a biological control for the European brown snail *(Helix aspersa),* which for many years has been a serious pest in California orange groves. Both species had been originally introduced from the Mediterranean basin but prior to Dr. Fischer's experiments they had never encountered one another in the same climatic zone in North America.

◀ European land snails are preyed on by many different bird and mammal species (thrushes, blackbirds, gulls and various others shorebirds, shrews, hedgehogs, badgers, and fieldmice) as well as frogs, toads, and snakes; thrushes, for example, have acquired the trick of dropping snails from a great height in order to break their shells, and even set aside particular rocks, known to country people as "thrushes' anvils," that they habitually use for this purpose. *Rumina decollata,* like the more familiar European "escargot" (usually *Helix pomatia),* has a relatively thick shell and accordingly less to fear from these predators, though it may be attacked by insects, particularly in the early stages of life. The eggs of land snails, which are deposited on the ground, are devoured by numerous invertebrate predators, including beetle and fly larvae, sowbugs, slugs, and nematode worms.

Rumina decollata is the only example of the genus; other members of the same family, collectively known as door snails, are herbivorous for the most part. One of the *Rumina's* more distant relatives, the African land snail *(Achatina fulica),* has spread from its original home to the Far East and Oceania (where it was deliberately introduced by the Japanese during World War II as a potential food source for their garrisons on remote island outposts) and has recently appeared in Florida, where it has already become a serious agricultural pest. In addition to having a insatiable appetite for growing plants, *Achatina fulica* also feeds on organic debris and other snails. H.C.

S

377
Sub.: Arthropoda
Cl.: Insecta
O.: Orthoptera
S.o.: Ensifera
F.: Ephippigeridae

SADDLE CRICKET *Ephippiger ephippiger* 377

This flightless Old World grasshopper is found in brushlands and forest clearings and in stands of low-lying shrubs and bushes (including vineyards), though its "song"—a short, two-note stridulation that is produced by both sexes—is heard more often than the creature itself is seen; its local name in Burgundy is *"tizi,"* in imitation of its song. Its body is green, for the most part, and it is difficult to distinguish the larval and the adult forms on account of the adult's rudimentary elytra (brachyelytra), which are shaped something like the bowl of a spoon and partially concealed beneath the protuberant, saddle-shaped posterior section of the pyothorax (hence the genus and species names, meaning "saddle-bearer").

▶ The saddle cricket is an opportunistic forager, feeding primarily on leaves and buds (and in a terrarium, it can be raised on bits of fruit and cake crumbs). It also seizes insects with its forelegs when they comes within reach, though probably a fair proportion of these have already been crushed or injured in some other way. It is said to be especially fond of the eggs of moths (noctua, bombyx, pine processionary), which are attached to the undersides of leaves and twigs in long tubelike clusters.

◀ The saddleback grasshopper is the favorite prey of the Languedocian sand wasp *(Palmodes occitanicus)*. The wasp paralyzes the grasshopper by inserting its stinger into the latter's spinal ganglia, then drags it off to its burrow, using its long antennae as a sort of towrope. The wasp deposits its eggs inside the grasshopper's body, which is devoured piecemeal by the developing larvae. The praying mantis also feeds avidly on the larvae of the saddleback grasshopper. R.C.

Saga pedo 378

378
Sub.: Arthropoda
Cl.: Insecta
O.: Orthoptera
F.: Sagidae

This rare species of carnivorous grasshopper is found in the Mediterranean Basin and generally goes unnoticed because of its small size (70–80 mm). The females are wingless, making them even less conspicuous, and since this species generally reproduces parthenogenetically, the winged males are absent from a large part of its range; the elytra (wing covers) of the males are also somewhat atrophied. The slender, elongated body of *Saga pedo* is green with pinkish-white lateral stripes; the head is elongated as well. The larva, which is similar in form and color to the adult, hatches out in early May, undergoes 8 larval moults, and completes its development in July. *Saga pedo* is found in sunny, open terrain (heath, garrigue), generally on the ground, sometimes among the branches of a bush.

▶ The most striking feature of this species are its well-developed forelegs, which, like those of the preying mantis, are provided with powerful adhesive pads on the tarsi; the femur and tibia are also studded with sharp spines, which are also useful in seizing and grasping prey. (Because these formidable appendanges appear to be pointing, *Saga* is named for a Hindu goddess who points out the way to lost travelers.) *Saga pedo* moves both by jumping and walking along the ground, and both the larvae and the adults feed almost exclusively on locusts, which they catch by surprise in the grass, grasping and restraining the prey with the adhesive pads on their forelegs, and eating them alive in short order. *Saga pedo* can be kept in a terrarium, and need not be fed on live locusts, since it will also subsist quite contentedly on cherries, plums, and other sugary fruits.

◀ The predators of this species, if any, are unknown. R.C.

SAND SHRIMP *Crangon crangon* 379

379
Sub.: Arthropoda
Cl.: Crustacea
O.: Decapoda
S.o.: Caridea
F.: Crangonidae

This is the common shrimp of European coastal waters from the Gulf of Finland to the Mediterranean. It is normally found in regions of low salinity (e.g. near the mouth of a river) on a sandy, muddy, or mixed substrate at depths between 0 and 50 m; during certain stages of its life cycle, it migrates into estuaries and brackish lagoons. The eggs of this species are between 0.35 and 0.61 mm in diameter, and hatch in 3 (at 18° C) to 10 weeks (at 10° C). The planktonic larvae initially measure 1.8 mm and grow to a length of 4.7 mm in their first 5 weeks of life. The males develop more slowly than the females and attain a maximum size of 75 mm, as opposed to 95 mm in the case of the females. Lifespan seems to vary between 2 and 5 years, depending on climate, water temperature, etc.

▶ The sand shrimp is an opportunistic predator that will feed on virtually any organism of appropriate size that comes within its grasp, notably free-swimming nematodes and *Nereis*, amphipods *(Gammarus)*, copepods, juvenile mussels, acorn barnacle *(Balanus)* larvae, and Mysidacea (small shrimplike crustaceans); in the North

Sea, each of these prey animals becomes seasonally abundant, more or less in the order mentioned, during the course of the year. The sand shrimp also feeds on small mollusks, notably cockles and sunset shells *(Tellina).* In the absence of smaller prey, the sand shrimp may attempt to catch small fish or others of its own species; in more normal circumstances, it competes for food with swimming crabs of the genus *Carcinus* and *Liocarcinus.*

The sand shrimp detects its prey with the aid of olfactory receptors located on the whiplike stalks of its first pair of antennae and on its walking legs; in the aquarium, the shrimp may be observed as it hunts from ambush, burying itself in the sand and springing out to grasp its prey in the powerful subcheliform pincers at the end of its first pair of walking legs. *(Subcheliform* means that the first joint of the appendage is doubled over, like a nutcracker, to enable the organism to grasp its prey, but it has not developed into a true claw, or *chela).* The prey is torn apart by the shrimp's maxillipeds before being ingested.

◀ The sand shrimp is the primary source of food for a number of bottom-dwelling (benthic) species in the North Sea and is frequently found among the stomach contents of the sea snail (a small fish of the family Liparidae), sandgoby, sea poacher, anglerfish, whiting, smelt, flounder, bib *(Gadus luscus),* thornback ray *(Raja clavata),* sole, and plaice. Studies carried out along the North Sea coast of Germany have suggested that 145 billion sand shrimp are eaten by predators every year in this region alone. Apart from the species of fish just mentioned, these predators include the shrimp *Pandalus montagui,* which feeds on juvenile sand shrimp, the prawn *Palaemon adspersus,* cephalopods, and cormorants. The sand shrimp's only defense is concealment, either by burying itself in the sand or by means of protective mimicry of the underlying sediment.

The annual catch for the West German fishery is between 20,000 and 25,000 metric tons, and in some parts of the Atlantic, the Channel, and the North Sea, between one fifth and half as many sand shrimp are caught by human fishermen as by marine predators. A special dragnet is used for shrimp and other creatures, such as crabs, that can bury themselves in the superficial sediment layer. In this case, the trawl is equipped with a double pocket, the purpose of the extra pocket being to shoo away as many as possible of the numerous small fish that congregate on the bottom. The North Sea shrimping season lasts from July till October.

The members of the family Crangonidae are found in a variety of habitats, from coastal estuaries to abyssal plains. In European waters, *Crangon allmani* is found at a slightly greater depth and on a muddier substrate. *Philocheras trispinosus* is found in coastal waters with a higher salinity than those inhabited by the sand shrimp, due to a smaller admixture of fresh-water outflow. *Crangon septenspinosus* occupies essentially the same ecological niche as the sand shrimp on the other side of the Atlantic (and these two are regarded by some authorities as different varieties of the same species). All of these coastal shrimp feed primarily on polychaetes, ostracods, copepods, and small mollusks, and possibly on decaying flesh as well. In Europe, an alternate name for the sand shrimp is *brown shrimp,* a term that in North America is more frequently applied to the Brazilian shrimp *(Penaeus aztecus),* also quite common in the Gulf of Mexico and thus very familiar to North American seafood fanciers.

J.-P. L.

SAND WASP *Ammophila sabulosa* 380

380
Sub.: Arthropoda
Cl.: Insecta
O.: Hymenoptera
S.o.: Apocrita (= Petiolata)
G.: Aculeata
F.: Sphegidae

Ammophila is a genus of solitary digger wasps best known, along with the mason wasps of the genus *Eumenes,* as a relentless destroyer of cutworms and other small caterpillars. *Ammophila* seeks its prey in the sand along the roadside; it also feeds on nectar and pollen. Most species attain a length of 20 to 25 mm, a very respectable size for a wasp; its coloration is vivid (orange-yellow or red mixed with black), its body contours are extremely svelte, with a narrow abdomen attached by a long pedicel ("wasp waist") of 5 or 6 mm. Sand wasps of the genus *Ammophila* are found in virtually all tropical and temperate regions, with at least a dozen species being represented in Western Europe alone.

▶ *Ammophila* prefers an exposed, sunny terrain with sparse vegetation and friable, sandy soil in which the female digs its burrow. This is generally a vertical tunnel about 8 to 10 cm long terminating in an oval egg chamber. The dirt excavated from the tunnel is borne aloft and scattered in the air, and the entrance is sealed with a pebble and camouflaged with grains of soil when the female goes off in search of a nice fat cutworm (the common name for the larva of the noctua moth, in this case primarily *Agrotis segetum* and *Agrotis exclamationis).*

Cutworms are nocturnal, concealing themselves in tufts of grass during the day; the female *Ammophila* examines each potential place of concealment in the vicinity of its burrow, its antennae pointing directly in front of it and continually vibrating. When the prey has been located, the female straddles the cutworm's upper body (the cutworm being generally somewhat larger than *Ammophila sabulosa),* grasps the cutworm with its mandibles, and bends its abdomen forward until the stinger can be inserted into the segment right behind the cutworm's head. Then, *Ammophila* gradually works its way down the cutworm's body, successively passing its stinger all the way through the cutworm's body and into each of the ventral nerve ganglia until it has reached the ninth or tenth segment, at which point the cu tworm will be completely paralyzed.

Then the wasp once again grasps one of the cutworm's anterior body segments with its mandibles and, still straddling its prey, begins to drag it toward its burrow. When it reaches the burrow, it rolls back the pebble at the entrance and begins to crawl down the tunnel, its head still facing the entrance, dragging its prey behind it. The cutworm is placed in the egg chamber, where the wasp also deposits its egg, then affixes it to the cutworm's body by means of a sticky mucous secretion. The female wasp crawls back down the tunnel, and the entrance is permanently sealed; each female *Ammophila* lays several eggs, and constructs and provisions several nests in this manner. When the larva emerges, it feeds on the body of the cutworm—which initially at least is paralyzed but living—until its development is complete. All *Ammophila* species provision their nests with cutworms or other hairless caterpillars; some of them, notably *A. heydeni,* provide several caterpillars for each nest.

◀ Though the system just described is admirable in theory, its every stage is fraught with perils both for the immature and the adult *Ammophila.* Velvet ants (actually a kind of small parasitic wasp), notably *Dasylabris maura* L., have been recovered from the burrows of *Ammophila heydeni,* presumably having devoured the sand wasp larvae in the egg chamber. The strepsipteran *Eupathocera sphecidarum* (a kind of beetlelike parasite) deposits its eggs in the abdomen of the female Ammophila, and the growth of the larva eventually sterilizes and debilitates its host. The sarcophagid fly *Metopia leucocephela* Rossi deposits its eggs inside the body of the captive cutworm during the course of its journey to the egg chamber; opportunistic foragers, such as ants *(Formica cinerea)* and ant lion larvae (genus *Myrmeleon),* sometimes make off with the carcass altogether during the brief interval in which the female *Ammophila* is obliged to let go of it in order to unseal the entrance to its burrow.

R.C.

SAVI'S PYGMY SHREW *Suncus etruscus* 381

381
Sub.: Vertebrata
Cl.: Mammalia
O.: Insectivora
F.: Soricidae

Savi's pygmy shrew is believed to be the world's smallest mammal. The adult weighs only 2 g on the average (3.5 g in the case of a pregnant female about to drop a litter of 5 young) and measures between 40 and 50 mm from its snout to the base of its tail (65–80 mm altogether). Average longevity is about 16–18 months in the wild and 30–32 months in captivity. This tiny predator is fairly common in Spain, southern France (particularly along the coast and northward along the slopes of the Charentes and the Rhone Valley), Italy, and Greece. It is interesting to note that its range corresponds very closely to that of the ilex tree, *Quercus ilex*—which is to say that the two species have similar climatic requirements, not that the presence of one is in any way essential to the other.

Savi's pygmy shrew is almost invariably found among ruins, in the crevices of low masonry walls, or in gravel pits and similar terrain. It is especially prevalent in the countryside around Banyuls-sur-Mer, on the Mediterranean coast near the Franco-Spanish border, where the terraced fields are separated by low, unmortared retaining walls. This habitat provides the shrew with an abundance of insect prey, mild and relatively uniform climatic conditions, and a labyrinth of interconnecting fissures and tunnels that it would not be incapable of excavating for itself. It is most often found in abandoned terrace fields as well as in olive groves and uncultivated heath and brushland (garrigue).

▶ We have no definitive information on the diet of this species in nature, since the stomach contents of the specimens we have been able to examine consisted of nothing more than an undifferentiated whitish pulp. Most captive specimens expressed a preference for insects, arachnids, myriapods, and gastropods obtained from collecting boxes that had been placed in their original habitats. (The captive shrews showed no interest in the bodies of dead mammals; one shrew appeared to be terrified when confronted with a live mouse or a young sparrow, and made no attempt to attack either of these creatures.)

At any rate, we were able to conclude from these experiments that Savi's pgymy shrew will feed on virtually every small organism that it encounters in its usual habitat, provided that these are of suitable size (no larger than a large grasshopper or a young ocellated lizard) and unprotected either by chemical weaponry—which excludes certain millipedes, oil beetles *(Meloe)*, and the hemipteran *Graphosoma*. Tough chitinous armor excludes numerous ground beetles and the myriapod *Glomeris*. Like other shrews (notably *Crocidura russula* and *C. suaveolens)*, *Suncus etruscus* is quick to turn cannibal when its usual prey is unavailable (a phenomenon observed by us in cases where several individuals had been caught in the same trap).

In observing the captive shrews, we were struck not only by their feverish and unremitting activity but also by the difficulty they seemed to have in locating prey animals that were more than a few centimeters away. Savi's pygmy shrew has very poor eyesight, and its sense of smell is not acute beyond a distance of 15–20 cm. It may be able to sniff out a motionless prey animal from only a short distance away, but even then, the shrew appears to hunt down its quarry with a series of hesitant, random motions and will not attack until it has actually brushed up against the other animal with its sensitive whiskers, or vibrissae. Its keen sense of hearing and its ability to detect even the slightest vibrations in the ground may also come into play during this final tracking phase.

Once the prey is successfully located, the captive shrews proved themselves to be agile and relentless hunters; once a single prey animal had been caught and killed, the captive shrew went into a sort of killing frenzy that compelled it to hunt down every other living creature in the terrarium within a matter of minutes. A single bite from a shrew's sharp teeth is enough to dispatch an adult cricket, though judging by the rapidity with which the cricket's normal defense and flight responses were extinguished (which seemed disproprotionate to the strictly physical effects of the bite), it seems likely that the saliva of Savi's pygmy shrew, like that of the water shrew *Neomys fodiens*, contains a neurotoxin of some sort.

Savi's pygmy shrew has the highest metabolic rate of any mammal—more specifically, the hemoglobin in its blood transports a larger volume of oxygen, and the number of red blood cells per microliter of blood is extremely high (18 million per μl as against 5 million in humans). In addition, its heart is twice as large as other mammals' (in proportion to their other bodily dimensions), and its heart pumps blood at the astonishing rate of 900–1400 beats per minute. This explains the deadly urgency of the shrew's continual search for food; simply in order to live at such a pace, this "stomach equipped with teeth," as one writer has called the pygmy shrew, is obliged to consume the equivalent of twice its own weight every day. (Thus, it is hardly surprising that some of our captive specimens rejected their more accustomed prey in favor of a high-energy diet of beef spleen).

Shrews have generally been represented in the literature as solitary, irascible creatures, and even accidental encounters between them were generally thought to be fatal. More recent investigations have suggested that hostilities may be confined to an exchange of shrill cries and a few harmless feints and jostlings. The captive specimens we observed tried to avoid all contact with one another except during the breeding season. Then, after a few preliminary faceoffs of this kind, a sufficiently cohesive bond was finally formed, and the male and female along with several successive litters were able live together in a single nest. (It is true, however, that such a harmonious family situation could not endure for very long if food was in short supply, since one of the adults would invariably eat the others.)

◀ Though our information on this subject is also incomplete, it seems likely that weasels, martens, genets, and other small carnivores of the Mediterranean Basin feed occasionally on *Suncus etruscus* and other shrews. The screech owl and the tawny owl probably take a heavier toll of this particular species, since the bones of shrews are regularly found in the excremental pellets of these birds of prey. The ladderback snake *(Elaphe scalaris)* and the Montpellier snake *(Malpolon monspessulanus)* are both well represented in the region around Banyuls-sur-Mer; these colubrids, along with full-grown ocellated lizards (measuring 40–60 cm), should also probably be included among the predators of Savi's pygmy shrew. The adults, and more frequently the young ones in the nest, may also fall victim to marauding shrews of other species.

Until recently, Savi's pygmy shrew, too small to be eaten or hunted for sport and a tireless destroyer of insect pests, has been fortunate in its relations with human beings, and the architectural and agricultural techniques of former times (terraced fields and mortarless drywall construction) have provided it with an ideal habitat. However, the agricultural techniques of the present day (notably the use of chemical herbicides and pesticides) as well as grassfires and brushfires, which have become endemic in the south of France during the past few decades, threaten the destruction both of its habitat and of the insects, small reptiles, and invertebrates on which this species feeds. R.F.

SAWFISH *Pristis pristis* 382

382
Sub.: Vertebrata
Cl.: Chondrichthyes
O.: Hypothremes
F.: Pristidae

The sawfish typically measures between 4.5 and 6 m, though roughly a third of this length is taken up by the sawfish's saw, a long flattened beak that has a row of about 20 teeth extending laterally from each side. Though the rest of the sawfish's body looks something like that of a shark or a dogfish, it is more closely related to the rays and skates; its pectoral fins, which are attached behind its head, are not as well developed as the swept-back "wings" of the ray, and it mainly relies on its caudal fins to propel it through the water.

The sawfish is a coastal bottom-dwelling (benthic) fish, found in the Mediterranean and the Eastern Atlantic, roughly from Portugal to Angola; some species are also found in coastal estuaries. The female sawfish gives birth to about 20 live young at a time—though as long as the developing embryos are still in the female's genital tract, their beaks remain soft and flexible and their teeth are covered with a sheath of conjunctive tissue. The saw begins to stiffen and the protective sheath sloughs off within a few hours of birth, however, so that the newborn sawfish are quickly able to fend for themselves.

▶ The sawfish preys on small benthic fish, crustaceans, mollusks, and worms, and uses its toothed saw either as a rake or a saber—raking mollusks and other burrowing creatures out of the sand or mud on the bottom, or, in open water, hacking its way through a school of small fish with powerful saber strokes, very much in the manner of the swordfish. It swallows a certain number of the casualties on the spot and leaves the rest to be devoured by tuna and other predators. The sawfish's more conventionally fishlike proportions, in contrast to the batlike, laterally elongated contours of the rays, allow it to maneuver more freely while pursuing a school of fish in open water, whereas the rays are obliged to seek their prey almost exclusively along the bottom.

◀ The sawfish's toothed beak is an effective deterrent even against much larger marine predators, with the exception of the sperm whale. As far as human fishermen are concerned, the sawfish is pursued as a trophy fish, as well as for its pectoral fins, which are locally consumed by man, and for its oil-rich liver. F.M.

SCALARE *Pterophyllum scalare* 383

383
Sub.: Vertebrata
Cl.: Osteichthyes
S.cl.: Teleostei
O.: Perciformes
F.: Cichlidae

This tropical freshwater fish measures between 20 and 25 cm and is found in the "black waters" of the Amazon Basin (regions subject to annual flooding, where little sunlight penetrates the forest canopy and the high concentrations of tannins and humins released by decaying vegetation turns the water the color of strong tea). The female attaches its eggs to a plant stalk or other vertical surface, which it carefully cleans off beforehand to make sure that the eggs will adhere; both parents stand guard over the eggs, aggressively defending them against predators, and the vividly contrasting silver-and-black coloration of the adults is intensified during the incubation period. The hatchlings are similarly kept safe inside a burrow that is excavated out of the bottom shortly before the eggs hatch and is provided with several escape hatches.

The scalare hunts by day and often seeks its prey along the bottom or amid the dead branches and drowned vegetation along the bank; in open water, it swims in schools, sometimes in association with another cichlid, *Cichlasoma severum*. Though it will accept nonliving prey in the aquarium, in nature its predatory reflexes are triggered only by living prey animals that exhibit some degree of motion. The scalare feeds primarily, and voraciously, on tetras and other characids, young cichlids, and aquatic insects; a small individual (15 cm) can consume 7 or 8 smaller fish (3–4 cm) in a few minutes.

The scalare's small, delicate-looking mouth can be opened very wide to create a powerful suction, so that the prey animal is immediately aspirated into the oral cavity of the predator; the scalare's teeth erupt from the pharyngeal bones and are thus used exclusively for chewing rather than for prehension. While feeding along the bottom it uses the same predatory mechanism in reverse, expelling a stream of water from its mouth to dislodge bottom-dwelling creatures from their burrows in the mud and immediately sucking them into its mouth them when they float up into the water.

The scalare hunts by sight, slowly approaching within a few centimeters of its victim before attempting to ingest it, and actively giving chase if the quarry is alerted by its approach. Its laterally flattened body enables to pass very easily through a bed of aquatic plants, the branches of a fallen tree, or other underwater obstruction, and it sometimes tilts over on its side when in pursuit of a hatchling or an insect that is poised just below the surface, thus placing its mouth in the closest possible proximity to the prey. It swims by moving its unpaired fins in order to avoid creating vibrations in the water and, especially, ripples on the surface, which might attract the attention of an aerial predator.

◀ Piranhas, catfish, larger cichlids, watersnakes, and aquatic birds are the chief predators of the scalare. When it first catches sight of a potential predator, the scalare executes a quarter-turn, thus presenting the predator with an unrevealing frontal view of its slender disk-shaped body and the filaments of its pelvic fins (somewhat resembling an elongated capital A). In its normal vertical orientation, the scalare is also scarcely visible from above. A predator launching a broadside attack without warning may be confused by the hallucinatory display of wavy silver-and-black stripes presented by an entire school of scalares, particularly since the quarry will turn aside at the last moment, like duellists, to provide their attacker with the slenderest possible target.

Schooling fish benefit from the fact that the predator is likely to make a sort of random lunge at the entire school rather than a concerted attack on a particular individual. When pursued, the scalare will attempt to retreat, headfirst, into a crevice in the bank or some similar refuge; even if its caudal fin is left protruding, it can quickly be regenerated if the stalk is not too badly damaged in a predator attack.

The common name "scalare" may refer either to *Pterophyllum scalare* or to a smaller congener, *P. einekei*, both of which are very popular aquarium fish. Commercial stocks are generally bred in captivity—including several distinctive strains that have been produced by breeders and are not found in nature—so that the popularity of these two species with tropical-fish fanciers has had little effect on populations on the wild. *P. scalare* is locally consumed by nomadic hunters in the Amazon Basin but is not fished on any sort of intensive or commercial scale. M.D.

SCARLET-CHESTED SUNBIRD
Nectarinia senegalensis **384**

384
Sub.: Vertebrata
Cl.: Aves
O.: Passeriformes
F.: Nectariniidae

This species measures 13 cm in length and is found in scrub forests and woodlands (though not on subdesert steppes or in dense rainforests) all across tropical Africa, from Senegal to Kenya. The sunbirds are a large family of Old World passerine birds that occupy more or less the same ecological niche as the New World hummingbirds, both being closely associated with flowering plants and playing an important role in pollination. In contrast to the often irascibly solitary hummingbirds, the mating bond is very strong among sunbirds, though they often congregate in larger groups outside the mating season while gathering nectar from flowering trees and shrubs; on these occasions, they may often be seen chasing each other and squaking shrilly, but their intentions appear to be playful rather than truly aggressive. Sunbirds often build their nests in the vicinity of wasp's nests, for reasons that are not well understood.

▶ The sunbirds feed on nectar and fruit pulp as well as on insects and other small arthropods. Those species that feed primarily on nectar can be identified by their broad curving beaks, whereas the beaks of primarily insectivorous species tend to be narrower and straighter; the sunbird's very long protractile tongue is well adapted for sipping nectar as well as retrieving insects from inside the tubular bells of flowers. Insects may be caught in flight or while alighting on a petal or a branch, and sunbirds frequently scavenge dead insects from spider webs.

◀ Several species of African cuckoo, including the emerald cuckoo *(Chrysococcyx cupreus)*, Klaas's cuckoo *(C. klaas)*, and the bronze cuckoo *(Chalcites)*, are small and agile enough to lay their eggs in the sunbird's enclosed hanging nest.

The 150 sunbird species are mainly found in Africa, though they also range as far east as Indonesia and Australia. Unlike the hummingbirds, they are not strong fliers and are incapable of long migrations. The spider hunters *(Arachnotera)*, found in the Indo-Malaysian archipelago, are more sturdily built and less brightly colored; all have very long straight beaks (most notable in the case of the yellow-eared spider hunter, *A. chrysogenys)* and are thus well equipped for catching spiders outright rather than merely robbing their webs.

J.-F. T.

SCHELTOPUSIK *Ophisaurus apodus* 385

385
Sub.: Vertebrata
Cl.: Reptilia
O.: Squamata
S.o.: Sauria
F.: Anguidae

This large legless Eurasian lizard typically measures between 80 and 120 cm, though it may occasionally attain a length of up to 150 cm. It is found in treeless, open countryside, including moors and open fields, and is particularly common in stone walls and hedges between cultivated fields. The scheltopusik ranges from Dalamatia in the west to Turkestan in the east, by way of Greece, the Crimea, Turkey, Syria, Iraq, and Iran. Its name (sometime seen as "sheltopusik" or "sheltopusick") represents a Germanic spelling of the Russian word *zhyoltopusik*, or "yellowbelly." The scheltopusik is diurnal in its habits, though it sometimes seeks shelter during the hottest part of the day, and its spends a period of winter dormancy (varying in duration with the region) at a end of a long underground burrow. Its lifespan in captivity may exceed 20 years.

▶ The scheltopusik feeds on large insects, including beetles, butterflies, and moths, and snails; it crushes the shells of snails with its broad, blunt teeth, and sometimes feeds opportunistically on other creatures, including small rodents, birds' eggs and nestlings, other lizards, and small snakes. It is a marauding hunter, though it pauses frequently to await the approach of a potential prey animal. It will actively pursue an insect or other prey, raising up the forepart of its body so it can seize the prey it its jaws and sometimes striking they prey against the ground if the pressure of its teeth along is not sufficient to cause death. Like many lizards, the scheltopusik generally pushes a prey animal into its mouth by pressing it against a rock or against the ground.

◀ Unlike other legless lizards, the scheltopusik can move along the ground very rapidly and is quick to seek cover at the approach of danger. The adults rarely show themselves in open country and have no particular predators or competitors, since they are the only reptiles in the habitat that feed on snails and hard-shelled insects; the younger lizards are more vulnerable to attack and are thus even more discreet in their habits. The scheltopusik is not readily mistaken for a snake of any kind and, since it preys on harmful insects in cultivated fields, is regarded as a useful ally by farmers in the Crimea; elsewhere in the Soviet Union, this species has been shown to play a dominant role in the biological control of certain borers that feed on cereal crops.

Congeners, collectively known as alligator lizards, are found in the Eastern United States (where they are sometimes known as glass snakes), North Africa, and Southeast Asia and Indonesia; all are smaller than the scheltopusik, timid in their habits, rarely encountered by man.

D.H.

386
Sub.: Arthropoda
Cl.: Chilopoda
O.: Scolopendromorpha
F.: Scolopendridae

Scolopendra cingulata 386

This Old World centipede typically measures 60–80 mm, though some tropical congeners (notably *S. subspinipes* and *S. gigantea)* may attain lengths of up to 25 cm. *S. cingulata* is found throughout much of the Mediterranean Basin, including Southern Europe, the Balkans, North Africa, Syria, and the Black Sea coast. In Southern Europe, it is most prevalent in dry, open country (heath, garrigue, and upland meadows) and is sometimes encountered while turning over the soil in a garden or in leafmold, though it is not normally found in wooded areas. Specific data is lacking, but typical longevity for this species is on the order of several years.
▶ Like other centipedes, *S. cingulata* catches its prey in its curved venomous pincers and will feed on virtually any small invertebrate that is not protected by a stout carapace or toxic chemical weaponry (e.g. scorpions and large spiders). Some tropical species are large enough to prey on snakes and lizards, toads, birds, and rodents, and their venom has even been reported to have caused death in humans (in a few very exceptional cases). *S. cingulata* has a somewhat more sinister reputation than it deserves, since it is not capable of inflicting any permanent harm on its occasional human victims. Its venom is considerably more potent in the spring than in the winter; in general, its bite is accompanied by acute pain and localized swelling, which subsides after a few days.
◀ There are a few species of scorpion (including the Languedocian scorpion, *Buthus occitanus),* perhaps a few species of beetles or even ants, that prey opportunistically on S. cingulata from time to time, though in view of the difficulties involved in subduing such a quarry, the word "opportunistic" may not seem very well chosen. These encounters are probably very rare, however, since these predators naturally tend to concentrate on less challenging prey; only a few vertebrate species (including raptors and certain primates) have ever been reported to prey on centipedes, and the rare travelers' accounts of human beings feeding on centipedes were probably inspired by individual acts of bravado rather than customary or habitual behavior.

Most centipedes are found in tropical or subtropical latitudes; other scopendromorphs found in the same regions as *S. cingulata* belong to the genus *Cryptops* for the most part and are considerably smaller, and so little is known of their ecological role as predators that it would be difficult to identify any two species as being competitive with one another. Such species as *C. parisi* and *C. hortensi,* for example, may coexist in the same habitat, often in conjunction with centipedes belonging to the order Lithiobiomorpha (see discussion at the end of the entry for *Lithobius forficatus.)*

J.-J. G.

387
Sub.: Arthropoda
Cl.: Insecta
S.cl.: Mecopteroidea
O.: Mecoptera
F.: Bittacidae

SCORPION FLY *Bittacus tipularius* 387

This rather ungainly-looking insect lives in bushes along riverbanks or other low-lying, humid areas; it is about 18 mm long with a wingspan of 40 mm. Its wings are narrow, transparent, and yellow in color, without any darker pigmented patches. The scorpion fly is airborne for a brief period at dusk, then spends the night clinging to the underside of a twig, somewhat in the manner of a long-legged cranefly (which is *tipula* in Latin, hence *tipularius).* The name "scorpion fly" refers to the fact that the male habitually arches up its greatly enlarged genital organs in much the same way that a scorpion does its sting.

The adults emerge in May and are particularly active from July to September. During courtship, the male presents the female with a little food parcel or an entire prey animal. Copulation, which lasts for 2 hours, takes place with both partners dangling from a twig by their forelegs while finishing off the food parcel; the male maneuvers his body around until its ventral surface is adjacent to the ventral surface of the female's. The male dies shortly afterward.

The female digs a small egg chamber in the ground; the eggs are white at first, later greenish gray, and oddly polyhedral in shape, something like misshapen dice. The eggs will hatch only in the presence of moisture. The larval stage lasts from the middle of summer until (probably) the following spring, when the larva excavates a little tunnel for itself fitted with a protective lid, in which it undergoes nymphosis.

▶ The adult is equipped with an elongated beak formed from the clypeus (the sclerotized plate on the front of an insect's head, directly behind the "lips" and mouth parts) and buccal plates. The forelegs have only a single claw; the scorpion fly seizes its prey with its posterior pair of legs, the last 3 segments of which are folded back to make a set of pincers. The scorpion fly stations itself on a twig in a shady spot, not too far above the ground, and awaits the approach of a gnat or a spider. It seizes the prey with its posterior pair of legs (which dangle toward the ground when not in use) and pushes forward until its beak can be inserted and the prey drained of fluids, after which the scorpion fly makes a fastidious toilette.

The larvae are about 15 or 20 mm long and look something like caterpillars, with 8 fleshy ventral pads and a sort of staple (called a pygopod) on the tenth abdominal segment, which permit the larva to propel itself along in a series of laborious arcing motions like a measuring worm. It crawls over the topsoil, between the leaves and blades of grass, and feeds on dead insects and other organic debris and occasionally on other small creatures even less mobile than itself, which it captures with its forelegs. R.C.

Scutigera coleoptrata 388

388
Sub.: Arthropoda
Cl.: Chilopoda
O.: Scutigeromorpha
F.: Scutigeridae

The adults grow to a length of 30 mm, and the postembryonic development of this species is comparable to that of *Lithobius forficatus* (q.v.). The scutigeromorphs are largely confined to tropical regions, but this species is found as far north as Great Britain, North America, Central Europe, and southern Russia, and more frequently in the Mediterranean Basin, North Africa, and the South Atlantic island of St. Helena. *S. coleopterata* is a thermophilic creature, often found in warm, dry, relatively open terrain (including fields, lawns, gardens, and parks); it is also the common house centipede of Europe and North America, and is thus most frequently seen scuttling along the floors and walls of human dwellings as fast as its long bristly legs will carry it. Longevity of the scutigeromorphs is on the order of 3–4 years.

▶ Like the lithiobiomorphs, the scutigeromorphs are predators rather than scavengers, and *S. coleopterata* is remarkably quick in its movements, catching flies (which, along with various other insects, appear to comprise the greater part of its diet) the moment they alight on a horizontal surface and immoblizing them with its venomous hooked pincers. *S. coleopterata's* vision is keen enough to enable it to detect its prey partially by means of visual cues, though it also makes use of its tactile and olfactory organs.

A fairly complex sequence of predatory responses in this species have been observed in nature and the laboratory; interestingly, predation and aggression, as well as any alteration in the environment, are all associated with grooming behavior in *S. coleopterata*. In the terrarium, males and females have sometimes been observed to prey on one another after mating.

◀ Ants have occasionally been observed carrying off the bodies of scutigeromorphs, sometimes in fragmentary form, though it is not clear that they are capable of subduing a living centipede. As noted elsewhere, there is a fairly limited number of amphibians, reptiles, and insectivorous mammals that feed on centipedes, though we have no specific information on the predators of *S. coleopterata* in particular. *S. coleopterata* is the only scutigeromorph in Western Europe, so it seems probable that its competitors are to be sought among the larger arachnids and predacious insects, though, once again, we have no definite information on this subject. J.-J. G.

SEA CUCUMBER *Holothuria tubulosa* 389

389
Sub.: Echinoderma
C.: Holothuroidea
O.: Aspidochirota
F.: Holothuyriidae

This particular species of holothurian, or sea cucumber, is common, even abundant, along the shores of the Mediterranean, less so along the Atlantic coast from the Bay of Biscay to Portugal. It lives in shallow water, frequently in the vicinity of patches of *Posidonia* or other aquatic plants. It displays a negative reaction to light (negative phototropism) and often covers itself with coarse sand. Thus, it is rarely found below a depth of 30 m, where the substrate becomes increasingly muddy. The sexes are separate (in contrast to many marine invertebrates, which are hermaphroditic), and reproduction occurs throughout the year with an annual peak in August and September. The free-swimming (planktonic) larva passes through 3 morphologically distinct stages, known as auricularia (lobate), doliolaria (barrel-shaped), and pentactula (five-tentacled).

▶ The sea cucumber feeds in much the same manner as an earthworm, extracting bacteria and other nutrients from the sand that it shovels into its mouth with its spadelike buccal tentacles. It feeds continuously as it creeps along the sandy bottom, though there is a question (which many would probably regard as moot) as to whether it can more properly be regarded as a predator or a scavenger, feeding exclusively on the bacterial layer on the surface of the sand and on dead invertebrates and other organic debris found beneath it.

In any event, the sea cucumber is not an indiscriminate feeder, since it appears to be able to pick out patches of sand that are especially rich in these nutrients. Depending on local conditions, the digestive contents of *Holothuria tubulosa* might typically include sandworms, nematodes, crustaceans (usually digested) as well as such inorganic debris as mollusk shells, the spines of sea urchins, and a large proportion of sand. In certain coastal areas, the presence of a dense population of holothurians results in a significant rerrangement and alteration of the composition of the seabed in much the same way that earthworms "till" the soil on land.

◀ This species has no known marine predators, probably because the sea cucumber's outer teguments are extremely tough and its body can be readily contracted. Other holothurians have also evolved 2 remarkable defensive mechanisms not present in *Holothuria tubulosa*. The first of these is Cuvier's organ, which secretes adhesive mucus filaments that the creature expels from its anus and in which a pursuing predator may become stickily entangled. Many holothurians can also detach their internal organs from the body cavity and expel them in a similar manner, perhaps to decoy predators away from the organism itself or perhaps to serve some physiological purpose that is not yet understood. Whatever the reason, the holothurian's internal organs are speedily regenerated. Unlike several of its congenerics (along with certain species of the genera *Microthele, Actinopyga,* and *Thelenota*), *Holothuria tubulosa* is not normally consumed by man.

Other holothurian species found in European waters include *H. stallati, H. mammata, H. forskali,* and *H. impatiens. H. impatiens* is also quite common along the coasts of Indonesia, New Guinea, and Melanesia, where, along with a number of other species found in association with coral reefs, it serves as an important article of commerce under the name *trepang, shekin,* or *bêche-de-mer*. Sea cucumbers are harvested by divers or by fishermen with dragnets and then processed for export on the beach, i.e. the sea cucumbers are sorted and graded (those with the thickest outer teguments being especially prized), then cleaned, boiled, smoked, and dried.

Trepang, which is used in the preparation of soups and other dishes, is a staple item in Chinese cuisine especially; in Hong Kong and Singapore, the principal centers of the trade, about 1500 metric tons are sold each year. The total annual catch throughout the Pacific (including Japan and Korea) is currently estimated at 30,000 metric tons. Trepang played such an essential role in the early economic development of Papua and Micronesia that the coastal variety of Pidigin English spoken there came to be known as Bêche-de-Mer (or more accurately, Beachy-la-Mar).

A.G.

SEA ELEPHANT *Mirounga leonina* 390

390
Sub.: Vertebrata
Cl.: Mammalia
O.: Pinnipedia
F.: Phocidae

The sea elephant is the largest member of the family Phocidae (seals and sea lions). The bulls may weigh as much as 3 or 4 metric tons and attain a length of 4.5–4.8

m, though the cows are considerably smaller (400–900 kg, 2.6–2.8 m). The species owes its name to the bulbous mass of erectile tissue that is situated just in front of the nostrils and which, especially in the case of the bulls, inevitably suggests an elephant's trunk. The elephant seal reaches sexual maturity at 3 to 5 years; maximum life expectancy for both sexes is about 20 years. This species is found only on 3 subantarctic island groups: South Georgia (with a population of 300,000 individuals), Kerguelen (290,000), and Macquarie (100,000), with some variation in average size and weight among these 3 different populations.

The elephant seal comes ashore only for two months out of the year: October, which is its mating season, and November (the beginning of summer), when it sheds its winter coat; it eats nothing during either of these periods. During the mating season, the cows are rounded up into harems by the dominant bulls, whereas they assemble in single-sex groups during the November molt; individual elephant seals appear to have little contact with one another during the rest of the year. Each year, the cow elephant seal gives birth to a single pup, which weighs 60 kg at birth and will have put on an additional 90 kg by the time it is 23 days old, by which time even the very substantial fat reserves that are accumulated by these creatures during their many months at sea will be virtually depleted.

▶ No direct observations of the submarine feeding habits of the elephant seal have ever been made, though in a very few instances, the stomach contents of non-fasting elephants that were captured on land have been analyzed. These consisted of the beaks of cephalopods and the remains of fish. In view of the pronounced sexual dimorphism displayed by this species, it seems likely that the diet of the bulls and the cows is very different. We know that the cows tend to remain in shallow water, where they presumably feed on cuttlefish and smaller fish, whereas the males are known to be in the habit of making very long ocean voyages (up to 3000 km) and to be capable of remaining submerged for as long as 45 minutes at a time; hence it seems likely that they feed on larger cephalopods from the ocean depths. In either case, the elephant seal must consume about 6 percent of its body weight every day in order to sustain itself through its two- or three-month fast on shore.

◀ The killer whale is the only native of these latitudes that is even capable of preying on the elephant seal, though the pups might conceivably be in some danger from the leopard seal. In any case, the toll that is taken by these predators (if any) does not appear to be very considerable. Before the turn of the century, however, the elephant seal was extensively hunted by man for its blubber—understandly so, since the carcass of a smallish male elephant seal (3 m long) might yield up to 350 liters of oil.

M.P.

SEA MOUSE *Aphrodite aculeata* 391

391
Sub.: Annelida
Cl.: Polychaeta
F.: Aphroditidae

The sea mouse is a marine polychaete worm with a broad flattened body, about 10 to 20 cm long and 4 to 5 cm wide. As with all annelids (segmented worms), its back is covered with a row of overlapping scales called *elytra* (a term more frequently used to describe the sclerotized wing covers of beetles and other flying insects), though these are largely concealed by a dense covering of bristles. *Aphrodite* owes its flattering scientific name to the shimmering, iridescent coloring along the sides of its body and its more prosaic common names (*sea mouse* in English, *taupe de mer*, or "sea mole," in French), first bestowed by sailors and fishermen, to the dark-gray color of its back. The sea mouse lives on muddy or sandy bottoms, though it does not dig tunnels like many marine worms; it is rarely washed up on shore but frequently caught in fisherman's dragnets all along the coast of Europe.

▶ Most early twentieth-century authorities, following Darboux (1899), have acknowledged that *Aphrodite* is a predator; later researchers have discovered the remnants of other polychaete worms, both sedentary and free-swimming, as well as crustaceans in its digestive tube. Recent experiments in the laboratory have shown that the sea mouse will take prey only when it is in the vicinity of a sandy substrate in which it can readily conceal itself. It will not hesitate to attack creatures that are somewhat larger that itself, notably the polychaeate *Nereis vireus,* which it swallows headfirst, a procedure that requires a convulsive effort on the part of its entire body and is accompanied by rustling and flapping of its dorsal scales. Though the sea mouse is equipped with 4 jaws, these are used only to secure the prey, which is subsequently swallowed whole.

◀ Like its terrestrial namesake, the sea mouse furnishes one of the lower links in a number of different food chains; it is preyed on by numerous bottom-dwellers, including the larger marine worm species, starfish, hermit crabs, sea snails, and many different kinds of fish.

Harmothoë imbricata is about the only one of the Aphroditae whose feeding habits have been studied in some detail. It only eats live prey, which it detects initially by vibration and subsequently by chemical stimuli; it strikes quickly, abruptly protruding its proboscis (which in this case is the retractable anterior tip of its digestive tube), firmly grasping the prey with its jaws, and then ingesting it by contracting its muscular pharynx. The planktonic larva of *Harmothoë* and related species feed on diatoms and stored food reserves during its initial free-floating (pelagic) phase; later, when it descends to the bottom (benthic phase), the proboscis is already functioning, and the sea mouse is launched on its career as a predator. C.M.

392
Sub.: Vertebrata
C.: Mammalia
O.: Carnivora
F.: Mustelidae

SEA OTTER *Enhydra lutris* 392

The sea otter is considerably larger than the European common otter, attaining a length of 1.2–1.3 m (though its tail is only about 30 cm long) and weighing as much as 25 or 30 kg. Like the other members of the subfamily Lutrinae, this species displays a number of adaptations to an aquatic mode of life, most notably its webbed toes and thick, water-resistant fur, which ranges in color from reddish-brown to black. The sea otter is currently found on both sides of the North Pacific, all along the Kamchatka Peninsula and much of the coast of California, though its range was formerly much more extensive. It spends most of its time in the water; in summer, it wraps itself in long strands of kelp to prevent being swept out to sea while it sleeps, and in winter, after the kelp beds have been swept away by the tides and heavy seas, the sea otter tends to stay close to the rocky shoreline, though it never ventures more than about 10 m inland.

▶ The sea otter exploits virtually all the possibilities of its marine habitat, feeding on mollusks, gastropods (notably abalone), and large quantities of sea urchins; during the summer, its diet includes a larger proportion of fish and crustaceans. The sea otter dives to an average depth of 5 m (but occasionally as far down as 50 m) in search of its prey, remaining submerged for only about a minute as a rule, though it is able to hold its breath for as long as 3 minutes. It locates its prey esentially by sight, and after it has picked up a mollusk or other marine organism, it tucks it into one of the two deep folds on either side of its chest and swims back up to the surface, where it devours its prey while floating comfortably on its back.

The sea otter preys avidly on sea urchins, breaking off their spines with its front paws, then pressing hard against the shell and making a hole through which the urchin's soft body parts can be extracted. The sea otter's method of opening scallops, Pismo clams, and other bivalves constitutes a rare instance of tool-using behavior in the mammalian world: Along with the mollusk, the otter brings up a small, flat stone from the bottom; while floating on the surface, stretched out at full length on its back, it places the stone on its chest and uses it as a sort of anvil, cracking open the mollusk by pounding it on the stone. Other mollusks with thinner shells, crustaceans, and fish are simply ground up by the otter's powerful molars before being swallowed.

◀ The pelt of the sea otter was an important article of commerce for Russian fur trappers and Yankee sea captains engaged in the China trade during the 18th and 19th centuries, which is the main reason that this species no longer occupies much of its former range along the shores of the Northern Pacific. Though protective legislation has been in effect since 1911 in certain coastal regions where the sea otter is still prevalent, the clandestine persecution of the California sea otter by abalone fishermen has continued to be a source of controversy and public concern in recent years.

The sea otter conceals itself in thick offshore kelp beds partially in order to protect itself from sharks and killer whalers, its main natural enemies other than man. When the kelp beds disappear in winter, the sea otter is vulnerable to attack by these predators, which is why, with the additional dangers of winter gales and heavy seas, it prudently hugs the shore or even remains on land during the winter.

The pelt of the Northern fur seal (q.v.) was known as "sea otter" or "loutre de mer" to the European fur trade, understandably resulting in some confusion between these two species, which are not at all closely related. P.A.

SECRETARY BIRD *Sagittarius serpentarius* 393

393
Sub.: Vertebrata
Cl.: Aves
O.: Falconiformes
F.: Sagittariidae

This large terrestrial raptor measures between 112 and 127 cm and weighs between 3.8 and 4.1 kg. It is found in regions of low-lying ground cover (<50 cm in height) on the open savananahs and arid or subdesert steppes of eastern and southern Africa. It builds its nest in a thornbush, and avoids both deserts and wooded areas. Depending on the local abundance of prey animals, an individual territory may cover an area of as much as 10,000 ha, though males and females sometimes hunt in pairs. The quill-like feathers at the back of the bird's head suggest a resemblance to a busy 19th-century scribe with a sheaf of quill pens tucked behind his ear, hence "secretary bird."

▶ The secretary bird patrols its extensive hunting grounds on long storklike legs rather than from the air, though in other respects its morphology is quite comparable to that of the other raptors. It feeds primarily on rodents and other small mammals weighing less than 500 g, grasshoppers and locusts, and both venomous and nonvenomous snakes. It seeks its prey by striding across the savannah, its eyes fixed on the ground, pausing to snatch up an insect in its beak or pin a rodent to the ground with its talons.

When it comes upon a snake, it begins to dance around it in a sort of frenzy while flapping its outstretched wings, then snatches the snake with its long talons (its legs are protected against snakebite by a thick layer of scales) and tosses it up in the air, repeating this performance until the snake loses consciousness. Since the secretary bird feeds primarily on rodents and insects, however, these spectacular encounters occur rather rarely.

◀ The adult secretary bird has virtually no predators, though the nestlings may occasionally fall prey to kites and corvids. The pale chanting goshawk *(Melierax musicus)* as well as various large bustards and storks all pursue similar predatory strategies, though competition with humans is limited since the secretary bird tends to live in unpopulated or thinly settled areas. This species may have suffered some loss of habitat as grazing lands are increasingly brought under cultivation, though the secretary bird is generally regarded as a useful control on snakes, insects, and rodents, and thus is protected from deliberate molestation by religious taboos and other customary safeguards.

The secretary bird is the sole representative of the family Sagittaridae. The cariamas, though assigned to the order Gruiformes (cranes, storks, etc.), occupy the same ecological niche on the pampas of Argentina, and the crested cariama *(Cariama cristata)* is almost identical in appearance to the secretary bird, even to the bristly topknot on its head. J.-F. T.

SERVAL *Leptailurus serval* 394

394
Sub.: Vertebrata
C.: Mammalia
O.: Carnivora
F.: Felidae

This small, long-legged African felid generally measures about 1 m in length exclusive of the tail, which is comparatively short (30–40 cm). It has a narrow head and broad, elongated ears; its coat is spotted with black, and its eyes are yellowish. The serval is found throughout most of the nondesert regions of Africa, with the several subspecies being distinguished by the distribution of their markings as well as their various habitats: savannah, arid steppes, forests, etc. In general, however, the serval is fond of thick brushwood or tall grass where it can easily conceal itself and is most often found at the edge of a marsh or along the banks of a river or stream. The serval is a solitary hunter, and the size of its territory varies substantially with the nature of the terrain.

▶ The serval preys on rodents, hares, and the young of various larger mammals as well as on birds, and it seems to particularly enjoy catching fish, which it flips out of the water with a backhanded swipe of its paw. Its long forelegs are useful for investigating the burrows of rodents, and the occupants are sometimes killed on the spot with a blow from one paw, sometimes hooked by the serval's claws and dragged out into the open, then killed with a bite to the back of the neck. The serval catches young antelopes by stalking very cautiously through the grass, like a miniature version of the cheetah, then making a short dash and pouncing on the prey as soon it draws within range.

◀ The serval is quick and agile enough to escape from larger carnivores and appears to have no natural enemies; at any rate, attacks on this species by other felids or canids have never been reported.

P.A.

395
Sub.: Arthropoda
Cl.: Insecta
O.: Coleoptera
F.: Coccinellidae

SEVEN-SPOTTED LADYBUG
Coccinela septempunctata **395**

This ladybug species, one of the most common in Western Europe, is found throughout the Palaearctic zone as well as in India. Its body is smooth, globular, highly convex, and measures about 5–8 mm. The elytra are bright red, with a large wedge-shaped black spot at the junction of the 2 elytra, as well as 3 smaller black spots farther down on each side. The female lays about 50 eggs per day, or a lifetime total of between 500 and 1000. The larva's elongated pale-gray body is marked with 4 orange spots; its relatively long legs give it a ponderous, trundling appearance, which has reminded some observers of a very small alligator. The body of the nymph is globular, like the adult's; the adult may survive for up to 12 months.

▶ The seven-spotted ladybug is stenophagous, which is to say that it feeds on only a few different kinds of food; for the most part, adult and the larva prey actively on aphids, especially those species of the genus *Aphis* that are associated with gramineous plants (grasses). The aphid is swallowed whole by the adult and the more advanced larvae; only the larvae in the initial stages of development digest their prey externally by means of an infusion of salivary secretions. An adult ladybug may eat as many as 100 aphids in a day, but when aphids are scarce, the larvae may begin to feed on one another and the adults to rely on emergency rations of nectar and plant sap in order to assure their own survival if not the emergence of future generations.

◀ The larvae, nymphs, and adults (though not the eggs) of this species are preyed on by a variety of insectivorous birds, rodents, and spiders, and systematically hunted down by the ants that exploit aphid colonies as a source of honeydew. The seven-spotted ladybug may also be parasitized by chalcid or braconid wasps, tachnia flies or *Phoridae,* mites (acarians), nematodes, fungi, and even protozoans *(Gregarina).* J.B. and G.I

SHARP-SNOUTED SNAKE LIZARD
Lialis burtonis **396**

Like all the members of its family, *Lialis burtonis* is a snakelike legless lizard; its body is roughly cylindrical, its snout highly elongated, and there is no precise demarcation between its head and neck. The largest specimens may attain a length of 50 cm (almost half of which is accounted for by the lizard's tail), and coloration varies from beige to dark brown. *Lialis burtonis* is found in southeastern New Guinea and all but the southeasternmost parts of Australia. It is a very adaptable species, found in moist tropical forests, deserts, and cultivated regions, though most frequently in scrub or second-growth forest with herbaceous ground cover.

▶ In contrast to its fellow pygopodids, *Lialis burtonis* feeds on other reptiles, primarily on other lizards and occasionally on snakes as well. The stomach contents of 26 adult specimens examined by Kluge (1974) included 26 skinks, 3 agamids, 1 gecko, 1 pygopodid, and 1 snake. These figures seem to reflect a genuine dietary preference on the part of *Lialis* rather than the relative abundance of these reptiles in the environment, since *Lialis* exhibits a similar preference for the Scincidae when offered a choice of prey animals in captivity.

Lialis may be active at any hour of the day, particularly after nightfall. While searching for prey, it raises the forepart of its body off the ground and keeps its head held high, so that it presumably detects its prey primarily by sight and only secondarily by olfaction. It systematically investigates the clumps of grass and foliage in which smaller lizards might have taken refuge and also crawls into their burrows; in the latter case, of course, it detects its prey exclusively by olfaction. It immobilizes the prey animal by clamping its jaws around the body, then works its jaws up to the level of the animal's lungs and squeezes until it dies of suffocation. Then *Lialis* advances its jaws up to the animal's snout and begins to swallow it headfirst.

Its jaws are not flexible enough to enable to engulf its prey by means of lateral chewing motions, as snakes do; instead, it braces the prey animal's body against the ground and pushes it slowly down its gullet. In certain cases, *Lialis* will apparently try to to provoke another lizard into shedding its tail by autotomy, which it immediately devours while allowing the lizard itself to escape. The reasons for this behavior are not well understood; the size of the prey animal is not the only factor involved, since *Lialis* sometimes eats only the tails of lizards that are small enough to be swallowed whole. The only saurian prey that *Lialis* rejects outright are geckos of the genus *Diplodactylus,* whose skin exudes a viscous and no doubt repulsive liquid. *Lialis* probably feeds on insects when more substantial prey is not available.

◀ Monitor lizards, numerous species of snakes (most of them venomous), birds of prey, and even carnivorous marsupials are all known to prey on *Lialis burtonis* and other lizards of comparable size. *Lialis* has no way of defending itself except by taking refuge in its burrow and biting vigorously when cornered; neither of these of these tactics is of much use against the snakes, which can easily pursue it into its burrow and cut short its struggles by injecting it with venom.

The New Guinean snake lizard *(Lialis jicari),* found exclusively on that island, is the only other representative of the genus. It is a little larger than *L. burtonis* but very similar in other respects. The other pygopodids, which exhibit a variety of behaviors intermediate between terrestrial hunters such as *Lialis* and strictly subterranean burrowers such as *Aprasia,* tend to be rather small and feed primarily on invertebrates.

H.S.G.

396
Sub.: Vertebrata
Cl.: Reptilia
O.: Squamata
S.o.: Sauria
F.: Pygopodidae

SHEATFISH *Silurus glanis* 397

397
Sub.: Vertebrata
Cl.: Osteichthyes
S.cl.: Teleostei
O.: Siluriformes
F.: Siluridae

Also known by its German name, wels, this European catfish is among the most massive of all freshwater fish; it typically measures between 2 and 3 m (with a recorded maximum length of 5 m) and weighs between 200 and 300 kg. The sheatfish is found in Alpine lakes, slow-moving streams, and the backwaters of larger rivers in Central and Eastern Europe (including the Danube and its tributaries). The female lays its eggs in a nest of aquatic plants that is constructed by the male, and the male also stands guard over the hatchlings for several days.

▶ The sheatfish spends much of its time in a sheltered spot along the bottom or near the bank of a stream, and often hunts from ambush by concealing itself among the submerged roots of a tree. It will attack virtually anything that moves through the water or on the surface—including ducks and other waterfowl, which it catches by the legs and drowns—though it is more likely to feed on frogs and fish, occasionally on small semiaquatic mammals.

The sheatfish is generally attracted by vibrations in the water; its sense of hearing as well as its lateral line detectors play a role in long-distance prey location, and its sense of sight comes into play only at the final moment before it sizes the prey animal in its jaws; it does not necessarily aim for the head, since it generally suffices to close its jaws on any part of the prey animal's body. The sheatfish sometimes stations itself outside the underwater entrance to the burrow of a small mammal or in a spot frequented by a flock of waterfowl and hovers almost motionless in the water by paddling very gently with its fins while it awaits the return of its quarry.

The sheatfish is not fond of direct sunlight; it appears on the surface only at dusk, and hunts along the bottom or conceals itself in ambush beneath a ledge or in a cavity in the bank during the day. Like the other silurids, it can detect the presence of frogs or worms buried in the mud with its sensitive barbels, and it invariably swallows prey animals whole (including those up to half its own size).

◀ This species is well protected by its dark gray color, its reclusive habits, and its enormous bulk. Other than man, the bear is the only creature that is capable of landing a full-grown adult sheatfish, though ospreys and pike may sometimes feed on the juveniles. The sheatfish is avidly consumed by man, though the very largest specimens have recently become rather rare in Central Europe; the problem seems to be that, unlike the sheatfish itself, the aquatic creatures it feeds on tend to be susceptible to organic pollution. Canalization of streams and similar engineering projects have also deprived this species of its favored habitat. The sheatfish has been raised commercially on fish farms in Hungary, though not on a very substantial scale. M.D.

398
Sub.: Mollusca
Cl.: Gasteropoda
O.: Stylommatophora
F.: Testacellidae

SHELLED SLUG *Testacella haliotidea* 398

This sluglike European land snail has only a rudimentary shell and otherwise looks very much like a small dead leaf; it generally only measures 4 or 5 cm, though it is capable of stretching out to a length of 10 cm. Its shell was thought to resemble a miniature abalone *(Haliotis)* shell, which is the origin of the species name, *haliotidea.* The shelled slug lives in friable soil, under rocks, or in rotten logs, and is sometimes encountered on the surface at night. It is thought to be able to displace particles of soil by expelling air from its pulmonary orifice and thus slowly tunnel its way through the ground.

The shelled slug remains buried throughout the winter or for the duration of a prolonged dry spell; it envelops itself in a sort of earthen cocoon bound with mucous secretions. This species originated in Southern Europe, but has since become acclimated in Central Europe and North America.

▶ Like other gastropods, the shelled slug ingests its food with the aid of a flexible chitinous belt called the radula, which is studded with numerous sharp toothlike projections and essentially serves the functions of both teeth and tongue. In this species, the radula also serves as an organ of prehension, since the radular teeth take the form of minute grappling hooks and the radula itself can be projected forward to snag a prey animal (generally an earthworm) and pull it back toward the snail's mouth. The shelled slug can devour an earthworm in several minutes, though it generally only feeds every 2 or 3 days; it also feeds on slugs and other snails (including members of its own species), and the droppings of the shelled slug typically contain radular teeth as well as fragments of cuticle and pieces of grit from the crops of earthworms.

◀ The mole is probably the chief predator of the shelled slug, which may also be preyed on by hedgehogs, toads, and shrews. As with many invertebrates that live in the topsoil layer, populations of this species have been adversely affected by the widespread use of malacides (pesticides that are specially formulated to be toxic to snails), as well as other chemical pesticides, herbicides, and chemical fertilizers.

Several *Testacella* species are found in Western Europe, of which *T. haliotidea* is the most common; the others, including *T. maugei* and *T. scutula,* can only be distinguished by minute differences in the structure of their shells and certain features of their reproductive systems. There are a number of New World tropical snails with simple fusiform shells that are similar in their feeding habits, including *Glandina, Englandina, Poiretia* (family Spiraxidae). The former two are found exclusively in tropical America; *Poiretia* is also found in the Mediterranean Basin.

Englandina rosea has been introduced into Florida as a biological control on the giant land snail *(Achatina achatina),* another introduced species that is very destructive to crops (see the note at the end of the entry for *Rumina decollata).* This experiment has not been regarded as a success, however, since *Englandina rosea* has begun to prey indiscriminately on indigenous snails that the sponsors of the program had also hoped to protect both by avoiding the use of chemical pesticides and by selectively eliminating *Achatina* as a competitor. H.C.

SHINING GROUND BEETLE *Chrysocarabus rutilans* 399

399
Sub.: Arthropoda
Cl.: Insecta
S.cl.: Coleopteroidea
O.: Coleoptera
F.: Carabidae

A handsome carabid beetle is found primarily in the beech forests and upland meadows on both the French and Spanish slopes of the eastern Pyrenees (including the principality of Andorra). It measures between 29 and 35 mm, and its body is entirely golden or coppery-red (in Latin, *rutilans*), with the exception of certain subspecies on the southern slopes in which the head and pronotum are green. It has 3 deep grooves running along its elytra (wing covers) and is completely flightless, lacking even vestigial wings.

▶ The long-legged *Chrysocarabus rutilans* can move quite nimbly through the litter on the forest floor and actively seeks out its prey, which largely consists of snails and other mollusks, earthworms, and insect larvae. It hunts at night, locating its prey by touch, apparently, rather than by sight, and spends the daylight hours concealed in the forest litter, under a rock, or in a tree trunk. Like most predacious insects, it digests its food "preorally" by injecting digestive enzymes into the tissues of its prey as it masticates them with its powerful mandibles.

The larvae are less conspicuous than the adults; they lead a largely subterranean existence, and their preference for snails is even more pronounced than in the adult. The larva's long mandibles are well suited to the task of puncturing the snail's shell and extracting its contents, which it devours completely.

◀ Like other carabids, *Carabus rutilans* is sometimes preyed on by insectivorous birds, including small nocturnal raptors, as wells as frogs and even foxes; it defends itself against these predators in the same manner as the bombardier beetle—with a fusillade of superheated butyric acid. Adult beetles may be infected by a fungus, *Beauveria bassiana*, during their winter dormancy, and the larvae are occasionally parasitized by a species of tachnia fly, *Vivania cinerea*. As is generally the case with the larger, showier European ground beetles, amateur naturalists, dealers, and collectors have also taken their toll of this species; increased use of the upland forests for commercial and recreational purposes have also contributed to the decline of local populations.

Other members of the genus *Chrysocarabus*, notably *C. hispanus* and *C. splendens*, also tend to feed primarily on snails rather than insects (as with ground beetles of the genus *Carabus*). J.-C. M.

SHOEBILL *Balaeniceps rex* 400

400
Sub.: Vertebrata
Cl.: Aves
O.: Ciconiiformes
F.: Balaenicipitidae

About 100 cm long with a wingspread of 130 to 150 cm, the shoebill is a long-legged storklike bird with a short neck, a ridge running down its back that ends up in a little hook, dull-gray plumage with patches of greenish down, and a very distinctive bill that is shaped like an elongated wooden shoe. It is found only in the almost impenetrable marshes of the Upper Nile and the Bahr-el-Ghazal as well as the Ubangi-Chari in Central Africa, perhaps in a few other areas as well; it lives in papyrus swamps, thickets of tall grass, and on floating mats of vegetation.

It rarely takes to the air except when frightened, when it may attempt a short, low-altitude flight into a nearby tree with its great bill tucked back against its crop. The shoebill lives alone, in pairs, or in small groups and is very similar in its behavior to the herons of temperate latitudes, spending hours in the water, perched on one leg, immobile, with the lower part of its body barely immersed and its head poised just above the surface.

▶ The shoebill feeds on frogs, toads, fish, and even baby crocodiles, which it catches in its bill and swallows without further delay; it also uses its bill to probe the muddy bottom in search of turtles and lungfish. It hunts alone or in groups of two or three, taking great strides through the deeper water with its wings partly open in order to drive its prospective prey into the shallows. The shoebill sometimes feeds on the floating carrion that it encounters in the course of its hunting sweeps.

◀ Because of the shoebill's diffident nature and the inaccessibility of its habitat, very little else is known of its behavior in the wild. It is frequently seen in zoos, however, since in addition to its picturesque appearance, it is hardy and adapts easily to captivity and is eagerly sought after by animal collectors.

The taxonomic family Balaenicipitidae was created expressly to accommodate the shoebill. J.-P. R.

SHORT-FINNED SQUID *Illex coindeti* 401

401
Sub.: Mollusca
Cl.: Cephalopoda
O.: Decapoda
F.: Omnastrephidae

Squids are basically classified into two groups—those whose eyes are covered by a film of skin and those (suborder OEgopsida) whose eyes are surrounded by a flexible lens; the short-finned squid is of the latter type. It is found in the Atlantic, the North Sea, and the Mediterranean, though rarely in coastal waters, at a depth of 50 to 200 m.

The young are only about 2 mm long when they emerge from the egg. Their fins are rudimentary, and a form of jet propulsion (created by expelling a pressurized jet of water from the mantle) is their only means of locomotion. The so-called brachial crown at the base of the mantle, still not very developed at this stage, is already provided with tentacles with which the young squid catches its prey; these tentacles later fuse along the medial plane into a single organ of predation called the *rhynchoteuthion*, the tip of which is studded with adhesive suckers. In time, this organ is redifferentiated into separate tentacles. Very little is known about the adult squid's migration patterns and mating habits; lifetime does not seem to exceed 3 or 4 years. In the aquarium, the short-finned squid prefers to remain on the bottom with its tentacles spread out around it and, though it is quite a good swimmer, it seems to be basically sedentary by disposition.

▶ Very young squid, already equipped with adhesive suckers on their tentacles, probably feed on copepods and crustacean larvae. Older squid, including the adults, feed primarily on fish, especially anchovies of the genus *Engraulis*, along with crustaceans and other small cephalopods. The squid's eyes are comparable to those of mammals in their complexity, and the excellent eyesight of the short-finned squid is presumably a great help in locating its prey in the murky depths below 50 m. In other respects, its hunting technique is probably similar to that of the common European squid, *Loligo vulgaris* (q.v.).

◀ The short-finned squid is preyed on by whales and other ceteceans, probably by certain pelagic fish as well. It is caught in dragnets or with baited drag lines and though its flesh is not considered as tasty as that of *Loligo vulgaris*, it is still consumed in substantial quantities by man.

Three related species are found along the Atlantic coast of the Americas: *Illex illecebrosus* in the North Atlantic, *I. oxygonus* off the Southeastern U.S. (including the Gulf of Mexico) and in the Caribbean, its range partly overlapping with that of *I. coindeti*, and *I. argentinus* off the coast of Argentina. Not much more is known of the hunting behavior of other families of deep-sea squid; certain families have evolved morphological adaptations that may assist in predation—some of the suckers on their tentacles have assumed the form of hooks or claws, for example. Other species have developed luminous organs that (presumably) act as searchlights and help them to locate their prey in the sunless depths of the ocean; in others, bioluminescence appears to serve a purely defensive function, obscuring the silhouette of the squid's body at depths at which there is very little incident light. It also seems that the predatory species of the abyssal depths are attracted by the bioluminiscent displays of other creatures, notably certain fish and crustaceans.

S.v.B.

SHORT-NOSED ECHIDNA *Tachyglossus aculeatus* 402

402
Sub.: Vertebrata
Cl.: Mammalia
O.: Monotremata
F.: Tachyglossidae

Along with the Australian duckbilled platypus, the echidnas, or spiny anteaters, of Australia and New Guinea are the only surviving representatives of the monotremes, or egg-laying mammals. The echidna, very similar to a hedgehog in appearance, is about 45 cm long and weighs between 2 and 6 kg. It back is covered with sharp spines as much as 8 cm long; its belly is furry rather than spiny, and its head is small with a long, tapering muzzle (somewhat shorter in several of the Australian species). The second digit of its hind feet is equipped with a very long claw that enables it to scratch its back—and thus rid itself of the parasites that live among its spines—without doing itself an injury.

There are several different species, sufficiently different in their morphology to be classified into three different families. The echidna ranges widely throughout Australia and New Guinea and is particularly prevalent on rocky, brush-covered slopes in mountainous country, though it is also found in forests and on the flatlands, and even in the vicinity of human settlements. Individual echidnas have been known to survive for as long as 50 years

▶ The echidna is nocturnal in its habits and conceals itself during the day in a crevice in the rocks or in the abandoned burrow of some other creature. It has weak eyes,but is very sensitive to subterranean vibrations, and it wanders about at night in search of beetles, termites and earthworms. It catches its prey with its tongue, which is coated with a viscous, adhesive substance; the echidna has no teeth, and it chews its food with the help of keratinous pads along its gums, taking in a certain amount of topsoil with every mouthful.

◀ The echidna is a fast and skillful burrower, though it exercises this talent only when it senses the approach of danger, whereupon it quickly digs itself into a shallow trench with only its spines protruding. It has few predators, with the occasional exception of the dingo, and since its flesh is tasteless and it does no damage to crops, henyards, etc., it is not persecuted by man. In addition, it enjoys the protection of Australian law. P.A.

SHORT-TOED EAGLE *Circaetus gallicus* 403

The short-toed eagle, or harrier eagle, is a large Old World raptor that is found in warm temperate regions between sea level and an altitude of 1600–2000 m, or wherever surface conditions are agreeable to reptiles, including marshes, overgrown croplands, waste-and scrublands of various sorts, throughout Southern Europe and parts of Siberia, North Africa, and Turkey. The limits of its summer range may vary in accordance with short-term climatic fluctuations that are injurious to reptiles—as for example in France, where the short-toed eagle has recently been retreating southward in response to a series of very wet springs. It spends the winters on the savannahs of the African Sahel, from Senegal to Ethiopia.

The genus *Circaetus* is intermediate between the eagles and the buzzards, and the short-toed eagle is perhaps more reminiscent of a buzzard or a harrier in its behavior: patrol flights along the perimeter or over the interior of its territory are interspersed by prolonged rest stops at a perch from which it can survey at least some part of its domain. The short-toed eagle is 64–72 cm long with a wingspread of 1.6–1.9 m; it weighs between 1.5 and 2 kg. It reaches sexual maturity by the age of 3 or 4; maximum longevity is about 17 years.

▶ The short-toed eagle preys on various species of colubrids (the dark-green snake, smooth snake, ringed snake, Montpellier snake), vipers (the asp and the pelias viper), and lizards (the ocellated green lizard and the slow-worm), and in exceptional circumstances on moles and small rodents, frogs, toads, goldfinches, turtledoves, and, while in its African winter home, on guineafowl.

The short-toed eagle's strong talons are equipped with protective scales, and its vision is very keen; its eyes are as large as an owl's. It hovers motionless in midair until it has a firm binocular fix on a reptile or other prey animal, then plummets downward from a height of between 20 and 50 m, talons outstretched, wings folded back in the customary "divebomber" posture. The stiff bristles (vibrissae) around its face provide some protection against snakebite, but it is wary of venomous species and takes care to crush a reptile's head as soon as it catches it.

The short-toed eagle begins to ingest its prey in midair, though it generally leaves a short length of a reptile's tail dangling from its beak in order to facilitate the midair transfer to its mate or its single chick in midair—behavior that is also reminiscent of the harrier. The chick will typically require between 120 and 150 g, or the

403
Sub.: Vertebrata
Cl.: Aves
O.: Accipitriformes
F.: Accipitridae

equivalent of 2 or 3 small grass snakes, per day.

◀ The short-toed eagle is not subject to predation, though it is ocasionally persecuted by man, and its habitat has been increasingly restricted by land-clearing and other human activities. In spite of the fact that an individual eagle controls a very extensive territory (about 4000 hectares), encroachments by other members of its own species will be immediately repelled with warning cries and a sequence of threatening aerobatic displays.

Six other *Circaetus* species make their permanent home on the African savannah, though Beaudoin's eagle *(Circaetus beaudoini)* is regarded by some authorities as a geographical race of the short-toed eagle. The subfamily Circaetinae also includes several other genera of snake-eating eagles, including the elegant bateleur ("tumbler") eagle *(Terathopiis ecaudatus)*, which preys on ground-dwelling birds (including guineafowl) and small mammals as well as reptiles. J.-F. T.

404
Sub.: Vertebrata
Cl.: Mammalia
O.: Carnivora
F.: Felidae

SIBERIAN TIGER *Panthera tigris altaica* 404

The Siberian tiger, and not the African lion, is the largest member of the cat family, sometimes attaining a length of up to 2.8 m, with its tail measuring an additional 60–90 cm. It is more heavily built than the Bengal tiger—its entire body conveys the impression of tremendous power—and its yellowish-or reddish-brown coat is considerably longer and thicker; like the more familiar tropical subspecies, it has a white underbelly and a pattern of transverse black stripes running all along its body.

The Siberian tiger is now restricted in its range to the Amur-Ussuri river basin, which extends into Manchuria and the maritime region of Siberia, as well as a few isolated areas in North Korea, though it is found in a variety of terrain types, from tall grass to coastal rain forest, that afford sufficient opportunity for concealment. The size of the individual hunting territory also varies considerably in accordance with the local abundance of game, from about 40 sq km in Korea to as much as 4000 sq km in parts of the Amur region.

▶ The Siberian tiger preys on virtually all large mammals with which it shares its habitat, including deer, wild boar, saiga, lynxes, badgers, hares, and even occasionally on bears. Various studies have determined that this subspecies requires about 9 kg of fresh meat per day, on the average. Since it is likely to go for several days without encountering a prey animal of sufficient size to fulfill these requirements, it is fully capable of consuming up to 50 kg of meat at a single feeding.

The Siberian tiger prefers to hunt at night, stalking noiselessly to within a few meters of the quarry, though if cover is less than optimal, it is often sighted by the quarry before it is close enough to pounce. If not, it lunges out at the prey animal and throws itself on top of it. Smaller animals are killed with a bite to the back of the neck; larger ones are caught by the throat and suffocated. The quarry is dragged or carried to a source of fresh water, and the tiger begins its meal with the tenderest parts of the carcass (generally the thighs and belly of a ruminant).

◀ As with the Bengal subspecies, older individuals prefer to feed on less elusive quarry, which sometimes includes livestock and, less frequently, human beings—which is the principal reason why no more than 150–300 of these magnificent creatures are still at large. The Siberian tiger has no natural enemies other than man.

Apart from the Siberian tiger and the Bengal tiger *(P. t. tigris)*, there are five other subspecies of *Panthera tigris:* the Indochinese tiger *(P. t. corbetti)*, the Sumatra tiger *(P. t. sumatrae)*, the Java tiger *(P. t. sondaica)*, the Caspian tiger *(P. t. virgata)*, and the Chinese tiger *(P. t. amoyensis)*, the last 3 of which are barely surviving, if not already extinct, in the wild. P.A.

405
Sub.: Arthropoda
Cl.: Crustacea
O.: Amphipoda
F.: Caprellidae

SKELETON SHRIMP *Caprella spp.* 405

This small, inconspicuous filiform (thread-shaped) marine crustacean is between 1 and 3 cm long and is found in shallow and moderately deep waters all over the world, especially in beds of seaweed or sponges, or among colonies of alcyonarians, hydrozoans, bryozoans, tunicates, and other soft-bodied sessile (fixed) invertebrates. It develops directly, without undergoing a planktonic phase, and probably lives for about a year.

▶ The skeleton shrimp is a ferocious little creature that feeds on many varieties of vagile (self-propelled) microfauna as well as sessile invertebrates of the type just mentioned. Sensory organs on its antennae can detect minute perturbations in the current, warning it of the approach of a potential prey animal. It stands upright, brandishing its second pair of legs, which are equipped with hooklike pincers; in appearance, these are rather like a rudimentary version of the long, stalklike arms of the praying mantis. If the prey ventures within a very short distance, the shrimp seizes and immobilizes its prey with this second pair of legs, then tears off bite-size fragments and conveys them to its tiny mouth with its forelegs.

The free-swimming or -floating prey of the skeleton shrimp consists primarily of other small crustaceans—copepods, isopods, amphipods (including *Corophium,* gammarids, and other skeleton shrimp)—as well as protozoans, diatoms, and foraminifera. The feeding habits of the skeleton shrimp can be described as opportunistic, since it will devour almost any small creature that the current brings within its reach; it is possible that it feeds on these free-swimming organisms during the day and on its sessile or nearly immobile invertebrate neighbors at night.

◀ It has been established that one skeleton shrimp species, *Caprella acanthifera,* is sometimes caught by such larger members of the family Caprellidae as *Phthsica marina,* and it seems likely that skeleton shrimp are sometimes eaten by fish and other crustaceans, but no particular predator appears to prey on them regularly or systematically.

One of the skeleton shrimp's close relatives is *Cyamus,* the whale louse, described as a degenerate amphipod that is parasitic on whales and other ceteceans. (It is perhaps worth noting that *degenerate* in this sense merely refers to the extreme simplification of *Cyamus'* morphology in response to the undemanding circumstances of its parasitic mode of life.)

P.N.

SKUNK *Mephitis mephitis* 406

406
Sub.: Vertebrata
Cl.: Mammalia
O.: Carnivora
F.: Mustelidae

This familiar North American carnivore measures between 30 and 45 cm and weighs between 1 and 2.5 kg. Its thick, bushy tail measures 18–25 cm, and, in addition to the usually white tip, is lighter in color than the fur on its body; the distinctive white stripes running either side of the skunk's body converge at the forehead and the base of the tail. The skunk is more prevalent in open fields and prairies but is also frequently encountered in suburban and even urban areas. Families of skunks live in burrows, which may be shared with a variety of other creatures, including opossums, rabbits, foxes, badgers, and armadillos. The males are solitary during the warmer months but will rejoin their consorts in the burrow for their winter hibernation. The entrance to the burrow is marked by the scats (feces) of its residents.

▶ The skunk is an omnivore, feeding on bees and wasps, earthworms, fish, crustaceans, snakes, toads, and small mammals (no larger than a rabbit) as well as bird's eggs, honey, and vegetables. Small rodents are its favorite prey; it hunts for them at night, locates prey animals with the aid of its sense of smell, and dispatches them with a bite to the throat or the nape of the neck. After it catches a bee or a wasp, it dashes it to the ground and rolls it around with its paw until the stinger has broken off before attempting to swallow it. The skunk is protected from snakebite more by the thickness of its fur than by its quickness or agility, though it is quick enough to catch fish and crustaceans by stationing itself on the bank and flipping them out of the water with a backhand stroke of its paw.

◀ The skunk is best known, of course, for its ability to repel predators by squirting an oily foul-smelling liquid (with remarkable accuracy) from the glands at the base of its tail. A puma, coyote, or other carnivore that has had a previous encounter with a skunk will take flight as soon as the latter adopts its preliminary defensive posture—lifting its tail and turning its back toward the interloper. If this fails to evoke the desired response, the skunk will turn around to face the other animal, growl threateningly, and then continue its intimidatory display before it actually unleashes its chemical battery.

The spotted skunk *(Spilogale putorius)*, found in some regions of North America, is primarily arboreal and makes its home in a hollow tree. P.A.

407
Sub.: Vertebrata
Cl.: Aves
O.: Passeriformes
F.: Alaudidae

SKYLARK *Alauda arvensis* 407

The skylark is a small Old World songbird (suborder Passeres), brown and gray in color with black bands along its back, a crest on its head, a conical beak, long, slender wings, and a white border around its tail; it typically measures 16.7 or 17 cm from beak to tail and weighs 22 to 47 g with a wingspan of 34 to 36 cm. It is found throughout Europe and Asia (including Japan) and the northwest corner of Africa and has also been introduced into Australia and New Zealand by homesick Britons, as well as Vancouver Island and Long Island (unsuccessfully in the latter case).

In Western Europe, the skylark prefers sunny, open country and is quick to take up residence in areas that have recently been cleared of trees and brush. The skylark is a gregarious bird, living in small bands or sometimes in fairly large ones. In autumn, the skylark population of the Mediterranean countries is augmented by seasonal migrants from Scandinavia, including members of a related species, the horned lark *(Eremophila alpestris)*, noted for its "horns" or "eyebrows" (a tuft of erect black feathers above the eyes), also as the only lark species that is indigenous to the New World.

The skylark makes its nest in a bed of dry grass; the female sits on her clutch of 3 to 5 eggs for about 12 days while the male provides all her food. A pair of skylarks frequently raise two broods in a season, sometimes more. In poetry and folklore, the skylark is best known for its startling, nearly vertical flight and its melodious whistling song, which is frequently imitated or alluded to in the folk music of Great Britain and Ireland (e.g., "The Lark in the Morning," which refers to both these characteristics), Hungary, Rumania, and a number of other European countries.

▶ When not eating or sleeping, the skylark spends most of its time in the air. During the winter it lives primarily on seeds and grains, and with the coming of spring, it switches over to a diet of beetles, aphids, grasshoppers (which it catches on the ground, for the most part), insect larvae, spiders, sowbugs (myriapods), snails, and earthworms.

◀ The flesh of the skylark is greatly esteemed by a number of small birds of prey, including the European sparrowhawk, the peregrine falcon, the merlin (another falcon), the long-eared owl, the screech owl, and the gray shrike. Small carnivorous mammals may also catch skylarks on occasion.

Unlike many songbirds, the skylark actually benefits from the clearing of land for agriculture, though it will not live in cornfields and is sometimes locally threatened by the use of pesticides; in balance, the skylark population of Western Europe has increased in recent years. Though protected by law from human predators, the skylark is also considered a great delicacy by countryfolk, particularly in Occitania (southwestern France). Lark-hunting with nets is still a popular pastime when the migratory flocks arrive in this region in the spring and fall; the populations of entire towns and villages are recruited to act as beaters, and between 200,000 and 500,000 skylarks are thought to be taken in this manner every year.

Since skylarks normally prefer to fly straight up into the air to escape from predators, a device called a "larking glass," or *miroir aux alouettes,* equipped with movable mirrors to reflect the rays of the sun, has traditionally been used to lure them into the poacher's net when the services of an entire village are not available. In French the phrase *miroir aux alouettes* has taken on the additional significance of a flashy or showy trinket that is likely to attract the attention of the less discriminating consumer. (On the other hand, such English expressions as "skylarking" and "what larks!" probably derive from the North English dialect "lake," meaning "game" or "sport," and have nothing to do with the family Alaudidae).

Apart from *Eremophila alpestris* mentioned above, there are several other horned lark species, all adapted to life in cold or

mountainous regions. Oher notable Old World larks include the calandra, or calender lark, which is found in Central Asia and throughout the Mediterranean basin, and several species of crested lark, found in Southern Europe and North Africa; the genus *Mirafra* is represented in Africa and Madagascar, Malaysia, and Australia.
J.-P. R

SLOANE'S VIPERFISH *Chauliodus sloani* 408

This common bathypelagic fish typically measures between 15 and 20 cm and sometimes attains a length of 30 cm. It is found throughout the warm seas of the world, including the Mediterranean, at a depth of 100–3000 m, since it remains in very deep water during the day and comes up toward the surface at night. The viperfish has a large head and an enormous mouth with very sharp teeth, so that in spite of its modest size, its appearance is still rather terrifying. Its body is studded with luminescent organs (photophores), and it is a very poor swimmer.

▶ The viperfish feeds on a variety of marine organisms, some of them of substantial size, including shrimp, copepods, and other crustaceans, cephalopods, lanternfish, hatchetfish, and other bathypelagic species. It always hunts in darkness, whether in deeper water by day or closer to the surface at night, and its predatory technique is quite similar to that of the anglerfishes: the viperfish moves through the water with its head upward and its body tilted at an angle of about 45°. The flexible second ray of the dorsal fin has developed into a long, whiplike forward-pointing stalk, so that the luminous organ at its tip is left dangling directly over the viperfish's mouth.

Bathypelagic organisms with large, light-sensitive eyes are readily attracted by this lure, and the viperfish has only to open its mouth in order to swallow them; it is able to do so without entirely sacrificing the element of surprise, since its mouth its also lined with luminous photophores. Unlike the anglerfishes, however, the viperfish does not swallow its prey by aspiration. Its cervical vertebrae are very flexible, and a set of muscles anchored to the first cervical vertebrae jerks the viperfish's head back and pushes the lower jaw forward, so that its head seems to be about to detach itself from the rest of its body when its enormous gaping mouth is opened wide.

The viperfish lunges forward as it opens its mouth, so that the prey is speedily conveyed into its gullet, and at the same time, the viperfish's gills and heart, which are placed well forward, just behind the lower jaws, are rapidly shifted out of the way to prevent these organs from being damaged as the prey struggles to free itself. The bony girdle to which the heart is attached slides backward, and the gill arches are pushed aside like swinging doors and flattened against the ribs; after the prey has been ingested, the viperfish's vital organs are restored to their normal positions. The stomachs of viperfish that have been netted at sea are generally empty, which implies that digestion takes place very quickly, though it is also possible, given the viperfish's other anatomical peculiarities, that its stomach contents might be disgorged immediately when it is frightened or severely stressed.

◀ Slow and clumsy in its movements, the viperfish is not well equipped for flight or for defense. The snake mackerel *(Nesiarchus nasutus)*, which measures up to 1.2 m and is found at depths of up to 1200 m, is known to be a predator of the viperfish.

Apart from *Chauliodus sloani* , there are 2 other species of viperfish *(C. danae, C. barbatus)*, and 5 subspecies of of *C. sloani* are also recognized. There may be some overlap in the ranges of, hence competition among, different species and subspecies, but though their ranges may be contiguous or even superimposed (i.e. two different varieties may be found at different depths in the same oceanic region), they never actually overlap.

The Stomiidae (scaly dragonfishes) are a related family of about 10 species that are quite similar in their habits and appearance. *Stomias boa*, which is fairly common in the Atlantic and the Mediterranean, measures about 20 cm and has a chin barbel (rather than a dorsal fin ray) tipped with a luminous organ to attract its prey as well as an array of photophores along its flanks. It also carries out a nightly migration to within 100 m of the surface.

408
Sub.: Vertebrata
Cl.: Osteichthyes
S.cl.: Teleostei
O.: Salmoniformes
F.: Chauliodontidae

Another dragonfish family, the Malacostidae (loosejaws), even outdoes the viperfish in the realm of conspicuous consumption: *Malacosteus niger* is also common at depths of 0–2500 m in the Atlantic; the degenerate state of its cervical vertebrae enables it to jerk its head and upper jaw straight upward at an angle of 90° to the rest of its body while alowing its lower jaw to drop, so that it is able to swallow prey animals that are truly enormous with respect to its own modest length of 20 cm.

F.M.

409
Sub.: Vertebrata
Cl.: Chondrichthyes
O.: Galeiformes
F.: Scylliorhinae

SMALL-SPOTTED DOGFISH *Scylliorhinus canicula* 409

This small dogfish typically measures between 60 and 70 cm in length, and the largest specimen on record measured no more than 1 m. It is generally found at depths between 10 and 85 m, though it occasionally ventures down to about 400 m, and on a variety of substrates, though most frequently on sandbanks interspersed with beds of seaweed. This species is particularly common in the Western Mediterranean, though it is also found along the Atlantic coasts of Europe and Africa.

▶ The small-spotted dogfish feeds on small benthic fish (e.g. gobies, blennies) as well as crustaceans, mollusks, and marine worms. It is a nocturnal hunter, though it locates its prey by sight, and the ventral location of its mouth generally obliges it to take its prey from the bottom, from an outcropping of rock, or some other horizontal surface; occasionally, however, it will pursue a small fish into open water and still contrive to swallow it by positioning its mouth directly over the moving prey.

◀ The sinuous shape and sandy color of the small-spotted dogfish enable it to conceal itself quite well along the bottom, and its thick skin is provided with a rough, abrasive coating of "sharkskin" denticles, which means, in effect, that the small-spotted dogfish insists on being swallowed whole or not at all. As a result, its typical predators are all quite a bit larger than itself, including sharks, porpoises, and groupers. This species is intensively fished with both trawls and longlines in the Mediterranean , and (once the head, skin, and other identifying marks have been removed) is often sold in French fishmarkets under the name *saumonette* ("little salmon").

J.-Y. S.

410
Sub.: Arthropoda
Cl.: Crustacea
O.: Stomatopoda
F.: Gonodactylidae

SMASHER SHRIMP *Gonodactylus chiragra* 410

The Gonodactylidae, a family of marine crustaceans, are collectively known as mantis shrimps (mantis prawns) or squillas. They are on the order of 10 cm long, and are generally found in the shallow coastal waters of the Indian and Pacific Oceans. Active by night, like the majority of stomatopod crustaceans, the mantis shrimp takes refuge by day in a cavity or recess in a coral reef or a submerged rock.

▶ The smasher shrimp gets its name from the form and function of its two second maxillipeds, each of which is shaped like a hammer rather than the pincerlike claws with which the other mantis shrimps are equipped. The dactylopodite (the segment at the tip of the maxilliped) and the propodite (the adjacent segment) are smooth or nearly so, with just a few serrations, and the proximal portion of the dactylopodite is greatly thickened and enlarged. The smasher shrimp specializes in prey that is well protected but not especially mobile—bivalves, gastropods, hermit crabs, and, to a lesser extent, crabs and brittle stars. Within this grouping, however, it tends to prefer prey animal that require a reasonable small initial effort—the smaller individuals, for example, rather than the larger—though it is capable of dealing out hundreds of smashing blows every day with no apparent signs of fatigue or of damage to its hammer claws. The impact of these repeated blows would be sufficient to break the glass in an ordinary home aquarium.

The smasher shrimp locates its prey by sight. In the case of a crab or other vagile prey animal, it will attempt to stun the prey animal with its initial onslaught and then go on to shatter the claws, walking legs, and carapace of its victim. Once the prey has been disabled, the smasher carries it off to its lair or some other sheltered spot and devours it; the distal portion of the smasher's dactylopodite terminates in a sharp point, enabling it to pick the flesh of its victims out of the broken shell.

If the smasher shrimp encounters a prey animal that might be able to give a better account of itself in combat—a hermit crab, for example—it immediately carries it off to its lair, where the hermit's efforts to defend itself would be greatly hampered, and then smashes its borrowed shell by wedging it up against a rock or coral wall and holding it in place with its third, fourth, and fifth pairs of maxillipeds. Like a blackbird, the smasher shrimp may have a favorite rock that it uses like an anvil to break up the shells of its prey.

◀ The stomatopods are all extremely agile in their movements and not especially vulnerable to predation. They defend their territories against intruders of their own species by exchanging ritual taps rather than lethal hammerblows, though a combatant may occasionally have one of its hammer claws shattered; the damage will be repaired, however, after two or three molts. The smasher shrimp can also defend itself by rolling up in a ball, thus presenting its spiny caudal fan (telson) to its adversary; this armored appendage is used to block the entrance of the smasher's burrow and secure it against attack. As a final defensive measure, the meropodite (fourth segment, counting the dactylopodite as the first) of the hammer claw often bears several pigmented "eyes," which—in conjunction with a series of threatening postures assumed by the smasher—may also be of use in discouraging a potential predator.

There are several other species of smasher shrimp, also found in the vicinity of coral reefs and either too small or too uncommon to be economically exploited by man. *Odontodactylus scyllarus,* found in the Indo-Pacific region, grows to a length of about 20 cm; a blow from the hammer claw of*Hemisquilla ensigera,* 25 cm long, is equivalent to the impact of a low-caliber bullet.

P.N.

SMOOTH CARRION BEETLE *Ablattaria laevigata* 411

411
Sub.: Arthropoda
Cl.: Insecta
O.: Coleoptera
F.: Silphidae

This species is so called, first of all, because its shell is almost perfectly smooth (there are no elytral ridges) and secondly because the members of its family (Silphidae) are collectively referred to as carrion beetles, even though the smooth carrion beetle and the black carrion beetle *(Phosphuga atrata)* feed exclusively on land snails and slugs and never on carrion. Both species measure between 12 and 15 mm, and *Ablattaria* (as the genus name implies) might conceivably be confused with a cockroach, though the resemblance is not very pronounced. Both species have narrow heads with long mandibles (as do the snail-eating tiger beetles of the genus *Cychrus),* and *Ablattaria's* thorax is somewhat narrower toward the front; *Phosphuga's* shell is also solid black, but not as smooth. Both species are found in Western Europe, including the British Isles.

▶ *Ablattaria* feeds on small snails or the juveniles of larger species. It begins by biting the snail's foot with its mandibles and spraying it with the secretions of its anal glands, both of which activities naturally cause the snail to retreat into its shell. The snail defends itself by producing a plug of sticky froth to cover its retreat, and *Ablattaria* prefers to wait until this begins to subside before inserting its narrow head into the opening of the snail's shell and continuing to nip at the snail's body with its mandibles. These assaults are accompanied by a continuous flow of toxic salvia, which initially serves to immobilize the snail and ultimately to dissolve its flesh into a sort of thick snail chowder, which is sopped up by the brushlike filaments on *Ablattaria's* mandibles and transferred to its mouth.

◀ The diet and feeding habits of *Ablattaria laevigata* are basically identical to those of the nocturnal *Phosphuga atrata,* mentioned above. *Ablattaria* has no known predators.

R.C.

SMOOTH SNAKE *Coronella austriaca* 412

412
Sub.: Vertebrata
Cl.: Reptilia
O.: Squamata
S.o.: Ophidia
F.: Colubridae

This small European colubrid is about 60 or 70 cm long. Its body is variable in color—frequently grayish, sometimes with a tinge of pale pink—and it is more readily identifiable by the black bar markings around the eyes. The smooth snake is active during daylight hours and prefers a warm, dry climate, wooded terrain or brushland, hedgerows, stony waste-ground, disused quarries, and the like; it is found in many parts of Europe, from sea level up to an altitude of 1800–2000 m. The smooth snake spends the fall and winter (from October to the beginning of April) in hibernation. Mating begins in May; this species is ovoviviparous (which means that the eggs are incubated inside the female's body), and the young are born sometime between late August and early October.

▶ Sixty-five percent of the diet of the smooth snake consists of lizards; since the smooth snake's mouth is relatively small, it must restrict itself to prey about the size of the average wall lizard *(Lacerta muralis)* or smaller, including green lizards and slow-worms. In habitats that it shares with voles and other snakes, it may feed on these as well, particularly the meadow vole, the "terrestrial" water vole (*Arvicola terrestris*), and young asp vipers *(Vipera aspis)*. On occasion, it may also feed on nestlings, insects, and earthworms.

The smooth snake may attack from ambush, but normally pursues its prey more actively. In the manner of most colubrids, it coils the forepart of its body into an S-curve preparatory to striking out abruptly and seizing some part of the prey animal's body in its teeth. In many cases, the smooth snakes coils itself partially around the prey and attempts to suffocate it by constriction, though of course this technique is considerably less decisive when administered by the smooth snake than by the python or the anaconda.

The prey animal is often swallowed while still alive; the smooth snake is an aglyphous (solid-toothed, nonvenomous) species, but it is thought that its saliva contains toxins powerful enough to dispatch a vole, though it is unlikely to have any effect on a human being or other large mammal. (The venom glands of the cobra and other snakes that inject their venom into their prey are actually heavily modified salivary glands.)

◀ The predators of the smooth snake include diurnal raptors as well as ordinary barnyard fowl; shrews and other small insectivores sometimes feed on young smooth snakes, and the eggs, young, and adults are all consumed by larger colubrids, notably the Aesculapian snake and the dark-green snake *(Coluber viridiflavus)*. The smooth snake is relatively aggressive toward human beings and will attempt to strike if approached too closely; as with other colubrids, however, its second line of defense is the foul-smelling liquid secreted by its cloacal gland.

A similar species, the southern smooth snake *(Coronella girondica)*, is found in southwestern France. It is nocturnal or crepuscular (active at dusk) and feeds on wall lizards, geckos, smaller snakes, and insects. In contrast to the smooth snake, *C. girondica* is timid and relatively docile and will generally submit to being handled without attempting to strike. J.C.

SNAKEBIRD *Anhinga rufa* 413

413
Sub.: Vertebrata
Cl.: Aves
O.: Pelecaniformes
F.: Anhingidae

The African snakebird, or darter, is found in marshes and along the shores of rivers and lakes in sub-Saharan Africa and on Madagascar. It measures between 70 and 80 cm, with a wingspread of 118–125 cm, and weighs between 1. 3 and 1. 5 kg. The snakebird, with its webbed feet and its legs set well to the rear of its body, like a rudder, is a first-class swimmer; it sometimes swims on the surface, but spends much of its time with its body completely submerged and only its long, sinuous neck out of the water. The snakebird is a gregarious species, nesting in colonies in a grove of trees or bushes overhanging the water. It is also a strong flier and sometimes sets off on long migrations in search of a suitable watercourse during the dry season.

▶ The snakebird feeds on fish, amphibians, crustaceans, and aquatic insects, impaling the larger prey animals with its long pointed beak and catching the smaller ones in its open mandibles. Due to the extreme muscular tension in its neck and a sort of spring-bolt mechanism that abruptly releases this tension, the snakebird can thrust its beak forward with enormous force. It always catches its prey underwater but is obliged to come up to the surface it to ingest it, flipping the prey animal into the air so that it can swallow it lengthwise; larger prey animals that have been impaled on the snakebird's beak may have to be brought ashore before they can be pried loose and swallowed. The nestlings are fed by regurgitation of smaller prey animals.

◀ The snakebird is directly competitive with the common cormorant in some areas, though it has no known predators. During the 19th century, the long glistening plumes from the snakebird's back were in great demand for the millinery trade, and certain African populations were considerably reduced by plume hunters during this period.

The snakebirds (including *Anhinga melanogaster* of India and Southeast Asia and *A. novaehollandiae* of Australia) and the New World anhinga *(A. anhinga)* form a single genus and are closely related to the cormorant. All the snakebirds lack the uropygial gland that is located directly above the anus of other waterbirds and whose oily secretions help to make their plumage water resistant; consequently, after the snakebird emerges from the water, it is obliged to sit on a branch with its wings outstretched until its feathers have dried off. J.P.R.

SNAKE FLY *Raphidia cognata* 414

Insects of this family are known collectively as "snake flies" in English and "camel-neck flies" *(Kamelhalsfliegen)* in German because of the abrupt angle at which the elongated prothorax is tilted away from the rest of the insect's body, suggesting (if not a camel) a snake that has raised its head to strike. *Raphidia* does not exceed 20 mm in length. Its body is black, the abdomen flecked with small yellow spots, and its wings are long and transparent; wingspread varies between 16 and 20 mm. Its forelegs, which are adapted for walking, are attached toward the rear of the prothorax rather than in front, as is the case with the Mantispidae. The female's body terminates in a long, black, slightly curved ovipositor *(raphis* means'needle' in Greek).

The larva and the nymph are both very similar to the adult. Nymphosis takes place in April or May, and at first the nymph is whitish in color and only assumes its adult pigmentation after about 15 days. Its body is slightly curled and generally lies concealed in a crevice in the ground or a natural cavity of some kind, since it is not protected by a cocoon. The adult can be observed (though generally with difficulty because of its dark color) alighting on the leaves and branches of shrubs and fruit trees or flying along the edges of small groves of trees and forest clearings.

▶ The *Raphidia* larva is often found in the subcortical tunnels dug by bark beetles (Scolytidae), more often concealed under shreds of partly detached bark or patches of lichen and moss on the outer trunk of a tree. It occurs in essentially the same sort of terrain as the adult. The tip of the larva's abdomen is provided with an extrudable elastic rim that functions like a suction cup, thus enabling it to attach itself to a particular spot, then to detach itself and pull itself up to a new resting place on the bark of a tree. The larva is especially active by night, when it feeds avidly on a variety of small insects, including gnats, deathwatch beetles (Anobiidae), and mosquitoes at rest; it catches its prey and slices it into pieces with its sharp mandibles.

The adult *Raphidia* is not a very avid hunter, and sometimes is content to feed on the bodies of insects that have been killed or crushed by crushed by some other agency. It locates its prey by sight, rearing back its long prothorax and rapidly thrusting its head forward to catch a small insect (most often of the species just mentioned) with its mouthparts.

◀ The snake fly larva is sometimes parasitized by an ichneumonid wasp.

R.C.

414
Sub.: Arthropoda
Cl.: Insecta
O.: Raphidioptera
F.: Raphidiidae

SNAPPING TURTLE *Chelydra serpentina* 415

415
Sub.: Vertebrata
Cl.: Reptilia
O.: Chelonia
S.o.: Cryptodira
F.: Chelydridae

The overall length of this species is about 100 cm; the carapace alone measures about 30–45 cm, and weight varies between 15 and 30 kg, though it may be as much as 45 kg in exceptional cases. The snapping turtle's head is massive and its neck very long, enabling it to strike like a snake; the tail is slightly prehensile and surmounted by a ridge of bony knobs, and the shell is relatively reduced in size, rounded, and notched in back. This species is one of the most numerous and most widely distributed of all turtles, ranging from North America down to northwestern South America, and is found in a variety of aquatic habitats, including brackish water; in general, it prefers a muddy bottom where vegetation, submerged roots, or dead branches provide opportunity for concealment.

The snapping turtle generally remains motionless and submerged thoughout the day, though it occasionally comes out to bask in the sun; it generally stays in water that it is just deep enough for it to push its nostrils above the surface when its neck is fully outstretched. It is more active at night, particularly during the warmer months, and generally moves by clumping along the bottom, though it can swim fairly quickly if necessary. The female scoops out a nest about 10–17 cm deep and deposits in it several clutches of round eggs measuring 24–33 mm in diameter, numbering between 20 and 50 (sometimes as many as 80) altogether. The newly hatched snapping turtles measure about 25–30 cm in length (35 cm including the tail); longevity is in excess of 20 years.

▶ The adult snapping turtle generally hunts from ambush, waiting for a suitable prey animal to venture within striking distance of its lair, though the younger snappers (and the hungrier adults) are entirely capable of hunting for food along the bottom in a more aggressive manner. The snapping turtle is omnivorous, and its diet consists of 30–50 percent vegetable matter by volume; animal prey includes small mammals and birds, amphibians and their larvae, snakes, turtles, fish (about 30 percent of the remainder), worms, mollusks, crustaceans (10–30 percent), insects, and carrion (as much as 20 percent).

The snapping turtle detects its prey by smell as well as by sight; it can stretch out both its legs and its neck very rapidly indeed, enabling it to strike at a prey animal from a surprising distance away; birds and small mammals are often seized in this manner, dragged beneath the surface, and drowned. Smaller prey animals are swallowed immediately; larger ones will be picked to pieces by the snapping turtle's powerful beak and claws.

◀ Foxes, skunks, raccoons, and crows are among the many predators that have been known to dig up and devour the snapping turtle's eggs. The young turtles may also fall victim to raptors or wading birds as well as certain fish, bullfrogs, and snakes, and the alligator almost certainly preys on the adults as well. Adults snapping turtles will defend themselves quite vigorously against less formidable predators, scratching with their claws and striking with their snakelike necks; predators may also be deterred by the snapping turtle's musk glands, which exude a pungent and extremely unplesasant aroma.

Snapping turtles are caught by humans for a variety of reasons, first of all perhaps because they are very easy to catch, second because they feed on food fish, ducklings, and other desirable species (though the snapper's reputation for voracity and destructiveness may be somewhat exaggerated), third because they can readily be substituted for a number of less readily available species in recipes for turtle soup, "terrapin," and other traditional dishes. Sometimes the demand for this unprepossessing gourmet item may exceed the supply in the wild, and raising snapping turtles in captivity has become a minor cottage industry in certain areas of the United States. Snapping turtle eggs are also consumed locally, and the stuffed shells of this species are often used as ceremonial objects by American Indian tribes. Because of its fecundity, the snapping turtle is not threatened in the wild and continues to play an important role as a scavenger and a natural regulator of population levels in freshwater habitats.

The alligator snapping turtle (*Macroclemys temminckii;* q.v.) is larger than the snapping turtle, more specialized in its feeding habits, and more restricted in its range. The big-headed turtle (*Platysternon megacephalum*) of Southeast Asia, in spite of its exotic appearance—in addition to its disproportionately large head, it also has an unusually flat shell—more closely resembles the majority of aquatic turtles in its behavior and feeding habits.

R. Bo.

SONG THRUSH *Turdus philomelos* or *ericetorum* 416

Sub.: Vertebrata
Cl.: Aves
O.: Passeriformes
F.: Muscicapidae

The song thrush, or throstle, measures 20–21 cm, with a wingspread of 34–35 cm, and weighs between 50 and 95 g. It is found in woodlands and parklands, in wilderness as well as populated regions and at various altitudes, from Western Europe to Western Asia, though it rarely ranges south of a line extending from northern Spain to northern Iran. Some European populations are sedentary; those that spend the spring and summer in more northerly areas often migrate to North Africa or Southern Europe for the winter.

In the forest, the song thrush is very shy; it rarely emerges from cover and flies only for short distances, though it is much more willing to show when it frequents city parks and suburban lawns. During its migrations, its flight is straight and direct, its wingbeats rapid and interrupted at regular intervals. The song of *T. philomelos* is sonorous and pleasing to the ear and is frequently interspersed with imitations of other birds. It build its nest relatively close to the ground, and during the nesting period, it is tolerant of the presence of other birds and vigorously aggressive toward members of its own species.

▶ The diet of the song thrush is highly varied; it feeds on fruits and berries in their season and is particularly fond of grapes. It forages through the leafmold layer for earthworms, mollusks, insects, spiders, and other small creatures , and is noted for its habit of grasping the outer rim of the shell of a snail and other mollusks with its beak and hammering it against a rock until it breaks. It feeds on both freshwater and marine mollusks, including periwinkles and *Lymnaea,* in this fashion, and the appropriately sized "anvil" rocks in its territory are often surrounded by the broken shards of snail shells.

The nestlings are fed on beakfuls of crushed insects that are brought back to them in the crops of their parents. The song thrush sometimes finds itself engaged in competition with the European blackbird *(Turdus merula),* with both parties tugging on the opposing ends of the same earthworm; the blackbird may even make off with the edible portion of a snail whose shell has just been laboriously cracked open by the thrush.

◀ The predators of the European song thrush include both diurnal raptors such as the European sparrow hawk and nocturnal raptors such as the long-eared owl; members of the family Corvidae such as jays and magpies prey on the eggs, the nestlings and the young, and the adult song thrush. In habitats that the song thrush shares with man, the domestic cat takes a considerable toll of its numbers as well. In France, the song thrush and the blackbird, as in medieval days ("four and twenty blackbirds" . . .), are still regarded as legitimate gamebirds, and according to the French National Fish and Game Bureau (Bureau de la Chasse), almost 26 million birds of both species were taken by hunters during the 1974–5 season, with the aid of fowlers' nets, nooses and snares, and firearms of various kinds.

Apart from migratory thrushes from North America and Siberia, three congeneric species are found in Western Europe. The fieldfare thrush *(Turdus pilaris)* is a noisy, gregarious bird that lives in open spaces (its common name is derived from the Old English words for "field dweller"), feeds on vegetable matter in the spring summer and on insects during the winter. It will not hesitate to attack, either singly or in groups, any predator that approaches its nesting site. The mistle thrush *(T. viscivorus)* is similar to the fieldfare thrush in its diet and its habits; it is so called because it often feeds on mistletoe berries, and thus plays an important role in the propagation of the seeds. The redwing *(T. musicus),* somewhat smaller than the two preceding species and by no means to be confused with the North American red-winged blackbird (*Agelaus phoenicius),* is often found in humid meadows.

J.-P. R.

SOUTH AMERICAN OVENBIRD
Furnarius rufus **417**

417
Sub.: Vertebrata
Cl.: Aves
O.: Passeriformes
F.: Furnariidae

The ovenbird, or rufous hornero, measures about 20 cm from beak to tail and is found in cultivated fields or parklands interspersed with groves of trees, including city parks, in the tropical and temperate regions of South America. The reddish plumage is identical in both sexes, and average longevity is probably 2 or 3 years, on the same order as that of a blackbird or a thrush. The ovenbird frequently builds its nest in the vicinity of human dwellings. It is monogamous and will defend its territory (generally less than a hectare) against intruders of its own species. The ovenbird (in Spanish, *el hornero,* "the baker") is so called because its large, roughly dome-shaped hanging nest, made of adobe and about the size of a small child's head, was thought to resemble an old-fashioned baker's oven. The nest has an entrance on one side and is divided into two separate chambers; the inner chamber contains the nest and the eggs or nestlings.

▶ The ovenbird searches for food by stabbing its beak into the ground, somewhat in the manner of a starling, in the hope of unearthing an insect or larva, spider, worm , or snail.

◀ The ovenbird's adobe nest will resist the onslaughts of most predators, but the guira cuckoo and the ajaja hawk are both strong enough to break into the egg chamber and make off with the eggs or nestlings. The adults are sometimes caught by the sparrow hawk or the aplomado falcon *(Falco fuscocaerulescens)*. The ovenbird, the national bird of Argentina, enjoys generally friendly relations with human beings.

The family Furnariidae comprises 219 species of small passerine birds; their plumage, in various shades of gray or maroon, is inconspicuous and identical in both sexes. All are found in South America, from sea level to the highest peaks of the Andes, and in a variety of habitats, from deserts to swamps and tropical forests. All are insectivorous, though they may seek their prey in the trees, on the ground, or, like *Furnarius rufus*, in the topsoil layer. Miners *(Geositta, Upucerthia)* build their nests at the end of deep underground burrows; the subfamily Synallaxinae includes numerous species that nest communally in large piles of brushwood, which—in the case of the rufous-fronted thornbird *(Phacellodomus rubifrons)*, for example—may reach a height of 3 m. Some ovenbird species *(Cinclodes)* occupy the same ecological niche and even display the same fidgety mannerisms as the European wagtails. The name "ovenbird" (see North American Ovenbird) is also commonly applied to a North American warbler, *Seiurus aurocapilus* (family Parulidae), and to avoid confusion with this species, the Furnaridae are sometimes referred to as horneros by ornithologists. J.-F. T.

418
Sub.: Vertebrata
Cl.: Reptilia
O.: Squamata
S.o.: Sauria
F.: Anguidae

SOUTHERN ALLIGATOR LIZARD
Gerrhonotus multicarinatus **418**

This large North American lizard has an elongated head, short, sturdy legs, and a more or less prehensile tail; coloration varies among different populations and individuals from gray to reddish-brown to yellowish. Its back and flanks are covered with regularly shaped spots, darker on the flanks and often having a white border. Its broad rectangular scales are arranged in parallel rows (hence, "alligator lizard"), and it varies between 20 and 40 (occasionally 50) cm in length, slightly more than half of which is accounted for by the tail. Longevity may exceed 12 years.

This species is found along the Pacific coast of North America, from southern Washington to Baja California, and in a variety of different biotopes: prairies and open rangeland, brush, chapparal, scrub vegetation, thinly forested plains and hillsides, and sand dunes in coastal and semidesert regions, particularly in the vicinity of a watercourse. The southern alligator lizard requires some ground cover, however scanty, and especially prefers rocky slopes strewn with boulders, beneath which it often seeks refuge. It is primarily active during the day or at dusk, depending on the season and the temperature.

▶ In spring and early summer, the alligator lizard leaves its burrow, which is often located between the roots of a bush or shrub, warms itself for a moment in the sun, and goes off to forage in the grass, which is still green and swarming with young grasshoppers. At the height of summer when the grass is dry and brown, the adult grasshoppers and other insects take refuge in the bushes, where the alligator lizard goes to hunt, beginning in the late afternoon and particularly after sunset. In this season, the lizard rarely leaves its burrow in the morning, but in autumn, it comes out at midday to warm itself in sunny patches among the vegetation.

The younger lizards remain active later in the year than the adults, but in general this species undergoes a period of latency between October and March, which is occasionally interrupted on warmer days in the southern parts of its range. Though the alligator lizard may emerge from its burrow occasionally, it does not resume feeding, continuing to live on the fatty reserves stored in its tail. During the remainder of the year, it feeds primarily on beetles and caterpillars as well as grasshoppers, as well as wasps, Mormon crickets, butterflies, ants, termites, true bugs (hemipterans), potato bugs *(Stenopelmatus)*, spiders, scorpions, and sowbugs.

Occasionally, but much more rarely, it

may also feed on slugs, smaller lizards *(Eumeces, Sceloporus)*, bird's eggs and nestlings, even on young rodents. It is not very susceptible to the venom of scorpions and spiders (including the black widow, *Lacrodectes)*, which it feeds on with impunity. The alligator lizard is a marauding hunter, exploring twigs and branches and the undersides of rocks; it detects moving prey by sight, sniffs out prey that is motionless or concealed, digging in the ground or in vegetable litter to retrieve it. When it spies a mobile prey animal, it slowly begins to crawl in its direction, pausing ocasionally and tilting its head up or down or to the right or left, presumably to verify its initial identification as well as to get a fix on the quarry from various angles and thus to estimate the distance that still separates them.

When the prey animal is 10–30 cm away, the alligator lizard pauses once again; it moves its hindlegs forward, pressed firmly against its body, then pushes off with them and propels itself forward to seize the prey animal in its jaws. The prey is masticated (more or less thoroughly, depending on its size) and immediately swallowed, the lizard contorting its neck to assist the passage down the esophagus. Young lizards and other small vertebrates are caught by the back of the neck, crushed, and swallowed headfirst; larger prey animals are shaken from side to side, then pressed against the ground to make them easier to swallow. The alligator lizard will often attack a creature that is too large for it to swallow, even after trying for an hour or more, and will then have relinquish its prey altogether or content itself with whatever part of its victim (a leg, for example) it might have succeeded in detaching.

◀ The natural enemies of the alligator lizard includes numerous snakes (the northern black racer, the rubber boa, various kingsnakes and milksnakes, garter snakes, and rattlesnakes) and several predatory birds (shrike, red-tailed hawk, sparrowhawk) as well as various small carnivores that forage under rocks or among the vegetable litter on the surface of the soil. Domestic cats also catch adult alligator lizards, and other creatures that are normally insectivorous (birds and small lizards) will sometimes prey on the young. Many alligator lizards are run over on the roads or killed by prairie and brush fires.

The southern alligator lizard has a large repertory of evasive tactics when pursued by a predator. It may try to reach shelter in a burrow or a patch of dry leaves by slithering rapidly along the ground, undulating its tail and body from side to side. Like most lizards, it is capable of shedding its tail at will (autotomy); the tail keeps quivering even after it is detached from the lizard's body, thus providing an additional bit of distraction for the bemused attacker. The number of specimens captured in the wild whose tails have been regenerated suggest that this is an especially useful ploy in the case of the alligator lizard.

The lizard may try to conceal itself among the leaves of a plant or bush, and may even take to the water if necessary. It sometimes crawls into the thick branches of a shrub, loops its tail around a branch, and then clamps its jaws down on its tail, making it very difficult to dislodge. In lieu of detaching its tail, it may simply void its bowels, in the hope that its foul-smelling excrement will deter a predator with senstitive nostrils from pursuing the matter any further. If even this fails to deter, the alligator lizard coils its body into a circle, then suddenly lunges at its pursuer, jaws agape and ready to bite.

The southern alligator lizard's principal competitors include a number of skink species as well as its close relative, *Gerrhonotus coeruleus,* a more northerly species that is found primarily in the pine forests of British Columbia but whose range still overlaps to some extent with that of *Gerrhonotus multicarinatus.* An important difference between the two species is that *coeruleus* is ovoviparous (it incubates its eggs internally) whereas *multicarinatus* is viviparous and does not incubate its eggs with its own body heat, so that *coeruleus* is able to bear its young in a colder climate and at a higher altitude (up to 3000 m) than *multicarinatus. Coeruleus* also seems to be less dependent on fresh water and is probably more effective in exploiting the resources of a given biotype—for example, it is larger and thus capable of capturing (and swallowing) a larger assortment of prey animals.

Other *Gerrhonotus* species, found in the southwestern United States and Latin America, like *Gerrhontous coeruleus* give birth to live young and thus enjoy an adaptive advantage over oviparous species at higher altitudes. D.H.

419
Sub.: Vertebrata
Cl.: Mammalia
O.: Cetacea
F.: Physeteridae

SPERM WHALE *Physeter macrocephalus* 419

The sperm whale is the largest of the toothed (as opposed to baleen) whales. Adult males may attain a length of 18 m (average length is closer to 15 m) and a weight of 70 metric tons, females a length of 12 or 13 m (average length, 10 m) and a weight of 35 tons. It is found in all the world's oceans, particularly between the fortieth and sixtieth parallels in both Hemispheres, as well as in the Mediterranean during the winter. The female sperm whale is sexually mature at the age of 4, the male at the age of 5. The female bears a single calf at the end of a gestation period of 16 or 17 months and continues to nurse for about a year thereafter. The sperm whale's maximum longevity is probably about 40 years.

▶ Apart from all other considerations, commercial whaling operations have at least provided us with some very comprehensive data on the feeding habits of the sperm whale. With rare exceptions, it feeds on other creatures on the ocean depths, and available information suggests that these whales can dive to a depth of at least 1000 m and remain submerged for up to 90 minutes. For the most part, the sperm whale catches only squid and octopus (normally in a ratio of about 4 to 1, though the precise proportion depends on the age and sex of the animal in question, the season, location, etc.) that range in size between 0.5 and 3 m. Giant squid of the genus *Architeuthis*—which, like the sperm whale, sometimes attain a length of 18 m—have sometimes been found among the stomach contents of the latter, and the heads of sperm whales often bear impressive scars inflicted by the squid's beak and tentacles, strongly suggesting the possibility of titanic contests between these two enormous creatures, evenly matched in size if not in bulk.

Sperm whales also feed on pelagic bony fish (cod, herring, tuna) as well as on sharks and rays, though in a much smaller proportion; the size of these prey varies between 0.2 and 2 m. The least significant component of the sperm whale diet consists of crustaceans as well as small pinniped mammals (seals and walruses). The quantity of nourishment required by a sperm whale in a single day is estimated at 1 metric ton. Very little is known about its hunting technique apart from the fact that it finds its prey by echolocation; its eyes are very weak and play absolutely no part in seeking out prey at a depth where the sunlight never penetrates. The sperm whale has a row of conical teeth along its lower jaw, though not the upper, which makes them relatively useless for chewing; they are used exclusively to seize and hold prey animals, which are subsequently swallowed whole.

◀ With the possible exception of the killer whale, the sperm whale has no natural enemies other than man. (The giant squid does not attempt to prey on the sperm whale, and it is thought that the whale almost always emerges the victor in these encounters.) As readers of *Moby Dick* may recall, the sperm whale was particularly sought after for its spermaceti, a waxy subtance that is still used in the manufacture of perfume and as a light lubricating oil. Another semilegendary sperm whale product is ambergris, which accumulates in the whale's intestines but is more frequently (though still very rarely) found floating on the ocean surface; it also has important applications in the perfume industry.

The sperm whale's closest living relative is the pygmy sperm whale *(Kogia breviceps)*, which because of its comparatively small size (less than 4.5 m) is more like a dolphin from an ecological and behavioral standpoint. V. de B.

420
Sub.: Arthropoda
Cl.: Insecta
O.: Hymenoptera
G.: Aculeata
F.: Sphegidae

Sphex albisectus 420

The smallest of the 10 *Sphex* species found in France, *S. albisectus* measures between 12 and 18 mm in length. The pedicel ("wasp waist") is extremely slender—the Sphecidae are sometimes known as thread-waisted wasps—and the species name *albisectus* refers to the fact that each body segment is tipped with a cream-colored band. The rims of its large eyes are contiguous with the joint of the mandibles; the spiny legs are strong and well adapted for digging and earth-moving. This species is also found in Southern Europe, North Africa, and the Sahel.

▶ As with other digger wasps, the larvae of this species are fed on insects that are caught by the female wasp, paralyzed, and dragged off to the wasp's underground nest, which consists of a straight vertical shaft with an enlarged brood chamber at the bottom. Before the female *Sphex* sets off in search of insect prey, it blocks the entrance to the nest with a small flat pebble camouflaged with a sprinkling of dust.

The female wasp begins by locating a medium-sized locust (generally *Oedipoda, Chortippus,* or *Calliptamus),* which is generally to be found within a short distance of the nest, then hops up on its back, and overcomes the wild bucking of the locust's strong jumping legs by injecting its venom into a neural ganglion just behind the locust's head. After a few convulsive shudders, the locust becomes completely paralyzed within seconds. The wasp seizes one of the locust's antennae in its mandibles and drags it back toward its nest. A short distance away from the entrance, it releases its hold on the quarry, unblocks the entrance to the shaft, and leaves its burden on the surface while it crawls down to inspect the interior.

When the wasp returns to the surface, it drags the locust a little closer to the head of the shaft, often leaving it with its antennae dangling over the edge, and goes down for a second time to inspect the chamber. In most cases, it decides to continue with the operation; it grasps one of the locust's antennae in its mandibles and backs down the shaft, finally leaving the locust of the floor of the chamber, which is clearly of substantial size relative to the diameter of the shaft.

The female *Sphex* deposits its egg somewhere on the upper part of the locust's body, most frequently near the membrane that covers the "hipjoint" of one of the jumping legs. The wasp crawls back up the shaft, blocks the entrance to the nest with a single stone, and covers the stone with dirt. Each female *Sphex albisectus* will dig and stock a number of nests in this manner, though it is not known how many.

◀ The tachinid fly *Apodacra plumipes* sometimes manages to secrete its own egg on the body of the paralyzed locust while it is still in transit to the nest, with the result that the tachnid larva, which is the first to hatch out, is able to devour both the locust and the larva of *S. albisectus.*

Sphex, which is the Greek word for wasp, is a very large genus, consisting of about 300 species worldwide, which are found in all but the polar regions of the earth.

R.C.

SPINY LOBSTER *Palinurus elephas* 421

421
Sub.: Arthropoda
Cl.: Crustacea
O.: Decapoda
S.o.: Macrura Reptantia
F.: Palinuridae

This Old World crustacean measures about 50 cm and weighs up to 4 kg. The larger individuals have probably lived for 8 or 10 years, and the female produces 100–200,000 eggs, in a single session. The larval stage lasts for 10–12 months. The common spiny lobster, sometimes called the European rock lobster, spiny crayfish, or langouste (especially in a culinary context), lives in clear, calm water at a depth of 20–80 m; it will not live in muddy or brackish water, and is generally found on submerged rocks, reefs, and other hard surfaces; it ranges along the Atlantic coasts of Europe from Norway (south of the Arctic Circle) to Cape Bojador (West Africa, 26° N) and throughout the Mediterranean. The spiny lobster is migratory, gregarious, and primarily active at night, taking shelter during the day in a cavity or crevice in or under a boulder or coral block, etc., or even inside a submerged wreck.

▶ The planktonic larva of the spiny lobster, or phyllosome (so called because its body is flattened and shaped something like a leaf), is a ferocious carnivore, feeding on copepods, scyphomedusans and hydromedusans, and the larvae of fish, crustaceans, mollusks, ascidians, and polychaetes, which it locates by sight. The juvenile and adult spiny lobsters include a certain amount of seaweed in their diet, posibly as an aid to digestion (see below), and track their animal prey both by sight and with the aid of an array of chemoreceptors located on the antennules and antennae and the first four pairs of walking legs; nevertheless, the spiny lobster's olfactory sense is not very finely tuned.

The spiny lobster prefers to feed on creatures with highly calicfied shells, notably echinoderms—starfish and brittle stars (a hungry spiny lobster can eat up to 30 of the latter in a single session), regular and irregular sea urchins, and sometimes holothurians and crinoids. Echinoderms make up about 50 percent of the spiny lobster's diet, mollusks another 25 percent, including mussels, scallops, limpets, trochid sea snails *(Gibbula)*, and basket whelks. Spiny lobsters also feed occasionally on ascidians and avidly, though more rarely, on bryozoans, alcyonarians, and even on sponges (though probably for the sake of the algae and other organisms that they contain, since the food value of the sponge itself is virtually nil). They feed on fish and other crustaceans when circumstances permit; they are generally too slow-moving to catch mobile prey animals, though a spiny lobster sometimes succeds in prying a hermit crab out of its borrowed shell.

The spiny lobster seizes its prey with its first 3 pairs of legs (less frequently with the fourth and never with the fifth). It begins by gnawing at the edges of a mollusk's shell with its powerful mandibles; it will eat a whelk, for example, by breaking off bits of the shell and swallowing them with the whelk's flesh as it goes along, in more or less way the same way that one generally eats an ice cream cone. It will tip a sea urchin over on its back and gnaw a circular hole around the urchin's mouth, through which it extracts all the edible parts of the animal. It browses on sessile coelenterates, sponges, and the like by biting off a little piece at a time, like a a goat nibbling on a hedge.

The spiny lobster does not emerge from concealment and thus does not eat at all during the week following its molt (which occurs at least twice a year in the case of the juveniles, only once a year in sexually mature individuals). Unlike many crustaceans, the spiny lobster is not a cannibal; it naturally chooses the largest prey animals that are available to it, though individual spiny lobsters will never attempt to open a shell that would be too thick for them to crack. It frequently grazes on patches of *Desmarestia*, a kind of seaweed with a very high acidity that may be of use in digesting seashells and other calcareous material.

◀ The planktonic larvae of the spiny lobster are devoured by pelagic fish (notably tuna and mackerel), and the bottom-dwelling juveniles and adult are preyed on by a variety of other creatures, including octopuses *(Eledone, Octopus)*, congers, rays, large gadids, anglerfish, and sharks. A shark was discovered to have 16 spiny lobsters in its stomach (which may not seem excessive for a creature weighing 160 kg), and seals and various seabirds occasionally dine on them as well. Spiny lobsters are caught with dragnets as well as lobster pots, and the human fishery accounts for millions of spiny lobsters every year in European waters (1500 metric tons in 1966). Such intensive (not to say excessive) fishing often produces a population boom in species such as the common starfish on which the spiny lobster normally feeds, which in turn has an adverse efffect on the harvesting of mussels and scallops.

The spiny lobster has no large catching or crushing claws with which to defend itself, like the lobster *(Homarus)*, and is much more vulnerable to predators. The juveniles rarely leave the crevices in the rocks where they are obliged to find their food as well as shelter (which is why they rarely turn up in lobster pots). The adults will turn to face a predator, then very quickly propel themselves backward by flicking their tails back and forth; this potential for evasive action gives them a certain freedom of movement, and they sometimes wander as far as 50 m from the nearest shelter.

If approached, the spiny lobster stands its ground and begins to stridulate, warning the other lobsters to get back into their holes; this may also attract the attention of a large predatory fish, thus distracting the original adversary and giving the spiny lobster an opportunity to make its escape. Finally, the spiny lobster is able to shed its limbs and antennae by autotomy if they are seized by a predator and to later regenerate them in one or two molts. The spiny lobster's defensive spines not only make it a less tempting target when it is out in the open, they also make it very dificult to dislodge from its crevice in the rock.

A number of other varieties of spiny lobster are found in the warm and temperate waters of the world. The pink spiny lobster *(P. mauritanicus)* also prefers echinoderms and mollusks, though it feeds on faster-moving prey, including fish, crustaceans, and cephalopods, as well as on the eggs of fish and other creatures. The green spiny lobster *(P. regius)* is less partial to echinoderms and also feeds on relatively mobile prey animals. The Cape crawfish *(Jasus)* feeds on sea urchins by scooping out the flesh through the peristome without cracking their shells and is the object of a commercially important fishery (31,000 metric tons in 1966); it is merchandised under the name "South African rock lobster." P.N.

SPONGICOLOUS SHRIMP *Typton spongicola* 422

422
Sub.: Arthropoda
Cl.: Crustacea
O.: Decapoda
S.o.: Caridae
F.: Palaemonidae

This small sedentary shrimp (15–25 mm) is found in the Eastern Atlantic from the British Isles to the Cape Verde Islands and throughout the Mediterranean; it lives at moderate depths, between 15 and 100 m, most frequently between 60 and 90 m, and its body is neutral or yellowish in color. *Typton spongicola* is invariably found in association with a sponge, which provides the shrimp with both its permanent residence and its exclusive source of food.

▶ *Typton* generally selects a large or medium-sized sponge with relatively large oscula (external pores), which make it easy to penetrate into the interior. In the Channel, this species is found in association with the sponge *Desmacidon fruticosum;* it feeds on (or inhabits) many different varieties of sponge in the Mediterranean. There is record of an individual *Mycale syrinx* that was inhabited by no fewer than 27 spongicolous shrimp. The manner in which the sponge's tissues are actually ingested by the shrimp remains unknown, though it seems quite clear that this happens.

◀ The spongicolous shrimp has no known predators; the nutritive value of the sponge is very low, so the chances of *Typton's* inadvertently being consumed by a predator seem slight. There are a number of other species of spongicolous shrimps in the tropical and temperate waters of the world, but we know even less about them than we do about *Typton spongicola*.

P.N.

SPOONBILL *Platelea leucoridia* 423

423
Sub.: Vertebrata
Cl.: Aves
O.: Ciconiiformes
F.: Threskiornithidae

This Old World member of the ibis family measures 80–90 cm in length, with a wingspan of 120–130 cm, and weighs 1700–1800 g. It prefers to live in broad expanses of shallow water, including lakes, river deltas, estuaries, and tidal mudflats in both tropical and temperate regions. The spoonbill builds its nest in reedbeds, groves of trees, and directly on the ground on small , unfrequented islands, and often forms nesting colonies in association with other species, including herons, egrets, cormorants, and ibises. There are currently thought to be only about 1000–1500 nesting pairs in Europe (primarily in the Netherlands, the Danube delta, and in Spain), though this species is also found in significant numbers in North China and Central Asia. European populations spend the winter in sub-Saharan Africa or along the shores of the Red Sea.

▶ The spoonbill's tough, flattened, broad-tipped beak is used both for detecting and for catching prey animals. The spoonbill probes in the mud with its beak and will feed on virtually every small aquatic creature it is likely to encounter in this manner, though most frequently on aquatic insects and their larvae, worms, leeches, mollusks, crustaceans, frogs, small fish, and reptiles. Spoonbills often hunt in groups, advancing slowly while hunched over the water with their beaks immersed and the mandibles held slightly apart. The spoonbill moves its beak from side to side in a methodical sweep of the strip of muddy water in front of it, snaps its mandibles shut the instant it encounters a prey animal, then tilts up its beak and swallows it.

◀ Like most large wading birds, the adult spoonbill has no natural enemies. Its foraging technique is sufficiently specialized to avoid competition with other species, and its habit of nesting in colonies affords a measure of protection for the eggs and nestlings. On the other hand, this species has been adversely affected by the draining of marshlands and the disappearance of suitable habitat since even before the beginning the modern era; it has not nested in France in appreciable numbers since the 16th century, for example. The millinery craze of the Victorian and Edwardian eras resulted in a widespread massacre of many remaining European populations. Though the spoonbill is currently protected throughout Europe, it still faces a number of serious threats to its survival, including the development of wetlands, pesticide and PCB residues, and other forms of chemical pollution.

The African spoonbill *(Platea alba)* is very similar to the European spoonbill; 4 other *Platea* species are found in Africa and Asia, and the roseate spoonbill *(Ajaja ajaja)* of tropical America, now very seriously endangered, has been assigned to a separate genus. J.-F. T.

424
Sub.: Vertebrata
Cl.: Aves
O.: Passeriformes
F.: Muscicapidae

SPOTTED FLYCATCHER *Muscicapa striata* 424

This small migratory bird, 12–13.5 cm long, with a wingspread of 23–24 cm, and weighing between 13 and 20 g, is found in flat, wooded terrain (second-growth forests, parklands, lawns and gardens) from Western Europe to Baluchistan, though never above an altitude of 155 m. It is very common in Western Europe, where it spends the spring and summer (May—early September), the remainder of the year in subsaharan Africa.

Solitary and inconspicuous in its habits, the appearance of the spotted flycatcher may not be familiar even to those who occasionally share its habitat (in city parks, for example). It is a graceful flier, though it sometimes appears to suffer from fits of nervous twitching when it is sitting on its perch. It rarely alights on the ground and nests in climbing plants such as ivy or Virginia creeper *(Ampelopsis).* It actively defends its territory against other members of its own species.

▶ The spotted flycatcher feeds not just on flies but on a variety of flying insects, including butterflies, bees and wasps, beetles, dragonflies, etc. It stations itself on a perch and flies off in pursuit of flying insects, which it catches on the wing but does not swallow until it has returned to its perch.

◀ The eggs or nestlings of the spotted flycatcher are often devoured by jays, magpies, domestic cats, and even squirrels. This species is not directly affected by human beings or their activities.

A great many species of the family Muscicapidae are found throughout the tropical and temperate regions of the Old World. The paradise flycatchers *(Tchitrea)* of tropical Africa are noted for the spectacular development of their tailfeathers, sometimes four or five times as long as their bodies. The black flycatcher and the collared flycatcher *(Ficedula albicollis)* are also found in Western Europe. New World flycatchers, also called tyrant flycatchers, belong to the family Tyrannidae.

J.-P. R.

SPOTTED HYENA *Crocuta crocuta* 425

With its long neck, stocky body, powerful, grinning jaws, and its shrill, demonic laugh, the spotted hyena has always been regarded as one of the repulsive creatures in nature. Nicknamed "Satan's mount" by African Muslims, it was thought to subsist exclusively as a ghoul and graverobber because it hunted by night and fed on corpses that had been buried in shallow graves on the outskirts of a village or had been left unburied on the field of battle. The hyena is about 1.6 m long with a short, shaggy tail; its weight varies between 50 and 90 kg, and it lives for about 25 years. It no longer found in North Africa but ranges all over the rest of the continent. A pack of hyenas, about 30 in number, never wanders far from its communal burrow, which is always one that has been appropriated from (or abandoned by) some other animal. The size of the hyena's hunting grounds varies with the local abundance of prey animals, and its boundaries are always scrupulously delimited by various olfactory markings.

▶ The spotted hyena has the most powerful jaws of any terrestrial carnivore, with powerful shearing fangs and sturdy back teeth that can crack the hardest bones. The massive development of the jaw muscles also extends to the shoulders and forequarters, which is what makes the spotted hyena seem particularly stocky and its hindquarters disproportionately frail. It has excellent night vision, which gives it a clear advantage over the herbivores on which it preys, which see very poorly at night. The hyena does not follow the migrations of the great herds, and when wildebeest and other large migratory herbivores are not to be found, it preys on smaller, sedentary creatures.

During the day, hyenas patrol their hunting grounds singly and in pairs, feeding exclusively on small animals—often newborn or young herbivores—which they attack with a sudden rush and with a bite to the throat or the neck, snapping the cervical vertebrae. Hyenas are frequently unsuccessful during these daytime forays, and it is then that they devote themselves to their secondary task (in competition with jackals and vultures) of ridding the savannah of carcasses; however, they feed primarily on live prey, and carrion represents only a small proportion of their total diet.

At night, the entire pack participates in the hunt, often led by a single dominant individual; the pack generally selects a weak or an isolated herd animal as its quarry, and at night the hyena's succcess rate as a predator increases dramatically. Once the quarry animal takes flight, one of the pack will try to slow it down by nipping at its legs and underbody. The hyena has no claws and is not strong enough to knock a large antelope or a zebra off its feet; the pack dispatches its prey in much the same way as the Cape hunting dog, by tearing off strips of hide and flesh and attempting to break through to the animal's viscera while it is still on its feet and running. As cruel and grotesque as this may seem, it appears that the hyena actually makes its kill more quickly and efficiently than the large hunting cats, which suffocate their prey by clamping their jaws over the animal's mouth and nostrils.

◀ Like the Cape hunting dog, the hyena has a number of large and powerful competitors, the lion most prominent among them. When the hyena pack makes its kill, lions and other large cats are often attracted by the hyenas' yipping cries of excitement. (When experimenters played a tape recording of the hyena's excited "laughter" in the Serengeti game parks, bands of lions began trotting up immediately in hopes of snatching the carcass). It was formerly supposed that the hyena pack gathered around a lion's kill in order to be the first to dispose of the leftovers, but it is statistically more probable that it was the lions that had robbed the hyenas of their lawful prey. This misconception has often been cited in recent years to illustrate how our observations of animal behavior may be influenced by subjective, and often mistaken assumptions derived from folklore and what we believe to be the "character" or "nature" of the animal in question. It does sometimes happen, on the other hand, that a solitary lion will be unable to defend its kill from a pack of hyenas and will be forced to relinquish it to them. No other creature ever attempts to prey directly on the spotted hyena, and when pursued by human hunters, a hyena will sometimes lie down and play dead if no other means of escape are available.

The two other extant hyena species, the striped hyena and the brown hyena, are both assigned to the genus *Hyena* rather than *Crocuta;* the brown hyena, found in southern Africa, is now very rare.

P.A.

425
Sub.: Vertebrata
Cl.: Mammalia
O.: Carnivora
F.: Hyenidae

STARFISH *Asterias rubens* 426

426
Sub.: Echinoderma
Cl.: Asteroidea
O.: Forcepulata
F.: Asteriidae

This common starfish species is found in the eastern coastal waters of the Atlantic, from Senegal to the White Sea, primarily on fine sand and in the vicinity of beds of wrack grass or rocky outcroppings. *Asterias* lays its eggs in the spring, and a single female can release as many as 2.5 million eggs into the water in an hour. The planktonic stage of development lasts for 3 months, after which the larvae, now at least 5 mm long, spend the remainder of their first year of life in the intertidal zone, feeding on the slick of algae that covers the stones and boulders along the shore, where they also will be safe be from the violence of the surf and the ocean swell. In their second year, they move out into deeper water, particularly toward the undersea rocks with their beds of oysters, mussels, and acornshells and the sands where the clams and other lamellibranchs lie buried.

▶ *Asterias rubens* is a relentless predator and, as noted, especially fond of mollusks, which makes it a serious hazard in oyster, mussel, and scallop beds and other commercial operations of this kind; it also feeds on other invertebrates, including marine worms, crustaceans, gastropods, and sea urchins. Three techniques by which a starfish could force open the shells of bivalves have been suggested in the literature: first, by secreting toxins to paralyze the muscles that keep the bivalve's shell clamped shut; second, simply by prying the shell open by main force; and third, by extruding the lobes of its stomach and introducing it into the joint between the two halves of the shell.

As far as the first technique is concerned, *Asterias* does produce hemolytic toxins called saponins, but though these have been shown to evoke a flight response in mollusks, they do not seem to play a direct role in predation. Starfish will occasionally attack small mollusks (about half the diameter of the starfish's central disk) by prying them open with their arms and podia (tube feet), but the third technique is the one most commonly used by *Asterias* in dealing with lamellibranch mollusks: Once the prey has been located, the starfish grasps it with its arms and hunches up the dorsal surface of its central disk (a maneuver referred to as the "big back" reflex); this reflex invariably precedes the extrusion of the stomach lobes, in which only the ventral surface of the disk is physically involved.

The extruded stomach looks something like a small balloon and is filled with digestive fluids under high pressure; it is this internal pressures that enables the starfish's extruded stomach to slowly insinuate itself into the joint between the valves of the mollusks's shell, whereupon it drapes itself very neatly over the mollusk's internal organs like a latex mold. This entire process of external digestion probably takes about 5 or 6 hours on the average, though this varies considerably in relation to the size or the starfish and its prey. In the case of soft-bodied prey animals, *Asterias* also extrudes its stomach, consumes part of the prey, and then transfers the products of digestion to its digestive tube (equivalent of the intestines); this is the same feeding technique employed by many other starfish species, including those belonging to other families.

◀ Another starfish species, *Crossaster papposus*, found in the same biotopes, is a member of the family Solasteridae, which prey avidly on other echinoderms (and on other creatures as well). *Crossaster* is both predator and competitor as far as *Asterias* is concerned, and, in effect, the two species have worked out a bizarre sort of damage- limiting compromise. *Crossaster* pursues *Asterias* and grabs hold of one or more of its arms, which are immediately detached, by the process known as autotomy, whereby echinoderms and other invertebrates are able to shed their limbs in order to escape from predators. *Crossaster* feasts on only an expendable portion of *Asterias,* which is able to make its escape and, in time, to regenerate its missing arms.

The spider crab *Hyas araneus* and the hermit *Lithodes maia* have also been observed to prey on *Asterias* in oyster beds off the coasts of Great Britain. Here as well, the crab will lock its pincers on the tip of one of *Asterias's* arms, and the arm is immediately detached. Since the wounding attacks of these predators do little or nothing to control *Asterias's* total numbers, periodic population booms in various regions of the globe have resulted in intensive human counterattacks on behalf of their imperilled oyster beds, etc.

Asterias forbesi and *A. vulgaris* are the most common starfish species on the North American coast of the Atlantic; their feeding habits, as with virtually all the members of the family Asteriidae, are very similar to those described above. It is interesting to note that fossil starfish have been discovered in the typical "big back" position in conjunction with fossil bivalves, thus preserving a graphic record of animal behavior from many millions of years ago.

A.G.

STARLING *Sturnus vulgaris* 427

427
Sub.: Vertebrata
Cl.: Aves
O.: Passeriformes
F.: Sturnidae

This starling's total body length is 17–18 cm with a maximum of 37 cm; it weighs between 76 and 80 g. Average life expectancy at birth is only 14–17 months, but the record longevity for this species is 19 years. The female reaches sexual maturity after its first year, the male after its second. The starling is a native of Europe but has been introduced into North America and many other regions of the globe by settlers and colonists. It is found in cultivated fields, mixed or deciduous forests, parks and gardens, and, increasingly, in cities and towns as well.

The starling's most notable characteristic is its extreme gregariousness. Normally, thousands of starlings roost together in clamorous dormitories, though mating pairs prefer the privacy of a hollow tree or a hole in a wall in which to build their nest. Most of the starlings of Western Europe are sedentary, but their numbers are greatly increased when they are joined in winter, particularly along the coasts of the Mediterranean, by large migratory flocks from Central and Eastern Europe.

▶ The starling is an omnivore that generally seeks its insect prey in loose, friable soil, bobbing its head back and forth as it does so in distinctive fashion. It feeds on cranefly and cockchafer larvae, Colorado potato beetles, ants, libellulids, grasshoppers, leaf rollers (tortricid moth caterpillars, especially the larvae of the pea-green oak twist, *Tortrix viridana).* Like the cuckoo, the starling is undeterred by the urticant (stinging) hairs with which many caterpillars protect themselves. Other prey may include earthworms, snails, leeches, and spiders, and in exceptional cases, the starling may feed on lizards and small fish.

It is even more eclectic in its choice of vegetable food, including (to restrict ourselves to the category of fruits and grains) cherries, elderberries, gooseberries, raspberries, pears (ripe), apples (rotten), grapes, ripe olives, dates, cereal grains, peas, lettuce and other leafy vegetables. The starling generally moves along the ground, often in large flocks, stabbing its beak into the loose soil and opening it up to make a little pit in which it searches for its prey; this hunting behavior is innate and can also be observed in captive birds. The starling also hunts for prey on sandbanks and mudbanks and turns over clumps of seaweed in search of larvae and other tiny creatures. It also stations itself on an elevated perch and catches flying insects on the wing like a swallow, and it even sometimes hitches rides on the backs of cattle and other livestock in order to peck for ticks and parasite larvae.

◀ The starling is notorious for its proficiency in sweeping all ecological competitors from the field, and rollers and other birds that nest in caves are frequently evicted by starlings. Starlings also furnish a significant fraction of the diet of some birds of prey, including goshawks, sparrow hawks, and peregrine falcons (as much as 5 percent in the case of the latter), though a large flock of starlings is able to defend itself quite adequately by contracting into a very dense formation (the ornithologist's term for this is "mobbing") that few predators will attempt to penetrate. At night, however, screech owls and other owls, stone martens, pine martens, and weasels may have some success in raiding the starling's dormitories.

Human beings, especially Europeans, have played the role of the sorcerer's apprentice in contributing to the starling's conquest of a considerable portion of the globe. The species was introduced into North America, Australia, and South Africa during the nineteenth century, generally with catastrophic results, and many indigenous species have subsequently been displaced by these vigorous and prolific invaders. In Northern Europe, on the other hand, the starling's presence is regarded as desirable because it preys on cranefly larvae, and nesting boxes have been set out to attract it in certain areas.

Starlings have adapted very well to modern life, particularly the heated dormitories and abundant food resources that are provided by our larger cities. European starling migrations are accompanied by a great deal of damage to cherry orchards and vineyards, particularly in southern France, and to olive groves in Tunisia. As in North America, various attempts to deflect these migrations or to destroy starling dormitories, involving nets, poison, mechanical birds of prey, high-frequency transmissions and continuous tape recordings of starling distress calls, and even explosives, have so far proven themselves to be of limited utility.

Of the 21 starling species, only *Sturnus vulgaris* and *S. unicolor* are prevalent in Europe; the latter, very similar to the common European starling, is a native of the Iberian peninsula, Corsica, and the other large Mediterranean islands. The rose-colored starling *(Sturnus* syn. *Pastor roseus),* normally found in Northern India, sometimes ventures as far west as Central Europe, or even the British Isles, in pursuit of migratory flocks of locusts. The hill myna

(*Gracula religiosa*), also found in Northern India, has achieved some popularity as a household pet because of its ability to speak and is frequently encountered in petshops and aviaries, particularly in Europe.

J.-F. T.

428
Sub.: Vertebrata
Cl.: Mammalia
O.: Insectivora
F.: Talpidae

STAR-NOSED MOLE *Condylura cristata* 428

The star-nosed mole is common in eastern North America; it measures about 23 cm (including 11 cm for the tail) and is readily distinguished by the star-shaped corona of fleshy appendages, like little tentacles, at the tip of its muzzle; these are composed of no fewer than 22 bright-pink cartilaginous plates and are equipped with tactile and olfactory sensors. The star-nosed mole always locates its burrow near a pond or stream and is a very good swimmer.

▶ The star-nosed mole leaves its burrow only to hunt, and it seeks its prey primarily in the water. During the hunt, the 2 uppermost points of the star remain rigid and forward-pointing while the others are continually in motion. About 80% of its prey consists of leeches and aquatic worms and insects, though it also feeds on insect larvae, fingerlings, and crayfish, as well as on earthworms that it catches on land.

◀ The star-nosed mole is occasionally preyed on by feral cats and other small carnivores.

J.-J. B.

429
Sub.: Vertebrata
Cl.: Reptilia
O.: Squamata
S.o.: Sauria
F.: Gekkonidae

Stenodactylus petrii 429

This small gecko measures about 9–10 cm and is found in the high dunes (ergs) of the Sahara as well as in oases and palm groves and along the banks of intermittent watercourses (wadis). Its barely pigmented, almost transparent body is pinkish-beige in color, and, like its relatively long splindly legs, seems ill adapted to the rigors of desert life. Unlike most geckos, its toes are equipped with claws rather than adhesive pads. Its tail is slender and rather short, and the tail has two rows of 5–6 whitish tubercules along either side of its base. *S. petrii* is generally active from dusk until dawn, except when the nights are very cool, and remains dormant during the winter (October—March). This species has no common name in English, though in German it is known as "Petri's thin-fingered gecko" (*Petris Dünnfingergecko*) and in French as "Petri's stenodactyl" (*sténodactyle de Petri*).

▶ *S. petrii* is not a very agile predator, stalking slowly and awkwardly from one sparse tuft on vegetation to the next, though it is quick to pounce on and devour such small spiders and insects that it comes across. It feeds primarily on ants and termites as well as silverfish, small butterflies and moths, caterpillars, and young spiders. Its teeth are not strong enough to penetrate the tough chitinous shells of tenebrionids and other beetles, though it does feed on tenebrionid larvae and can be maintained quite easily in captivity on a diet of young cricket larvae.

◀ *S. petrii* protects itself from the heat of the morning sun by digging itself into the sand; it digs a fresh burrow every day and carefully seals itself in by flicking its tail back and forth until the walls of the burrow collapse behind it. Thus, the last external traces of its occupancy will be erased as soon as a light breeze springs up, as frequently happens just after sunrise on the desert. This turns out to be a sensible precaution, since the brown-necked raven (*Corvus ruficollis*) is in the habit of investigating all fresh burrows whose entrances are still visible on the surface in the early morning.

There are also a number of nocturnal snakes that feeds on this species, including the Arabian leaf-nosed snake (*Lytorhynchus diadema*), the back-fanged snake (*Malpolon moilensis*), sand snakes (*Psammophis sibilans* and *P. schoukari*), the sand viper (*Cerastes vipera*), and the horned viper (*C. cerastes*). *Stenodactylus petrii* and its congeners (along with many other gecko species) are feared and avoided by desert dwellers, who erroneously believe them to be poisonous, with the result that they are rarely molested and still more rarely consumed by human beings.

Two subspecies of *Stenodactylus sthenodactylus* and two *Tropiocolotes* species are found throughout the same geographic range and are similar in their habits—except that during the day they take shelter under rocks, since are are generally found in the arid, rock- or gravel-strewn terrain known as regs and hamadas rather than on the dunes (ergs). These geckos are also rather slow and clumsy in their movements and subsist essentially on small arthropods. R.V.

STONE CURLEW *Burhinus oedicnemus* 430

430
Sub.: Vertebrata
Cl.: Aves
O.: Charadriiformes
F.: Burhinidae

The stone curlew is so called because its piping cry resembles that of the common curlew and because it is generally found on arid or rocky ground. It typically measures 39 cm, with a wingspread of 76–85 cm, and weighs between 375 and 480 g. *Burhinus oedicnemus* is found in arid, sparsely vegetated regions of the Old World, especially in steppes or desert regions or (in Western Europe) in sand dunes, cultivated fields (provided the plants are widely spaced), and on chalk downs. The stone curlew is a sociable, monogamous bird that congregates in small flocks except during the mating season; populations may be either migratory or sedentary, depending on whether their habitat is temperate or tropical.

▶ The stone curlew is a nocturnal, ground-dwelling bird with large light-colored eyes and excellent night vision; it feeds primarily on crickets (including mole crickets), grasshoppers, locusts, beetles, spiders, caterpillars, mollusks, snails, slugs, earthworms, and the like. The stone curlew hunts at dusk or at night and sometimes preys opportunistically on somewhat larger creatures, including frogs, small rodents, lizards, and even small birds.

◀ The stone curlew is often found in the same biotope as the bustard, partridge, or the cream-colored courser *(Cursorius cursor)*, though its circadian rhythms and dietary preferences are sufficiently differentiated to avoid competition. The female lays a clutch of two eggs in a shallow depression in the soil; both parents guard the nest in relays, and the eggs (like the chicks and the adults) are drably colored and inconspicuous, though this does not prevent the nest from being plundered by crows on occasion. At the approach of danger, the adults flatten themselves against the ground or, if caught in the open, run off to take shelter behind a tussock or a tiny tuft of grass.

Nocturnal, extremely wary, and now legally protected in certain countries, the stone curlew is still indirectly threatened by the disappearance of its habitat (particularly sand dunes) and by the presence of chemical pesticide and fertilizer residues in the fallow fields in which it frequently makes its nest.

Nine species of stone curlew are found in the arid regions of tropical America and the Old World. J.-F. T.

STONEFISH *Synanceia verrycosa* 431

431
Sub.: Vertebrata
Cl.: Osteichthyes
S.cl.: Teleostei
O.: Scorpaeniformes
F.: Synancepidae

The adult measures up to 45 cm, but this species does not attain its distinctive adult form until it has reached a length of about 10 cm. The juvenile is pink with a traverse brown stripe and is shaped something like a Mediterranean rockfish (rascasse) or a sculpin. The head of the adult is broader than it is long, and its body is covered with warty excrescences and looks very much like a flat brown rock. The resemblance extends to its normal behavior as well as its color and texture, since it is among the most sedentary of all known fish. The stonefish is found in the calm coastal waters of the Pacific, particularly on coral reefs, and prefers a habitat that is strewn with rocks and broken coral.

▶ Like most rockfish (Scorpaeniformes), the stonefish hunts from ambush, in this case concealing itself on the bottom and often partially burying itself in sand and bits of coral; its gill slits are vented in such a way that this covering of sand will not be disturbed when water is expelled during respiration, and the backs of certain individuals may even be overgrown with seaweed. The stonefish's very wide slot-like mouth is located on the dorsal surface of its blunt snout, and when a curious reef fish swims close enough to investigate the rhythmic breathing motions of this otherwise unremarkable flat rock or the lumps and fringes growing along its back, the stonefish launches itself upward with an abrupt convulsive motion and swallows the victim in a single gulp. It also feeds on small crustaceans as well.

◀ The stonefish, which has no natural predators, has a venom gland attached to the middle of each of the spiny vanes of its dorsal fin. There are two grooves running from the venom glands to the tip of the spine, and the spine is normally covered with a sheath of skin, which is immediately retracted when the spine penetrates a human foot or the flesh of some other creature so that the venom can be pumped through these grooves and directly into the wound.

The stonefish's venom is very toxic to humans, even potentially fatal to those with existing cardiac problems, and the pain is so intense that a number of stonefish victims are reported to have killed themselves rather than endure it any longer. At the very least, the victim generally suffers from violent attacks of nausea and sometimes throws himself down and begins writhing around on the sand before losing consciousness altogether. The edematous swelling in the region of the wound generally does not subside for quite a while, and loss of feeling may persist for up to several months; in some cases, the site of the wound will be marked by a permanent scar. An antivenom manufactured in Australia has been available for several years and must be administered by injection within the hour.

Synanceia torrida is found along the coasts of Australia, and the two remaining species are also found in the Pacific; all are dangerous to man, and their feeding habits are essentially as described above.

J.-M. R.

432
Sub.: Vertebrata
Cl.: Mammalia
O.: Carnivora
F.: Mustelidae

STONE MARTEN *Martes foina* 432

The stone marten, or beech marten, is an Old World member of the weasel family. It is about 40–50 cm long, excluding the tail, which counts for an additional 23–27 cm; its weight varies between 1.3 and 2.3 kg, the male being somewhat larger and heavier than the female. Its pelt is grayish brown with a grayish-white undercoat and long dark-brown guard hairs, and it has a white bib, generally bisected by a strip of brown, running down from the throat to the inside of the forelegs. In some individuals, the bib is solid or spotted or even yellow, and although the shape and color of the bib is sometimes cited as a means of distinguishing the stone marten from its relative the pine marten, this is not always a reliable indicator.

The stone marten ranges all over continental Europe (including the larger islands of the Western Mediterranean and excluding northern Russia and Scandinavia) as well as parts of Asia. It is generally found in rocky terrain, around cliffsides and boulder-strewn glacial moraines, as well as in quarries—since the stone marten does not avoid areas frequented by man and often seems to prefer them. In Alsace, for example, where the villages are built on rocky promontories in the midst of cultivated fields, the stone marten frequently takes up residence in farmhouses, barns, and other outbuildings; it especially appreciates the warmth of haystacks and haylofts and digs elaborate tunnels through the hay. It does not dig a burrow for itself in the ground, however, though it sometimes occupies those that have been abandoned by other creatures; otherwise it lives in hollow trees or in clefts or crevices in the rock.

◀ The stone marten is an opportunistic predator, and its diet varies greatly, according to the season, the type of terrain, and the opportunities available to the individual animal. It feeds on fruits and berries, which sometimes comprise a large proportion of its diet in summer, all kinds of insects, sparrows, starlings (which it catches at night in their treetop dormitories), as well as pigeons and domestic fowl when it can find its way into a dovecote or a henhouse. It is also very fond of bird's eggs, though it tends to feed primarily on rodents and other small mammals, including house mice, fieldmice, black rats, brown rats, voles (common, vole, vole rat, field vole), shrews (common shrew, white-toothed shrew), and moles.

Individual stone martens often rely almost exclusively on some particularly abundant local food resource, such as a cherry orchard or a cave inhabited by bats; one stone marten's den in a crevice among the rocks was found to contain no other postprandial refuse besides the wings of bats. Stone martens, like bears and raccoons, have gotten into the habit of investigating human garbage, and cheese wrappers and butcher paper occasionally figure in scholarly inventories of their caches. In its capacity as a predator rather than a scavenger, the marten relies primarily on its sense of sight, secondarily on its sense of smell. It does not stalk its prey slowly like a cat, but, according to Waechter (1975), concludes a rapid chase by inflicting a fatal bite at the nape of the neck that severs the spinal column of its victim; death occurs within several tens of seconds.

A marten in a henhouse naturally provokes great excitement and consternation on the part of the hens; this in turn appears to send the marten into a killing frenzy, which will continue as long as any of the hens are still stirring. Räber cites an instance in which a stone marten had killed 34 out of 35 occupants of a henhouse. The thirty-fifth hen was sitting on a clutch of eggs and presumably refused to join in the general agitation; it was thus left undisturbed by the marten. This is an instance of what might literally be called wholesale butchery, since the marten is presumably intent on restocking one or more of the food caches that are scattered throughout its territory (Waechter, 1975). This would appear to be more practicable in the case of the marten's "natural" prey—songbirds and small rodents—rather than large domestic fowl, since the sort of aperture through which a marten might find its way into a henhouse will usually not accommodate a marten that is trying to drag off a plump chicken behind it.

Also, the marten is often obliged to make its escape after the surviving hens have sounded the alarm—leaving the slaughtered hens behind, uneaten, which has given rise to the popular belief that the marten drinks the blood of its prey or that it kills simply for pleasure. Waechter has provided a somewhat more satisfactory explanation: "The insatiable predatory behavior that is associated with the maintenance of a cache of excess food reserves appears to an adaptation to the irregular and uncertain availability of food sources." Certainly the stone marten is well aware of the location of its food caches and visits them quite frequently to replenish its supplies, or on days when the chase has been unsuccessful.

◀ There are very few creatures in Western Europe, the eagle owl among them, that prey on the stone marten. Though the marten itself feeds primarily on rodents, and henhouse massacres of the type just described are comparatively rare, it has been persecuted throughout Europe as an undesirable predator. It is now partially protected by law in France, though it is still trapped and hunted for its fur in many areas (notably the Balkans and the Urals) and has already has been exterminated in many others (particularly in Western Europe); martens that live in populated areas are often killed on the highways. F.T.

STORM PETREL *Hydrobates pelagicus* 433

433
Sub.: Vertebrata
Cl.: Aves
O.: Procellariiformes
F.: Hydrobatidae

This small pelagic seabird, also known as the stormy petrel, is found in the North Atlantic and the Mediterranean. It measures about 14–15 cm, has a wingspread of 31–34 cm, and weighs between 23 and 27 g. This species reaches sexual maturity at the age of 4 or 5; record longevity is 20 years. The storm petrel lives on the open sea, synchronizing its flight with the crests and troughs of the waves, occasionally fluttering its wings very quickly like a bat to increase its airspeed, sometimes swooping in low to pluck a prey animal out of the water or simply to come to rest on a calm sea.

This species is so called because its appearance was thought by early mariners to be a sign of heavy weather ahead, and the word *petrel* itself is thought to refer to the Biblical story of St. Peter walking on the Sea of Galilee. The storm petrel generally travels in small flocks, makes its nest in a burrow in the sand or in a crevice in the rocks, and assembles in enormous nesting colonies during the breeding season. During the day, the sitting birds and the nestlings in the colony seem to be continually humming to themselves, at night squabbling and complaining in loud, raucous voices. At any time of day, a petrel colony can be distinguished by the musky, penetrating aroma given off by its inhabitants.

▶ The storm petrel feeds on fish hatchlings and small fish, planktonic crustaceans, and the refuse discarded by boats. It hunts by sight and catches its prey while swimming on the surface as well as while in the air. Young petrels are fed by regurgitation and, in view of the logistic problems involved, on a fairly irregular schedule; sometimes 2 or 3 days may elapse between feedings.

◀ The petrel is vulnerable to predators mainly during the breeding season; rats and gulls prey on the fledglings when they emerge from their burrows and even on the adults when they return to the colony with food. Nesting colonies near well frequented beaches and seaside resorts have often suffered from human incursions, since the petrel's underground burrow, along with the eggs and chicks, can easily be trampled underfoot.

About 20 different species are referred to "storm petrels." Storm petrels that are found primarily in the Northern Hemisphere have short toes, long, pointed wings, and a forked tail (e.g. Leach's storm petrel, *Oceanodroma leucohoa);* those that live primarily in the Southern Hemisphere have longer toes, rounded wings, and a square tail (e.g., the soft-plumaged petrel, *Oceanodroma castne,* or Wilson's petrel, *Oceanites oceanicus).* Wilson's petrel nests along the coasts of Antarctica and the sub-Antarctic islands and spends the summer as far north as Great Britain and Labrador.

J.-P. R.

434
Sub.: Vertebrata
Cl.: Reptilia
O.: Squamata
S.o.: Sauria
F.: Scincidae

STUMPTAIL *Tiliqua rugosa* 434

As befits an Australian reptile, this species is well provided with nicknames, including bobtail, shingleback, and stumptail or stump-tailed lizard; the several *Tiliqua* species are also reefrred to collectively as blue-tongued skinks. The stumptail typically measures between 30 and 40 cm, sometimes attaining a length of 50 cm. It has a short, stocky body with a broad triangular head and a short tail, which is roughly the same size as its head. Its legs are short but sturdy, and its short toes are equipped with claws. Coloration varies from slate-gray to brown to black, depending on the region, and its back is covered with broad, corrugated scales (whence "shingleback").

This species is found in various habitats, ranging from second-growth forest to semidesert steppes, and in every region of Australia except for the northeast and the southeast coast. The stumptail is active by day, though rather sluggish in its movements. It sometimes spends the hottest part of a summer day in the abandoned burrow of a rabbit or a rodent. Captive specimens have been known to survive for at least 20 years.

▶ The stumptail feeds primarily on flowers, fruits, and berries (including cultivated species), occasionally supplementing this vegetarian diet with snails and insects, more rarely with small vertebrates, including other lizards, small snakes, birds' eggs and nestlings. The stumptail identifies an immobile prey animal (or vegetable food) by flicking it repeatedly with its tongue, though it also may locate a mobile prey animal by sight. It crushes snails, insects, and other hard-shelled creatures in its jaws before swallowing them. During the winter dry season, the stumptail feeds very little and lives primarily off its accumulated fat reserves; these are stored beneath the thick scales on its back as well as in its tail, so that its body and tail are roughly cylindrical at the beginning of this period and considerably flattened by the end.

◀ The chief predators of the stumptail include several python species and a number of other snakes as well as raptors, kookaburras, and gray kingfishers. When approached by a potential predator, the stumptail flops over on one side and arches its body, drawing its tail up toward its head and puffing up its torso while opening its mouth very wide. The stumptail's tongue is bright blue, which shows up very advantageously against the background of its pink gullet, and if the intruder tries to circle around, the stumptail wheels around in place, so that its open mouth is always facing toward the intruder.

This alarming display of pigmentation will often be sufficient to repel an aggressor, and if is actually attacked, the stumptail is capable of inflicting a painful bite and hanging on very stubbornly with its strong jaws. The same defensive tactics are exhibited by several congeners, which, though sometimes found in the same habitat, tend to rely more on animal prey and thus are not strictly competitive with *Tiliqua rugosa.*

Lizards belonging to different families and living in arid regions on other continents may be fairly similar in their habits, notably the iguanid *Sauromalus obesus* of the American Southwest and the agamid *Uromastix acanthinurus* of the Sahara.

D.H.

435
Sub.: Vertebrata
Cl.: Osteichthyes
O.: Acipenseriformes
F.: Acipenseridae

STURGEON *Acipenser sturio* 435

The sturgeon is an anadromous migratory fish, which is to say that it lives in salt water along the continental shelf and swims upstream to reproduce; the young sturgeons return to salt water almost immediately after hatching. The adult is usually between 2 and 3 m long and weighs between 150 and 200 kg, but the sturgeon grows very slowly; it may live longer than 50 years and attains a maximum size of 5 m and 500 kg. *Acipenser sturio* is found in the Mediterranean, the Black Sea, and on both sides of the North Atlantic, from Morocco to North Cape and from South Carolina to Hudson Bay.

▶ During the two stages of its cycle that are spent in fresh water, the sturgeon feeds on insect larvae (mayflies, libellulids, chironomids, mosquitoes), crustaceans, freshwater snails, and worms; in salt water, it catches crabs, shrimp, mollusks, marine worms, and small bottom-dwelling fish. In either case, its feeding methods are identical. Its tubular, protractile mouth is ventrally located on the sturgeon's long shovel-like snout; it has four long barbels just in front of its mouth that are richly supplied with sensory receptors like the catfish, equipped with tactile "whiskers" rather than olfactory organs, as is the case with other fish.

The sturgeon roots along the soft, muddy bottom with its snout, and as soon as one of the barbels comes into contact with a prey animal, the sturgeon ingests, or rather inhales, its prey with a reflexive sucking motion. The sturgeon's eyesight is very poor, but the barbels enable it to find prey that is buried in the mud or while swimming through the turbid waters of an estuary; the lining of the sturgeon's mouth is liberally supplied with papillae (tastebuds) to prevent it from ingesting too much inedible material, which is immediately expelled through the nostrils. The sturgeon eats slowly and absorbs a comparatively small amount of nourishment, considering the vast quantities of sea or river bottom that it processes in this manner. In short, the sturgeon's feeding strategy is not very efficient, and it must devote the greater portion of its long life to the search for food.

◀ The sturgeon has no defense against predators, apart from its sheer size and the fact that it lives in muddy water. Sharks, killer whales, seals, porpoises, and even certain seabirds feed on sturgeon in salt water, and they are caught by human fishermen with seines and gill nets during their final upstream migration as well as with dragnets on the open sea. Sturgeon cavair consists of mature oocytes, eggs that are not quite fully developed, and the harvesting of caviar does not necessarily result in the death of the female sturgeon, which can be "milked down" (by pressing along its abdomen) and relieved of its roe before being returned to the river, a technique that is obviously preferable from a conservationist standpoint. The flesh of the sturgeon is also a great delicacy of course, and a substance known as isinglass that is extracted from the swim bladder of the sturgeon and other large freshwater fish is used in clarifying wines and various food products.

There are about 20 extant species of sturgeon, notably *A. ruthenus*, found in the Danube, and *A. gueldenstaeti*, found in the Caspian and the Black Sea, as is the largest sturgeon species, the famous beluga (*Huso huso*), which may attain a length of 9 m , weigh as much a metric ton, and (though rarely) live to age of 75. J.-Y. S.

SURICATE *Suricata suricata* 436

436
Sub.: Vertebrata
Cl.: Mammalia
O.: Carnivora
F.: Viverridae

This small carnivore, also called the slender-tailed meerkat, measures 30–35 cm, with its tail contributing an additional 25 cm; the suricate's pelt is whitish gray, and it has a black mask around its eyes and sometimes black stripes along its back as well. Suricates can dig out extensive underground galleries with their powerful front paws, and a sizable number of individuals will often occupy the same burrow at night, which they sometimes share with ground squirrels and other small terrestrial mammals. When the food resources of a particular locality are exhausted, the suricate will move on to a new hunting ground 1 or 2 km further away from the communal burrow.

▶ The suricate is very catholic in its tastes, feeding on insects and other invertebrates, mice, birds, lizards, and snakes, and occasionally even on creatures slightly larger than itself; even after it has eaten its fill, a suricate will sometimes attempt to snatch food away from its denmates. The suricate hunts in small groups, which set out toward their hunting grounds at daylight; it locates its prey primarily by smell, scratching at the ground to locate a scent trail that it can follow. When it detects the scent of a small animal in its burrow, it begins to dig with its front paws until the prey has been entirely unearthed, then kills it with a bite to the back of the neck. If the prey animal is too large to be dispatched with a single bite, the suricate tramples it underfoot, then gives a little cry of triumph to summon its denmates to the kill.

◀ Vultures and other raptors are the main natural enemies of the suricate; the suricate's eyesight is very keen, however, and one member of the band is detailed to stand watch at the entrance to the burrow, scanning the skies and shrilling out a warning for its denmates when it sees the silhouette of a bird of prey. When actually attacked, the suricate throws itself down on its back and threatens the aggressor with its teeth and claws, then begins to shriek and makes a rapid dash toward the aggressor, with its tail bristling and erect.

Six subspecies of suricate, which is the only representative of the genus, are distributed throughout southern Africa. P.A.

437
Sub.: Vertebrata
Cl.: Aves
O.: Apodiformes
F.: Apodidae

SWIFT *Apus apus* 437

The swift measures about 16 cm, with a wingspread of 38–40 cm, and weighs between 29 and 50 g. It spends most of time in the air, and it has even been thought to sleep on the wing. The swift nests on plains as well as mountainous regions in Europe (except in Iceland and northern Scandinavia), the Middle East and Central Asia, and in North Africa; it spends the winter in more southerly regions. In Western Europe, the swift arrives at the end of April or the beginning of May and starts to fly south at the end of July.

▶ The swift remains active as long as the daylight lasts; as it name implies, it is a fast precision flier, skilled at aerobatics, and attains speeds of up to 200 kph, at least for short periods. The swift is more comfortable resting on the wall of a building or other vertical surface than on a branch, since all of its toes point forward, which makes it difficult for it to perch in the usual way. The swift nests in colonies and faithfully returns to its old nesting site every year, generally in a cavity in a high wall of a building, less frequently on a rock face or other natural surface.

The swift feeds on insects that it snaps up in midair, as well as on the tiny spiders that float through the air on gossamer parachutes. It seeks its prey in the built-up areas where it nests, as well as in the surrounding countryside, and may cover substantial distances in the course of a day's hunting. The parents feed their young with pellets of crushed insects held together with saliva, each of which consists of as many as 200–300 insects; normally, between 25 and 30 of these pellets disappear down the gullets of the nestlings every day, but the nestlings can also go for several days without food if weather conditions are unfavorable for flying. The fledglings begin to catch insects for themselves as soon as they leave the nest.

◀ The barn owl and the hobby (a small falcon) both prey avidly on the swift. Man, on the other hand, is the great benefactor of this species, providing it with shelter and (indirectly) with its food; the swift has adapted very well to an urban environment.

There are 20 species of swift, found in every region of the globe except for the poles. The chimney swift *(Chaetura pelagica)* nests in the eastern United States and spends the winter in South America; the cave swiftlet, or salangane *(Collocalia)*, is notable as the source of the main ingredient of bird's nest soup (actually the dried adhesive secretions with which the swiftlet fastens its nest to the wall of its cave, from which twigs and feathers have been removed, for the most part). J.-P. R.

438
Sub.: Vertebrata
Cl.: Osteichthyes
S.cl.: Teleostei
O.: Perciformes
F.: Xiphiidae

SWORDFISH *Xiphias gladius* 438

The swordfish is a migratory pelagic (deep-sea) fish. The adult is between 2 and 3.5 m long (about a third of this length consisting of its long flattened beak) and weighs between 60 and 150 kg; the largest specimen on record was 5 m long and weighed about 500 kg. The swordfish is very aggressive; swordfish attacks on fishing boats, in which the fish actually drive its sword into the hull, have frequently been reported. It is not known whether this represents an attempt on the part of the swordfish to repel an aggressor or to secure an enormous prey animal for itself (as the silhouette of a boat viewed from below and the vibrations created by the oars or the propellor might plausibly suggest). It prefers the warm waters between 45° N and S.

▶ The swordfish can swim as rapidly as the tuna and can dive to a depth of 800 m, which it sometimes does in order to prey on shoals of preabyssal fish; when it swims on the surface, the tips of its dorsal and caudal fins can be seen rising out of the water. The jaws of very young swordfish are already greatly elongated, and they are aggressive predators, creating general havoc among the zooplankton and the fry of other species. The adult swordfish makes use of its speed, strength, and endurance to catch substantial quantities of gregarious fish such as herrings, mackerel, and gar by means of a single concerted attack; it also feeds on squid (of which it is particularly fond), sauries *(Scombresox),* flying fish *(Exocetus),* and young tuna.

The upper part of the swordfish's body is dark and the lower part is a brilliant silvery color, which makes it easier for it to approach a school of fish either from above or below without being seen. When it comes within range, it begins to flail about with its sword—which is actually used more like a club—and the fish that are stunned or wounded in this bludgeoning attack are quickly consumed by the swordfish. The swordfish sometimes disables larger prey by impaling it with its sword.

◀ The giant squid is perhaps the only creature of the ocean depths that preys on the swordfish; the mako shark is the only other predator that is strong and swift enough to attack an adult swordfish, which it disables by attacking it from behind and biting off the flukes of its caudal fin, thus depriving it of its principal means of locomotion. A favorite quarry of deep-sea sports fishermen, the swordfish is also pursued by commercial fishermen with the aid of the harpoon, trawl, or floating line. The European fishery accounts for about 3000 metric tons per year, and in addition to fresh or frozen swordfish steaks, canned swordfish is eaten in Europe as well.

The feeding habits of the family Istiophoridae (sailfish and marlin) are similar to those of the swordfish; in these species, the elongated beak is rounded rather than flattened in cross-section. The sail of the Pacific sailfish *(Istiphorus orientalis)* is the most impresssive, though the black marlin *(Istiompax marline)* is the largest overall, sometimes exceeding 700 kg. Along with the latter, the striped marlin *(Makaira audax)* and the blue marlin *(M. ampla)* are also found in Indo-Pacific waters.

J. Y.-S.

T

TASMANIAN DEVIL *Sarcophilus harrisi* 439

439
Sub.: Vertebrata
Cl.: Mammalia
O.: Marsupiala
F.: Dasyuridae

This rather reclusive marsupial carnivore typically measures 50 cm exclusive of its tail, which measures 25 cm. It has short bow legs, its head is very large in proportion to its body, and there is little in its apearance that seems overtly diabolical; perhaps it is so called because its ears turn red when it is angry or distressed. In prehistoric times, it was found on the mainland of Australia, but has been confined to the large offshore island of Tasmania (where it is still fairly prevalent) since shortly after the arrival of the dingo. The Tasmanian devil is nocturnal, spending its days in a crevice in the rock or the abandoned burrow of some other creature, or curled up on a bed of dry leaves, though it sometimes comes out to warm itself in the sun.

▶ The devil's powerful jaws can be opened very wide (up to 90°), but its curved stumpy legs are not very well adapted for the chase, and it feeds primarily on carrion and small, slow-moving vertebrates. The Tasmanian devil is an inveterate henhouse robber, and sometimes preys on newborn lambs as well; it is one of the few carnivores that has acquired the habit of carrying uneaten food away with it in its mouth.

◀ Foxes and dingoes were never introduced into Tasmania, and, in spite of its penchant for feeding on livestock, the Tasmanian devil now enjoys the strict protection of the law, though it is still occasionally trapped or otherwise persecuted by indignant farmers. P.A.

440
Sub.: Vertebrata
Cl.: Mammalia
O.: Insectivora
F.: Tenrecidae

TENREC *Tenrec ecaudatus* 440

This insectivore measures about 38 cm in length. Its fur is coarse and interspersed with delicate guardhairs and a few brownish spines, like those of a hedgehog; its tail consists of no more than a rudimentary stub. Its original range was limited to Madagascar and the nearby Comoros, but it was also introduced onto Mauritius, Reunion, and the Seychelles during the 19th century. Tenrecs are active at dusk and at night, taking shelter during the day in a nest of dry leaves, under a rock, or in a hollow stump. During the winter, which is the dry season on Madagascar and also corresponds to the end of the tenrec's mating season, the tenrec retreats into its burrow, stops up the entrance with earth, and goes into true hibernation, which is characterized by a substantially diminished body temperature, heartbeat, and respiration rate.

▶ The tenrec feeds primarily on insects, though it seeems to include a larger proportion of plant material in its diet than do the more familiar insectivores (moles, shrews, and hedgehogs). The tenrec forages in the topsoil with its snout and front paws, scooping out conical pits some 2–5 cm deep. Like the SOLENODON, it is thought to detect its prey with the aid of a rudimentary system of echolocation, since it produces high-pitched (5–11 khz) but distinctly audible clicking sounds with its tongue. The tenrec crushes its insect prey in its very strong teeth before swallowing it; the dentition is even more highly developed in the related rice tenrecs *(Oryzorictes)*. Captive specimens can be maintained on fresh meat or the bodies of small birds.

The tenrec is the most prolific of all mammals; the female may give birth to a litter of as many as 21 offspring, each measuring 6–8 cm and weighing 10–18 g at birth. Consequently, tenrecs tend to forage in extended family groups of 20–30 individuals. The baby tenrecs will accept food from their mother's mouth when they are 3 days old, and in captivity they quickly learn to lap milk from a saucer.

◀ The tenrec seems to have few natural enemies other than man and the domestic dog; the tenrec is hunted for food (by both of these predators) on Madagascar, where the destruction of habitat has very critical consequences for a large number of unique insular species. J.-J. B.

441
Sub.: Vertebrata
Cl.: Reptilia
C.: Squamata
S.o.: Sauria
F.: Iguanidae

TEXAS HORNED TOAD *Phrynosoma cornutum* 441

The horned torn of North America, more accurately but much less commonly known as the horned lizard, is a miniature member of the iguana family with a flattened ovoid body studded with spiny tubercules and with long, sharp projections on its head reminiscent of the ceratopsian dinosaurs of the Mesozoic. Its back is gray or brown in color with a pattern of darker spots arranged symmetrically with respect to the mediodorsal line. Its tail is short and tapering, and its legs are attached almost horizontally with respect to its body, which is flattened both ventrally and dorsally. All of these features make it easy for the horned toad to bury itself in sandy or friable soil.

This particular species is found in open, arid terrain at an altitude ranging from 0 to 1800 m and from Kansas to Texas, as well as in an adjoining region extending from southern Arizona and New Mexico down into northeastern Mexico and including the large natural depression known as the Bolsón de Mapimí. Like certain skinks and vipers of the Sahara, the horned toad buries itself in the sand to insulate itself and thus can regulate its internal temperature by exposing more or less of its surface area to the rays of the sun—or it can bury itself entirely by tunneling headfirst into the subsoil and replacing the sand or earth above it with lateral movements of its body. The average longevity of this species probably exceeds 5 years.

▶ The horned toad feeds primarily on ants, which make up about 91% of its total intake in the Bolsón de Mapimí, the remainder consisting of about 6.6% mites and ticks and 2.1% beetles. In New Mexico, it feeds almost exclusively on harvester ants *(Pogonomyrex);* the stomach contents of individuals collected in southeast Texas consisted of beetles (38% by bulk), grasshoppers and other orthopterans (37%), hymenopterans (21%), and true bugs (4%). The relative proportions (except of course in the second case) may vary somewhat with the season.

In New Mexico, population density was calculated at 10 adult horned toads per hectare in an area that was also found to contain an average of 145 anthills. Assuming that each of these horned toads will consume an average of 71 ants per day during the warmer months (May—August), this suggests that a total of 85,200 worker ants, or an estimated 72% of the worker ant population, will be eaten by horned toads during this period; this would seem to impose an enormous burden on the reproductive capabilities of the harvester ants.

The horned toad travels an erratic zigzag course, occasionally stopping to feed in the vicinity of anthills for periods never exceeding 15 minutes, picking off a couple of dozen worker ants that pass within reach of its long adhesive tongue, and then setting off again, rarely in the same direction as before. It covers about 50 m per day in this manner, and its peak periods of predatory activity coincide with peak periods of activity for the worker ants.

◀ The predators of the Texas horned toad consist primarily of raptors and snakes. The area surveyed in the Bolsón de Mapimí, for example, consisted of of 20,000 hectares, in which there were 8 nesting pairs of red-tailed hawks and 11 nesting pairs of Swainson's hawks. The red-tailed hawks consumed an average of 2 horned toads per nest per day, comprising 24% of the biomass consumed by them; the Swainson's hawks consumed an average of 1 horned toad per nest every 2 days, or 4.3% of the biomass consumed. Roadrunners, northern shrikes, sparrowhawks, rattlesnakes, and various colubrids have also been observed to prey on horned toads.

When alarmed or excited, the horned toad is noted for its ability to discharge a fine stream of blood from the corner of its eye, up to a distance of 1 m—a phenomenon that appears to be brought about by the sudden rupture of the capillaries in the nictating membrane, or so-called third eyelid, which moves outward from the conjunctiva to cover the entire eyeball. Only the larger and more vigorous individuals appear to possess this bizarre defensive capability (if that is indeed what it is), which has also been observed in 2 congeneric species, *Ph. coronatum* and *Ph. solare.* The spiny excrescences on the horned toad's back do not appear to serve any defensive function, though the horned toad's habitual immobility and protective coloration probably protect it to some extent from aerial attacks by buzzards and other birds of prey.

Seven species of *Phrynosoma* have been recognized in North America; all of them are found in arid, though not necessarily desert, regions of the West. *Ph. douglassi,* for example, may range as far north as British Columbia; it is also the only horned toad species that gives birth to live young.

C.G.

Thanasimus formicarius 442

442
Sub.: Arthropoda
Cl.: Insecta
O.: Coleoptera
F.: Cleridae

The adult of this species of checkered beetle is 7–10 mm long; its head is black, the corselet (upper part of the thorax) and the base of the elytra are orange-red; the posterior part of the elytra is black with two wavy yellowish stripes (the first of which has a thin black stripe inside it), the underparts are entirely red, and the feet black. *Thanasimus formicarius* may be found on the bark of pine trees that are infested with bark beeetles of the family Scolytidae; it occurs all over Europe and is active between spring and fall.

▶ *T. formicarius* is so called not because it eats ants (though it may occasionally do so) but because it scurries along the bark of trees in the manner of a carpenter ant and because its long legs give it a somewhat antlike silhouette. Both the larva and the adult *formicarius* feed voraciously on bark beetles, particularly *Ips sexdentatus*, *Ips acuminatus*, and *Tomicus piniperda*. The larva makes its way through the galleries excavated by these and other species devouring more or less everything in its path, including the eggs, nymphs, and imagos of the bark beetle as well as on other members of its own species. The adult seeks its prey on the surface and may sometimes be seen on fenceposts from which the bark has not been stripped; it catches adult scolytid beetles as they emerge from the entrance to their galleries as well as any other small insects that it might encounter.

◀ The adult may be seized in flight by insectivorous birds, including several varieties of Eurasian woodpecker (the green, black, and great spotted woodpeckers), which may come upon on both the larva and adult checkered beetle in the course of their excavations.

Thanasimus formicarius looks very much like its relative *Thanasimus rufipes*, which never exceeds 7 mm in length and lacks the black border on its elytral stripe; in addition, the second stripe is placed closer to the apex of the elytra (i.e. further back). The family resemblance is not quite so pronounced in the case of *Allonyx quadrimaculatus*, which is is shiny black except that each of the elytra has 2 spots or stripes covered with fine white hairs and the pronotum (snout) and the antennae are red. A.D.

443
Sub.: Vertebrata
C.: Chondrochthyes
S.cl.: Elasmobranchii
O.: Rajiformes
F.: Rajidae

THORNBACK RAY *Raja clavata* 443

This species measures between 90 cm and 1.25 m and is found along sandy or muddy substrates in the coastal waters of the Atlantic and Mediterranean, sometimes out to a depth of 500 m. The thornback ray is a graceful swimmer, propelling itself through the water with deep, winglike strokes of its broad pectoral fins, which extend laterally from just behind its head. While at rest, it supports its weight on the edges of its pectoral fins rather than lying directly on the bottom like a flatfish.

The "thorns" of the thornback ray are actually long curved spines attached to a bony baseplate; these spines, which are found in various sizes and distributions in other species of ray, are homologous to the placoid scale of a shark and are particularly well developed in the case of *Raja clavata*. The electric organs located along the side of the thornback's tail (unlike those of the electric rays, or Torpedinidae) are capable of delivering a very feeble electric shock, and may possibly play a role in prey detection or electronavigation, though this has not been established for certain.

▶ The thornback ray hunts from ambush, and each of its dorsally mounted eyes is located on a raised moundlike protuberance, so that it can continue to survey its environment while the rest of its body is concealed beneath a sprinkling of sand or mud. When a fish or crustacean ventures within striking distance, the thornback surges forward and traps the prey animal underneath its broad, flat body, then carefully maneuvers it toward its wide, ventrally located mouth. The thornback's flat grinding teeth are also suitable for crushing the shells of mollusks, on which it also feeds in quantity.

◀ Though the thornback ray does not occupy a very high position in the benthic foodchain, it has no specific predators apart from man. The spotted camouflage pattern on its back makes it difficult to see as long as it remains poised over the the bottom, and the curved spines studded across its back and along its slender pointed tail also serve as an effective deterrent to predators. This species makes very good eating, and though it is often taken by coastal trawlers, it is not sold commercially, and thus is gratefully appropriated as a fringe benefit by their crews.

The flapper skate (*Raja batis*) reaches a length of up to 2.5 m and can weigh up to 100 kg. It is normally found at depths of 100–200 m, though it sometimes ventures out into deeper water (up to 600 m) during the winter months. This species is found all along the continental shelf of Europe, thus, in the North Sea and the Mediterranean (though rarely in the Baltic) as well as in the coastal waters of the Eastern Atlantic. It feeds on flatfish, crustaceans, and marine worms, though its teeth are not sturdy enough to deal with the shells of mollusks. Owing to its substantial size, the flapper skate is an important food fish and is caught with trawls as well as longlines baited with fish, meat, crabmeat, or shellfish; both the "cheeks"

and "wings" (pectoral fins) of the flapper skates are commonly eaten in Europe.
F.M.

THORNY DEVIL *Moloch horridus* 444

444
Sub.: Vertebrata
Cl.: Reptilia
O.: Squamata
S.o.: Sauria
F.: Agamidae

The bodies of both sexes of this small agamid lizard are covered with thorny spikes; there is a pronounced sexual dimorphism as far as body length and weight are concerned: The females measure between 16 and 21 cm and weigh between 35 and 90 g; the males measure between 10 and 18 cm and weigh between 10 and 40 g. The thorny devil is found almost exclusively in the vicinity of scrub vegetation on sand dunes and other sandy terrain in the deserts and semiarid grasslands of central, southern, and western Australia.

The devil is strictly sedentary throughout most of the year, venturing no more than a few tens of meters in any direction, and its territory always includes a bush or a clump of brushwood in the shade of which it can dig its burrow. It spends the night and the hottest part of the day in its burrow, venturing out shortly after sunrise to warm itself. When its body temperature has reached a comfortable level, it moves to the spot (always the same one and always a few meters away from its burrow) where it deposits its feces.

The thorny devil is heliophilic and most active during the autumn (March—May) and the spring (September—December). The breeding season is in August and September, when the devil may wander as far as several hundred meters from its home base in order to find a mate. It spends the entire winter (June and July) and summer (January and February) in its burrow.

▶ The thorny devil is a highly specialized predator, since it feeds exlcusively on 3 species of ants, all belonging to the genus *Iridomyrex*. Analyses of numerous stomach contents have also revealed the presence of bits of gravel, insect eggs, flower petals, and other vegetable debris, which were probably being transported by the ants at the moment they were devoured by *Moloch horribilis*.

The Australian outback is traversed by innumerable broad (2–3 cm) marching columns of ants, which makes it possible for the thorny devil to maintain itself in such a highly specialized niche, and the devil's territory always contains one or more pathways along which these columns invariably travel. In the early morning, the devil simply stations itself beside one of these pathways and starts lapping up ants with its tongue, at a rate of 35–40 ants per minute; the stomach of a single thorny devil was found to contain 2500 ants.

◀ The thorny devil's principal predators include snakes, bustards, and dirunal raptors; the devil may be able to escape from a bird of prey, if it catches sight of it overhead, by scuttling into its burrow. Other species that feed primarily on ants (lizards and insectivorous mammals) are found in different biotopes or have different circadian cycles or predatory techniques from the thorny devil.

Another agamid genus, *Phrynocephalus*, has adapted to a similar mode of life in the deserts of Central Asia, though it is not as highly specialized as *Moloch* and feeds on all sorts of small arthropods. The horned toads (q.v.), or horned lizards, found in arid regions of the American Southwest *(Phrynosoma)* are iguanids, though their morphology and appearance (notably their flattened bodies and thorny spines) as well as their feeding habits are remarkably similar to those of the agamid *Moloch horribilis*. This is a striking instance of evolutionary convergence, the exigencies of a similar habitat and an almost identical mode of life having given rise to very similar adaptations in 2 different saurian families.
D. H.

THREE-SPINED STICKLEBACK *Gasterosteus aculeatus* 445

445
Sub.: Vertebrata
Cl.: Osteichthyes
S.cl.: Teleostei
O.: Gasterosteiformes
F.: Gasterosteodae

The three-spined stickleback is about 10 cm long and weighs about 30 g; it is found in fresh, brackish, and salt water, along the coast as well as in the open sea, and all throughout the cold and temperate regions of the Northern Hemisphere. It lives in schools, and the males and females pair off only for the purposes of reproduction; the three spines (Old English, *stichel)* for which the species is named are located just in front of the dorsal fin.

▶ The stickleback is notorious for its gluttony , and it will devour almost any creature of the appropriate size that it encounters, though it feeds primarily on worms, copepods, insect larvae, fish eggs, and young fry; not surprisingly, it is considered an undesirable intruder on commercial fish farms. The stickleback hunts by sight, sometimes actively patrolling its territory of several square meters, sometimes concealed in the vegetation and darting out at its prey from ambush; in either case,the prey is consumed while the stickleback is still in motion.

◀ In salt or fresh water, the stickleback may fall victim to a variety of different predators, including the codfish, salmon, eel, pike, and perch, heron, great black-backed or herring gull and the otter. When it is approached by a predator, it erects its three dorsal spines and locks them into place; it then remains perfectly immobile, so that most potential predators, especially the smaller ones, find it impossible to close their jaws without impaling themselves on its spines and are finally compelled to relinquish their prey. Many learn a lesson from this and are reluctant to attack the stickleback in subsequent encounters. The stickleback's protective coloration also makes it difficult to distinguish from the vegetation on the bottom.

The stickleback is not a food fish in the usual sense though it is valued a source of fish meal in some parts of the world. It adapts very easily to aquarium life and is thus sought after not only by hobbyists but by laboratories as well for use in aquatic toxicity studies.

The nine-spined stickleback *(Pongitius pongitius)* is found in more northerly regions, the fifteen-spined stickleback *(Spinachia spinachia)* off the coasts of the British Isles and in the North Sea, and the four-spined stickleback *(Apeltes quadracus)* off the eastern coast of North America.

J.-Y.S.

THRESHER SHARK *Alopias vulpinus* 446

446
Sub.: Vertebrata
Cl.: Chondrichthyes
O.: Galeiformes
F.: Alopiidae

The thresher shark, or fox shark, is found primarily on the opean ocean, occasionally in coastal waters, between 45° N. lat. and 45 ° S. lat., provided the water temperature remains between 15 and 25°C. It can readily be distinguished by the remarkably well developed dorsal fluke of its caudal fin (thought to resemble the blade of a thresher's scythe), which is fully as long as the rest of its body and enables it to attain very high speeds when swimming just below the surface, generally with the tip of its dorsal fluke protruding. This species is not dangerous to man, though it may inflict some damage with its flailing caudal fin when landed on the deck of a fishing boat.

▶ The thresher shark feeds primarily on mackerel, herring, and other gregarious fish, occasionally on squid as well. It hunts singly or in small packs, swimming rapidly around a school of fish in gradually tighter circles and churning up the water with its caudal fin (a technique that is also employed by tuna and killer whales). Unlike other pelagic sharks, which attack larger prey animals and devour them piecemeal, the thresher shark swallows its prey whole. Like other sharks, it is extremely sensitive to changes in frequency of the wave patterns generated by prey animals (as well as by other members of the shark pack) as they move through the water.

◀ The adult thresher shark is unlikely to be overtaken by any predator, though their remains are occasionally recovered from the stomachs of killer whales and sperm whales. The thresher shark is not fished commercially, though it is a highly combative species and thus of some interest to deep-sea sports fishermen.

The common names "thresher shark" or "fox shark" are also applied to 3 rarer congeners: *Alopias superciliosus, A. profundus* (found at depths in excess of 300 m), and *A. pelagicus.* J.-Y. S.

TIGER BEETLE *Cicindela hybrida* 447

447
Sub.: Arthropoda
Cl.: Insecta
O.: Coleoptera
F.: Cicindelidae

The adult tiger beetle (so called because of its voracious appetite) can often be seen scurrying over bare patches of ground and sunken country roads in the bright April sunshine; when alarmed by the approach of a human, the tiger beetle takes to the air as easily as a mosquito and alights on a safer spot a few meters away. As soon as the sun goes down or simply moves behind a cloud, the tiger beetle disappears into a hole or crevice in the ground, sometimes into a short tunnel that it digs for itself with its forelegs.

▶ The adult tiger beetle uses its large prominent compound eyes to locate its prey at a distance. It is an active hunter and a redoubtable predator, seizing its prey in its powerful curved mandibles and devouring it on the spot. The larva of the tiger beetle is quite distinctive in appearance: Its head bulges out on top and is scalloped or recessed on the bottom, and it has no fewer than 12 single-lensed eyes (stemmata) that enable it to look in all directions at once, though only up to a distance of 4 to 6 cm.

It digs a sloping vertical tunnel for itself in the sandy soil at the side of a road that is about 3–5 mm in diameter and may ultimately attain a length of 15–45 cm. It bends its body into an S curve and wedges itself into the entrance to this tunnel with its feet and with the additional aid of two little hooks on the dorsal surface of its fifth abdominal segment. The head and thorax of the larva form a disk that fits very snugly over the entrance to the tunnel, and there, with its mandibles spread wide, it awaits the approach of its prey. The tiger beetle larva is quick to retreat into its lair at the sight of any danger, but when an insect approaches, it seizes it in its curved mandibles and drags it down into its tunnel. The tunnel, which is barely wide enough to accommodate the larva itself, is scrupulously maintained, and the larva never permits its feces or the inedible portions of its victims to accumulate in the interior.

The larva seals the entrance to its tunnel at the approach of winter and in anticipation of each of its several larval molts; it grows slowly, and like many predacious flying insects, it takes 2 or 3 seasons to complete its larval development, though it remains in the same burrow during this entire time.

◀ Hister beetles of the genus *Saprinus* sometimes attack the tiger beetle larva in its tunnel, but the larva's foremost natural enemy is a species of solitary wasp, *Methoca ichneumonides*. The male is about 10 mm long, the wingless female between 4 and 8 mm; the female charges boldly into the tiger beetle's burrow and paralyzes it by stinging it in the neck, then deposits a single egg between the forelegs of the tiger beetle larva and blocks up the entrance to the tunnel. It seems curious that the normal flight reaction of the tiger beetle larva is somehow inhibited in the presence of the female wasp, but this nonethless appears to be the case. As is usual with parasitic solitary wasps, the larva is assured an ample food supply, and the helpless tiger beetle larva is devoured piecemeal by the developing parasite.

The North American bee fly *Spogostylum anale* parasitizes the larva of *Cincindela scutellaris* in a slightly different manner. The female bee fly locates the entrance to the larva's tunnel by aerial reconnaissance, then alights nearby and thrusts its ovipositor into the dirt around the tunnel. The newly hatched bee fly larvae make their way into the tunnel and affix themselves between the legs of the tiger beetle larvae—out of reach of the mandibles out their host, which also prevents the tiger beetle from dislodging them by rubbing against the walls of the tunnel.

Between 1500 and 2000 tiger beetle species have been classified thus far, including one Indonesian species that lives in trees and bushes and lays its eggs underneath their bark; the larva drills out a tunnel for itself in the wood of one of the branches. R.C.

TOCO TOUCAN *Ramphastos toco* 448

448
Sub.: Vertebrata
Cl.: Aves
O.: Piciformes
F.: Ramphastidae

This large tropical bird measures about 60 cm in length and is found in groves of coconut palms and other dense, low-lying forest terrain from the Guianas down to Brazil and northern Argentina. Toucans congregate in noisy family groups or in larger bands of up to a dozen individuals and build their nests in abandoned woodpecker holes or natural cavities in a hollow treetrunk.

▶ The toucan's enormous, brightly colored beak has evolved as an intraspecific recognition signal rather than as a food-gathering device. The toucan has become very skillful at picking berries with the tips of its mandibles and then flicking them into its gullet by snapping its head back very quickly. Toucans also feed on termites and other insects, spiders, and other small invertebrates, and are also in the habit of raiding the nests of tyrants and other smaller birds, feeding on the nestlings as well as on the eggs; they are also reported to catch fish on occasion.

◀ The toucan is certainly quite conspicuous in its appearance and its habits, and undoubtedly attracts the attention of numerous avian predators. The toucan's gregarious behavior is useful not only in locating food sources but also in warning of the approach of danger. Its impressive beak may serve as a deterrent to the smaller raptors; the ornate hawk eagle *(Spizaetus ornatus)* and the white hawk *(Leucopternis schistacea)* are known to prey occasionally on the larger toucan species. The toucan has always been pursued by South American Indians for its ornamental tailfeathers, and the nestlings are frequently taken from the nest and raised as pets by jungle-dwellers and settlers alike, since they become very tame and even affectionate in captivity—though clearly the second of these practises is no less detrimental to the survival of a species in the wild than the first.

The toucan family (Rhamphastidae) comprises 7 genera and 40 species altogether, which collectively range from southern Mexico down to the northern limits of the grasslands in Argentina. The beaks of the aracaris *(Pteroglossus)* are markedly serrated, and the fiery-billed aracari *(P. frantzii)* is noted for its habit of using the nesting cavity as a year-round dormitory for as many as 5 or 6 individuals, which provides an excellent defense against predators. J.-F. T.

449
Sub.: Vertebrata
Cl.: Aves
O.: Passeriformes
F.: Certhidae

TREE CREEPER *Certhia familiaris* 449

This small forest-dwelling bird is 11–13 cm long, with a wingspread of 19–20 cm, and weighs 7–10 g. It is found in mountain pine forests throughout the Northern Hemisphere. It gets about by hopping along a treetrunk—and is equally adept at hopping up or down the trunk, headfirst in either case—or from branch to branch; it moves from tree to tree in swift, short flights. It builds its nest in a hollow tree, in a pile of brushwood, or beneath a slab of loose bark. The tree creeper is sedentary, discreet, and essentially solitary, though it often mingles with flocks of titmice.

▶ The tree creeper uses its thin, curved beak to catch insects and their larva, sowbugs, spiders, snails, and other small creatures that it dislodges from the crevices in the bark of a conifer. It works its way up (or down) the trunk in a methodical spiral, pausing about halfway up or at the height of the first small branches (in the former case) and swooping down to the bottom of a neighboring tree to begin again. It sometimes takes to the air, briefly, in pursuit of flying insects that have fled at its approach.

◀ The tree creeper is fortunate in that it has neither predators nor competitors, nor is it directly affected by man or his activities.

A kindred species, the short-toed tree creeper *(Certhia brachydactyla)*, can be readily distinguished from the tree creeper only by its song and is found in groves or forests of deciduous trees, most conspicuously in public parks and gardens. The wall creeper *(Tichodroma muraria)*, or tichodrome, lives on vertical rock faces in mountainous areas, moves about in a series of short hops, like the tree creeper, and occasionally stops and flutters its wings in an abrupt, nervous fashion. It is a handsome bird, its wings a brilliant carmine, and it ranges from Spain to Western China. The creepers (Certhidae) are closely related to the nuthatches (Sittidae).

J.P.R.

Trichogramma maidis 450

Tricho means "hair" in Greek and *gramma* means "line," and the wings of this family (Trichogrammatidae) of small parasitic wasps can be readily distinguished by the fanlike array of straight bristles that covers much of their surface. *Trichogramma* often measures less than 1 mm and varies considerably in color, from oyster-white to black; its globular compound eyes are invariably vermillion-red. Reproduction is arrhenotokous, which means that the fertilized eggs produce female offspring, the unfertilized eggs produce males by parthenogenesis. Sexual differentiation is pronounced, particularly as regards the female's knobbed clublike antennae and the whiplike antennae of the males, which are also covered with bristles in the latter case.

▶ *Trichogramma* is of considerable economic importance to man, since its parasitoidal larva feeds on the eggs of the

European corn borer moth *(Ostrinia nubilalis),* a species that is so called because its larvae (rather than the adult moths) feed on cornstalks and are in fact one of the world's most destructive insect pests. The adult *Trichogramma* feeds on honeydew, nectar, and other sugary substances, but the female wasp is always alert for chemical traces (kairomones) of the adult corn borer moth—specifically a volatile compound that is present in the scales sloughed off by the female moth. These olfactory cues permit *Trichogramma* to retrace the moth's steps of back to the spot where it has laid its eggs. (In general, the term *pheromone* refers to a chemical attractant that is recognized by a member of the same species, *kairomone* to a chemical signal that enables the organism to be identified by a member of a different species).

Trichogramma examines each of the eggs by brushing it with the tips of its antennae, searching for traces of other volatile chemical markers that have been deposited by the tarsi or ovipositor of a female of its own species (possibly including itself). If these are discovered and the egg has already been parasitized, it moves on the next one. If not, it pierces the chorion (egg case) with its ovipositor and injects a dose of venom that kills the corn borer embryo before it inserts its own egg. The female *Trichogramma* generally lays between 50 and 100 eggs, depending on the host species with which it is associated.

In the laboratory, *Trichogramma* will insert as many as 3 or 4 of its own eggs into the egg of the European corn borer moth and only one of its own eggs into the eggs of the meal snout mouth *(Pyralis farinalis),* which belongs to the same family as the corn borer. In nature, parasitic specificity is maintained from one generation to the next, but when an attempt is being made to propagate large numbers of trichogrammatids in the laboratory, the female can readily be induced to cooperate with this scheme by parasitizing the eggs of a related moth species.

Very early on, the *Trichogramma* embryo surrounds itself with a membrane of specialized cells that can extract nutrients from the vitellus (equivalent to the yolk sac) of the host through the chorion of its own egg. The chorion eventually splits open some time before the embryo finally detaches itself from this membrane and completes the transition to the larval stage, so that it is referred to as a free-living larva during this interim period. Even after the digestive membrane is abandoned by the larva, it continues to perform its original function of "metabolizing" the remaining contents of the corn borer's egg, all of which will eventually be consumed by the *Trichogramma* larva.

Toward the end of *Trichogramma* 's final larval stage, the corn borer eggs that have parasitized can by distinguished by the darker color of *Trichogramma's* nymphal cocooon. If the external temperature drops below a certain level, *Trichogramma's* development may be arrested at the larval stage for over a year, but normally the larva undergoes nymphosis in due season. The emerging adult finally cuts its way through the chorion of the corn borer's egg with its mandibles and will soon be ready to mate; the females will also sniff out the track of the corn borer moth.

◀ Parasitized eggs of the corn borer moth are introduced into an artificial casing that protects them from crickets, lacewings, cochineal insects, and other creatures that feed on insect's eggs, and are made available to farmers by government agricultural agents or commercial distributors. These capsules are sometimes broadcast by hand or spread by tractor; they may also be dropped over the fields by crop-dusting planes. The adult *Trichogramma maisis* is a strong flier and can be relied upon to disperse itself over a radius of several kilometers, and either the eggs or the adults of this species are released over more than a million hectares of cornfields every year in Europe, the United States, and China.

There are roughly 50 known *Trichogramma* species and many undoubtedly remain to be identified. Morphologically they are almost identical and are found in a variety of habitats, including the banks of streams and ponds, low-lying foliage, orchards, and woodlands. They have also been introduced into an estimated 15 million ha of cropland as a biological control on borers as well as torticid and noctua caterpillars (leaf-rollers and cutworms). Wasps of the genus *Trichogrammatoidea* have also been widely used as a biological control on caterpillars, especially in Africa.

J.V.

450
Sub.: Arthropoda
Cl.: Insecta
S.cl.: Hymenoptera
F.: Trichogrammatidae

451
Sub.: Vertebrata
Cl.: Osteichthyes
S.cl.: Holostei
O.: Amiiformes
F.: Lepisosteidae

TROPICAL GAR *Lepisosteus tristoechus* 451

The gars, or Lepisosteidae, are a family of large freshwater fish, weighing as much as 150 kg in the case of the tropical, or giant, gar, which which is found in Mexico, Cuba, and Central America. North American species include the longnose gar *(L. osseus)*, which is the most common, the alligator gar *(L. platystonus* syn. *spatula)*, and the shortnose gar *(L. tristoechus)*; a fourth species, *L. sinensis*, is found only in China. All the gars are sluggish fish that spend much of their time basking in shallow, often brackish water that is rich in vegetation; they approach the bank only during the spawning season. The gar's compartmentalized swim bladder serves as an auxiliary organ of respiration, and it often gulps air directly from above the surface. The gar resembles the pike is its morphology and predatory behavior, and a number of European and Australian fish belonging to different families are also referred to as gars, garfish, or garpike (e.g., the NEEDLEFISH).

▶ The gar is quite voracious, feeding on aquatic birds as well as other fish. The hatchlings are carnivorous from the outset, preying on insect (especially mosquito) larvae. The young gar quickly begins to take an interest in larger prey, and by the time it is 5 cm long, it will be feeding on small Cyprinidae (catfish and their relatives) in batches of 15 or so at a time. The larger individuals feed on the very largest prey and disdain such small fry as these; the gar has very irregular feeding habits and a relatively slow digestive rate (24 hours from ingestion to excretion) in comparison to other freshwater fish, though its rate of growth is nevertheless quite rapid.

The gar's hunting behavior presents a striking contrast to its otherwise placid demeanor; while at rest, the young gars resemble nothing so much as floating chips of wood. The gar detects its prey by sight and, as it draws closer, verifies its location with the aid of the sensory organs along the lateral line. It gives chase very abruptly and is capable of literally bounding along the surface of the water in pursuit of its prey. Like the crocodile, the gar attacks its prey with a broadside lunge and can inflict serious injury with its long rows of needle-like teeth; once the prey has been immobilized, the gar works it around in its mouth, still grasping it very firmly, until it is in position to be swallowed headfirst.

◀ The gar is protected from predators by a cuirass of tightly articulated thick bony (ganoid) scales, strong enough—at least in the case of the dorsal armor of the older individuals—to withstand a spear fired from a skindiver's speargun. The tropical gar is an important food fish, and its congeners are sometimes, though ever more rarely, consumed by man (notably by the Seminoles in Florida). The gar's armored scales are also occasionally made into souvenir costume jewelry in Mexico and Louisiana. F.M.

452
Sub.: Vertebrata
Cl.: Aves
O.: Anseriformes
F.: Anatidae

TUFTED DUCK *Aythya fuligula* 452

This Old World diving duck, also called the tufted pochard, is 40–47 cm long, with a wingspread of 67–73 cm, and weighs between 0.65 and 1.1 kg. It is found at relatively low altitudes and prefers the placid waters of shallow lakes, including artificial lakes, and reservoirs (less than 15 m). Its numbers are currently on the increase in Western Europe, with a winter population (primarily in Northwestern Europe) estimated at 525,000 individuals; it is also found in parts of Asia. A banded bird has survived past the age of 14, but the average life expectancy for the adult in the wild is probably 1 to 7 years, with an average annual adult mortality estimated at 46 percent.

▶ The tufted duck not only tolerates but has learned to exploit the proximity of human beings and is one of several species that may be seen accepting tribute from its admirers in the parks of Geneva and other European cities, where it also explores the sewer system in search of prey. It is diurnal, gregarious, and omnivorous, feeding primarily on mussels (especially the zebra mussel, *Dreissena polymorpha)* and other mollusks *(Cardium, Littorina, Hybridia, Pisidium, Anodonta, Lymnaea)*, which may comprise as much as 90 percent of its diet. The remainder consists of gammarids and other small crustaceans, insect larvae (chironomids, backswimmers, caddis flies, diving beetles, libellulids), and vegetable material such as pondweed, bulrushes, and the seeds and foliage of other aquatic plants (notably *Potamogeton)*. The tufted duck will dive down to a depth of about 8 m in search of prey, propelling itself through the water with its webbed feet and rummmaging through the sandy or muddy bottom with its bill; the duration of this entire procedure varies between 10 and 40 seconds.

◀ Like other ducks, the tufted duck builds its nest on the ground, so that the eggs and ducklings are vulnerable to predators—foxes, polecats, crows, marsh harriers, or gulls in this case. A weak or injured adult may sometimes be finished off by a goshawk or a white-tailed eagle, but the peregrine falcon is the only bird of prey that is swift enough to catch a tufted duck in flight. While human hunters continue to take a substantial toll of this species every year, the tufted duck has at least been able to profit from certain alterations of the environment made by man—the creation of artificial lakes and reservoirs, for example, or the eutrophication of Lake Leman (among others), which has resulted in an enormous increase in the number of zebra mussels along the lakeshore. In areas where hunting is prohibited—city parks, wildlife sanctuaries, and the like—the tufted duck has become a familiar visitor.

The duck family (Anatidae) comprises 145 species altogether and is classified into several subfamilies, which are in turn divided into tribes. The tribe Aythyini includes 15 species of diving ducks grouped into 2 genera, *Aytha* (pollards and scaups, 12 species) and *Netta* (crested pochards, 3 species); the red-crested pochard *(Netta rufina)*, a very handsome bird, is a characteristic inhabitant of the Camargue marshlands of Southern France. J.-F. T.

TUPAIA *Tupaia spp.* 453

453
Sub.: Vertebrata
Cl.: Mammalia
O.: Insectivora
F.: Tupaiidae

The family Tupaiadae comprises about 15 species of small squirrel-like mammals, sometimes known as tree shrews, that are thought to be the only surviving representatives of a group intermediate between the insectivores and the prosimians (lemurs, tarsiers, bush babies). In the past, the tupaias have been included, though with significant reservations, either among the insectivores or the primates, though now it is more usual to assign them to a separate order of their own. Tupaias are generally solitary, though they are sometimes seen in small family groups; the more than 10 species belonging to the genus *Tupaia* collectively range from India and China down through Southeast Asia into Indonesia and the Philippines.

Tupaias are basically arboreal though they spend a substantial portion of their time on the ground, especially when hunting for food. Some species sleep curled up among the surface roots of trees; all tupaias can run along branches with impressive speed and agility, and one species, the feather-tailed tree shrew *(Ptilocercus lowii)*, also moves along the ground by hopping like a gerbil or a kangaroo rat.

▶ Tupaias are omnivorous, feeding on insects, worms, lizards, mice, and bird's eggs as well as on leaves, fruit pulp, and seeds, though the relative proportions of animal and vegetable food may vary from one species to the next. They use their long incisors to catch and immobilize their prey, then, gripping the prey animal firmly with their front paws, sit back on their haunches to devour it. All the *Tupaia* species (excluding *Ptilocercus)* are active by day; they have excellent vision (their retinas consist entirely of cones), and their senses of smell and hearing are also very acute.

◀ Tupaias are not regarded as edible by any predator and have no natural enemies, though they have a large repertory of vocalizations that they emit (including snuffling, squealing, chirping, and shrieking) when they are startled or threatened, or imagine themselves to be so. Each tupaia species has acquired a rodent "double" by evolutionary convergence—a species of squirrel that has become almost identical in appearance to the tupaia and thus shares its immunity from predation, which is an excellent example of so-called Batesian mimicry. J.-J. B.

TWIG SNAKE *Thelotornis kirtlandi* 454

454
Sub.: Vertebrata
Cl.: Reptilia
O.: Squamata
S.o.: Ophidia
F.: Colubridae

This venomous arboreal colubrid typically measures between 100 and 140 cm, depending on the locality. Its belly is white, its upper body is gray or brownish with spots in a contrasting color that may be more or less pronounced, and its scales are very rough, similar in texture to the bark of the dead branches on which this species is frequently encountered. It is found throughout the wooded regions of sub-Saharan Africa, though particularly in clearings, at the edge of the forest, and in other comparatively sunlit areas, occasionally in brush or tall grass, though it prefers to live among the lower or middle branches of a tree. The twig snake is diurnal and rather sedentary, though it can move with equal facility along the ground, in the trees, or in the water when necessary; it is also somewhat gregarious, at least to the extent that 3 or 4 twig snakes are often encountered in a single tree.

▶ The twig snake feeds primarily on other arboreal reptiles, including chameleons, geckos, agamids, skinks, grass vipers, boomslangs *(Dispholidus typus)*, egg-eating snakes *(Dasypeltis scabra)*, and other twig snakes. It also feeds on tree frogs and on larger frogs and toads (notably *Bufo regularis)* that it finds along the ground in the latter case, as well as on bats, on flying termites during their nuptial flight, and on eggs and nestlings (as a result of which it is sometimes dispatched by the adult birds defending their nest). The twig snake actively seeks its prey for the most part, though it occasionally hunts from ambush, draped over a branch with the forepart of its body jutting off at an oblique angle and swaying slightly in a highly convincing imitation of a twig swaying in the wind.

The twig snake's eyes are frontally located, its pupils are horizontal, and its vision is very keen and probably binocular as well, since it is able to pick out a motionless prey animal from its background at a distance of 3 or 4 m. Irrespective of the speed and agility of the quarry, the twig snake's stalking behavior is invariable: It moves slowly along the branch, the forepart of its body curved into an S shape and still swaying from side to side in twiglike fashion. It pauses occasionally and protrudes its tongue, which is bright red along the blade and black at the tip, tilts it upward, and vibrates it for a substantial length of time—a maneuver that is not likely to be overlooked by even the most inattentive prey animal.

While it is not clear that the predatory reflexes of birds or lizards or other prey animals are actually triggered by this display (certainly the twig snake's tongue is not large or threatening enough to trigger a flight response in any case), it does seem to serve as an effective distraction, allowing the twig snake to continue its leisurely approach. It does not strike until it is in direct proximity to the quarry, and the quarry, now suspended in midair and gripped firmly in the twig snake's jaws, generally struggles vigorously.

The dentition of the twig snake is of the opisthoglyphic ("back-channeled") type: the venom glands are located at the roots of the modified back teeth, and the venom flows through a groove or channel into the wound. The twig snake partially opens and closes its its jaws to enlarge the wound and facilitate the flow of venom, and death generally ensues after several minutes of agony for the prey animal. The twig snake then proceeds to swallow its victim, tipping up the forepart of its body to secure the aid of gravity, and working its jaws from side to side to slowly advance the prey animal down its gullet.

◀ The long list of potential predators of the twig snake includes the short-toed eagle, the batteleur, the lizard buzzard, the long-crested eagle, and other birds of prey, the hornbill, the secretary bird, the marabout, and other long-legged terrestrial birds as well as large ophidophagous snakes of many species (colubrids, elapids), crocodiles, and carnivorous fish. In addition, rollers, magpies, shrikes, and other large passerine birds sometimes feed on young twig snakes, monitor lizards prey especially on the eggs and the young, and a number of carnivorous or omnivorous mammals sometimes feed on either the eggs or the adults, including genets, civets, mongooses, suricates, zorillas, ratels, warthogs and river hogs, baboons, and galagos. The newborn twig snakes may easily fall prey to large amphibians and arthropods (spiders, scolopendromorphs [large tropical centipedes], and beetles). Brushfires are an important (albeit indirect) cause of mortality among the adults, since most of the raptors and terrestrial birds listed above, including the hornbill, are in the habit of marauding in the path of an approaching brushfire and feeding on snakes, lizards, insects, and other refugees from the flames.

In ordinary circumstances, the twig snake's best defense against predators is its mimetic coloration, form, and even behavior (i.e. its characteristic twiglike swaying motion while it is draped across a branch). When it is startled or threatened, it flees rapidly from branch to branch or drops to the ground immediately. In general, it is only when lacking time to escape that it resorts to the intimidatory maneuver of inflating its throat pouch and displaying the brightly colored skin that lies beneath the scales as the latter are stretched apart.

Because its poison fangs are located in the back of its mouth, the twig snake does not pose a serious threat to humans or other large mammals. However, the toxicological effects of its venom are similar to those of another arboreal African colubrid, the boomslang, which has been held responsible for several human fatalities; the celebrated herpetologist Robert Mertens, for one, was fatally bitten while handling a captive twig snake. From an ecological standpoint, there are a number of other arboreal or semiarboreal snakes, including the colubrid *Philothamnus* and the mambas *(Dendroaspis)*, that are close competitors of the twin snake.

A congener, *Thelotornis capensis*, may attain a length of 195 cm. Other arboreal snakes belonging to the same subfamily (Boiginae) include the South American vine snakes *(Oxybelis)*, the South Asian *Dryophis*, and *Langaha nasuta* of Madagascar, all notable for their extremely slender and sinuous conformation, excellent day vision, and cryptic coloration (in addition to a possibly mimetic leaf-shaped projection on the nose of the female *Lanagaha nasuta)*.

The twig snake's characteristic maneuver of protruding its tongue slowly and vertically (whereas a snake normally "tastes" the air by darting out its tongue very rapidly and, of course, horizontally) has also been observed in a number of arboreal snakes that are not closely related, including several Asian tree vipers of the genus *Trimesurus* (see GREEN PIT VIPER) and the constrictor *Gonyosoma oxycephala*. D.H.

TWO-SPOTTED LADYBUG *Chilocorus bipustulatus* 455

455
Sub.: Arthropoda
Cl.: Insecta
O.: Cleoptera
F.: Coccinellidae

This is a very common species throughout the entire Palaearctic region (Europe and Northern Asia plus North Africa and much of the Middle East). Its smooth black body is highly convex and underslung; it measures from 3.5 to 4.5 mm overall. The black elytra are marked with 2 or 3 reddish spots beginning about two fifths of the way along their length, the borders of which are practically contiguous. This species is sedentary and multivoltine (i.e. it hatches out several clutches of eggs—in this case 3—in a single season); the adult lies dormant over the winter in a crack in a treetrunk or rolled up in a dried leaf that remains on the branch or has fallen to the ground. The larva can be recognized by a light-colored transverse stripe that runs across the dorsal side of the abdomen; the larva undergoes its nymphal molt and the adult emerges after about 8 or 9 days.

▶ The two-spotted ladybug is a prodigious destroyer of armored scale insects (diaspids) that feed on the leaves of trees. The ladybug larva frequently feeds on the mobile larvae of the scale insect, and the adult ladybug feeds almost exclusively on scales in every stage of their development, piercing the armored body shield and devouring their internal organs, including the eggs.

◀ Once hatched from the egg (which appears to be of no great interest to any other species), the two-spotted ladybug may encounter a variety of predators, including ants, spiders, lizards, and insectivorous birds—in addition to the more insidious threat posed by such parasites as the chalcid wasp, nematode worm, and *Gregarina* (a genus of protozoans that live in the alimentary canals of certain insects) as well as miscellaneous bacteria and fungi. J.B. and G.I.

V

456
Sub.: Vertebrata
Cl.: Mammalia
O.: Chiroptera

VAMPIRE BAT *Desmodontidae* 456

This notorious family comprises three monospecific genera of small tropical and subtropical bats, typically measuring only 9 cm in length (tail included) and ranging from Mexico down to Chile. The vampire bat's long incisors are needle-sharp, the upper canines are specialized for shearing, and even the molars of these blood-drinking bats have sharp cutting edges. Large colonies of vampire bats are often found in caves and deserted buildings, and the entrances to these daytime refuges can often be identified by the inmates' black tarry guano.

▶ Vampire bats are, of course, nocturnal predators that drink the blood of sleeping birds and mammals, including humans. The vampire begins by seeking out a spot that is only sparsely covered with hair or feathers; it can scuttle along very quickly and stealthily on its hands and wrists without allowing any other part of its body to touch the victim directly, and its heels are padded with a cushion of flesh as a further precaution against disturbing the victim's sleep.

The vampire makes a short incision in the victim's skin with its upper teeth, and as soon as the blood starts to flow, it curls up the edges of its tongue into a sort of hemispherical trough that, when pressed against its upper lip, forms a tube through which the bat begins to lap up the blood of it victim. The vampire bat's saliva contains an anticoagulant, and the rhythmic contractions of the muscles of its throat and chest begin to assist in this process as soon as the blood reaches its stomach.

The vampire bat is a voracious feeder, and will continue to drink the blood of its victims until its stomach is swollen into a ball and it is almost too heavy to fly away. The weight of the blood absorbed during a single such meal may be as much as 56 g, and Goodwin and Grennhall report that the average yearly consumption of a captive vampire bat was 23.4 liters. An individual vampire bat will return to the same victim night after night, if possible, even picking the same horse or cow out of the herd, and after it has drunk its fill, the vampire bat flies back to its cave or other refuge.

◀ The white-winged vampire bat *(Diaemus)* has two cylindrical scent glands located in the corners of its mouth that emit an extremely unpleasant odor when it is disturbed, with the result that this species at least is rarely troubled by predators. The chief danger to the vampire's victims is from infection rather than loss of blood, particularly since these species of bats sometimes transmit rabies, a form of trypanosomiasis that can be fatal in horses. In populated or stock-raising areas, vampire bats are smoked out of their lairs and killed whenever possible. J.-J. B.

W

457
Sub.: Vertebrata
Cl.: Mammalia
O.: Pinnipedia
F.: Odobenidae

WALRUS *Odobenus rosmarus* 457

The male walrus already weighs between 50 and 60 kg at birth and possibly as much as 1200–1600 kg when full grown (750–1250 kg in the case of the female). The walrus's upper canines continue to grow throughout its lifetime, resulting in spectacular curved tusks reminiscent of the saber-toothed cats of the Pleistocene. The walrus is found only in the Northern Hemisphere, with separate populations ranging eastward from the east coast of Greenland to Spitzbergen and westward from the eastern Canadian Arctic to the Bering Straits; the total population consists of about 155,000 individuals. Walruses are generally found in shallow water and frequently set out to sea on an iceberg or a slab of floating pack ice. The female is sexually mature at the age of 6 or 7, the male at 8–10. After a gestation of 15 months, the female gives birth to a single pup, which it continues to nurse for up to two years.

▶ The walrus feeds on 65 different species of marine invertebrates; analyses of stomach contents suggest that mollusks and echinoderms predominate, followed by tunicates and crustaceans. The walrus uses its mighty tusks for the utilitarian task of foraging in the sand or mud and digging these creatures out; it cracks their shells with its powerful flat-topped molars, which are shaped something like cobblestones.

◀ An adult walrus may occasionally be attacked by a killer whale, though all authorities agree that the extent of this predation is very limited and has little or no effect on population levels. The bearded seal *(Erignatus barbatus)* feeds on many of the same marine invertebrates as the walrus, but it remains to be seen whether this potential competition actually has any effect on the feeding habits or the population dynamics of either species. M.P.

WANDERING ALBATROSS *Diomedea exulans* 458

458
Sub.: Vertebrata
Cl.: Aves
O.: Procellariiformes
F.: Diomedeidae

One of the largest of living seabirds, the wandering albatross has a wingspan of 280 to 320 cm, measures between 11 and 135 cm from the tip of its long, hooked beak to the tip of its tail, and weighs from 7 to 12 kg; its has white feathers with handsome black plumes, once much sought after by the millinery trade. Lifespans of 30 and 53 years have been reported for related species (black-browed and Laysan albatross, respectively). The wandering albatross may be found in all the oceans of the Southern Hemisphere and is occasionally, though rarely, lured past the equator by freak weather conditions. It lives (and sleeps) on the open ocean for months on end and only comes ashore during the nesting season on the remote South Atlantic islands of South Georgia and Tristan da Cunha.

▶ The wandering albatross has a strong preference for cuttlefish, though it also feeds on fish and crustaceans, which it captures with its hooked beak, as well as on carrion and refuse floating in the wake of vessels. Its relatively large bulk (for a seabird) leads us to supppose that it has a voracious appetite. The albatross has been the object of superstitious interest on the part of mariners for centuries, primarily because it is said to tag along in the wake of vessels for many days at a time; actually it appears that individual albatrosses, like mariners, are relieved by their watchmates after a stint of several hours (and certainly the albatross's reasons for following ships are far more prosaic than nautical mythology and *The Rime of the Ancient Mariner* might indicate). This is not to suggest, on the other hand, that the wandering albatross is not a prodigious traveler; according to Austin, some individuals may circumnavigate the globe *twice* in the interval between breeding seasons.

Thanks to its long, narrow wings, the albatross is a formidable flier, or rather glider, since it can cruise for hours at sea with its wings outstretched and almost motionless, buoyed up by the lightest breeze. Because of its substantial body weight, the albatross does have some difficulty in taking off either from land or water without a strong following wind. While ashore, the wandering albatross is very gregarious, living in large breeding colonies on flat or slightly sloping terrain. Like many avian inhabitants of small islands that have only been settled by man in recent times, the wandering albatross has no natural fear of humans.

◀ A predatory gull called the skua, or "seahawk," will sometimes attack a sick or wounded adult albatross. Albatross chicks are frequently taken by skuas, still more frequently by adults of their own species. The wandering albatross once suffered greatly from the activities of plume hunters but has since been reprieved, not so much by protective legislation as by the vagrant winds of fashion.

There are 13 albatross species, 9 ranging over the southern oceans, 4 in North Pacific, all similar in their feeding and living habits to the species described.

J.-P. R.

459
Sub.: Arthropoda
Cl.: Insecta
S.cl.: Orthoptera
O.: Ensifera
F.: Tettigoniidae

WART BITER *Decticus verrucivorus* 459

This Old World relative of the North American katydid (a name that is sometimes applied to the entire family of Tettigoniidae, or long-horned grasshoppers) is sometimes assigned to a different genus and thus referred to as *Tettigonia verrucivora*. A denizen of the "grass forest," this big-bodied grasshopper measures 30–38 mm is brownish or greenish in color, has a powerful head with a broad forehead, and may be observed between July and September in open fields, plots of waste ground, and similar terrain. The relatively short elytra, barely longer than the body, are marked with dark spots.

The female is somewhat longer than the male (30–45 mm), its abdomen terminating in a long curved ovipositor, popularly known as a "saber." The male stridulates during the day, producing a measured series of clicks that occasionally increase in tempo and duration. The larva is simply a smaller version of the imago (adult), and, as is invariably the case with insect species that develop continuously and thus do not undergo a metamorphosis, the wings are initially present as a pair of exterior sheaths that grow progressively larger and assume their adult form at the time of the final molt. The wart biter *(Warzenweisser, mange-verrues)* is so called in recognition of a widespread folk belief that the mildly corrosive expectorations produced by this species will cause warts to disappear—a course of treatment that was reportedly still being prescribed by naturopathic healers as recently as 1940.

▶ This species is fairly common in Western Europe, including mountainous regions; it can inflict a painful bite on a human finger with its powerful mandibles (hence *Decticus,* "biting"), though other grasshoppers, especially the handsome red- or blue-winged OEdipodinae, comprise its more usual diet. Its seizes its prey slowly but firmly with the aid of prehensile suction cups on its forelegs and dispatches it by biting through the central ganglia at the back of the grasshopper's neck, then devours it except for the wings and elytra. This species also feeds on seeds and grains to some extent; following copulation, which lasts for several hours, the female seizes and eats the male in the manner just described as soon as the partners have separated.

The larva, though unable to fly and generally too small to catch grasshoppers, pursues and feeds on less substantial prey in a similar manner.

◀ *Decticus verrucivorus* has no systematic predators that we know of, though it is possible that the larvae are sometimes parasitized by sand wasps of the genus *Sphex.*

Among many related species of similar habits we might mention *Decticus albifrons* F., the white-fronted bush cricket, which was extensively studied by the great naturalist J.-H. Fabre.

R.C.

460
Sub.: Vertebrata
Cl.: Reptilia
O.: Squamata
S.o.: Ophidia
F.: Viperidae

WATER MOCCASIN *Agkistrodon piscicvorus* 460

The water moccasin, or cottonmouth, is a large, thick-bodied viper that typically measures between 70 amd 140 cm; the largest reported specimen measured 188 cm. It is the only semiaquatic viper, though it displays no special adaptations to life in the water. It is found in lakes, streams, rivers, swamps, and marshy regions in the southeastern United States, from Virginia to Florida and westward as far as Missouri, Oklahoma, and Texas. Longevity in captivity may exceed 20 years.

▶ A close relative of the rattlesnake, the water moccasin is also equipped with solenoglyphic dentition and heat-sensitive pits on its snout that are used in detecting prey (see WESTERN DIAMONDBACK for additional details). It is eclectic in its feeding habits, preying on small mammals and birds, snakes, young turtles, and especially frogs, toads, salamanders, and fish—as well as the occasional baby alligator in the case of the larger individuals. Young water moccasins feed primarily on amphibians and their larvae, as well as on fish and invertebrates.

The water moccasin hunts from ambush, lying coiled up and partially hidden from view on the bank, the top of a stump, along a branch overhanging the water, or on a hummock or a drift of vegetable debris along the shore. It is a strong swimmer, but does not actively pursue its prey through the water. Instead, it simply waits for a prey animal to come within range, then lunges forward and strikes with its venomous fangs, keeping its teeth clenched in the wound, since its venom is hemotoxic and fairly slow-acting on the warm-blooded animals that constitute the bulk of its prey.

The water moccasin is diurnal throughout most of the year though it is usually active after sundown during the summer. It hibernates during the winter, and though it may venture out to warm itself in a patch of sunlight from time to time, it never feeds during these occasional outings.

◀ The length and bulk of the adult water moccasin are considerable enough to protect it from most predators—with the notable exceptions of the alligator (insofar as it still exists in the wild), the snapping turtle *(Chelydra serpentina),* the alligator snapping turtle *(Macroclemys temmincki),* and a few other large snakes and fish ; it may occasionally fall prey to a large wading bird or raptor as well. The moccasin may attempt to frighten off an aggressor by opening its jaws very wide and displaying the spotless white lining of its mouth (hence "cottonmouth"), which presents a striking contrast with the black or dark-brown color of its exterior.

Like the rattlesnake, the water moccasin may attempt to warn off trespassers by thumping the tip of its tail, which, although not equipped with rattles, makes a distinctly audible sound when it strikes against the ground or dry leaves. This behavior (also found in the asp, the death adder, and the rat snakes) may have evolved as a means of attracting prey by mimicking the faint rustling sounds made by a worm or an insect. Young moccasins still vibrate the tips of their tails to attract prey animals, but "rattling" serves an exclusively defensive function in the adult.

Though the moccasin's venom is comparatively weak, it can inject enough of it into the wound to kill a large mammal. It is reluctant to strike and will generally try to escape by swimming if given an opportunity to do so; if it is cornered or surprised, however, it defends itself by striking violently and repeatedly. Like the rattlesnake, it will try to prevent itself from being caught by a marauding kingsnake by executing a curious evasive maneuver—pressing its head and tail down flat while arching the rest of its body off the ground and wriggling it in a series of wavelike muscular contractions.

The moccasin is at least partially competitive with a number of more specialized predators, mainly turtles and colubrids, that prefer to feed either on invertebrates or ectothermic (warm-blooded) vertebrates.

A close relative, the copperhead *(Agkistrodon contortrix),* is found in wooded or mountainous regions of the eastern United States (including the Southeast), though the two species rarely occur in the same biotope. The copperhead is noted for its broken camouflage pattern, which provides it with perfect concealment on a leaf-strewn, sun-dappled forest floor. There is one other North American species, and 9 more species are found in northern Asia, where the genus is thought to have originated, ranging from the fringes of Europe to Japan. (A new classification system has recently been proposed in which these Old World species would be assigned to a separate genus; their venom, at least in certain cases, is much more potent than that of the moccasins or copperhead, and their bite is often fatal to horses and livestock). The Halys viper *(Agkistrodon halys)* of Northern and Central Asia is consumed by man in certain areas, and its flesh and blood have been used in shamanistic rites as well as for medicinal purposes.

D.H.

WATER SHREW *Neomys fodiens* 461

461
Sub.: Vertebrata
Cl.: Mammalia
O.: Insectivora
F.: Soricidae

This is the largest European shrew, though it weighs no more than 15 g and measures between 80 and 100 mm (plus an additional 55–65 mm for the tail); both sexes are about the same size. The water shrew's upper body is basically mole-colored, varying from dark gray to black; the underbody is white or yellowish, and the demarcation between the two color areas is generally sharp. The tail is brownish and has a collection of stiff bristles along the underside that serve as a paddle or rudder when the animal is in the water;

similar bristles are found on the feet, which are larger than those of the common Eurasian shrew. The eyes are very small, the ears practically hidden by the animal's fur. The points of the water shrew's 30 teeth are reddish-pink in color, a peculiarity that is more pronounced in young individuals.

This species ranges throughout most of mainland Europe (excluding southern and central Spain and portions of the Lower Danubian plain); in the Mediterranean basin, it is more prevalent in upland regions than along the coast and may be found at altitudes up to 2000 m in the Pyrenees and up to 2500 m in the Alps. It is normally found along the banks of watercourses, small ponds, and marshes, though it is not as closely associated with this semiaquatic environment as was once assumed and sometimes ventures rather far afield. Maximum longevity for this species is probably in the vicinity of 32 months.

▶ The water shrew is found in habitats that are abundantly provided with small crustaceans and aquatic larvae. According to some authorities, it feeds primarily on freshwater shrimps, gammarids, and gastropods as well as on the mayfly, caddis fly, and dragonfly (libellulid) larvae, all of which it catches in the water. It catches earthworms and various insects on land, and in captivity can frequently be induced to feed on adult amphibians or their larvae (efts, tadpoles) and mealworm larvae, though unlike some of its relatives, it does not prey on voles or other small mammals.

The water shrew is commonly regarded by fishermen as a "destructive" (i.e. competitive) species, but we know that it is often found along the banks of streams that have no fish in them, and in general, it seems more likely that a water shrew will be swallowed up by a hungry pike, eel, or trout than the other way around. It is possible that shrews feed on hatchlings and young fry, though we have no definite information on this point. Even if the occasional healthy fish or crayfish finds its way into a shrew's stomach, it seems likely that these tiny hunters serve the usual benign predatory function of culling out the sick and the disabled.

There have been several published reports of "game larders," caches of partially eaten frogs, secreted by shrews along riverbanks, but it seems more likely that these authors have stumbled upon the shrew's refectory rather than its larder and that these are simply the remains of a rather extravagant gourmet meal. When prey is abundant, the water shrew tends to eat only its favorite parts of the animal (namely, the internal organs) and discard the rest; the same behavior has been observed among captive shrews that are provided with an ample supply of earthworms.

The water shrew hunts primarily at dawn and at night. It seeks its prey by diving into a stream and running along the streambed, investigating the sediment and even turning over pebbles in search of insect and amphibian larvae, small crustaceans, etc. Both the richness of its diet and the alacrity with which it hunts for food may be explained by the fact that its extremely rapid metabolism requires it to consume almost the equivalent of its own weight (12–15 g) every 24 hours. Like other shrews, the water shrew's submaxillary glands secrete a venomous saliva that is potent enough to kill an animal much larger than those it customarily preys on; a few tenths of a milligram of the extract from these glands will almost immediately prove fatal when injected into a vole weighing 20 g.

◀ The stoat, the fisher, the marten, the barn owl, and the tawny owl all feed avidly on the water shrew, as occasionally do the heron and the large fish (pike, eel, trout) mentioned earlier. Since the water shrew is never found in the vicinity of streams that have been polluted, dredged, deepened, or "scoured" (the way that canals in Europe are emptied and then cleared of algae and other organic material), the presence of large numbers of water shrews along a riverbank is likely to be a reliable sign of good fishing, though fishermen are unlikely to be convinced of this until all the water shrews (and all the fish) are gone. R.F.

WATER STICK INSECT *Ranatra linearis* 462

462
Sub.: Arthropoda
Cl.: Insecta
O.: Hemiptera
F.: Nepidae

This variety of water scorpion measures between 60 and 70 mm; its body is cylindrical, greatly elongated in the manner of its namesake, and tipped with a very long breathing tube (30–37 mm), which serves as a sort of snorkel and provides *Ranatra* with a continuous air supply while it stations itself in a clump of aquatic plants a short distance below the surface and awaits the approach of a prey animal. Its head is triangular, and both the beak and the antennae are rather short and divided into 3 jointed segments. *Ranatra's* highly developed forelegs are used for grasping and prehension (see below); its second and third pairs of legs are not modified for swimming in any way, so that, like the other water scorpions (Nepidae), it normally crawls along the bottom or clambers up the stalks of aquatic plants. *Ranatra linearis* is principally found in pools, ditches, ponds, and similar bodies of stagnant water, occasionally in streams where the current is very slow, and is widely distributed throughout Europe and Asia.

▶ Both the larvae and the adults feed on the larvae of mosquitoes, dragonflies, diving beetles, true bugs (Hemiptera), and other insects, as well as on crustaceans and their eggs, fish hatchlings, and roe. *Ranatra* relies primarily on concealment (namely the close resemblance between its sticklike body and the adjoining plant stalks) to catch its prey, then reaches out abruptly with its long forelegs to seize and grasp the prey while it inserts its beak and devours all but the tough outer tegument in a matter of seconds.

◀ Backswimmers (Notonectidae), diving beetles (Dytiscidae), and other water scorpions (Nepidae) are among the more mobile aquatic insects that are known to prey on *Ranatra;* other predators include crayfish, some species of fish, and aquatic birds.

The family Nepidae consists of about 170 species worldwide. *Nepa* is a typical genus of water scorpions; they are also equipped with long grasping forelegs and snorkel tubes that enable them to breathe while underwater. J.-C. D.

WATER STRIDER *Gerris lacustris* 463

463
Sub.: Arthropoda
Cl.: Insecta
S.cl.: Pterygota
O.: Hemiptera
F.: Gerridae

This is one of the more common Western European species, which is also found in North Africa and parts of Asia. Water striders, also called pond skaters or water spiders, are perhaps the most familiar of all aquatic insects. They have 2 long antennae and a long beak (or rostrum) composed of 4 jointed sections; the body is dark in color, narrow, and elongated, something like a spider's. Their thin outrigger legs and the water-repellent bristles on the tarsi enable them to skate or skip over the surface of the water, sometimes very quickly. For obvious reasons, they prefer stagnant or slow-moving water and are usually found in groups, sometimes in concert with other species of the family Gerridae. The larva is quite similar in appearance to the adult, though the body is proportionately less elongated, particularly in very young individuals.

▶ The water strider supports itself on its medial and posterior sets of legs and uses the anterior set, much shorter than the others, to grasp its prey, though it fact the strider is a very opportunistic predator, subsisting mainly on dead or disabled insects that it finds floating on the surface or on the leaves of aquatic plants. These normally include chironomids, mosquitoes, psychodids (tiny sand flies), caddis flies, odonata, and various lepidoptera, among others. Once the strider has its prey in hand, it inserts its rostrum and injects its digestive juices into the captive insect, whose tissues liquefy very quickly and are ingested by the strider.

◀ It seems reasonable to suppose that certain aquatic birds habitually prey on *Gerris* and its relatives, but we have no definite information on the subject. The strider repels intruders of its own species by drumming on the surface of the water with its feet. The family Gerridae numbers about 200 species worldwide, including several that live in (or rather on) salt water. J.-L. D.

WEASEL *Mustela nivalis* 464

464
Sub.: Vertebrata
Cl.: Mammalia
O.: Carnivora
F.: Mustelidae

Just 16 to 23 cm long with an additional 4 to 7 cm for the tail and weighing only 35 to 90 g, the weasel is Europe's smallest carnivore (the North American variety of *Mustela nivalis,* called the least weasel, is the smallest carnivore in the world). It is very similar to the stoat, or ermine, with which it is often confused. The ermine's body is longer (though no thicker) than the weasel's; the weasel's tail is shorter and less bushy, and has no black hairs at the tip. The weasel's back is brown or light brown, its underparts white; the line of demarcation between the two color areas is usually sinuous or irregular. In Northern Europe and in mountainous regions, some weasels grow a white winter coat like the ermine.

The Old World variety of *Mustela nivela* is found throughout Europe (including the larger islands in the Mediterreanean but excluding Ireland) and in North Africa. It lives in open, forested, or mixed terrain and in the Alps has been found at altitudes as high as 2700 m (though it generally gives way to the ermine at higher altitudes). The weasel also does not climb or swim as well as the ermine and is not as frequently found around water. It makes its home in a rockpile or stone wall, frequently in a burrow dug by a mole or a vole, or perhaps even in a farmer's hayloft.

The weasel is less conspicuous than the ermine in its habits, since it spends most of its life underground, scurrying through the narrow tunnels dug by the rodents on which it feeds; a weasel can squeeze through an opening as little as 2.5 cm in diameter. The weasel frequents these tunnels not only in search of food but also as its normal means of getting from place to place; it uses them as a refuge from its numerous enemies and the elements and as a nursery for its young.

▶ The weasel eats insects, mollusks, lizards, and certain small birds, as well as fieldmice, rats, voles, moles, and shrews; it is also partial to eggs. As noted, rodents form the greater part of its diet; average daily consumption is about 33 g in the case of the slightly larger male, 23 g in the case of the female (a substantial fraction of the weasel's body weight in either case). The weasel methodically explores all the holes, crevices, burrows, woodpiles, and underground tunnels in its domain, a hunting technique that it also shares with the ermine. It takes most of its prey underground rather than on the surface; hearing, sight, and sense of smell all come into play. Aboveground, it can usually catch up with a rodent in a few rapid bounds; underground, the extreme suppleness and flexibility of its body is a considerable asset. Not surprisingly, the weasel has been shown to be a powerful check on the rapid, lemminglike expansion of the vole population that occurs in the upland fields and pastures of Europe.

As noted earlier in connection with the ermine (q.v.), the weasel does not suck the blood of its prey, a misconception that was probably inspired by the monomaniacal tenacity with which the weasel clings to a fresh kill. Marquart observes of a weasel at the Hague zoo that it "had just caught a brown rat and continued to stand mesmerized over its prey, clutching it so intently that the rat's body could be lifted off the ground without inducing the weasel to let go of it." Like certain other mustellids (members of a family that includes the skunk, the mink, the otter, and the wolverine), the weasel maintains a cache of emergency rations in some sequestered spot; Hainard mentions one such cache discovered by a forester in a poplar trunk that contained 44 freshly killed mice, 2 wagtails, and a large quantity of acorns.

◀ Among the larger carnivores that sometimes prey on the weasel we might mention the fox, the European wild cat *(Felis silvestris)*, the feral cat, the marten, the buzzard, the vulture, the buzzard, the eagle, the screech owl, the wood owl, the eagle owl, and even the viper. The weasel is protected in some countries but is still unjustly stigmatized by farmers as a raider of chicken coops; this may be true of some of its larger relatives such as the polecat, but the weasel, though admittedly often found skulking about the henyard, is interested exclusively in the rats that feed on the grain. F.T.

465
Sub.: Vertebrata
Cl.: Osteichthyes
O.: Ceratodiformes
F.: Lepidosirenidae

WEST AFRICAN LUNGFISH
Protopterus annectens **465**

This species measures between 50 and 70 cm (occasionally as much as a meter) in length, and is found in the tropical regions of West Africa, particularly on the muddy bottom of large pools and waterholes, flooded marshes, and other bodies of stagnant water rich in vegetation. *Protopterus's* body is elongated, its paired fins are long and ribbonlike (filiform), and in addition to its gills, it is equipped with a pair of auxiliary lungs that supply a large percentage of *Protopterus'* s oxygen requirements , so that it frequently comes up to breathe on the surface.

Protopterus is likely to spend the dry season buried in the mud at the bottom of a dried-up pool or waterhole; it can survive for quite lengthy periods by encasing itself in a sort of adobe shell (a mixture of mud and mucus) and breathing air directly through a tube that protrudes through the layer of dried mud that protects it from the equatorial sun. *Protopterus* is a solitary creature that (when circumstances permit) defends a circular territory about 100 sq m in area, and the intrusion of another lungfish into this territory elicits the same aggressive response as the appearance of a prey animal.

▶ Until it has attained a length of about 35 cm, the juvenile *Protopterus* feeds exclusively on insect larvae; older juveniles (35–70 cm) feed on larvae, snails, and occasional small fish (cichlids). The full-grown *Protopterus* (70 cm) feeds on frogs, small fish, and crustaceans, as well as occasionally on snails. It begins to search for food at dusk and continues into the night; its eyes are small and probably of little use in prey detection. The lateral line is well developed, however, and *Protopterus's* long trailing pelvic and pectoral fins are well provided with sensory organs (tastebuds in effect) that can immediately identify a prey animal that is lying cocnealed along the bottom. The fish that *Protopterus* preys on are all mediocre swimmers, and though it may pursue them for a short distance along the bottom, it never attempts to give chase in open water.

◀ The chief natural enemies of the adult are the crocodile and the African fish eagle *(Haliaeëtus vocifer)*, since *Protopterus* is most vulnerable to attack when it comes up to the surface for air. It may occasionally be disinterred from its refuge during the dry season by a pack of jackals or Cape hunting dogs as well.

There are 3 other *Protopterus* species besides *P. annectens* : *P. amphibius* (30–40 cm) is also found in West Africa, *P. dolloi* (85 cm—1 m) in the Congo Basin. *P. aethiopicus*, sometimes referred to as the East African lungfish (up to 1.4 m), is found in the great lakes of Central and East Africa, and thus usually in clearer water than the other three species. The South American lungfish *(Lepidosiren paradoxa)* is found in the Amazon Basin and can attain a length of up to 1.25 m; it is basically similar in its habits to *Protopterus*, and, like *Protopterus*, is locally consumed by man.

The Australian lungfish *(Neoceratodus forsteri)* attains a length of up to 1.25 m and can weigh up to 50 kg; longevity is in excess of 20 years. It has only a single lung and depends on its gills to a larger extent that the other *Dipnoi*, though it also comes up to the surface quite frequently for a breath of air. It is a sedentary creature, feeding on worms, crustaceans, mollusks, and aquatic plants; during the early part of this century, its range was restricted to two rivers in Queensland, but since then it has been successfully introduced into a number of other lakes and rivers throughout the state. F.M.

WESTERN DIAMONDBACK *Crotalus atrox* 466

466
Sub.: Vertebrata
Cl.: Reptilia
O.: Squamata
S.o.: Ophidia
F.: Viperidae

The western diamondback is one of the largest of the New World rattlesnakes and may in exceptional cases attain a length of over 220 cm and a weight of 7 kg. Average size is probably between 75 and 100 cm. Longevity may theoretically exceed 25 years, though the diamondback is rarely permitted to reach this age in nature. This species is widely distributed throughout the American Southwest (California, Arizona, New Mexico, Texas, Oklahoma, and Arkansas) and northern Mexico, and isolated colonies are scattered throughout the southern states as well. The western diamondback is found in a variety of different habitats, though generally in arid terrain: deserts, plains and scrubland, dry subtropical forests, canyons, and mountains (up to an altitude of 2300 m).

▶ The dentition of the western diamondback is of the solenoglyphic type that is characteristic of all pit vipers—the long hollow poison fangs, normally pointed toward the back of the rattlesnake's mouth, are attached to its short, highly flexible jawbone (maxilla). When the snake opens its mouth and prepares to strike, the fangs spring into position, something like the blade of a gravity knife, by means of the pivoting action of the jawbone; as a result, the rattlesnake's fangs (and consequently its venom) can penetrate deep into the tissues of its victim. The Western diamondback and the Eastern diamondback

(C. adamanteus), which is found primarily in the southern states, are the largest and most dangerous of North American rattlesnakes.

The rattlesnake's fangs wear out rather quickly and are replaced several times a year; behind each fang, there is an entire row of replacements in various stages of development, and one of these will erupt and become fully functional as soon as (or even before) its predecessor has loosened and fallen out. The worn-out fangs are often left buried in the flesh of a prey animal and subsequently swallowed, and eventually excreted in perfectly intact condition, by the rattlesnake.

The western diamondback is crepuscular or nocturnal in its habits when the weather permits, though when it is too cold during the evenings for it to remain active, it begins to hunt during the day as well. Its prey consists largely of small rodents, squirrels, and rabbits, to which an occasional bird or lizard (the dietary staples of the juvenile diamondback) may be added. Rattlesnakes actively search out their prey or hunt from ambush. In either case, the prey is detected partially by its smell though primarily by sight and by thermoreception.

The facial pits (*fossae* is the technical term) of rattlesnakes and other pit vipers actually consist of an inner and outer chamber separated by a membrane that is richly supplied with nerve endings. These are capable of registering a temperature difference of several thousandths of a degree between the two chambers, and with the aid of these remarkably sensitive heat-seeking sensors, the rattlesnake is able to home in on the position of any object whose temperature is greater than that of the surrounding air. This method of prey location is not uniquely characteristic of the pit vipers; the boids (boas and their relatives) have labial fossae that serve a similar purpose, though the neurological mechanism involved is somewhat more primitive.

It is the movement of the prey animal that causes the rattlesnake to strike, however. In the case of a smaller animal, the snake will continue to grasp it with its jaws until the animal is dead, then proceed to swallow it whole, beginning with whichever part of the body the venom was injected into. In the case of slightly larger prey, the rattlesnake will strike, with its fangs erected, but will not attempt to seize the prey immediately. It returns to its original position and waits for some time before it begins to seek out the carcass of its victim, darting its tongue in and out as it samples the olfactory trail that the dying creature has left along the ground.

The rattlesnake's venom is hemolytic, which means that it destroys red blood cells, and it causes death quite rapidly in a small warm-blooded animal. Once the snake has successfully tracked its prey, it begins to swallow it, usually headfirst, though it will not hesitate to disgorge its meal if it is attacked or interrupted either while ingesting a prey animal or shortly afterward. The process of digestion lasts from several days to 2 or 3 weeks, depending on the external temperature and the size of the prey; younger rattlesnakes digest their food more quickly than adults. It has been suggested that an adult rattlesnake probably feeds no more than 5–15 times a year, primarily during its period of greatest activity, which lasts from April to July; it eats nothing during the winter. Females tend to eat more than males during the spring, though this situation is reversed during the summer.

▶ The most frequent mammalian predators of the western diamondback include foxes, coyotes, badgers, skunks, lynxes, and bobcats, though this and other rattlesnake species comprise a larger proportion of the diet of eagles, buzzards, and roadrunners (especially *Geococcyx californenesis*). The western diamondback is also subject to predation by domestic fowl and other gallinaceous birds, wading birds, and crows and their relatives (Corvidae); there are a number of colubrids that feed on rattlesnakes as well, notably the common king snake *(Lampropeltis getulus)*, which appears to be at least partially immune to the rattlesnakes's venom.

The rattlesnake's famous rattle consists of horny segments of skin that are loosely nested together like a stack of baskets and that emit an unmistakable percussive warning call, clearly audible at a distance of several tens of meters, when the rattlesnake vibrates its tail. When threatened or alarmed, the rattlesnake always gives this warning (when time permits) before it strikes. Several theories have been announced to account for this phenomenon, perhaps the most persuasive of which contends that the rattlesnake evolved on the plains of North America and was thus constantly at risk of being trampled by vast herds of ungulates, notably the ancestors of the North American bison. The rattlesnake's venom would not have been strong enough to bring down a creature as large as a bison, but the association of the warning rattle with the rattlesnake's painful bite might have had a suitably chastening effect and assured the ancestral rattlesnake an appropriate measure of respect on the primeval prairie. If the rattlesnake's warning is ineffective, however, it will strike, or attempt to strike, several times before eventually fleeing.

When confronted with a king snake, however, which it detects by odor, the rattlesnake assumes a very different posture, pressing its head and tail firmly against the ground, arching its body upwards and flopping it back and forth in an effort to strike the king snake (not with its fangs, however, but with the upraised portion of its body). These contortions on the part of the rattlesnake presumably make it more difficult for the king snake to immobilize and devour it.

There are more than 30 species of rattlesnake, comprising the genera *Crotalus* and *Sistrurus,* ranging from Canada to Argentina, most of them found in the arid regions of Mexico and the American West. The smaller species tend to feed on lizards, and probably the best known of these is the horned rattlesnake, or sidewinder *(Crotalus cerastes),* which, like a number of Old World vipers, has evolved a distinctive method of locomotion that involves only 3 points of contact between the animal's ventral surface and the hot desert sands.

The rattlesnake's seasonal pattern of activity is naturally much less pronounced in those species that are found in tropical regions, notably the cascabel *(Crotalus durissus)* of Central and South America. These tropical rattlesnakes seem somewhat more aggressive—their temperaments are usually described as "vicious" or "irascible"—and their venom contains a higher proportion of neurotoxins than that of their northern counterparts. Rattlesnake bites only very rarely prove fatal in humans, but even in temperate regions, the larger *Crotalus* species may still represent a significant hazard. Rattlesnakes are of limited economic importance to man, except insofar as they play a role in controlling the population of rodents in semiarid regions.

D.H.

WESTERN WHIPTAIL *Cnemidophorus tigris* 467

467
Sub.: Vertebrata
Cl.: Reptilia
O.: Squamata
S.o.: Sauria
F.: Teiidae

This teiid species is common throughout the semidesert regions of the southwestern United States, particularly in open country. Five subspecies are recognized, their collective ranges extending from Baja California up to eastern Oregon and southern Idaho and from South and West Texas down through northern Mexico. The teiids occupy roughly the same niche as the Old World lacertids, and the whiptails are very similar to lacertids found in arid regions, notably *Acanthodactylus.* The tail and body of the western whiptail are elongated, and its legs are and sturdy and well equipped for running, which is its principal means of defense against predators. Total body length may be up to 30.5 cm.

Population density in a given locality varies greatly in acordance with seasonal and annual fluctuations in rainfall and available food resources—from just a few individuals per hectare to as many as 70. In the northerly parts of its range, the western whiptail emerges from its winter latency in May. The adults generally spend the summer months in estivation, though some individuals in more southerly regions remain continuously active from April to the end of August. The circadian cycle is dependent on soil temperature, so that the northern subspecies tend to become active at a later hour in the morning. The circadian cycle of the species as a whole is said to be bimodal and discontinuous, which simply means that its daily period of activity is interrupted by a lengthy siesta at midday. Maximal longevity is 4–5 years, but only about 30 percent of western whiptails survive their first year of life.

▶ The western whiptail is insectivorous, scurrying rapidly from bush to bush and foraging vigorously underneath each one. It scratches up the ground and turns over bits of bark and dead twigs with its snout, locating its prey by sight, by smell, and even by taste—namely, with the aid of its long, forked protractile tongue. The northern subspecies feed primarily on termites, the southern subspecies on beetles, grasshoppers, and crickets.

◀ When pursued by a predator, the western whiptail prefers to take flight across open ground rather than seeking refuge in its burrow, and when it gets up to speed, it lifts its forelegs off the ground and continues along very rapidly on its hind legs like a biped. The numerous predators of this species include the roadrunner, the sparrowhawk and a number of other raptors, the shrike, the collared lizard, the leopard lizard, and most of the larger snakes of the region, including the western diamondback, Mojave rattlesnake, prairie rattlesnake, whipsnakes , bull snakes , and the western patch-nosed snake *(Salvadora hexalepis)*.

Sixteen *Cnemidophorus* species are found in the United States, many of which reproduce parthenogenetically and may thus consist of populations made up entirely of females (Mayhew, 1968); in such a case, a great many young lizards hatch out of a single fertile but unfertilized egg laid by a mature female. These species include the aptly named *C. perplexus* and *C. uniparens* as well as at least 7 others. A less striking anomaly was reported by Schall (1977), concerning as many as 5 sympatric *Cnemidophorus* species (3 bisexual and 2 parthenogenetic) in one area in southwestern Texas. Normally so many different species belonging to the same genus would not be found in the same habitat. In this case, though their reproductive strategies are basically similar, species differentiation is maintained by a variety of other factors—differences in size, choice of hunting grounds, microhabitats, prey animals, hunting behavior, circadian rhythms, and methods of thermoregulation. (In any case, the question of interbreeding would not ordinarily arise with the parthenogenetic species). C.G.

WHEATEAR *OEnanthe oenanthe* 468

468
Sub.: Vertebrata
Cl.: Aves
O.: Passeriformes
F.: Muscicapidae

This small thrushlike bird measures 14 cm in length, with a wingspread of 24 cm, and weighs between 21 and 35 g. It is found among shortgrass vegetation interspersed with boulders and rocky outcroppings, which category may include a wide variety of habitats, notably coastal dunes, alpine meadows and karst depressions (up to an altitude of more than 200 m), Mediterranean garrigue, and Arctic tundra. In Western Europe, the wheatear is largely confined to coastal and mountainous regions. This species ranges all across Eurasia to western Alaska and the coasts of Greenland and Labrador, and the European, Siberian, and North American populations all spend the winter on the arid plains of the Sahel.

The wheatear is a shy, animated, and rather fidgety-seeming bird with contrasting black, white, and gray plumage. It adapts poorly to cultivated land or any other manmade habitat, and its common name originally had nothing to do with wheat, "wheatear" being a genteel substitution for the Middle English equivalent of "white arse." Life expectancy is probably several years.

▶ The wheatear feeds on beetles, butterflies and caterpillars, locusts, crickets, and ants as well as mollusks and spiders. Its seeks its prey along the bare ground or in shortgrass vegetation, sometimes swooping down from its perch on a boulder to intercept an insect in flight.

◀ The wheatear conceals its nest in a heap of stones, a crevice between two boulders, or in a disused burrow, and the cryptic plumage of the fledglings and their thrushlike habit of flattening themselves out against the ground also helps to protect them against predators. Ermine, colubrids, and hawks (including harriers and, in more northerly regions, merlins) are all known to prey on this species in its spring and summer range; the lanner falcon and the red-headed falcon are the wheatear's chief predators in its winter range.

The male wheatear defends a territory of 1 ha or more not only against trespassers of its own species but other small birds as well, and intraspecific battles over turf (quite literally) are fairly frequent. The wheatear is one of numerous species of small migratory birds that is still being trapped and netted in substantial numbers during its migratory flights over the Mediterranean Basin; populations have declined in other parts of Europe as a result of the clearing of wastelands for agriculture and the development of seashore recreational areas.

The genus OEnanthe is a homogeneous grouping of about 20 species, most of which are found in the steppes and desert regions of the Old World and all of which are essentially similar in appearance as well as in their feeding and nesting behavior. *OEnanthe oenanthe* is the only one that ranges as far afield as North America; the other two species that are most frequently encountered in Europe are the black-eared wheatear *(OE. hispanica)*, which ranges from the Iberian peninsula as far east as Iran, and the black wheatear *(OE. leucura)*, which is also found in Southern Europe and North Africa.

The wheatears and their relatives are collectively known as chats. The plumage of the whinchat *(Saxicola rubetra)* and the stonechat *(S. torquata)* is more colorful than that of the wheatears, and both are found in grassy meadows interspersed with bushes and shrubs. The stonechat spends the winter on the savannahs of Central and East Africa whereas the whinchat ventures no further south than the Mediterranean Basin. J.-P. R.

WHELK *Buccinum undulatum* 469

The whelk is a seasnail with an elongated conical shell that may attain a length of 12 cm, though 8 or 9 cm is closer to the average. The shell is built up of 7 or 8 convex courses, somewhat suggestive of a spiral ramp or an ancient Mesopotamian ziggurat, with a series of wavelike (=*undatus)* ridges running along its length. It is covered with a chitinous membrane called the periostracum, which is covered in turn with tiny, backward-pointing protuberances (papillae). The shell's anterior edge is notched to accommodate the whelk's breathing tube, or siphon, a rolled-up extrusion of the mantle that pumps water into the branchial cavity, permitting respiration when the whelk is partially buried in sediment; the siphon, like the rest of the whelk's body, is grayish-white with black spots.

The whelk prefers cool or cold water and is found between the tidal mark and a depth of 200 m on both sides of the North Atlantic along the Arctic coasts of Russia and Siberia, and down into the Bering Sea and the Sea of Okhotsk. Unlike other gastropods (e.g., the winkle and periwinkle) that are quite at home in the intertidal zone and sensibly bury themselves in the sand and batten down their opercula when stranded by an outgoing tide, the whelk becomes totally disoriented, sometimes crawling still further up the beach rather than back down toward the water. It thus perishes in large numbers during spring tides or in areas, such as Canada's Bay of Fundy with its famous tidal bore, where the tidal variation is especially great. Under normal circumstances, the whelk is found in a variety of marine environments—coarse sand, gravel, pebbles mixed with seashells, or wrack grass.

▶ The whelk is exclusively carnivorous and feeds primarily on carrion, provided that decomposition is not too far advanced. It is especially fond of crustaceans and can consume every morsel of flesh inside the shell of a dead crab by piercing the shell around one of the joints and inserting its long, extensible pink trunk. The whelk's mouth is located at the tip of the transparent trunk, so that you can, if you choose, see a fragment of bright-yellow digestive gland sliding along its length when you hold it up to the light.

469
Sub.: Mollusca
Cl.: Gasteropoda
O.: Monotocardia
Sup.f.: Buccinacea
F.: Buccinidae

The whelk scouts out its prey by continually changing the orientation of its siphon as it pumps water in and out of its branchial cavity, which is outfitted with sensitive chemoreceptors. Whelks are very adept at following this molecular trail back to its source; on one occasion, a whelk in an aquarium was separated from a dying sea cucumber (holothurian) by a kind of wooden shutter. The whelk was able to extrude its trunk through a gap in the shutter and across the intervening distance of 18 cm (more than twice the length of its shell) in order to feast on this succulent prize. The whelk also uses this technique to feed on prey that is wedged into a narrow crack, under a rock, or buried in sediment to a depth of several centimeters.

The whelk also attacks living prey, primarily bivalves, that are immobile by nature or immobilized by circumstance. Instead of boring a hole in the shell (the method favored by other carnivorous gastropods such as *Murex* and *Urosalpinx),* the whelk stations itself alongside or perches on top of its prospective victim and waits for it to open its shell; then it inserts the wedged-shaped anterior edge of its own shell into the aperture and extends its trunk to attack the adductor muscles that normally clamp the shell closed. It uses this technique to open scallops, cockles, oysters, mussels, and steamer clams (which may attempt to escape by burrowing into the substrate but which are vulnerable in that they are incapable of closing their shells all the way).

The cockle *(Cardium)* is the whelk's favorite live prey; it can open and eat one in less than an hour, but will be perfectly satisfied with only two a week, even in April, when its appetite appears to be at its heartiest. It continues to hunt all year round, however, unless the water temperature drops below 5°C. Whelks will also attack fish that are caught in a net, leaving nothing but the skin and bones; Danish fishermen report hauling in plaice (a kind of flounder) that were being devoured by as many as 20 whelks simultaneously.

◀ Man has used the whelk as fishbait (particularly in the cod fishery), as food for fish in aquaculture facilities, and as food for himself in several regions of Europe, notably northwest France, Belgium, and the southeast of England. Social history records that 8,000 whelks were consumed in a feast celebrating the installation of an archbishop at Canterbury in 1504, and the whelk stall is a traditional fixture at popular festivities and seaside resorts in Flanders and Great Britain. Whelks are sometimes caught with dragnets but more frequently in traps, something like crab or lobster pots, that are usually baited with crabs.

The cod, the wolffish *(Anarichas lupus),* the cat shark (genus *Scyliorhinus),* and certain rays are the whelk's most important aquatic predators, several of which are adroit enough to eat the whelk's body without swallowing the shell. The hermit crab *Cancer pagarus,* which specializes in gastropods and other creatures protected by a calcified shell, can crack even the toughest whelk in its powerful pincers; crabs of the genus *Telmessus* feed on whelks in the North Pacific. Lobsters eat whelks, and the spiny lobster *Palinurus elephas* breaks off the projecting edge of the whelk's shell in its powerful mandibles and eats its flesh in much the same determined manner as a child finishing off an ice cream cone—though a lobster may take as long as 20 minutes to eat half of a large whelk. The whelk is also attacked by starfish *(Marthasterias glacialis* and *Asterias rubens),* but it may be able to defend itself against slow-moving predators of this kind by digging in with its fleshy foot and abruptly spinning its shell around, thereby propelling itself for some distance and perhaps also driving away the aggressor by rebounding against it smartly with its shell.

Whelks of the genus *Buccinum* are widely distributed throughout the temperate zones of both hemispheres as well as the Arctic and Antarctic Oceans; whelks of the genera *Colus* and *Neptunea* are prevalent in Northern European waters and similar in their habits, though they lack the free-swimming larval stage that is characteristic of the genus *Buccinum. Neptunea antiqua,* for instance, sometimes turns up on English whelk stalls and in fishmarkets, and the foot of the large *N. polycostata* of southern Japan is a prime ingredient in sushi. In addition, the empty shell of *N. antiqua* was formerly used as a lamp on the Faeroes, Orkneys, and Shetlands when fueled with seabird's fat; the wick was inserted into the siphon notch in the shell.

As shallow-water carrion-eaters, the Buccinidae are in direct competition with many fish and crustacean species, notably hermit crabs of the genus *Pagarus.* In northern waters, incidentally, the whelk's is the only shell that can accommodate the larger hermits, and on a sandy or muddy bottom, it may also provide the only solid surface for an anemone to adhere to. The original tenant may still be in possession or its empty shell may have been reoccupied by a hermit crab. J.V.

WHIRLIGIG BEETLE *Gyrinus substriatus* 470

470
Sub.: Arthropoda
Cl.: Insecta
S.cl.: Pterygota
O.: Coleoptera
S.o.: Adephaga
F.: Gyrinidae

The adult of this aquatic beetle measures between 5 and 7 mm. Like the other Gyrinidae, its body is ovoid and dorso-ventrally flattened, black on top with metallic highlights, black on the bottom with rust-colored spots; it has two very short antennae and two pair of eyes (a characteristic that is unique to the Coleoptera). Life expectancy for the adult is no more than a year. The whitish, highly elongated larva measures between 18 and 20 mm bythe time its development is complete; its head is small and flat, its mouth has not yet developed, and, as with the larva of the great diving beetle *(Dytiscus marginalis),* it ingests its food through a tube that runs the length of its hooklike mandibles. Each of its abdominal segments is equipped with two ciliated appendages, as long as its swimming legs; these are tracheo-branchia, which enable the larva to breathe underwater and are also of some use in swimming.

This particular species is common in Europe and is also found in North Africa and Western Asia, generally in stagnant water, though small colonies can often be observed gyrating about the surface of a sluggish stream or an irrigation ditch.

▶ The forelegs of the adult are fairly long and tipped with powerful claws, and are used for seizing and gripping prey; in addition, the male has an adhesive "heel" on each of its tarsi. The second and third pairs of legs are very short and flattened, and used exclusively for swimming. The whirligig is able to swim underwater for fairly long periods, and sometimes does so to escape from preadtors, but is normally to be found sculling about the surface, sometimes very rapidly, describing a series of eccentric loops and spirals for which the whirligigs (Gyridae) are named. The whirligig is active during the day, particularly so in good weather, and is an active predator and scavenger, feeding on small insects that it traps with its long prehensile forelegs, reduces to small fragments with its grinding mouthparts, and then conveys to its mouth.

It also feeds on flying insects, dead or alive, that have fallen onto the surface, including flies and mosquitoes, butterflies and moths, hellgrammites, dragonflies, damselflies, etc. Its normal prey consists primarily of mosquitoes (Culicidae), chironomids, blackflies (Simuliidae), and other small flies that are found in great numbers along the banks of ponds and streams (Empididae, Psychodidae, and others). After a very brief period of development, the whirligig larva takes up residence in deeper water, where it probably feeds on chironomid and mosquito larvae for the most part; homophagia (cannibalism) is probably quite prevalent as well. As noted earlier, it catches, externally digests, and finally ingests its prey with the aid of its hollow pincerlike mandibles.

◀ Both the larvae and adult whirligigs are eaten by fish. The larvae, for all their ferocity, are highly vulnerable to predators and are frequently devoured by other aquatic larvae, including diving beetles, hellgrammites, dragonflies, as well as a variety of others.

The family Gyridae comprises almost 700 species,ofworldwidedistribution

J.-L. D.

WHITE-LIPPED TREE VIPER
Trimeresurus albolabirs **471**

471
Sub.: Vertebrata
Cl.: Mammalia
O.: Carnivora
F.: Procyonidae

The white-lipped tree viper is a thick-bodied snake, covered with carinate (ridged or keel-shaped) scales; its head is broad and clearly distinct from its neck. It generally grows to a length of 70 cm, occasionally 1 m. This species is arboreal and is found in dense forests as well as in relatively open terrain (notably in hedges bordering gardens and rice paddies and in shrubs and bushes of all kinds) throughout Southeast Asia, from the Himalayas to Indonesia. Collectively the pit vipers (including the rattlesnakes) comprise the subfamily Crotalinae of the family Viperidae, and the arboreal "lance-headed snakes" of the genera *Bothrops* and *Trimeresurus* are also known as fer-de-lances after the most notorious member of this group, the highly venomous *Bothrops atrox* of South America.

▶ The dentition of the white-lipped tree viper, and indeed of all vipers, is of the type called solenoglyphic, which means that its poisonous fangs, normally pointed toward the rear, are tilted upright as the viper is about to strike by means of a pivoting motion of the maxilla. *Trimeresurus* seems to feed rather opportunistically on small lizards and salientians (frogs and toads); it is nocturnal, and hunts either by concealing itself in the foliage and attacking its prey from ambush or by flushing out and pursuing small animals along the ground. It detects potential prey animals by sight, odor (by means of olfactory organs located on the tongue and in Jacobson's organ), and thermoreception (the pit that gives these reptiles their name is actually an indentation between the eye and the nostril that is sensitive to infrared radiation).

Once the prey has been located, *Trimeresurus* anchors itself firmly by wrapping its prehensile tail around a branch, then it slowly moves closer, the forepart of its body compressed into an S-curve, darting its tongue to distract the prey, and then abruptly straightening out its body to seize the animal in its jaws. The prey, dangling from the viper's jaw in empty air, naturally begins to struggle, which only serves to accelerate the effects of the viper's venom; the venom is hemolytic—that is, it breaks down red blood corpuscles—and otherwise rather slow-acting. As soon as the prey animal is dead, the viper starts to swallow it, beginning, if feasible, at the site where the venom was injected, otherwise head first.

◀ The predators of the white-lipped tree viper probably include birds of prey as well as other snakes. Newborn vipers are quite tiny and are preyed on by many creatures that normally feed only on invertebrates: small birds, batrachians, and even centipedes and other large arthropods. The adult *Trimeresurus* is virtually invisible so long as it remains motionless against a background of green leaves. When threatened, it strikes out in the direction of the intruder, mouth open, even before the latter has come within range of its fangs. Its bite can be very painful but is not usually fatal to humans.

A number of other species of "tree viper" are found in Southeast Asia; those that are basically green in color were formally grouped into a single species, *Trimeresurus gramineus.* Others have brown or gray as the background color with spots that may be more or less pronounced, notably *T. purpureomaculatus,* which is primarily found in rocky terrain. The habu *(T. flavoviridis)* of the Ryukyu Islands is essentially terrestrial and may attain a length of 1.6 m; its venom is much more potent than that of its arboreal congeners. Another insular species, *T. okinawensis,* is noted for its habit of restraining its prey by wrapping a single body coil around it, though it is does not kill by constriction. Wagler's palm viper *(T. wagleri)* is maintained as a pensioner at certain religious establishments in Malaysia—notably the temple of the sorcerer-demigod Zho Sukung in Penang—and is accordingly known as the temple viper.

Tree vipers are represented in Africa by the genus *Atheris,* fairly small snakes in which the carinate formation of the scales is quite pronounced. The bush viper *Atheris squamigera* is between 50 and 70 cm long, ranges from Ghana to Kenya, and feeds primarily on tree frogs. In the tropical rain forests of Central and South America, a comparable niche is occupied by certain arboreal crotalids, notably the eyelash viper *(Bothrops schlegelli).* These arboreal vipers sometimes find their way into shipments of agricultural produce originating in tropical countries and are thus familiarly known as "banana snakes." D.H.

WHITE-NOSED COATI *Nasua narica* 472

472
Sub.: Vertebrata
C.: Mammalia
O.: Carnivora
F.: Procyonidae

This rather comical-looking member of the raccoon family has short, inconspicuous ears and an elongated muzzle with a flexible, trunklike snout; its coat is thick and may be dark brown, yellowish-gray, or light gray in color. It measures between 75 and 135 cm (plus an additional 36–68 cm for its stiff, bushy tail) and weighs between 3 and 6 kg; maximum longevity is probably 14 years. Collectively the 4 coati, or coatimundi, species range from the Southwestern U.S. down through Mesoamerica, Columbia and Venezuela, the Andes, and Argentina; they are found in a variety of habitats including jungle and highland forest, grasslands, mountains, and even on the fringes of the desert.

▶ The coati spends its night stretched out on the branch of a tree and its days in search of food; coatis live and forage in small bands, typically covering 2 or 3 km a day in search of food and ranging over a territory of about 40 hectares or so. The band makes no attempt to defend its territory against interlopers, and although the territories of different bands may overlap to some extent, instances of intraspecific aggression are virtually unknown.

The coati feeds (in order of preference)

on insects and worms, lizards, mice and other small mammals, and birds, as well as on fruits and tubers. The coati band explores its territory calmly and systematically, with some of its members pulling down dead branches in an attempt to dislodge rodents and lizards from their holes, or digging up the topsoil layers with their snouts and powerful forelegs in search of insects, while others set off in pursuit of any small animals that might have been flushed from cover. The coati will also climb trees to eat tropical fruit and always holds its head back while doing so, presumably so as not to miss out on any of the juice.

◀ The coati frequently falls prey to the large hunting cats of Mesoamerica (notably the jaguar and the puma) as well as anacondas and other large constrictors. When confronted with a predator, the coati takes refuge in the trees or tries to flee along the ground in a series of short, ungainly bounds. The coati's foraging technique is clearly destructive to crops, and the coati is sometimes persecuted by man as well as hunted for its flesh, considered to be something of a delicacy; however, since the coati lives for the most part in sparsely settled areas, its survival does not seem problematical at the moment.

There are 3 other coati species—Nelson's coati *(Nasua nelsoni)*, the South American coati *(Nasua nasua)*, and the mountain coati *(Nasuella olivacea)* of the Andes. P.A.

WHITE PELICAN *Pelecanus onocrotalus* 473

This Old World pelican (sometimes called the eastern white pelican to distinguish it from a closely related New World species) measures between 140 and 175 cm, has a wingspread of 270 cm, and weighs about 10 kg. It reaches sexual maturity at the age of 3 or 4; life expectancy in the wild is unknown, though pelicans are noted for their longevity. This particular specie is quite demanding in its choice of habitat—it will nest only in inaccessible reedbeds surrounded by vast, undisturbed marshlands—with the result that only a few small colonies remain in Europe, consisting of about a thousand pairs altogether. Most of them are found in protected habitats in the estuaries of the Danube or the Volga, or along the marshy shores of several remote mountain lakes in northern Greece. The white pelican spends the winter in the Aegean and Eastern Mediterranean or on the east coast or in the interior of Africa, sometimes as far south as the Cape of Good Hope.

▶ The white pelican's fishing grounds are often quite far away from its nesting site, and, all appearances to the contrary, this species is a powerful flier as well as an excellent swimmer. White pelicans fly in formation between their fishing and nesting grounds, sometimes (as in the case of the remnant populations in northern Greece) negotiating mountain passes as high as 1000 m above sea level. In Europe the white pelican feeds on carp, bream, perch, and pike, in Africa on tilapia and other local species—exclusively on fish in either case. Average daily consumption is about 1 kg, and an adult pelican can carry a load of up to 3 kg in its capacious throat pouch; a carp weighing 2 kg is the largest individual catch on record.

Once they have arrived at their fishing grounds, a formation of white pelicans continues to behave cooperatively, arranging themselves in a semicircle and slowly driving their quarry into the shallows, using their bills to strain the fish out of the water. The sight of pelican chicks sticking their heads into their parents' gaping maws gave rise to the medieval legend that pelicans feed their chicks on their own blood when more conventional food is unavailable.

◀ The white pelican's size protects it from predators; it is often found in association with a related species, the Dalmatian pelican *(Pelecanus crispus)*, though the two are not direct competitors, since the Dalmatian pelican is a solitary hunter that dives for fish in deeper water. This species is also very wary of human beings, and will not remain for long in any area where its nesting site is likely to be disturbed. The range of the Dalmatian pelican extends into Central Asia and as far as Northern China, but in Europe, the disappearance of suitable habitat for both species has been greatly accelerated in recent years by land reclamation and the drainage of marshlands.

There are 8 pelican species worldwide. The brown pelican *(P. occidentalis)*, found

473
Sub.: Vertebrata
Cl.: Aves
O.: Pelecaniformes
F.: Pelecanidae

on both the Atalantic and Pacific coasts of the New World, is noted for its spectacular fishing technique—with its wings tucked back and its body fully extended, it plummets into the water from a height of up to 20 m, breaking the surface with such an impact that fish are stunned by the concussion. J.-F. T.

474
Sub.: Vertebrata
Cl.: Aves
O.: Ciconiiformes
F.: Ciconiidae

WHITE STORK *Ciconia ciconia* 474

The white stork has a number of picturesque folkloric associations (builds its nests on the chimneys of thatched cottages, brings good luck as well as newborn human infants) but is now very rare in Western Europe, though still common in Spain and Southeastern Europe. It prefers a warm, dry climate and a flat, sparsely wooded terrain—wet meadows, shallow ponds (unless overgrown with aquatic vegetation), cultivated fields with irrigation ditches. The stork populations of Europe and the Mediterranean basin spend the winter in subsaharan Africa; the white stork is a strong flier, but European storks prefer to "ford" the Mediterranean at the Bosporus or Gilbraltar.

The white stork measures between 100 and 115 cm with a wingspread of 155–165 cm and weighs between 2.6 and 4.4 kg; it begins to nest between the ages of 3 and 5. One individual is known to have survived to the age of 26, undoubtedly a record for this species. The stork is a gergarious bird, and nesting pairs frequently band together in small colonies where the terrain permits. Even though they may have good reason to be suspicious of man, storks continue to build their nests on the roofs and elsewhere in the vicinity of human dwellings, even in the midst of villages or small towns.

▶ The stork seeks its prey both in the water and on dry land, striding along with its beak pointed downward, darting out with its long neck to catch insects and small rodents, frogs, reptiles, and earthworms (on land) as well as frogs, fish, aquatic insects, and grass-snakes (in muddy water). In its African winter home, storks are cooperative hunters, feeding primarily on a species of migratory locusts *(Schistocera peregrina)*; individual storks have been known to devour as many as 30 of these insects per minute. The white stork feeds its nestlings with a regurgitated pulp composed of insects, worms, and other invertebrates until they grow big enough to cope with more substantial prey— frogs, small fish, etc.

◀ The white stork implacably defends its nesting area against intruders of its own species. It occupies a lofty position in the food chain, since it has no predators. It lacks any competitors to speak of, and though numerous other species depend on the same food sources, including raptors, wading birds, and passerines, they use different hunting strategies, hunt at different times of day or in slightly different terrains. The stork has been held in reverence by virtually all cultures; in addition to the familiar notions that storks bring babies and/or good fortune to the house on which they build their nest, it was also believed that it was the storks' annual migration that perpetuated the cycle of the season, rather than the other way around.

Unlike owls, hawks, and other large raptors, the white stork can hardly said to have benefited from the fact that the widespread popular myths and superstitions concerning it have largely lost their hold over the human imagination in recent years. Traditional taboos against killing a white stork in East Africa were formerly more effective in the service of conservation than today's protective regulations; the stork is a conspicuous, slow-moving bird, and the hunters are now equipped with guns. Lebanon—which lies squarely on the white stork's migratory route, both outward and homeward bound—is of course one area where high-powered rifles have become especially plentiful in recent years. In Western Europe, the stork population has declined to such an extent that current conservation efforts are more or less cancelled out by the annual depredations of gun-toting vandals; in several areas, notably in Alsace, an attempt has been made to reacclimate individuals bred in captivity to life in the wild.

The family Ciconidae consist of 18 different species, found in Europe, Asia, Africa, America, and Australia. The true storks comprise a numerous subfamily, the more prominent members of which include the European black stork *(Ciconia nigra)*, which nests in wooded areas and is slightly smaller and rarer than the white stork, as well as Abdim's stork *(Sphenorhyncus abdimii)* of Africa, the maguari *(Exenura galatea)* of South America, and the open-billed storks of Africa and Southern Asia (genus *Anastomus).*

Classified in separate taxa within the family Ciconidae are the wood storks, saddlebills, marabous, and adjutant storks. The wood stork subfamily consists of only two species, the American wood stork *(Mycteria americana),* which ranges between the Southeastern U.S. and Tierra del Fuego, and the Indian painted stork *(Ibis leucocephalus).* The West African saddlebill *Ephippiorhyncus senegalensis* is the best known of several African saddlebill species, which are sometimes called jabirus (a name that is also rather loosely and confusingly applied to the maguari stork of South America). Finally, we have the carrion-eating marabous and adjutant storks of Africa and India, of which the African marabou *Leptoptilos crumeniferus* is probably the best known. J.-F. T.

WHITETHROAT *Sylvia communis* 475

475
Sub.: Vertebrata
Cl.: Aves
O.: Passeriformes
F.: Muscicapidae

This small migratory bird weighs between 11 and 20 g; it has a wingspread of 19–22 cm and an overall length of 13.5–14.5 cm. It ranges from Western Europe to eastern Siberia as well as throughout the arid, subdesert regions of Africa and Asia Minor as far east as Afghanistan. In Europe, it is commonly found in fields and uncultivated grasslands, especially in the vicinity of brushwood or thick shrubbery, up to an altitude of 1000–1500 m. It arrives in Western Europe in April, and in October sets off on its return migration to its winter home in the Sahel.

▶ Nervous, frenetic, and very shy, the whitethroat is difficult to observe, since it spends much of its time in bushes and thick brushwood, and is adept at flying through tight, obstructed passages; even when flying between bushes, in the open air, its flightpath remains erratic. It feeds on butterflies and caterpillars, beetles, ants, aphids, and other insects, snails, several varieties of spider—all of which prey it catches with its beak—as well as gooseberries, raspberries, and other fruit.

◀ There may be rivalry with other members of its own species during the nesting period, but otherwise the whitethroat has no particular competitors. Eleonora's falcon is its principal predator, though both the nestlings and the adults are probably taken from time to time by other diurnal raptors or terrestrial carnivores. In 1972, Moreau estimated that the migratory flocks that crossed the Mediterranean every autumn still numbered about 120 million individuals, but the whitethroat is one of several European birds—along with the woodchat shrike, the butcher-bird, redstart, and others—whose populations have sharply declined over the past few years. The reasons for this are not well understood, but several explanations have been proposed, including pesticide residues, global climatic shifts, and the disappearance or destruction of habitat in Europe and in the Sahel (perhaps including the widespread use of brushwood for cooking fires, which has caused extensive deforestation, even desertification, in many parts of Africa). J.-F.T.

WHITE-THROATED CAPUCHIN *Cebus capucinus* 476

476
Sub.: Vertebrata
Cl.: Mammalia
O.: Primates
F.: Cebidae

This medium-sized New World monkey weighs about 3 kg; its fur is black, though it has a white mask and a patch of white fur on top of its head, reminiscent of a monk's cowl ("Capuchin" being the name of an order of Franciscan friars). The capuchin has a prehensile tail, and its nasal septa are detached, so that the capuchin's nostrils are set very wide apart.

This species is commonly found in the forests of tropical America; it lives in bands of about 15 individuals, which occupy a territory of about 90 ha and take part in communal foraging expeditions of several km under the direction of a dominant male. The members of the band remain in vocal contact at all times, and are occasionally called upon to join forces to repel an incursion by a rival band of capuchins, though this happens rarely, since the territory occupied by each band is quite extensive (ø 90 ha).

▶ The capuchin band finds most of its food in the trees. The leaves and green shoots of trees provide them with most of their vegetable food (about 60 percent of its total intake), and the animal protein in their diet is chiefly obtained from xylophagous (wood-eating) insects and insect larvae. The capuchin also feeds on large orthopterans and other ground-dwelling insects, primarily those that are found among the forest litter. It tears off pieces of rotten wood with its paws to dislodge burrowing insects in the trees, and sometimes nibbles around the edges of inedible leaves in order to flush out the insect larvae that are hidden away in the center. It eats fruit very slowly and deliberately, like a cow chewing on a cud, in order to absorb all the nutrients, then spits out the masticated pulp.

◀ The ocelot is the most frequent predator of the capuchin, which is particularly at risk as soon as it touches down on the ground. The cohesion of the group is essential to the survival of the indiviudal capuchin, both in gathering food and in defending itself against predators. The long canines of the male may serve as an effective deterrent in the latter case, and even if not, the band generally throws itself into a kind of collective frenzy that is sufficiently violent and energetic to discourage most predators. In the forest canopy, the males have often been observed to deliberately detach dead branches from the trunk, thus causing them to fall for a considerable distance in order to frighten off a potential predator on a lower branch or on the ground directly underneath.

P.A.

477
Sub.: Vertebrata
Cl.: Reptilia
O.: Squamata
S.o.: Sauria
F.: Amphisbanidae

WIEGMANN'S WORM LIZARD
Trogonophis wiegmanni 477

Amphisbaenians are (almost always) legless burrowing lizards, well adapted to their subterranean mode of life and rarely encountered by man. Their collective relationship to other lizards is not very well understood; sometimes they are even classed as a separate suborder, Amphisbaenia, and thus upgraded to the same taxonomic level as Sauria (all other lizards) and Serpentes (snakes).

According to the classical bestiaries, the amphisbaena was a serpent with 2 heads, one at either end of its body. With no legs or external ears and only vestigial eyes, Wiegmann's worm lizard looks remarkably like a headless snake with 2 tails similarly situated. Its body is ringed and cylindrical, greenish-gray in color with rectangular black spots arranged in a checkerboard pattern. Its size varies between 12 and 26 cm, its weight between 5 and 10 g. *Trogonophis wiegmanni* is found in western Tunisia, northern Algeria (from the seacoast to the edges of the Sahara), and most of Morocco (north of the Atlas Mountains and the Plain of Sous); its preferred habitat includes forests of cork-oaks and "miracle berry" bushes *(Argania sideroxylon)*, grassy steppes, flat, rocky terrain, and sandy beaches, even if devoid of vegetation.

▶ *Trogonophis* feeds almost exclusively on ants and termites, perhaps supplemented by an occasional insect larva or mealworm (the larva of the tenebrionid beetles). The amphisbaenians are all very adept at detecting vibrations transmitted through the soil; once it has gotten an approximate fix by this method, *Trogonophis* uses its sense of smell to determine the insect's precise location. It is basically nocturnal, active from dusk onward, and frequently spends the day curled up under a flat stone that will transmit the sun's heat, sometimes emerging to warm itself in the direct sunlight, though only in the spring, when its heat is less intense. On these occasions, it often keeps its head or its tail buried in the sand, since, like its mythical namesake, it can slither off just as quickly in either direction.

Trogonophis digs an extensive network of shallow tunnels in sandy or dry and crumbly soil; its skin is very loosely attached, and it excavates its tunnels by rotating its body inside its skin like a drill bit and thrusting its thick-skinned, conical head through the soil in one direction and conveying the loosened soil particles in the other. It sometimes ventures up to the surface at night, particularly after a rain has flooded its tunnels. *Trogonophis* undergoes a period of more or less complete hibernation; its activity level is also greatly reduced during the months of July and August.

◀ *Trogonophis* is most frequently subject to predation by colubrids (members of a large family of nonvenomous snakes) like *Malpolon monspessulanus* and *Macroproto-*

don, and during its rare appearances on the surface, it may also attract the attention of hungry toads, rollers, magpies, shrikes, falcons, and other nocturnal raptors. If unable to make its escape, it adopts a curious defensive posture—rolling itself into a tight coil with its head and tail firmly tucked up against its body, which apparently makes it more difficult for it to be swallowed whole by its reptilian predators.

Trogonophis actively avoids contact with other members of its own species and may also find itself in direct competition with the other North African amphisbaenian, *Blanus cinerus,* in certain parts of its range. *Blanus* needs more moisture than *Trogonophis,* however, and prefers to dig its tunnels in firmer, more compact soil.

There are 130 amphisbaenian species distributed throughout the tropical and subtropical regions of the globe. Most are insectivores, but the largest species (up to 80 cm in length) also feed on reptiles and small mammals that they catch on the surface and drag down into their tunnels. When the prey animal is too large to be swallowed whole, the lizard clamps down its teeth and rotates its body in order to twist off a more manageable portion.

D.H.

WINTER WREN *Troglodytes troglodytes* 478

This little songbird measures 9–10 cm in length, with a wingspread of 14–15 cm, and weighs 14 g. It is found in dense, low-lying shrubs and thickets on lowland plains as well as in the mountains, woodlands, along the banks of watercourses, and along the coast. The more than 30 recognized subspecies of *Troglodytes troglodytes* are widely distributed throughout the North Temperate regions; the most wide-ranging of the Eurasian subspecies is found throughout most of Europe (excluding the northernmost parts of Scandinavia and European Russia) as well as in Central Asia.

The winter wren is a solitary bird, nervous and restless in its movements and rarely remaining stationary for long. It occasionally perches on an upper branch, the top of a fencepost, or a dead tree in order to proclaim the boundaries of its territory in song. Its nest is made of moss and is capacious, well-camouflaged, and generally located close to the ground. The Northern European populations spend the winter in Southern Europe; more southerly populations are sedentary, though small bands of winter wrens may sleep together in an abandoned nesting cavity in order to conserve body heat during the winter.

▶ Excluding occasional short, swift, straight-line flights, the winter wren patrols its territory on foot, continually searching for insects and spiders in the interior of shrubs and thickets, woodpiles, and hollow trees. It feeds primarily on butterflies and caterpillars, flies, mosquitoes, and their larvae, and true bugs, and catches its prey in its beak.

◀ Since the winter wren is small, inconspicuous, and rarely seen in the air or on open ground, it is unlikely to attract the attention of the sparrowhawk or other avian predators. The nestlings may occasionally fall prey to domestic cats and mustelids and—since this is one of the species whose nests are most commonly parasitized by the gray cuckoo—are perhaps more frequently ejected by their larger, more aggressive nestmates. The song of the winter wren is generally sufficient to discourage trespassers, and this species has no particular competitors, nor is it greatly affected by man or his activities.

The wren family (Troglodytidae) comprises about 60 different species, though the winter wren is the only one that is found in the Old World. The common house wren of the New World, which ranges from southern Canada down to the Falklands, is *Troglodytes aëdon.* J.-P. R.

478
Sub.: Vertebrata
Cl.: Aves
O.: Passeriformes
F.: Troglodytidae

WOLF SPIDER *Pardosa pullata* 479

479
Sub.: Arthropoda
Cl.: Arachnida
O.: Araneida
F.: Lycosidae

This small mycosid, which only measures 4–10 mm, is found all over the world, from the Equator to the polar regions; in the Temperate Zones, it is relatively common up to an altitude of 4000 m and in a variety of different habitats—meadows, forests, arid as well as humid coastal regions and the banks of watercourses. Two generations are hatched out every year; the female carries the egg cases around with it, still attached to its spinnerets, though the hatchlings, as is invariably true of spiders, are entirely on their own.

▶ The first-stage larvae, known as pulli, feed on gnats, midges, springtails (collembolans), and other very small insects. The adults and second-stage larvae are large enough to catch flies and mosquitoes as well as the emerging nymphs of other species, such as craneflies, that spend their larval stages underground. The wolf spider has excellent eyesight and will attack any moving creature small enough to overpower, though it appears to be drawn primarily by the motion of the insect, rather than by its shape, size, or any other aspect of its appearance. If food is in short supply, the larvae are able to go without food for long periods, though they have may to pass through a few additional molts later on, and thus postpone their arrival into adulthood, in order to to make up this dietary deficiency.

◀ Other spiders, pompilid wasps, and, in humid areas, frogs and toads are the principal predators of the wolf spider; homophagia is also not unknown. The adult *Pardosa pullata* is sometimes, though rarely, attacked by ichneumonids and other external parasites (cf. *Zatypota percontatorius*, below); the eggs are frequently consumed by various ectoparasitic fly and wasp larvae.

The genus *Pardosa* comprises several hundred species worldwide, of which about 50 are found in Western Europe.

J.-F. C.

480
Sub.: Vertebrata
Cl.: Mammalia
O.: Carnivora
F.: Mustelidae

WOLVERINE *Gulo gulo* 480

The Eurasian wolverine, or glutton, is a strong, stocky carnivore, the largest member of the weasel family; it is found in forests, marshes, and rocky terrain from northern Scandinavia (and formerly in northern Germany, Poland, and the Baltic states) to the Pacific island of Sakhalin. The North American wolverine, primarily found in Canada, is treated as a separate species *(Gulo luscus)* in some classifications, a subspecies in others. The Eurasian wolverine measures 70–82.5 cm (excluding its short tail, 12.5–15 cm) and weighs between 15 and 35 kg.

▶ The wolverine is a voracious predator, if perhaps not quite as insatiable as legend has made it out to be—forcing itself into the narrow space between two treetrunks, for example, and vomiting up the remains of a previous gluttonous meal in order to make room for a second. The wolverine is active both by day and by night and usually hunts from ambush, stationing itself on an overhanging branch and falling upon its prey from above. A wolverine will even attack a reindeer (most often a domesticated reindeer) in this fashion, setting off on a wild, galloping ride of 100 to 150 m, though it is sometimes thrown before it is able to bring down its prey.

It will also feed on lemmings and hares (it prefers the latter), bird's eggs, wasps' nests, bilberries, and carrion—sometimes making off with the severed head of a cow or other large animal and secreting in one of its caches in the trees, as much as 8 m from the ground. Hunters heap up cairns of boulders over their kills to protect them from wolverines, though this is not always effective, since the wolverine is capable of shifting very large stones if it is worth its while to do so. Wolverines, like bears, are also in the habit of breaking into isolated cabins and climbers' huts where provisions have been stored.

J.-J. B.

481
Sub.: Vertebrata
Cl.: Aves
O.: Charadriiformes
F.: Scolopacidae

WOODCOCK *Scolopax rusticola* 481

The woodcock is a medium-sized European gamebird, about 30 to 36 cm long with a wingspread of 60 to 6 cm and a weight of 200 to 400 g. It lives in wooded terrain (both decidiuous and coniferous forests) interspersed with wet meadows and ditches throughout the temperate regions of the Old World. It is a solitary bird, active at twilight, inconspicuous, and rarely encountered by humans (unless they are accompanied by dogs). Its plumage is mottled with various earth colors that provide an effective camouflage while nesting or roosting on the ground. The woodcock's nocturnal migrations have no fixed seasonal destination but occur on an irregular basis in response to local climatic variations

▶ The woodcock's long flexible beak is richly innervated and highly sensitive to the movements of earthworms and insect larvae that it probes for in the soil or the leafmold on the forest floor. But when the woodcock has its beak buried in the ground, literally inhaling earthworms, it can warily survey the skies overhead or the surrounding underbrush, since its eyes are placed all the way around at the back of its head. It may occasionally feed on small insects, spiders, and woodlice as well as earthworms; it is the only member of the order Charadriformes (which includes gulls, plovers, sandpipers, and other shorebirds) to have adapted to a woodland environment.

◀ The woodcock's principal defenses against predators are prudence, protective coloration, and the rapidity with which it takes flight from cover. It has no scent, which protects it from most carnivorous mammals, though the fact that it makes its nest on the ground still leaves the woodcock and its nestlings vulnerable to predation by wild boars and martens. The goshawk is another woodland hunter that may occasionally prey on the woodcock.

The woodcock is also hunted avidly by man, either on the ground with dogs or on the wing when the birds leave the shelter of the forest at twilight or during the males' nuptial flight in the spring— these latter two practices are technically prohibited in certain areas. According to statistics published by the French national fish and wildlife office (Bureau National de la Chasse), 1.5 million woodcock and 1.4 million snipe were shot by hunters during the 1974– 75 season in that country alone. The woodcock is remarkably prolific but not sufficiently so, it appears, to make good on such enormous losses year after year; collisions with high-tension lines, lighthouses, and other manmade obstacles during their nocturnal migrations are another notable cause of woodcock mortality.

The American woodcock *(Philohela minor)* is smaller than its European relative and differs in several other morphological details but is similar in its habits and also greatly prized as a gamebird. The family Scolopicidae includes 6 species of woodcock and 12 of snipe, a gamebird that lives in freshwater marshes in both the Old and New World, as well as more than 60 species of curlews, sandpipers, and other shorebirds, ranging in size from just a few cm to over 50 cm long. J.-F.T.

WOODPECKER FINCH *Cactospiza pallida* 482

482
Sub.: Vertebrata
Cl.: Aves
O.: Passeriformes
F.: Emberizidae

This species is found only on the arid uplands of the islands of Isabella, Fernandina, and Santa Cruz in the Galapagos. It is about the size of an ordinary finch, though somewhat heavily built and not a very skillful flier; its granivorous (seed-eating) relatives of the genus *Geospiza* often assemble in substantial flocks at the end of the breeding season, but the woodpecker finch is generally solitary.

▶ Woodpecker finches are omnivorous, feeding on insects for the most part as well as on fruit and mangrove leaves. The woodpecker finch is so called because it hunts for xylophagous (wood-eating) insects in rotten wood and under the bark of trees, and like the woodpecker, it locates its prey primarily with the aid of its sense of hearing. Instead of drilling with its beak, however, it uses the broken tip of a cactus spine as a combination surgical probe and bayonet, forcing it through the bark of tree or into a cavity in the trunk at the precise spot where a grub or borer is likely to be found; if all goes according to plan, the prey animal is impaled on the spine and immediately retrieved by the woodpecker finch.

This is one of roughly 4 or 5 known instances of tool-using behavior in the animal world, and *Cactospiza* has become a minor celebrity in the fields of ethology (animal behavior studies) and evolutionary biology. The basic skills involved are instinctive, though a period of apprenticeship is still required before the young woodpecker finches have thoroughly mastered their trade. The nestlings are fed exclusively on insects and insect larvae, and the woodpecker finch will not even attempt to make a nest and raise a brood in a very dry spring, when the supply of insects is certain to be insufficient.

◀ Potential predators of the woodpecker finch include the Galapagos hawk *(Buteo galapagoënsis),* the short-eared owl *(Asio flammeus),* the barn owl, and a species of viper *(Drominicus dorsalis).* All the Galapagos finches treat these predators with an innate fear, which varies somewhat in intensity from one island or one species to the next, but they remain unafraid of cats, humans, and other relative newcomers to the islands. A single subspecies of the sharp-beaked ground finch *(Geospiza difficilis)* has been exterminated by feral cats on Indefatigable Island, though it appears to have been the only such casualty so far.

Thirteen species of Galapagos finches, together with numerous subspecies, are distributed among the 24 different islands of the archipelago. During his visit to the Galapagos in 1835, Darwin was particularly troubled by the idea that their strikingly divergent physical characteristics (notably the size and shape of their beaks as well as their feeding habits) could best be explained by the hypothesis that they had developed in isolation from one another. At that time, however, Darwin was still a staunch creationist, who believed, in accordance with the current scientific orthodoxy, that the form of all animal species was fixed and unalterable, and thus could not be said to have "developed" at all. (The bayonet-wielding behavior of the woodpecker finch, by the way, had not yet been observed at the time of Darwin's visit to the Galapagos).

Darwin's theory of natural selection, which did not appear in print until almost 25 years later, suggested the mechanism by which physical differentiation among species descended from the same lineage and that shared the same habitat could occur—namely the pressure of competition for the same food source. The beaks of the ground finches *(Geospiza)*, which continued to feed on seeds and grains like their ancestors (probably a type of bunting) on the continent of South America, grew progressively shorter and thicker, for example, so they could eat seeds that were encased in tougher kernels.

The small ground finch *(G. fuliginosa)* occasionally feeds on insects and hunts for ticks on the backs of marine iguanas, and the short-beaked ground finch *(G. difficilis)* has learned to feed discreetly on the blood of the red-footed booby, jabbing its sharp beak into the bird's insensitive feet; it also robs the nests of seabirds. The insectivorous tree finch *(Camarhyncus)* feeds on seeds and grains as well as on insects, foraging on treetrunks or at the tips of branches, somewhat in the manner of a titmouse, or swooping down on ground-dwelling insects from above, in the manner of a shrike.

J.-F. T.

WORM LION *Vermileo degeeri* 483

483
Sub.: Arthropoda
Cl.: Insecta
O.: Diptera
F.: Rhagionidae

The adult insect is a little larger than a housefly, measuring 9–12 mm in length, and is reddish in color with brown stripes on the thorax, yellow legs, and highly iridescent wings. It occurs sporadically throughout Western Europe, and is generally seen on flowers and in hot, sandy areas in June and July. The firm, elastic skin of the larva is gathered into a series of transverse wrinkles and covered with stiff bristles. The body segments are retractile and highly mobile; the body is narrower toward the front, and the head is small and conical. There is a raised pad on the first abdominal sternite (one of the chitinous plates that covers that abdomen) that serves as a sheath for a retractile spine, and the posterior body segment is elaborated into a flattened, upraised pseudopod tipped with a four-pronged burrowing organ studded with numerous spiny projections. The larval stage lasts up to three years, since the larva grows very slowly and often goes without eating for several months during the winter; the free-living nymph remains buried in the sand and is not enclosed in a cocoon.

▶ Little is known of the feeding habits of the adult, though like the other rhagionids (snipe flies), it probably lives primarily on nectar and pollen, supplemented occasionally by smaller insects that it captures insides the bells of flowers. As suggested by its highly specialized morphology, the predatory strategy of the larva is somewhat more complex: it begins by digging a trap—alternately relaxing and then abruptly extending the coils of its springlike body and jabbing the various sharp appendages attached to its abdomen and posterior segment into the sand until it has excavated a conical pit, though not as big in diameter or as deep as the pit of an ant lion. The worm lion stretches out horizontally along the bottom of the pit and waits for insects to set foot on this slippery slope.

If is accurate to say that *Vermileo* uses its posterior segment like a four-pronged gardening fork, then it might also be said that it uses the tapering, elastic forepart of its body like a bullwhip, abruptly extending itself to its full length and coiling around the body of an ant (in France, generally

Aphaenogaster subterraneus or *Lasius brunnea)* or other insect that has stepped into its trap. The physicist and naturalist Réamur noted this behavior as early as 1753: "The worm lion outstretches itself, then violently recoils, thereby ensnaring its prey and dragging it back under ground to be devoured." The larva attacks its prey with its hooked mouthparts, which are jointed so that they can move freely in two different directions (i.e. both vertically and horizontally, an ability that is very unusual among insects), then discards the empty body case by tossing it up out of the pit after it has devoured the liquid contents.

The larva has no known predators, though it seems likely that the adults are sometimes devoured by crab spiders, which conceal themselves among the petals of flowers and feed on insects that come to drink the nectar. R.C.

WORM SNAIL *Vermetus triqueter* 484

The Vermetidae, or worm snails, are an unusual family of sessile gastropod mollusks, and this species, which is quite common in the Mediterranean, cements itself in large numbers to a flat expanse of offshore rock, so that a well established colony becomes inextricably commingled with a variety of other marine flora and fauna—calcium-encrusted seaweeds, sponges, annelids, tubeworms, sipunculids, probosicis worms, barnacles, sea urchins, and others—to form a sort of "wormshell conglomerate." *V. triqueter's* body is entirely enclosed by its planospiral shell (i.e. the shell is coiled in a single plane, in the manner of a planorbid's shell), one edge of which is permanently attached to the substrate.

In other vermetid species, the shell has generally lost its curl, so that it looks very much like the calcified test of a tubeworm. The snail's body, once removed from its shell, is also quite wormlike in appearance as well, and until the around the middle of the 18th century, it was commonly assumed by naturalists that the worm snail was a sessile ploychaete rather than a mollusk, which is the origin of both the common name *worm snail* and the family name Vermetidae. The cephalopodium—formed, as the name suggests, by the fusion of the head and the foot—is the only part of the snail's body that projects outside the shell. Its mouth is located between the two cephalic tentacles, and its eyes are located at the base of each of these tentacles. The snail's internal organs, collectively known as the visceral mass, are firmly attached to the inside of the shell by the columellary muscle.

As with most prosobranchs, the hindpart of the foot secretes a small operculum, which does not entirely obstruct the entrance when *V. triqueter* is fully retracted into its shell, and the sexes are differentiated. Fertilization takes places internally, though indirectly—the spermatozoids are released into the water and then gently wafted into the mantle cavity of the female by a ciliary current. Batches of 10–30 fertilized eggs are encased in protective capsules that are attached to the inside of the female's shell. The larva that emerges from the egg is already equipped with a velum (a sail-like organ of locomotion) and a spiral shell, and after a brief free-swimming period, it attaches itself permanently to the substrate.

484
Sub.: Mollusca
Cl.: Gastropoda
S.cl.: Prosobranchia
O.: Mesogastropoda
F.: Vermetidae

▶ The worm snail fishes for small planktonic organisms by releasing long strands of adhesive mucus into the water. The relatively large glandular organ (the pedal gland) that secretes this mucus strand is located in the interior of the snail's foot, the hindpart of the cephalopodium, which essentially serves no other purpose in this sessile species. In addition to the cephalic tentacles, mentioned above, there is another pair of long tentacles adjacent to the orifice of the pedal gland; the constant vibratory motion of these tentacles is imparted to the mucus strands with which *V. triqueter* ensnares it prey, thus increasing its effectiveness as a sort of one-dimensional plankton net. The prey consists of protozoans and unicellular algae, trocophores (polychaete larvae) as well as the larvae of mollusks and crustaceans, though one of the evolutionary shortcomings of this system is attested by the fact that the larval shells of *V. triqueter* are often found in the digestive tubes of the elder members of its own species.

◀ Apart from these instances of accidental cannibalism, *V. triqueter* has no natural enemies.

The Vermetidae are Mediterranean or tropical species represented by the genera *Vermetus* and *Serpulorbis* that share the following basic characteristics: shells more or less regularly spiral and firmly attached to the substrate by a large part of their surface, operculum reduced in size or absent, well developed pedal gland and tentacles, presence of a slit in the mantle of the female that allows the egg capsules to be removed from the mantle and attached to the inside of the shell.

The closely related Siquillaridae are a group of tropical species that are associated with coral reefs. In contrast to the first group, only a small part of the shell is cemented to the substrate (the reef in this case), and the operculum is well developed, sometimes covered with bristles, though the pedal gland and tentacles are correspondingly reduced in size or absent altogether. Instead of secreting a mucus "plankton net," these species feed in much the same way that clams and other lamellibranchs do, namely by creating a ciliary current that circulates water through their gill filaments. Finally, there is no slit in the female's mantle of the type described above, since the eggs are incubated inside the mantle and never removed. M.C.

485
Sub.: Vertebrata
Cl.: Aves
O.: Piciformes
F.: Picidae

WRYNECK *Jynx torquilla* 485

This Eurasian woodpecker measures 16–18 cm in length, with a wingspread of 27–28 cm, and weighs between 30 and 40 g. It is found in sunlit, sparsely wooded areas (including gardens, clearings and the edges of forests, and hedged fields) on plains and other low-lying terrain throughout most of Europe (excluding the British Isles and Scandivania) and eastward to the Pacific along a fairly narrow corridor extending through Central Russia, Manchuria, and the maritime regions of Siberia.

The wryneck is discreet in its habits and more often heard than seen. It hops agilely from branch to branch and moves along the ground in a series of longer leaps while keeping its tail tucked upward. Its flight pattern is undulating, like the woodpecker's, and it builds its nest in a natural cavity in a tree or in an abandoned woodpecker hole; it will also nest in a wooden birdhouse.

▶ The wryneck feeds primarily on ants and their larvae, which it searches for in the branches of trees as well as in anthills on the ground; it laps them up with its long slender tongue, which, like the woodpecker's, is coated with adhesive saliva. Unlike the woodpecker's, however, the wryneck's bill is too short and delicately made to batter through an obstruction of any kind; it can only feed on ant larvae in wet weather, for example, when the ants remove them from their nursery at the bottom of the nest to protect them from groundwater. The wryneck also feeds occasionally on beetles, butterflies, caterpillars, and spiders.

◀ The wryneck is named, and best known, for its gymastic intimidatory displays: It stretches out its neck like a serpent about to strike, then twists it from side to side (*wryneck* means "twisted neck" in Middle English) so that its head appears to have turned all the way around; at the same time, it fluffs up its plumage, closes its eyes, and hisses noisily, the sound increasing in volume until it abruptly smooths down its feathers and retreats.

The wryneck competes aggressively for nesting sites with titmice and other small passerine birds; when it finds a suitable nesting cavity that is already occupied by one of hese smaller species, it will eject the sitting tenant along with all eggs, nestlings, and nesting materials it finds there so the female wryneck can take possession. Substantial decreases in the wryneck populations of Central and Western Europe should probably be regarded as normal cyclical fluctuations rather than attributed to the use of chemical insecticides or the felling of dead trees.

The red-breasted wryneck *(Jynx ruficollis)* is found in central and southern Africa.

J.-P. R.

Y

YELLOW-AND-BLACK MUD DAUBER
Sceliphron spirifex 486

486
Sub.: Arthropoda
Cl.: Insecta
O.: Hymenoptera
F.: Sphegidae

This large solitary wasp is between 20 and 30 mm long; the first abdominal segment is attenuated into a yellow petiole that is almost as long as the rest of the abdomen, which is black. Its legs, the third pair especially, are extremely long and thin, and the overall effect is one of elegant emaciation. The female mud dauber, or potter wasp, builds its nest with mud retrieved from the banks of ponds or streambeds. The mud is kneaded between the mandibles, mixed with saliva, and rolled into a pellet about the size of a pea, which the wasp carries between the points of its mandibles as it flies back to its nesting site.

The pellets of mud are rolled out flat, this time using the mandibles as a trowel, to form the walls of the nest, which is placed in a sunny spot that is protected from the rain, often under the eaves of buildings, particularly against a wall of rough, undressed masonry, or attached to the side of a boulder or the branch of a tree. The nest consists of 5 or 6 vertically aligned cells, each about 3 cm long, which—once the eggs are in place and the cells provisioned with insect prey—are sealed with a final coating of mud; the finished nest has the appearance of a small block of adobe. This mud dauber species is found from the Cape of Good Hope as far north as the Loire Valley and the Dordogne, in Central France.

▶ The yellow-and-black mud dauber provisions its nest with small spiders (an average of 8 for each egg cell), particularly epeirids such as *Araneus cucurbitinus* and clubionids such as *Chiracanthium erraticum*. As soon as the nest is ready, the female flies off to search for spiders in the neighboring bushes and underbrush. As soon as it encounters a spider, it hurls itself upon it and quickly immobilizes it by stinging it several times in the cephalothorax. It carries its prey in its forelegs and mandibles as it flies back to the nest, then backs into the egg cell dragging the spider behind it, then switches positions and pushes the spider up against the rear wall of the cell.

It attaches its egg to the ventral part of the spider's abdomen, immediately posterior to the junction with the cephalothorax, and then flies off to collect a second spider (which it pushes ahead of it into the cell this time), and so on until the nest is fully provisioned. The entrance to the cell is sealed with a tough mortarlike substance that serves both to camouflage it and secure it against predators. The larva hatches a few days later and immediately begins to feed on the spiders; its larval growth is complete by August, but it does not undergo nymphosis immediately. Instead, it remains in the cell, its head pointed toward the sealed entrance, in a state of diapause, or suspended animation, and does not emerge until the following summer.

◀ In spite of all these precautions, the adult wasp that finally emerges from the cell may not be a yellow-and- black mud dauber. There is a species of cuckoo wasp (family Chrysidae), *Stilbum cyanarum*, that is able to penetrate the sun-baked adobe of the mud dauber's nest with its ovipositor and thus enables its own larva to feed not only on the spiders but the larva of *Sceliphron spirifex* as well. The cuckoo wasp also parasitizes other mason waps of the families Eumenidae and Chalicomenidae and, quite fittingly, is parasitized in turn by an ichneumon wasp, *Leptobatides abeillei*. R.C.

YELLOW-BILLED OXPECKER
Buphagus africanus 487

487
Sub.: Vertebrata
Cl.: Aves
O.: Passeriformes
F.: Sturnidae

This species, which attains a length of 22 cm, is found on the savannahs of sub-

Saharan Africa, from Senegal and Ethiopia down to the Cape, and always in association with wild or domestic herd animals.

▶ The oxpecker uses its short legs and curved claws to scale the sides of a cow, a Cape buffalo, an elephant, or a rhinceros in the much the same way that a woodpecker works its way up the sides of a tree; it uses its short, stiff tailfeathers as a sort of ratchet to keep from sliding back down again. The oxpecker's activities are generally of some benefit to its host—it feeds on ticks and fly larvae that are buried in the the animal's hide and sounds a warning cry at the approach of predators—but it sometimes acquires a taste for mammalian blood, and a few individuals seem to do serious harm to cattle by enlarging the wounds made by ticks, chiggers, etc.

◀ The oxpecker has no known predators or competitors.

A congener, the red-billed oxpecker *(Buphagus africanus)*, is found on the East African savannah, and a number of other species that are not closely related to the oxpeckers have adopted a similar mode of life. Both the oxpeckers and the anis *(Crotophagus anis)* of tropical America are collectively known as "tickbirds" (though the anis' Greco-Latin genus name discreetly proclaims the fact that it also hunts for insects in pats of cowdung). The cattle egret *(Ardeola ibis)*, which is found in Europe, Africa, South America, and now occasionally in North America, as well as the South American cattle tyrant *(Machetronis rixosa)* and the cowbird (q.v.), are often seen perched on the backs of cattle, though they normally hunt on the ground, pecking at insects that have been stirred up by the animals' hooves. J.-F. T.

YELLOW-HEADED ROCKFOWL *Picathartes gymnocephalus* 488

488
Sub.: Vertebrata
Cl.: Aves
O.: Passeriformes
F.: Picathartidae

This tropical passerine bird is more graceful in its movements than in its appearance: rather large (35 cm) and heavily built, it makes its way in short, surefooted hopping motions along the moss-covered walls of cliffside caverns, and is found exclusively in the dense mountain rainforests of West Africa. A small colony of rockfowl will take up residence in a cave of this sort, affixing their nests of mud and dried leaves to the vertical rock face.

▶ The yellow-headed rockfowl feeds primarily on insects, which it hunts along the ground, making its way in a series of abrupt, energetic leaps. It sometimes stations itself at the head of a column of driver ants and feeds on the insects that are flushed from cover by the approaching ants. It also hunts for frogs and snails by turning over the leaves of plants and probing among the litter on the forest floor. The rockfowl dispatches these larger prey by striking them smartly against the ground in attempting to swallow them; it feeds on fallen fruit as well as on animal prey.

◀ The yellow-headed rockfowl has few predators and no particular competitors. It is not a prolific species, and only its relatively long life expectancy (7 years is typical in captivity) enables it to maintain its numbers. It is credited with supernatural powers by the forest-dwelling peoples of the region, and thus is never hunted or otherwise molested by humans.

Picathartes auratus, which is found exclusively in Gabon and Cameroon, and *P. gymnocephalus*, described above, are the only representatives of an isolated lineage that is sometimes regarded as a separate family (Picathartidae), though nowadays more often included along with several other anomalous African species (notably the babblers) in the family Timaliidae.

J.-F. T.

YELLOW JACKET *Vespula germanica* 489

489
Sub.: Arthropoda
Cl.: Insecta
O.: Hymenoptera
F.: Vespidae

Small, primarily ground-nesting paper wasps of the genus *Vespula* are commonly called yellow jackets in English and, in Europe as well as in North America, are among the very small number of insect species that most of us can confidently identify. *Vespula germanica* is among the dozen most common wasp species found in Western Europe—and is probably the single species by which human beings, either through imprudence or sheer inadvertence, are most commonly stung. Fortunately, the yellow jacket's slim black body—spotted or striped with yellow and with the characteristic "wasp waist" indentation between thorax and abdomen—is instantly recognizable. Both yellow jackets and hornets build their nests out of "paper" or "cardboard," actually a mixture of chewed-up bark and plant fibers mixed with saliva. Réamur, the French physicist and naturalist, was the first to determine the composition of the paper wasp's nest, in 1720, when the manufacture of pulp paper (rather than rag paper) was still unknown in Europe.

The yellow jacket conceals its paper nest in a hole or an indentation in the ground, a crevice or a cavity in a wall, a burrow abandoned by a vole or other rodent, or an attic or an airspace under a roof. Yellow jacket colonies die off over the winter, except for a single female that has already been impregnated but remains in a state of ovarian diapause; it remains immobilized over the winter in a sheltered spot and is unable to lay its eggs until some time after it has begun to feed on nectar again in the spring. It is these females who build the egg cells that become the nucleus of the next year's colony, surrounding them with a more or less spherical protective paper wrapper. The cells are always open on the bottom, and the nest is suspended by a stalk of chewed paper pulp that has been stiffened with saliva and is sturdy enough to withstand a pull of several kilograms.

Until emergence of the first generation of workers (females that are sterilized by a process known as nutricial castration: ingesting a substance secreted by the original female that inhibits the development of their ovaries), it is the female that builds the egg cases, enlarges the protective cover of the nest, and catches insects with which to feed the larvae. Afterwards, these responsibilities are assumed by the workers, and the female devotes itself exclusively to the task of laying eggs.

▶ The insects that are brought back to the larvae have already been chewed up by the workers, and the resulting organic pap is fed to them a mouthful at a time, since the opening of the cell occupied by the yellow jacket larva faces downward and food cannot be left inside it. The adult yellow jackets, like other waps, feed on nectar, honeydew, and other sugary substances, which they sometimes find more convenient to collect from a windfall orchard or even an open jam jar than from the inner recesses of flowers. Yellow jackets are not as conspicuous as bees in their nectar-collecting activities, but they may often be seen filling up with water at the edge of a pond, puddle, fountain, trough, etc.

Similarly, the workers will sometimes gather tiny crumbs of meat from a butcher's block (or from a dinner plate) or even from the fresh carcass of an animal on which some of the flesh has been exposed. More often, however, they will forage vigilantly among the grass stems and other vegetation, especially in search of hairless moth caterpillars of the genus *Noctua*. The yellow jacket catches its prey with the forelegs, and if the caterpillar begins to struggle, the yellow jacket may subdue it by squeezing it between its mandibles before stinging to death. The body of the caterpillar is chewed up and mixed with saliva before being brought back to the nest, where it is regurgitated and fed to the larva in the manner described above.

◀ From an ecological standpoint, the yellow jacket's contribution as a destroyer of harmful insects (and as a source of food for a number of vertebrates species) clearly outweighs its more conspicuous activities as a disrupter of picnics and a despoiler of jam jars—even the occasional annoyance (and for some, the serious hazard) of getting stung. The European bee-eater *(Merops apiaster)* is one of several insectivorous birds that preys on *Vespula germanica* , and there are a number of staphyline wasps that feed on the eggs and larvae of social wasps (such as the yellow jacket) that build their nests underground. The nests of social wasps may also infested, or harmlessly colonized, by various species of parasites and scavengers.

Volucella zonaria is a large fly whose coloration mimics that of certain wasps, which enables it to attach its large (2.5 mm) white eggs to the outer wrapper of the nest, apparently under cover of darkness. The larvae were naturally assumed to be parasitic, but Fabre (with his usual patience and insight, plus a certain amount of courage in this case) was able to determine that they actually fill the role of scavengers, feeding on the bodies of dead larvae and workers that litter the bottom of the nest.

The activities of the ichneumon wasp *Sphecophaga vesparum* Curt., on the other hand, are of no discernible benefit to their hosts. The adult deposits one or more eggs on the wasp larva itself, inside its cell; the ichneumon larvae bury themselves inside the body of the yellow jacket larvae, though only one of the parasites survives. It feeds initially off the fatty tissue and the less essential organs of its host, saving the vital organs for the last few days of its larval phase, then hibernates inside the nymphal cocoon that was spun by the host.

The life cycle of the female stylops (order Strepsiptera) provides an extreme example of the phenomenon known as parasitic degeneration. The adult male is fully equipped with wings and legs, but the female, like the larva, has neither; its body remains embedded in that of its host, taking the form of a sort of herniated swelling on the wasp's intersegmental membrane, the inner portion of which corresponds to the cephalothorax of the stylops larva or of the adult female. Only the male undergoes nymphosis, and shortly after mating, little else of the female's body is visible except for a swollen abdominal sac containing thousands of eggs. The newly hatched larvae, known as triongulins, are very active and are soon making vigorous efforts to insinuate themselves into the bodies of other members of the wasp colony. The presence of the stylops larva causes sterility (so-called parasitic castration) but is not fatal to its host.

The triongulin of the rhipiphorid beetle *Metoecus paradoxus* feeds on the larva of social wasps that build their nests under ground. The adult beetle lays its eggs on flowers, and the triongulins attach themselves to visiting wasps as they gather nectar and return with them to the nest. The adult wasp, which is rarely seen and probably quite rare, is 8–12 mm long; this cycle is repeated twice a year. R.C.

Z

490
Sub.: Arthropoda
Cl.: Insecta
S.cl.: Hymenoptera
O.: Apocrita (Petiolata)
F.: Ichneumonidae
S.f.: Pimplinae

Zatypota percontatorius 490

This small ichneumon wasp never exceeds 12 mm in length. The female's ovopositor is non-retractible (and thus always visible), as is the case with *Rhyssa* and *Ephialtes*, two large genera that also belong to the subfamily Pimplinae, and the larva of *Zatypota* is ectoparasitic, which is not the case with the other species. The adults mate shortly after they emerge from the cocoon; the male pursues the female, clambers up on her back while grasping her thorax with his forelegs, then steps back, spreads his wings, and curves his abdomen forward to effect intromission. After this extremely brief coupling, the male will not mate with the same female again. The larva measures between 2.5 and 3 mm by the time it completes its development; initially cream-colored, it will have acquired a grayish tint by the time it is ready to spin its silken nymphal cocoon.

▶ The adult female wasp assures the survival of its larvae by attaching its eggs to the living bodies of small web-spinning spiders of the genus *Dictyna* (family Cribellatae), usually *D. puella* and *D. uncinata*, which are only about 3–4 mm long and generally found in the shallow creases in the leaves of various shrubs (including *Phillyrea* and *Cistus)* and ivy plants. The ichneumon touches down on the spider's web, which naturally brings the spider out to investigate; the spider starts to flee at the sight of the ichneumon and then remains motionless for a short time (as part of its instinctual response to danger). The ichneumon takes advantage of this opportunity and deposits its egg in the joint between spider's cephalothorax and abdomen. The spider is not paralyzed; it continues to go about its business as before, and there is no sign of the ovipositor's having pierced the skin.

The ichneumon larva grows slowly at first (especially if it hatches during the winter), clinging to the spider's curved abdomen, just at the edge of the cephalothorax, with its mandibles and ventral pads. In April, the larva undergoes its initial molt and then starts to grow very rapidly; it presses its mouthparts against the spider's abdomen and, without actually piercing the spider's body cavity, feeds on the fluids that flow out of the wound. In a day or two, the spider is drained dry. The larva spins out a continuous silk thread from its mouth and envelopes itself in a lacelike, conical cocoon, very wide at its base and about 5 or 6 mm long, beside the desiccated corpse of the spider; this process takes another 3 or 4 days. The threads are not continuously wrapped, so the larva can still be seen inside the cocoon; this also permits the larvae to eject the 18–20 pellets of excrement that it has been stockpiling inside its body case during this entire period. In 24 hours, the process of nymphosis is complete, and the adult wasp emerges after another 10 days. R.C.

ZORILLA *Ictonyx striatus* 491

491
Sub.: Vertebrata
C.: Mammalia
O.: Carnivora
F.: Mustelidae

All appearances to the contrary (striped black-and-white pelt and upraised bushy tail), this African weasel is not very closely related to the North American striped skunks. It measures between 30 and 40 cm, exclusive of its tail (20–30 cm). The zorilla (sometimes spelled zorille) ranges from the Sahel to the Cape of Good Hope; it digs its own burrow or occupies the disused burrow of another creature or a crevice in the rock.

▶ The zorilla is not a highly specialized predator; it prefers rodents and other small mammals, though it also feeds on frogs, birds, and insects. It is a solitary nocturnal hunter and detects its prey by olfaction. As soon as it has picked up the scent of a prey animal, it rushes to attack it; smaller creatures are consumed on the spot, larger ones are carried back to the zorilla's den.

◀ The zorilla also shares the quintessentially skunklike trait of defending itself by spraying a volatile and extremely foul-smelling liquid from its anal scent glands. When it is attacked, it turns its back and stiffens up its fur, a threat that is sufficient to deter most canine and feline aggressors. The raptors, which have no sense of smell, are a more serious threat to the zorilla, which is also persecuted by humans as a habitual chicken thief.

The Libyan striped weasel *(Poecilictis libyca)*, smaller and more northerly in its range, is also remarkably skunklike in its appearance. P.A.

INDEX

This index is a guide to all prey. All predators can be located in alphabetical order in the text of the book. This index is also a guide to the natural habitats of both prey and predators.

B

C

R

S